AF581004

Toxic Marine Phytoplankton

Toxic Marine Phytoplankton

Proceedings of the Fourth International Conference on Toxic Marine Phytoplankton, held June 26–30, 1989, in Lund, Sweden

Editors

Edna Granéli

Bo Sundström

Lars Edler

Department of Marine Ecology
University of Lund
Lund, Sweden

Donald M. Anderson

Biology Department
Woods Hole Oceanographic Institution
Woods Hole, Massachusetts, USA

Elsevier

New York • Amsterdam • London

Elsevier Science Publishing Co., Inc.
655 Avenue of the Americas
New York, New York 10010

Sole distributors outside the United States and Canada:
Elsevier Science Publishers B.V.
P.O. Box 211, 1000 AE Amsterdam, the Netherlands

This book is printed on acid-free paper.

Library of Congress Cataloging-in-Publication Data

International Conference on Toxic Marine Phytoplankton (4th : 1989 : Lund, Sweden)
Toxic marine phytoplankton : proceedings of the Fourth International Conference on Toxic Marine Phytoplankton held June 26–30 in Lund, Sweden / editors, Edna Granéli ... [et al.].
p. cm.
1. Marine phytoplankton—Toxicology—Congresses. 2. Water bloom-Toxicology—Congresses. 3. Marine toxins—Congresses.
I. Granéli, Edna. II. Title.
RA1242.M34I57 1989
615.9'5294—dc20 89-25737
ISBN 0-444-01523-X CIP

Current printing (last digit):
10 9 8 7 6 5 4 3 2 1

Manufactured in the United States of America

CONTENTS

V. PHYSIOLOGY AND BIOCHEMISTRY

VI. TOXIN CHEMISTRY

VII. REGIONAL OUTBREAKS AND MANAGEMENT ISSUES

VIII. WORKSHOPS

IX. CONFERENCE OVERVIEW

APPENDIX

FOURTH INTERNATIONAL CONFERENCE ON TOXIC MARINE PHYTOPLANKTON

JUNE 26-30, 1989
Lund, Sweden

HOST

Department of Marine Ecology
University of Lund
Sweden

MAIN SPONSORS

National Environmental Protection Board (SNV)
Swedish Natural Science Research Council (NFR)
The National Board of Fisheries
The Norwegian Research Council for
Science and the Humanities (NAVF)
Norwegian Fishery Research Council (NFFR)
Royal Norwegian Council for
Scientific and Industrial Research (NTNF)
Directorate for Nature Management, Norway
Intergovernmental Oceanographic Commission (IOC-UNESCO)
World Wildlife Fund (WWF)
Crafoord Foundation
The Swedish West Coast Foundation
University of Lund

MAIN SUPPORTER

ÅKERLUND & RAUSING

SPONSORS

Blekinge County Administration
Göteborg and Bohuslän County Administration
Kristianstad County Administration
Malmöhus County Administration
Göteborg and Bohuslän County Council
Halland County Council
Kristianstad County Council
Malmöhus County Council
Community of Lund

AB Ramlösa Hälsobrunn
Alfa-Laval Food Engineering AB
Ewos Aquaculture AB
ICA Eol AB
IDEON Research Park, Lund
Pripps Breweries
SAAB-SCANIA AB
SAS
Scandinavian Ferry Lines AB
The National Swedish Telecommunications
Administration (Malmö)
TT- Line AB

EDITORIAL COMMITTEE

Berit Sjöstedt Gustafsson
University of Lund
Sweden

Wilhelm Granéli
University of Lund
Sweden

CONFERENCE ORGANIZERS

Conveners

Edna Granéli
Lars Edler
Bo Sundström

Scandinavian committee

Einar Dahl, Norway
Matts Hageltorn, Sweden
Vagn Kr. H. Hansen, Denmark
Palle Krogh, Denmark
Odd Lindahl, Sweden
Karl Tangen, Norway

Local:
Lena Carlson
Per Carlsson
Nils Ekelund
Per Olsson
Berit S. Gustafsson
Rainer Weich

International Advisory Committee

D.M. Anderson, USA
J. Carreto, Argentina
M. Elbraechter, FRG
Y. Fukuyo, Japan
G.M. Hallegraeff, Australia
T. Okaichi, Japan
T.J. Smayda, USA
K.A. Steidinger, USA
S. Sudara, Thailand
F.J.R. Taylor, Canada
T. Wyatt, Spain
T. Yasumoto, Japan

A special prize - Swedish crystal - was awarded Professor Takashi Yasumoto during the Conference banquet. The Fourth International Conference on Toxic Marine Phytoplankton, in this way, officially recognized his outstanding achievements in the field of marine toxin chemistry.

PREFACE

In 1974, 1978 and 1985, International Conferences on Toxic Dinoflagellates were held in Boston, Massachusetts, USA, Key Biscane, Florida, USA and Saint Andrews, New Brunswick, Canada , respectively. An international symposium with a similar but broader scope was organized in 1987: the "First International Symposium on Red Tides", in Takamatsu, Kagawa Prefecture, Japan.

During the 1985 Saint Andrews conference it was decided that the Fourth International Conference on Toxic Dinoflagellates was to be held in Lund, Sweden in 1989. However, in the intervening years, a series of dramatic blooms of toxic marine phytoplankton species that were not dinoflagellates occurred. A devastating bloom of the prymnesiophyte *Chrysochromulina polylepis* affected the coastal waters of Sweden, Denmark and Norway from May to June 1988, killing benthic fauna and flora as well as pelagic organisms, and causing a loss of cultured salmon worth approximately 12 million US$. A bloom of the diatom *Nitzschia pungens* at King Edward Island, Canada (1987) poisoned 150 shellfish consumers, of which three died. These latter two algal species were not known to be toxic. New toxins were isolated and the shellfish poisoning associated with the *N. pungens* bloom was tentatively named "Amnesic Shellfish Poisoning" (ASP) after one of the main symptoms. To exclude the presentation of scientific information on these and other toxic non-dinoflagellate marine phytoplankton species from the pending and future conferences for purely taxonomic reasons was clearly unwise. The "new" toxic groups are studied with the same methods and often by the same scientists as the "traditional" toxic dinoflagellates. Therefore, the international advisory committee agreed to change the name of the Lund symposium to "International Conference on Toxic Marine Phytoplankton", but to consider it part of the original series of dinoflagellate conferences. However, since several phytoplankton species cause serious harm without actually being toxic, and since presentation of information on these was not excluded from the Lund conference and should not be excluded from future conferences, yet another change of name was proposed. It was thus suggested by several scientists that future conferences should be called "International Conference on Harmful Phytoplankton". The final decision in this matter will be taken by the newly appointed International Committee.

The Fourth International Conference on Toxic Marine Phytoplankton was held in Lund, Sweden, June 26-30, 1989. The local organizing committee consisted of members of the Plankton Ecology Research Group at the subdepartment of Limnology and Marine Ecology, department of Ecology, University of Lund.

The conference was attended by 250 scientists from 27 countries, representing North- and South America, Australasia, and Europe.

Scientific presentations were in the form of plenary lectures (8), oral presentations (44), posters (90) and workshops (6).

During the conference, Professor Takashi Yasumoto was presented a special award for his outstanding achievements in the field of marine toxin chemistry.

The difficulty that participants of international scientific conferences have in agreeing on anything but very general statements is well-known. At the Lund meeting, after long and quite-animated discussions, consensus was reached on the nomenclature of the *Alexandrium* - *Protogonyaulax* species complex, which has been debated by the scientific community for some time (see contributions by E. Balech and K. A. Steidinger). Throughout this volume, and hopefully in all journals covering this field of research, the genus name ***Alexandrium*** will be used.

After another long discussion, another consensus was obtained and a statement of concern was transmitted to the Intergovernmental Oceanographic Commission (IOC) in Paris concluding that there is a potential that global expansion of algal bloom problems is now occurring (see workshop; T. J. Smayda and A. W. White).

The role of bacteria in the production of shellfish toxins has long been a controversial issue. M. Kodama, in a plenary lecture, presented a review of several research programs that suggest that PSP-producing bacteria may play an important role in the toxicity of bivalves. The discussion that followed that presentation or the related poster session indicates that this topic remains highly controversial within this community and that it will surely be a central focus of future conferences.

Oral presentations and posters were treated equally with respect to publication in this Proceedings. Thus, each contribution (except plenary lectures, which were allowed up to 12 pages) was allotted six printed pages. Contributions were supposed to be delivered in final form (ready to print) to the editors a few weeks after the meeting. The scientific value of a conference volume lies principally in providing up-to-date information on a field of research, so prompt publication is necessary. This was also requested by the publishers, Elsevier, and the editor-in-chief Dr. Allan Ross. The editors thus had to process approximately 100 manuscripts during a period of two months. All manuscripts have been scrutinized by at least one referee (in the case of rejection, two) and by at least one of the Swedish editors (Granéli, Sundström and Edler). It is evident that the editing of such a large number of manuscripts in such a short period of time does not allow a final product equivalent to that found in international scientific journals. Many of the manuscripts arrived in a form far from "final" or camera-ready, and much time was spent correcting, rewriting, and in some cases retyping papers. The editors are therefore aware that linguistic, typing and even scientific errors may remain. We apologize to those that were late in submitting their manuscripts to us, as the time constraints discussed above and the strict page limit set by the publisher required that all late submissions be rejected.

Up to now, the conferences on Toxic Dinoflagellates (Toxic Marine Phytoplankton) have been organized by informally selected or self-elected groups of scientists. There has been no "International Society for Toxic

Dinoflagellates", and no official or permanent committee to decide on organizational matters. The conferences have thus not occurred at regular intervals. Participants at the Lund meeting decided by referendum that a suitable interval for future "International Conferences on Toxic Marine Phytoplankton" should be two years. This may seem a short period, but scientific development is rapid at present and the apparent increasing frequency of toxic blooms worldwide has made frequent international scientific communication and publication a necessity.

Representatives from five different countries offered to host the next conference. The offer by T. J. Smayda to arrange the 1991 meeting at Rhode Island, USA and the offer by P. Lassus to arrange the 1993 meeting in France were accepted by consensus. It was also decided that an international committee should be elected to support the local conference committees. Members should represent America, Australasia and Europe, as well as the most recent conference organizing committee and the organizers of the pending conference. The following scientists were elected: America: D. M. Anderson (USA), T. J. Smayda (USA), F. J. R. Taylor (Canada); Australasia: G. M. Hallegraeff (Australia), T. Okaichi (Japan), T. Yasumoto (Japan); Europe: E. Dahl (Norway), E. Granéli (Sweden), P. Lassus (France).

Finally, the editors wish to thank all members of the local organizing committee who, often without financial compensation, worked very hard to make the conference successful. We are also grateful for the very generous support given by research councils, county and community administrations and private companies. Without this personal and financial support the conference would not have been possible.

Edna Granéli Bo Sundström Lars Edler

University of Lund

D. M. Anderson

Woods Hole Oceanographic Institution

I HONORARY LECTURE IN MARINE TOXICOLOGY

MARINE MICROORGANISMS TOXINS - AN OVERVIEW

TAKESHI YASUMOTO
Faculty of Agriculture, Tohoku University, 1-1 Tsutsumidori Amamiyamachi, Sendai 981, Japan

ABSTRACT

Recent research progress made on chemical structures, analytical methods, and etiology of marine phytoplankton toxins is reviewed. Emphasis is laid on polyether toxins involved in diarrhetic shellfish poisoning, ciguatera, and on galactoglycerolipids with potent hemolytic and ichtyotoxic activities.

INTRODUCTION

Marine phytoplankton toxins are frequently involved in seafood-borne poisoning and massive fish kills, and thus are creating increasing public concern. In order to prevent or minimize the damages to public health, acquaculture, and marine ecosystems, the establishment of effective monitoring systems of toxins has been a prerequisite. Determining toxin structures provides not only the molecular basis for elucidation of the phenomena but also information useful for designing chemical assay methods to be used in monitoring. The structural studies of phytoplankton toxins have been, and still are, handicapped by the extremely limited availability of toxins. However, recent progress in spectroscopic techniques, especially NMR spectroscopy, has drastically reduced the necessary amount of samples and thus accelerated structural determination. Separation and analyses of complex toxin mixtures are also facilitated by the advances in high performance liquid chromatography (HPLC). Thus the number of toxins of which structures were determined has increased markedly in recent years. As toxins implicated in paralytic shellfish poisoning, amnesic shellfish poisoning, and brevetoxins will be described by other authors in these Proceedings, the author wishes to present results achieved in his laboratory on diarrhetic shellfish toxins, ciguatera and related toxins, and hemolytic and ichthyotoxic galactoglycerolipids.

Diarrhetic shellfish toxins

Structures and occurrence Toxin studies were carried out mainly with hepatopancreas of the scallop *Patinopecten yessoensis* collected in Mutsu Bay which is located at the northern tip of Honshu Island, Japan. Out of twelve toxins isolated so far, structures of nine toxins were determined as shown in Fig. 1. The toxins are separable into three groups according to their basic skeletons. The first group includes okadaic acid (OA), dinophysistoxin-1 (DTX1), and dinophysistoxin-3 (DTX3). The structure of OA was first determined by Tachibana *et al.* by an X-ray analysis of crystals obtained from sponges [1]. DTX1 and DTX3 were determined by comparison of spectral data with those of OA [2,3]. DTX3, which occurs in scallops but seldom in mussels, is a mixture of 7-*O*-acyl derivatives of DTX1 differing only in the acyl moiety. Biological activities of DTX3 vary according to degree of unsaturation in the acyls. A homolog having a saturated acyl, such as 7-*O*-palmitoylDTX3, has only one-tenth mouse lethality of DTX1, while a homolog having a highly unsaturated acyl retains one-third potency of DTX1 [3]. Hemisynthetic models of 7-*O*-acylOA indicate that effect of the degree of unsaturation in the acyls are less prominent on diarrheagenicity and cytotoxicity than on mouse lethality [4].

Toxic Marine Phytoplankton
Edna Graneli et al., Editors

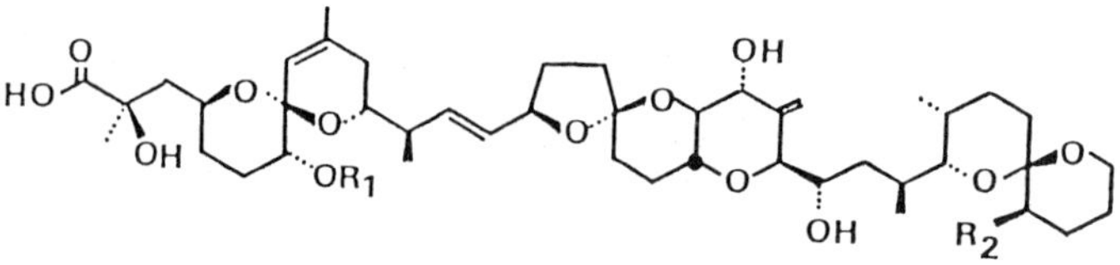

okadaic acid (OA): R_1 = H, R_2 = H
dinophysistoxin-1 (DTX1): R_1 = H, R_2 = CH_3
dinophysistoxin-3 (DTX3): R_1 = acyl, R_2 = CH_3

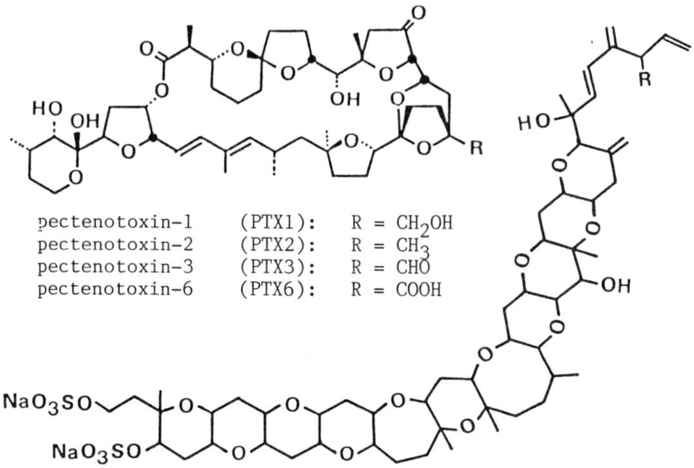

pectenotoxin-1 (PTX1): R = CH_2OH
pectenotoxin-2 (PTX2): R = CH_3
pectenotoxin-3 (PTX3): R = CHO
pectenotoxin-6 (PTX6): R = COOH

yessotoxin (YTX): R = H
45-hydroxyyessotoxin (45-OH YTX): R = OH

Fig. 1. Diarrhetic Shellfish Toxins

The second group comprises polyether macrolides named pectenotoxins (PTXs). The structure of PTX1 was determined by X-ray analysis, and those of PTX2, PTX3, and PTX6 by comparison of spectral data [3,5]. ^{1}H NMR data suggest that PTX4 and 7 are very close to, but not identical with, PTX1 and PTX6, respectively. Structural elucidation of PTX5 was unsuccessful due to the extremely small sample size. The occurrence of PTXs was proven in Japanese specimens but not in mussels from other countries.

The third group consists of yessotoxin (YTX) and 45-hydroxyYTX. The planar structure of YTX was determined mainly by modern NMR techniques [6]. Both YTX and 45-hydroxyYTX are characterized by the presence of two sulfate esters and by the resemblance with brevetoxins and ciguatoxins in having contiguously fused ether rings. The occurrence of YTX outside Japan was confirmed only in mussels collected in Sognefjord, Norway [7]. Histopathological effects of the toxins are described elsewhere in these Proceedings by Terao _et al._

The occurrence of diverse toxins is a characteristic feature of the scallops of Mutsu Bay. Yet toxin profiles fluctuate widely seasonally and annually. In contrast to Mutsu specimens, mussels from European countries were characterized by the predominance of OA [8]. Exceptionally mussels from Sognefjord, Norway, resembled Mutsu specimens in having DTX1 and YTX [7]. As will be discussed later, only DTX1 and PTX2 were found in a causative dinoflagellate _Dinophysis fortii_. It is presumed, therefore, that acylation of DTX1 to DTX3 takes place in scallops. It is also likely that 43-CH_3 of PTX2 undergoes a series of oxidation to yield CH_2OH (PTX1), CHO (PTX3) and finally COOH (PTX6).

Chemical analysis of DSP toxins Carboxylic acid toxins such as OA, DTX1, PTX6, and PTX7 can be derivatized into fluorescent esters by reaction with 9-anthryldiazomethane. Subsequent HPLC analysis allows us to determine the toxins quantitatively with a high sensitivity and specificity [9]. Fluorescence labelling of PTX4 and PTX7 is achievable by esterifying their 43-CH_2OH with 9-anthroylnitrile. The resultant fluorophores can be determined by HPLC [10]. Other toxins which do not possess functional groups suitable for derivatization have to be analyzed by monitoring UV absorptions of eluates from HPLC [10].

Toxin producing organisms Although _Dinophysis fortii_ and related species have been presumed to be the primary sources of toxins [11], confirmation of their toxigenicity has not been possible because of the difficulty in culturing these species. Aided by the high sensitivity and specificity of the fluorometric HPLC method, we were able to identify OA or DTX1 or both in 100 - 1000 cells of specimens of _Dinophysis_ collected under a microscope by capillary manipulation. The toxigenicity of the following seven species was confirmed: _Dinophysis acuminata, D. acuta, D. fortii, D. mitra, D. norvegica, D. rotundata,_ and _D. tripos_ [12]. The relative ratio of OA to DTX1 and the contents of OA and DTX1 in cells showed significant intraspecies variation depending on sampling sites and sampling season. The presence of DTX3 in _D. fortii_ was not confirmed. Tests on 10,000 cells of _D. fortii_ collected in Mutsu Bay proved the occurrence of PTX2 but no other PTXs [12]. As will be mentioned later, _Prorocentrum lima_ also produces OA and related compounds. However, the implication of this species in DSP has not been confirmed yet.

Ciguatera

In tropical and subtropical regions, many species of coral reef fish may bear toxins and thus cause intoxication named ciguatera when ingested. Several thousands of cases are officially reported annually. However, the actual number of cases may far exceed the reported figures. Previously the authors' group discovered an epiphytic dinoflagellate _Gambierdiscus toxicus_ as the primary source of ciguatoxin and maitotoxin [13]. The former toxin was isolated as the major toxin of ciguatera from the viscera of the moray eel _Gymnothorax javanicus_ by Scheuer's group in University of Hawaii [14,15]. However, structural studies were hampered by the difficulty of obtaining toxic fish, tedious isolation procedures due to extremely low concentration of the toxin in fish, and by the complex nature of the toxin. Recently in collaboration with Legrand _et al._, who succeeded in isolating 350 μg of pure ciguatoxin from 900 moray eels, we proposed the molecular formula of $C_{60}H_{86}NO_{19}$ for ciguatoxin on the basis of high resolution FAB mass spectra data [16]. A partial structure of ciguatoxin was determined by ^1H NMR measurements, but further elucidation of the whole structure was hindered by broadening and disappearance of certain signals due to molecular movements [17]. This problem was solved by measuring ^1H NMR at low temperature. Assignments of overlapping signals were helped by comparing spectra taken in different solvents. The planar structure shown in Fig. 2 was thus proposed for moray eel ciguatoxin [18]. From _G. toxicus_ collected in the Gambier Islands, a less polar analogue was obtained (Fig. 2)[18]. The characteristic feature of the structures is the presence of thirteen contiguously trans-fused ether rings of five through nine members. The structural resemblance of ciguatoxins with brevetoxins [19] and yessotoxin is thus pointed out. Comparison of structures of the two toxins indicates that oxidative modification of *G. toxicus* toxin takes place during the transmission through the food chain. Occurrence of analogous toxins in fish and in the dinoflagellate was also recognized [20]. The presence of primary alcohol at one terminal of moray eel ciguatoxin suggests the possibility of preparing *via* hemisuccinate a toxin-protein conjugate, which can be used for producing an anti-ciguatoxin antibody. Although a structural relationship between toxins in wild *G. toxicus* and in the moray eel was established, toxin production by cultured *G. toxicus* has been unsuccessful. Maitotoxin, a characteristic toxin of herbivorous ciguateric fish, was isolated from *G. toxicus* cultures and its chemical properties were partially defined [21].

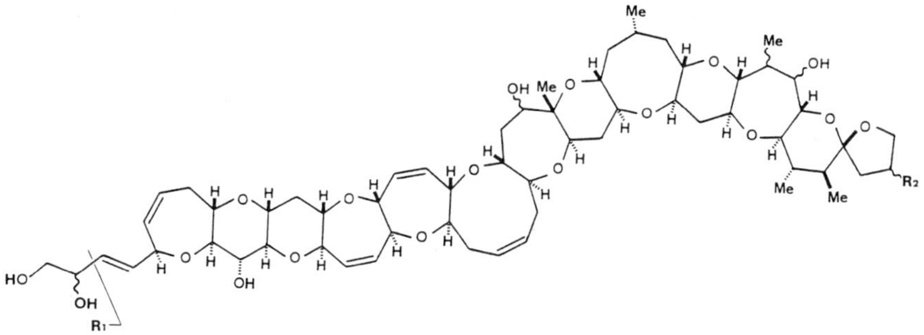

1: R_2 = OH ; 2: R_1 = CH_2=CH-, R_2 = H

Fig. 2. Ciguatoxin (1) from moray eels and its Congener (2) from *G. toxicus*

Toxins of Prorocentrum lima

During the ecological surveys on *G. toxicus* a number of other benthic dinoflagellates were found to produce toxins [22]. Among such benthic species, *P. lima* was found in abundance in coral reef areas [23]. Later, its presence in Vigo mussel farms was also noticed [12]. The organism produces in culture OA, diol esters of OA, DTX1, and a nitrogen-containing novel polyether macrolide named prorocentrolide (Fig. 3)[24-26]. Thus, *P. lima* is capable of producing toxins having entirely different skeletons, as is *D. fortii*. At present it is not certain whether or not OA and its derivatives of *P. lima* enter the food chain of ciguatera and diarrhetic shellfish poisoning.

Hemolysins of Amphidinium carteri

A. carteri, like *P. lima*, was collected during the survey on benthic dinoflagellates in coral reefs. The organism produced five hemolysins tentatively coded 1 through 5. Hemolysin-1 and 2 were determined to be *O*-β-D-galactopyranosyl-(1-1)-3-*O*-octadeca-tetraenoyl-D-glycerol and *O*-α-D-galactosyl-(1-6)-*O*-β-D-galactopyranosyl-(1-1)-3-*O*-octadecatetraenoyl-D-glycerol, respectively (Fig. 4)[25]. Hemolytic activities of hemolysin-1 and -2 were 80% and 25% that of commercial saponin (Merck). The minimum concentration of hemolysin-1 required to kill killifish was 7 μg/ml. Hemolysin-3 through -5 were far more potent than hemolysin-1 and -2, but their structures remain unknown. As will be discussed elsewhere in these Proceedings [Yasumoto *et al.*], hemolysin-1 and -2 are likely to be responsible for fish kills caused by red-tides of *Gyrodinium aureolum* and *Chrysochromulina polylepis*.

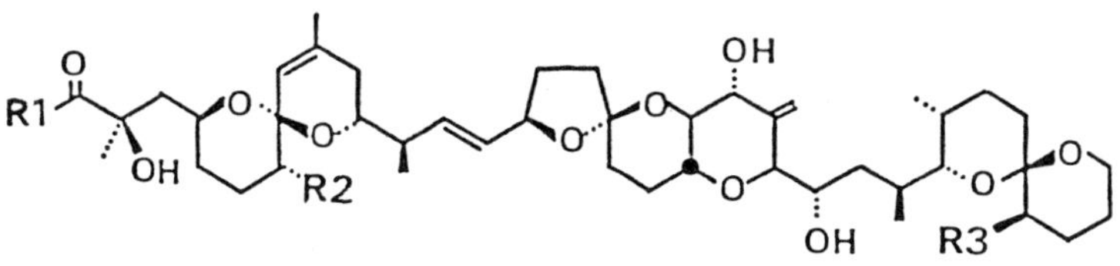

	R_1	R_2	R_3
okadaic acid	OH	OH	H
dinophysistoxin-1	OH	OH	CH_3
7-deoxyokadaic acid	OH	H	H
ethyl okadaate	CH_3CH_2O-	OH	H
diol esters	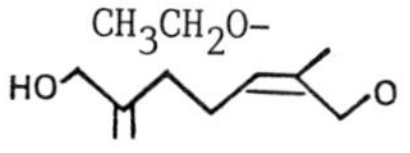	OH	H
	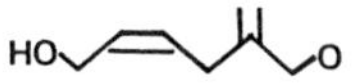	OH	H

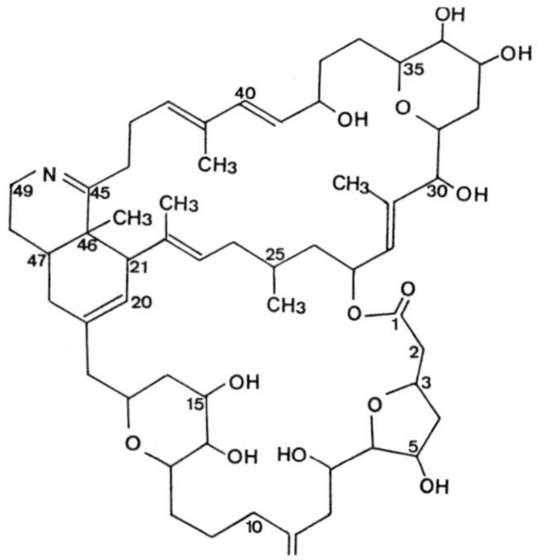

prorocentrolide

Fig. 3. Toxins of Prorocentrum lima

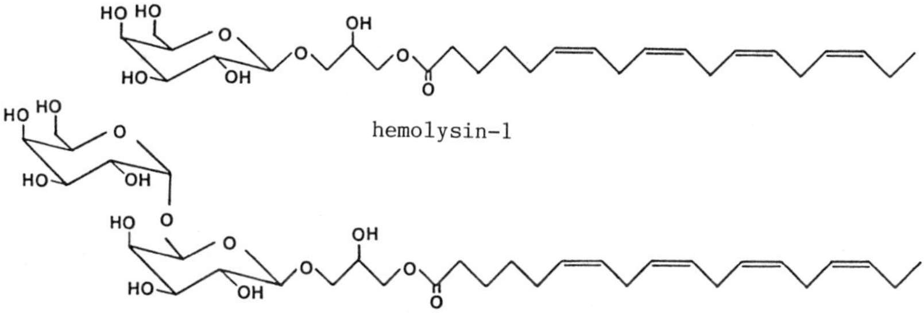

hemolysin-1

hemolysin-2

Fig. 4. Hemolysins of Amphidinium carteri

REFERENCES

1. T. Tachibana, P.J. Scheuer, Y. Tsukitani, H. Kikuchi, D.V. Engen, J. Clardy, Y. Gopichand, F.J. Schmitz, J. Am. Chem. Soc. 103 2469-2471 (1981).
2. M. Murata, M. Shimatani, H. Sugitani, Y. Oshima, T. Yasumoto, Nippon Suisan Gakkaishi 48 549-552 (1982).
3. T. Yasumoto, M. Murata, Y. Oshima, G.K. Matsumoto, J. Clardy, Tetrahedron 41 1019-1025 (1985).
4. T. Yanagi, M. Murata, K. Torigoe, T. Yasumoto, Agric. Biol. Chem. 53 525-529 (1989).
5. M. Murata, M. Sano, T. Iwashita, H. Naoki, T. Yasumoto, Agric. Biol. Chem. 50 2693-2695 (1986).
6. M. Murata, M. Kumagai, J.S. Lee, T. Yasumoto, Tetrahedron Lett., 28 5869-5872 (1987).
7. J.S. Lee, K. Tangen, E. Dahl, P. Hövgaard, T. Yasumoto, Nippon Suisan Gakkaisi 54 1953-1957 (1988).
8. M. Kumagai, T. Yanagi, M. Murata, T. Yasumoto, M. Kat, P. Lassus, J.A. Rodriguez-Vazquez, Agric. Biol. Chem. 50 2853-2857 (1986).
9. J.S. Lee, T. Yanagi, R. Kenma, T. Yasumoto, Agric. Biol. Chem. 51 877-881 (1987).
10. J.S. Lee, M. Murata, T. Yasumoto in: Mycotoxins and Phycotoxins 1988, S. Natori, K. Hashimoto, Y. Ueno eds. (Elsevier, Amsterdam 1989) pp. 327-334.
11. T. Yasumoto, Y. Oshima, W. Sugawara, Y. Fukuyo, H. Oguri, T. Igarashi, N. Fujita, Nippon Suisan Gakkaishi 46 1405-1411 (1980).
12. J.S. Lee, T. Igarashi, S. Fraga, E. Dahl, P. Hovgaard, T. Yasumoto, J. Appl. Phycol. 1, (1989) in press.
13. T. Yasumoto, I. Nakajima, R. Bagnis, R. Adachi, Nippon Suisan Gakkaishi 43 1021-1026 (1977).
14. M. Nukina, L.M. Koyanagi, P.J. Scheuer, Toxicon 22 169-176 (1984).
15. K. Tachibana, M. Nukina, Y.G. Joh, P.J. Scheuer, Biol. Bull. 172 122-127 (1987).
16. A.M. Legrand, M. Litaudon, J.N. Genthon, R. Bagnis, T. Yasumoto, J. Appl. Phycol. 1 (1989) in press.
17. M. Murata, A.M. Legrand, P. Cruchet,T. Yasumoto, Tetrahedron Let. (1989) in press.
18. M. Murata, A.M. Legrand, Y. Ishibashi, T. Yasumoto, J. Am. Chem. Soc. submitted.
19. Y. Shimizu, H.N. Chou, H. Bando, G. Van Duyne, J.C. Clardy, J. Am. Chem. Soc. 108, 514-515 (1986).
20. A.M. Legrand, P. Cruchet, R. Bagnis, M. Murata, Y. Ishibashi, T. Yasumoto, in this Proceedings.
21. A. Yokoyama, M. Murata, Y. Oshima, T. Iwashita, T. Yasumoto, J. Biochem. 104 184-187 (1988).
22. I. Nakajima, Y. Oshima, T. Yasumoto, Nippon Suisan Gakkaishi 47 1029-1033 (1981).
23. T. Yasumoto, U. Raj, R. Bagnis, Seafood Poisoning In Tropical Regions (Tohoku University, Sendai, 1984).
24. Y. Murakami, Y. Oshima, T. Yasumoto, Nippon Suisan Gakkaishi 48 69-72 (1982).
25. T. Yasumoto, N. Seino, Y. Murakami, M. Murata, Biol. Bull. 172 128-131 (1987).
26. K. Torigoe, M. Murata, T. Yasumoto, J. Am. Chem. Soc. 110 7876-7877 (1988).

II PLENARY LECTURES

SPECIES OF THE TAMARENSIS/CATENELLA GROUP OF GONYAULAX AND THE FUCOXANTHIN DERIVATIVE-CONTAINING GYMNODINIOIDS

Karen A. Steidinger
Florida Marine Research Institute, Department of Natural Resources, 100 Eighth Avenue S.E., St. Petersburg, Florida 33701-5095 USA

ABSTRACT

Based on Balech's analysis of Alexandrium [1, 2], including type locality material of A. minutum, and recent SEM analyses of thecal morphology and growth patterns, it is recommended that Alexandrium be accepted as the generic designation for the tamarensis/catenella group because it is the oldest available and valid generic name for this morphogroup. It is also recommended that the same conservative approach be taken with the fucoxanthin derivative-containing gymnodinioids, e. g., Gymnodinium (=Ptychodiscus) breve, G. nagasakiense, Gyrodinium aureolum, and others, and that their original names be retained until detailed studies are completed on existing available and valid genera, e. g., Gymnodinium, Gyrodinium, Ptychodiscus, Balechina, and Woloszynskia. These recommendations should alleviate the present confusion of simultaneous use of multiple valid and available species names.

INTRODUCTION

About 30 dinoflagellate species in the tamarensis or catenella group can be assigned to several different genera, e. g., Alexandrium, Gessnerium, and Protogonyaulax. Almost half of these species were previously in the genera Goniodoma and Gonyaulax; the remainder are new species. The group is characterized by a plate formula of Po, 4', 0a, 6'', 6c, 9-10s, 5''', 0p, and 2'''' (Kofoid nomenclature) although interpretation of plates can involve different plate assignments for the S.s.a., 1', and the 1''''. Typically, the species of this group share some common morphological characters, in addition to the plate formula, e. g., cingulum displaced less than 1.5x, thin cell walls, no spines or horns, +/- ventral pore between 1' and 4', characteristic apical pore plate, and smooth-walled resting cysts [3, 4].

Today, at least three valid and available generic names are being used interchangeably, based on author preference, for species of this group. The distinction between the genera, and ultimately the synonymy, involves the Po and 1' plates and whether they are separated or connected. Protogonyaulax was characterized by the Po and 1' being in contact with one another [5]. Alexandrium as originally illustrated (A. minutum) had variability in the Po and 1' contact [6], and the type for Gessnerium (=Gonyaulax monilata Howell) had a distinctly displaced 1' plate. In all probability, as can be inferred by recent scanning electron microscopy, the slightly separated Po and 1' represents plate rim overlaps in older cells and this could account for the variability of the Po and 1' contact as observed in A. minutum culture and field specimens by Hallegraeff et al. [7], Balech [2], and Montresor et al. [8]. If this is a valid conclusion, then Alexandrium and Protogonyaulax are synonymous and Alexandrium has priority. The distinctly separated Po and 1' or displaced 1', as in Gessnerium, may warrant a separate subgeneric status for species with this character.

Toxic Marine Phytoplankton
Edna Graneli et al., Editors

DISCUSSION

The taxonomy of dinoflagellates is confused by 1) lack of type specimens as reference material, 2) incomplete species decriptions or optically reversed illustrations, 3) recognition of conservative, although variable, morphological characters, 4) the use of two codes of nomenclature, and 5) unfamiliarity with existing taxa of limited distribution. This suite of circumstances has resulted in synonymy and homonymy and has caused confusion in the scientific literature as well as the lay press, particularly when the taxa involved were of medical or economic importance, e. g., toxic dinoflagellates.

The confusion surrounding about 30 species of Alexandrium involves all of the above problems as well as availability of type locality material to substitute for type specimens. Balech [2] recently obtained type locality material from Halim in Egypt and published a redescription of A. minutum.

The tamarensis/catenella group of Gonyaulax.

Historically, the tamarensis/catenella group originated with the description of Gonyaulax tamarensis Lebour [9] from England. Unfortunately, the epithecal plate pattern was optically reversed in the original illustrations. Optically inverted and reversed microscope images have, in several cases, led to erroneous species descriptions and illustrations. Later, Whedon and Kofoid [10] described two new, related species, G. catenella and G. acatenella from California waters. Braarud [11] named three varieties of G. tamarensis from Norwegian waters, one of which is synonymous with A. ostenfeldii, and although Braarud did not publish his variety globulosa with a ventral pore in the 1' plate, his original drawings depicted a pore [12]. Woloszynska and Conrad [13] found a species similar to G. tamarensis var. excavata (as described by Braarud), in Belgian waters, but thought it was a Pyrodinium and named it P. phoneus. Between 1953 and 1971, other researchers named new species that shared characteristics with Lebour and Whedon and Kofoid's species, e. g., G. monilata Howell [14]; G. conjuncta Wood [15]; G. fratercula Balech [16]; Alexandrium minutum Halim [6]; Gessnerium mochimaensis Halim [17]; G. balechii Steidinger [3]; and others. Howell [14] was one of the first scientists to recognize that one of these species was different from Gonyaulax sensu stricto.

Steidinger [3] and Taylor [4] were the first to characterize this group of dinoflagellates, many of which produce neurotoxins. Taylor [4] suggested that those with a displaced 1' plate (e. g., G. monilata and G. balechii) be transferred to Alexandrium. Later, the same author [18] suggested that those species with a displaced 1' plate be transferred to Pyrodinium but questioned that transfer if new information became available on cysts of the group. In 1979, Taylor left those with a displaced 1' plate in Gonyaulax and transferred nine species with the Po and 1' in direct contact to the new genus Protogonyaulax. At the same time, Loeblich and Loeblich [19] transferred nine species of this same group to Gessnerium. Balech and Tangen [12] re-evaluated Gonyaulax tamarensis and Goniodoma ostenfeldii and transferred 15 species to Alexandrium. Most recently, Balech [1] named six new Alexandrium species and recognized 22 species in the genus Alexandrium.

Balech's [1] criteria, and those of other taxonomists, for separating about 30 Alexandrium species depends on shape and position of the Po plate, shape and position of pores (foramen) in the Po and S.p. plates, shape of S.a., presence and size of the ventral pore, displacement of 1' plate, shape and size of the S.s.a., chain formation, shape and size of 6" plate, shape of cell, and cell dimensions. All these species were originally described in the following genera: Gonyaulax, Goniodoma, and Alexandrium. One new species, A. ibericum, is synonymous with A. minutum [2].

Balech [20] recognized those species of Alexandrium (≥6) with a displaced 1' plate as belonging to a subgenus of Alexandrium, i. e., Gessnerium. Since age of cells (not necessarily size of cell) can influence plate rim growth, the distinction between whether or not the Po and 1' plates are in contact or slightly displaced by an apparent thread (fig. 1) appears to be related to growth parameters and age of cells. Even the amount of theca reticulation and its development can be related to cell age and geographic area (see [2] and [8]).

The orientation of the Po plate and the suture line(s) or direct contact between the Po and 1' plates can be used to separate Alexandrium and its two subgenera, Alexandrium and Gessnerium, from two related genera Goniodoma and Pyrodinium. The latter two genera are thick walled and have prominent cingular lists. In the genus Alexandrium, the Po plate is longitudinally oriented and the relationship between the Po and 1' plates is either direct contact or one straight line from the Po plate to the 1' plate. In Goniodoma, the Po plate is oriented more horizontally and there are two lines oriented at almost right angles to one another. In Pyrodinium, the Po plate is broad ventrally and pointed dorsally, which is the reverse of Alexandrium and Goniodoma, and again there are two lines at almost right angles to one another. Figures 1-3 illustrate the differences between genera.

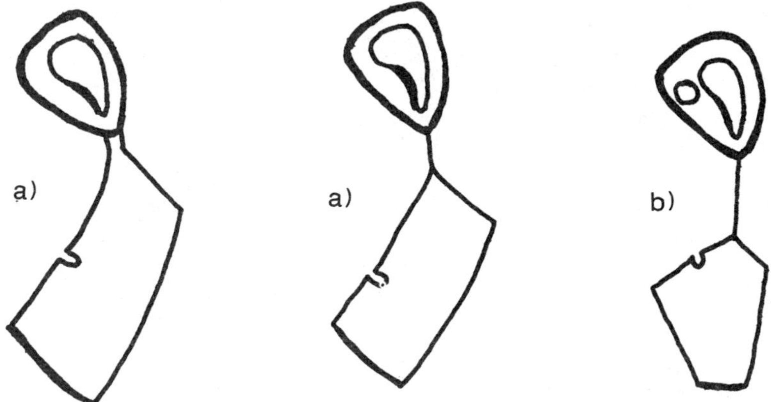

Figure 1. Schematic diagram of Po and 1' relationship of Alexandrium. a) subgenus Alexandrium, b) subgenus Gessnerium.

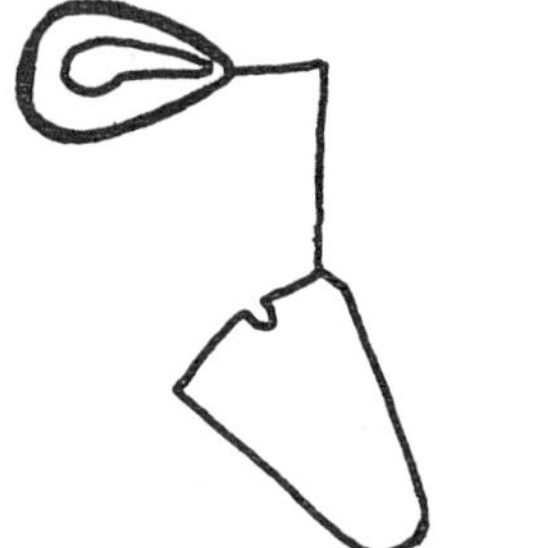

Figure 2. Goniodoma.

Figure 3. Pyrodinium.

The fucoxanthin derivative-containing gymnodinioids.

A portion of the following account for the gymnodinioids is taken from an Appendix in the Proceedings of the Taxonomy and Biology of Dinoflagellates and Other Harmful Phytoplankton Workshop, Sherkin Island Marine Laboratory, County Cork, Ireland, June 9-13, 1989 [21] as written by this author.

"Those photosynthetic gymnodinioids that are associated with toxic effects and have, or are presumed to have, fucoxanthin derivatives rather than peridinin as the dominant xanthophyll are currently in three genera: Gymnodinium, Gyrodinium, and Ptychodiscus. The species include Gymnodinium breve Davis, 1948 (=Ptychodiscus brevis); Gyrodinium aureolum Hulburt, 1957; Gymnodinium galatheanum Braarud, 1957; Gymnodinium veneficum Ballantine, 1956; and possibly, Gymnodinium mikimotoi Mikye and Kominani, 1935 and Spirodinium aureum Conrad, 1926. Another toxic Gymnodinium is G. catenatum, which does possess peridinin.

Gyrodinium aureolum was originally described from Massachusetts, U.S.A., and later identified as a fish-killing species in European waters. Whether the northern European G. aureolum is the species described by Hulburt is still in question, as discussed by Partensky and Sournia (1986). Partensky and Vaulot (1988) have suggested the designation G. cf. aureolum until certain morphological and cytological features are clarified, presumably by comparison with type locality material. Also, it would appear that Gymnodinium nagasakiense and Gyrodinium aureolum may be synonymous, based on morphological similarity. The former was described as having an elongated nucleus in the left half of the cell while the latter was described as having a round centrally-located nucleus. Comparison of several strains of these two species revealed that 1) cingulum displacement varied in each species (17% to 27%) to the extent that individual specimens of either species could be assigned to either Gymnodinium or Gyrodinium, and 2) the average chromosome number for both species was 117 + 3 S.E. (n=5, each strain). The major difference between these clonal strains (both species) was that northern European Gyrodinium aureolum contained 41% more DNA. Whether this difference in DNA content, and different population growth characters warrant separation of the northern European G. cf. aureolum and the Japanese Gymnodinium nagasakiense should await mtDNA, ctDNA, or DNA sequencing analyses and sexual compatability and viability experiments, although several scientists consider the two to be conspecific.

Gymnodinium breve, although similar to Gymnodinium nagasakiense and two other Japanese species, has a ventrally-directed apical carina, thecal ridges in the cingulum, 2-6 large pores in the lateral-dorsal left hypocone, can be preserved in 1% seawater-formalin or 2% glutaraldehyde, retains its cell shape and covering in clorox, and plasmolysis can leave a characteristic amphiesmal cell outline. G. breve, additionally, has two flagellar pores, a ventral ridge or ventral flange on the right sulcal extension on the epicone, an apical groove that extends onto the ventral and dorsal surfaces, and a displaced (<2x) cingulum. The Japanaese G. breve-like strains have distinct sulcal and apical groove junctures that separate them from G. breve; these strains have a short, ventral apical groove and an extended sulcal intrusion. G. nagasakiense and Gyrodinium aureolum also differ from the other species, in their greater cingular displacement, straight edge of the ventral flange, and in the apical groove and sulcal juncture.

In 1979, Steidinger transferred Gymnodinium breve to Ptychodiscus, on the basis of the ventrally-directed carina, thecal ridges in the cingulum, and resistant cell covering. The type species for Gymnodinium, the freshwater species G. fuscum, has membrane fragment templates in the amphiesmal vesicles, while G. breve does not. If amphiesmal structure is to be used for taxonomy on the genus level, then G. breve, and associated fucoxanthin derivative-containing species, belong in another genus. If Ptychodiscus has only one flagellar pore, then clearly G. breve does not belong in Ptychodiscus. The ultrastructure of the amphiesma of the type species of Ptychodiscus (P. noctiluca) is unknown. Since there are species such as Gymnodinium nagasakiense and Gyrodinium aureolum without a pronounced ventrally-directed apical carina and thecal ridges in the cingulum that are similar to G. breve and which contain fucoxanthin derivatives, the use of these features may not be reliable for generic distinction. Until type species of the genera Gymnodinium, Gyrodinium, and Ptychodiscus, as well as the related genera Balechina and Woloszynskia have been reinvestigated in detail, it is recommended to use the name Gymnodinium breve and not to introduce new nomenclatural combinations." [21]

Additional studies following a recent outbreak of Gyrodinium aureolum in Maine [22] may help clarify the taxonomy of the European G. cf. aureolum and the northeast United States G. aureolum, if the bloom recurs. Maine specimens and type locality material, if possible, from a Massachusetts salt pond should be studied as wild populations and cultured clones to assess morphology, cytology, and life history similarities or dissimilarities between G. aureolum, G. cf. aureolum, and Gymnodinium nagasakiense. The same material plus type locality material for the type species of Gymnodinium and Gyrodinium may also clarify the distinction between these two closely related genera and whether the range of cingulum displacement overlaps. Studies on the fucoxanthin derviative-containing gymnodinioids need to resolve 1) the external cell covering or amphiesma for pores, reticulate pattern, or other surface features, 2) contents of amphiesmal vesicles, 3) microtubular cytoskeleton (see [23]), 4) ventral ridge shape and variability, 5) apical groove shape, extent, and variability, 6) anterior sulcal intrusion and variability, 7) number of chromosomes and DNA content, 8) crossfertility of species and strains, and viability of daughter cells, 9) hypnozygotes, and 10) pigment composition.

At the recent Sherkin Island Marine Laboratory workshop [21], participants made the following recommendations to help alleviate some of the problems with taxonomy and systematics of toxic dinoflagellates. 1) Study type material or, if this is not available, collect material from the type locality if possible (e. g., Alexandrium minutum Halim [6], Gyrodinium aureolum Hulburt [24], Gyrodinium spirale (Bergh) Kofoid and Swezy [25], Gymnodinium breve Davis [26], G. fuscum (Ehrenberg) Stein [27], Ptychodiscus noctiluca Stein [28], Balechina pachydermata (Kofoid and Swezy) Loeblich and Loeblich [29], and others). This was done by Balech [2] for Alexandrium. 2) Establish and curate type specimen collections using prepared slides, photomicrographs, video tapes, living cultures, preserved specimens, and sediment samples for cysts. The collection should be in central repositories with ready access to scientists. 3) Incorporate cyst descriptions in species descriptions. 4) Incorporate biochemical characters (e. g., pigments, sterols, DNA, etc.) in species descriptions or characterizations. 5) Determine range of variability of "conservative" characters in vegetative cells. 6) Choose one code of nomenclature or establish a new code for unicells or flagellates. 7) Conduct mating experiments between morphospecies and determine interfertility and viability.

REFERENCES

1. E. Balech in: Toxic Dinoflagellates, D. Anderson, A. White, and D. Baden, eds., Elsevier, New York, pp. 33-38, (1985).
2. E. Balech, Phycologia 28(2):206-211, (1989).
3. K. A. Steidinger, Phycologia 10(2/3):183-187, (1971).
4. F. J. R. Taylor, Environ. Lett. 9:103-119, (1975).
5. F. J. R. Taylor in: Toxic Dinoflagellate Blooms, D. L. Taylor and H. H. Seliger, eds., Elsevier/North Holland, pp. 47-56, (1979).
6. Y. Halim, Vie et Milieu 11(1):102-105, (1960).
7. G. M. Hallegraeff, D. A. Steffensen, and R. Wetherbee, J. Plankton Res. 10(3):533-541, (1988).
8. M. Montresor, D. Marino, A. Zingone, and G. Dafnis in: 4th International Conference on Toxic Marine Phytoplankton E. Graneli et al., eds., (1989), in press.
9. M. V. Lebour, Mar. Biol. Assoc. U. K., Plymouth, (1925).
10. W. F. Whedon and C. A. Kofoid, Univ. Calif. Publ. Zool. 41:25-34, (1936).
11. T. Braarud, Avdh.ut.ov. Det. Norske Viden. Akad. Oslo. I Mat. Naturv. Klasse, 11:1-18, Pl. 1-4, (1945).
12. E. Balech and K. Tangen, Sarsia 70:333-343, (1985).
13. J. Woloszynska and W. Conrad, Bull. Mus. Hist. Nat. Belgique, 15(46); 1-5, (1939).
14. J. F. Howell, Trans Amer. Micros. Soc. 82(2):153-156, (1953).
15. E. F. J. Wood, Austr. J. Mar. Freshwat. Res. 5(2):171-351, (1954).
16. E. Balech, Bol. Inst. Biol. Mar. M. del Plata 4:1-49, (1964).
17. Y. Halim, Int. Rev. Ges. Hydrobiol. 52(5):701-755, (1967).
18. F. J. R. Taylor, Bibl. Bot. 132:1-234, (1976).
19. A. R. Loeblich III and L. A. Loeblich in: Toxic Dinoflagellate Blooms, D. L. Taylor and H. H. Seliger, eds, Elsevier/North Holland, pp. 147-156, (1979).
20. E. Balech, (1989), in prep.
21. Proceedings of Sherkin Island Workshop, (1989), in press.
22. S. Shumway, personal communication.
23. K. R. Roberts, M. A. Farmer, R. M. Schneider, and J. E. Lemoine, J. Phycol. 24:544-553, (1988).
24. E. M. Hulburt, Biol. Bull. 112(2):196-219, (1957).
25. C. A. Kofoid and O. Swezy, Mem. Univ. Calif. 5:viii + 538 pp., Pl. 1-12, (1921).
26. C. C. Davis, Bot. Gaz. 109(3):358-360, (1948).
27. F. R. von Stein, Wilhelm Engelmann, Leipzig, x + 154 pp., Pl. 1-24, (1878).
28. F. R. von Stein, Wilhelm Engelmann, Leipzig, 30 pp., Pl. 1-25, (1883).
29. A. R. Loeblich, Jr. and A. R. Loeblich III, J. Paleontol. 42(1):210-213, (1968).

NOVEL "BROWN TIDE" BLOOMS IN LONG ISLAND EMBAYMENTS: A SEARCH FOR THE CAUSES

ELIZABETH M. COSPER, CINDY LEE AND EDWARD J. CARPENTER

Marine Sciences Research Center, State University of New York
Stony Brook, New York 11794 USA

ABSTRACT

Unusual blooms of a previously unidentified chrysophyte, *Aureococcus anophagefferens*, have occurred in several coastal embayments along the northeast coast of the USA. The monospecific blooms were termed the "brown tide" due to the resulting water color. The first appearance of the "brown tide" occurred early in the summer of 1985 over a wide geographic range in non-contiguous bodies of water. The blooms did not return to some bays but, on Long Island, New York they recurred during the summer of 1986 and in diminishing densities during the summers of 1987 and 1988. Historically, a diverse group of small microalgal species dominate the phytoplankton biomass and productivity in Long Island bays during the summer. The continued dominance through several months at high cell densities (> 10^9 cells l^{-1}) of *A. anophagefferens* was the distinctive feature of consequence during these blooms.

Environmental variables which may be contributory to the occurrence of the "brown tide" include elevated salinities due to drought conditions, pulses of rainfall delivering organic and/or micronutrients to bay waters, reduced grazing and restricted flushing of bays. The "brown tide" species appears to be closely related to an open ocean chrysophyte, *Pelagococcus subviridis*, and possibly was seeded into northeast coastal bays from offshore when conditions during 1985 were particularly favorable for its growth. The ability of this species to maintain at least minimal populations during the winter months seems to allow for its recurrence during subsequent summers. Culture studies have shown that this species has growth requirements for trace elements, chelators and organic nutrients some of which are different from many common estuarine and coastal phytoplankton species. The competitive advantage of *A. anophagefferens* over other potentially co-occurring species probably relates to its heterotrophic and photoadaptive capabilities.

BLOOM OCCURRENCE AND DISTRIBUTION

Several coastal embayments along the northeast coast of the USA have recently experienced novel microalgal blooms for which there is no previous record. These monospecific blooms were popularly called the "brown tide" due to the resulting water color of golden-brown [1]. In the early summer of 1985 the first appearance of the "brown tide" occurred over a wide geographic range along the coast in non-contiguous bodies of water: Narragansett Bay, Rhode Island, and Long Island embayments in New York as well as Barnegat Bay in New Jersey (Fig. 1) [1,2,3,4,5,6]. The blooms were restricted to these coastal bay systems; blooms did not appear to follow a pattern of

Toxic Marine Phytoplankton
Edna Graneli et al., Editors

spreading from one bay system to the next. This suggests that the environmental factors contributing to these "brown tide" blooms were not just local conditions in a bay system but probably were more regional, e.g. involving meteorologically induced changes. The blooms on Long Island markedly reduced the extent of eelgrass (Zostera marina) beds because of increased light attenuation, and decimated populations of commercially valuable bay scallops (Argopecten irradians irradians) which were unable to graze adequately and starved to death [1,7,8]. Similarly in Narragansett Bay the mussels were unable to feed and populations were severely reduced [9].

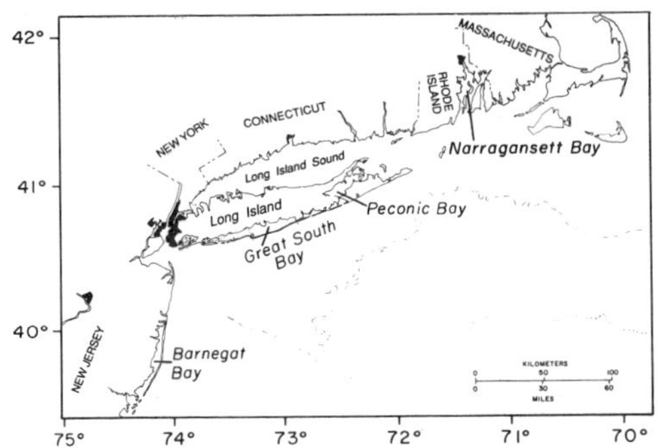

FIG. 1. Regional chart of northeast coastal bays affected by "brown tide".

In 1986 the blooms recurred throughout the summer months in the same Long Island embayments as previously. In Long Island and in Barnegat Bay, N.J. during the summers of 1987 and 1988 the "brown tide" blooms returned only in diminishing levels (Fig. 2) [10,2,3]. Since 1985 "brown tide" blooms have not returned to Narragansett Bay [4,6]. This paper documents and evaluates the findings to date of many scientific efforts to find the causes and factors which contribute to these unusual blooms. Details concerning any particular experimental methodology can be found in the cited references.

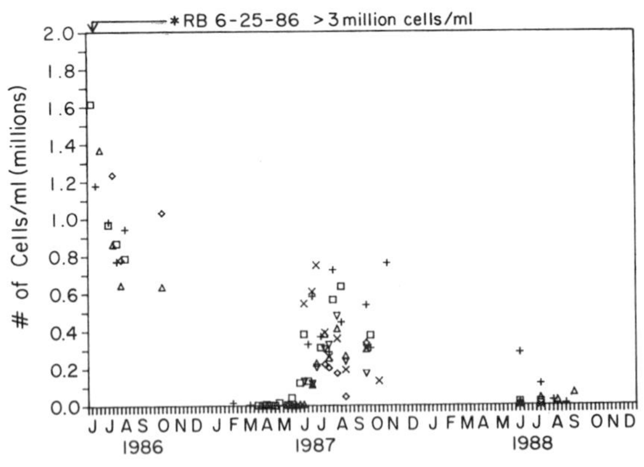

FIG. 2. Brown tide cell concentrations for sites sampled in Peconic and Great South Bays during 1986, 1987 & 1988.

BLOOM DYNAMICS ON LONG ISLAND

The "brown tide" species was dominant in terms of cell number and contributed more than 80% of total cellular phytoplankton volume throughout most of the bloom period during the summer months [1,10]. During the blooms phytoplankton biomass, as indicated by chlorophyll a levels, was not particularly elevated for Long Island bays, in comparison to other years, since concentrations reached levels less than 30 μg l^{-1} (Fig. 3 A), but chlorophyll a was concentrated in the smaller (less than 5 um) fraction (Fig. 3 B) [11,12,1,10].

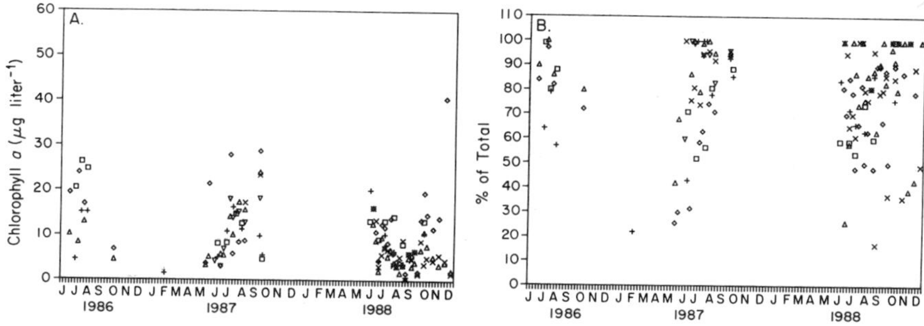

FIG. 3. (A) Chlorophyll a concentration in the < 5 μm fraction (B) % of total chlorophyll a in the < 5 μm fraction for sites sampled during 1986, 1987 & 1988.

Primary productivity levels were high but also were not different from pre-bloom years. The less than 10 μm fraction of the phytoplankton contributed more than 90% of the total photosynthetic activity throughout the bloom period; estimates of picoplankton carbon turnover were rapid, of the order of hours (10). Changes in inorganic nutrient levels, (nitrate, nitrite, phosphate and ammonium), again were not different from pre-bloom years [13,11,12]. Variations in inorganic macro- nutrients were not correlated with variations in the productivity of the "brown tide" (Fig. 4 A & B) and there is no evidence to support increased macro- nutrient loading as a cause of the blooms [10,14]. These findings are consistent with similar studies in Narragansett Bay, Rhode Island [6].

The continued dominance over extensive periods of time (~ 6 months) of a single, particularly small algal species was the distinctive feature during these blooms [13,11,12,10,14]. When rapid rates of growth were inferred from the rapid turnover of photic zone phytoplankton carbon or calculated using the frequency of dividing cell method, changes in phytoplankton biomass did not reflect such potential increases in populations [10]. Neither sinking of this small microalga nor flushing of the bays could account for such constant population densities, since daily sinking rates of such a small microalga would be small [15], and flushing of the bays is of the order of weeks [13,12]. Grazing by protozoans on the "brown tide" was evaluated both in culture experiments and during the summer of 1988 [16] and found to be a potential controlling factor of population densities.

Small microalgae known as "small forms" generally dominate the phytoplankton biomass and productivity in Long Island embayments such as Great South Bay [12] and Peconic Bay [11] during the summer. Usually the "small forms" are composed of many species of diatoms, chlorophytes, cryptomonads and small flagellates which vary in percent contribution to the total phytoplankton through the summer months. Similar to the "brown tide", dense blooms of two minute chlorophyte species occurred during the mid 1950's and were termed "small forms" and also affected oyster populations in Great South Bay [17,18].

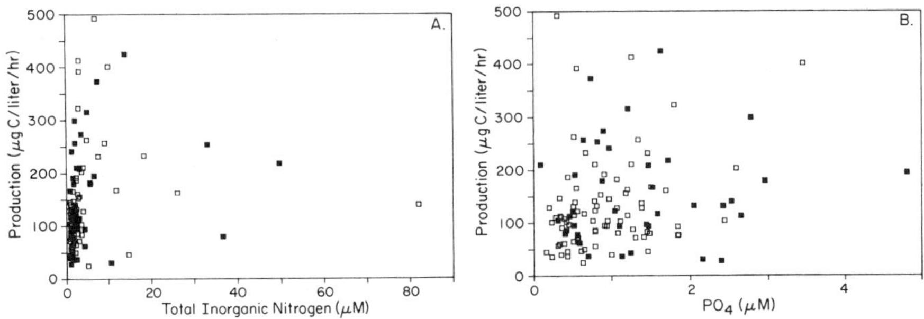

FIG. 4. Inorganic nutrients as μM of (A) total dissolved nitrogen (nitrate+nitrite, ammonium) and (B) orthophosphate (PO_4) versus primary productivity for sites in Peconic and Great South Bays from 1986 & 1987 (solid squares) & 1988 (open squares); r = 0.18 and 0.21, respectively.

BLOOM SPECIES

The "brown tide" microalga had not been previously identified and was recently named, Aureococcus anophagefferens [4]. A. anophagefferens was isolated into culture during the summer of 1986 [19] and again during the winter of 1987. The use of immunofluorescent detection techniques for A. anophagefferens [20] have already indicated the presence of A. anophagefferens in northeast coastal waters unaffected by the "brown tide" blooms. This implies that this species has been generally present in coastal embayments and that the blooms were caused by a unique set of environmental events in particular bay systems [14]. In contrast, however, other immunochemical analyses comparing A. anophagefferens to other ultraplankton [21] have, along with electron microscope analyses [4,5] and pigment analyses [22], demonstrated a close affinity between A. anophagefferens and a ubiquitous, open ocean microalga, Pelagococcus subviridis, isolated and identified from many areas in the Pacific Ocean [23,24] as well as from Norwegian waters [25]. The possibility that A. anophagefferens is an expatriate species only recently introduced into nearshore areas from offshore waters is an important issue which needs to be elucidated.

ENVIRONMENTAL CONDITIONS CONTRIBUTORY TO THE BLOOMS

Physical and Chemical Factors

The first year of the bloom, 1985, coincided with the start of a drought, and rainfall was the third lowest annual level in the last 37 years; the drought continued for several years thereafter (Fig. 5 A) [1]. The temporal pattern of rainfall appeared to be important to the formation of the blooms. Low rainfall during the winter and spring months of 1985 and 1986 were followed by abnormally large pulses of rain and potential freshwater inputs into the bays through runoff and/or ground water flow during the early summer, which coincided with the initiation of the "brown tide" blooms (Fig. 5 B). This pattern was not as definitive in 1987, and the blooms did not recur until later during the summer and never

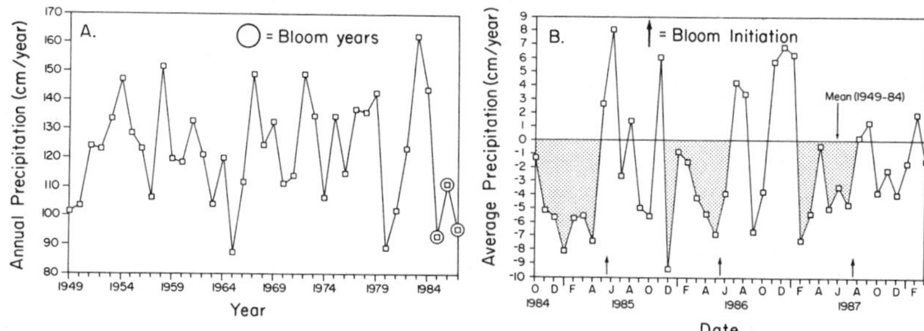

FIG. 5. (A) Total monthly precipitation for Brookhaven National Lab, Upton, Long Island, NY from 1949 to 1987, (B) Monthly precipitation from Oct. 1984 to March 1988 as deviations from the mean for each month from 1949 to 1984.

to the same degree, since densities reached only 10^8 cells l^{-1}. The drought led to elevated salinities in the bays; measurements during the summers since 1986 have been generally closer to 30 $^{o}/oo$ (Fig. 6 A), while previously they were typically about 25 $^{o}/oo$ [1,12]. During the summer of 1987, when salinities were initially lower, the "brown tide" first appeared at sites where salinities were still relatively high. As salinities increased during the summer and into the fall months, again the "brown tide" regained dominance. Thus, elevated salinities appear to be a significant contributing factor to the initial blooming of the "brown tide" species, A. anophagefferens. It is unclear, however, why the "brown tide" did not return to any significant extent during the summer of 1988, since salinities were again generally above 25 $^{o}/oo$. Possibly grazer control [16] had become effective in preventing the initiation of blooms. Temperature, in contrast, varied in a similar pattern in 1986, 1987 and 1988 (Fig. 6 B) and does not appear to be different from previous summers [11,12] nor to be contributory to the "brown tide".

Laboratory studies with cultures of <u>A. anophagefferens</u> adapted to growth at 30 ^{o}/oo indicate a severe reduction in

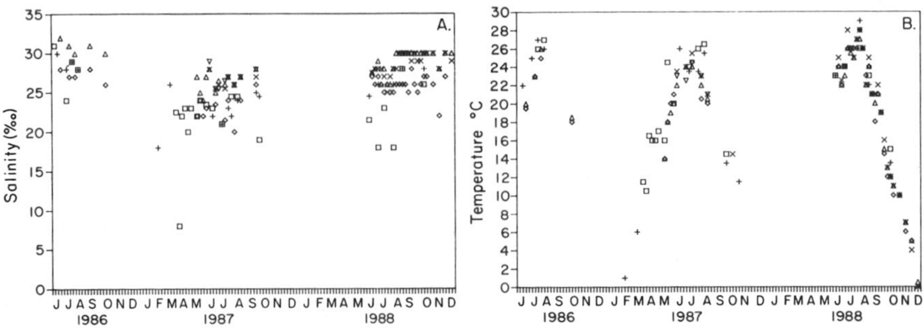

FIG. 6. (A) Salinity and (B) Temperature at stations sampled in Peconic and Great South Bays during 1986, 1987 & 1988.

growth rate below 28 ^{o}/oo as compared to good growth at 30 ^{o}/oo in standard Enriched Instant Ocean (EIO) f/2 media with inorganic phosphate (Pi) at f/20 (3.33 µM) [26] (Fig. 7). Growth in EIO f/2 with Pi at f/2 (33.3 µM) was never obtained at any salinity and is discussed in a later section. In the presence of glycerophosphate substituted for inorganic phosphate at f/2 and f/20 concentrations (f/2 GP, f/20 GP), growth of this alga was enhanced substantially at lower salinities (Fig. 7) when compared to growth in totally inorganic media [14]. Experiments to evaluate the ability of <u>A. anophagefferens</u> to grow at higher salinities representative of neritic, 32 ^{o}/oo, and oceanic, 35 ^{o}/oo, waters, (Fig.7) showed that this species could grow well at these higher salinities particularly in media with inorganic phosphate. Salinity is important to the growth of <u>A. anophagefferens</u> in Long Island Bays, but it must be evaluated in light of the presence of organic nutrients.

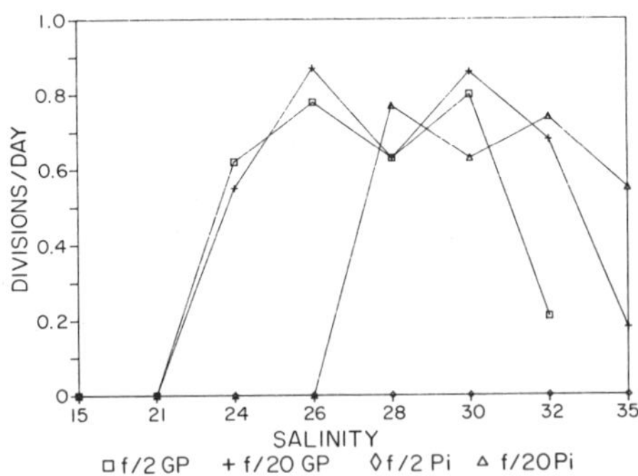

FIG. 7. Growth rate of <u>A. anophagefferens</u> versus salinity in EIO media with glycerophosphate, GP or inorganic phosphate, Pi.

Analysis of field samples obtained from helicopter flights throughout the Peconic Bays and Great South Bay on August 12, 1986 showed the reverse of what would be predicted from laboratory analyses and rainfall data; the greatest cellular concentrations were at 26 ^{o}/oo and the lowest at salinities above 30 ^{o}/oo [14]. The high inverse correlation (r= -0.94) between salinity and "brown tide" cell concentration found during August, 1986 was not found during June and July, 1987 (r= -0.64 and r= -0.46, respectively) when salinities were reduced relative to the previous summer, although the inverse trend still prevailed [14]. During the fall, on the October 1 trip, when the "brown tide" was reestablished and dominant, even if at reduced concentrations, the high inverse correlation (r= - 0.96) was again found [14].

Correlation analyses were performed between salinity and inorganic nutrient levels along transects sampled [14]. Nutrient distributions (when significant) were positively correlated with salinity, indicating higher levels in more offshore areas or in areas where coastal waters had mixed into the bays. The distribution of "brown tide" cells relative to salinity cannot simply be explained by variations in inorganic macro-nutrient concentrations. In addition, no correlation between primary productivity of "brown tide" bloom waters and inorganic macro-nutrient levels can be demonstrated (Fig. 4 A & B) [10].

An evaluation of temperature indicated that the best rate of growth was at 20 and 25 ^{o}C both for the summer and winter clonal isolates indicating that _A. anophagefferens_ is a warm water species consistent with its formation of summer blooms [14]. Growth, however, over a wide temperature range can be obtained, if given enough time to adapt, so that even at 5 ^{o}C doubling times of 10 days are realized. Such growth rates at low temperatures would be enough to maintain populations during the winter months in the minimally flushed bays on Long Island [14,2] and could allow populations to reseed the waters the following spring and summer.

Comparative growth experiments

Since the addition of organic phosphorus compounds such as glycerophosphate and fructose-1,6-diphosphate [29] to artificial seawater media affected the growth of _A. anophagefferens_ relative to growth on inorganic phosphorus (Fig. 7) [14], consideration was given to the idea that the alga may benefit from these compounds serving as chelating agents [27,28]. If so, substitute chelators might have the same effect as the addition of organic phosphate. The addition of nitrilotriacetic acid (NTA) and citric acid (CA), chelators commonly used in culture media [26], resulted in growth on inorganic phosphate (Pi f/2) which was higher than that on EDTA (Fig. 8)[29]. Thus, NTA and CA may be better chelators than EDTA for growth factors important to _A. anophagefferens_.

Brand [30] hypothesized that high phosphate concentrations could inhibit uptake of arsenate and vanadate from solution, particularly for open ocean phytoplankters. When 10 nM additions of arsenate and vanadate were added to the EIO medium, a significant but small increase of growth in the presence of EDTA occurred with each of these trace elements [29]. Selenium was also added to media based on observations made by Keller and Guillard [31], that some species of oceanic plankton require selenium for growth. Additional selenium as selenite did slightly enhance growth of _A. anophagefferens_ with EDTA but not with CA or NTA [29].

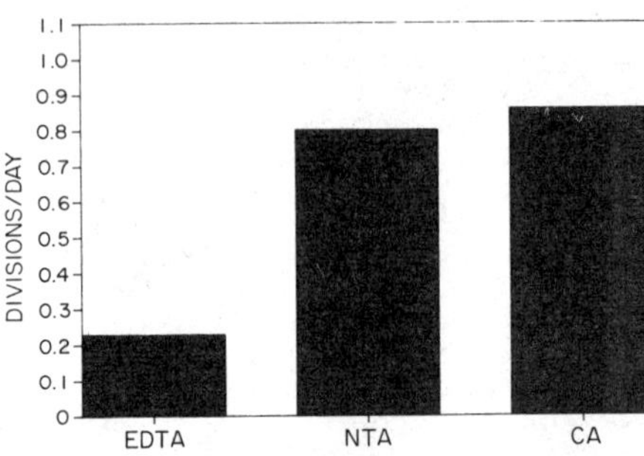

FIG. 8. Growth of A. anophagefferens as a function of chelator, EDTA, NTA and citric acid, CA; all chelators at f concenration.

NTA and citrate have, in part, replaced phosphates in commercial soaps and detergents in an effort to slow eutrophication of coastal bays. Ryther and Dunstan [32] suggested that this type of replacement might actually contribute to eutrophication processes. Once in seawater, NTA may undergo biological degradation to glycine or glycolic acid [33], two compounds which may be used directly by algae [34]. If anthropogenic sources of NTA are reaching the bays, organisms such as A. anophagefferens may use it or its breakdown products as chelators or as an energy or carbon source to enhance growth.

Heterotrophic uptake experiments

A. anophagefferens can grow on urea as the sole nitrogen source at rates comparable to growth on nitrate; growth on ammonium was obtained only after a long adaptive period (Fig.9). Urea can contribute substantially to the total pool of available nitrogen in Long Island bays [12]. Growth on

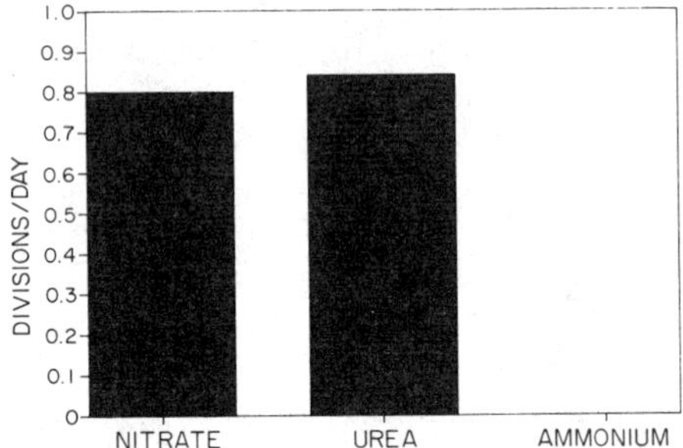

FIG. 9. Growth of A. anophagefferens on different nitrogen sources, nitrate, urea, ammonium, in EIO-GP media.

glutamic acid as the sole nitrogen source has also been obtained [29], and A. anophagefferens has shown significantly higher uptake rate constants per unit cell volume for glutamic acid than other potentially co-occuring species tested (Fig. 10 A) [29]. Similarly, glucose uptake rate constants on a per unit cell volume basis showed that A. anophagefferens would have a significant uptake advantage over Nannochloris sp. and Minutocellus polymorphus (Fig. 10 B) [29]. The ability of A. anophagefferens to effectively compete for and utilize organics for growth in laboratory experiments may potentially enhance its ability to maintain blooms in Long Island coastal bays.

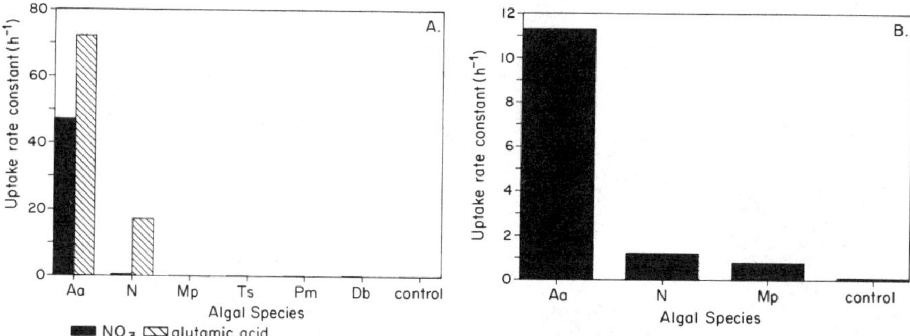

FIG. 10. (A) Glutamic acid (glu) uptake and (B) glucose uptake rate constants per unit cell volume for algae grown with f/2 nitrate, NO_3 or 10 µM glu as sole nitrogen source; species A. anophagefferens (Aa), Nannochloris sp. (N), Minutocellus polymorphus (Mp), Thalassiosira pseudonana (Ts), Prorocentrum minimum (Pm) and Ditylum brightwellii Db).

Biological Interactions

It does not appear that A. anophagefferens excretes compounds that inhibit the growth of other phytoplankton commonly found in Long Island embayments. Filtrates of media used to grow the "brown tide" alga, obtained from all stages of the growth cycle of the "brown tide" (early, mid, and late exponential phase), were added at concentrations from 0.1% to 100% to fresh enriched media. For all four species, Thalassiosira pseudonana (3H), Prorocentrum minimum, Ditylum brightwellii, Nannochloris sp. (WNB 7/22), tested, the growth was either enhanced or there was little effect [14]. At only 10% of its own filtrate taken from a senescent culture, A. anophagefferens was growth inhibited.

CONCLUSIONS

"Brown tide" blooms may have resulted from several factors 1) higher than average salinities in the bays during the early summers of 1985 and 1986, 2) freshwater runoff or groundwater inputs of organic and possibly certain inorganic micro- nutrients, compounds which may be essential to the rapid growth of the A. anophagefferens, 3) restricted flushing by coastal waters of the Long Island bays which results in long residence times for water on the order of weeks [13,34], allowing for the retention and maintenance of large populations of "brown tide" cells within these embayments (Fig. 11).

Further reduction in flushing of the bays by coastal waters because of changes in sea level [35] could have also been contributory. Grazing pressure during the early stages of bloom initiation was possibly also reduced [36,16].

The "brown tide" bloom scenario has some similarity to the "green tide" blooms of the 1950's in Great South Bay [17,18]. During the early fifties a lowering of salinity selected for two estuarine species, Nannochloris sp. and

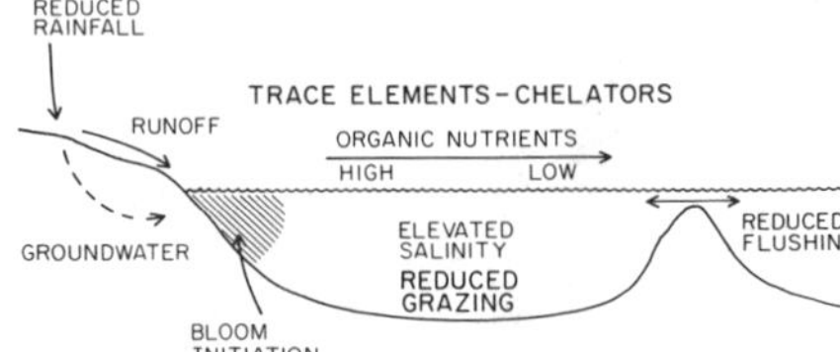

FIG. 11. Hypothetical model depicting the conditions conducive to initiation of "brown tide" blooms.

Stichococcus sp., with a salinity of 17 $^{o}/oo$ optimal for growth. The recurrence of the "green tides" for several summers afterwards appeared to depend on the restricted circulation of the inshore bays and the overwintering of large enough seed populations to initiate the next summer's growth. Effluents from duck farms, which flowed into Great South Bay through creeks, were found to be supplying nitrogeneous nutrients and promoting the growth of these two species of microalgae and these effluents were subsequently restricted [17,18].

The ability of A. anophagefferens to outcompete all other phytoplankton species and maintain dominance throughout the summer possibly relates to its specific micro-nutrient needs, heterotrophic capabilities and photoadaptive capabilities. The photosynthetic characteristics of A. anophagefferens relative to other species under severe light limitation [1] might be particularly important since recent evidence indicates that light absorption characteristics and pigments of this species are more characteristic of a deep- dwelling oceanic species than a coastal form [37,22]. The ability of A. anophagefferens to dominate the phytoplankton community over extensive periods of time during the summer months does not appear to relate to the excretion of substances which could inhibit the growth of other potentially co- occurring species. Laboratory experiments, however, evaluating such allelopathic interactions between species must be evaluated cautiously [38,39].

Other unusual phytoplankton blooms both past and present have resulted from subtle, but long term anthropogenic eutrophication combined with altered environmental conditions [14]; examples of such are: the sewering of Providence, R.I. and the red tide of 1898 [40], duck farm effluents and the green tides of the 1950's [17,18], nutrient loading and acid rain leachates in Scandinavian waters associated with red tides and unusual Chrysochromulina blooms of 1988 [41,42,43], and unusual Gulf Stream meanders seeding Ptychodiscus brevis into enriched coastal waters off North Carolina [44].

Since A. anophagefferens is a species not previously known to cause blooms, environmental conditions contributory to the blooming could in part relate to new anthropogenic influences in these bays such as different chelators in detergents [29] or new lawn treatments [45]. Drought conditions, elevated salinities, pulses of rain delivering specific nutrients to the bay waters, along with restricted flushing of bay waters [35] set the scenario for the formation of a phytoplankton bloom [46]. The selection of the particular "brown tide" species would then have related both to specific chemical conditions and any selective grazing pressures [16,36] extant during the early summer of 1985.

ACKNOWLEDGMENTS

Research supported by Suffolk County, New York; The Living Marine Resources Institute of the State Of New York; and United States Environmental Protection Agency. We thank Saou-Lien Wong, Allen Milligan, Kristen Drewes, Valerie Philbrick and Jim Carretta for excellent technical assistance. This research represents in part the Master theses of Susan Dzurica and Mathew Cottrell. Support for C. Lee and for HPLC analyses was also provided by the Office of Naval Research. Special thanks go to the New York Air National Guard, 106 ARRG, Westhampton Beach, Long Island for making the aerial surveys possible. Contribution No. 654 of the Marine Sciences Research Center of the State University of New York.

REFERENCES

1. E.M. Cosper, W.C. Dennison, E.J. Carpenter, V.M. Bricelj, J.G. Mitchell, S.H. Kuenstner, D.C. Colflesh, and M. Dewey, Estuaries 10, 284-290 (1987).
2. R. Nuzzi, and R.M. Waters, in: Novel Phytoplankton Blooms: Causes and Impacts of Recurrent Brown Tides and Other Unusual Blooms, E.M. Cosper, E.J. Carpenter and V.M. Bricelj, eds. (Springer-Verlag, Berlin 1989).
3. P. Olsen, ibid.
4. J.McN. Sieburth, P.W. Johnson and P.E. Hargraves, J. Phycol. 24, 416-425 (1988).
5. J.McN. Sieburth, and P.W. Johnson, in: Novel Phytoplankton Blooms: Causes and Impacts of Recurrent Brown Tides and Other Unusual Blooms, E.M. Cosper, E.J. Carpenter and V.M. Bricelj, eds. (Springer-Verlag, Berlin 1989).
6. T.J. Smayda,and T.A. Villareal, ibid.
7. M. Bricelj, and S. Kuenstner, ibid.
8. W.C. Dennison, G.J. Marshall and C. Wigand, ibid.
9. G.A. Tracey, R.L. Steele, J. Gatzke, D.K. Phelps, R. Nuzzi, M. Waters and D.M. Anderson, ibid.
10. E.M. Cosper, E.J. Carpenter and M. Cottrell, ibid.
11. S.F. Bruno, R.D. Staker, G.M. Sharma, and J.T. Turner, Estuaries 6, 200-211 (1983).
12. J.S. Lively, Z. Kaufman, and E.J. Carpenter, Est. Coast. Shelf Sci. 16, 51-68 (1983).
13. C.D. Hardy, Marine Sciences Research Center Special Report, 3, State University of New York, Stony Brook, N. Y. (1976).
14. E.M. Cosper, W. Dennison, A. Milligan, E.J. Carpenter, C. Lee, J. Holzapfel and L. Milanese, in: Novel Phytoplankton Blooms: Causes and Impacts of Recurrent Brown Tides and Other Unusual Blooms, E.M. Cosper, E.J. Carpenter and V.M. Bricelj, eds. (Springer-Verlag, Berlin 1989).
15. T.J. Smayda, Marine Geol. 11, 105-122 (1971).
16. D.A. Caron, E.L. Lim, H. Kunze, E.M. Cosper and D.M. Anderson, in: Novel Phytoplankton Blooms: Causes and Impacts of Recurrent Brown Tides and Other Unusual Blooms, E.M. Cosper, E.J. Carpenter and V.M. Bricelj, eds. (Springer-Verlag, Berlin 1989).
17. J.H. Ryther, Biol. Bull. 106, 198-209 (1954).
18. J.H. Ryther, in: Novel Phytoplankton Blooms: Causes and Impacts of Recurrent Brown Tides and Other Unusual Blooms,

E.M. Cosper, E.J. Carpenter and V.M. Bricelj, eds. (Springer-Verlag, Berlin 1989).
19. E.M. Cosper, Appl. Phycol. Forum 4, 3-5 (1987).
20. D.M. Anderson, D.M. Kulis and E.M. Cosper, in: Novel Phytoplankton Blooms: Causes and Impacts of Recurrent Brown Tides and Other Unusual Blooms, E.M. Cosper, E.J. Carpenter and V.M. Bricelj, eds. (Springer-Verlag, Berlin 1989).
21. L. Campbell, L.P. Shapiro, E.M. Haugen and L. Morris, ibid.
22. R.R. Bidigare, ibid.
23. J. Lewin, R.E. Norris, S.W. Jeffrey and B.E. Pearson, J. Phycol. 13, 259- 266 (1977).
24. M. Vesk, and S.W. Jeffrey, J. Phycol. 23, 322-336 (1987).
25. J. Throndsen, and S. Kristiansen, Abstrs, second International Phycol. Congr., Copenhagen p. 160 (1985).
26. J. McLachlan, in: Handbook of Phycological Methods, Culture Methods and Growth Measurements, J.R. Stein, ed. (Cambridge University press, Cambridge 1973) pp. 25-51.
27. L.G. Sillen, and A.E. Martell, Stability Constants of Metal- Ion Complexes (Burlington House. London 1964).
28. F.M.M. Morel, Principles of Aquatic Chemistry (John Wiley & Sons. New York 1983).
29. S. Dzurica, C. Lee, E.M. Cosper and E.J. Carpenter, in: Novel Phytoplankton Blooms: Causes and Impacts of Recurrent Brown Tides and Other Unusual Blooms, E.M. Cosper, E.J. Carpenter and V.M. Bricelj, eds. (Springer-Verlag, Berlin 1989).
30. L.E. Brand, in: Photosynthetic Picoplankton, T. Platt and W.K. Li, eds., Can. Bull. Fish. Aquat. Sci. 214, 205-233 (1986).
31. M.D. Keller, and R.R.L. Guillard, in: Toxic Dinoflagellates, Anderson, White and Baden, eds (Elsevier, Amsterdam 1985) pp.113-116.
32. J.H. Ryther, and W.M. Dunstan, Science 171, 1008-1013 (1971).
33. A.V. Palumbo, F.K. Pfaender and H.W. Paerl, Environ. Toxicol. Chem. 7, 573-585 (1988).
34. D.W. Pritchard, and E. Gomez-Reyes, Marine Sciences Research Center, SUNY, Stony Brook, N.Y., Special Report No. 70 (1986).
35. M. Vieira, in: Novel Phytoplankton Blooms: Causes and Impacts of Recurrent Brown Tides and Other Unusual Blooms, E.M. Cosper, E.J. Carpenter and V.M. Bricelj, eds. (Springer-Verlag, Berlin 1989).
36. L.E. Duguay, D.M. Monteleone and C. Quaglietta, ibid.
37. C.S. Yentsch, D.A. Phinney and L. Shapiro, ibid.
38. P.E. Hargraves, R.D. Vaillancourt and G.A. Jolly, ibid.
39. S.Y. Maestrini, and D.J. Bonin, Can. Bull. Fish. Aquat. Sci. 210, 323-338 (1981).
40. S.W. Nixon, in: Novel Phytoplankton Blooms: Causes and Impacts of Recurrent Brown Tides and Other Unusual Blooms, E.M. Cosper, E.J. Carpenter and V.M. Bricelj, eds. (Springer-Verlag, Berlin 1989).
41. E. Graneli, P. Olsson, P. Carlsson, B. Sundstrom and O. Lindahl, ibid.
42. E. Dahl, O. Lindahl, E. Paasche and J. Throndsen, ibid.
43. T.J. Smayda, ibid.
44. P.A. Tester, P.K. Fowler and J.T. Turner, ibid.
45. J.McN. Sieburth, ibid.
46. L.B. Slobodkin, ibid.

NOVEL AND NUISANCE PHYTOPLANKTON BLOOMS IN THE SEA: EVIDENCE FOR A GLOBAL EPIDEMIC

THEODORE J. SMAYDA*
***Graduate School of Oceanography, University of Rhode Island, Kingston, Rhode Island 02881**

ABSTRACT

Evidence is presented from the Baltic Sea, Kattegat, Skagerrak, Dutch Wadden Sea, North Sea, Black Sea, Tolo Harbour (Hong Kong) and Seto Inland Sea that a long-term trend in increased frequency and dynamics of novel phytoplankton blooms of indigenous species, both benign and harmful ones, has accompanied nutrient enrichment of coastal waters and inland seas on a global scale. Anthropogenic enrichment of N and P has led to long-term declines in the Si:N and Si:P ratios. It is suggested that the decline in Si:P ratio has particularly favored non-diatom blooms in response to nutrification, and such ionic ratio regulation within nutrification is a key factor associated with the global epidemic of novel phytoplankton blooms and phylogenetic shifts in biomass predominance in the sea. A feature independent of the global bloom epidemic is the apparent global spreading of certain species, some of which exhibit blooms. Local stimulation of the indigenous "hidden flora" to detectable levels may also be increasing.

INTRODUCTION

A fundamental cellular characteristic of the micro-algae is their vegetative growth cycle: cell division occurs through binary fission. This has led to a fundamental ecological characteristic of the phytoplankton: population growth expressed as cell numbers follows the geometric progression 2^n, where n represents the number of cell divisions. Increased nutrients in the nutritionally-dilute sea; changed water quality; modified grazing pressure, and/or improved growth conditions generally can stimulate growth leading to blooms. Blooms, in fact, are not only a common characteristic of the phytoplankton at the cellular, population and community levels, but an ancient trait (phytoplankton date back at least 3 billion years) as well, upon which marine food-web dynamics and energy-transfer linkages have evolved. An unfortunate conceptual and methodological approach traditionally applied to phytoplankton bloom events has been to restrict the term bloom to an outburst of growth which results in high biomass, such as the spring bloom [1]. Tett [2] suggests a bloom occurs when the sea surface is discolored, or when chlorophyll concentrations exceed about 10 mg m^{-3}. This definition is too restrictive. Blooms of individual species are continuous in the sea; this drives succession [3]. One or more species is always in bloom at any given time in any given watermass, region or season. For some species, maximal abundance during its bloom may be only 10^4 cells $liter^{-1}$ or less; for others, 10^7 cells $liter^{-1}$. The local carrying capacity, intrinsic cellular properties and competition coefficients determine the magnitude of each species' bloom before its demise. The dual question before us is whether a global increase in *novel* phytoplankton blooms is in progress and, if so, whether this has an anthropogenic linkage. Novel blooms refer collectively to bloom events considered to be unusual, anomalous, exceptional, increasingly harmful, toxic or a nuisance relative to the long-term bloom patterns, types and frequencies established regionally and for a given location. Evaluation of natural phenomena to assess the occurrence of long-term trends and patterns is fraught with difficulty. Inadequate regional data bases, different methodologies, inadequate time-series, *etc.* compromise such efforts. I must therefore emphasize that the present contribution is to be considered as the preliminary findings of an ongoing analysis. Two convergent global aspects will be evaluated: the species-spreading phenomenon and frequency of novel bloom occurrence. (Text figures referenced are located at end of this article.)

Toxic Marine Phytoplankton
Edna Graneli et al., Editors

EVIDENCE FOR A GLOBAL SPREADING OF BLOOM SPECIES

The first report of the dinoflagellate *Gyrodinium aureolum* in European waters, a species known previously only from the northeast coast of the U.S. was spectacular. Its mass occurrence along the Norwegian coast from the outer Oslofjord westward to Bergen during October - November 1966 discolored the surface waters, accompanied by sea trout mortality [4]. Since 1966, *G. aureolum* has spread throughout North European waters (Fig. 1), its mass occurrences frequently discoloring surface waters yellow-brown to coffee-brown, and accompanied by massive fish mortality [5]. Spreading of the dinoflagellate *Gymnodinium catenatum*, toxic in shellfish rearing areas in Spain, Japan and Tasmania, has also been described [6]; this species was previously known only from southern Californian waters. A remarkable Indo-Pacific spreading of the toxic dinoflagellate *Pyrodinium bahamense* var. *compressa* between 1972 -1984 has been documented [7]. For Japanese waters between 1978 and 1982 [8], PSP-producing species *Alexandrium tamarense* and *Gymnodinium catenatum* spread into eight new areas, accompanied by shellfish poisoning; a spreading of diarrhetic shellfish poisoning (DSP) into 17 new areas attributable to *Dinophysis fortii* and *Dinophysis acuminata* also occurred. *Ptychodiscus brevis* has spread from the Gulf of Mexico to the North Carolina coast, accompanied by shellfish toxicity in October 1987 [9].

These and other examples suggest that certain toxic species are spreading globally, both on a trans-oceanic scale and regionally within a given geographic region, some of which exhibit toxic bloom events following their apparent introduction into new regions. Regional spreading of non-toxic species, such as diatoms, has also been documented [10, 11]. The spreading and dispersal mechanisms remain enigmatic, however. A transport of cyst stages in the ballast water of large cargo ships leading to geographic spreading of certain species has been proposed as one method [6]; transport in altered current patterns is also a possibility. However, these dispersal mechanisms do not fully explain the apparent range extensions of all immigrant species. This suggests that in at least some cases the apparent global and regional spreading phenomena are illusory. Such occurrences may not represent new introductions, but rather the proliferation to detectable levels of indigenous populations which previously were a component of the "hidden flora" and escaped detection. Should (and where) this be the case spawns a new question, however: Why is there an apparent global and regional increase in unusual blooms of previously unimportant species, many of which are toxic?

EVIDENCE FOR INCREASED NOVEL BLOOM FREQUENCY

At the meeting on Exceptional Plankton Blooms convened by ICES in Copenhagen in October 1984, some investigators concluded that there was little supporting evidence for an increased frequency of exceptional blooms, at least in Atlantic waters [12,13,14]. Three years later at the Red-tide Symposium in Takamatsu, Japan, Anderson [15] concluded that there has indeed been a major global expansion "--- of the geographical extent, frequency, magnitude, and species complexity of red tides throughout the world." The 1988 *Chrysochromulina polylepis* outbreak in Scandinavian waters [16]; the 1987 toxic bloom of the diatom *Nitzschia pungens* f. *multiseries* off eastern Prince Edward Island, Canada [17], and toxic occurrence of *Ptychodiscus brevis* off the N. Carolina coast [9] since the Takamatsu meeting have a significance which transcends their localized occurrences. They need to be viewed also as possible local manifestations of, and further evidence for an apparent global epidemic of novel phytoplankton blooms.

The view that there is currently a global bloom epidemic has been legitimately challenged by arguments that it is merely an artifact of increased monitoring; improved analytical techniques, or increased awareness of ichthytoxic and PSP outbreaks accompanying increased mariculture of finfish and shellfish. Resolution of this is hampered by the lack of a suitable number of long-term data sets. Among the data sets suitable for such analyses, however, there is clear evidence for a long-term increase in the annual frequency in red-tide blooms at a given locality, including Hong Kong [18] and South Africa [19]. Continuous monitoring of PSP toxicity in mussels since 1959 at Lepreau, Bay of Fundy, has shown a marked increase in mean monthly toxicity since 1968 [20]. The Tolo Harbour results [19] (Fig. 2) are particularly provocative. An 8-fold increase in number of red-tide outbreaks per year occurred between 1976 and 1986. Most interesting, this epidemic has been accompanied by a 6-fold increase in human population level and a concurrent 2.5-fold increase in nutrient loading associated with industrial and domestic waste. These parallel increases implicate an anthropogenic component in the form of nutrient enrichment as contributing to the increased red-tide outbreaks.

The *Gyrodinium aureolum* distributional pattern in North European waters (Fig. 1) exhibits a conspicuous absence along the Dutch and German coasts. (I accept this apparent exclusion as real, given the intense, continuous scrutiny of phytoplankton growth in these waters by Dutch and German scientists.) A significant feature of this region is that it is the recipient of numerous, major European rivers well-known to be chemically-laden with industrial and domestic effluent. Is the apparent, conspicuous exclusion of *Gyrodinium aureolum* from these waters attributable to chemical modification by riverine inputs? In this instance, chemical modification may be preventing red-tide outbreaks of *Gyrodinium aureolum*, whereas in Tolo Harbour chemical enrichment may be stimulatory to such blooms. In both cases, a modified chemical environment resulting from anthropogenic activities is implicated as a factor influencing exceptional phytoplankton blooms. Clearly, this cannot be established from the data at hand, but this situation in the Wadden Sea and the Tolo Harbour results prompted further assessment of a possible connection between unusual phytoplankton bloom events, anthropogenic modification of coastal waters, and long-term trends in primary production.

BLOOMS IN THE SKAGERRAK, KATTEGAT AND BALTIC SEA

Several long-term data sets are available for this region. A clear, long-term (1971 - 1982) increase in N and P levels (January - February) prior to the spring bloom, both in the outflowing surface waters of the Kattegat and in the deeper waters flowing inwards from the Skagerrak has been reported by Andersson and Rydberg [21]. In the surface waters, total N levels increased 3.5-fold over this 12-year period; total P levels about 1.7-fold. For the Öresund, PO_4 concentrations have increased 9-fold since 1940 [22]. Primary production levels increased concurrently at various locations within this region [23, 24]. A notable long-term increase is evident for Halskov Rev (Fig. 3), particularly during the summer. Average daily production rates between 1966 and 1981 increased about 2.5-fold. Annual production rates increased about 2-fold between 1966 and 1977, remaining relatively constant between 1953 - 1966 [23]. Nielsen and Aertebjerg [24] established a hyperbolic correlation between annual summer production levels and inorganic N levels in bottom water (available for vertical diffusion); production rates were maximal at about 7 μmol L^{-1} NH_4+NO_2+NO_3 (Fig. 3). Clearly, long-term increases in N and P levels within this region have been accompanied by a long-term increase in primary production, although regional variations occur [23, 25].

For the Gulf of Finland, where surface layer phosphorus concentrations increased 40% between 1962 - 1978 [26], a similar correlation exists between increased annual primary production rates and P concentrations (Fig. 3; [27]). Data presented by [28] indicate a strong positive correlation between interannual variations in mean summer total P concentrations and primary production rates in the Gulf of Bothnia and Gulf of Finland between 1958 - 1983. In the central and southern regions of the Baltic Sea, winter surface layer concentrations of PO_4 and NO_3 increased about 4- to 8-fold [29]. Some of this euphotic zone enrichment can be attributed to long-term hydrographic trends favoring stronger vertical mixing and fluxes of nutrients from deep water. But Nehring [29] concluded that these long-term nutrient increases show "-- that eutrophication in the Baltic proper continues unabated." Long-term increases in phytoplankton biomass and primary production apparently have accompanied these increased nutrient levels. For the southern Baltic Sea, mean summer chlorophyll biomass during 1977 - 1981 was approximately double the 1968 - 1976 mean [30]; mean annual primary production during 1981 - 1985 was about 25% greater than 1971 - 1974 levels [31]. Although the Baltic Sea exhibits regional differences in the magnitude of the nutrient loadings and associated long-term increases in primary production, Wulff *et al.* [32] concluded that many coastal areas of this sea exhibited "-- severe signs of eutrophication, with toxic phytoplankton blooms and anoxic bottoms in the late 1970s and early 1980s".

Within this Nordic region, a progressive nutrient enrichment accompanied by an altered productivity appears to have been in progress for about two decades. The corollary of this is that changes in phytoplankton niche structure can be expected [33]. This prospect has been evaluated generally, based on reports of novel, unusual, or otherwise noteworthy blooms relative to the normal phytoplankton successional cycle and dynamics established for this region. Fig. 4 (right panel) presents some of the reports of unusual phytoplankton bloom episodes between 1964 and 1988 in southern Scandinavian waters. Two general response categories are evident: new species' introductions, some of which then spread regionally exhibiting toxic bloom events; and indigenous species exhibiting unusual, toxic bloom-events. In 1964, the rhaphidophycean *Olisthodiscus luteus* exhibited an unprecedented appearance discoloring the polluted surface waters of the inner Oslofjord [4]. The first appearances of *Gyrodinium aureolum* in 1966 and *Prorocentrum minimum* in 1979 in Norwegian waters were followed by their subsequent spreading, with first occurrences of *Gyrodinium aureolum* and *Prorocentrum minimum* in the Kattegat reported in 1981 [24], and with continued spreading and toxic bloom episodes of *Prorocentrum minimum* thereafter through 1983 in the western Baltic Sea, Danish coastal waters and Kiel Bay [34, 35]. Within the Oslofjord, *P. minimum* has become a regular component of the phytoplankton since its 1979 appearance, forming large blooms [36]. The diatom *Thalassiosira punctigera* first appeared in 1979, after being first recorded in North European waters in the western English Channel in 1978 (Fig. 4) and three years later off the Dutch coast [10]. Should *Chrysochromulina polylepis* not have been a rarified component of the indigenous community, it is another example of a novel species introduction accompanied by a toxic bloom event.

Exceptional bloom episodes of indigenous species co-occurred with bloom events of immigrant species. *Ceratium* species, commonplace bloom organisms within this region [37,38], exhibited an unusual bloom episode in 1980 accompanied by fish mortality [39], leading Lindahl and Hernroth [22] to comment that "-- mass occurrences of *Ceratium* spp. are a new phenomenon in the area --". The silicoflagellate *Distephanus speculum* produced a toxic bloom (to fish) in 1983 [24], and the PSP-producing *Alexandrium tamarense*, recorded for the first time in Limfjord in 1983, exhibited a PSP outbreak in 1987 [40]. The cyanobacterium *Nodularia spumigena*, toxic to mammals, bloomed extensively in the Kattegat and Belt Sea in 1975 and 1976 [24]; exhibited widespread, sustained blooms in the Baltic Sea between 1969 - 1973 [41] and, together with *Aphanizomenon flos-aquae*, bloomed at the entrance into the Gulf of Finland during the 1978 - 1981 summers [42]. Toxic occurrences lethal to dogs occurred for the first time in 1984 [32]. Assessment of the long-term patterns in bloom episodes within the Baltic Sea is complicated by long-term trends and regional variations in changing salinity [43]. (The associated osmotic effect influences species' distributions and occurrences in these brackish waters.) For the Gulf of Finland, a conspicuous 6-fold increase in maximal bloom concentrations of the dinoflagellate *Gonyaulax catenata* is evident between 1968 - 1975, followed by a progressive decrease thereafter through 1981 [42]. The chlorophyte *Monoraphidium contortum* became more significant at high phosphorus levels [43].

Over the 25-year period from 1964 - 1988, then, the waters of the Skagerrak, Kattegat and Baltic Sea generally have exhibited unusual blooms involving at least 11 different taxa (Fig. 4), characterized also by the wide-spread dispersal and establishment within the local community of several, apparently newly introduced, aggressive, toxic bloom-producing species. This trend may have accelerated since 1979.

BLOOMS IN THE SOUTHERN NORTH SEA

A similar pattern of a long-term increase in nutrient loading accompanied by increased primary production and occurrences of novel, unusual phytoplankton blooms are also evident for the southern North Sea, notably the Dutch Wadden Sea. Numerous, nutrient-rich rivers discharge into the Wadden Sea and German Bight (Fig. 1). Since 1930, riverine nutrient inputs of P and N into Dutch coastal waters have increased 7- and 5-fold, respectively [44, 45]. The long-term increases in mean NH_4, NO_3 and PO_4 concentrations in the Rhine River illustrate the importance of such riverine "nutrient-pumps" in this region (Fig. 5). In the two decades since 1955, PO_4 levels in Rhine River water at Lobith increased 7.5-fold; NO_3 about 3-fold. In contrast, long-term mean reactive silicate concentrations have remained relatively constant at about 135 μM during the winter and 50 μM during the summer [44].

The regional impact of these long-term, riverine nutrient discharge patterns is very clear. A conspicuous, long-term increase in winter (= pre-spring bloom), nearshore PO_4 concentrations has resulted: 1978 levels were about 3-fold greater than in 1961 [44]. A pronounced onshore-offshore PO_4 gradient has developed; nearshore, winter PO_4 levels (> 3 μM) currently

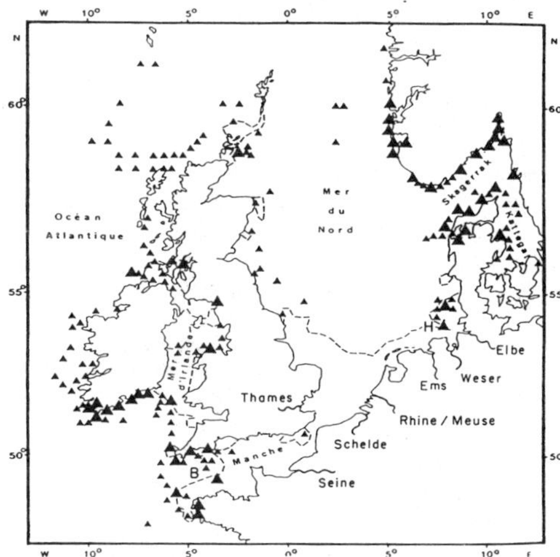

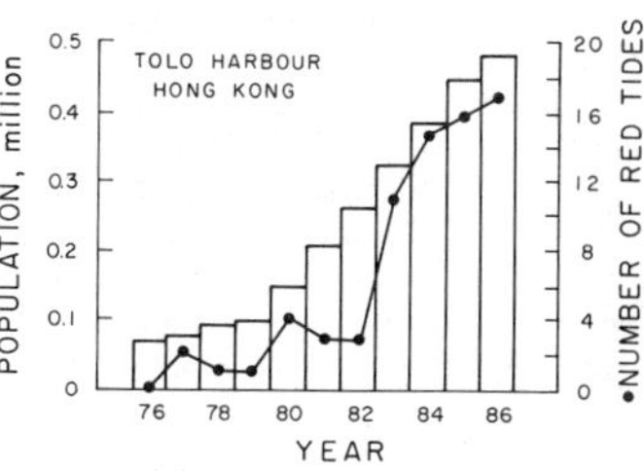

Fig. 2. Annual number of red-tides in Tolo Harbour, Hong Kong, and human population levels in Hong Kong from 1976 - 1986 (from Lam and Ho [18]).

Fig. 1. Distribution and abundance of *Gyrodinium aureolum* in the North Sea, Skagerrak and Kattegat. Small triangles: cell concentrations $<10^6$ L^{-1}; large triangles $>10^6$ L^{-1} and discolored water; hatched lines show position of frontal systems (modified from Fig. 15 in Partensky and Sournia [72]). H = Helgoland; B = Boalch's production station.

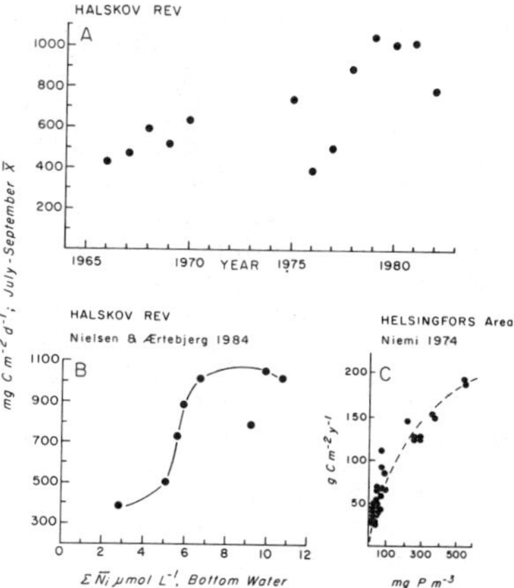

Fig. 3. (A). Average daily summer carbon production at Halskov Rev during 1963 - 1982 and (B) as a function of inorganic nitrogen levels in the bottom waters (> 30 m) during that period. (C). Annual carbon production related to average annual phosphorus concentrations in the Gulf of Finland. A,B modified from Nielsen and Aertebjerg [24]; C from Niemi [27].

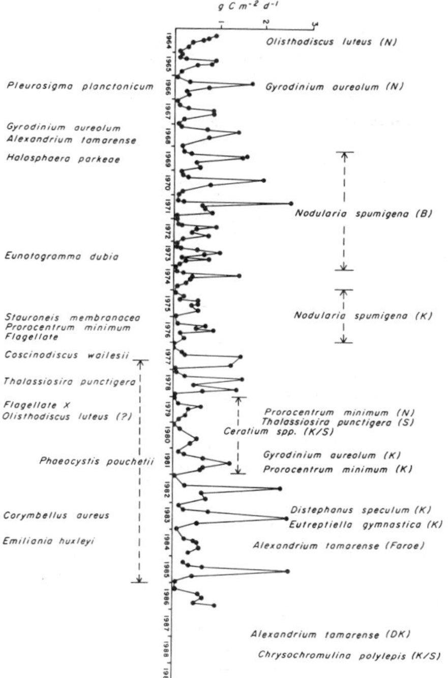

Fig. 4. Right Panel: first occurrences and/or persistence of some novel, exceptional phytoplankton species' blooms in Norwegian (N) coastal waters, Danish fjords (DK), Kattegat (K), Skagerrak (S) and in the Baltic Sea (B) during 1964 - 1988. Left Panel: As for right panel, except annual calendar records species' occurrences in the Dutch Wadden Sea and southern North Sea. Middle Panel: monthly daily primary production rates in western English Channel at International Hydrographic Station E1 (50°02'N, 4°22'W) during 1964-1988 (modified from Boalch *et al.* [59]; Boalch [11]). Data presented in right and left panels from various sources.

exceed concentrations offshore by about 4-fold. In 1961, differences along this onshore-offshore gradient were modest. Silicate concentrations, in contrast, did not reveal any long-term changes based on regional surveys carried out in 1962 and 1974 [46]. This is consistent with the relatively constant long-term reactive silicate concentrations in Rhine River water [44]. Significant, long-term increases in primary production have accompanied these elevated nutrient levels. Fransz and Verhagen's [47] simulation model for 1980 yields a conspicuous, progressive onshore increase in annual primary production for that year along a 143 km offshore-onshore nutrient gradient. Estimated production increased along this gradient by about 12-fold, from ca. 60 to 700 g C m^{-2} y^{-1}. In contrast, simulated year 1930 gradient rates increased only about 1.5-fold. Direct measurements within the shallower, Wadden Sea (Fig. 6A) confirm the coincident long-term increases in primary production and PO_4 concentrations [48, 49]. Between 1950 and 1981, winter PO_4 levels increased about 5-fold; late summer values by 7-fold. Year-round concentrations also progressively increased over this time span (Fig. 6B). Primary production of the pelagic phytoplankton and tidal-flat microphytobenthos increased by 2- and 3-fold, respectively (Figs. 6C, E), and microphytobenthic biomass trebled between 1967 - 1981 (Fig. 6D).

Increased phytoplankton numerical abundance, shifts in phylogenetic predominance and novel bloom episodes also accompanied the long-term increases in nutrients - biomass - primary production in Dutch coastal waters. For the Marsdiep area, total phytoplankton cell numbers increased over the 17-year period from 1969 - 1985 [50], during which diatoms decreased initially (1969 - 1974), then increased markedly thereafter; flagellate abundance increased continuously over this 17-year period. The number of days per year when flagellates exceeded 10^6 cells L^{-1} increased from < 10 in 1969 - 1972 to about 300 days in 1985. Simulation models suggest summer flagellate biomass levels increased 2- to 4-fold above 1930 levels in the more open, progressively nutrified Dutch coastal waters [45]. This significant phylogenetic shift is due principally to the remarkable response of the haptophycean *Phaeocystis pouchetii*. Mass occurrences began in 1977 [49, 50, 51]. Since 1976, its mean annual abundance increased about 8-fold; the number of days per year that its abundance exceeded 10^6 cells L^{-1} progressively increased to about 100 days. Thus, between 1973 - 1985 an epidemic of *Phaeocystis pouchetii* blooms has characterized Dutch coastal waters. During this period, its abundance, duration of its spring bloom and period of occurrence within the annual cycle have increased significantly [51]. Other novel species' occurrences or blooms have accompanied this *Phaeocystis* epidemic (Fig. 4, left panel). *Prorocentrum minimum* and *Prorocentrum balticum* recorded for the first time in 1976 and 1978, respectively, have exhibited exceptional blooms; in contrast, *Ceratium fusus* and *Ceratium lineatum* have virtually disappeared since 1974 and 1976, respectively [52, 53]. Novel diatom behavior includes the 1972 bloom episode of *Eunotogramma dubia* and 1974 occurrence of *Pleurosigma planctonicum*; both species then disappeared. Similar to the Kattegat and Baltic Sea, then, there is clear evidence that increased primary production has accompanied increased nutrient loading in Dutch coastal waters, accompanied by modified phytoplankton behavior, including exceptional blooms.

NUTRIENT - PRODUCTION - BLOOM PATTERNS ELSEWHERE IN THE NORTH SEA

Evaluation of whether a similar linkage between nutrients, primary production and modified phytoplankton bloom and species behavior occurs elsewhere in the North Sea is frustrated by the apparent lack of similar, concurrent long-term measurements. Nonetheless, the more subjective trend analyses when combined into a regional overview suggest a long-term nutrient-production-bloom pattern for much of the North Sea analogous to that described for the Dutch and Nordic coastal waters.

Measurements of nutrients and phytoplankton biomass have been made several times weekly over a 23-year period (1962 - 1984) at Helgoland located in the German Bight (Fig. 1; [54]). Although Rhine River water is transported into the German Bight, the collective nutrient discharge from the Elbe, Ems and Weser Rivers into this region is more important [55]. Helgoland lies within a region where watermasses from the central North Sea and Elbe River-discharge admix (Fig. 1). Significant long-term increases in PO_4 and NO_3 concentrations occurred in both Elbe-influenced water and the central North Sea watermass over the 23-year period. Mean winter NO_3 concentrations increased by about 4-fold; PO_4 concentrations 1.5-

fold. Mean annual NO3 concentrations in the Elbe River-influenced water (16.26 μM) were 2.3-fold greater than in the saltier, central watermass. Reactive silicate concentrations, in contrast, decreased by about 5-fold in the Elbe River-influenced watermass, but remained constant in the central watermass. A pronounced increase in phytoplankton biomass and shift in phylogenetic group predominance occurred in both watermass types. In the Elbe-influenced region, flagellate biomass (as carbon) increased 16-fold in 23 years; but diatom biomass remained constant. In the saltier, central watermass, flagellate biomass increased about 6-fold, whereas diatom biomass *decreased* 2-fold! Primary production measurements are unavailable. However, the observed long-term increases in biomass most likely were a consequence of increased primary production accompanying increased nutrient levels, given the relationships between these variables found for the comparison regions (Figs. 3, 5, 6), rather than to long-term reductions in grazing pressure. The long-term phylogenetic shift in biomass predominance from diatoms to flagellates in the German Bight is remarkable. Not only is it similar to that in Dutch coastal waters, a similar long-term increase in, and shift to flagellate biomass predominance appear to characterize much of the North Sea. Based on changes in the green coloration of Continuous Plankton Recorder (CPR) silks, Gieskes and Kraay [56] reported phytoplankton biomass increased throughout the North Sea between 1956 - 1975, which they attributed to increased microflagellate abundance. This qualitative index of phytoplankton abundance continued to increase after 1975 [14]. In May 1983, an extraordinary bloom (9 x 10^6 cells L^{-1}) of the colonial microflagellate *Corymbellus aureus* occurred in the open North Sea near Fladen Grounds [57]. This prymnesiophyte had not been recorded since its description 10 years earlier. Other notable long-term phytoplankton changes recorded for the North Sea include exceptionally early blooms of *Ceratium* spp. which characterized the Southern Bight of the North Sea during the early 1970s [58]. For the North Sea generally, "-- blooms of diatoms and *Ceratium* spp. have declined over the last 26 years [1958 - 1983] ---" [14]. Collectively, these floristic shifts reflect a general long-term phenomenon which appears to be ongoing in the North Sea: a pattern within which flagellates have increased, certain dinoflagellates have decreased, diatoms have either decreased or remained constant, and new species invasions have occurred. These are in addition to the remarkable *Gyrodinium aureolum* spreading and bloom phenomena (Fig. 1).

A long-term production time-series (1964 - 1986) is available [11, 59] for the western English Channel (Fig. 4, middle panel) along with tow-net observations of phytoplankton species [11,12]. Significant interannual differences in seasonal coverage and in number of annual production measurements (n = 4 to 10) characterize this time-series; these preclude rigorous trend analysis. Several production cycles may have occurred during this 23-year period, possibly with a long-term tendency for increasing maximal summer production rates. Progressive increases in maximal monthly production rates between 1964 - 1971 and between 1979 - 1985 are suggested; the 1972 - 1978 pattern is less clear. Based on CPR "color", biomass in this region (CPR Area D3) doubled between the early 1950s and 1970s [56]. This suggests a long-term increase in nutrients, but long-term nutrient data sets are scarce for English coastal waters where riverine discharges can be rich in nutrients [55]. North of the Channel in the Thames Estuary (Fig. 1), winter phosphate concentrations increased from 1.2- to 1.7-fold between 1962 - 1974 [46].

It is unresolved whether a long-term increase in nutrient levels stimulated the reported increased biomass in the English Channel similar to that in the North Sea, or contributed to the primary production cycles. A major similarity with the North Sea is in the occurrence of novel, unusual, even toxic occurrences and blooms of phytoplankton (Fig. 4, left panel). Following the 1966 occurrence of *Gyrodinium aureolum* in Norwegian coastal waters, it was recorded off Plymouth in 1968, and has since spread throughout the North Sea (Fig. 1), often forming toxic blooms. Two major phytoplankton shifts have occurred within the North Sea since 1960, in addition to increased flagellate biomass levels and a reduction in diatom:flagellate biomass ratio. In the well-mixed, enriched eastern regions, notably Dutch coastal waters, *Phaeocystis pouchetii* has increased to predominance. In the stratified and frontal zone regions, the immigrant species *Gyrodinium aureolum* has established itself, causing recurrent toxic bloom epidemics. Accompanying this remarkable regional spreading and toxic blooming of *Gyrodinium aureolum*, novel introductions of diatoms (*Pleurosigma planctonicum*, *Stauroneis membranacea*, *Coscinodiscus wailesii* and *Thalassiosira punctigera*) were commonplace. Dense blooms of *Coscinodiscus wailesii*, first noticed off Plymouth in 1977 and now established in the phytoplankton community, clogged fishing nets interfering with trawling [11]. *Skeletonema costatum* which "--- used to be considered the weed of the phytoplankton --- (through) the 1970s ---" has almost disappeared in recent years. Interpretation of these novel phytoplankton species events is compromised by 50-year Russell Cycle patterns [12]. Nonetheless, these species patterns parallel similar behavior for the North Sea proper, Dutch coastal waters, Kattegat, Skagerrak and Baltic Sea. For the North Sea generally, at least 14 different taxa have exhibited novel, extraordinary blooms or occurrences, paralleling similar events of the 11 different taxa discussed earlier for the Kattegat and Skagerrak.

In summary, there is considerable evidence that significant changes in phytoplankton species occurrences, biomass and productivity; novel species occurrences; exceptional or unusual blooms; phylogenetic shifts in predominance, and spreading phenomena have occurred within various regions of the North Sea generally and in the Skagerrak, Kattegat and Baltic Sea. In fact, a regional disequilibrium in phytoplankton community structure may have resulted. Collectively, these events and those in Tolo Harbour (Fig. 2) support the hypothesis that novel phytoplankton blooms, both harmful and benign, are increasing in the sea, and are linked to, and a consequence of long-term increases in coastal nutrient levels, accompanied also by increased primary productivity.

BLACK SEA AND SETO INLAND SEA

The long-term, coincident increases (1957 - 1984) in mean annual PO_4 concentrations and maximal abundance of the red-tide, anoxia-producing [61, 62] dinoflagellate *Exuviaella cordata* along the Romanian coast of the Black Sea provide further, regional support for this global hypothesis (Fig. 7). Maximal bloom abundance of *Exuviaella cordata* increased about 100-fold and mean annual PO_4 concentrations about 10-fold over this 28-year period. Between 1959 - 1967 there was a highly significant linear relationship (log-log plot) between these variables (r^2 = +0.87); for the 1959 - 1984 period, the relationship is hyperbolic (r^2 = +0.77). (The Black Sea data set, including sources, will be discussed in greater detail elsewhere.) If increased bloom frequency is indeed associated with an increased nutrient buildup, then a reduction in nutrient concentrations should be accompanied by a decrease (not necessarily in 1:1 ratio) in novel bloom frequency. This appears to have occurred in the Seto Inland Sea (Fig. 8; [63, 64, 65]). Between 1965 - 1976, the number of confirmed red-tide outbreaks increased progressively by about 7-fold, from 44 to 326 per year. This was accompanied by increased nutrient enrichment; for example, between 1962 - 1969 the COD loading increased 2-fold [64]. Effluent controls were initiated in the mid-1970's to reduce COD loading; between 1976 - 1985, the number of annual red-tide outbreaks decreased by about 2-fold.

NUTRIENT RATIO HYPOTHESIS

A regionally persistent observation has emerged: an apparent long-term, increased frequency of novel bloom episodes accompanies coastal nutrient enrichment. This association is detectable on a global basis: Baltic Sea, Kattegat, Dutch Wadden Sea, Black Sea, Hong Kong, Seto Inland Sea. Available long-term production measurements from the Baltic Sea, Kattegat and Dutch Wadden Sea indicate primary production increases with enrichment. Productivity *per se* can not be the determinant inducing novel floristic changes and blooms. More likely, selection of the bloom-species is associated with some aspect of the increased nutrient levels. Increased nutrient concentrations elicit yield-dose responses, but such reactions are an unlikely explanation for the observed novel species occurrences, bloom events and phylogenetic shifts. Such yield-dose responses would primarily influence the magnitude of the species' blooms, Michaelis-Menten kinetics notwithstanding. Changes in the ionic ratios of essential nutrients accompany nutrient loading. The N:P and Si:P ratios in the long-term trend (1955 - 1975) in mean chemical composition of Rhine River water, decreased curvilinearly by about 6-fold over the 20-year period (Fig. 5). Obviously, the specific patterns and trends in nutrient concentrations and ratios will vary in the different fluvial-estuarine systems. The significance of such changing ratios is that riverine nutrient delivery not only influences yield-dose responses. Redfield Ratio and resource-competition effects also accompany increased nutrification of coastal waters. There is abundant evidence that phytoplankton species are sensitive to nutrient ratios. Numerous investigators have focused on the N:P ratio to evaluate whether N or P was the more limiting to total community production. The more relevant issue

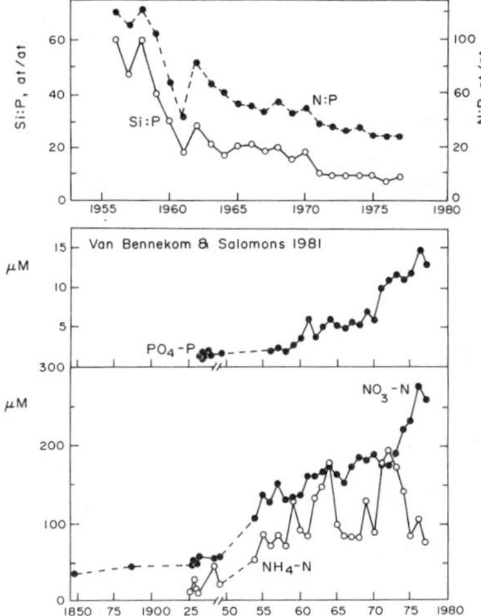

Fig. 5. Long-term trends in the mean annual concentrations of NH_4, NO_3 and PO_4 in the Rhine River at Lobith and in atomic ratios of N:P and Si:P (modified from van Bennekom and Salomons [44]).

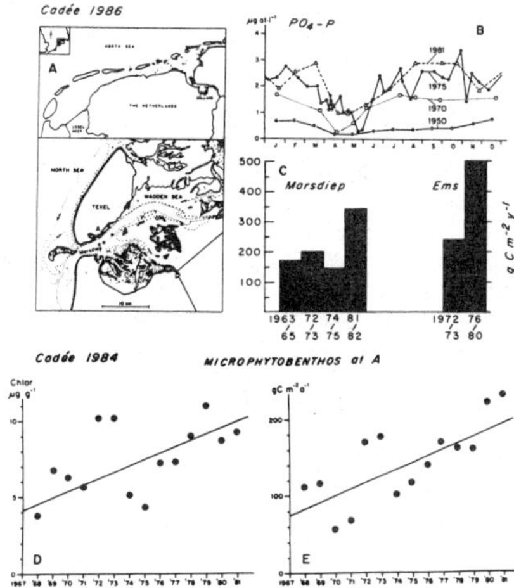

Fig. 6. A. Dutch Wadden Sea; B. Annual PO4 cycles between 1950 and 1981 at Marsdiep inlet; C. Interannual variations in pelagic primary production at Marsdiep and Ems estuary (Dollard); D. Long-term (1967 - 1981) patterns in mean annual chlorophyll biomass; E. primary production of the microphytobenthos near Marsdiep. (A-C modified from Cadée [49]; D, E from Cadée [48].)

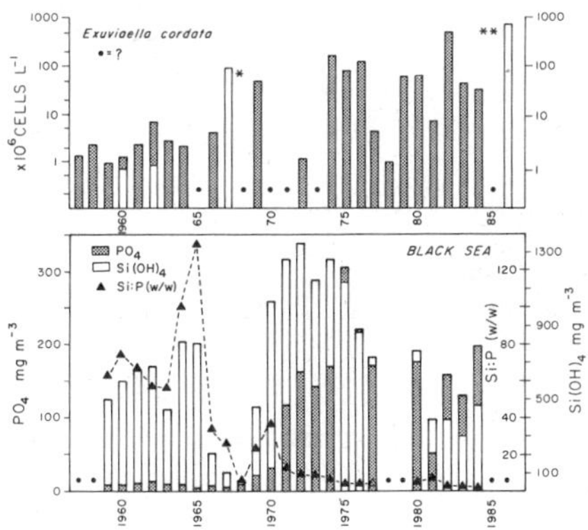

Fig. 7. Upper Panel: maximal annual abundance of *Exuviaella cordata* in Black Sea near Constanta along Romanian coast, except for observations at l'Etang Mangalia (*) and off Bulgarian coast (**). Histograms when both open and hatched indicate differences in maximal abundance reported for that year. Lower Panel: Mean annual PO_4 and $Si(OH)_4$ concentrations and Si:P ratio at Constanta. (Figure based on various sources.)

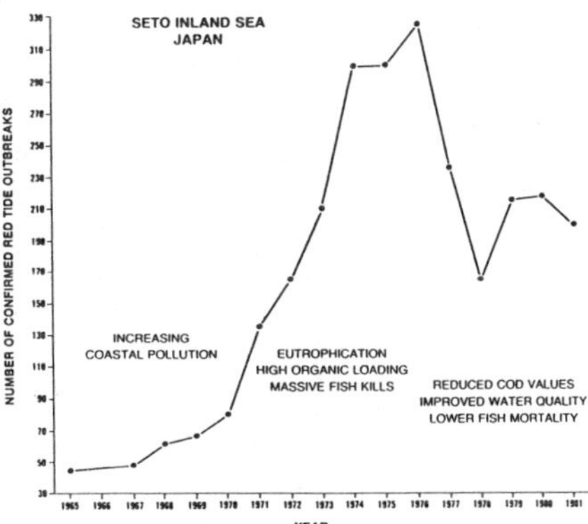

Fig. 8. Trends in red-tide outbreaks from 1965 - 1981 and accompanying nutrient loadings and fish mortality in Seto Inland Sea (From Prakash [65].

before us, however, is whether altered nutrient ratios accompanying nutrification could be an important species-selection mechanism involved in the associated linkage of increased novel species bloom events. Experimental data on the influence of nutrient ratios as regulators of inter-specific competition among marine phytoplankton are virtually non-existent, in contrast to that available for limnetic species [66]. Given the fundamental sub-division of the phytoplankton into those species which require Si and those which do not, the *in situ* ionic relationships of Si to N and to P, and associated species bloom events will be evaluated. In practical terms, this simplifies evaluation of the regional field results to the issue of diatom blooms *vs.* non-diatom blooms *vs.* ionic ratios.

The long-term decline in Si:P ratio of the nutrifying Rhine River (Fig. 5) and in Dutch coastal waters generally [44, 47, 49] has been pointed out. For the German Bight, long-term (23-year) increases in PO_4 concentration and decreasing, or constant reactive Si concentrations in the watermasses around Helgoland [54] have also led to a long-term decrease in the Si:P ratio. Within the Baltic Sea near Tvärminne, the Si:P ratio (by weight) progressively decreased from 53:1 to 21:1 between 1969 - 1984 [67]. Long-term reactive Si data do not appear to be available for the Skagerrak and Kattegat. However, in the contiguous waters of the Belt Sea near Kiel (Fig. 1) winter, pre-bloom Si concentrations progressively decreased from 27 µM to 16 µM over a 13-year period (1973 - 1985) [68]; inorganic N levels were constant at about 13 µM, while PO_4 decreased initially (1969 - 1977), then increased thereafter. A significant reduction in the Si:N ratio between 1973 - 1984 and, since 1977, in the Si:P ratio has occurred. Similar, if not greater long-term nutrient ratio shifts are expected to have occurred in the increasingly enriched, contiguous Kattegat. A similar linkage occurs between a long-term decrease in the Si:P ratio in the Black Sea and an increased number, intensity and duration of *Exuviaella cordata* red-tide blooms (Fig. 7). Bodeanu [69] reported the Si:P ratio (w/w) in Romanian coastal waters decreased from about 98:1 during 1960 - 1970 to about 10:1 during 1971 - 1975 and to about 4:1 during 1976 - 1980.

During these periods of long-term decreases in the Si:P ratios in the various regions of the Baltic Sea, Kattegat, Dutch coastal waters, certain regions of the North Sea, and Black Sea, significant blooms of non-silicon requiring groups and species have emerged with increased frequency, persistence and/or replaced diatoms as the dominant biomass group. The remarkable long-term emergence and increased predominance of *Phaeocystis pouchetii* in Dutch coastal waters [51] is one example. Fransz and Verhagen [47] have stated that in these waters "--- the increase in N and P and essentially constant discharge of reactive Si [have changed the] nutrient first depleted by phytoplankton from N or P some 50 years ago to Si nowadays --" and that diatom bloom events are currently regulated by silicate availability. The 23-year decrease in Si:P ratio near Helgoland has been accompanied by a long-term increase in flagellate carbon-biomass, constant diatom biomass, and progressive 10-fold decrease in diatom:flagellate biomass ratio [54]. The increased number and spreading of flagellate blooms in the Kattegat (Fig. 4) has occurred during the period when in contiguous waters pre-spring bloom reactive Si concentrations decreased by about 2-fold, accompanied by declining Si:N and Si:P ratios [68]. Extensive blooms of the blue-green alga *Nodularia spumigena* have occurred in the Baltic Sea (Fig. 4), and vigorous annual blooms of several other nitrogen-fixing species [70], particularly in nutrient-enriched regions, further demonstrate the importance of changing nutrient ratios as regulators of species occurrences and blooms. Their emergence in the Baltic Sea has paralleled a decreased ratio of N:P accompanying P enrichment [70]. Heterocystic blue-green algae have high P requirements, and through their N-fixing capability become excellent competitors for N [66]. The brackish Baltic Sea is inherently favorable to the occurrence of blue-green algae and green (Chlorophycean) microalgae [42, 43]. Chlorophycean phytoplankters, however, are mediocre competitors for P and have high N demands [66]. Thus, just as the ratio of Si:P and its changes appear to gate the competition between diatoms and non-diatoms for nutrients, the N:P ratio regulates the competition between, and bloom events of the blue-green algae and chlorophytes in the Baltic Sea in response to regional differences in nutrient enrichment.

It appears, then, that long-term decreases in Si:P ratios in these different regions and changes in the N:P ratio in the Baltic Sea are accompanied by increased, often recurrent bloom episodes of non-siliceous phytoplanktonic groups. This suggests that nutrient ratios are indeed involved in the linkage between the increase in novel blooms and increased nutrient levels based on regional comparisons. Obviously, other factors will be involved in regulating bloom events, but such nutrient ratio effects are considered to be of major importance. Their role as bloom regulators within a given region over several annual cycles of variable nutrient loading is considered elsewhere [60]. The field results implicating the role of Si in bloom phenomena confirm and amplify the views of Officer and Ryther [71] on the potential importance of Si in regulating the responses of phytoplankton to coastal nutrient enrichment. The events and paradigms developed in this paper will be more fully evaluated and expanded upon elsewhere.

ACKNOWLEDGEMENTS

I wish to thank Dr. Edna Granéli for her invitation to present this paper at the conference and for provision of a travel grant. Dr. Pia Elena Mihnea and Dr. Nicolae Bodeanu kindly provided data and information for the Black Sea. Ms. Blanche Coyne drafted the figures and typed the manuscript.

REFERENCES

1. M. Parker and P. Tett, Rapp. P.-v Réun. Cons. int. Explor. Mer 187, 5-8 (1987).
2. P. Tett, Rapp. P.-v. Réun. Cons. int. Explor. Mer 187, 47-60 (1987).
3. T.J. Smayda in: The Physiological Ecology of Phytoplankton, I. Morris, ed. (Blackwell Scientific Publ., Boston 1980) pp. 493-570.
4. T. Braarud and B. Heimdal, Nytt Mag. Bot. 17, 91-97 (1970).
5. K. Tangen, Sarsia 63, 123-133 (1977).
6. G.M. Hallegraeff, D.A. Steffensen and R. Wetherbee, J. Plank. Res. 10, 533-541 (1988).
7. J.L. McLean in: Toxic Red Tides and Shellfish Toxicity in Southeast Asia, A.W. White, M. Anraku and K.K. Hooi, eds. (Southeast Asian Fisheries Development Center, Singapore 1984) pp. 92-97.
8. M. Anraku in: Toxic Red Tides and Shellfish Toxicity in Southeast Asia, A.W. White, M. Anraku and K.K. Hooi, eds. (Southeast Asian Fisheries Development Center, Singapore 1984) pp. 105-109.
9. P.A. Tester, P.K. Fowler and J.T. Turner in: Novel Phytoplankton Blooms. Causes and Impacts of Recurrent Brown Tides and Other Unusual Blooms, E.M. Cosper, E.J. Carpenter and V.M. Bricelj, eds. (Lecture Notes on Coastal and Estuarine Studies, Springer-Verlag, Berlin 1989).
10. G.R. Hasle, Nord. J. Bot. 3, 593-608 (1983).
11. G.T. Boalch, Br. phycol. J. 22, 225-235 (1987).
12. G.T. Boalch, Rapp. P.-v. Réun. Cons. int. Explor. Mer 187, 94-97 (1987).
13. M. Parker, Rapp. P.-v. Réun. Cons. int. Explor. Mer 187, 108-114 (1987).
14. P.C. Reid, G.A. Robinson and H.G. Hunt, Rapp. P.-v. Réun. Cons. int. Explor. Mer 187, 27-37 (1987).
15. D.M. Anderson in: Red Tides: Biology, Environmental Science and Toxicology, T. Okaichi, D.M. Anderson and T. Nemoto, eds. (Elsevier Science Publishing Co., N.Y. 1989) pp. 11-21.
16. R. Rosenberg, O. Lindahl and H. Blanck, Ambio 17, 289-290 (1988).
17. S.S. Bates and 16 co-authors, Can. J. Fish. Aquat. Sci. 46, 1203-1215 (1989).
18. C.W.Y. Lam and K.C. Ho in: Red Tides: Biology, Environmental Science and Toxicology, T. Okaichi, D.M. Anderson and T. Nemoto, eds. (Elsevier Science Publishing Co., N.Y. 1989) pp. 49-52.
19. D.A. Horstman, Fish. Bull. S. Afr. 15, 71-88 (1981).
20. A.W. White, Rapp. P.-v. Réun. Cons. Int. Explor. Mer 187, 38-46 (1987).
21. L. Andersson and L. Rydberg, Est. Coast. Shelf Sci. 26, 559-579 (1988).
22. O. Lindahl and L. Hernroth, Mar. Ecol. Prog. Ser. 10, 119-126 (1983).
23. E. Gargas, S. Mortensen and G. Aertebjerg Nielsen, Ophelia, Suppl. 1, 123-144 (1980).
24. A. Nielsen and G. Aertebjerg, Ophelia, Suppl. 3, 181-188 (1984).
25. L. Edler in: Gödning av havsområden kring Sverige. En Kunskapsöversikt., R. Rosenberg, ed. (Statens Natur-vårdsverk PM 1808 1984) pp. 74-140.
26. M. Perttilä, P. Tulkki and S. Pietikäinen, Finn. Mar. Res. 247, 38-50 (1980).
27. Å. Niemi in: Nionde Nordiska Symposiet om Vattenforskning, Trondheim 27-29 June 1973. Nordforsk Miljövårdssekretariatet Publ. 1974 4, 173-188 (1974).
28. J. Lassig, J-M. Leppänen, Å. Niemi and G. Tamelander, Finn. Mar. Res. 244, 101-115 (1978).
29. D. Nehring, Ophelia, Suppl. 3, 167-179 (1984).

30. D. Nehring, S. Schulz and W. Kaiser, Rapp. P.-v. Réun. Cons. int. Explor. Mer 183, 193-203 (1984).
31. H. Renk, J. Nakonieczny and S. Ochocki, Kieler Meeresforsch., Sonderh. 6, 203-209 (1988).
32. F. Wulff, G. Aertebjerg, G. Nicolaus, Å. Niemi, P. Cizewski, S. Schulz and W. Kaiser, Ophelia, Suppl. 4, 299-319 (1986).
33. T.J. Smayda and T.A. Villareal in: Novel Phytoplankton Blooms: Causes and Impacts of Recurrent Brown Tides and Other Unusual Blooms, E.M. Cosper, E.J. Carpenter and V.M. Bricelj, eds. (Lecture Notes on Coastal and Estuarine Studies, Springer-Verlag, Berlin 1989).
34. J.P. Mommaerts, Ann. Biol. 41, 82-84 (1986).
35. E. Granéli in: Dinoflagellat blomningar. Förekomst, orsaker och konsekvenser i marin miljö: En kunskapsöversikt. Statens Naturvårdsverk Rapport 3293, 1-133 (1987).
36. K. Tangen, Sarsia 68, 1-7 (1983).
37. K. Tangen, Blyttia 38, 145-158 (1980).
38. T. Noji, U. Passow and V. Smetacek, Ophelia 26, 333-349 (1986).
39. L. Edler, Limnologica 15, 353-357 (1984).
40. O. Moestrup and P.J. Hansen, Ophelia 28, 195-213 (1988).
41. U. Horstmann, Merentutkimuslait. Julk./Havforsknings. Inst. Skr. 239, 83-90 (1975).
42. K. Kononen and Å. Niemi, Ophelia, Suppl. 3, 101-110 (1984).
43. K. Kononen, Kieler Meeresforsch., Sonderh. 6, 281-294 (1988).
44. A.J. van Bennekom and W. Salomons in: River Inputs to Ocean Systems. UNEP/UNESCO, 1981, pp. 33-51.
45. H.G. Fransz in: The Role of Freshwater Outflow in Coastal Marine Ecosystems, S. Skreslet, ed. (Springer-Verlag, Berlin 1986) pp. 241-249.
46. A.R. Folkard and P.G.W. Jones, Mar. Poll. Bull. 5, 181-185 (1974).
47. H.G. Fransz and J.H.G. Verhagen, Neth. J. Sea Res. 19, 241-250 (1985).
48. G.C. Cadée,.Neth. Inst. Sea Res. Publ. Ser. 10, 71-82 (1984).
49. G.C. Cadée, Neth. J. Sea Res. 20, 285-290 (1986).
50. G.C. Cadée, Mar. Biol. 93, 281-289 (1986).
51. G.C. Cadée and J. Hegeman, Neth. J. Sea Res. 20, 29-36 (1986).
52. M. Kat, ICES C.M. 1982/L: 22 (1982).
53. M. Kat, ICES C.M. 1988/L: 5 (1985).
54. J. Berg and G. Radach, ICES C.M. 1985/L:2 (1985).
55. U. Brockmann, G. Billen and W.W.C. Gieskes in: Pollution of the North Sea. An Assessment, W. Salomons, B.L. Bayne, E.K. Duursma and U. Förstner, eds. (Springer-Verlag, Berlin 1988) pp. 348-389.
56. W.W. Gieskes and G.W. Kraay, Neth. J. Sea Res. 11, 334-364 (1977).
57. W.W. Gieskes and G.W. Kraay, Mar. Biol. 92, 45-52 (1986).
58. R.R. Dickson and P.C. Reid, J. Plankton Res. 5, 441-455 (1983).
59. G.T. Boalch, D.S. Harbour and E.I. Butler, J. mar. biol. Ass. U.K. 58, 943-953 (1978).
60. T.J. Smayda in: Novel Phytoplankton Blooms: Causes and Impacts of Recurrent Brown Tides and Other Unusual Blooms, E.M. Cosper, E.J. Carpenter and V.M. Bricelj, eds. (Lecture Notes on Coastal and Estuarine Studies, Springer-Verlag, Berlin 1989).
61. N. Bodeanu and M. Usurelu in: Toxic Dinoflagellate Blooms, D.L. Taylor and H.H. Seliger, eds. (Elsevier North Holland, Inc. N.Y. 1979) pp. 151-154.
62. P.E. Mihnea in: Toxic Dinoflagellate Blooms, D.L. Taylor and H.H. Seliger, eds. (Elsevier North Holland, Inc. N.Y. 1979) pp. 77-82.
63. T. Okaichi in: Science for Better Environment. Proceedings International Congress on the Human Environment, Kyoto, Pergamon Press 1975) pp. 455-460.
64. T. Yanagi, Mar. Poll. Bull. 19, 51-53 (1988).
65. A. Prakash, Rapp. P.-v. Réun. Cons. Int. Explor. Mer 187, 61-65 (1987).
66. U. Sommer, Prog. Phycol. Res. 5, 124-178 (1987).
67. Å. Niemi and A.M. Åstrom, Ann. Bot. Fennica 24, 333-352 (1987).
68. B. von Bodungen, Ophelia 26, 91-107 (1986).
69. N. Bodeanu, Trav. Mus. Hist. Nat. "Gr. Antipa" Bucuresti 26, 69-83 (1984).
70. I. Rinne, T. Melvasalo, Å. Niemi and L. Niemistö, Finn. Mar. Res. 248, 117-127 (1981).
71. C.B. Officer and J.H. Ryther, Mar. Ecol. Prog. Ser. 3, 83-91 (1980).
72. F. Partensky and A. Sournia, Cryptogamie Algologie 7, 251-276 (1986).

TOXIN VARIABILITY IN *Alexandrium* SPECIES

DONALD M. ANDERSON*
*Biology Department, Woods Hole Oceanographic Institution, Woods Hole, MA 02543 USA

INTRODUCTION

One of the more important yet least understood aspects of the ecology and physiology of the dinoflagellate *Alexandrium* is that toxicity is highly variable both between isolates of the same species and in single isolates grown under different conditions [1-9]. The following summary focuses on some of the more important aspects of toxin variability, not in the form of a comprehensive review, but rather as an indication of the types of insights such variability provides on important issues in dinoflagellate taxonomy, population dynamics, toxin biosynthesis, and general physiology.

VARIABILITY BETWEEN POPULATIONS

In regions subject to recurrent episodes of Paralytic Shellfish Poisoning (PSP), it is now clear that toxicity is not caused by one widespread, homogeneous species but instead by a variety of different subpopulations that may be varieties of one species or a group of closely-related species. Two regional populations of *Alexandrium* have been well characterized in this respect, one on the northeast [6,16] and the other on the northwest [5,7,15] coast of North America. Despite the inherent limitations of the mouse bioassay for saxitoxin, a clear indication of the extent of *Alexandrium* population heterogeneity over large geographic areas resulted from a study that quantified the toxicity of 34 strains isolated from the region between New York and Nova Scotia [6]. This study is noteworthy because it not only documented a two order of magnitude variation in toxicity among isolates, but it also revealed that this toxicity followed a decreasing trend from north to south.

Toxicity was expressed as toxin content (μmouse units per cell), so the differences between isolates could have been due to variability in the total number of moles of toxin contained in a cell or in the mixture of toxins (toxin composition) of variable potency that each isolate produced [10,11]. Although the former possibility remains to be rigorously tested, extensive differences in toxin composition have since been demonstrated in this collection of isolates using HPLC [12]. It is now clear that the high toxicity of the northern isolates reflects their production of the highly potent carbamate toxins (e.g. saxitoxin (STX) and gonyautoxins I-IV (GTX I-IV)), whereas southern isolates typically contain the weaker N-sulfocarbamoyl toxins (e.g. B1,B2, C1-4). Although the toxin content data suggest a continuous north to south trend, cluster analysis (based on toxin composition differences) suggests the existence of three related groups within this regional population - a distinct cluster of

Toxic Marine Phytoplankton
Edna Graneli et al., Editors

northern isolates and two more geographically-diverse clusters of southern isolates [12]. The observed north to south trend in toxin content could thus result from latitudinal differences in environmental parameters (such as temperature) and their influence on the establishment of genotypically different blooms. Alternatively, the clustering of toxin composition types might reflect species dispersal from several origins or spreading centers.

On the west coast of North America, studies by Hall [5] and Cembella et al., [7] also document considerable variability in toxin composition between regional populations of Alexandrium. Some regional populations were found to be more genetically diverse than others, possibly reflecting the extent of advection and mixing of populations in those areas. Most importantly, there was no correlation between toxin composition and the morphological characteristics typically used to separate tamarensoid from catenelloid morphotypes [7]. A similar finding resulted from toxin composition analyses of the east coast Alexandrium populations [12]. Some strains identified as A. fundyense had toxin profiles that were more closely related to those of A. tamarense isolates than to other A. fundyense strains. These two studies, both based on toxin composition differences, have important implications relative to the species concept in dinoflagellates [13]. Morphologists have consistently had difficulties discriminating between species in the "tamarensis complex" [13,14], and biochemical information from isozyme electrophoresis and toxin composition [15,16] should therefore be useful in resolving the confusion. Since isolates were grown under identical conditions in each of the toxin composition studies described above, the observed differences are presumed to be genetic. To some, such differences should be sufficient for the definition of separate species, yet others argue that such distinctions are too restrictive and that a better alternative is to recognize the existence of physiological races or varieties within one "supraspecies" [13]. Resolution of this important taxonomic controversy will take time, but it is clear that toxin variability will be a major factor in the debate.

VARIABILITY IN INDIVIDUAL ISOLATES

Toxin Composition. Variability in the toxicity of a single isolate is generally attributed to differences in the rate of toxin production or accumulation under different growth conditions and not to any differences in toxin composition. Until recently, those studies that have used chromatographic methods to quantify saxitoxin and its derivatives have all concluded that toxin composition is a relatively stable or conservative property in Alexandrium species [5,7-9]. It is now clear, however, that toxin composition changes in a single isolate do occur under different growth conditions. This was first noted by Boczar et al., [17] using batch cultures of A. tamarense and A. catenella. In one strain that produced only neosaxitoxin (NEO) and STX, the mole % of NEO increased from 8 to 44% as the culture aged, with STX decreasing proportionally. Another culture that produced a more diverse array of toxins also showed variability through time, with GTX I and IV increasing as GTX II and III decreased. These changes were only significant 10 days or more after cell division had ceased - the toxin composition remained relatively constant during exponential growth. The cultures were nutrient replete, so growth presumably ceased due to CO_2 depletion.

In a more recent study of A. fundyense in both batch and semi-continuous cultures where growth was limited by factors other than CO_2, these same trends and some new ones were verified [12]. Nitrogen limitation in semi-continuous culture favored the production of toxins C1,2 and GTX I,IV, whereas phosphorus limitation produced cells with high relative abundance of GTX II,III. As batch cultures aged, the interconversion between STX and NEO was again observed. In addition, the accumulation of GTX II,III in cells that had ceased dividing due to phosphorus depletion was noted [12]. In all cases, STX reached its highest relative abundance when growth was most rapid.

These are all relatively new data, so their implications are only now being considered. At first glance, these results raise doubts as to the fundamental validity of using toxin composition as a chemotaxonomic marker in the context described above for the east and west coast North American populations of Alexandrium. However, on closer examination, it is clear that the changes in batch culture occur only after exponential growth has stopped and the cells become senescent. Replicate cultures assayed at the same stage of growth have essentially identical toxin profiles. Since toxin composition comparisons have always been made between cultures grown under identical conditions and harvested at the same stage of exponential growth [7,12], the conclusion that the toxin differences have a genetic basis is still correct and the clustering analyses and other comparisons still valid. It should also be noted that it was the relative abundance (mole %) of the different toxins that changed - the specific array of toxins produced did not vary in these studies.

From a chemical standpoint, most of the compositional changes observed in the semi-continuous cultures [12] are reasonable, in that as growth rate increased, a decreasing trend in one toxin was typically associated with an increase in another toxin that could result from a relatively simple chemical transformation. For example, with nitrogen-limited growth, a decreasing trend in the relative abundance of GTX I,IV and C1,2 as growth rate increased was associated with an increasing trend in STX, NEO and GTX II,III. We know that C1 and C2 are easily hydrolyzed to GTX II and GTX III in vitro. Likewise, oxidation at the N-1 position converts STX to NEO, and conversion of GTX I and GTX IV to GTX II and GTX III requires only the cleavage of the N-1-hydroxyl group. We must be careful here, as these statements imply interconversions between toxins (as might occur in a batch culture approaching stationary phase), whereas each toxin profile at a given growth rate in the semi-continuous culture study [12] was from a culture that was essentially in steady state. The more correct way to view the compositional trends would be to recognize that differences in cell physiology under the growth conditions of each culture altered either the rate of biosynthesis of each toxin or the extent of conversions between toxins at equilibrium. Given our relative ignorance of the pathways involved in synthesis of saxitoxin and its derivatives [18], it is not yet possible to explain these compositional changes on the basis of specific biochemical reactions that might be expected under different growth or physiological conditions (i.e. why would a cell growing slowly due to lack of phosphorus produce predominantly 11-hydroxysulfate toxins?). We are thus in the awkward position of knowing that the compositional changes described above will someday be logical given a more complete knowledge of biosynthetic pathways and cell physiology during growth, but at at present, this background knowledge is lacking and we are left with unexplained observations.

Toxin Content and Net Toxin Production Rates. The toxin content of individual isolates has been shown to vary with temperature, salinity, irradiance, degree of nutrient limitation, and with growth stage in culture,[1-9,17,19,20]. In an attempt to synthesize these diverse observations, all of which are from batch cultures, it is first necessary to recognize the dynamic nature of these batch cultures. Growth conditions are initially optimal, but then rapidly change as one nutrient is depleted or when growth is limited by other variables such as light or CO_2 depletion. Generally, toxin content is highest in mid-exponential growth, decreasing thereafter as the cells enter stationary phase [1,3,5,8,17,19]. This means that during the early stage of growth with optimum conditions with no nutrient limitation or stresses, toxin accumulates because it is produced faster than it is transferred to daughter cells during division. The optimum conditions are short-lived, however, and the cells soon experience nutrient limitation or the effects of CO_2 depletion (which presumably causes the eventual cessation of growth in nutrient-replete cultures). The subsequent decrease in toxin content that is often observed indicates that the rate of toxin production decreases even though cell division continues. This occurs with both nitrogen limitation and CO_2 limitation [1,3,5,8,19], but when phosphorus is limiting, toxin content continues to rise, even after the cells have stopped dividing [5,18,19]. These changes in late stages of growth indicate several important points: (a) the pathways of toxin synthesis are more easily inhibited by nitrogen limitation or CO_2 depletion (pH or CO_2 limitation) than are those involved in cell replication; and (b) cell division is more sensitive to phosphorus limitation than is toxin synthesis.

Cultures grown under sub-optimal temperatures show the same convex toxin content curve through time described above for many batch cultures, but the entire curve is elevated. This type of enhanced toxicity superimposed on the growth stage variability described above may also be the case with light limitation, although published data are not sufficiently detailed in time [9] for this to be verified.

The first attempt to synthesize these different observations was by Proctor et al., [1] and Ogata et al., [9] who argued that toxin content was inversely proportional to growth rate. This conclusion was based on the high toxin content of cells grown at sub-optimal temperatures and light. It is now clear that this is not a general relationship that applies to all growth conditions, since recent studies using nitrogen-limited semi-continuous cultures of Alexandrium revealed a direct proportionality between growth rate and toxin content [19], the opposite of the hypothesized inverse relationship. Phosphorus limited semi-continuous cultures did follow an inverse relationship, however, so at least three variables (temperature, light, phosphorus) seem to affect growth rate and toxin accumulation in a similar manner. This relationship is only relevant in comparisons between growth treatments, such as with cultures grown at different temperatures. Within one culture, the opposite generally applies, namely that toxin content is highest when the cells are growing fastest.

But why would cells growing slowly due to sub-optimal temperatures contain more toxin than faster growing ones? Are they producing toxin at the same rates, but retaining more due to a lower rate of cell division? How and why does the rate of toxin production vary with growth rate under different conditions? To address these questions it is necessary to move beyond toxin content as a measure of toxicity. Toxin content reflects the difference between toxin production and a series of loss terms that include catabolism, leakage into the medium, and most importantly, toxin transferred to new cells during division. A more useful measure for comparing the dynamics of toxin accumulation under different conditions would be based on the rate of production (fmol toxin $cell^{-1}$ day^{-1}). This approach was taken in an extensive study of several Alexandrium isolates to be published elsewhere [19]. A few conclusions from that study should be emphasized. In batch culture, there is a direct proportionality between growth rate and net toxin production rate (i.e. as growth rate increases, so does the rate of toxin production). This is not a 1:1 relationship, however, as toxin is consistently produced in excess of the amount transferred to daughter cells when nutrients are abundant and growth rapid. As CO_2 depletion or nitrogen limitation occurs in late exponential growth, toxin is "lost" to daughter cells faster than it is produced. These toxin production excesses or deficits are generally within a factor of 2 of levels needed for balanced growth, but the differences are sufficient to produce the typical convex toxin content curve through time in batch culture. In the approximate steady state of semi-continuous cultures, the net toxin production rate is also directly proportional to growth rate, again indicating that the faster the cells grow, the faster they produce toxin. Since there are conditions under which toxin synthesis is blocked but cell division continues and other conditions where the opposite occurs, there is no simple relationship between growth rate and toxin content.

By themselves, these observations tell us little about the regulation of toxin production. However, in conjunction with a series of concurrent physiological measurements, some useful relationships emerge [19]. For example, in most of the batch cultures, free arginine (the amino acid precursor to saxitoxin [18]) varied as the mirror image of toxin content, being very low when toxin content peaked and increasing rapidly when toxicity declined. Arginine synthesis thus continued and accumulated in the cell under conditions that inhibited toxin synthesis. In contrast, arginine pools remained low in the P limited culture that increased so dramatically in toxin content during stationary phase. Phosphorus limitation thus blocked pathways necessary for cell division but did not affect toxin synthesis. Interestingly, the low temperature culture that had elevated toxin content contained high arginine levels during early exponential growth. Since the protein level of these cells was low, it may be that the low temperature inhibited protein synthesis (and thus growth rate), resulting in a surplus of arginine within the cell that could be used for toxin synthesis. The general hypothesis that emerged was that enhanced toxin production is associated with increased availability of arginine in the cell at times when the toxin biosynthetic pathway is still functional [19]. Arginine can be made more available through the blockage of competing pathways (e.g. for general cell division or protein synthesis) or from a reorganization of N metabolism that results in greater activity of the arginine synthetic pathway. The latter has been suggested as an explanation for enhanced _de novo_ synthesis and accumulation of arginine in P-deficient citrus leaves, possibly as a mechanism for detoxifying excess ammonia [21]. If this mechanism holds for phosphorus deficient _Alexandrium_ cells, arginine would not accumulate due to the existence of an alternative pathway that requires arginine (toxin synthesis).

The variable toxin contents of cells at different stages of growth and under different growth conditions clearly reflect a number of different mechanisms and physiological conditions. The discussion above emphasizes the complexity of the toxin biosynthetic process and stresses the need for concurrent studies of the pool sizes and regulation of arginine and other important metabolites.

CELL CYCLE VARIABILITY

All studies of toxin variability in _Alexandrium_ species have examined toxin changes on time scales from days to weeks. The discussion above demonstrates how toxin production and growth can be closely coupled. Without knowledge of the metabolic role of the saxitoxins in the dinoflagellate, however, such data once again remain observations without physiological explanations. Reasoning that we need to know more about the finer details of toxin synthesis as cells divide, we initiated a cell cycle study of one _A. fundyense_ isolate for which the preliminary data can be summarized here.

The generalized cell division cycle for eukaryotic cells is usually depicted with 4 discrete sequential intervals called G_1, S, G_2, and M (Fig. 1). G_1 and G_2 are the gaps separating DNA synthesis (S) and cell division or mitosis (M). The time it takes to traverse one complete cell division is the generation time of the cell; the duration of each cell cycle stage varies with cell type and with different growth conditions.

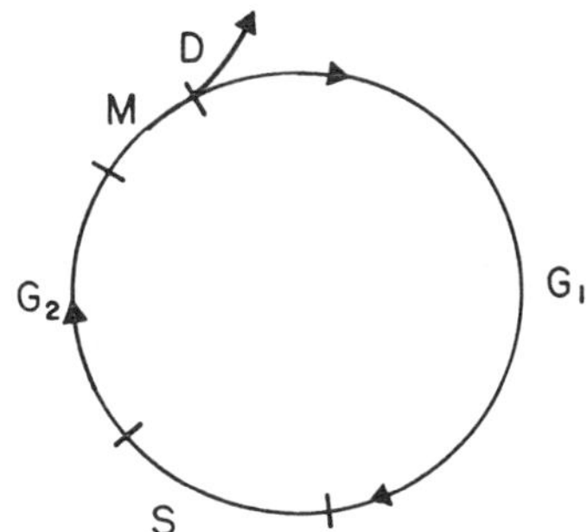

FIG. 1. Generalized cell cycle diagram. S corresponds to DNA synthesis, M to mitosis, D to cytokinesis and daughter cell separation. G_1 and G_2 are gaps between these events.

The objective of our study, was to see whether toxin synthesis is continuous during the cell cycle, or occurs instead during discrete intervals. In order to conduct a study of this type, the culture had to be synchronized such that the divisions of individual cells in a population were aligned as tightly as possible to result in simultaneous division. In essence, the objective was to induce an entire culture to express the behavior of a single cell. This required prolonged storage in constant darkness, after which the culture was released to a normal 14:10 L:D cycle. Cell concentration, cell volume, toxicity, and some physiological parameters were measured at hourly intervals. The number of cells in each cell cycle stage was also determined using propidium iodide stain and flow cytometry. As seen in Fig. 2A, a small amount of division occurred 15-20 hrs after the normal photocycle was re-established, but the major division burst caused the cell concentration to double over a 6 hr interval beginning 45 hours after the lights were first turned on. As seen in Figure 2C, toxin analyses are not yet complete, but those samples that have been run suggest some important relationships. Figure 2C shows that toxin content began to increase slightly before the increase in DNA synthesis that was first apparent at hr 34 and peaked at hr 40. The subsequent decrease in toxin content through hr 48 coincided with the division burst. Simple book-keeping of the total toxin in the culture shows that toxin synthesis occurred during the first half of the S phase, but then completely stopped for more than 10 hrs as the cells went through mitosis and divided at the end of our experiment. Toxin synthesis was thus not continuous throughout the cell cycle, but was temporally linked to the pulse of DNA synthesis. Other

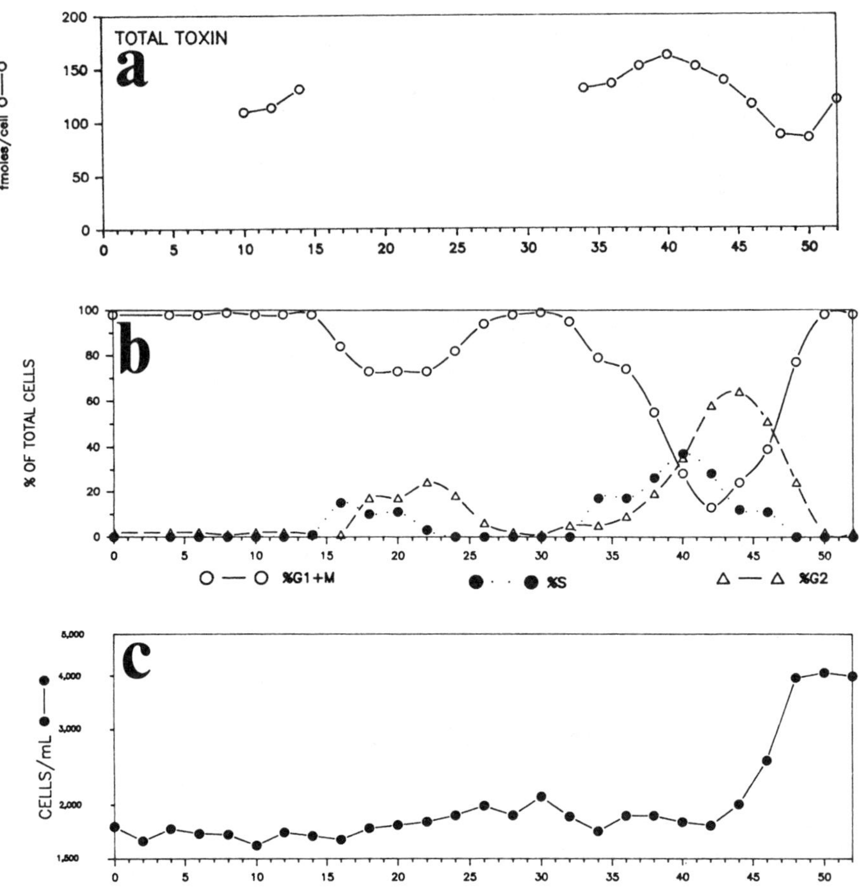

FIG. 2. Toxin production during the Alexandrium cell cycle. Starting at time 0, the cells experienced a regular 14:10 L:D photocycle. (a) Total toxin (fmoles/cell) (missing points represent samples not yet analyzed due to a shortage of standards). (b) Percentage of cells in each cell cycle stage. O = G_1+M, ● = S, and Δ = G_2. Note that prior to the major division burst, DNA synthesis began at hour 34. (c) Alexandrium fundyense cell concentration. Note the major division burst beginning hour 44.

experiments with two complete division bursts are underway to test the general validity of this important observation.

We now can hypothesize that some of the variability in toxicity under different growth conditions may be due to cell cycle variability. For example, Olson et al., [22] showed that low temperatures caused an increase (up to 5 fold) in the duration of all cell cycle phases of a diatom and a coccolithophore, but increases in only G_1 when nitrogen was limiting. If these relationships are valid for Alexandrium, we could explain the elevation in toxin content observed at low temperatures [5,9,19,20]. Specifically, when low temperature slows growth, the S phase will be proportionately longer, allowing toxin to be produced for a longer time between divisions and resulting in an elevated toxin content. (Recall that batch culture experiments showed that toxin synthesis could proceed at high rates even at low temperatures that inhibited protein synthesis and growth [19]). There are no published data showing the effect of P limitation on cell cycle stage durations in phytoplankton, but it is possible that the S phase might be lengthened (and possibly other stages as well) due to a shortage of P for energy and for nucleic acid synthesis. Thus the enhanced toxicity in P-limited Alexandrium might also be cell cycle related. These are clearly speculations based on a rather limited dataset, but they do demonstrate the level of knowledge that we must have if we are to understand toxin variability over small time scales. Cell cycle studies will not answer all of our questions about toxin variability, but they should provide insights relevant to many of our presently unexplained observations.

OVERVIEW

Toxin variability in Alexandrium isolates takes a variety of forms. Differences in total potency or toxin content between populations can now be attributed in part to heterogeneity in toxin composition, but explanations for those differences and for geographic trends in toxicity remain out of reach. Such trends may reflect environmental selection of certain genotypes, but it's too soon to even guess what role the different toxins might play (if any) in that selection. Alternatively, geographic patterns in toxin content or composition might simply reflect dispersal patterns. For individual isolates, we now see that the commonly accepted paradigm of invariant toxin composition is wrong and that dramatic changes are possible with different growth conditions. Here again, explanations for the observed compositional changes await more knowledge of organism physiology and biochemistry. In addition to toxin compositional changes, the number of moles of toxin in individual isolates (toxin content) can vary with growth conditions. This variability takes two forms in batch culture. One is a growth stage effect that reflects the balance between variable rates of toxin synthesis and the dilution of that toxin through cell division. A direct proportionality between growth rate and the rate of toxin production is generally observed, even though the rate of toxin production can vary by more than two orders of magnitude under different growth conditions. The second type of toxin variability in batch culture is a general enhancement of toxicity due to temperature or light effects - an enhancement seen in addition to the growth stage variability described above. There are no physiological explanations for this enhancement, but one possibility is that variability in the duration of the different cell cycle stages might regulate the amount of toxin produced in each division cycle, and thus determine the toxin content. Alternatively, the combined effects of these environmental variables on the pool size of free arginine and the activity of the toxin biosynthetic pathway could determine the level of toxin accumulation. As we learn more about toxin variability, these mechanisms will undoubtedly become clearer, and hopefully will enable us to address the fundamental question of whether the saxitoxins are important in Alexandrium metabolism and cell replication or are instead only secondary metabolites with no essential functional role.

ACKNOWLEDGEMENTS

This work was supported in part by the National Science Foundation (OCE-8614210), by the Office of Sea Grant in the National Oceanic and Atmospheric Administration through grant NA86AA-D-SG0090 (R/B-76) and by the Donaldson Charitable Trust. Contribution No.7158 from the Woods Hole Oceanographic Institution.

References

1. N.H. Proctor, S.L. Chan, and A.J. Trevor, Toxicon 13, 1-9 (1975).
2. A.W. White, J. Phycol. 14, 475-479 (1978).
3. A.W. White, and L. Maranda, J. Fish. Res. Bd. Can. 35, 397-402 (1978).
4. Y. Shimizu, in: Toxic Dinoflagellate Blooms, D.L. Taylor and H.H. Seliger, eds. (Elsevier, New York 1979) pp. 321-326.
5. S. Hall, Ph.D. Thesis, Univ. of Alaska, Fairbanks, AK. (1982).
6. L. Maranda, D.M. Anderson and Y. Shimizu, Est. Coast. Shelf Sci. 21, 401-410 (1985).
7. A.D. Cembella, J.J. Sullivan, G.L. Boyer, F.J.R. Taylor and R.J. Anderson, Biochem. System. Ecol. 15, 171-186 (1987).
8. G.L. Boyer, J.J. Sullivan, R.J. Anderson, P.J. Harrison, and F.J.R. Taylor, Mar. Biol. 96, 123-128 (1987).
9. T. Ogata, T. Ishimaru, and M. Kodama, Mar. Biol. 95, 217-220 (1987).
10. A.A. Genenah and Y. Shimizu, J. Ag. Food Chem. 29, 1289-1291 (1981).
11. S. Hall and P.B. Reichardt, in: Seafood Toxins, E.P. Ragelis (ed.) (Amer. Chem. Soc., Washington, D. C. (1984) pp. 113-123.

12. D.M. Anderson - unpublished data.
13. F.J.R. Taylor, in: Toxic Dinoflagellates, D. M. Anderson, A. W. White, and D. G. Baden, (eds.). (Elsevier, New York, 1985) pp. 11-26.
14. F.J.R. Taylor, Environ. Letters. 9, 103 (1975).
15. A.D. Cembella, and F.J.R. Taylor, Biochem. System. Ecol. 14, 311 (1986).
16. B.A. Hayhome, D.M. Anderson, D.M. Kulis and D.J. Whitten. Mar. Biol. 100, in press. (1989).
17. B.A. Boczar, M.A. Beitler, J. Liston, J.J. Sullivan, and R.A. Cattolico, Plant Physiol. 88, 1285 (1988).
18. Y. Shimizu, M. Norte, A. Hori, A. Genenah, and M. Kobayashi, J. Amer. Chem. Soc. 106, 6433, (1984).
19. D.M. Anderson, D.M. Kulis, J.J. Sullivan, S. Hall, and C. Lee, Mar. Biol. (in press).
20. A.W. White, Toxicon 24, 605 (1986).
21. E. Rabe, and C.J. Lovatt, Plant Physiol. 81, 774 (1986).
22. R.J. Olson, D. Vaulot and S.W. Chisholm, Plant Physiol. 80, 918 (1986).

POSSIBLE LINKS BETWEEN BACTERIA AND TOXIN PRODUCTION IN ALGAL BLOOMS

MASAAKI KODAMA
Laboratory of Marine Biological Chemistry, School of Fisheries Sciences, Kitasato University, Sanriku, Iwate 022-01, Japan

ABSTRACT

Bacterial production of paralytic shellfish toxins (PSP toxins) was reviewed. Several species of marine bacteria which were isolated from toxic dinoflagellates produced PSP toxins. Toxin productivity of these bacteria increased when they were grown under starved conditions. Main toxin components of bacteria grown under these conditions were gonyautoxins which were major toxins of dinoflagellates, suggesting the possible association of these bacteria with toxin production of dinoflagellates. It was also suggested that bivalves become toxic by ingesting these bacteria.

INTRODUCTION

Paralytic shellfish toxins (PSP toxins) are potent neurotoxins produced by several species of dinoflagellates such as *Protogonyaulax tamarensis* etc. [1]. In the bloom of these species, bivalves become toxic by ingesting them. Contamination of PSP toxins to bivalves poses a severe problem to their culture industry as well as public health. Therefore, many studies on the causative dinoflagellates and the toxins they produce have been carried out in the countries where this form of poisoning has been known to occur, e.g. Canada, U.S.A. and Japan. At present, taxonomy and ecology of the organisms, as well as chemical nature of the toxins, are well elucidated. However, little is known about biosynthesis or physiological role of toxins in these dinoflagellates. Recently, we have found that PSP toxins are produced by bacteria which is coexisting with toxic dinoflagellates [2,3]. In this article, topics on bacterial production of PSP toxins are reviewed, especially in relation to toxin production of dinoflagellates.

TOXIN PRODUCTION OF *P. TAMARENSIS* IN VARIOUS CONDITIONS

Toxin production of *P. tamarensis* has been suggested to be affected by various physiological or environmental factors: the toxicity of *P. tamarensis* is different in its various growth stages [4-8]; salinity affects the toxin production of *G. excavata* (*P. tamarensis*)[9]; toxin production of *P. tamarensis* changes during its cell cycle [10]. In a field survey, it was also suggested that the toxicity of *P. tamarensis* which occurs under a lower temperature is greater than that which occurs under a higher temperature [11]. Recently, photosynthesis is reported to be essential for toxin production of *P. tamarensis* [12].

On the other hand, there are various data on the toxicity of different strains of *P. tamarensis* isolated from various regions of the world [6-10, 12-14]. As the toxicity of *P. tamarensis* is known to be affected by physiological or environmental factors, these data cannot be directly compared. However, the toxicities of these strains seem to differ from each other. White [15] reported the high toxin content of *Gonyaulax excavata* (*P. tamarensis*) in nature. He also demonstrated that the toxicity of the cell in nature fluctuated during the bloom, suggesting that the toxicity of the cells in different water masses was different, even if they

Toxic Marine Phytoplankton
Edna Graneli et al., Editors

bloomed in the same bay during the same period. These data suggest that the toxicity of different clones differs from one another. We examined the toxicity of various clones isolated from the same area during the same season, by culturing under the same conditions [16]. In Fig. 1 are plotted the toxicity of various clones of P. tamarensis against their growth rate (μ_2). No clear correlation was obtained. Maximum differences in toxicity and growth rate were about 100 and 4 times, respectively. These results

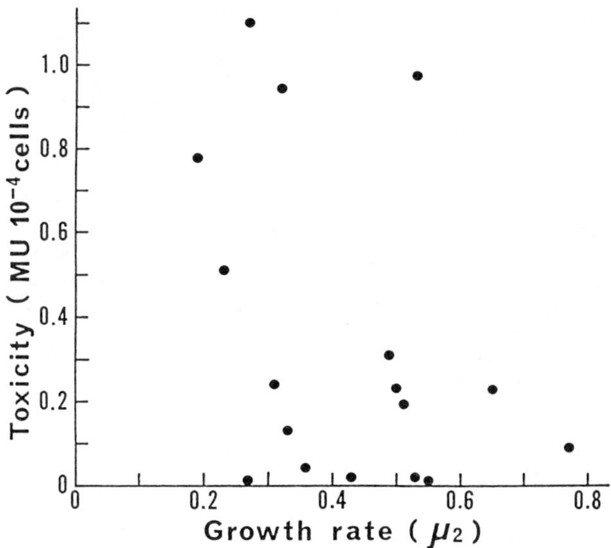

FIG. 1. Toxicity of various clones of P. tamarensis as a function of growth rate [16].

showed that growth characteristics and toxin production of many of the clones were different from each other. The large difference of toxicity among the various clones might be explained by the heterothalism of P. tamarensis [17]. The occurrence of nontoxic strains [18-20] would support this idea. However, the toxicities of various subclones from a single clonal strain are also different from each other (maximum difference is about 20 times). The cultures of these subclones are prepared from a single cell isolated from a clonal culture. Thus, genetic type of each subclone is identical. From these, we concluded that the toxin production of P. tamarensis is not a hereditary characteristic, that is, the information on toxin production is not coded for in the genes of P. tamarensis.

INTRACELLULAR BACTERIA IN P. TAMARENSIS

Toxin production in P. tamarensis is suggested to be an acquired characteristic. Association of external bacteria in the toxin production of P. tamarensis can be considered plausible. However, confirmation of toxin production of P. tamarensis under axenic state [7] denies this hypothesis. On the other hand, some bacteria are known to occur in the inside of the cell in some species of phytoplankton [21]. Intracellular bacteria are also observed in P. tamarensis [22]. Based on this observation, Silva [23] presented a hypothesis that P. tamarensis toxins are not produced by P. tamarensis itself, but by the intracellular bacteria or their association with P. tamarensis. However, efforts to find the intracellular bacteria in P. tamarensis by another workers resulted in no significant finding [24]. We examined sections of various clones of P. tamarensis under the electron microscope [25]. Some of the examples are

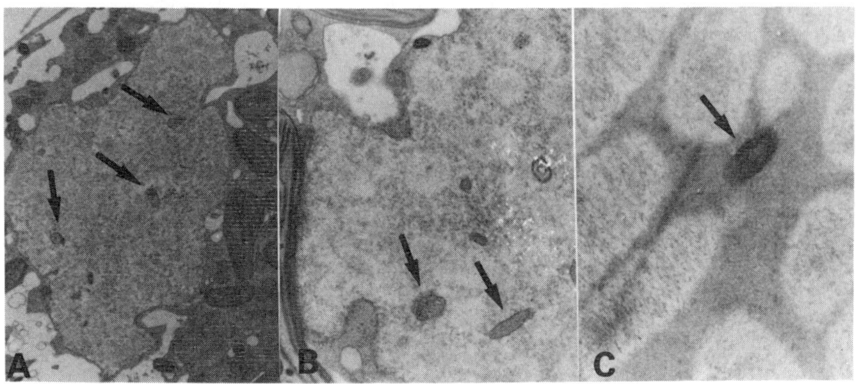

FIG. 2. Electron microscopic observation of intracellular bacteria in P. tamarensis [25]. Bacteria are indicated by arrows.

shown in Fig. 2. In these cases, the outline of the chromosomes is not clear (Fig. 2-A and B). However, cells showing clear chromosomes but having bacteria were also found (Fig. 2-C).

Based on this observation, isolation of bacteria from P. tamarensis was attempted. The clone having intracellular bacteria was cultured in the media containing antibiotics [3]. The cells showed normal growth, indicating that antibiotics did not affect the growth. The harvested cells at the late-exponential phase were washed thoroughly with sterilized sea water and homogenized. The homogenates were inoculated to nutrient agar plates and incubated for 24 hr. Colonies of bacteria appeared on every plate. All the colonies showed the same shape and color, indicating that these are single species. According to the criteria proposed by Shimidu

TABLE I. Relationship between mounts of bacteria derived from P. tamarensis and toxicity of P. tamarensis [2].

Culture*	Number of colonies from 10^4 cells	Toxicity of the cell MU/10^4 cells
A	15,900	0.578
B	1,500	0.097
C	330	0.057

*Culture was conducted in the medium without antibiotics (A), containing gentamicin sulfate, sodium ampicillin and sodium cloxacillin in concentrations of 5, 50, 50 mg/L (B) and 15, 150, 150 mg/L (C).

[26], the isolate was assigned to Moraxella sp. Table I shows the relationship between toxicity of P. tamarensis and number of bacterial colonies derived from P. tamarensis [3]. The number of bacterial colonies derived from P. tamarensis decreased, as the concentration of antibiotics mixture which added to the medium of P. tamarensis increased. In contrast, the toxicity of P. tamarensis decreased with increase of antibiotics mixture concentration, indicating the positive relationship between the toxicity of P. tamarensis and the number of colonies derived from P. tamarensis. In addition, bacteria were obtained from the cultures of only toxic strains of dinoflagellates [27]. These results also support Silva's hypothesis [23].

PRODUCTION OF PARALYTIC SHELLFISH TOXINS BY BACTERIA ISOLATED FROM TOXIC DINOFLAGELLATES

A bacterium Moraxella sp. was isolated from P. tamarensis. This bacterium produced saxitoxin (STX) when cultured in a normal nutrient medium, although toxin productivity was very low [2,3]. Under the conditions applied, gonyautoxins (GTXs), major toxins in P. tamarensis were detected neither in the cultured cells nor cell-free medium. Generally, production of substances by bacteria depends on the culture conditions. Thus toxin production of Moraxella sp. grown under various conditions was examined.

Fig. 3 shows the toxicity change of Moraxella sp. cell grown under various culture conditions in commercially available media [28]. No significant toxicity was observed in all the cultures during 64 hr. Only the toxicity of the cells grown in Heart Infusion (Difco) by standing

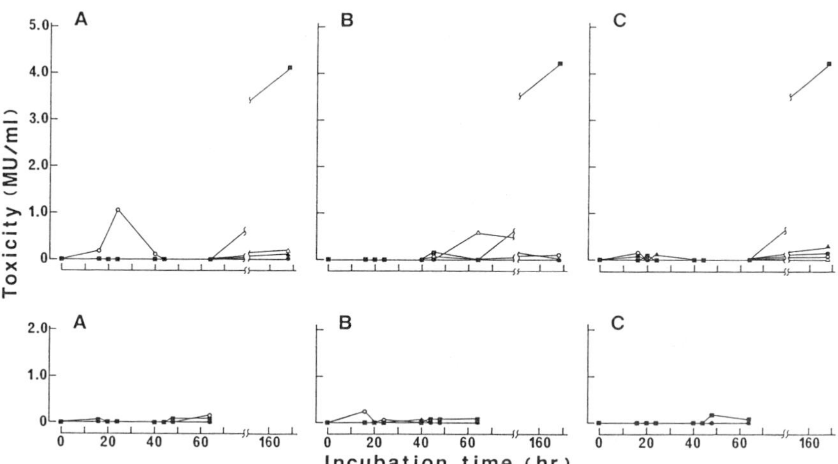

FIG. 3. Time course of toxin production by Moraxella sp. under various conditions [28].
The toxicity of cells grown under various conditions was assayed by mouse neuroblast cell culture method [31,32].
(○): Heart Infusion, (●): Brain Heart Infusion, (△): Tryptic-Soy Broth, (▲): CAYE medium containing 1% Casamic acid and 1% yeast extracts, (■): Marine Broth; A: standing, B: shaking, C: rotatory culture methods.
Upper: 25°C, Lower: 37°C, A: standing culture, B: shaking culture, C: rotatory culture.

culture at 25 C increased during 24 hr, whereas it decreased after 24 hr. The cells in Tryptic-Soy Broth (Difco) by shaking culture at 25°C also showed weak toxicity at 64 hr which disappeared after 168 hr incubation. On the other hand, Moraxella sp. showed considerable toxicity in all the culture methods when cultured in Marine Broth (Difco) at 25°C for 168 hr, suggesting that the toxin production increases under starved condition. In order to culture Moraxella sp. in starved condition, it was inoculated to extremely diluted medium (1% Marine Broth, diluted with seawater) or natural seawater [29]. The cell growth in these media was too poor to be traced by turbidity. However, slight turbidity was observed after 10 days incubation, showing that it did grow slightly in these media. Fig. 4 illustrates a HPLC chromatogram of the extract of Moraxella sp. cells grown in the seawater for 10 days. Clear peaks corresponding to GTX 1 and 4 followed by those of GTX 2 and 3 appeared. Faint peak of neoSTX was also observed. These results show that main toxin components of Moraxella sp. grown in these conditions are GTXs. Fig. 5 and Table II show the changes of total toxin amounts per 1 L of culture and proportion of toxin components which were calculated from the results of HPLC-fluorometric analysis, respectively. In the seawater, Moraxella sp. produced PSP toxins within 24 hr, but the amount was low during a week. However, it increased considerably at 10th day. The pattern of toxicity change in 1% Marine Broth was similar to that in the seawater, although the toxicity did not increased as in the seawater. In both cases, the growth was too poor to be

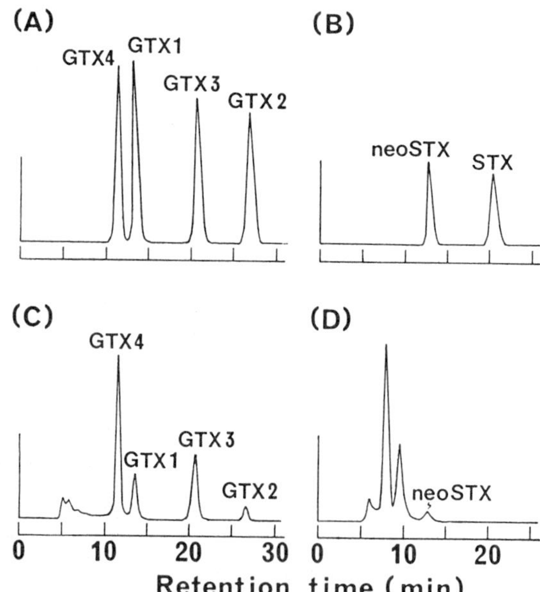

FIG. 4. HPLC-fluorometric analysis of the extract from Moraxella sp. cells cultured in natural seawater for 10 days at 20°C [28].
Mobile phase for GTX fractions, (A),(C): 2 mM 1-heptanesulfonic acid/10 mM phosphoric acid (pH 7.2); mobile phase for STX fractions, (B),(D): 2 mM 1-heptanesulfonic acid/10 mM phosphoric acid/10% acetonitrile (pH 7.2).
(A): GTX standards, GTX1 (5 nmole), GTX2 (1.3 nmole), GTX3 (0.5 nmole), GTX4 (12 nmole); (B): neoSTX and STX standards, neoSTX (9 nmole), STX (5 nmole); (C) and (D): extract of Moraxella sp.

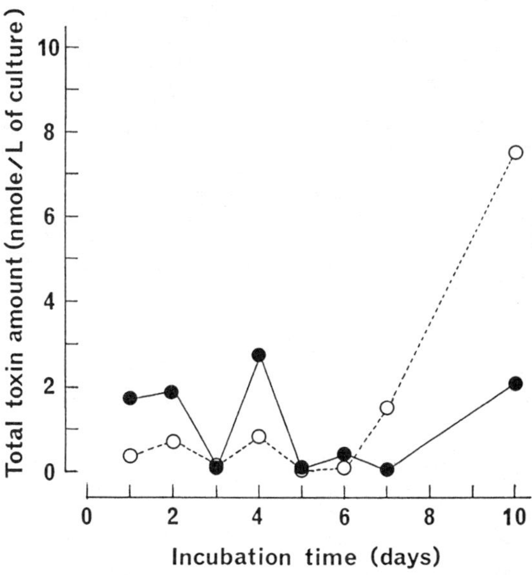

FIG. 5. Changes of total toxin amount during culture [28].
●: culture in 1% Marine Broth; ○: culture in natural seawater.

TABLE II. Toxin composition of Moraxella sp. cultured in natural seawater and 1% Marine Broth [28].

Culture medium	Incubation time (day)	Toxin composition (mol%) GTX1	GTX2	GTX3	GTX4	neoSTX	STX
	1	41.0	11.8	2.7	44.5	-	-
	2	25.6	8.0	2.2	64.2	-	-
	3	100.0	-	-	-	-	-
Natural seawater	4	50.5	8.7	1.8	28.7	-	10.3
	5	100.0	-	-	-	-	-
	6	70.5	23.7	5.8	-	-	-
	7	56.7	-	-	43.3	-	-
	10	10.0	0.5	1.4	87.3	0.7	-
	1	28.1	-	2.5	69.4	-	-
	2	20.3	6.2	1.8	71.7	-	-
	3	37.0	-	-	63.0	-	-
1% Marine Broth	4	1.0	0.1	3.4	95.5	-	-
	5	-	-	-	100.0	-	-
	6	34.8	4.4	-	60.8	-	-
	7	90.7	9.3	-	-	-	-
	10	40.7	10.4	1.6	47.3	-	-

traced by the turbidity, showing that the number of harvested cells was much smaller than that grown in normal media. This fact indicates that toxin content per cell grown in nutrient deficient conditions is much higher than that in commercial media, and supports the results described above that Moraxella sp. produces toxin in starved conditions. Toxin profile in both cases changed throughout the incubation period (Table II). However, main components were always GTXs, especially GTX 1 and 4. In the extreme cases, most of the toxin was GTX 1 or 4. STX and neoSTX were scarcely observed. The toxin profile obtained here is comparable to that of P. tamarensis from which Moraxella sp. was isolated, showing that PSP toxins Moraxella sp. produces depend on the culture conditions. As described above, other species of bacteria were isolated from different strain of P. tamarensis or other toxic species of dinoflagellates than P. tamarensis [27]. When the culture conditions described above were applied to some of them, high productivity of PSP toxins was also observed [27]. These results show that toxic dinoflagellates possess bacteria with toxin productivity which seem to be linked with toxin production of dinoflagellates.

MARINE BACTERIA, POSSIBLE CAUSATIVE ORGANISMS OF PARALYTIC SHELLFISH POISONING

It has been confirmed by many authors that bivalves become toxic during blooms of toxic dinoflagellates. Needler [30] monitored the abundance of G. tamarensis and the toxicity of mussels in the Bay of Fundy and reported that both parameters correlated well. We have also monitored these parameters in detail at Ofunato Bay, Iwate Prefecture, Japan since 1981 [11, 31]. The toxicity of scallop well correlated with the abundance of P. tamarensis. However, the scallop toxicity increased often after complete disappearance of P. tamarensis, suggesting the presence of unknown causative organism(s) than dinoflagellates. In 1988, this phenomenon was also observed (Fig. 6). Therefore, a 100 L of seawater was collected in July and December of 1988 from the station. The seawater samples were filtered through the sieve with an opening size of 20 µm and membrane filters with pore sizes of 5 and 0.45 µm, successively. The toxin in the particles trapped by each sieve were analyzed by HPLC-fluorometric analysis.

The extracts from all the fractions of the waters collected at July showed peaks of PSP toxins in HPLC-fluorometric analysis. Fig. 7 shows the chromatograms of each fraction collected at July 24 [29]. Most of the toxicity was concentrated into 0.45 - 5 µm fraction. A large peak corresponding to GTX 1 was detected in the 0.45 - 5 µm fraction with small peaks corresponding to GTX 2,3,4. Trace amounts of STX and neoSTX were detected. Calculated toxin amounts in 0.45-5 µm fraction was 61.8 MU per 100 L, which is likely to be responsible for increase of scallop toxicity. This fraction showed sodium channel blocking activity in neuroblast cell culture bioassay [31]. Assumed toxicity from the standard curve of STX in this method [32] was comparable to that from HPLC analysis (about 50 MU/100 L). In contrast, toxin contents of the same fraction collected at December when no P. tamarensis was observed and scallop toxicity was decreasing, was negligible (0.1 MU). PSP toxins were also detected in the same particle fraction prepared from cell-free cultured medium of P. tamarensis. In HPLC-fluorometric analysis, a clear peak of GTX 4 followed by a large peak of GTX 1 appeared (Fig. 8)[28]. The faint peaks corresponding to GTX 2,3, neoSTX and STX were also detected. This fraction also showed sodium channel blocking activity in neuroblast cell culture assay.

In the seawater or cell-free cultured medium of P. tamarensis in which PSP toxins were detected, bacteria reacted with anti-Moraxella sp. serum were found, suggesting that PSP toxins in these samples originate from Moraxella sp.

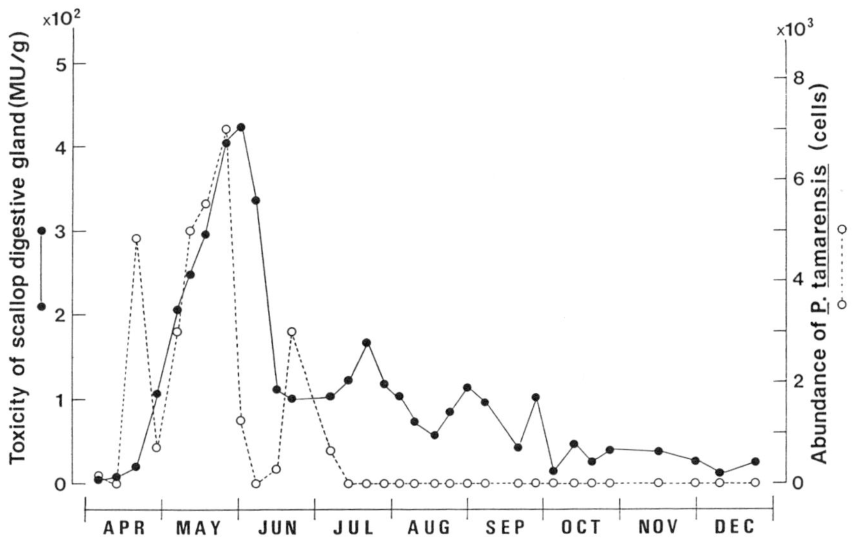

FIG. 6. Seasonal variation of the scallop toxicity and the abundance of Protogonyaulax tamarensis at Ofunato Bay in 1988 [29].
Abundance of P. tamarensis is expressed as sum of cell number observed in 1 liter of each 2 m layer from surface to bottom.

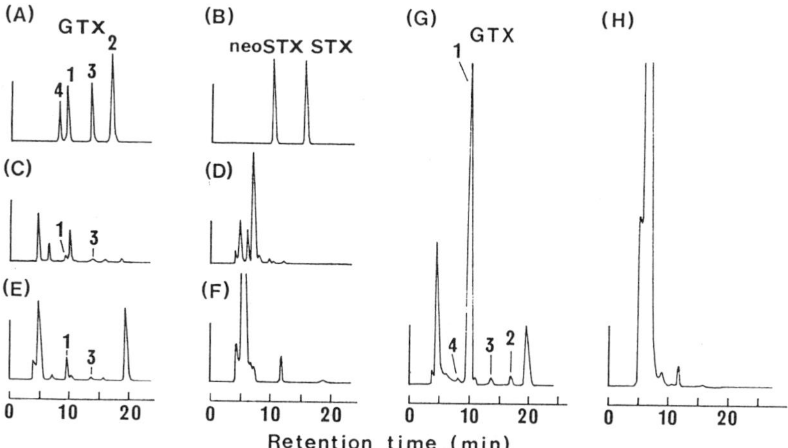

FIG. 7. HPLC-fluorometric analysis of the extract of each particle fraction from seawater [29].
Mobile phase for GTX fractions, (A),(C),(E),(G),and mobile phase for STX fractions, (B),(D),(F),(H): refer to Fig. 4.
(A): GTX standards; (B): neoSTX and STX standards; (C) and (D): over 20 μm fraction; (E) and (F): 5 - 20 μm fraction; (G) and (H): 0.45 - 5 μm fraction.

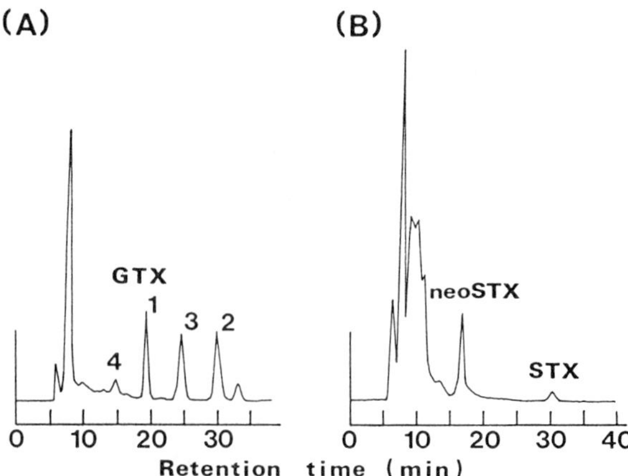

FIG. 8. HPLC-fluorometric analysis of the extract of 0.45 - 5 μm fraction from cultured medium of P. tamarensis [29].
Mobile phase for GTX fraction (A) and STX fraction (B): refer to Fig. 4.

Nelinda et al. [24] isolated extracellular bacteria from the cultured medium of various strains of P. tamarensis, cultured them in nutrient rich media, and examined their toxin productivity. However, they did not detect any PSP toxins in the bacterial extracts. Toxin production of their isolates grown in nutrient rich conditions might be too small to be detected.

It has been a general belief that dinoflagellates are the primary producers of PSP toxins and responsible for the shellfish toxicity. However, the present data suggest that PSP toxins-producing bacteria as well as toxic dinoflagellates are responsible for the toxicity of bivalves. Interestingly, no increase of scallop toxicity was observed without bloom of P. tamarensis during monitoring survey from 1981-1988 [33]. This fact suggests the close relationship between P. tamarensis and toxin-producing bacteria.

REFERENCES

1. E.J. Schantz, Annals N.Y. Aca. Sci. 479, 15-23 (1986).
2. M. Kodama, T. Ogata, and S. Sato, Agric. biol. Chem. 52: 1075-1077 (1988).
3. M. Kodama in: Mycotoxins and Phycotoxins '88, A Collection of Invited Papers Presented at the Seventh International IUPAC Symposium on Mycotoxins and Phycotoxins, Tokyo, Japan, 16-19 August 1988, S. Natori, ed. (Elsevier, Amsterdam) pp. 391-398.
4. A. Prakash, J. Fish. Res. Bd. Canada 24, 1589-1606 (1967).
5. A.W. White and L. Maranda, 35, 397-402 (1978).
6. Y. Oshima and T. Yasumoto, in: Toxic Dinoflagellate Blooms, D.L. Taylor and H.H. Seliger, eds. (Elsevier, Amsterdam 1979) pp, 377-380.
7. H.T. Singh, Y. Oshima, and T. Yasumoto, Nippon Suisan Gakkaishi 48, 1341-1343 (1982).
8. G.L. Boyer, J.J. Sullivan, R.J. Andersen, P.J. Harrison, and F.J.R. Taylor, in: Toxic Dinoflagellates, D.M. Anderson, A.W. White, and D.G.

Baden, eds. (Elsevier, Amsterdam 1985) pp. 281-286.
9. A.W. White, J. Phycol. 14, 475-479 (1978).
10. M. Kodama, Y. Fukuyo, T. Ogata, T. Igarashi, H. Kamiya, and F. Matsuura, Nippon Suisan Gakkaishi 48, 567-571 (1982).
11. T. Ogata, M. Kodama, Y. Fukuyo, T. Inoue, H. Kamiya, F. Matsuura, K. Sekiguchi, and S. Watanabe, Nippon Suisan Gakkaishi 48, 563-566 (1982).
12. T. Ogata, T. Ishimaru, and M. Kodama, Mar. Biol. 95, 217-220 (1987).
13. A.W. White and L. Maranda, J. Fish. Res. Bd. Can. 35, 397-402 (1978).
14. A. Prakash, J.C. Medcof, and A.D. Tennant, Bull. Fish. Res. Bd. Can. No. 177, 1-87 (1971).
15. A.W. White, Toxicon 24, 605-610 (1986).
16. T. Ogata, M. Kodama, T. Ishimaru, Toxicon 25, 923-928 (1987).
17. D.H. Turpin, P.E.R. Dobell, and F.J.R. Taylor, J. Phycol. 14, 35-38 (1978).
18. C.M. Yentsch, B. Dale, and J.W. Hurst, J. Phycol. 14, 330-332 (1978).
19. R.J. Schmidt and A.R. Loeblich III, J. Mar. Biol. Ass. U.K. 59, 479-487 (1979).
20. M. Kodama, T. Ogata, Y. Fukuyo, T. Ishimaru, P. Pholpunthin, S. Wisessang, K. Saitanu, V. Panikiyakarn, and T. Piyakarnchana, Nippon Suisan Gakkaishi 53, 1491 (1987).
21. G.F. Leedale, Osterr. Bot. Z. 116, 279-294 (1969).
22. E.S. Silva, in: Proceedings of International IUPAC Symposium on Mycotoxins and Phycotoxins, Lausane (Pahotox Publication, 1979).
23. E.S. Silva, Mar. Algae Pharm. Sci. 2, 269-288 (1982).
24. M. Nelinda, V. Dimanlig, and F.J.R. Taylor, in: Toxic Dinoflagellates, D.M. Anderson, A.W. White and D.G. Baden eds. (Elsevier, Amsterdam, 1985) pp. 103-108 .
25. M. Kodama and T. Ogata, Mar. Asia Pac. J. Pharmacol. 3, 99-109 (1988).
26. U. Shimidu, in: Methods for investigation of marine microorganisms, H. Kadota and N. Taga eds. (Japan Scientific Societies Press, Tokyo, 1985)(in Japanese) pp 228-233.
27. T. Ogata, M. Kodama, S. Sakamoto, S. Sato, and U. Shimidu, in this book.
28. M. Kodama, T. Ogata, S. Sakamoto, S. Sato, T. Honda, and T. Miwatani, submitted to Toxicon.
29. M. Kodama, T. Ogata, S. Sato and S. Sakamoto, submitted to Mar. Ecol. - Progress Series.
30. A.B. Needler, J. Fish. Res. Bd. Can. 7, 490-504 (1949).
31. K. Kogure, M.L. Tamplin, U. Simidu, and R.R. Colwell, Toxicon 26, 191-197 (1988).
32. S. Sato, T. Ogata, M. Kodama, and K. Kogure, in: Proceedings of the Japanese Association of Mycotoxicology, Supplement No.1, IUPAC'88 and ICPP'88, Mycotoxins and Phycotoxins, K. Aibara, S. Kumagai, K. Ohtsubo and T. Yoshizawa, eds. (The Japanese Association of Mycotoxicology, Tokyo, 1988) pp. 3-4.
33. K. Sekiguchi, N. Inoguchi, M. Shimizu, S. Saito, S. Watanabe, T. Ogata, M. Kodama, and Y. Fukuyo, in: Red Tides: Biology, Environmental Science, and Toxicology, T. Okaichi, D. Anderson, and T. Nemoto, eds. (Elsevier, Amsterdam, 1989), pp. 399-402.

BIOSYNTHESIS OF DINOFLAGELLATE TOXINS

YUZURU SHIMIZU,* SANDEEP GUPTA, AND A. V. KRISHNA PRASAD
Department of Pharmacognosy and Environmental Health Sciences, College of Pharmacy, The University of Rhode Island, Kingston, Rhode Island 02881, U.S.A.

ABSTRACT

Toxigenicity of dinoflagellates varies even in an identical species from proximate waters. This variation raises an interesting question about the toxin production or regulation mechanism in the organisms. As the first step to investigate the problem, we have been conducting the study of biosynthetic sequence of two representative classes of dinoflagellate toxins: saxitoxin and brevetoxin analogues. The molecules of saxitoxin analogues were shown to be derived from arginine, acetate, and methionine methyl. Our recent works using stable isotope markers, ^{13}C, ^{2}H and ^{15}N have further elucidated intricate mechanisms involved in the construction of the molecules. Brevetoxins produced by *Gymnodinium breve* were speculated to be unusual polyketides formed from dicarboxylic acid precursors. The hypothesis has been tested by using various precursors, such as ^{13}C-labelled succinate, propionate, glycerol, pyruvate, malonate, and acetoacetate.

INTRODUCTION

The production of so-called red tide toxins is surrounded with mysteries. Despite their notoriety, many organisms involved in red tides actually do not produce confirmed biotoxins. Even among those well-known toxic organisms such as *Alexandrium* and *Gymnodinium* spp., many species and strains do not produce toxins at least in culture. Varied toxin productivity in the strains of *Alexandrium* (=*Protogonyaulax*, *Gonyaulax*) has been documented by various researchers. For example, in a study on *Alexandrium* species found along the U.S. east coast, the toxicity levels were found to vary widely from non-toxic to extreme virulency [1]. This variation is also seen in other species such as *Pyrodinium bahamense* var. *compressa*, whose Caribbean sibblings are generally known to be non-toxic. *Gymnodinium breve* (=*Ptychodiscus brevis*) is currently only one brevetoxin producing species, although a number of closely related species are known to cause deleterious red tides. On the other hand, a number of distantly related organisms have been found to produce or contain saxitoxin, neosaxitoxin and gonyautoxins (Fig. 1). Okadaic acid derivatives responsible for so-called diarrhetic shellfish poisoning (DSP) are found in sponges, *Dinophysis* spp. and *Prorocentrum* spp.
The occurrence of such unique structures in taxonomically remote species led to various speculations regarding the mechanism of the toxin production in the red tide organisms. A popular thought is to set endosymbiotic or associated bacteria as common vectors. The theory has support by microscopic observation of certain organism cells [2]. In fact, Kodama, *et al.* reported the isolation of such an endosymbiotic bacterium from a Japanese *Alexandrium* species and identification of saxitoxin in its culture medium [3]. In the past decade, we have been attempting similar experiments with *Alexandrium tamarense* from the U.S. east coast and a

Toxic Marine Phytoplankton
Edna Graneli et al., Editors

neosaxitoxin and saxitoxin producing cyanobacterium, *Aphanizomenon flos-aquae.*

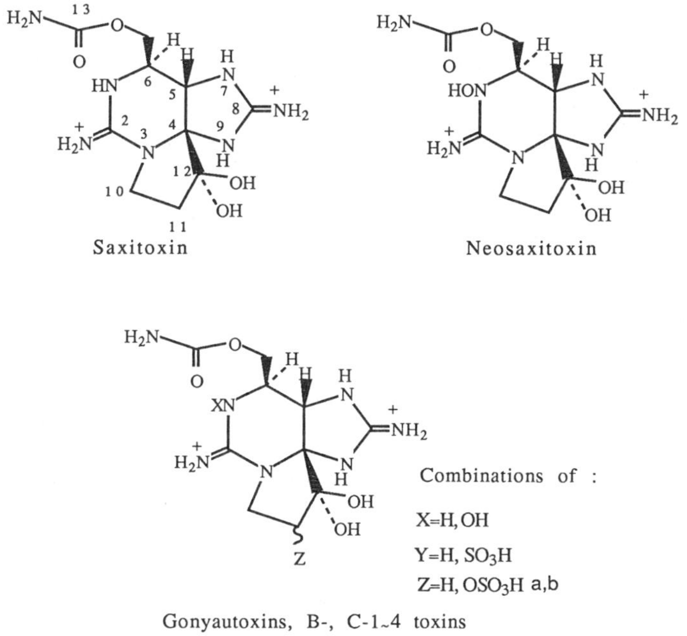

Saxitoxin

Neosaxitoxin

Combinations of :

X=H, OH

Y=H, SO_3H

Z=H, OSO_3H a,b

Gonyautoxins, B-, C-1~4 toxins

FIG. 1. Structures of paralytic shellfish toxins.

We have observed mouse-toxicity in some culture broths, but have been unable to identify saxitoxin or any other related toxins [4]. Another plausible theory is to assume the presence of a plasmid or transposable element responsible for the toxigenicity. However, our efforts to isolate plasmids from the toxic strains of the organisms have been so far in vain. We have not seen much difference in the DNA patterns of the toxic and nontoxic strains. Similar attempts made with *Gymnodinium breve* have also been unsuccessful.

In order to investigate determinants of the toxigenicity of the organisms, we need a good understanding of the metabolic steps involved in the formation of the toxin molecules. Therefore, parallel efforts have been made to study the biosynthesis of two representative groups of toxins: saxitoxin analogues and brevetoxins. The feeding studies with these autotrophic dinoflagellates were found to be extremely difficult. The organisms either refuse the uptake of exogenous organic compounds or utilize them only after metabolizing to smaller fragments, resulting in randomization. Such conversions can occur inside the algal cells and/or outside by the associated bacteria. Despite these difficulties, we have been able to elucidate the basic pathway of the biosynthesis of saxitoxin analogues, and also unravel a portion of the very unusual pathway of brevetoxin biosynthesis.

THE BIOSYNTHESIS OF SAXITOXIN ANALOGUES

The origin of the unique perhydrosaxitoxin skeleton of saxitoxin analogues was a subject of various speculations. In the early stage of the study, such speculative precursors as ordinary purine derivatives or C_7 aminosugars were ruled out [5]. Instead, α-ketoglutarate or arginine was implicated to form the straight carbon chain of the toxin skeleton. However, repeated feeding of [1-^{13}C]arginine could not effect the expected enrichment of C-5 of gonyautoxin-II or neosaxitoxin. The mystery was solved later, when it was revealed that C-1 of arginine is actually lost in the biosynthesis. The revised pathway involves the condensation of acetate to the α-carbon of arginine followed by decarboxylation and introduction of C_1 unit from S-adenosylmethionine (SAM) to form the C-13 side chain [6] (Fig 2), The pathway was conclusively proved by feeding [2-^{13}C, 2-^{15}N]arginine and [*methyl*-^{13}C]methionine to the culture of toxic *Aphanizomenon flos-aquae* strain [7] (Fig 3). We further investigated the mechanism of the introduction of the side chain carbon.

First , we designed an experiment to determine how many of the original three hydrogens on the methyl group of S-adenosylmethionine will be retained in the toxin

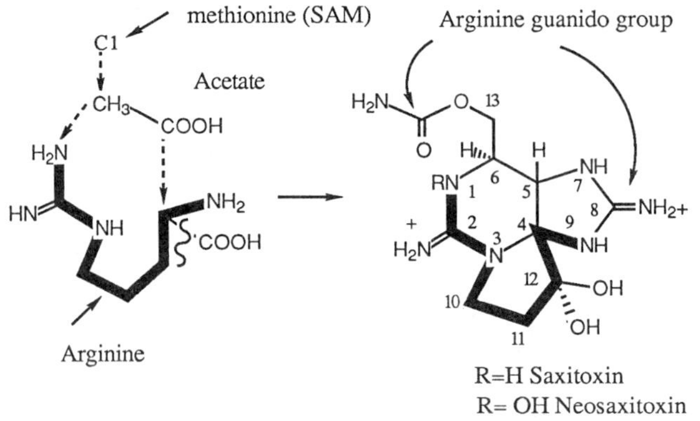

FIG. 2. Building blocks of saxitoxin analogues as determined by precursor feeding studies.

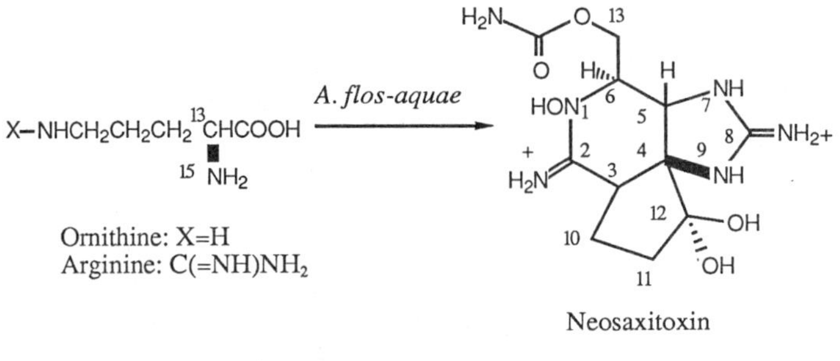

FIG. 3. Incorporation pattern of [2-^{13}C, 2-^{15}N]ornithine into neosaxitoxin. The dark line indicates the presence of spin-spin coupling in NMR.

molecule. The double-labelled precursor, [*methyl*-^{13}C, $^2H_3$]methionine was fed to

A. flos-aquae, and then isolated neosaxitoxin was examined by ^{13}C NMR. As previously observed [6], methionine methyl group was efficiently incorporated in C-13 of neosaxitoxin. Here, the signal for C-13 was composed of a singlet for a natural abundance ^{13}C-peak, and a triplet of introduced ^{13}C bearing deuterium, which is slightly shifted upfield due to a deuterium isotope effect. Since the spin number of deuterium is 1, the appearance of a triplet signifies that only one deuterium atom was left on the methylene carbon [8]. The cleanliness of the signal indicates that the loss was not due to the partial washout by exchange (Fig 4a).

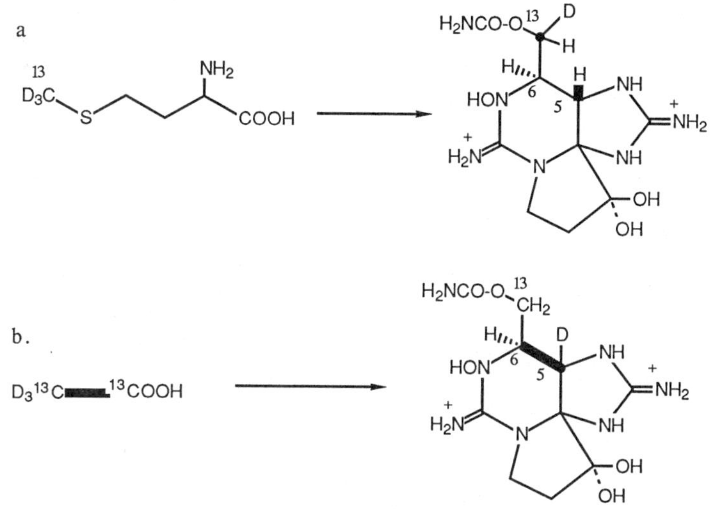

FIG. 4. Incorporation patterns of ^{2}H, ^{13}C-double labelled precursors into neosaxitoxin.

The transfer of methyl group from SAM usually undergoes by an electrophilic attack by methyl cation in SN_2 mode, i.e., a nucleophilic substitution from SAM's side. The reaction often accompanies elimination of proton and/or migration of hydride ion.

In an earlier experiment [7], we confirmed that labelled acetate will be incorporated into C-5 and C-6 of the toxin molecule as a unit. Accordingly, we decided to feed [1,2-^{13}C, 2-$^{2}H_3$]acetate . If the deuterium from the doubly-labelled acetate remains on one of the incorporated ^{13}C carbons, the NMR signal of that particular carbon should collapse or split by a spin-spin coupling with the quadrupole deuterium.

The initial feeding of the double-labelled acetate, however, resulted in the isolation of neosaxitoxin, whose ^{13}C NMR spectrum showed no trace of deuterium. We were, at that time, unable to determine if this absence was a result of washout due to a quick exchange of methyl hydrogens of acetate as often seen in biosynthetic studies of fungal metabolites, or due to a truly mechanistic reason. However, after repeated attempts, we succeeded in the isolation of neosaxitoxin, which still retained portion of deuterium (ca. 40 %) in the molecule. Interestingly, the deuterium was attached to C-5, and its signal appeared as a composite of a natural abundance singlet peak, a doublet of incorporated ^{13}C with no deuterium, and a multiplet of incorporated ^{13}C carrying deuterium. The chemical shifts of these signals were spaced as expected from the known isotope shifts of ^{13}C and ^{2}H. Since C-5 of neosaxitoxin is derived from the carboxyl carbon of acetate, its deuterium must be derived from the neighbouring C-6 of methyl origin by migration [8] (Fig.4b). These findings resulted in the proposal of a pathway, which involves the formation of an exomethylene epoxide and aldehyde (Fig. 5).

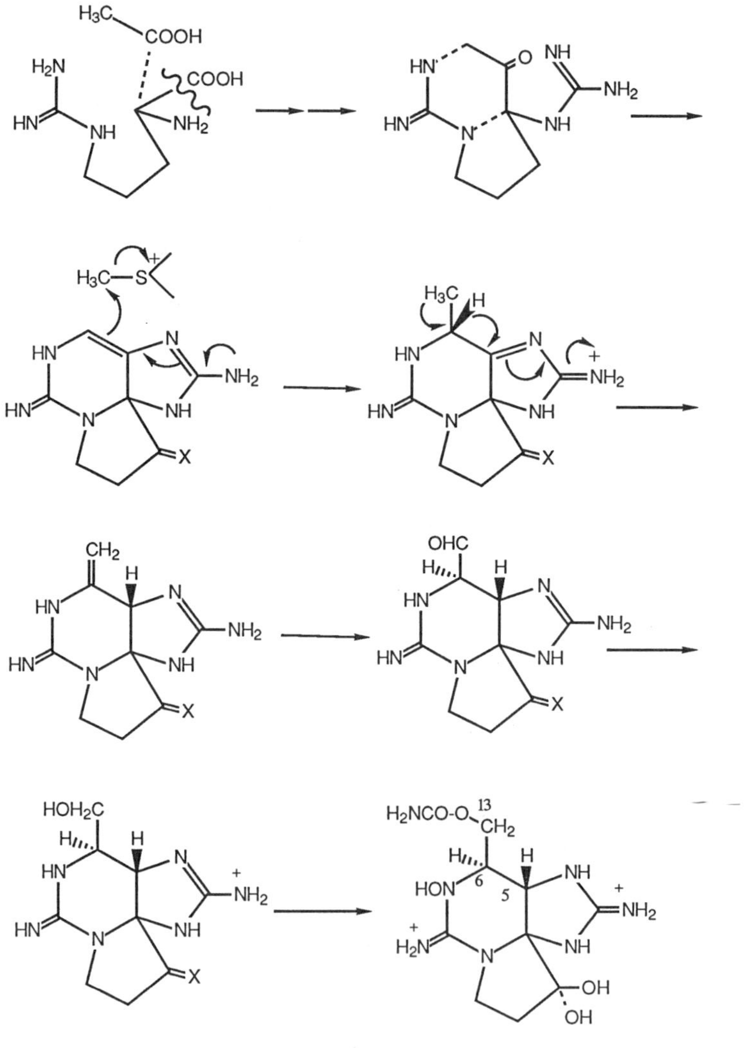

FIG. 5. Biosynthetic pathway of neosaxitoxin as judged by various feeding studies.

Here, an important question was whether one of the lost methionine methyl hydrogen atoms migrates to C-6 of neosaxitoxin. Such migration seemed quite possible in a mechanism as shown in Fig. 6a, especially since it was shown that the C-6 hydrogen is not derived from acetate methyl hydrogens. To solve this question, we considered a sptepwise feeding of [*methyl*-^{13}C]acetate and [*methyl*-^{13}C, $^{2}H_3$]methionine. If successful, the experiment would bring forth the molecular species with a combination of ^{13}C-6-^{13}C-13 in the same molecule, which should show a spin-spin couplng between the two nuclei. Moreover, if C-6 receives a deuterium atom from C-13, the C-6 signal should be split further by the deuterium (Fig. 6b).

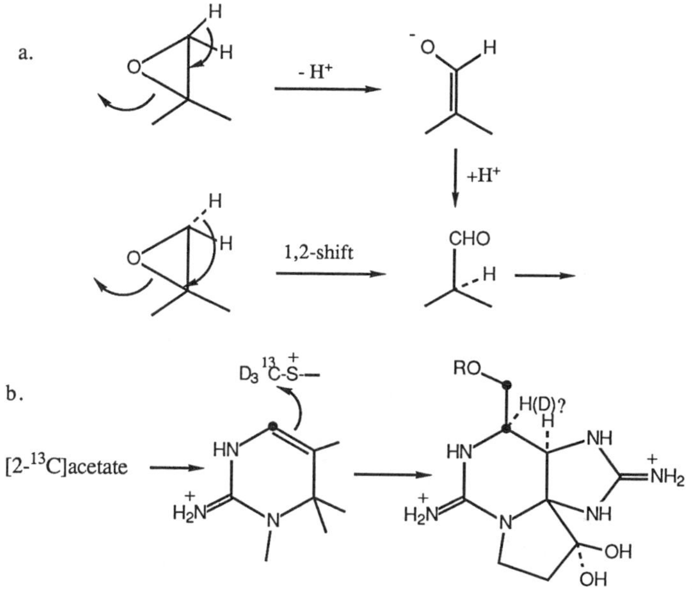

a.

- H^+

+H^+

1,2-shift

CHO

b.

$D_3{}^{13}C$-$\overset{+}{S}$—

[2-^{13}C]acetate

RO

H(D)?

FIG. 6. Stepwise feeding of [2-^{13}C]acetate and [methyl-^{13}C, $^{2}H_3$]methionine.

After repeated trials under different feeding conditions, we were able to isolate neosaxitoxin, whose NMR spectrum showed a doublet slightly upfield of the uncoupled ^{13}C signal for C-6. There was no sign of splitting of the signal due to deuterium, thus excluding the possibility of hydride ion transfer.

BIOSYNTHESIS OF BREVETOXINS

Brevetoxins also have unique structural features. There are two basic skeletons: brevetoxin-A type and brevetoxin-B type (Fig. 7). Both of them are composed of all *trans-syn* linear polycyclic ether rings. At the first glance they appeared as typical acetogenins or polyketides derived from acetate. Methyl branches could come from either methylation with SAM or replacement of acetate units with propionate. Epoxidation of *trans*-double bonds to *trans*-epoxides followed by a cascade of epoxide opening would result in the linear cycles with required stereochemistry.

We had started to study the incorporation of ^{13}C-labelled acetate into brevetoxin-B, long before its structure was established, hoping that the incorporation pattern would provide clues for the structure of this apparent acetogenin. The results were, however, very perplexing; the acetate units were incorporated in an unexplicable manner, and considerable randomization was observed [9] (Fig. 8A) . Lee, *et al.* also reported similar results with no explanation [10].

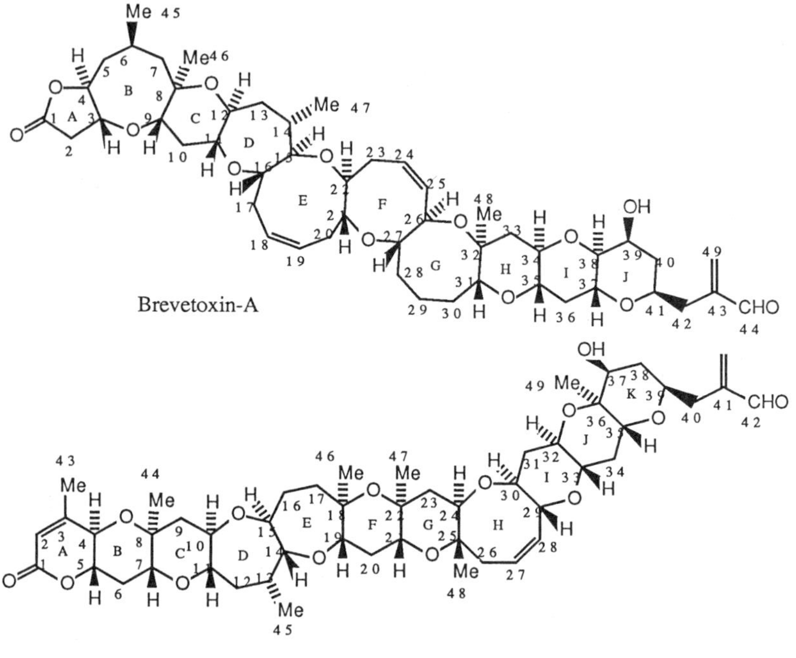

Brevetoxin-A

Brevetoxin-B

FIG. 7. Structures of two representative toxins from *Gymnodinium breve*.

After careful examination of the NMR spectra of enriched brevetoxin-B, we reached a conclusion that the compound is not an ordinary acetogenin, and most acetate units are incorporated only after they are converted to other metabolites, such as succinate and α-ketoglutarate [9]. Pulse feeding of [2-^{13}C]acetate afforded brevetoxin-B, the ^{13}C NMR of which showed spin-spin couplings only between certain carbons. These couplings reflect new carbon connectivities created by linkage of two acetate methyl-originated carbons. A well-known example of such reactions is the condensation of acetate to oxaloacetate to form citrate. Citrate is then converted to α-ketoglutarate, succinate and oxaloacetate to complete the TCA cycle. Asymmetricity of the molecules is lost at the succinate step, and the α-carbon of oxaloacetate carries a half of acetate methyl carbon incorporated in the previous condensation. In the next condensation, another connectivity will be formed (Fig. 9**A**).

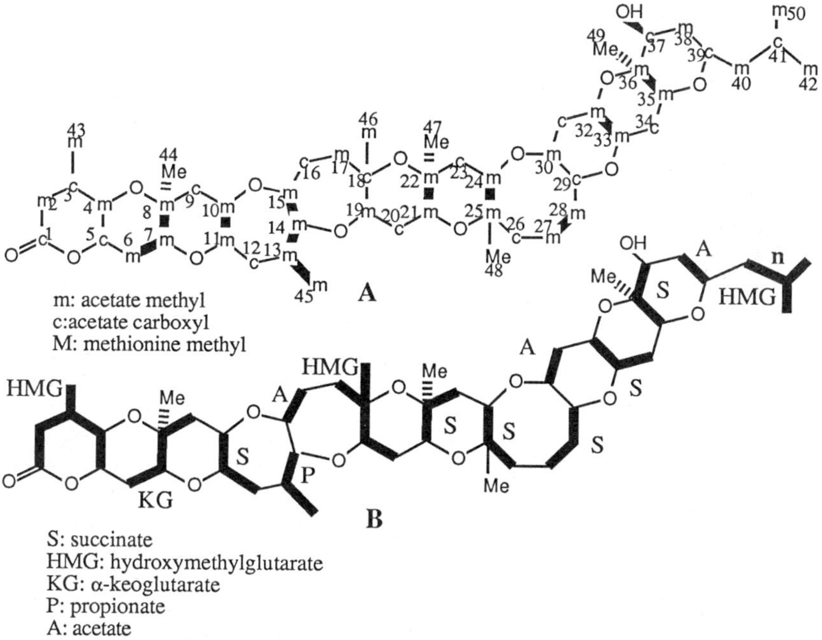

FIG. 8. Incorporation patterns of labelled acetate and methionine into brevetoxin-B (**A**) and putative building blocks (**B**).

We speculate that brevetoxins are a new type of mixed polyketides, which are formed by condensation of dicarboxylic acids, acetate and propionate (Fig. 8**B**). The condensation of the dicarboxylic acids occur in such a manner that one end of carboxyl group undergoes Claisen-type condensation, whereas the other end receives condensation on the α-methylene moiety (possibly after activation by carboxylation) followed by a loss of carboxyl group (Fig. 9b). Using this hypothesis, we were able to explain most building blocks of the brevetoxin-B and their labelling pattern [9].

The involvement of succinate or other dicarboxylic acids are known in the biosynthesis of certain microbial metabolites, but the condensation on the α-methylene is unprecedented. In attempts to confirm this hypothesis, such putative precursors as [2,3-$^{13}C_2$]succinate, [1,4-$^{13}C_2$]succinate, [1-^{13}C]propionate, [2-^{13}C] propionate, [1,2,3-^{13}C3]glycerol, [1,2,3,4-$^{13}C_4$]acetoactate, [2-^{13}C]malonate and [2-^{13}C]pyruvate were fed to the organism. Either randomization or no significant incorporation was noticed in most of the above cases.

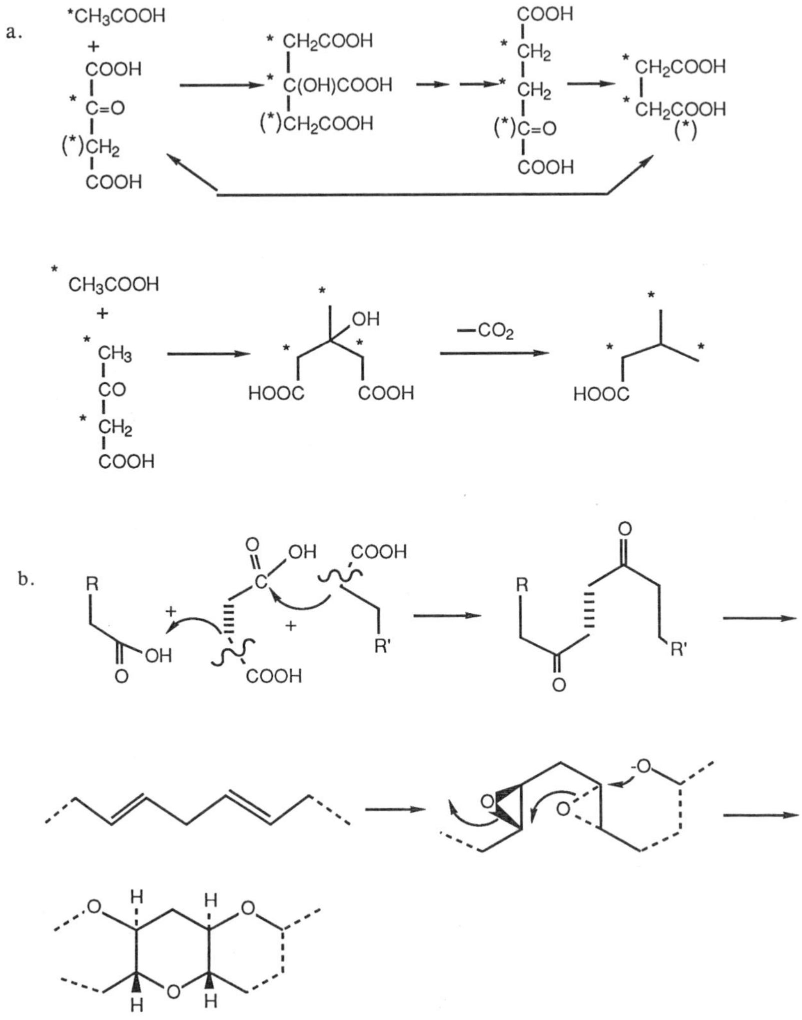

FIG. 9. The fate of acetate labels in putative building blocks (a) and possible mechanism of carbon chain formation in the biosynthesis of brevetoxins (b).

The mechanism of the polycyclic ether formation is of great interest. If the concomitant ring opening of *trans*-epoxides is proved to be true, it will be one of the most spectacular biochemical reactions. There are basically two ways to start the epoxide opening. One is to start from the left terminus by protonation on the carbonyl function, which will become the lactone ring of the toxin molecule [11, 12] (Fig. 10**A**). The other way is to start by the opening of *cis*-epoxide on the right terminal ring [12] (Fig. 10**B**). This idea was derived to explain the presence of 6β-

hydroxyl group in all brevetoxins so far isolated.

We have recently elucidated the structure of hemibrevetoxin-B (GB-N) [13]. The compound is one of the "smaller" (about half the size of other brevetoxins) toxins recognized earlier and known under such names as GB-4, and GB-M. Significantly, the structure of hemibrevetoxin-B is roughly the right half of brevetoxins, and has an α-methylene aldehyde group and is identical with the two terminal rings of brevetoxins. The presence of such compounds in the same organism clearly rules out the cyclization from the left side of the molecule (Fig. 10**A**). Rather it fits the aforementioned alternative mechanism involving the opening of *cis*-epoxide at the right terminus followed by hydride migration and a cascade of epoxide openings (Fig. 10**B**)

CONCLUSION

Our studies on the biosynthesis of saxitoxin derivatives and brevetoxins have proved that the biosynthetic pathways of red tide toxins are very unique and rather extraordinary. However, despite progress made in the recent years, the level of our understanding of the biosynthesis of marine toxins in general is still substantially below that of the terrestrial plant or fungal metabolites. At this point, we have no knowledge of enzymatic systems involved in each step of the biosynthesis. We have also no knowledge of the trigger or regulation mechanism of the toxin production. Attempts to find common vectors for the toxin production are also inconclusive at best.

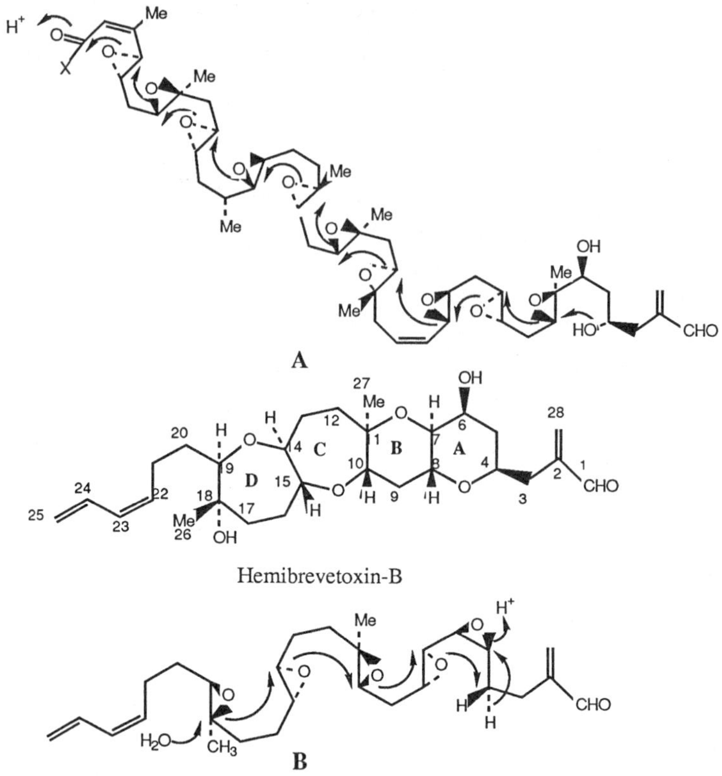

FIG. 10. Structure of hemibrevetoxin-B isolated from *Gymnodinium breve* and two possible cyclization mechanisms (**A**) and (**B**).

ACKNOWLEDGEMENT

The work described in this paper was supported by NIH grants, GM24425 and GM28754, which is greatly appreciated.

REFERENCES

1. L. Maranda, D. M. Anderson and Y. Shimizu, Estuarine, Coastal and Shelf Science 21, 401-410 (1985).
2. E. S. Silva IUPAC Symposium on Mycotoxin and Phycotoxin, Lausanne: Pahotox publication, 1979.
3. M. Kodama, T. Ogata and S. Sato, Agric. Biol. Chem. 52, 1075-1077 (1988).
4. C. K. Walker and Y. Shimizu, unpublished (to be included in C.K.W.'s Ph.D.

thesis, University of Rhode Island).

5. Y. Shimizu, M. Kobayashi, A. Genenah and N. Ichihara in: Seafood Toxins - ACS Symposium Series 262, E. R. Ragelis ed. (American Chemical Society, Washington, DC, 1984) pp. 151-160.
6. Y. Shimizu, S. Gupta, M. Norte, A. Hori and A. Genenah in: Toxic Dinoflagellates, D.M. Anderson, A. White and G.D. Baden eds. (Elsevier, New York 1985) p. 271.
7. Y. Shimizu, M. Norte, A. Hori and M. Kobayashi, J. Am. Chem. Soc. 106, 6433-6434 (1984).
8. S. Gupta, M. Norte and Y. Shimizu, J. Chem. Soc., Chem. Commun. accepted.
9. H. N. Chou and Y. Shimizu, J. Am. Chem. Soc. 109, 2184-2185 (1987).
10. M. S. Lee, D. S. Repeta, K. Nakanishi and M. G. Zagorski, J. Am. Chem. Soc. 108, 7855-7856 (1986).
11. K. Nakanishi, Toxicon 23, 473-479 (1985).
12. Y. Shimizu in: Natural Toxins: Animal, Plant and Microbial, J. B. Harris ed. (Clarendon Press, Oxford 1986) p. 123.
13. A. V. K. Prasad and Y. Shimizu, J. Am. Chem. Soc. accepted.

III ORGANISMS

A SHORT DIAGNOSTIC DESCRIPTION OF ALEXANDRIUM

E. BALECH
C.C. 64, 7630 Necochea, Argentina

Alexandrium Halim 1960, emend. Balech - Gonyaulacaceae (Gonyaulacidae) with the following tabular formula, expressed in the Kofoidean system as modified for the genus by Balech and Tangen (1985): Po, 4', 6'', 6c, 5''', 2'''' and 10 S (± 1). Cingulum descending generally about one cingular width. Neither horns nor spines present. Lists very poorly developed. Po-plate large, with a main large comma shaped pore, rarely narrow (oval). First apical plate connected or disconnected with Po, with or without ventral pore. Thecal surface usually almost smooth, however, sculptured in a few species. The sulcus consists of the following plates: the anterior sulcal plate (s.a.), the posterior sulcal plate (s.p.); two pairs of lateral plates: the right anterior, the right posterior, the left anterior, and the left posterior; two small plates between the lateral plates: the median plates (anterior, s.m.a., and posterior, s.m.p), and finally two tiny accessory plates.

Chloroplasts usually present. Nucleus transversely elongated.

Type species: Alexandrium minutum Halim 1960, p. 102.
Type locality: Harbour of Alexandria.

It is proposed to subdivide the genus into 2 subgenera: the nominative Alexandrium and the subgenus Gessnerium Halim (type species Alexandrium (Gessnerium) monilatum (Howell)).

In the subgenus Alexandrium the plate 1' is always rhomboidal and connected (directly or indirectly) to Po. In Gessnerium the plates are truly disconnected and differently shaped. The most closely related genera are Pyrodinium and Goniodoma. They differ from Alexandrium by having a thick theca, which is strongly porulated and with much developed lists (sulcal and cingular in the former genus, cingular in the latter). In both genera the sulcus has a somewhat different organization, with a slightly smaller number of plates. In addition Pyrodinium develops spectacular lists and spines in the hypotheca and produces cysts of a different kind.

In Goniodoma the Po-plate is of a different type (foramen not well developed), with the main axis transversally orientated. Thus the suture starting from the acute pole of Po is directed to the left of the cell instead of straight down as in Alexandrium. To reach the plate 1' following plate borders it is necessary to follow 2 borders in Goniodoma but only one in the subgenus Gessnerium, which is the only one likely to be confused with Goniodoma.

Published 1990 by Elsevier Science Publishing Co., Inc.
Toxic Marine Phytoplankton
Edna Graneli et al., Editors

SOME COMMENTS ON THE USE OF THE GENERIC NAMES PTYCHODISCUS AND ALEXANDRIUM

Ø. MOESTRUP AND J. LARSEN
Institut for Sporeplanter, Øster Farimagsgade 2D,
DK-1353 København K, Denmark

ABSTRACT

The generic status of Ptychodiscus brevis (Davis) Steidinger is discussed and it is argued that this species should be retained in the genus Gymnodinium. Use of the name Alexandrium for the PSP-producing species originally included in Gonyaulax is advocated in preference to either Gonyaulax or Protogonyaulax.

INTRODUCTION

Increasing problems with toxic algae in Danish and adjacent waters have accentuated the need for a plankton guide to be used for identification of toxic marine algae. A guide has therefore been prepared (1), mainly as a handbook for people working in environmental monitoring, but hopefully also useful to other persons interested in or working on marine plankton.

Several taxonomic problems or difficulties are pointed out in the guide, but two issues deserve particular attention and are here discussed in more detail.

The taxonomic position of Ptychodiscus brevis (Davis) Steidinger

Recent investigations have shown that Gyrodinium aureolum Hulburt and Gymnodinium nagasakiense Takayama & Adachi are very closely related (2), perhaps even conspecific. On basis of ultrastructural information we shall here argue that Ptychodiscus brevis should be given the same generic status as either of these two species.

The genus Ptychodiscus was described by Stein (3), and later reinvestigated by Boalch (4). He reports that P. noctiluca, the type species, has a tough structureless theca and the cells are compressed apical-antapically. The presence of chloroplasts is not mentioned.

Gymnodinium breve Davis was tranferred to Ptychodiscus by Steidinger (5), mainly based on the presence of an apical crest, which is also found in Ptychodiscus. Gymnodinium breve does not possess a tough theca, but only a delicate pellicular layer (Steidinger, communicated in Taylor (6)). The transfer of the species to Ptychodiscus was questioned by Taylor (6), and Sournia (7) considered it unjustified.

The ultrastructure of G. breve was investigated by Steidinger et al. (8). In their interpretation, the theca 'consists of four membranes: an outer membrane, two membranes of appressed flattened vesicles, the plasmalemma. An inner membrane is located beneath the plasmalemma'. This interpretation implies that there is no electron dense material in

Toxic Marine Phytoplankton
Edna Graneli et al., Editors

the lumen of the thecal vesicles, and the innermost membrane is apparently not considered part of the theca.

The theca of G. nagasakiense is shown in Figs 1,2. It consists of an outer continuous membrane, the plasmalemma, covered by a diffuse electron opaque layer. Beneath the plasmalemma are flattened membrane vesicles which contain fine plate material. Large vacuoles are located immediately

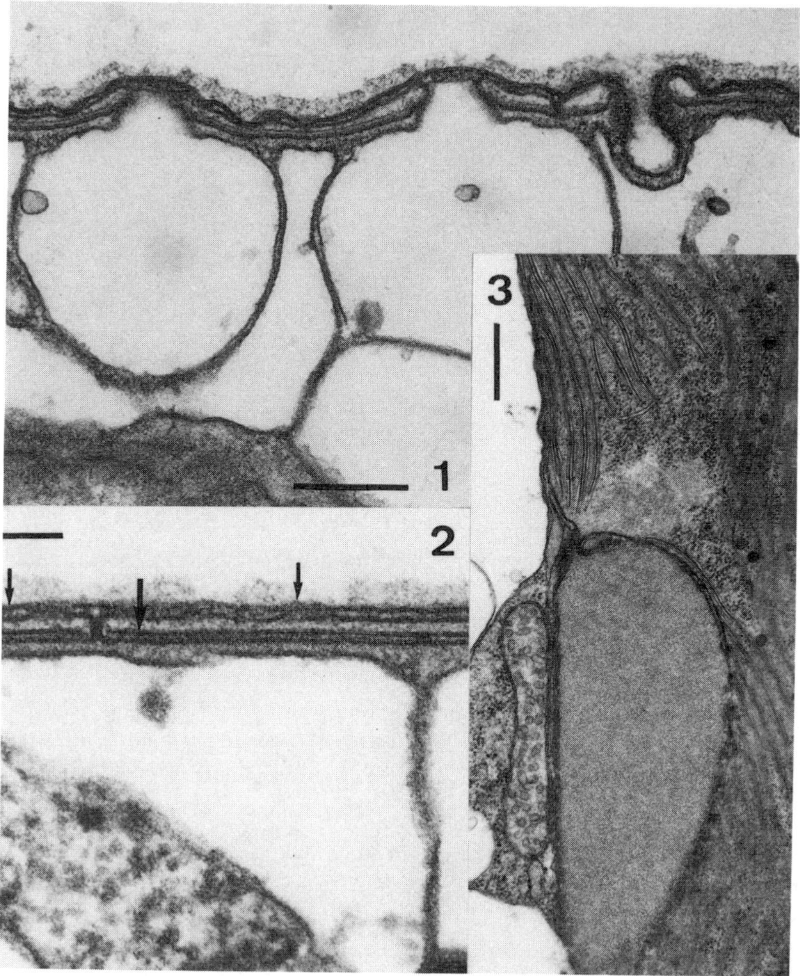

FIG.1. The theca of G. nagasakiense; scale bar = 0.25 µm. FIG.2. Detail showing the plasmalemma (small arrows) and the thecal vesicles with the fine plate material (arrow), the plate material is not in contact with the thecal vesicles; scale bar = 0.1 µm. FIG.3. Chloroplast and pyrenoid of G. nagasakiense; scale bar = 0.5 µm. Culture material isolated by J. Larsen from Port Phillip Bay near Melbourne, Australia.

beneath the theca and towards the exterior their membranes are situated adjacent to the inner membranes of the thecal vesicles.

A re-examination of the EM photographs published by Steidinger et al. (8 - Figs 3,4,7,8,9) suggests that the ultrastructure of the theca in G. breve may well be similar to that of G. nagasakiense. The outermost membrane of G. breve appears to be covered by a diffuse layer and, more important, what Steidinger et al. interpret as the inner membrane of the thecal vesicles may in fact be equivalent to the plate material found in G. nagasakiense, compare Figs 1,2 with (8: Fig.3, lower left corner, and Figs 8,9); and the innermost membrane may be part of the vacuolar system also in G. breve (8: Fig.4). Accepting this interpretation, the ultrastructure of the theca of G. breve is similar to what is found in G. nagasakiense, and clearly different to what can be expected in Ptychodiscus noctiluca with its tough theca.

Also the structure of the chloroplasts and pyrenoids in G. breve appears to be similar to those of G. nagasakiense and G. aureolum, compare Fig.3 with (8: Figs 12,17) and (12).

In addition, the pigment contents suggest a close relationship between Gyrodinium aureolum, G. nagasakiense and G. breve. They all lack peridinin which is otherwise characteristic of photosynthetic dinoflagellates, and they contain fucoxantin or fucoxantin derivatives (9,10 and pers. obs. on G. nagasakiense).

Based on these considerations we suggest that the combination Gymnodinium breve is retained until such information is available that a major revision of the Gymnodinium/ Gyrodinium complex can take place.

On the naming of the 'Gonyaulax tamarensis' complex

Naming of the PSP producers originally included in the genus Gonyaulax continues to be difficult, but the question of which generic name is most correct has recently become clearer. Two approaches may be considered, a broad or a narrow generic concept. In a broad concept all species are retained in the same genus, using the name Gonyaulax. In a narrow they are referred to different genera, one of which contains the PSP producing species. The former solution has at the present time few followers, despite the fact that detailed studies on ultrastructure have never been performed on any of the involved species, toxic or non-toxic. Thus details of flagellar structure, which in other groups of flagellates are being used increasingly for taxonomic and phylogenetic purposes, are entirely lacking in the Gonyaulax complex.

Somewhat ironically, however, the name Gonyaulax may actually be illegitimate. The type species, G. spinifera, produces cysts of a type for which the name Spiniferites was applied by Marshall in 1850, i.e. 16 years before Diesing created the name Gonyaulax. If Spiniferites is part of the life cycle of G. spinifera, the correct name of the genus is Spiniferites according to the Botanical Code of Nomenclature. The situation has been further complicated, however, by the study of Hall and Dale (11) who reported that several different types of cysts, belonging to several different genera, all yield G. spinifera-like cells. This obviously questions the identity of G. spinifera, which is in great need of careful and critical studies.

The narrow generic concept, i.e. splitting the genus Gonyaulax into 2 or more genera, one containing the PSP producing species, has presently many advocates. This appears the most satisfactory approach at the moment, considering the pronounced similarity in plate pattern and morphology of the PSP producing species. The correct name for these species is Alexandrium, a conclusion confirmed by re-examination of material of the type species, A. minutum, from the type locality, Alexandria Harbour, Egypt (13). Gessnerium is a later synonym of Alexandrium. Taylor has suggested to establish a separate genus, Protogonyaulax, for tamarensis and its close allies. Alexandrium minutum has now, however, been found to be one of these allies. It is a PSP producer which has lately caused mussel poisoning in France (14) and South Australia (15). Taylor's claim that contact versus non-contact between the first apical plate in the theca and the pore plate is a generic character was refuted by Balech and Tangen (16) and Balech (13) who found it to be a variable feature in both A. minutum and A. tamarense. Protogonyaulax is therefore a later synonym of Alexandrium and its use should be discontinued in order to avoid further nomenclatural confusion and create a more stable taxonomy.

REFERENCES

1. J. Larsen and Ø. Moestrup, Toxic marine algae from Danish waters (Fiskeriministeriets Industritilsyn, København 1989).
2. F. Partensky, D. Vaulot, A. Couté and A. Sournia, J. Phycol. 24, 408-415 (1988).
3. F.R.von Stein, Der Organismus der Infusionstiere, Abt. III, Hft.II, Arthrodelen Flagellaten (Leipzig 1883).
4. G. Boalch, J. mar. biol. Ass. U.K. 49, 781-784 (1969).
5. K.A. Steidinger in: Toxic dinoflagellates, D.M. Anderson, A.W. White and D.G. Baden, eds. (Elsevier, Amsterdam 1985) pp. 435-442.
6. F.J.R. Taylor in Toxic dinoflagellates, D.M. Anderson, A. W. White and D.G. Baden, eds. (Elsevier, Amsterdam 1985) pp. 11-26.
7. A. Sournia, Atlas du phytoplancton marin 1 (Éditions du Centre National de la Recherche Scientifique, Paris).
8. K.A. Steidinger, E.W. Truby and C.J. Dawes, J. Phycol. 14, 72-79 (1979).
9. K. Tangen and T. Bjørnland, J. Plankton Res. 3, 389-401 (1981).
10. T. Bjørnland, F. Pennington, F.T. Haxo and S. Liaaen-Jensen, 7th Int. IUPAC Symposium on Carotenoids, Munich (1984).
11 D. Wall and B. Dale, Micropaleontology 14, 265-304 (1968).
12. G.C. Kite and J.D. Dodge, J. Phycol. 21, 50-56 (1985).
13. E. Balech, Phycologia 28, 206-211 (1989).
14. E. Nezan and M. Ledoux, Red Tide Newsletter 2, No. 1: 2-3. Sherkin Island Marine Station, Co. Cork, Ireland (1989).
15. D.A. Steffensen, Third International Phycological Congress, Abstracts, 41 (1988).
16. E. Balech and K. Tangen, Sarsia 70, 333-343 (1985).

THREE *ALEXANDRIUM* SPECIES FROM COASTAL TYRRHENIAN WATERS (MEDITERRANEAN SEA)

MARINA MONTRESOR, DONATO MARINO, ADRIANA ZINGONE AND GIORGIO DAFNIS
Stazione Zoologica "A. Dohrn" - Villa Comunale, 80121 Napoli, Italy

ABSTRACT

Information on morphological characteristics and variability of *Alexandrium balechii*, *A. tamarense* and *A. minutum* from coastal Tyrrhenian waters is given. The first finding of *A. balechii* outside the type locality is reported. The variability in the ornamentation of hypothecal and sulcal plates in *A. balechii* and *A. minutum* was considerable. In the latter species, Po and 1' plates are always connected, despite the fact that the intercalary bands of the neighbouring plates sometimes mask this connection.

INTRODUCTION

The genus *Alexandrium* was created by Halim [1] who described the type species *A. minutum*. Several species previously described as *Gonyaulax* were transferred to this genus [2], due to morphological differences from the type species of *Gonyaulax*. The genus, which has also been considered as subgenus of *Gonyaulax* [3], includes the *Protogonyaulax* species of Taylor [4] and Fukuyo [5].

Red tide outbreaks, at times toxic, caused by *Alexandrium* species, have frequently been observed in different geographic areas. As for the Mediterranean Sea, information on red tides is scarce and poorly documented. After the review of Jacques and Sournia [6], several dinoflagellate red tides have been described from coastal areas of the Adriatic Sea [7, 8, 9, 10, 11, 12, 13], the Tyrrhenian Sea [14] and the Levantine Basin [15].

Only a few *Alexandrium* species have been reported from the Mediterranean Sea: *A. minutum* Halim [1, 15, 16, 17, 18], *A. tamarense* (Lebour) Balech [12, 19, 20, 21, 22] and *A. catenella* (Whedon & Kofoid) Balech [22].

In this paper we report the presence of three *Alexandrium* species in the Gulfs of Naples and Salerno (southern Tyrrhenian Sea).

MATERIALS AND METHODS

Surface phytoplankton samples were collected from the inner part of the Gulfs of Naples and Salerno with a Niskin bottle and/or a 20 µm mesh net and preserved with formaldehyde to a final concentration of 1.5%. Clonal cultures were established from the natural samples and maintained in F/2 Guillard medium without silicate addition: *A. balechii*: Clone 16 (1985), SA1 and SA2 (1988); *A. tamarense*: clones 19, 20 and 22 (1988); *A. minutum*: clone MM3B6 (1986). Samples were stained with the chloral hydrate-hydriodic acid technique [23] for plate pattern analysis. Fixed natural samples and cultures were dehydrated in increasing concentrations of ethanol, critical point dried, coated with gold and examined with a Philips EM2 scanning electron microscope.

DESCRIPTION

Alexandrium balechii (Steidinger) Balech was described as *Gonyaulax balechii* from coastal waters of Florida, where it reached high concen-

Toxic Marine Phytoplankton
Edna Graneli et al., Editors

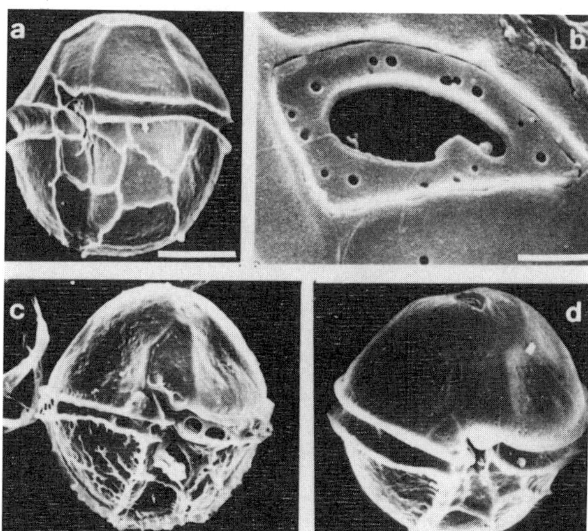

FIG. 1. SEM-graphs of A. balechii. a), c) and d) ventral view. Scale bar = 10 µm. b) apical pore plate. Scale bar = 1 µm.

trations [24]. To date, this is the first finding of the species outside the type locality.

Our specimens (Figs 1a-d) are rounded in shape, 31.3-41.2 µm in length and 28.8-41.3 µm in width. The nucleus is U-shaped and lies in the equatorial plane of the cell. The plate formula is: Po, 4', 6'', 6C, 9S, 5''', 2''''. The pore plate (Fig. 1b), which lacks an attachment pore, shows a quite large, drop-shaped apical pore with no evident callus. Plate 1', which lacks a ventral pore, is not in contact with Po. The cingulum is displaced 1/2 to 3/4 one cingulum width. Pores are scattered on the entire theca. In addition to these, hypothecal and sulcal plates show a great variability of sculptures in both natural and culture samples. Some specimens have only pores on these plates (Fig. 1a), others have vermiculations (Fig. 1c) while still others show a heavy reticulation which is clearly discernible with the light microscope or ridges along the sutures between hypothecal plates (Fig. 1d). Our clones are bioluminescent.

Alexandrium tamarense, described as Gonyaulax tamarensis by Lebour [25], includes A. excavatum (Braarud) Balech & Tangen [26]. The species has an ample geographic distribution and has been reported to cause toxic outbreaks along Norwegian [27], Danish, Swedish [26], English [28] and West Atlantic coasts [29, 30].

Specimens from the Gulf of Naples (Figs 2a, b) are rounded in shape, not compressed dorso-ventrally, 31.1-43.8 µm in length and 31.3-42.5 µm in width. Pairs of cells or chains of 4 individuals have often been observed in exponentially growing cultures, although attachment pores were not observed. The nucleus, that is U-shaped, is located in the equatorial plane of the cell. The epitheca is more or less hemispherical with shoulders sometimes evident. The hypotheca is slightly asymmetrical with the left half longer than the right one. The cingulum is excavated, left descending, displaced about one cingulum width and with ridges along

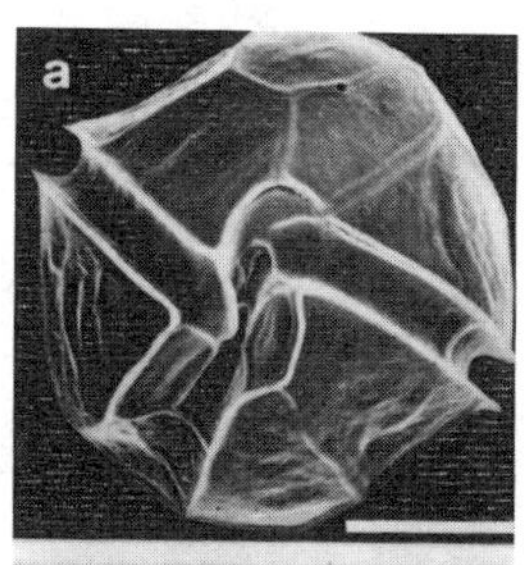

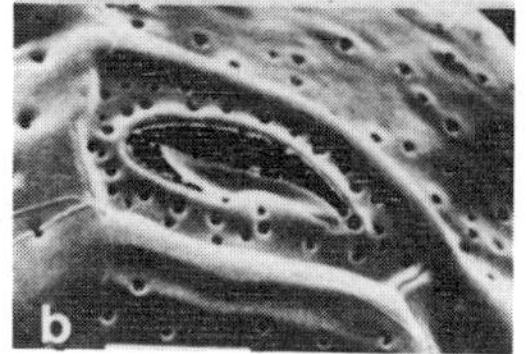

FIG. 2. SEM-graphs of A. tamarense: a) ventral view. Scale bar = 10 µm. b) apical pore plate. Scale bar = 2 µm.

the sutures with pre- and post-cingular plates. The sulcus is excavated and delimited by small lists along the borders with the hypothecal plates. An antapical depression is visible in ventral view. The plate formula is: Po, 4', 6'', 6C, 9S, 5''', 2''''. The S.ac.p. plate was not observed. The pore plate (Fig. 2b) is in contact with the 1' plate and is oval, with small scattered pores; the central pore is comma-shaped. The ventral pore is always present. All of the plates are smooth and covered with numerous scattered pores. No variability either in plate morphology or in ornamentation was observed. Resting cysts were not observed either in nutrient depleted clonal cultures or in clone cross experiments carried out using both depleted and standard medium. Towards the end of the log phase, several pellicle cysts were observed. Our clones are bioluminescent.

Alexandrium minutum Halim was reported to cause blooms in the harbours of Alexandria (Egypt) [1, 17], Adelaide (Australia) [31] and Izmir (Turkey) [15]. Recently, *A. ibericum* Balech, which was described from the coast of western Spain and Portugal [2], has been included in this species as synonymous [31]. The species has also been recorded from coastal waters of France [31, 32], United States [31] and Ireland [33].

Our specimens (Figs 3a-e, 4a-e) are rounded to oval in shape, slightly dorso-ventrally compressed, 21.3-30.0 µm in length and 18.8-23.8 µm in width. The nucleus is U-shaped. The left part of the hypotheca is often prominent anteriorly. The cingulum is deeply excavated and displaced from 1 to 1.5 times its width. Lists are present along the sutures between cingular and epithecal and hypothecal plates. The sulcus, which is bordered by small lists, is deeply excavated and reaches the antapex, where a slight depression is occasionally evident in ventral view. The plate formula is: Po, 4', 6', 6C, 9S, 5''', 2''''. The pore plate is drop-shaped, at times with a straight dorsal side. Numerous pores are scattered on this plate, the callus is well developed; no attachment pore is present. Po and 1' are connected through a narrow apical extension of 1' plate (Fig. 3c), but appear to be disconnected in some specimens due to the development and overlapping of the ventral intercalary bands of plates 2' and 4' (Fig. 3d). Plate 1' is asymmetrically rhomboidal in shape, with a ventral pore along the suture with 4'. The height of the 6'' plate is always greater than the width. The S.p. is quadrangular to heart-shaped. The S.d.p. is often longer than the S.s.p.. S.m.a. and S.m.p. are partially covered by the fins of S.d.a. and S.s.a.. The S.a. plate is wider than long with a slanting plica on its apical part.

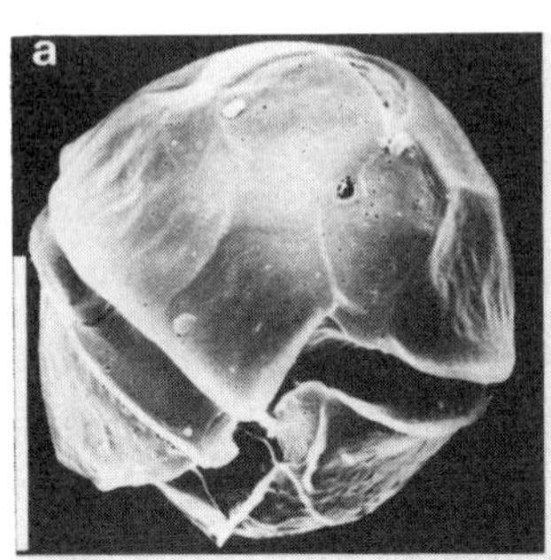

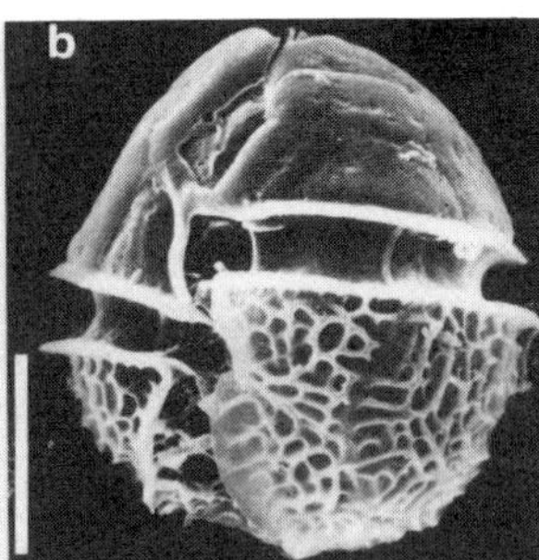

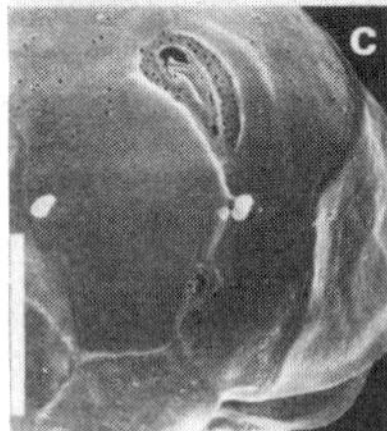

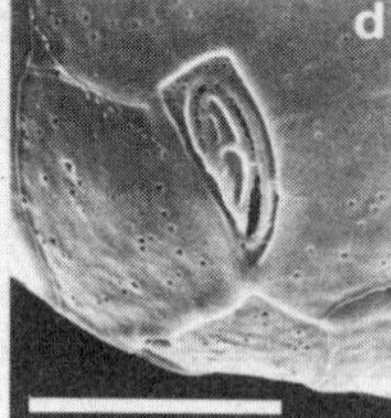

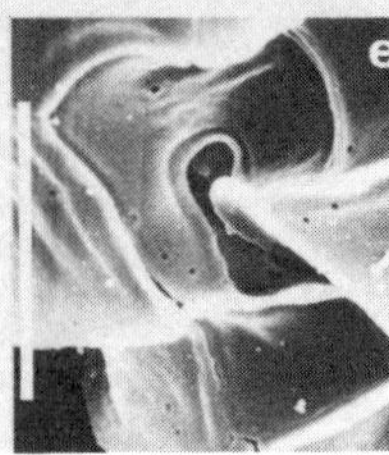

FIG. 3. SEM-graphs of *A. minutum*: a) and b) ventral view. Scale bar = 10 µm. c) and d) apical pore plate. e) detail of the sulcus. Scale bar = 5 µm.

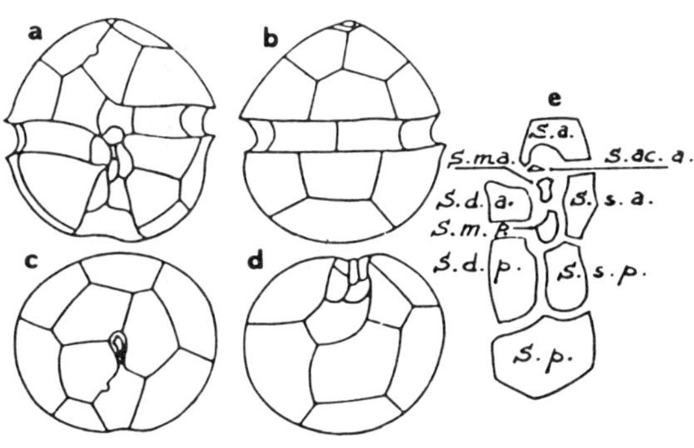

FIG. 4. A. minutum. Schematic drawings. a) ventral view. b) dorsal wiew. c) apical view. d) antapical view. e) sulcal plates.

A small triangular S.ac.a. platelet is present near the flagellar pore (Figs 3e, 4e). A S.ac.p. was not obseved in our samples. Plate 1'''' is small and pentagonal and is completely displaced on the left part of the hypotheca. Plate 2'''', which is very wide and pentagonal, is shifted to the left. There is a high variability in the appearance of the thecal plate surface in both natural and culture samples. Some specimens are smooth with scattered pores, others show a wide degree of ornamentation in the hypothecal and sulcal plates, ranging from a faint vermiculation (Fig. 3a) to a heavy reticulation, which in some cases hides plate sutures (Fig. 3b). The epitheca is always smooth. Our clone is non-bioluminescent.

The original description [1] is incomplete as regards some features such as the morphology of Po, the sulcal platelets and the presence of the ventral pore. In the plate notation followed by Balech [2], the intercalary plate 2p corresponds to S.p., whereas the smallest cingular platelet could be the S.a.. Some differences between specimens from different geographical areas are summarized in the following table.

	EGYPT [1]	SPAIN [2]	AUSTRALIA [31]	ITALY
LENGTH	16-23.2	22-29	20-24	21.3-30.0
WIDTH	13-20.3	15-22	20-24	18.8-23.8
Po-1'	disconnected	variable	variable	variable
ORNAMENTATION	smooth	faint areolae	smooth	smooth to heavy hypothecal reticulations
TOXICITY	probably non toxic	?	toxic	?

OCCURRENCE

The Gulfs of Naples and Salerno are adjacent embayments along the southern Tyrrhenian coast. Both gulfs are characterized by the rapid transition from eutrophic coastal waters and oligotrophic Tyrrhenian offshore waters. This is particularly evident in the Gulf of Naples during summer, when chlorophyll a concentrations of up to 46 mg/m^3 were recorded in coastal waters [34]. In contrast, surface chlorophyll a concentrations in offshore waters ranged from 0.01 to 0.15 mg/m^3 [35]. Summer phytoplankton populations in coastal waters are generally dominated by small diatoms and phytoflagellates whereas dinoflagellates never exceed 35% of the population [36]. Red tides caused by dinoflagellates have not been recorded. A. minutum and A. tamarense were found in net samples during the summer of different years (1985-86 and 1988 respectively) in proximity to the Naples harbour (14°16.0' E; 40°49.0' N). A. minutum was dominant, although no water discolouration was observed.

The Gulf of Salerno showed less marked gradients for hydrological and biological parameters as compared to the Gulf of Naples, with chlorophyll a values ranging from 0.02 to 4.33 mg/m^3 in surface coastal waters [37]. A. balechii was found in the inner part of this Gulf (14°44.1' E; 40°40.1' N), where it produced blooms in the summer period of several years. In August 1980 A. balechii reached a concentration of 5.5x10^6 cells/l, constituting more than 95% of the phytoplankton population. In July 1981 the same species, misidentified as Gonyaulax cf. grindleyi [37], reached a concentration of 4x10^7 cells/l. We observed less intense blooms in the same area in summer 1985 and 1988. A. balechii blooms were concentrated in the first centimeters depth during the day, giving the water an oily, yellowish-brown aspect. Intense bioluminescence was evident at night. No PSP case was recorded during the outbreaks, in spite of the fact that mussels growing on the rocks are usually eaten by the local population. In 1985, A. minutum was also found in samples collected during the A. balechii bloom.

CONCLUDING REMARKS

Our observations on Alexandrium species from coastal Tyrrhenian waters demonstrate a high variability for two of the morphological characters which are considered significant in the taxonomy of the genus. The appearance of the Po-1' configuration in A. minutum varied in relation to the extension of the intercalary bands of the neighbouring plates and hence to the age of the cells. This confirms that Po-1' appearance cannot be used as a distinctive character at the genus level [27, 31], or to separate different Alexandrium species.

Another character which showed great variability was the thecal plate ornamentation in A. balechii and A. minutum. In both species the hypothecal and sulcal plates were at times smooth, but more often showed more or less marked ornamentations. Plate ornamentations in dinoflagellates have been studied in detail by several authors [38, 39], but their relation to age and environmental conditions, as well as to the phylogeny of the species, is still an open question. It is worth pointing out that the absence of evident thecal plate ornamentation is not a rule in the genus Alexandrium. The ornamented species could be considered closer to the genus Gonyaulax, where thecal ornamentations reach a high level of development. No Gonyaulax species is however known to have ornamentations only on hypothecal and sulcal plates of mature cells.

AKNOWLEDGMENTS

The authors wish to thank Dr. K.A. Steidinger for helpful discussion, and G. Gargiulo and F. Esposito for their technical assistance.

REFERENCES

1. Y. Halim, Vie Milieu 11, 102-105 (1960).
2. E. Balech in: Toxic Dinoflagellates, D.M. Anderson, A.W. White and D.G. Baden, eds. (Elsevier, Amsterdam 1985) pp. 33-38.
3. E. Balech, Publ. Espec. Inst. Esp. Oceanogr. 1, 1-310 (1988).
4. F.J.R. Taylor in: Toxic Dinoflagellate Blooms, D.L. Taylor and H.H. Seliger, eds. (Elsevier, Amsterdam 1979) pp. 47-56.
5. Y. Fukuyo, K. Yoshida and H. Inoue in: Toxic Dinoflagellates, D.M. Anderson, A.W. White and D.G. Baden, eds. (Elsevier, Amsterdam 1985) pp.27-32.
6. G. Jacques and A. Sournia, Vie Milieu 28-29, 175-178 (1978-79).
7. I. Marasović and I. Vukadin, Biljeske-Notes, Split 49, 1-7 (1982).
8. S. Fonda Umani, Oebalia 11, 141-147 (1985).
9. I. Marasović, Estuar. Coastal Shelf Sci. 28, 35-41 (1989).

10. I. Marasović, Rapp. Comm. int. Mer Médit. 30, 186 (1986).
11. L. Boni, M. Pompei and M. Reti, Giorn Bot. Ital. 117, 115-120 (1983).
12. L. Boni, M. Pompei and M. Reti, Nova Thalassia 8 (Suppl. 3), 237-245 (1986).
13. A. Artegiani, R. Azzolini, M. Marzocchi, M. Morbidoni, A. Solazzi and F. Cavolo, Nova Thalassia 7 (suppl. 3), 137-142 (1985).
14. G.C. Carrada, R. Casotti and V. Saggiomo, Rapp. Comm. int. Mer Médit. 31, 61 (1988).
15. T. Koray and B. Buyukisik, Rev. Int. Océanogr. Méd. 91-92, 25-42 (1988)
16. N.M. Dowidar, Bull. Inst. Ocean. & Fish., A.R.E., 4, 320-344 (1974).
17. Y. Halim, Acta Adriatica 18, 31-38 (1976).
18. B.R. Heimdal, J.P. Taasen and M. Elbrächter, Sarsia 63, 75-83 (1977).
19. R. Margalef, Inv. Pesq. 8, 89-95 (1957).
20. R. Margalef, Inv. Pesq. 33, 345-380 (1969).
21. R. Margalef, Pubbl. Staz. Zool. Napoli 37 suppl., 40-61 (1969).
22. R. Margalef and M. Estrada, Inv. Pesq. 51, 121-140 (1987).
23. R.J. Schmidt, V.D. Gooch, A.R. Loeblich III and J.W. Hastings, J. Phycol. 14, 5-9 (1978).
24. K.A. Steidinger, Phycologia 10, 183-187 (1971).
25. M.V. Lebour, The dinoflagellates of the northern seas (Plymouth 1925)
26. O. Moestrup and P.J. Hansen, Ophelia 28, 195-213 (1988).
27. E. Balech and K. Tangen, Sarsia 70, 333-343 (1985).
28. A.M. Mortensen in: Toxic Dinoflagellates, D.M. Anderson, A.W. White and D.G. Baden, eds. (Elsevier, Amsterdam 1985) pp. 165-170.
29. H.R. Benavides, R.M. Negri and J.I. Carreto, Physis A 41, 135-142 (1983).
30. A. White, Rapp. P.-v. Réun. Cons. int. Explor. Mer 187, 38-46 (1987).
31. G.M. Hallegraeff, D.A. Steffensen and R. Wetherbee, J. Plankton Res. 10, 533-541 (1988).
32. E. Nezan, C. Beln, P. Lassus, G. Piclet and J.P. Berthome. This volume.
33. K. Duncan, C. Holland, P. Smith and A. Stewart, Red Tide Newsletter 1, 5-6 (1988).
34. M. Modigh, M. Ribera d'Alcalà, V. Saggiomo, G. Forlani and E. Tosti, Rapp. Comm. int. Mer Médit. 29, 109-110 (1985).
35. M. Ribera d'Alcalà, G.C. Carrada, R. Casotti, M. Modigh, V. Saggiomo and S. Vescia, Nova Thalassia 7 (Suppl. 3), 129-135 (1985).
36. A. Zingone, M. Montresor and D. Marino, Mar. Ecol. PSZN I (in press).
37. D. Marino, M. Modigh and A. Zingone in: Marine Phytoplankton and Productivity, O. Holm-Hansen, O. Bolis and R. Gilles, eds. (Coastal and Estuar. Stud., Springer-Verlag, Berlin 1984) pp. 89-100.
38. J.D. Dodge, J. Plankton Res. 5, 119-127 (1983).
39. C. Andreis, M.D. Ciapi and G. Rodondi, Bot. Mar. 25, 225-236 (1982)

THE USE OF OPTICAL PATTERN RECOGNITION IN DINOFLAGELLATE TAXONOMY

KAREN A. STEIDINGER, CHRISTINA CHASE, JULIE GARRETT, BEHZAD MAHMOUDI
BEVERLY ROBERTS, CARM TOMAS AND EARNEST TRUBY
Florida Marine Research Institute, 100 Eighth Avenue S.E., St. Petersburg, Florida 33701-5095, USA

ABSTRACT

The use of optical pattern recognition systems (OPRS) is being pursued as a tool for numerical taxonomy. Several *Prorocentrum*, *Ceratium*, *Protoperidinium*, *Gymnodinium/Gyrodinium* and *Ornithocercus* species, among others, have been separated by size and shape characteristics. With OPRS that use circularity/rectangularity for generating ratios and Fourier analyses for generating harmonic series, within species shape/size variation can be quantified and used to compare knowns, as well as unknowns. Image data can be compared for similarity by cluster analyses, discriminant function analysis and other methods. Sample size is a critical consideration because of variation and skewness in some samples.

INTRODUCTION

Cell size and shape are variable yet conservative morphological characters. Often taxonomists identify species based on shape and size or image recognition without looking at fine detail. With OPRS that use Fourier analyses, image recognition can be quantified without bias for use in identification of toxic and non-toxic dinoflagellates.

RESULTS AND DISCUSSION

In comparing within *Prorocentrum lima* and *P. concavum* variation and between *P. lima* and *P. concavum* variation, using OPRS circularity and rectangularity, *P. lima* and *P. concavum* could be separated by an ANOVA [1] using Notched Box-and-Whisker plots (Figs. 1 and 2). These plots are derived from statistical software [2] and display summary statistics graphically. The box depicts the middle 50% of the data values; the vertical lines ("whiskers") represent the range of data values that are within 1.5 times the interquartile range. The horizontal line through the box is the median. The indented sides (notches) represent the 95% confidence interval of the median. If the two boxes do not have

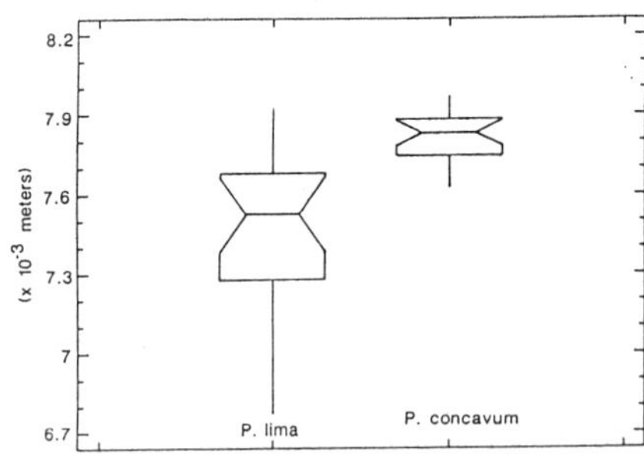

FIG. 1. Notched Box-and-Whisker plot for rectangularity of *Prorocentrum lima* and *P. concavum* (shape's area/area of its minimum closing rectangle).

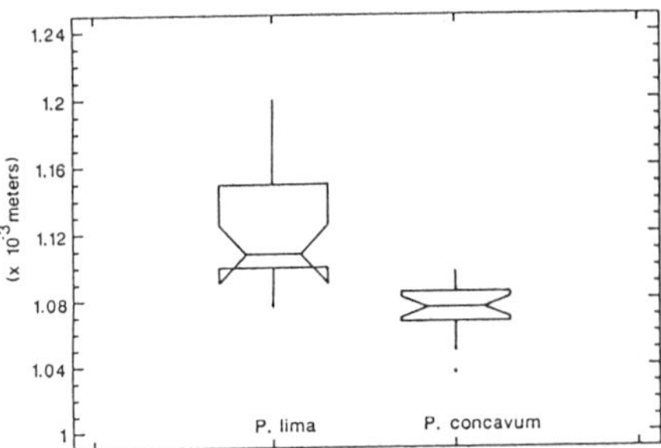

FIG. 2. Notched Box-and-Whisker plot for circularity of *Prorocentrum lima* and *P. concavum* (perimeter squared/area).

Toxic Marine Phytoplankton
Edna Graneli et al., Editors

overlapping confidence intervals, the groups are considered significantly different. If the confidence interval extends beyond the end of the box (ends of box fold over), statistics are suspect. Although the age of cells can cause wide variation in shape/size due to the growth margin of older cells, quantitative differences between the two *Prorocentrum* species were still resolved [1].

Ceratium hircus has been classified as *C. furca* var. *hircus*, but this taxon is consistently distinct in its robustness and the curvature and almost equal length of its antapical horns. OPRS using Fourier descriptors and discriminant analysis has separated the two species by morphometrics (Fig. 3). *C. hircus* is principally an estuarine species with a narrower temperature range than *C. furca*. *C. hircus* and *C. furca* can co-occur in warm water coastal areas.

Ornithocercus steinii and *O. thumii* are often mis-identified because of their overlapping size ranges and general similarities. OPRS using Fourier descriptors and discriminant analysis has separated the two species, even though the sample size was too small (Fig. 4).

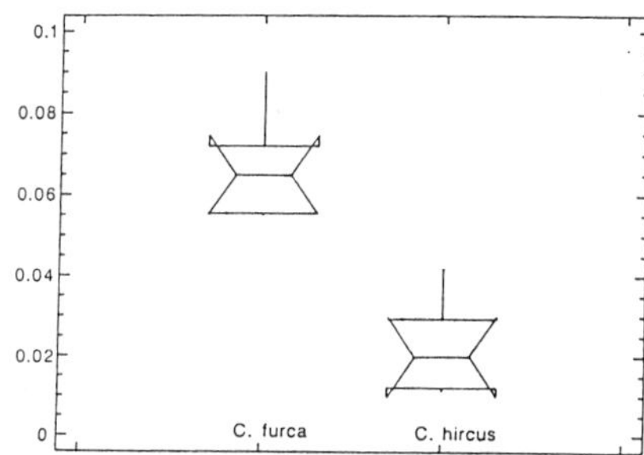

FIG. 3. Notched Box-and-Whisker plot of harmonic coefficient for *Ceratium furca* and *C. hircus.*

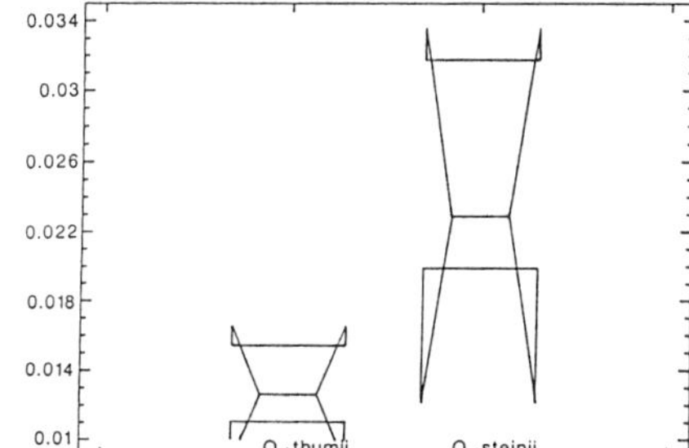

FIG. 4. Notched Box-and-Whisker plot of harmonic coefficient for *Ornithocercus steinii* and O *thumii.*

Statistical analysis [2, 3] and results associated with an optical pattern recognition system (OPRS) can be significantly improved by processing an adequate sample size for each population or class. Various statistical methods can be applied to determine number for specimens needed to detect a true difference between means. For example, in the case of *Ornithocercus steinii* and *O. thumii* based on three samples from each population, the coefficient of variation (CV) was estimated to be 16.9% and 24.8% respectively, and the CV for pooled data was 21.8%. Using appropriate formulae given by Sokal and Rolph [4], the number of specimens needed to be measured from each population at 80% certainty of detecting either a 5% difference or a 20% difference between two means at a 1% level of significance was N=380 and N=25 respectively. Therefore, the sample size for *Ornithocercus* was inadequate; however, the sample sizes for *Prorocentrum* and *Ceratium* were adequate. In addition to sample size, random sampling of different geographic areas should be incorporated, to assess geographic variation.

REFERENCES

1. K. Steidinger, C. Babock, B. Mahmoudi, C. Tomas, and E. Truby in: Red Tides: Biology, Environmental Science, and Toxicology, T. Okaichi, D. Anderson and T. Nemoto, eds. (Elsevier, New York, 1989), pp. 285-288.
2. Statigraphics: statistical graphic system, STSC. INC., (1986).
3. SAS User's Guide: STATISTICS, SAS Institute, (1985).
4. R.R. Sokal and F.J. Rolph, Biometry. The Principles of Statistics in Biological Research, 2nd ed. (W.H. Freeman, New York, 1981).

IV ECOLOGY AND BLOOM DYNAMICS

CHEMICAL TYPOLOGY OF COASTAL SEDIMENTS IN RELATION TO RED TIDES

G. ARZUL and P. GENTIEN
IFREMER - Centre de Brest - DERO/EL - B.P. 70 - 29283 Plouzané - France

ABSTRACT

The study of different chemical parameters characterizing the coastal sediments showed possible relationships between high levels of organic matter and total copper concentrations, and "red tide" occurrences. Sediment extracts had toxic effects on the growth of Alexandrium tamarensis in April, however, it was not observed in sediment elutriates in July.

INTRODUCTION

Red tides seem to induce along French coasts more and more nuisances (1). However, very localized areas appear to remain free of these outbreaks. The global mechanisms usually considered as basic factors do not give real insights on this geographical distribution (2). Local environmental conditions may play a role in the development of these red tides. Effect of sediments in coastal zones on the development of dinoflagellate blooms has been pointed out not only by seeding of populations from cysts (3) but also by its influence on the pelagic system as storage compartment with release of phosphate in anoxic conditions (4), as source of copper ions (5, 6) or as possible source of inhibitory substances for dinoflagellate growth (7, 8).

Coastal zones are highly variable environments and the sediments in shallow areas may be considered as an integrator of the different influences exerted on the ecosystem. We examined the superficial sediment (mud fraction) in different sites selected from 5 years results of the French red tides monitoring program (Fig. 1) (9).

FIG. 1. Sampling stations along the French coast were : Harbours (H), Agricultural (A) and Shellfish farm (S) areas. Observations of red tides, (-) and ban of shellfish harvest areas (...) reported in 1985 (1).

We measured at all sites nitrogenous and phosphorous nutrients released by the sediments, total organic carbon, nitrogen, iron and copper concentrations, and estimated the effects of sediment water extracts on A. tamarensis growth. The sampling was done twice, one in April before the ocurrence of the spring bloom, and the other in early July after sedimentation of the spring bloom's organic matter, but before the second red tide period.

MATERIALS AND METHODS

Sampling

Silty surface sediments were hand collected with polythene corers at low tide. Stations in Fig. 1 were sampled three to four times in an area of 1m^2. Samples were immediately frozen at - 20°C until treatment in the laboratory. Only surface, clear layer of sediments were considered in this study, cores of maximum 2 cm deep.

Water exchangeable nutrients

After thawing the sediment samples, exchangeable nutrients were estimated by

Toxic Marine Phytoplankton
Edna Graneli et al., Editors

water extraction of the surface layer for 2 hours, using 5 g of sediment per 100 ml of 40 g/l KCl dissolved in bidistilled water. Extractions were performed in aerobic conditions at non controlled pH, 8 to 9. After filtration of sediments on Whatman GF/C filters, the extracts were collected and frozen until analysis. Dissolved mineral nitrogen (NO_2 + NO_3) and phosphorus concentrations were determined by Treguer and Le Corre methods (10). In a few cases, elutriate was used for toxicity tests of A. tamarensis. This procedure allows an estimation of the exchangeable nutrient content designated here as interstitial water.

Organic matter

Total organic carbon and total organic nitrogen were measured on crushed dried samples (40°C - 48 hours), with a Carlo Erba Auto-analyser 1106. Carbonate was estimated by calcimetry (11).

Metals

Total iron and copper were measured by Atomic Absorption spectrometry after acid digestion (12).

Toxic effects on growth of Alexandrium tamarensis

Various volumes of aqueous sediment extracts were added to 250 ml of A. tamarensis cultures in Provasoli medium grown at 20°C and 50 $\mu E \cdot m^{-2} \cdot s^{-1}$ light intensity. In each flask cell populations were counted every day until final yield to estimate growth.

RESULTS

In coastal zones, sediments reflect mainly telluric origin, their composition providing good indication of coastal activities. Values reported in this paper are averages of 7 to 18 nutrient concentrations found in each type of zones. We present here the results of three types of zones chosen according to the economical activities of the coastal neighbourhood : shellfish farming, agriculture and breeding, harbours and cities.

Dissolved nutrients

Average values in nitrite plus nitrate concentrations (Fig. 2) varied from 586 ± 111 µmol N/l of porewater in samples from a little fishing harbour in an unaffected area, to 257 ± 223 µmol N/l in shellfish farm vicinities where red tides were usually reported. Inorganic nitrogen (NO_2 + NO_3) concentrations were higher in July than in April. Phosphorus variations followed the same pattern, highest concentrations (258 ± 251 µmol P/l) were found in harbour areas (Fig. 2).

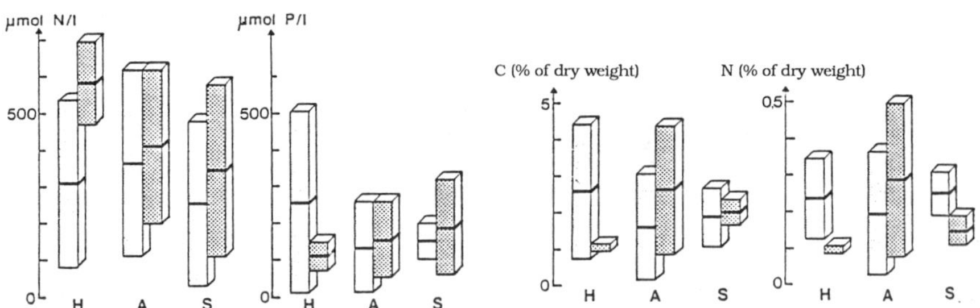

FIG. 2. Concentrations in the interstitial water for nitrogen (NO_2 + NO_3) phosphorus (PO_4) in samples coming from harbours (H), agricultural (A) and shellfish farms (S) areas.
affected
unaffected

FIG. 3. Organic carbon and nitrogen contents in samples coming from harbours (H), agricultural (A) and shellfish farms (S) areas. Legend as in Fig. 2.

Organic matter

Total organic matter shown in Fig. 3, includes dissolved and particulate forms originating from both marine and terrestrial inputs. Lowest values were found near extensive agricultural areas (0.22 ± 0.24 % C ; 0.03 ± 0.01 % N). In harbours and urban neighbourhoods, organic matter was higher, up to 2.54 ± 1.82 % C and 0.23 ± 0.11 % N, reflecting contamination by industrial and domestic discharges (13). C/N ratios varied around 10 but no relationship appears to be present between remineralization estimated by C/N, and nuisance blooms (Fig. 3).

Metals

Iron concentrations were usually high in coastal sediments and their relatively low. Average iron concentrations varied from 0.31 ± 0.14 mmol Fe/g of dry sediments in shellfish farm areas unaffected by red tide, to 0.58 ± 0.28 mmol Fe/g in agricultural areas with red tide occurrences (Fig. 4).

Copper concentrations were a thousand times lower than iron concentrations, with higher concentrations in harbour sediments, average concentration was 1.38 ± 2.17 µmol Cu/g sediment. Near agricultural and shellfish farming areas, average concentrations were 0.15 ± 0.04 to 0.36 ± 0.29 µmol Cu/g sediment, and copper concentrations varied according to organic matter contents indicating possible metal complexation (14, 15). Copper was highly correlated with organic matter as estimated by total nitrogen content (Fig. 5). At the bay of Bourgneuf, which has never been affected by red tides, the sediment contained low levels of organic nitrogen and copper. At Aber Ildut area no red tides occurred until 1986, when a Prorocentrum micans bloom took place. In 1988 an outbreak of Alexandrium minutum occurred.

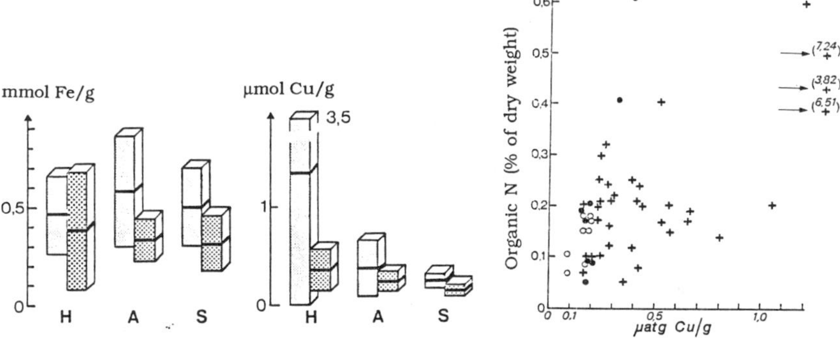

FIG. 4. Total iron and copper concentrations in samples from harbours (H), agricultural (A) and shell-fish farms (S) areas. Legend as in Fig. 2.

FIG. 5. Total copper concentrations versus total organic nitrogen in April and July (1985) samples:
+ red tides usually observed
• red tides observed since 1985 at Aber Ildut site
○ unaffected zone at bay of Bourgneuf site.

Biological effect of sediment elutriates

In order to indicate a possible relationship between environmental quality and dinoflagellate development, Alexandrium tamarensis cultures were grown with different dilutions of sediment elutriates. Increased concentrations of elutriates in April lead to lower final growth yields (Fig. 6). A comparison of control and test cultures showed a higher number of thecae in the test growth medium, at any given time. Thecae can be used as tracer of cellular division or encystement. It has been shown (16) that large amount of thecae in the culture can only result from a stimulation of cellular division, leading to an increased death of cells. The effect of sediment extracts on growth was still measurable at low concentrations corresponding to washing out of 1 cm layer of surface sediment distributed equally in a 10 m water column. In contrast, the same experiments performed with July sediment elutriates from the same area showed that toxic effects of sediment disappeared, allowing normal dinoflagellate growth. Therefore spring bloom sedimentation possibly leads in some areas to environmental conditions fit for dinoflagellate growth.

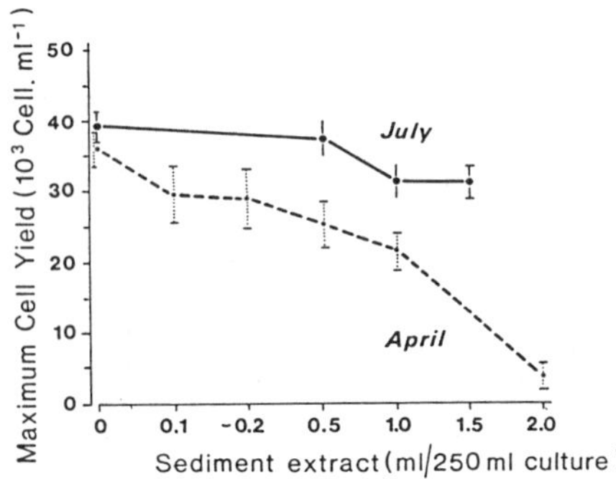

FIG. 6. Effect of sediment extracts on Alexandrium tamarensis growth.

DISCUSSION

Sediment enrichment of nutrients results from bacterial degradation of organic matter in oxidized layers (17). Particularly high levels of nutrients indicates exogenous supplies, as nitrogenous salts in agricultural vicinity, or phosphorus in urban neighbourhood. The position of areas unaffected by red tides (Fig. 2) shows that dissolved nutrients cannot be used as a discriminatory criterion.

Organic C and N concentrations varied irregularly from April to July, generally decreasing with bacterial degradation in summer (18). They are enhanced by polluted waste leading to anoxic environment. As reported by Uyeno and Nagai in (19), anoxia leads to favourable conditions for red tide blooms. They found that stimulating effect of water extracts of bottom sediments on Heterosigma inlandica growth was greater in hypoxia periods.

High concentrations of copper in harbours are probably due to the use of antifouling paint and brass material of boats.

Stations with high organic matter and high copper concentrations in sediments were always affected by recurrent nuisance blooms. Areas unaffected by red tides had the lowest organic matter and copper sediment contents. Nevertheless, no quantitative relation appeared between red tides frequencies and the levels of these two parameters which can only be considered as discriminatory criteria. Low levels of organic matter and copper did not allow a discrimination between stations, showing that sediment composition is not the only source of variation. However, this approach proved its usefulness in pointing out the need for monitoring at Aber Ildut (Fig. 5).

Although it is quite impossible to decide of causal relationships between the sediment compositions and red tide occurrences, it is to be noted that high levels of organic matter may ensure favourable conditions for all phytoplankton blooms (20). On the other hand, influence of copper has also been reported affecting phytoplankton, particularly dinoflagellate growth (21).

Growth inhibition of an aqueous extract of sediment of a station affected by nuisance blooms and the neutralization of this effect in July suggest important modifications of the environment resulting from bloom sedimentation (23). In the same areas, phytoplankton growth regulation by environmental constituents have been largely documented (21, 22, 23, 24). It suggests that spring blooms in coastal waters, after sedimentation of organic matter on the bottom, act as a favourable enhancing red tide factor.

ACKNOWLEDGEMENTS

We gratefully thank A. Youenou, A. Derrien, M.P. Crassous, M. Lunven for their technical assistance all along sampling, analyses and bioassays. We also acknowledge the secretary work of J. Huguen and P. Bodénes. Support for this work was provided by the IFREMER Red Tides Program.

REFERENCES

1. J.P. Berthome and C. Belin. Rapport DRV-88.005-CSRU/NTES, IFREMER, Nantes, France (1988).
2. K. Steidinger and K. Haddad. Bioscience 31, 814-819 (1981).
3. D.M. Anderson end F.M.M. Morel. Est. Coast. Mar. Sci. 8, 279-293 (1979).
4. S. Ishio and K. Kondo, Bull. Jap. Soc. Sci. Fish. 46, 977- 989, (1980).
5. D.M. Anderson end F.M.M. Morel. Limnol. Oceanogr. 23, 283-295, (1978).
6. N. Fisher. Limnol. Oceanogr. 31, 443-449 (1986).
7. P.J. Hannan and C.E. Patouillet. N.R.L. Memorandum, Report 3952 , (1979).
8. S. Ishio, T. Nishimoto and H. Nakagawa. Nippon Suisan Gakkaishi 54, 1175-1181 (1988).
9. P. Lassus, M. Bardouil, J.P. Berthome, P. Maggi, P. Truquet and L.Le Dean. Aquatic living resources, Vol 1, N° 1-2, 3-4, 155-164 (1988).
10. P. Treguer and P. Le Corre, Manuel d'analyse des sels nutritifs dans l'eau de mer. 2nd ed., UBO, Brest, France (1975).
11. A. Vatan. Manuel de Sédimentologie. Ed. Techniq. Paris, 383-386 (1967).
12. J.L. Charlou et M. Joanny. Manuel des analyses chimiques en milieu marin. CNEXO documentation, 285-295, IFREMER, Brest, France (1983).
13. J.V. Klump and C.S. Martens. Geochim. Cosmochim. Acta 45, 101-121 (1981).
14. J.G. Parker. Est. Coast. Shelf Sci. 15, 373-384 (1982).
15. D.T. Rudnick and C.A. Oviatt. J. Mar. Res. 44, 815-837 (1986).
16. N. Fukazawa, T. Ishimaru, M. Takahashi and Y. Fujita. Mar. Ecol. Prog. Ser. 3, 217-222 (1980).
17. D.R. Ackroyd, A.J. Bale, R.J.M. Howland, S. Knox, G.E. Millward and A.W. Morris. Est. Coast. Shelf Sci. 23, 621-640 (1986).
18. R. Rosental, G.A. Eagle and M.J. Orren. Est. Coast. Shelf Sci. 22, 303-324 (1986).
19. P. Gentien and G. Arzul. Rapport 88-07-EL - IFREMER (1987).
20. H. Iwasaki, J. Ishii, and S. Ueda. Bull. Jap. Soc. Sci. Fish. 53, 1065-1072 (1987).
21. E. Granéli and W. Granéli. International Symposium on utilization of coastal ecosystems : planning pollution and productivity. Rio Grande, Atlantica, Vol. 2, 51, special issue, (1982).
22. E. Granéli and H. Persson. International Symposium on utilization of coastal ecosystems : planning pollution and productivity. Rio Grande, Atlantica, Vol. 2, 51, special issue, (1982).
23. A. Prakash. Rapp. P.V. Reun. CIEM. Exceptional plankton blooms, 187, 61-65 (1987).
24. K.E. Cooksey and B. Cooksey. J. Phycol., 14, 347-352, (1978).

FOLLOW-UP OF A BLOOM OF THE TOXIC DINOFLAGELLATE GAMBIERDISCUS TOXICUS ON A FRINGING REEF OF TAHITI

RAYMOND BAGNIS*, ANNE-MARIE LEGRAND** AND AKIO INOUE*
*Institut Territorial de Recherches Médicales Louis Malardé, BP 30, Papeete Tahiti, French Polynesia; **Research Center for the South Pacific, Kagoshima University, 21-24 Korimoto, 1-Chome, Kagoshima, Japan

ABSTRACT

The population densities, per gram algae, of the toxic benthic dinoflagellate *Gambierdiscus toxicus* (GTD) on the Hitiaa fringing reef in Tahiti, have been followed since the end of 1976. At that time, during two months, GTD varied between 100 and 800. From mid 1977 to the end of November 1987, it stayed permanently under 50, most of the time under 10. In December 1988 and January 1989, GTD increased from 50 to 13,000 in 3 weeks. It decreased to ca 3,000 one week later and to under 100 one month later. The population densities (OLD) of another benthic dinoflagellate, *Ostreopsis lenticularis,* frequently associated to *G. toxicus* in ciguateric areas, but not found to be toxic in French Polynesia, were also high during the same period.
The toxin production of *G. toxicus* was checked at various stages of the bloom.

INTRODUCTION

Since the finding of the benthic dinoflagellate *Gambierdiscus toxicus* (GT) Adachi and Fukuyo (1) as a causative agent of ciguatera (2,3), a follow-up of its population and of that of *Ostreopsis lenticularis* (OL) Fukuyo, another benthic dinoflagellate frequently associated to it, has been initiated on a fringing reef of Tahiti island, sporadically exposed to ciguateric risk. Samples were taken weekly in 1977-1978, monthly from 1979 to 1985, quarterly from 1986 to 1988. The available data show that except in early 1977, when GTD rose to close to 800 cells per gram of the macroalga (g.a.) *Turbinaria ornata* (4), the population has stayed below 25 per g.a. The density of *O. lenticularis* (OLD) has grown sporadically to close to 1,000, but most of the time has stayed below 50 (5). The occurrence in December 1988 of an OLD > 5,000 and of a GTD close to 50 urged us to pay special attention, and gave us the opportunity to follow the evolution of a bloom of *G. toxicus* and *O. lenticularis* under the scopes of density, toxin production and transmission to the commonest first link of the ciguateric food chain, i.e. the local surgeonfish maito (*Ctenochaetus striatus*).

MATERIAL AND METHODS

The study was carried out from December 12, 1988 to March 29, 1989, on the Hitiaa fringing reef (Fig. 1), from where no clinical outbreaks of maito-poisoning had been known over the last 3 or 4 years. The sampling of the host macroalgae to evaluate GTD and OLD took place weekly and was made at the 3 to 6 m depth on the oceanic reef slope in the median part of the 250 m long reef. The monitoring of *G. toxicus* toxicity was carried out about weekly and that of *C. striatus* ciguatoxicity about monthly, starting at beginning of January 1989.

Approximately 50 g of algal tufts of *Jania* sp. were collected. The

Toxic Marine Phytoplankton
Edna Graneli et al., Editors

presence and the number of GT and OL cells were checked as previously described (5). The GTD and OLD values are given per gram of wet algae. From January 1989, larger samples of *Jania* sp. (750 to 7,000 g, depending on the GTD level) were collected. After centrifugation at 2,500 g, epiphytic sediments were homogenized twice in methanol (50 ml/10 g) at room temperature during 1 min, and then centrifuged at 1,000 g. The combined supernatants were concentrated under vacuum. The residue was suspended in water and the toxins extracted three times with twice its volume of butanol. After concentration under vacuum of the butanol fraction, the residue was suspended in acetone, left at -20°C for at least 4 h and filtered at 0°C. The acetone filtrate was concentrated under vacuum. The precipitate was suspended in methanol, concentrated and kept separately. Intraperitoneal acute toxicity of the acetone filtrate and precipitate was evaluated on Swiss mice (OFI, IOPS CAW, IFFA-CREDO, France) weighing 18-20 g. A small fraction of each sample was diluted in 0.5 ml of saline solution. Groups of three animals were administered with one dose evaluated according to the number of GT cells. Symptoms and survival time were noted. Toxicity was determined from standard dose-response curves preliminarily estahlished (unpublished data) and expressed in mouse-units (MU) per g of mouse (6).

A sample of 4 g of ground raw flesh of *C. striatus* was homogenized twice in 5x its volume of acetone and centrifuged at 1,000 g. The supernatants were concentrated. The dried acetonic residue was dissolved in 3x5 ml of $CHCl_3$ and adsorbed on a microcolumn of 1 g silica. The ciguatoxin-like subtances were eluted with $CHCl_3$-MeOH (90:10) and evaporated under nitrogen flux at 40-50°C. The toxicity of this fast extraction residue was checked by the mosquito-bioassay and expressed in equivalent gram of fish (fge). Three classes of fish were differentiated: toxic ≤ 7 fge; borderline > 7 to ≤ 10 fge, safe: > 10 fge (7).

Toxicity results were expressed as a mean value plus or minus the mean standard error. Comparison between several batches was performed by the unpaired Student t-test. A value (probability of error) less than 6% was considered significant.

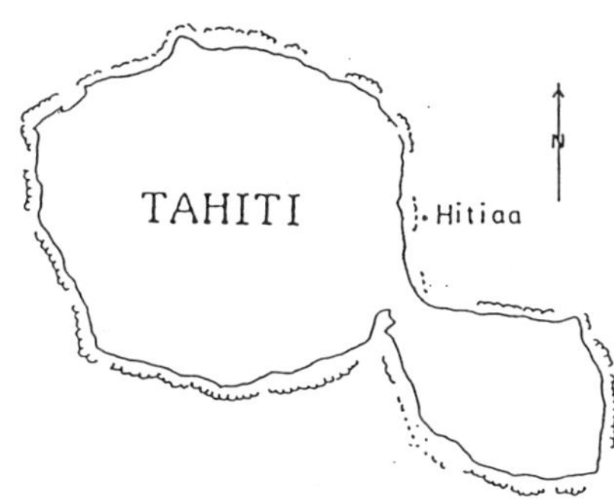

Fig. 1. Island of Tahiti with the Hitiaa fringing reef

RESULTS AND DISCUSSION

Epiphytic dinoflagellate populations (Fig. 2)

The described bloom involved both *G. toxicus* and *O. lenticularis*. It was first detected on December 12, 1988. Presented data result from 15 samplings from then until March 29, 1989. During three weeks, the GTD level increased gradually from about 50 to ca 13,000 on January 4 - most likely the peak of the bloom. Then it decreased slowly to under 1,000 at the end of the month and to under 100 (with a mean value of 60 ± 36) in February-March. The last value of the series was 25 (end of March). GTD levels were below 20 in April and May. The OLD level increased steadily from 5,800 on December 12 to culminate at 15,600 on December 27. During January it slowly dropped from 10,000 to 2,000 before rapidly decreasing to under 50 (Jan. 27 – Feb. 6). It remained at this level during February and half of March, to decrease to a value of 15 at the end of March. In April and May OLD levels stayed below 5.

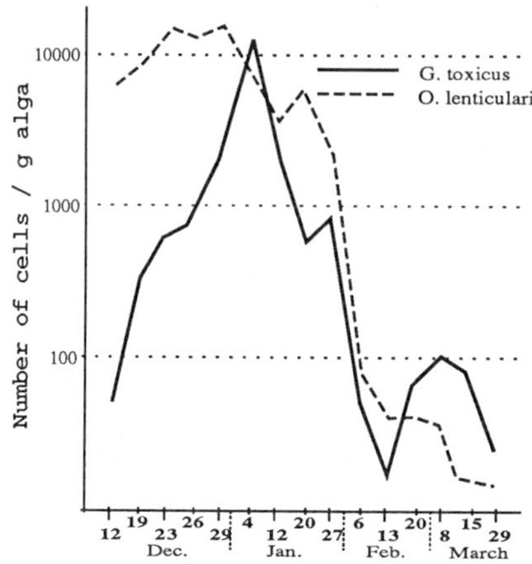

Fig. 2. Population density of *G. toxicus* (GTD) and *O. lenticularis* (OLD) on Hitiaa reef

The occurence of relatively large populations of both *O. lenticularis* and *G. toxicus* at almost the same time and in the same location, is not a frequent event in French Polynesia. Moreover, this was the first time that we had the opportunity to follow the onset, the developlment and the end of a bloom of *G. toxicus*. It is also interesting to note that this bloom occurred at the same period of the year as a smaller peak of *G. toxicus* that occurred 12 years ago on the Hitiaa reef (4,5). But that "bloom" was well in progress when detected and was monitored on the macroalga *T. ornata,* whereas the present bloom was studied on *Jania* sp., which was very abundant this year on the outer slope of the Hitiaa reef. The unusual abundance of *Jania* sp. could be one factor contributing to the bloom of *G. toxicus,* when taking into account its affinity for this macrophyte (8).

Toxicity of sediments associated to *Jania* sp.

Though the dinoflagellate *O. lenticularis* has several times been suggested as toxic and able to produce water-soluble and fat-soluble toxins (9,10,11), no correlation has been found in French Polynesia between the abundance of *O. lenticularis* and the presence of ciguateric fish. In 1980, when a bloom of *O. lenticularis* occurred on the Papeete reef, with OLD fluctuating between 1,000 and 18,000 during three months, the maito (*C. striatus*) remained non toxic. Moreover, a recent study on the toxicity of two samples of respectively 9.5 and 5.8 x 10^6 cultured OL cells, originating from the Hitiaa fringing reef, showed a total absence of ciguatoxin-like (OTX) and maitotoxin-like (MTX) substances, whereas cultured GT cells from the same area produced small amounts of each (12). Therefore, the amounts of these compounds in sediments was only compared to the number of GT cells (Table I). For 1,000 GT cells of each sample, the values ranged from 0.05 to 1.5 MU for CTX and from 1 to 10 MU for MTX with mean values of 0.59 ± 0.16 MU and 3.45 ± 1.02 MU respectively.

Although the GTD values registered in Hitiaa were much lower than those registered in Gambier Is. during the most important bloom ever reported (2,10), the toxin production of the wild dinoflagellates is comparable (6). Thus, the mean number of GT cells needed to produce 1 MU of toxin in Hitiaa and Gambier samples was 5,688 ± 2,223 and 2,393 ± 762 respectively for CTX, and 558 ± 91 and 334 ± 74 respectively for MTX, values that not differ significantly.

Recently in Puerto Rico, *O. lenticularis* and its associated microflora have been considered as a potential primary ciguateric vector (9). As for *G. toxicus,* which can be abundant without ciguatera (12,13,14), the occurence of non toxic strains or the absence of an appropriate associated microflora may explain the apparent difference between Caribbean islands and French Polynesia when it comes to *O. lenticularis* toxicity.

TABLE I. Toxicity of sediments, associated to *Gambierdiscus toxicus* (GT) in Hitiaa fringing reef

Date of sampling	Number of GT cells	Amount of toxin in MU		Number of cells to produce 1 MU of	
		CTX like	MTX like	CTX like	MTX like
04.01.89	$8x10^6$	1360	10000	5882	800
	$10x10^6$	10000	10000	1000	800
12.01.89	$4.2x10^6$	4200	7350	1000	571
20.01.89	$4.5x10^5$	450	675	1000	667
27.01.89	$6x10^6$	9000	45000	667	133
	$4.8x10^6$	240	6000	20000	800
06.02.89	$3.2x10^5$	20	400	16667	1000
13.02.89	$8x10^4$	Not determined	135	-	588
20.02.89	$3.5x10^5$	55	440	6667	800
08.03.89	$2.7x10^5$	Not determined	2650	-	100
15.03.89	$5.2x10^5$	260	5220	2000	100
25.03.89	$1.7x10^5$	90	525	2000	333

Toxicity of *Ctenochaetus striatus* surgeonfish (maito)

Three samples of five maito were captured and tested by the mosquito bioassay: M1 (105±16 g) in early January (at the peak of the GT bloom), M2 (125±14 g) at the end of February and M3 (152±16 g) at the end of March. All the fishes were found toxic (Table II). But the mean toxicity (in fge) of 2.07±0.55 for the M1 sample, 1.85±1.3 for the M2 sample and 3.67±1.24 for the M3 sample, did not differ significantly. The data point out a significant change in the ciguateric risk on the Hitiaa reef, since all the quarterly mosquito bioassays performed during the years 1986 - 1988 on comparable samples of maito were negative (fge > 10).

TABLE II. Toxicity of *C. striatus*, expressed in equivalent g of fish (fge)

Date of capture	Weight (g)	Toxicity (fge)
04.01.89 (M1)	106	4.20
	130	2.82
	90	0.15
	95	1.76
	107	1.72
	92	1.76
15.12.89 (M2)	110	0.15
	136	2.37
	143	6.73
	114	0.00
	120	0.00
28.03.89 (M3)	156	1.72
	176	5.64
	150	4.30
	142	6.69
	135	0.00

These results also show that when a bloom of *G. toxicus* occurs on a reef, the maito living in the area become very rapidly toxic, and may rapidly transfer the ciguatoxicity to the higher trophic levels of the food chain, which can explain sudden and patchy outbreaks of ciguatera in the human population, in areas previously safe over a long period of time.

ACKNOWLEDGMENTS

This study was jointly supported by the Government and Territory Assembly of French Polynesia and by the French Ministry of Research, through the Pasteur Institute of Paris. We want to express our gratitude to them, to the divers of Malardé lnstitute, Mr J. BENNETT and Mr M. BARSINAS and to Dr Y. OSHIMA, Associate Professor, Tohoku University, Japan for his advice.

REFERENCES

1. R. Adachi and Y. Fukuyo, Bull. Jpn. Soc. Sci. Fish. 45, 67-71 (1979).
2. T. Yasumoto, I. Nakajima, S. Bagnis and S. Adachi, Bull. Jpn. Soc. Sci. Fish. 43, 1021-1026 (1977).
3. R. Bagnis, S. Chanteau and T. Yasumoto, C.R. Acad. Sci. 28, 105-108 (1977).
4. T. Yasumoto, A. Inoue, R. Bagnis and M. Garçon, Bull. Jpn. Soc. Sci. Fish. 45, 395-399 (1979).
5. R. Bagnis, J. Bennett, C. Prieur and A.M, Legrand in: Toxic Dinoflagellates, D.M. Anderson, A.W. White and D.G. Baden, eds. (Elsevier, New York 1985) pp. 177-182.
6. R. Bagnis, S. Chanteau, E. Chungue, J.M. Hurtel, T. Yasumoto and A. Inoue, Toxicon, 18, 199-208 (1980).
7. A, Pompon, E. Chungue, I. Chazelet and R. Bagnis, Bull. OMS. 62, 639-645 (1984).

8. M. Chaine, Etude du Micro et du Meiobenthos Algal Associés au Dinoflagellé *Gambierdiscus toxicus* Adachi et Fukuyo Agent Causal Princeps de la Ciguatera par la Méthode des Substrats Neufs Artificiels (Atoll de Mururoa, Polynésie Française), PhD Thesis, University of Sciences and Technics of Languedoc, France 1987.
9. T.R. Tosteson, D.L. Ballantine, C.G. Tosteson, V. Hensley and A.T. Bardales, Appl. and Env. Microbiol. 155, 137-141 (1989).
10. J.W. Bomber, Ecological Studies of Benthic Dinoflagellates Associated with Ciguatera from the Florida Keys. M.S Thesis, Florida Institute of Technology, Melbourne, Florida 1985.
11. E.G. Besada, L.A. Loeblich and A.R. Loeblich, Bull. Mar, Sci. 32, 723-735 (1982).
12. N. Gillespie, R. Lewis, J. Burke and M. Holmes, Proc. of the Fifth Int. Coral Reef Congress, 437-441 (1985).
13. F.J.R. Taylor and M.S. Gustavson, An underwater survey of the organism chiefly responsible for ciguatera fish poisoning in the eastern Carribean region: the benthic dinoflagellate *Gambierdiscus toxicus*. in: A. Stefanon and N.J. Flemming (eds), Proc. VIIth Int. Diving Sci. Symp., Padova, Italy (1983).
14. R. Bagnis, J. Bennett, M. Barsinas, M. Chebret, G. Jacquet, I. Lechat, Y. Mitermite, P. Pérolat and S. Rongeras, Proc. Fifth Int. Coral Reef, Tahiti, 475-482 (1985).

DISTRIBUTION OF ORGANIC COMPOUNDS DURING A BLOOM OF CHRYSOCHROMULINA POLYLEPIS IN THE SKAGERRAK

U. BROCKMANN* AND E. DAHL**
*Institut für Biochemie und Lebensmittelchemie der Universität Hamburg, Martin-Luther-King-Platz 6, D-2000 Hamburg 13, F.R.G.;
**Statens Biologiske Stasjon Flødevigen, N-4817 His, Norway

ABSTRACT

Nutrient elements, bound in dissolved organic substances, can become a major controlling factor for algal growth when inorganic nutrients are depleted. During the late phase of the extended bloom of *Chrysochromulina polylepis* in the Skagerrak/Kattegat during spring 1988, concentrations of particulate and dissolved organic substances (C,N, P) were estimated together with hydrographic measurements, and nutrient and phytoplankton analyses. Along a profile through the Skagerrak between Skagen and Hällö a stable and most biologically rich pycnocline at 7-25 m depth separated two distinct chemical regimes: (i) a surface layer where nitrate (<0.2 µg at N dm^{-3}) and phosphate (<0.02 µg at P dm^{-3}) were nearly depleted and dissolved organic compounds (DOM) reached maximum concentrations (>10 µg at N dm^{-3} and >0.05 µg at P dm^{-3}), (ii) lower layers with higher inorganic nutrients and lower DOM. As a result of this, the pycnocline exhibited a minimum concentration of total dissolved nitrogen (<10 µg at N dm^{-3}) together with maximum cell densities (>$16 \cdot 10^6$ cells dm^{-3}) of *C. polylepis*. Because of these vertical nutrient profiles, the cells could utilize both inorganic and organic bound nutrient elements from the adjacent layers. The main N-source was however probably nitrate as seen from mixing curves. High concentrations of dissolved carbohydrates (>3 µmol Glc-eq dm^{-3}) in the mixed layer, significantly negative correlated with salinity, indicate that the dominating source for DOM was the Baltic outflow.

INTRODUCTION

A bloom of *Chrysochromulina polylepis* (Prymnesiophyceae) in the Skagerrak/Kattegat lasted from early May until mid of June in 1988. As the bloom was most extensive, it covered an area of 60.000 km^2 [1]. Among the questions asked during this large bloom was: What are the ecological reasons for development and stabilization of such a bloom? Important conditions are stratification, light climate and supply of nutrients. In addition to inorganic nutrients organic bound nutrient elements can be utilized by phytoplankton species [2]. Depletion of nutrients like phosphate can increase the toxicity of flagellates [3]. The distribution of dissolved organic substances can be used as indication for release processes. Hence inorganic nutrients as well as dissolved organic nitrogen and carbon compounds were

Toxic Marine Phytoplankton
Edna Graneli et al., Editors

analysed. The particulate organic fraction was used together with dissolved nutrients to characterize the physiological state of the bloom related to the nutrient status and to pin point the limiting nutrition element.

METHODS AND MATERIALS

The study was carried out during the RV "G.M. DANNEVIG" cruise on May 30th, 1988. Locations of stations are given in FIG. 1. Salinity and temperature were measured by probes. Water samples were taken by Niskin bottles and filtered through glass fiber filters (GF/C) by vacuum suction. Filtrates were partly fixed with $HgCl_2$ (0.1 % w/w) for dissolved nutrient analysis using an autoanalyzer. Subsamples were analysed directly for nutrients by colorimetric methods. Dissolved organic nitrogen (DON) and phosphorus (DOP) were determined following combustion with peroxidisulphate. Dissolved carbohydrates (DCH) were analysed as glucose equivalents (Glc-eq dm^{-3}), using the l-tryptophane/sulfuric acid reaction adapted to an autoanalyzer [4]. Following hydrolysis, particulate phosphorus (POP) and carbohydrates (PCH) were analysed by methods described for dissolved constituents, particulate carbon and nitrogen with a CHN-Analyzer. Cell numbers were counted on board directly in a Palmer-Maloney slide from acetic acid iodine fixed samples.

RESULTS AND DISCUSSION

During the late phase of *Chrysochromulina polylepis* bloom [5] along the transect from Skagen towards the Gullmar fjord (Isle of Hällö), there was a shallow mixed layer stabilized by a narrow density gradient of $\Delta\sigma_t$ 3-4 between 5 and 10 m. In the mixed layer, temperature was above 11 °C and salinity below 30 ‰. Near the Swedish coast the outflowing Baltic water reduced the surface salinity to less than 20 ‰, affecting still central positions (up to St.249). The lower part of the densicline had a weaker gradient of $\Delta\sigma_t$ 2 between 10 and 25 m, covering a central layer (30-100 m) with temperatures below 7 °C and salinities above 34 ‰. Near Skagen downwelling was evident by a deep pycnocline as well as by low nutrient concentrations at 30m depth. Vertical temperature profiles indicated partial upwelling at St.247 and 249. Here temperatures of less than 7 °C were detected at 10 m depth.

Main ranges of nutrient concentrations for the different layers are presented in TABLE I. Due to remineralization processes, intermediate ammonium maxima appeared with concentrations between 1-2 µg at N dm^{-3} in around 30 m depth, especially in the eastern part of the transect. Released ammonium was permanently converted by nitrification, indicated by increased nitrite concentrations (0.1-0.3 µg at dm^{-3}) coupled with ammonium maxima. The atomic ratios of nitrate/phosphate were between 20 and 50 within the mixed layer and below 10 in the densicline. Hence if one of these nutrients can be assumed to be growth-limiting, then it was phosphate within the mixed layer and nitrate within the densicline. The ratios found in the deeper water masses were between 10 and 16. They were in accordance with findings on May 11, south of Christiansand and Arendal [6]. The cells of *C.polylepis* occurred at this time in the lower pycnocline at depths between 10 and 25 m (FIG. 1), reaching cell numbers of more than $10 \cdot 10^6$ dm^{-3}.

TABLE I. Nutrient concentrations at a transect Skagen-Hällö on May 30th, 1988.

	PO_4^{3-}	SiO_2	NH_4^+	NO_2^-	NO_3^-	DON	DOP
				µg at dm^{-3}			
mixed layer (0-7 m)	0-0.02	0.5-1	0.2-0.4	0-0.04	0.1-0.2	10-15	0.05-0.16
pycnocline (7-25 m)	0.02-0.3	0.3-1	0.4-2	0.02-0.2	0.2-3	7-10	0.05-0.10
central layer (30-100 m)	0.5-0.7	1-4	0.5-2	0.05-0.3	5-10	4-7.5	0-0.05
bottom layer (> 100 m)	0.7-0.9	4-7	0.4-0.8	0.01-0.2	10-12	4-6	0-0.1

Due to exhaustion of some of the inorganic nutrients, also organic compounds could serve as nutritial basis of the bloom [2]. The vertical distribution of total dissolved nitrogen shows a layer of minimum concentrations (below 10 µg at dm^{-3}) (FIG. 2) where C. polylepis appeared with maximum cell numbers. This intermediate minimum-layer was mainly caused by nitrate decreasing within the pycnocline towards the mixed layer and DON decreasing towards the lower layer (FIG. 3). There was a significant negative correlation between DON and salinity (FIG. 4) with an intermediate DON maximum around S = 32 ‰. This was within the lower pycnocline, the maximum algal layer. The nitrate/S mixing diagram revealed two completely separated water masses: (i) above 33 ‰, nitrate concentrations were mostly below 0.2 µg at dm^{-3}, (ii) at 33 ‰ there was an abrupt nitrate increase to 5 - 10 µg at dm^{-3}. From the absence of any linear mixing curve it can be concluded that C. polylepis mainly utilized nitrate as nitrogen source from the lower pycnocline.

Dissolved organic substances DON, DOP and DCH showed highest concentrations in the mixed layer (TABLE I). DCH reached here 3-4 µmol Glc-eq dm^{-3}, below 30 m only 0.5-1 µmol Glc-eq dm^{-3}. Maximum values were detected near the coasts. Since all of the analysed dissolved compounds showed significant negative correlations with salinity, these compounds were probably discharged within the Baltic outflow. At a salinity of 32 ‰ where a DON-maximum in the DON/S mixing diagram occurred, similar maxima were found for DCH with an increase of 1.5 µmol Glc-eq dm^{-3}. Phytoplankton-release of dissolved organic substances, such as carbohydrates, increases especially when nutrients are exhausted [7]. Increased concentrations of DOM in the maximum cell layer can therefore be interpreted as a consequence of increased release activity. However, since the cells of C. polylepis can be very fragile [1], part of DON and DCH might have been lost from the cells during filtration. The ratios of DON/DOP (µg at) were mainly between 50 and 150, showing no clear vertical gradients, but indicating again P-deficiency.

The distribution of particulate organic material was similar to that of cell concentrations of C. polylepis (FIG. 3). Concentrations of POC varied between 2 and 100 μg at C dm^{-3}. C/N ratios were mainly between 6 and 14 and around 10 - 14 in the layer of C. polylepis. N/P ratios were here in the range of 20-40 and C/P ratios above 300. PCH concentrations were always above 0.5 μmol Glc-eq, reaching partly more than 2 μmol. In the central and deeper water the concentrations dropped below 0.2 μmol Glc-eq dm^{-3}. Particulate organic compounds were significantly correlated with cell numbers allowing a rough estimation

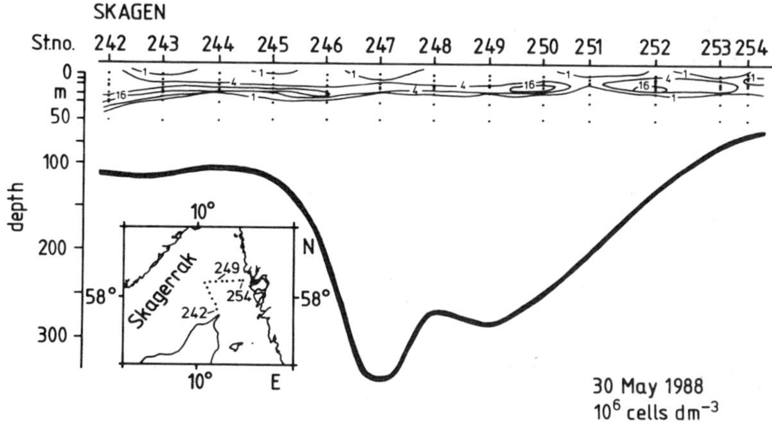

FIG. 1. Cell numbers of Chrysochromulina polylepis along a transect through the Skagerrak.

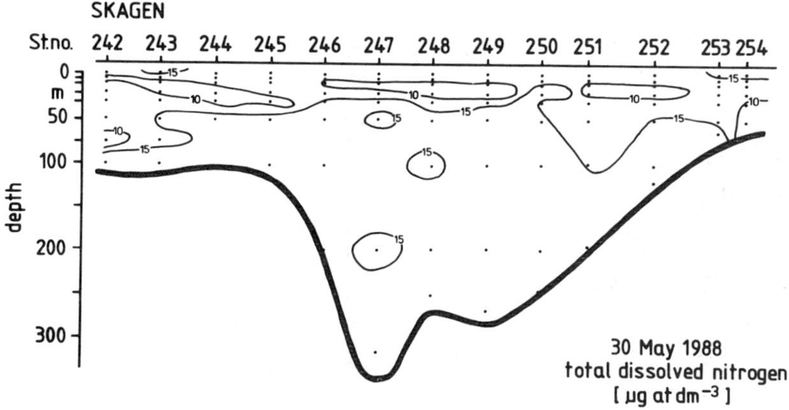

FIG. 2. Total dissolved nitrogen compounds along a transect through the Skagerrak.

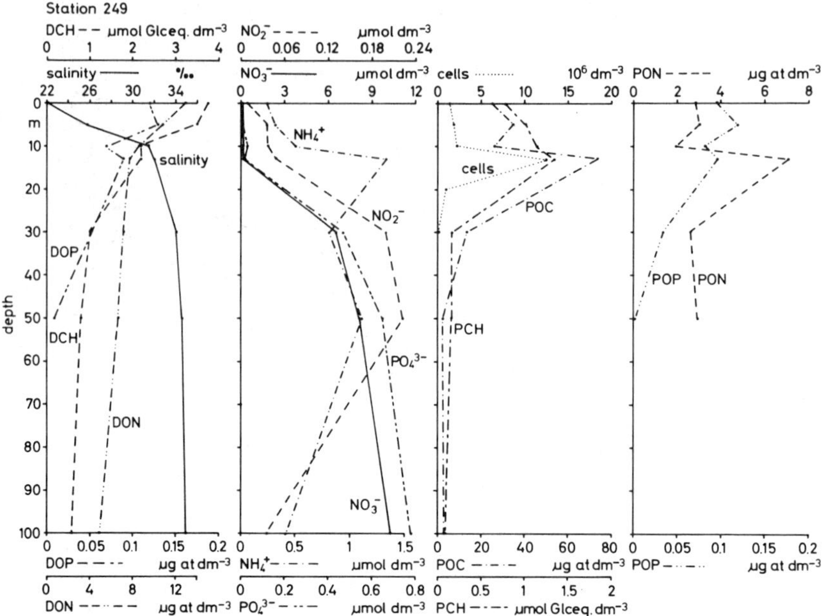

FIG. 3. Vertical profiles of dissolved and particulate organic substances at a central position (St.249) in the Skagerrak on 30 May, 1988 (see FIG. 1).

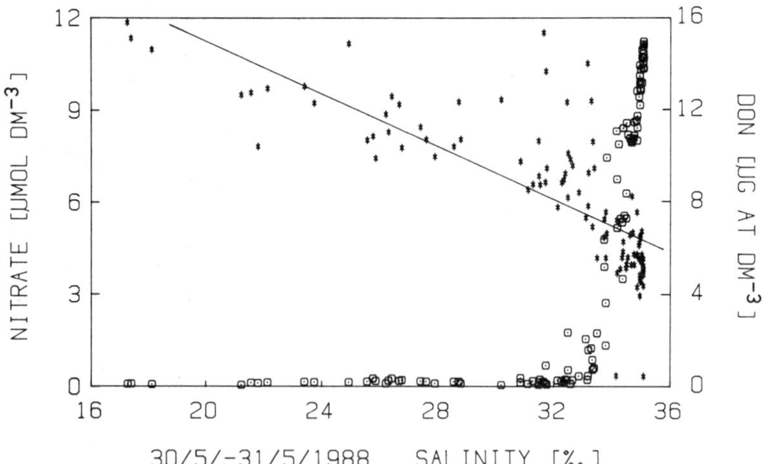

FIG. 4. Mixing diagram of DON (✱) and nitrate (⊡) with salinity. S/DON correlation: r = -0.78; n = 101.

of cell contents (TABLE II). Comparison with culture experiments [1] revealed that our findings were higher for C (1 pg at $cell^{-1}$) N (0.1 pg at $cell^{-1}$) and lower for P (0.01 pg at $cell^{-1}$), reaching only half of the culture values. This, together with the atomic ratios, suggests that P-limitation occurred which can increase the toxicity of the cells as found for other species [3]. The concentrations of PCH and DCH were significantly correlated suggesting close connections between the dissolved and particulate pools. The fraction of DCH surpassed twice the particulate fraction.

TABLE II. Estimated cell contents of Chrysochromulina polylepis

	µg at/10^6 cells*	r	n
C	3.4 - 4.3	0.83	80
N	0.32 - 0.42	0.78	80
P	0.010 - 0.013	0.75	82
	µmol/10^6 cells*		
PCH	0.108 - 0.132	0.78	82

*calculated for (15 and 30) 10^6 cells dm^{-3}
r = correlation coefficient, n = number of data points

CONCLUSIONS

In spite of the abundance of DON (originating from the Baltic outflow) in the mixed layer, nitrate was the dominating nitrogen source for C. polylepis. The P-deficiency which was found for cell contents, was only reflected in the nitrate/phosphate ratios of the mixed layer from where the cells originated a couple of days before [5], but in the pycnocline where maximum cell numbers of the late bloom stage were found, nitrate/phosphate ratios were below 10, indicating N-limitation. Increased concentrations of dissolved organic substances in the layer with maximum cell numbers can be related to release processes. The prolonged stabilization of the bloom within the same area may be explained by repeated local downwelling, advection and upwelling causing transient nutrient enrichment and seeding of surface water.

REFERENCES

1. E. Dahl, O. Lindahl, E. Paasche and J. Throndsen in: A novel phytoplankton bloom. Causes and impacts of recurrent brown tides E.M. Cosper et al., eds. (Springer Lecture Notes on Coastal and Estuarine Studies 1988), 24 pp.
2. A.B.J. Sepers, Hydrobiologia 52, 39-54 (1977).
3. G.L. Boyer, J.J. Sullivan, R.J. Andersen, P.J. Harrison and F.J.R. Taylor in: Toxic dinoflagellates D.M. Anderson, A.W. White and D.G. Baden, eds. (Elsevier, 1985) pp. 281-286.
4. K. Eberlein and M. Schütt, Fresenius Z.Anal Chem 323, 47-49 (1986).
5. O. Lindahl and E. Dahl, this volume.
6. E. Dahl, Vann 3B, 512-523 (1988).
7. U. Brockmann, V. Ittekkot, G. Kattner, K. Eberlein, K.D. Hammer in: North Sea dynamics J. Sündermann, W. Lenz, eds. (Springer, New York 1983) pp. 530-548.

DEVELOPMENT AND DISPERSAL OF RED TIDES IN THE PORT RIVER, SOUTH AUSTRALIA

JEAN A. CANNON
University of Adelaide, Department of Botany, Box 498. G.P.O. Adelaide, South Australia, 5001.

ABSTRACT

The development of three major blooms was monitored in the Port River (Adelaide, South Australia) during spring 1988. These were mixed blooms of *Prorocentrum micans* with approximately one fifth *Alexandrium minutum*, *Gymnodinium sanguineum* with *A. minutum*, and a nearly monospecific bloom of the toxic species, *Alexandrium minutum*.

The dinoflagellate blooms develop upstream in bottom layers at 4-5 metres depth close to a sewage effluent outfall. These underwater blooms move down the river with the tidal flow and surface when there is minimal water movement (when calm weather coincides with neap tides). Large increases in cell numbers (up to $9x10^8L^{-1}$) coincide with minimum water movement. Dinoflagellates form a dense band from the surface to 4 metres deep and show diel migration in the water column. The blooms break up when the water becomes destratified. Aerial photos show the bloom confined to the edges of the river rather than in the centre where wind, shipping and tidal flow appear to be the major destratifiers. Salinity stratification and tidal amplitude appear to be important factors in bloom development

Culture experiments with natural sediment samples produced *Alexandrium minutum* and *Prorocentrum micans* cultures. Blooms of these species appear to be initiated from cysts.

INTRODUCTION

The Port River is an estuarine waterway connected with a man made marine lake (West Lakes) with a salinity range from 30-37.8°/oo. The staggered tidal flow through a series of one way gates results in a one way tidal flushing arrangement such that water enters West Lakes from the sea and leaves via the Port River. Stormwater from many major drains as well as effluent discharge from the Port Adelaide Sewage Works result in eutrophic conditions especially in the Port River (Fig. 1). Thirty to forty mega litres per day of sewage effluent result in total P concentration which exceeds $1mgL^{-1}$ and total N which exceeds $3mgL^{-1}$ [1].

Patches of discoloured water have occurred intermittently in the Port River for many years and have been reported by the South Australian Engineering and Water Supply Department since 1981 [1]. Over 20 different species of dinoflagellate have been recorded. One of these is the toxic species *Alexandrium minutum* which has resulted in the Port River being closed for collection of shellfish in 1987 [2] and in 1988.

There are concerns about the effect toxic blooms might have on public health and planned housing developments along the Port River as well as the possibilities of the blooms spreading to fish nursery areas in Barker Inlet and the quarantine station. There is also potential for their spread both locally and internationally in ballast water as this is a busy dock area.

There have been few cases where events preceding a red tide have been investigated, on the other hand numerous studies have been undertaken once red patches are observed [3]. This paper reports on events before a bloom and the detection of underwater blooms before they became evident on the surface.

This paper aims to provide an understanding of the development and dispersal of dinoflagellate blooms in the Port River.

Toxic Marine Phytoplankton
Edna Graneli et al., Editors

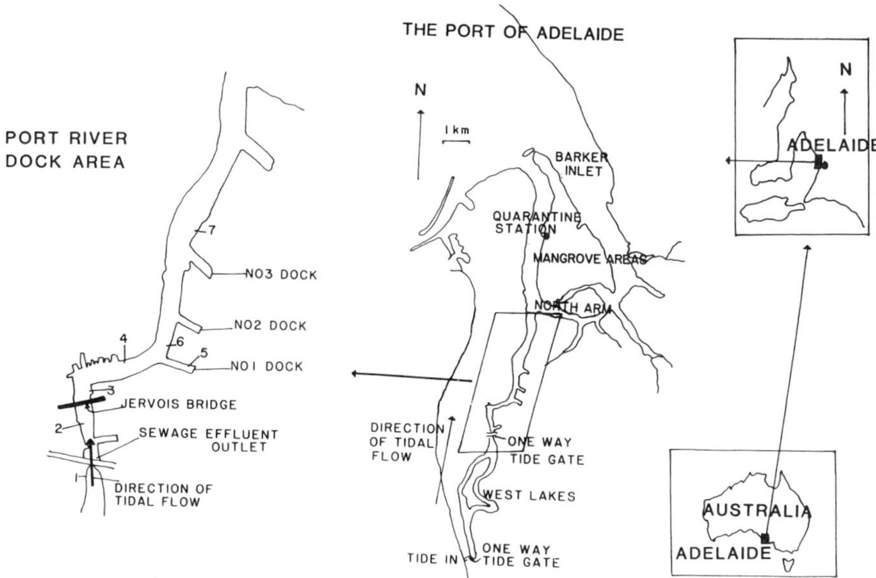

FIG. 1. **Maps showing the location of the Dock Area and the Port of Adelaide.**

MATERIALS AND METHODS

Seven sites were sampled, six along the river and one in West Lakes (Fig. 1). Surface sampling was with a bucket with subsamples in a 1 litre bottle. Vertical net pulls indicated whether there were dinoflagellates in deeper water and when this was the case, samples were collected at one metre depth intervals using a 1 litre Freidinger bottle. Samples were preserved in formalin and subsamples of 100 ml were used for counting. Normally samples were concentrated 10x before counting, in Sedgewick Rafter cells, but during blooms samples were not concentrated and at times had to be diluted 10x to enable the material to be counted.

Regular collections were at the same time of day to take account of diel migrations of dinoflagellates. The diel migrations were recorded at Site 3 by collections at various depths at 5 times during a 24 hour period from 2300hrs on 3/10/88 to 1800hrs on 4/10/88.

Salinity/temperature measurements were taken using a TPS LC181 conductivity meter at 1 metre depth intervals.

Preliminary analysis of the data indicated that blooms develop when calm weather conditions coincided with minimum tidal movement especially around the Jervois Bridge (Fig. 1). An index of stability was defined such that the change in salinity with depth at each location was divided by the tidal change for that time (in m) and this (Δ salinity/Δ tide) was called Stability ($^{o}/_{oo}$/m); this was then graphed against time and compared with summed cell numbers (for 0, 1, 2, 3 & 4m) for the same day.

Sediments from the bottom of the river were obtained using a grab sampler and placed in the dark in a refrigerator until they could be sieved. Samples were sonicated and sieved so that the fraction between 20 and 80μm was examined wet in counting cells, using a compound microscope. Isolated single cysts were placed in G.P. culture medium [4] in 2ml culture trays. and incubated at 17°C in a 12/12 L/D regime.

RESULTS AND DISCUSSION

The bloom species of dinoflagellates appear first in 3-4m depth (near the bottom) by the Jervois Bridge (Site 3, Fig.1). Blooms can be detected underwater at sites downstream between 1 and 3 days after their initial upstream detection and the dinoflagellates appear to move down the river with the largely one way tidal flow at a depth of three to four metres. Whilst it is probable that these cells act as "seed" [6] for the rest of the river some additional initiation may also occur downstream. The term "bloom" is defined in this study as > 10^5 cells L^{-1}. The blooms can be detected first near the Jervois Bridge suggesting that this is the major site for initiation of these blooms.

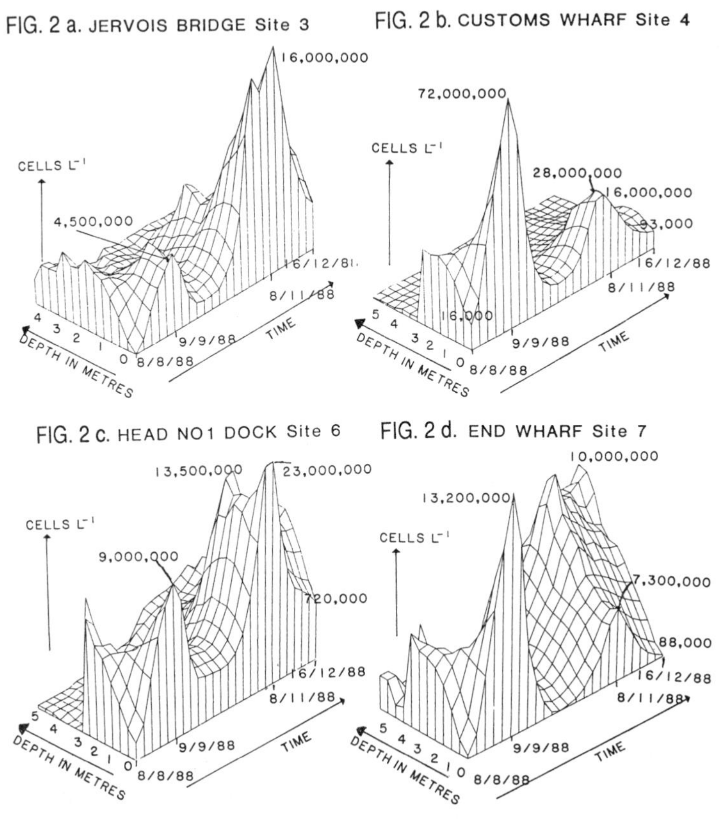

FIG. 2. Cell numbers (L^{-1}) plotted against time and depth at 4 locations in the Port River (Adelaide, South Australia). 2a illustrates the initiation of the blooms on the bottom near the Jervois Bridge in contrast to 2b & 2c where the flat area of the graph illustrates very low cell numbers on the bottom with "seeding" occurring downstream at a depth of 3-4 metres. Fig 2d, which is in deeper water, shows some spreading of the depth of the bloom.

High numbers of dinoflagellates may be found in 3-4m depth for two or three weeks while weather and tidal conditions maintain a mixed rather than stratified water column. When the water column stabilized the blooms developed along the entire dock area on the same day and the increase in cell numbers appears to occur over the full depth of the bloom. This is illustrated in Figs. 2a-d. These are descriptive graphs which illustrate the trends in the data and there is a rounding factor in the computation which prevents false zeroing between the discrete samples. The vertical scales vary from site to site.

Blooms form a dense band several metres thick. They are patchy and are most dense along the sides of the river, accumulating in bays and docks rather than in the centre where tidal flow and shipping movement are greatest.

Diel movement of the dinoflagellate blooms occurred and is illustrated in (Fig. 3).

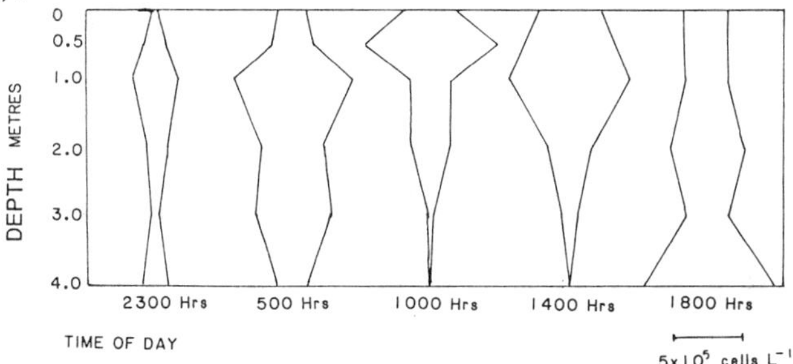

FIG. 3. **Diel Migration of Dinoflagellates at Site 3 on 3 & 4 October, 1988.**

Blooms coincide with stability but not all stable water conditions result in blooms. On some occasions dinoflagellates were only present in very small numbers, i.e. initiation had not occurred. This can be seen in Fig. 4 where a high index of stability on the 22nd September (day 45) occurred shortly after a large (up to $9x10^8$cells L^{-1}) bloom of *Prorocentrum micans*.

Stability of the water column appears to be the major factor controlling bloom development although nutrient levels are extremely high decreasing exponentially (for N, r^2=.93 & for P, r^2=.89 for 6 points) with distance down the river. The main weather component in Stability is wind which results in a mixed water column and a decrease in cell numbers. The sewage effluent which is released into the Port River near the West Lakes outlet (Fig. 1) contributes both nutrients and fresh water (30-40ML per day) to the system. The fresh water wedge increases the stratification and thus the greatest Stability and also greatest nutrient levels occur at the Jervois Bridge.

Blooms disperse at the End Wharf unless a very large bloom (Fig. 2d)has built up along the entire river. The November bloom spread down river to the mangrove areas by the Quarantine Station and into the North Arm (Fig. 1) but dispersed rapidly coinciding with a northerly gale on 8/11/88 leaving only isolated discoloured patches which gradually diminished as they moved downstream into more open water. The End Wharf shows much less stability (Fig. 5). This is partially because there is no freshwater wedge present but also because the river is wider, in line with prevailing wind conditions and has much more shipping movement. Nutrient levels are lower near the End Wharf. Diatoms are the most frequent phytoplankton found at this location.

No1 dock (Site 5) differs from the other sites. It is where the greatest numbers of dinoflagellates are found during every bloom. Blooms accumulate in No1 dock and peak about three days after the rest of river.

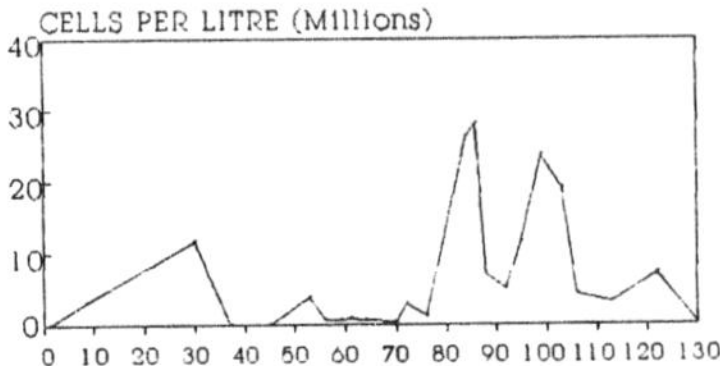

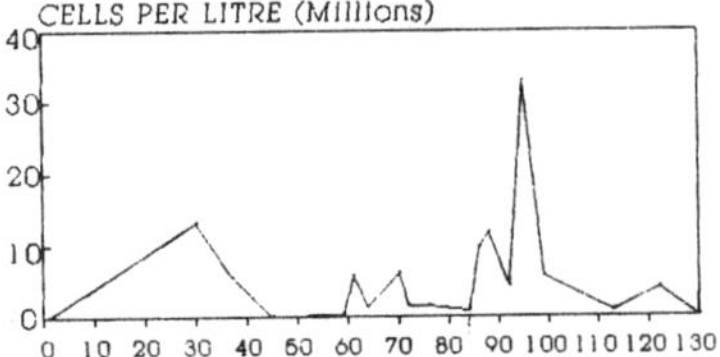

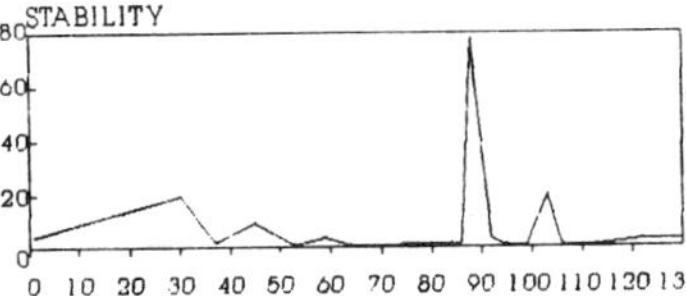

FIG. 4.

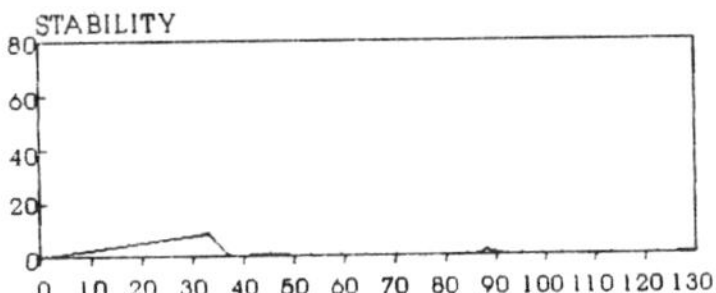

FIG. 5.

Stability (°/oo/m) graphed against time and compared with summed cell numbers (for 0, 1, 2, 3 & 4m) for the same day, at the Jervois Bridge and at the End Wharf.

(Fig. 6). It should be considered as an offshoot off the main river. The same applies but to a lesser extent in No2 and No3 docks which are further down the river (Fig. 1), not at the bend where the bloom sweeps around and are subject to more shipping movement. Table 1 illustrates the high cell numbers and the time lag after stable water conditions.

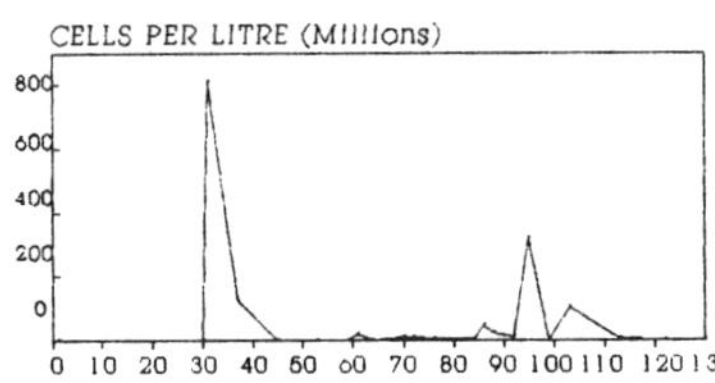

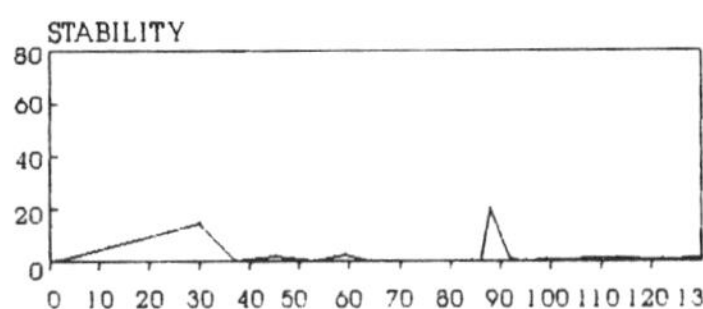

FIG. 6.

Stability (°/oo/m) graphed against time and compared with summed cell numbers for the same day in No1 Dock. Note different scale.

TABLE I
High dinoflagellate numbers in No1 Dock showing a 3 to 4 day time lag after minimal tidal movement

Species	neap tide dates	date	cell nos (L^{-1})
Alexandrium minutum	27,28,29/Sept.[2]	01/10/86	10^5
Alexandrium minutum	16,17,18/Oct. [2]	21/10/87	$4.8x10^8$
Alexandrium minutum	5,6/June	08/06/88	$1.1x10^7$
Prorocentrum micans	5,6,7/Sept.	08/09/88	$8x10^8$
Gymnodinium sanguineum	4,5,6/Oct.	08/10/88	$8x10^5$
Alexandrium minutum	1,2,3/Nov.	02/11/88	$1x10^6$
Alexandrium minutum	1,2,3/Nov.	04/11/88	$7x10^6$
Alexandrium minutum	1,2,3/Nov.	08/11/88	$3x10^7$
Alexandrium minutum	built up from 04/Nov.	11/11/88	$3x10^8$

Culture experiments with natural sediment samples produced *Alexandrium minutum* and *Prorocentrum micans* cultures. Blooms of these species appear to be initiated from cysts. Further work on this is in progress. *P. micans* was hatched in culture medium from isolated single cysts but this has not yet been achieved with *A. minutum* which was hatched from a small amount of sieved sediment containing cysts in a petri dish of Port River water.

SUMMARY

Dinoflagellate blooms in the Port River, South Australia, appear to be initiated underwater from cysts mainly around the Jervois Bridge. These cells move downstream at 3-4m depth and act as "seed" for the river. Cell numbers increase over a wide depth range and surface when there is a stable water column. Collection of bottom water samples from the Jervois Bridge region has enabled early detection of blooms and prediction of the last two major dinoflagellate blooms.

Stability of the water column appears to be the major controlling factor for bloom development in this high nutrient environment. Salinity stratification and tidal amplitude appear to be very important factors in determining stability.

Dinoflagellates are exploiters of calm stratified conditions and are unable to maintain their position in mixed water [6]. Dinoflagellates survive best in conditions of low turbulence and high nutrients and the End Wharf where the blooms tend to break up has high turbulence and lower nutrients.

ACKNOWLEDGEMENTS

I thank Dr A. Cheshire, (Dept. of Botany, Uni. of Adelaide) for assistance with analysis of data, Drs G. Hallegraeff and S. Blackburn (CSIRO, Hobart) for helpful discussions, Dr A. Butler (Dept of Zoology, Uni. of Adelaide) for the loan of the conductivity meter and Dr. D. Steffensen and the E&WS Dept for logistic support.

REFERENCES

1. D.A. Steffensen. Engineering and Water Supply Department. Pers. comm.
2. G.M. Hallegraeff, D.A. Steffensen and R. Wetherbee. J. Plankton Res. 10: 533 - 541. (1988).
3. D. Blasco. Toxic Dinoflagellate Blooms. Elsevier Science Publishing Co. Inc. Pp. 209-214. (1979).
4. A.R. Loeblich and V.E. Smith. Lipids 3:5-13 (1968).
5. K.A. Steidinger. Progress in Phycological Research 2: Elsevier Science Publishing Co. Inc. Pp. 147-188. (1983).
6. F.J.R. Taylor. in The Biology of Dinoflagellates. Blackwell Scientific Publications. Pp. 404. (1987).

GRAZER ELIMINATION THROUGH POISONING: ONE OF THE MECHANISMS BEHIND CHRYSOCHROMULINA POLYLEPIS BLOOMS?

PER CARLSSON*, EDNA GRANÉLI*, PER OLSSON*

*Dept. of Marine Ecology, Univ.of Lund, Box 124, S-221 00 Lund, Sweden

ABSTRACT

The tintinnid Favella ehrenbergii (Claparède & Lachmann) was incubated together with Heterocapsa triquetra (Ehrenberg) and the toxic Prymnesiophycean Chrysochromulina polylepis (Manton & Parke) in order to study wether F. ehrenbergii was affected by C. polylepis toxin. Both growth rate of F. ehrenbergii and ingestion rate were negatively affected by the addition of C. polylepis. Addition of high concentrations ($60 \cdot 10^3$ cells·ml^{-1}) of C. polylepis caused death of F. ehrenbergii at the same rate as when it was incubated in seawater without food. C. polylepis that had been grown under phosphorus limitation caused higher negative growth rates of F. ehrenbergii than C. polylepis grown in complete f/2 medium, suggesting that P-deficiency stimulated toxin production in C. polylepis. Feeding activity of F. ehrenbergii stopped when C. polylepis grown under phosphorus limitation were added at densities above $1.5 \cdot 10^3$ cells·ml^{-1}. The results show that C. polylepis affects feeding and growth rate of a protozoan grazer negatively. This limitation of grazing may contribute to the ability of the alga to form blooms.

INTRODUCTION

In May 1988, a large area of Kattegat and Skagerrak was affected by an algal bloom consisting of the prymnesiophycean Chrysochromulina polylepis. The bloom of C. polylepis did not consist of any unnaturally high biomass and did not cause any elevated sedimentation rates or high oxygen consumption rates (1, 2). The drastic effects of the bloom were instead coupled to the high toxicity of the algae. The 1988 C. polylepis-bloom was considered to be an ecological disaster because of the large economic losses (3, 4). C. polylepis has been found in these waters before, but never before at such high densities (up to $60 \cdot 10^3$ cells·ml^{-1}) and has not until now been known to produce toxins (3). Suggested explanations of the C. polylepis-bloom have been many, among them eutrophication, increase in trace metal concentrations etc.

As such a diversity of organisms (both plants and animals) were killed during the C. polylepis-bloom, we have put forward the hypothesis that one of the mechanisms behind the (success of this organism in producing a) bloom (in 1988) was the ability of C. polylepis to grow without any elimination of cells by suitable predators. In this paper we attempt to test this hypothesis by culturing the tintinnid Favella ehrenbergii together with C. polylepis at different cell concentrations. The tintinnids were fed the dinoflagellate Heterocapsa triquetra on which they normally grow well so that they would not starve (to death) for lack of food if it did not graze C. polylepis.

We also address the question of whether toxin production of C. polylepis increases when the alga is grown in a medium with an excess of nitrogen in relation to phosphorus.

Toxic Marine Phytoplankton
Edna Graneli et al., Editors

MATERIALS AND METHODS

The study was based on one tintinnid ciliate (Favella ehrenbergii), one dinoflagellate (Heterocapsa triquetra), used by us as the normal food source when culturing F. ehrenbergii, and the toxic prymnesiophycean Chrysochromulina polylepis. F. ehrenbergii was isolated from surface water samples from the Limfjord, Denmark in September 1988 and in August 1989. H. triquetra was isolated from the Öresund, Denmark. C. polylepis was isolated from a water sample from Kattegat during the C. polylepis bloom in May 1988. The algae were grown in non-axenic batch cultures in f/2 medium (5). F. ehrenbergii was cultured in Bibby 25 ml Tissue Culture Flasks and carefully rinsed cells were transferred weekly to new medium with concentrations of H. triquetra of 10^3-10^4 cells·ml^{-1}. All cultures were maintained at 15°C and a constant light intensity of 80 $\mu E \cdot m^{-2} s^{-1}$ (measured with a spherical light meter from Biospherical Instruments).

The ciliate growth experiments were performed in Nunclon Multidishes (6 x 4 = 24 wells). One dish with initially one ciliate in each well (= 24 ciliates) and 2 ml of algal suspension was used for each treatment: 1) Only medium (without algal cells), 2) Only H. triquetra ($10 \cdot 10^3$ cells·ml^{-1}) H. triquetra ($10 \cdot 10^3$ cells·ml^{-1}) together with 2, 5, 10, 30 or $60 \cdot 10^3$ cells·ml^{-1} of C. polylepis. At least every 24 hours the numbers of ciliates were counted in each Multidish and the ciliates were carefully transferred with a micropipette to newly prepared algal cultures (with the same initial cell concentrations in order to keep the algal concentrations approximately constant). This experiment was performed both for C. polylepis grown in full f/2 medium and C. polylepis grown in f/2 medium where PO_4^{3-} was omitted (PO_4^{3-} concentration in the water from which the medium was prepared was 2.0 μM). As a criterion for phosphate-deficiency, alkaline phosphatase activity (APA) was measured according to (6) in the C. polylepis stock culture.

The experiments concerning the ingestion rate of F. ehrenbergii, fed H. triquetra with or without C. polylepis in the same cultures, were also performed in Nunclon Multidishes. During the first ciliate ingestion rate experiments, two ciliates were incubated in each 2 ml well in 4 replicates together with different concentrations of H. triquetra (1 - $18 \cdot 10^3$ cells·ml^{-1}.) for 4 hours. The second ciliate ingestion experiment consisted of a constant concentration of H. triquetra ($10 \cdot 10^3$ cells·ml-1) in all wells (4 replicates) and different concentrations of C. polylepis (0, 1.5, 3, 7.5, 24 and $55 \cdot 10^3$ cells·ml-1). Algal cell concentrations in the wells were determined initially and finally. Algal cells were fixed in Lugol´s solution and at least 200 cells of each sample were counted in a 0.1 ml Palmer-Maloney chamber. The equations of Frost (7) were used to calculate the ingestion rate of the ciliates.

In order to study whether C. polylepis toxin production affected other planktonic algae, growth experiments were performed with P-limited C. polylepis and H. triquetra. The experimental design consisted of 3 replicates each of: 1) H. triquetra alone ($3 \cdot 10^3$ cells·ml^{-1}), 2) H. triquetra + $2 \cdot 10^3$ C. polylepis·ml^{-1}, 3) H. triquetra + $7 \cdot 10^3$ C. polylepis·ml^{-1} and 4) H. triquetra + $35 \cdot 10^3$ C. polylepis·ml^{-1}.

RESULTS

When fed only H. triquetra ($10 \cdot 10^3$ cells·ml^{-1}), F. ehrenbergii had growth rates during the first 48 hours of 0.44 and 0.58 divisions·day^{-1}, respectively (the higher value for the experiment with C. polylepis grown in full f/2 medium and the lower when C. polylepis was grown in f/2 medium with P omitted) (Fig. 1).

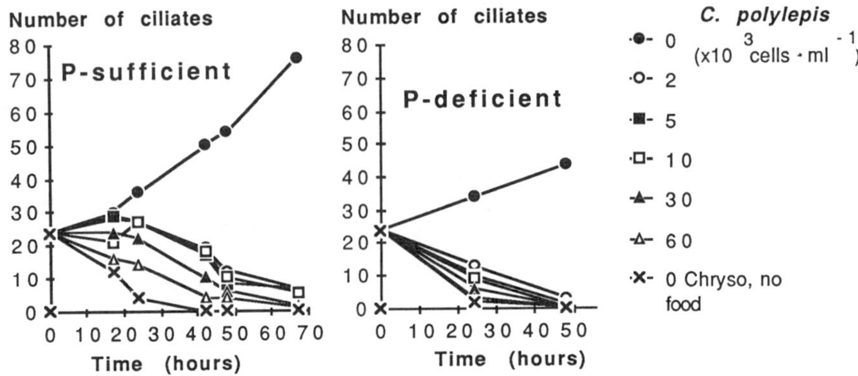

FIG. 1. Growth curves for F. ehrenbergii when fed H. triquetra alone (10^3cells·ml-1) or H. triquetra with different concentrations of C. polylepis present. (C. polylepis was cultured either in complete f/2 medium (P-sufficient) or with P omitted (P-deficient)).

Both growth rate of F. ehrenbergii and ingestion rate were negatively affected by the addition of C. polylepis. The growth rate of F. ehrenbergii was negative even at the lowest concentration of C. polylepis used ($2 \cdot 10^3$ cells·ml^{-1}) both when C. polylepis had been grown in full f/2 medium and medium with P omitted. Addition of higher concentrations (up to $60 \cdot 10^3$ cells ml^{-1}) of C. polylepis caused death of F. ehrenbergii at higher rates and at the highest concentration, the ciliates died at the same rate as when they were incubated in seawater without food (Fig. 1). C. polylepis that had been grown under phosphorus limitation (PO_4^{3-} concentration less than 0.1 μM and an alkaline phosphatase activity between 5-20 nM P·min^{-1} in the used C. polylepis cultures) caused higher negative growth rates of F. ehrenbergii than C. polylepis grown in complete f/2 medium, suggesting that P-deficiency stimulated toxin production in C. polylepis (Fig. 2).

In the absence of C. polylepis, F. ehrenbergii ingested H. triquetra according to a normal saturation curve, with saturation occurring between 10 and $15 \cdot 10^3$ cells·ml^{-1} of H. triquetra (Fig. 3). Ingestion rate for F. ehrenbergii was negatively affected at the lowest density of C. polylepis that was used ($1.5 \cdot 10^3$ cells·ml^{-1}) (Table I). At a concentration of $3 \cdot 10^3$ cells·ml^{-1} of C. polylepis, F. ehrenbergii completely stopped feeding. The negative values of the ingestion rate are an artefact due to better growth of H. triquetra in the presence of F. ehrenbergii than when H. triquetra was alone in the controls.

Growth rate of H. triquetra was not affected by the presence of C. polylepis up to a concentration of $10 \cdot 10^3$ cells·ml^{-1}. However, at a concentration of $60 \cdot 10^3$ cells·ml^{-1}, the growth rate of H. triquetra decreased markedly (Table II).

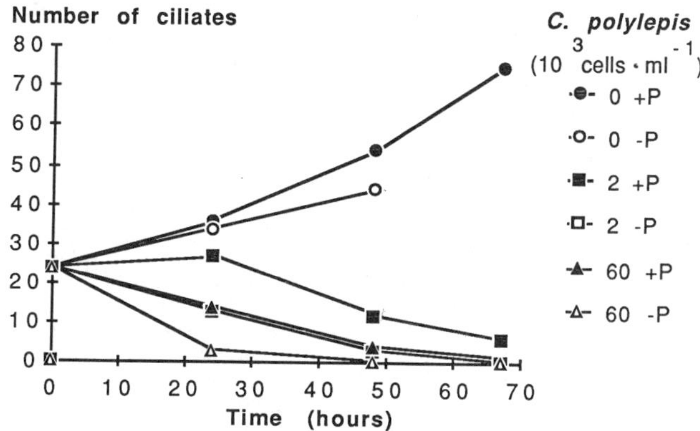

FIG. 2. Comparison of F. ehrenbergii's growth rate at lowest ($2 \cdot 10^3$ cells·ml^{-1}) and highest ($60 \cdot 10^3$ cells·ml^{-1}) concentration of C. polylepis cultured either in complete f/2 medium (+P) or with P omitted (-P).

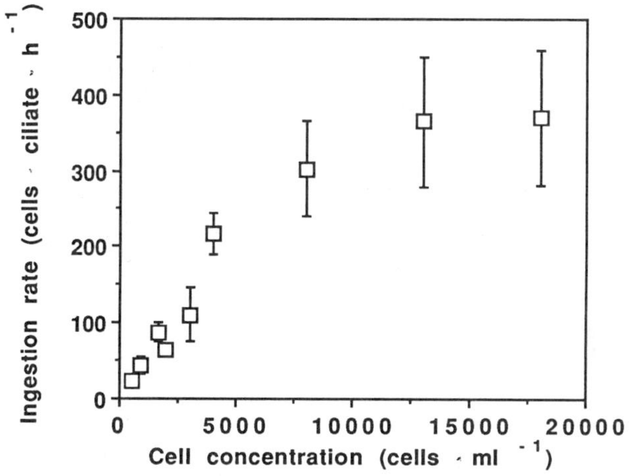

FIG. 3. Ingestion rate for F. ehrenbergii eating H. triquetra at different concentrations of the alga.

DISCUSSION

Tintinnids are suggested to have an upper limit for food particle size around 41-45% of the oral diameter of the lorica (8, 9). This means that the upper size of food particles that can be ingested by F. ehrenbergii is around 30-40 µm. However, unpublished data by Helene Munk Sørensen (in: 10) suggests a value of around 15 µm for the lower size limit. Visual observation by us showed that the ciliate was not able to retain C. polylepis by the membranelles and that the ciliate did not show any feeding reaction when it encountered C. polylepis cells. The suggestion that C. polylepis may be too small to be ingested to any large extent by this tintinnid is, however, not crucial for the interpretation of our results, since we supplied F. ehrenbergii with H. triquetra as a suitable food source during all experiments. Obviously not only was C. polylepis unsuitable as a food source for F. ehrenbergii, but this alga also prevented the ingestion of other algae, normally suitable as food for the ciliate. However, death rate of F. ehrenbergii in the presence of C. polylepis was never greater than the rate under complete starvation (no food no C. polylepis, see Fig. 1). This suggests that the toxin produced by C. polylepis does not kill F. ehrenbergii directly, but only through an effect on food intake. If C. polylepis is not eaten by the ciliate and incorporated into food vacoules, this means that the toxin produced by the algae must be excreted into the water, affecting the ciliate from outside or that direct contact between F. ehrenbergii and C. polylepis causes the former to stop feeding.

The ingestion rates for F. ehrenbergii in our experiments seem to be higher than previously reported values: Rassoulzadegan (11) found an ingestion rate of 1140 pg C·tintinnid^{-1}·h^{-1} (for F. ehrenbergii eating natural particles sized 1.3-30 µm). This is equivalent to an ingestion rate of only 7 cells of H. triquetra per ciliate per hour. However, Taniguchi and Takeda (12) found that Favella taraikaensis (which is smaller than F. ehrenbergii) could eat 82 cells per hour of Prorocentrum minimum (18 µm length) which is in the same size range as H. triquetra.

TABLE I. Ingestion rate (n=4, ± SE) for F. ehrenbergii eating H. triquetra in cultures with only H. *triquetra* (10^3 cells·ml^{-1}) or H. triquetra + C. polylepis (1.5-55·10^3 cells·ml^{-1}).

Number of C. polylepis (·10^3 cells·ml^{-1})	Ingestion rate (cells·ciliate·h^{-1})
0	637 ± 36
1.5	261 ± 160
3.0	-80 ± 65
7.5	-84 ± 60
24	-162 ± 180
55	-106 ± 70

C. polylepis has not, until the bloom in May 1988, been reported to be toxic. The closely related algae Prymnesium parvum, however, becomes toxic when it is grown in medium with a low phosphorus content (13). Dahl et al. (4) and Edvardsen et al. (14) have also shown that C. polylepis becomes toxic when grown in P-deficient medium. The phosphate concentration during the bloom of C. polylepis was below 0.1 µM which may have contributed to the toxicity.

The higher death rates of F. ehrenbergii when it was presented P-limited C. polylepis suggest that the algae increase the toxin production at a high

N/P ratio. It has been shown by Boyer et al. (15) that toxin production per cell in the dinoflagellate P. tamarensis increased dramatically in phosphorus limited cultures. Wyatt and Reguera (16) suggested that toxin production could be a mechanism for detoxification in a nitrogen-enriched environment. The increased nitrogen load to the Swedish west coast, which is increasing the N/P ratio in the water, may have induced a higher toxin production by C. polylepis during the 1988 bloom.

The strong negative effect observed on the tintinnid in our experiments might have had large ecological implications for the formation and persistance of the C. polylepis bloom in Kattegat and Skagerrak during May 1988. Measurements of cell numbers, primary production and cell carbon content on May 22, gave a doubling time of 1.2 days for the exponentially growing C. polylepis population (1, 4). If the algae were able to grow at this rate and were not subjected to any large loss factors such as grazing from microzooplankton, an initial hypothetical concentration of $1 \cdot 10^3$ cells·ml^{-1} of C. polylepis may attain a population density of more than $60 \cdot 10^3$ cells·ml^{-1} in one week.

TABLE II. Growth rate (doublings per day) ± SE (triplicates) for H. triquetra alone and with 2, 7 and $35 \cdot 10^3$ cells·ml^{-1} of C. polylepis grown in P-sufficient medium (+P) and P-deficient medium (-P).

C. polylepis	0	2	7	$35 \cdot 10^3$
H. triquetra	0.39±0.09			
H. triquetra +*C. polylepis* (+P)		0.32±0.09	0.49±0.04	0.14±0.11
H. triquetra +*C. polylepis* (-P)		0.31±0.07	0.35±0.02	0.27±0.06

Conclusions: 1) C. polylepis affects microzooplankton ingestion rate and growth rate negatively due to its toxicity, 2) Growth rate of C. polylepis is therefore probably unaffected by microzooplankton grazing, 3) With sufficient supply of nutrients, the avoidance of grazing helps C. polylepis to attain bloom concentrations, 4) A high N/P ratio in the medium increases the toxin production of C. polylepis.

ACKNOWLEDGEMENTS

This study was supported by the Swedish Environmental Protection Board, contract no: 531 2118-2 and 5332059-4, The Swedish Natural Research Council, contract Dnr: B-BU 4970-302, The Crafoord Foundation, The Futura Foundation and The Royal Physiographic Society, Lund. We would also like to thank Eva Schöllhorn and Birgitta Gisby for technical assistance.

REFERENCES

1. O. Lindahl in: The Chrysochromulina polylepis algal bloom along the Swedish west coast 1988, O. Lindahl, ed. (SNV Report 3602) pp. 49-54. (In Swedish).
2. E. Granéli, P. Carlsson, P. Olsson, B. Sundström, W. Granéli, and O. Lindahl in: Novel phytoplankton blooms. Causes and impacts of recurrent brown tides. E.M. Cosper , E. J.Carpenter, M. Bricelj, eds. (Springer-Verlag, Berlin in press).
3. R. Rosenberg, O. Lindahl, and H. Blanck, Ambio 17, 289-290 (1988).
4. E. Dahl, O. Lindahl, E. Paasche, and J. Throndsen in: Novel phytoplankton blooms. Causes and impacts of recurrent brown tides. E. M. Cosper , E. J. Carpenter, M. Bricelj, eds. (Springer-Verlag, Berlin in press).
5. R. L. L. Guillard, and J. H. Ryther, Can. J. Microbiol. 8, 229-239 (1962).
6. K. Petterson, Int. Rev. ges Hydrobiol. 64, 585-607 (1979).
7. B. W. Frost, Limnol. Oceanogr. 17, 805-815 (1972).
8. J. F. Heinbokel, Mar. Biol. 47, 191-197 (1978).
9. P. Spittler, Oikos Suppl 15, 128-132 (1973).
10. T. Fenchel, Ecology of protozoa (Springer-Verlag, Berlin1988).
11. F. Rassoulzadegan, Ann. Inst. Oceanogr, Paris, 54,17-24 (1978).
12. A. Taniguchi and Y. Takeda, Mar. Microb. Food Webs. 3, 21-34 (1988).
13. M. Shilo in: The Water Environment, W. W. Carmichael, ed. (Plenum Press, 1982) pp. 37-47.
14. B. Edvardsen, F. Moy and E. Paasche in: Toxic Marine Phytoplankton, E. Granéli, B. Sundström, L. Edler and D. M. Anderson, eds. (In this book)
15. G. L. Boyer, J. J. Sullivan, R. J. Anderson, P. J. Harrison, and F. J. R. Taylor, Mar. Biol. 96, 123-128 (1987).
16. T. Wyatt, and B. Reguera in: Red Tides: Biology, Environmental Science and Toxicology, T. Okaichi, D. M. Anderson, and T. Nemoto, eds. (Elsevier, New York 1989) pp.33-36.

GYRODINIUM AUREOLUM BLOOM ALONG THE NORWEGIAN COAST IN 1988

Einar Dahl* and Karl Tangen**
*Flødevigen Biological Station, N-4817 His, Norway,
**OCEANOR, Pirsenteret, N-7005 Trondheim, Norway.

ABSTRACT

A bloom of the dinoflagellate *Gyrodinium aureolum* along the Norwegian coast in the autumn 1988 was monitored by a water sampling programme including fish farm sites and Skagerrak waters, and by an underwater fibre optical probe deposited in coastal waters on the south coast, transmitting real time data through a satellite communication system. The bloom lasted from the beginning of August to January and covered the whole coastline from the Swedish coast in the south (58°N) to the Trondheim area in the north (65°N), thus affecting a much larger area than previous blooms in Norwegian waters. The 1988 bloom developed in inshore and nearshore coastal waters, and not in the open Skagerrak, in contrast to other *G. aureolum* blooms in this area. Observations during the bloom confirmed previously known characteristics of this species: surface or subsurface population maxima, a patchy horizontal distribution with brown water discolouration, and accumulation of cells in specific localities due to wind-driven surface currents. In the whole bloom area *G. aureolum* was accompanied by the dinoflagellate *Ceratium furca*, which in some localities was the dominant species, in terms of cell numbers. Fish mortality was recorded in a large number of fish farms, affecting Atlantic salmon, rainbow trout and cod, although the total loss was relatively small, probably not exceeding 100 tons.

INTRODUCTION

Gyrodinium aureolum Hulburt was first recorded along the Norwegian coast in the autumn 1966 as a bloom causing fish mortalities [1]. Since, harmful blooms of this species along the Norwegian coast has occurred in 1976, 1981, 1982 and 1985 [2,3]. The season for these blooms have been August to November, and the geographical area mainly along the south and south-west coast of Norway except for one bloom in northern Norway in June 1982.

Large blooms of *G. aureolum* along the southern coast have to a great extent been due to advection and concentration of off-shore populations of the alga [4], and corresponding, the blooms along the south-west coast have been due to along-shore advection and concentration by the Norwegian Coastal Current. However, in 1988 a large bloom of *G. aureolum* developed differently both in geographical spreading and in duration. The present investigation describes and documents new aspects of the occurrence of *G. aureolum* in these waters and relates it to mortality in fish stocks kept in net cages.

MATERIALS AND METHODS

The data on the 1988 bloom was partly collected through a monitoring programme [3] and partly by information and samples from fish farmers along the coast. In addition information was gained from a new fibre optical sensor for detection of algae in the field, deposited in coastal waters outside Arendal with real time monitoring and data transfer through a satellite system [6].

Toxic Marine Phytoplankton
Edna Graneli et al., Editors

RESULTS AND DISCUSSION

During the first weeks of August 1988 increasing concentrations of *Gyrodinium aureolum* were recorded along the southern coast of Norway. By 20 August brown patches with large amounts of this dinoflagellate were reported from different sites along the coast from the outer Oslofjord to the southernmost part of Norway (Fig. 1), and mortalities among encaged fish exposed to such brownish water were recorded.

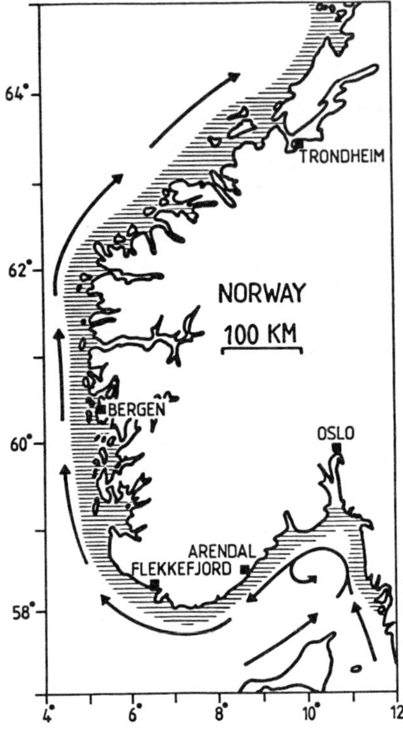

FIG. 1. The shaded area shows where along the Norwegian coast *Gyrodinium aureolum* occurred in the autumn 1988. The arrows indicate the surface currents.

During the next two weeks the bloom spread westwards, with records of fish mortality in fish farms in the Flekkefjord area. Especially rainbow trout and cod were affected. The following weeks and months the bloom extended northwards along the west coast of Norway and eventually terminated north of Trondheim in November (Fig. 1).

The occurrence of *G. aureolum* as a mean concentration for the upper 5 m water column at the monitoring station at Flødevigen nearby Arendal on the south coast of Norway showed large day to day changes (Fig 2). In the period from August to November *G. aureolum* was abundant essentially in two periods, one in August-September the other in the end of October (Fig. 2). Correspondingly, fish kills were observed mainly in the same two periods when *G. aureolum* populations were highest. In January 1989 still more than 1 million *G. aureolum* cells per litre were present in the Flødevigen/ Arendal area.

During the bloom a real time monitoring system with continuous transmission of data through satellite communication to the land station in Trondheim was operative. An underwater fibre optical probe measured transparency in three different wavelengths at 3 m depth [6] and detected

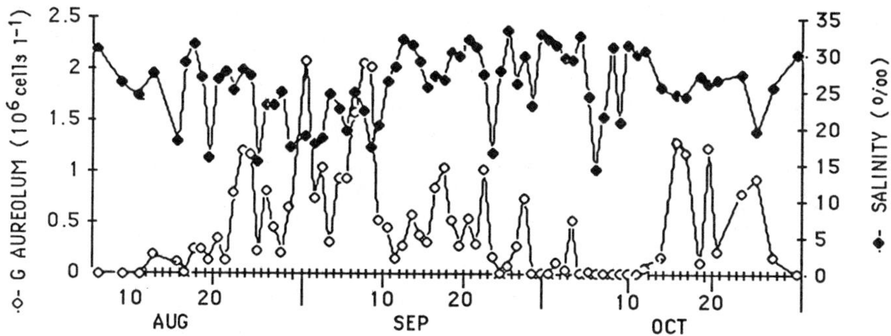

FIG. 2. The occurrence of *Gyrodinium* (cells · l^{-1} in a 0-5 m integrated sample),and the salinities measured at 1 m depth on the Flødevigen bay for the period August - November 1988.

the early bloom stages from the beginning of August. Fig. 3 shows diurnal variations in relative attenuation in all three wavelengths, corresponding to higher concentrations of *G. aureolum* at the probe depth during day-time than during night.

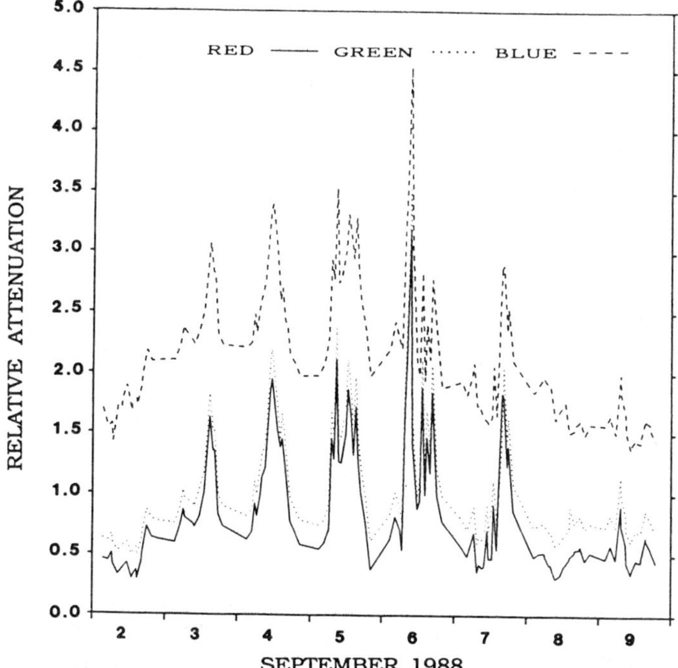

FIG. 3. Measured light attenuation in the period between September 2 and September 9 1988 according to the *in situ* optical sensors located at 3 m depth to the west of Arendal (58°18'N, 8°45'E), indicating a diurnal fluctuation in cell concentration at 3 m depth.

Initial populations of *Gyrodinium* as subsurface maxima in the central parts of Skagerrak were not recorded in the summer 1988, although the off-shore monitoring programme [3] was intensified as a result of the *Chrysochromulina* bloom in May [5]. Thus the bloom along the Norwegian coast in the autumn 1988 was due to growth in coastal waters rather than to advection and concentration of off-shore populations. This change in bloom pattern may complicate monitoring and early warning possibilities of *Gyrodinium*-blooms in the future. However, consistent with previous observations, the bloom in 1988 along the Norwegian coast also formed distinct subsurface maxima in the early bloom development stage (Aug. 15 - Sep. 5) (Fig. 4).

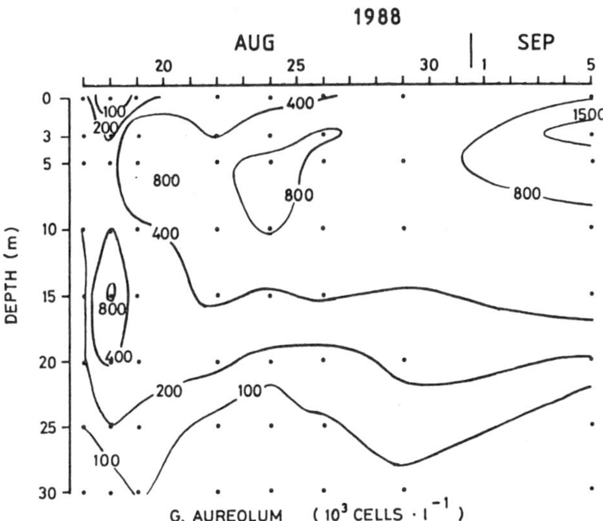

FIG. 4. The distribution of *Gyrodinium aureolum* at Flødevigen, Arendal during the early stage of the bloom.

During the whole bloom period and along the same geographical area the *Gyrodinium* populations were accompanied by *Ceratium furca* which in many localities was more abundant than *G. aureolum*. Brown patches of the two species were usually more or less monospecific, although there were also examples of mixed populations (see Table I). It is noteworthy that there were examples of a nearly complete vertical separation of the two species, with a *Gyrodinium* population below a surface *Ceratium* population.

TABLE I. *G. aureolum* and *C. furca* in mixed populations recorded during the bloom, examples of cell concentrations (10^6 cells · 1^{-1}).

Area and date	*G. aureolum*	*C. furca*
Bergen area 22 Aug	0.3	0.45
- " - 6 Sep	10.5	1.6
- " - 6 Sep	4.0	10.3
Oslofjord 9 Sep	7.1	<0.1
Flekkefjord 6 Sep	30	0.5
Trondheim area 25 Sep	8	0.8
- " - 27 Sep	0.5	3.2

Observations in 1988 were consistent with previous records on the behaviour and distribution of the *Gyrodinium* populations along the Norwegian coast in some respects: harmful concentrations were mainly restricted to the upper 2-3 m during day-time, and the distribution was patchy and cells were accumulated in bays due to wind-driven surface currents. The 1988 autumn bloom described in this paper differed, however, from previous autumn blooms in other respects : by extensive growth along the coast rather than advection of off-shore populations, by greater northward extension along the west coast (Fig. 1), and by greater persistence in the Skagerrak coastal waters.

There is no quantitative information on the total loss of farmed fish, because the individual losses on each farm were generally small and not covered by insurance. To our knowledge a few farms and a ship transport with living rainbow trout lost about 10 tons each during the bloom period, whereas a relatively large number of farms lost from a hundred to one thousand fish specimens of varying size, of Atlantic salmon, rainbow trout and cod. The fish behaviour differed, from Atlantic salmon which tried to avoid dense algae populations by swimming deeper, to rainbow trout which continued to swim in the surface, thus being exposed to higher algae concentrations. Mainly Atlantic salmon, by far the most common species in Norwegian fish farming, was, however, lost. It is believed that the losses were kept at a minimum due to a rapid warning and advisory system, in combination with local contingency plans.

REFERENCES

1. T. Braarud and B.R. Heimdal, Nytt Mag. Bot. 17, 91-97 (1970).
2. K. Tangen, Sarsia 63, 123-133 (1977).
3. E. Dahl in: Aquaculture - A Biotechnology in Progress, N. De Pauw, E. Jaspers, H. Ackefors and N. Wilkins, eds. (European Aquaculture Society, Bredene, Belgium 1989) in press.
4. O. Lindahl, Sarsia 71, 27-33 (1986).
5. O. Lindahl and E. Dahl, this volume.
6. F. Volent and K. Tangen. ARGOS Newsletter 36, 12-13 (1989).

EFFECTS OF *AUREOCOCCUS ANOPHAGEFFERENS* ("BROWN TIDE") ON THE LATERAL CILIA OF 5 SPECIES OF BIVALVE MOLLUSCS

CHRISTY DRAPER,[1] LOUIS GAINEY, [1] SANDRA SHUMWAY,[2,3] AND LYNDA SHAPIRO[3]
[1] Department of Biological Sciences, University of Southern Maine, Portland, ME 04103; [2] Dept. of Marine Resources, W. Boothbay Harbor, ME 04575; [3] Bigelow Laboratory for Ocean Sciences, W. Boothbay Harbor, ME 04575

ABSTRACT

Aureococcus anophagefferens had no effect on the lateral ciliary activity of *Mya arenaria*, *Geukensia demissa*, and *Argopecten irradians*. *Aureococcus* caused a significant decrease in the lateral ciliary activity in isolated gills of *Mercenaria mercenaria*, and *Mytilus edulis*. The lateral cilia of the species that were unaffected by *Aureococcus* were also unaffected by the neurotransmitter dopamine, while the lateral cilia of the species inhibited by *Aureococcus* were also inhibited by dopamine.

INTRODUCTION

During the summers of 1985 and 1986, several concurrent blooms of a previously undescribed chrysophycean alga were reported from Long Island embayments, Narragansett Bay, Rhode Island and Barnegat Bay, New Jersey [1,2,3,4]. The organism was subsequently identified as a chrysophycean alga, *Aureococcus anophagefferens* [2,5]. This species ranges in size from 1.5-2.0 µm in diameter and occurred in concentrations as high as 10^6 cells ml^{-1} which resulted in discoloration of the waters or "brown tide" [1,3].

The effects of these "brown tides" on shellfish populations were immediately evident and, in some cases, catastrophic. Populations of the mussel, *Mytilus edulis*, experienced mortalities of 95% during the bloom and the bay scallop fishery was virtually eliminated by two successive "brown tides" during the summers of 1985 and 1986 in New York; however, the mechanisms responsible for these mortalities remained unclear [6,7].

The efficiency with which bivalve molluscs can retain particles filtered from seawater is known to vary between species of shellfish and is, in most cases, directly related to the size of the algal cells with small cells generally retained less efficiently than larger cells [8,9]. The very small size of *A. anophagefferens* led investigators to determine the effects of this alga on feeding rates in several species of bivalve molluscs. The gills of scallops, *Argopecten irradians*, and mussels, *Mytilus edulis*, captured cells with only 36 and 59% efficiency respectively [7]. Furthermore, feeding rates in both species were significantly depressed when fed cultured *Aureococcus* at bloom densities. Reduced feeding rates have been observed in *M. edulis* and the quahog, *Mercenaria mercenaria*, in the presence of *A. anophagefferens*; reductions in clearance rates by the mussels were independent of particle size and substances dissolved in bloom waters had no effect on clearance rates [10].

Although the small size and high density of *Aureococcus* had detrimental effects on bay scallops, these effects were not enough to account for the starvation experienced by bivalves in the field; these results might be ascribed to toxic effects following more prolonged exposure to bivalves to *Aureococcus* cells [7]. Other authors have recently suggested that chronic toxicity of *A. anophagefferens* cells at high densities might be responsible for the detrimental effects observed in

Toxic Marine Phytoplankton
Edna Graneli et al., Editors

bivalves [10].

In the present study, the effects of exposure to both pure cultures of *A. anophagefferens* on ciliary activity in several species of bivalve molluscs have been examined using isolated gills. Activity of the lateral cilia has been monitored in an effort to assess possible toxic effects of the alga on the feeding mechanisms of bivalve molluscs.

MATERIALS AND METHODS

Original cultures of *Aureococcus anophagefferens* were supplied by E. M. Cosper (MSRC, State University of New York, Stony Brook). The bivalves used in the experiments were obtained from the Maine Department of Marine Resources, Boothbay Harbor, ME. with the exception of *Argopecten irradians* which were supplied by S. Tettlebach (Southampton College, Southampton, N.Y.). Animals were kept at 10~ C on a 12 hr light/dark cycle prior to use. The bivalve gills were removed and placed in bowls with 10 ml of artificial sea water; the gills were used after the appearance of metachronal waves on the lateral cilia, usually within 1 to 12 hrs depending upon the species. All of the gills were observed with a compound microscope; a strobe was used to determine the frequency of ciliary beating. Gills were first observed in artificial sea water; for the treatments, the artificial sea water was replaced with bloom concentrations of brown tide, (approximately 10^6 cells/ml). For each species of bivalve, 15 treatments and 8 controls were used. Readings were taken every hour for 6 hrs and again at 24 hrs. All experiments were performed at 10°C.

Two way repeated measures analysis of variance was used, with time, treatment, and animal nested within treatment as factors. All tests were at the .05 level of significance. Errors are 1 standard error.

RESULTS AND DISCUSSION

Aureococcus had no significant effect on the activity of lateral cilia of the following species of bivalve molluscs: *Mya arenaria*, *Geukensia demissa*, and *Argopecten irradians*. However, *Aureococcus* inhibited the lateral cilia of both *Mercenaria mercenaria*, and *Mytilus edulis*. When *Mercenaria* gills were exposed to *Aureococcus*, the percent original ciliary rate began to decline within an hour of exposure, and reached a minimum of 32% (± 15) 6 hr after exposure. Twenty four hours after exposure the mean ciliary rate had returned to 85% (± 6.5) (Fig. 1). When *Mytilus* gills were exposed to *Aureococcus*, the percent original ciliary rate began to decline within 1 hr and reached a minimum of 48% (± 10) 3 hr after exposure. Twenty four hours after exposure, the mean ciliary rate had returned to 74% (± 15) (Fig. 1). When *Mytilus* gills were exposed to water from which *Aureococcus* had been removed by filtration, there was no effect upon lateral ciliary activity (Fig. 2).

Aureococcus had no effect on lateral ciliary activity in either *Mya arenaria*, *Argopecten irradians*, or *Geukensia demissa*. However, *Aureococcus* inhibited lateral ciliary activity of *Mytilus* and *Mercenaria*. Exogenously applied dopamine had no effect on lateral ciliary activity in *Geukensia demissa* [*11*]. We have found similar results for both *Mya* and *Argopecten* (unpublished results). Dopamine is known to inhibit lateral ciliary acitivity of *Mytilus* [12]; moreover, dopamine as well as the enzymes for its synthesis and degradation are present in the gill [13]. We have also found that exogenously applied dopamine inhibits the lateral ciliary activity of *Mercenaria*. However, water from which *Aureococcus* had been removed by filtration had no effect on clearance rates of *Mytilus*, ruling out the possibility that a brown tide exudate that mimics dopamine is responsible for ciliary inhibition. In addition, dopamine inhibition occurs within minutes of application, not several hours, as seen with *Aureococcus* inhibition. The gills of

Mytilus are known to secrete both chymotrypsin and alpha amylase [14] and this enzyme activity could account for the time lag for inhibition if the gills are digesting part of the cell wall of *Aureococcus* and releasing a dopamine agonist. Further studies with dopamine antagonists and enzyme inhibitors will test this hypothesis. The discrepancy between our results on isolated gills of *Argopecten* and those of Bricelj and Kuentsner [7], who found reduced clearance rates in intact scallops, can be explained by assuming that the compound released from *Aureococcus* has its effects on the nervous system of the animal rather than on the gills, which are unaffected by dopamine.

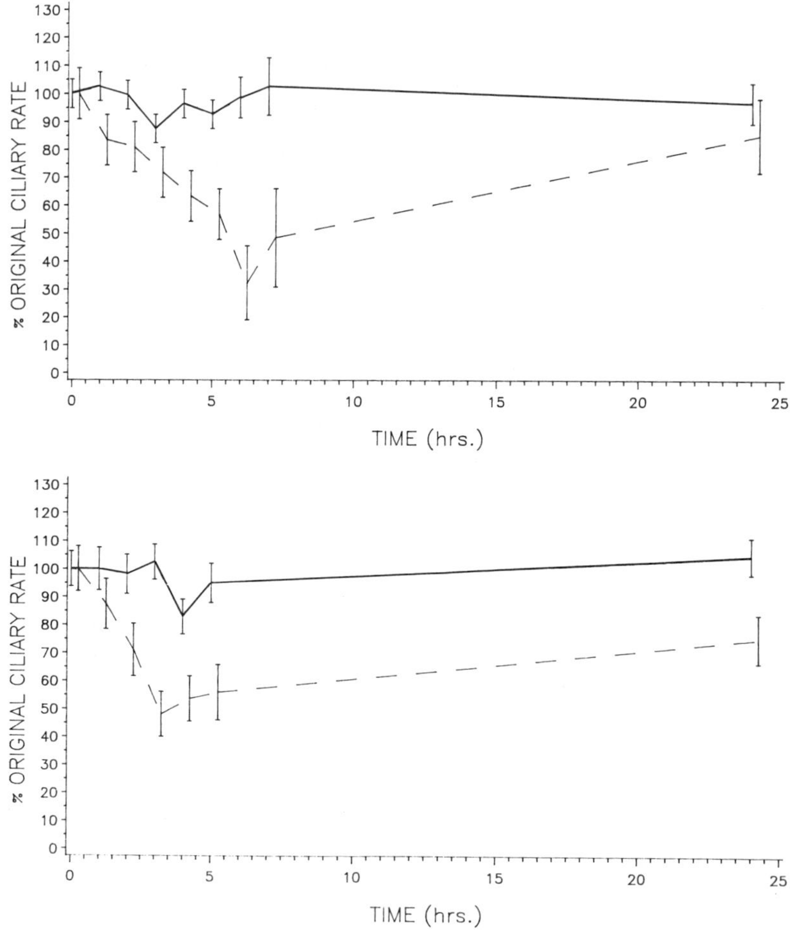

FIG. 1. Percent original ciliary rate of *Mercenaria mercenaria* (top), and *Mytilus edulis* (bottom) exposed to *Aureococcus*. Error bars are ± 1 standard error. Solid line: Control; Dashed line: Exposed to *Aureococcus*.

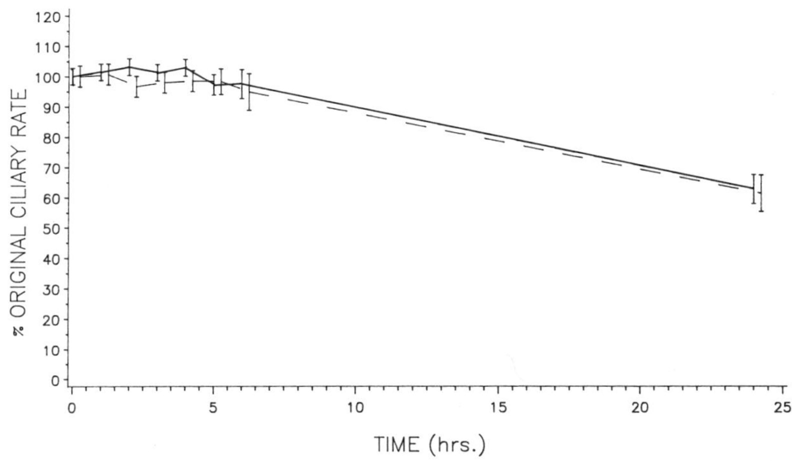

FIG. 2. Percent original ciliary rate of *Mytilus edulis* exposed to water from which *Aureococcus* had been removed with a .8 micron filter. Error bars are ± 1 standard error. Solid line: Control; Dashed line: Exposed to *Aureococcus* "water".

REFERENCES

1. E. M. Cosper, W. C. Dennison, E. J. Carpenter, V. M. Bricelj, J. G. Mitchell, S. H. Kuenstner, D. Colflesh, and M. Dewey, Estuaries *10*, 284-290 (1987).
2. J. McN. Seiburth, P. W. Johnson, and P. E. Hargreaves in: Proc. Emergency Conference on "Brown Tide." (State Dept., New York State, Albany, N.Y., 1986).
3. G. A. Tracey, P. W. Johnson, R. W. Steele, P. E. Hargreaves, and J. McN. Sieburth, J. Shellfish Res. 7, 671-675 (1988).
4. P. Olsen, in: Proc. Emergency Conference on "Brown Tide." (State Dept. New York State, Albany, N.Y., 1986).
5. J. McN. Sieburth, P. W. Johnson, and P.E. Hargreaves, J. Phycol. *24*, 416-425 (1988).
6. G. A. Tracey, Trans. Amer. Geophys. Union *66*, 1303 (1985).
7. V. M. Bricelj, and S. H. Kuenstner in: Notes on Coastal and Estuarine Studies. E. M. Cosper et al, eds. Springer-Verlag, N.Y., (1989).
8. F. Mohlenberg, and H. U. Riisgard, Ophelia *17*, 239-246 (1978).
9. B. L. Bayne, and R. C. Newell, in: The Mollusca v. IV, A.S.M. Saleuddin and K. M. Wilbur, eds. Academic Press, N.Y. (1983) pp. 407-515.
10. G. A. Tracey, Mar. Ecol. Prog. Ser. *50*, 73-81 (1988).
11. E. J. Catapane, Comp. Biochem. Physiol. *75C*, 403-405 (1983).
12. A Paparo, and E. Aiello, Comp. Gen. Pharmacol. *1*, 241-250 (1970).
13. E. J. Catapane, G. B. Stefano, and E. Aiello. J. exp. Biol. *83*, 315-323 (1979).
14. E. Pequignat, Mar. Biol. *19*, 227-244 (1973).

EFFECTS OF GYRODINIUM CF. AUREOLUM ON PECTEN MAXIMUS (POST LARVAE, JUVENILES AND ADULTS)

E. ERARD-LE DENN *, M. MORLAIX * and J.C. DAO **
*IFREMER - Centre de Brest - * DESO/EL - ** DRV/LPMDC - B.P. 70 - 29263 Plouzané - France

ABSTRACT

Since 1976, several outbreaks of G. cf. aureolum have been reported along the french coasts. This dinoflagellate was suspected to affect all the life stages of the scallop P. maximus in the Bay of Brest (FR). The effects appeared as a recent constraint in the management of shellfish resources i the Bay of Brest and its vicinity. In vitro, some preliminary bioassays have been designed to evaluate the toxicity of two strains of G. cf. aureolum. These experiments demonstrated that, following nine months of continuous culturing, specific toxicity of G. cf. aureolum strain of the Bay of Brest regressed considerably.

INTRODUCTION

Since its first description by Hulburt (1957) (1) the eastern coast of the USA, numerous minor or major red tides outbreaks of Gyrodinium cf. aureolum (*) have been reported along the continental shelf of the N.E. Atlantic ocean. The most conspicuous results of such occurrences are dead fishes, bivalves and lugworms (2, 3, 4, 5).

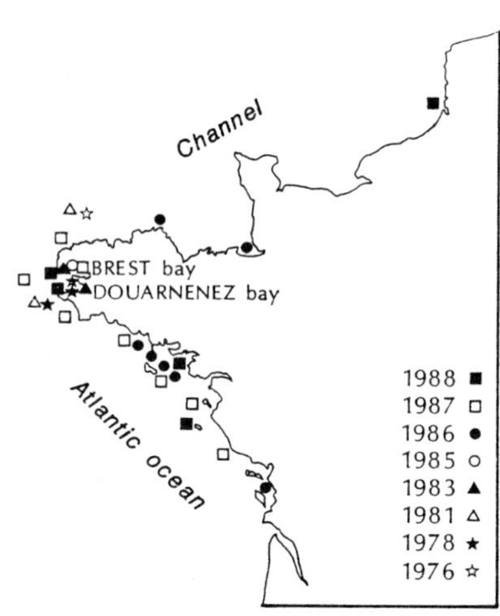

FIG.1. Sitings of G. cf. aureolum blooms along the English Channel and Atlantic coasts from 1976 to 1988 (7).

Since 1976, several outbreaks of G. cf. aureolum have been reported along the North West coast of France and more especially in the bays of Brest and Douarnenez (Fig.1). In these areas, deaths of bivalves and growth inhibition of scallops (Pecten maximus) have been observed which are, possibly, related to G. cf. aureolum bloom.

This paper describes and discusses ths effect of the red tides caused by G. cf. aureolum in the bays of Brest and Dournanenez. It examines also the susceptibility to G. cf. aureolum toxins on P. maximus which have been reported dead during red tides, and presents some preliminary bioassays designed to evaluate the toxicity of G. cf. aureolum.

(*) The specific epithet aureolum is retained, however, the use of the designation G. cf. aureolum is suggested for European specimens until firm taxonomic feature are found (6).

GYRODINIUM CF. AUREOLUM OUTBREAKS IN BREST BAY

G. cf. aureolum was first associated with problems on shellfish beds in July 1978. During this period important mortalities in mussels beds of Douarnenez Bay have been reported and slight mortalities of two-year scallops (Pecten maximus) in the Tinduff nursery ponds (Fig. 2A) were noted. Surviving wild Pecten maximus showed a "stress ring" on shells clearly indicating growth inhibition.

Toxic Marine Phytoplankton
Edna Graneli et al., Editors

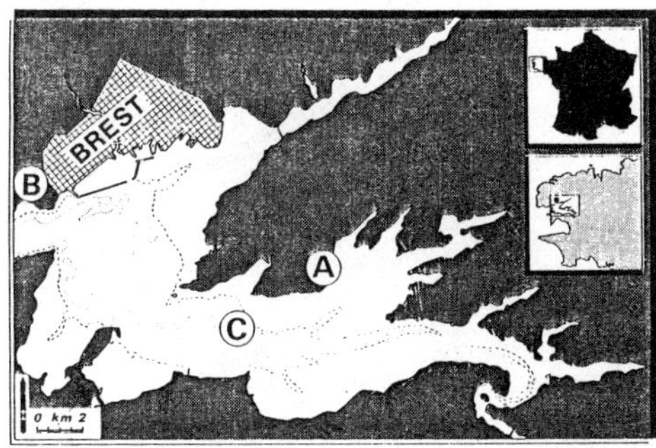

Fig. 2. Brest Bay scallops study sites.
A) post larvae in Tinduff nursery. B) spat in trays at Ste Anne du Portzic. C) adults at bottom station.

During September 1980, 50 to 70% of the mussel population have been destroyed in Douarnenez Bay.

In mid July 1983, at site A, some mortalities occurred among the 2.5 mm post larvae and later on numerous 1 mm post larvae died. At another site, site B, surviving post larvae showed later "stress rings" which further have been used for dating growth inhibition periods, G. cf. aureolum in July-August 1983 reached densities of 200 000 cell·l^{-1} and induced the mortality of all the stock of 550 000 post larvae in a few days at site A. At site B, only 150 000 post larvae survived from an initial stock of 1 000 000 individuals.

During summer 1985 in the bay of Brest, some disturbances occurred on all the breeding stages of Pecten maximus. A bloom of G. cf. aureolum lasted several weeks from the end of July at densities of several hundred thousand cells per liter (maximum 800 000). Mass mortalities of the post larvae stage (0.25-3 mm) caused an abnormal loss of 4 000 000 individuals in the nursery and in the trays (Fig. 3). During this event, growth of the juvenile stages (5-30 mm) stopped completely for one month and resumed on the 26th of August when the red tide vanished (Fig. 4).

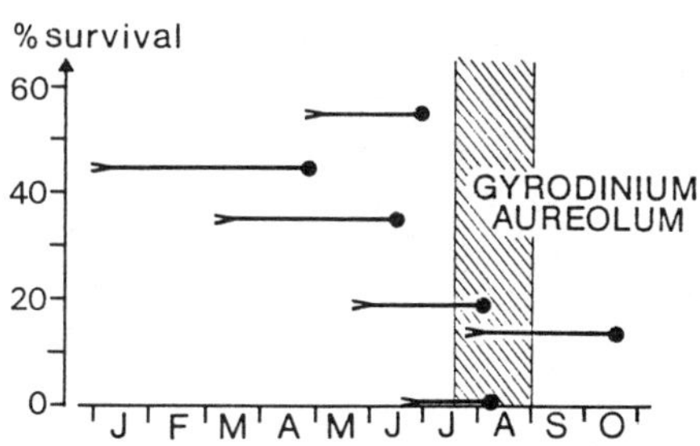

FIG. 3. Spat survival in trays for different batches in 1985. > Beginning of the culture spat size - 3 mm. - duration of the culture • % survival at the first grading spat size - 15 mm.

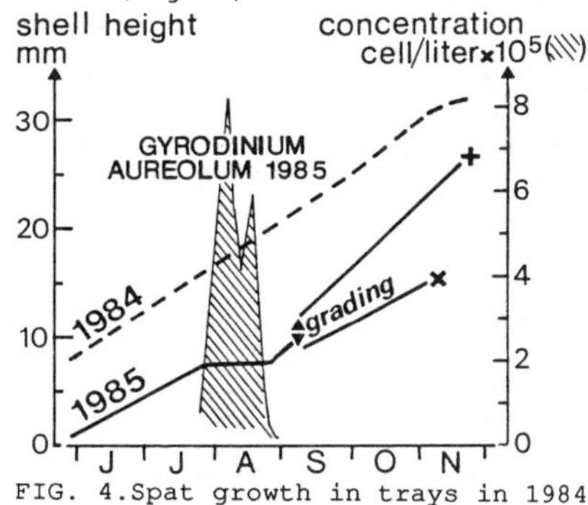

FIG. 4.Spat growth in trays in 1984 and 1985. + greater spat -25%, x smaller spat deformed - 75%

After recovery, a "disturbance ring" has been observed on every shell (figure 5). Most of them were bent out of shape (Fig. 6). These shell abnormalities induced further mortalities later on, probably caused by predation.

On adult shells (> 30 mm) released on the bottom (site. C), a characteristic disturbance growth ring was also observed without bendings (Fig. 7) . The reproduction cycle was modified with abnormally low values of the gonad index (Fig. 8).

FIG. 5. Shell disturbance ring (summer 1985). Concentric growth ridges crowded together forming a distinct ring on the shell. From August 26th the width of the ridges formed with a daily periodicity became much larger .

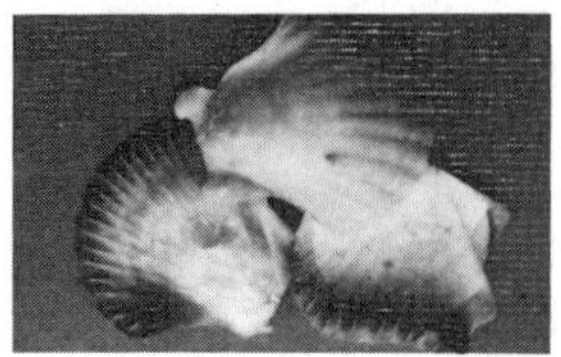

FIG.6. Shell bending affects up to 75 % of spat.

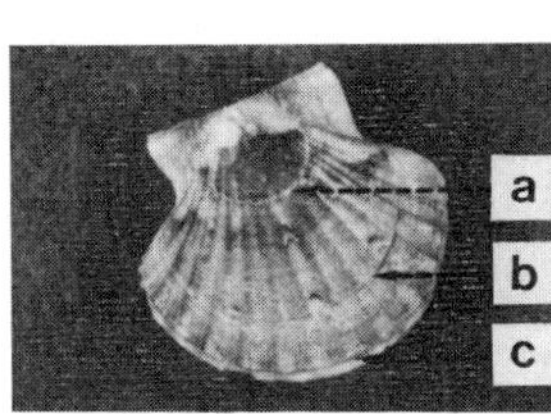

FIG. 7. Wild adult shells showed a characteristic prominent ring (b).
a) autumn 1984 showing ring.
b) disturbance ring August 1985.
c) December 1985 ring.

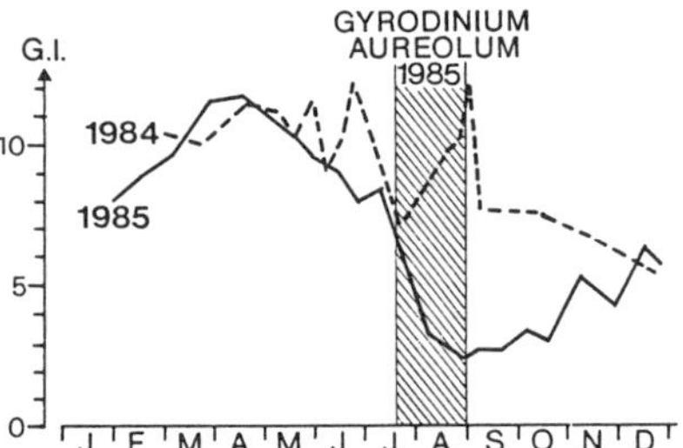

FIG. 8. Evolution of the gonad index in 1984 and 1985.

$$G.I = \frac{\text{wet weight of gonad (dg.)}}{(\text{shell height mm})^3} \times 1000$$

No G. aureolum bloom was noted in 1986 as in 1984, but, in July 1987 another outbreak was associated with growth inhibition for Pecten maximus larvae as in 1985. In 1988, although maximum densities of G. aureoulum reached only 160 000 cells$\cdot l^{-1}$, shellfish recruitment was as low as during the years 1983. 1985 and 1987.

Observations reported above suggest that there probably exists a simultaneity between the occurrence of G. cf. aureolum in the Bay of Brest and alterations caused to P. maximus. In order to verify this hypothesis, the effect of G. cf. aureolum on P. maximus has been examined in laboratory studies using G. cf. aureolum cultures.

TOXICITY OF G. CF. AUREOLUM : PRELIMINARY TESTS

MATERIALS AND METHODS

Scallops. **Post-larvae of P. maximus** (2 mm in size) were obtained from the Tinduff nursery pond (Bay of Brest). Post-larvae were acclimated at least two days before each study began. Test animals were fixed (50 individuals per batch) seived (500 µm meshes) and laid into 2 liters tanks of filtered water (0.2 µm) without air supply. Tests were conducted in a temperature controlled room (≈ 14°C). **Juveniles of P. maximus** (8-19 mm in size) were collected in the trays of the shellfish farm of Ste Anne du Portzic (Bay of Brest). Before exposure, test animals were acclimated to confined conditions, during 2 days. After this period, they were removed and placed (70 juveniles per batch) into 10 liter tanks of filtered water

(0.2 µm) provided with air supply. Tests were conducted in a temperature controlled room (≈ 18°C). Salinity and pH varied little during each test. Animals were fed with forage algae (see below). Every two days, the sea water was renewed.

Each test comprised 4 batches (4 replicates) and experiments were conducted as follows for post-larvae of P. maximus.

- 1 control test with post-larvae and forage algae
- 1 test with post-larvae + forage algae + G. aureolum culture 500 000 cells$\cdot l^{-1}$.
- 1 test with post-larvae + forage algae + G. aureolum culture 1 500 000 cells$\cdot l^{-1}$.
- 1 control test post-larvae not fed.

In addition, periodic observations on the different algae culture were conducted in order to control the filtration rate. Unusual behaviour was described and death times recorded if death occurred during observation periods. After 4 days tests were terminated because control post-larvae fed with algae died. Similar experiments were carried out with juveniles but the duration of the test was as long as 50 days, but results only corresponding to a test duration of 20 days havs been retained (mortality rate of control organisms ≤5%).

Algae. The forage algae were composed of 3 species of unicellular algae : Pavlova lutheri, Chaetoceros calcitrans and Isochrysis aff. galbana. Density of 10^8 cells per liter has been chosen as a suitable amount of food. G. cf. aureolum strains were PML-497 A strain isolated from the Plymouth (UK) area and provided by Dr. G.C. Green (Plymouth Marine Laboratory). Strain Tinduff was isolated in July 1987 from Tinduff nursery ponds by Dr. F. Partensky (Roscoff Marine Laboratory). The cultures of these species were grown on F/2 medium (8) under a 12.12 h L:D cycle light. Light was provided by fluorescent tubes (50 µmoles$\cdot m^{-2} \cdot s^{-1}$) with a temperature of 18±1°C. Cell counts were made on iodine-preserved samples with a Malassez or Nageotte slide.

RESULTS AND DISCUSSION

From this preliminary test on lethal effect (Tab. 1), Plymouth strain appeared not to be very toxic for the post-larvae. This slight avtivity was detected at a density of $6.2 \cdot 10^6$ cells$\cdot l^{-1}$. A four month culture at a density of $8 \cdot 10^5$ cells$\cdot l^{-1}$ (Tinduff strain) exhibited, on the second day, a more intense toxicity. Similar rates of mortality,have been induced with a nine month culture at a density of $1.5 \cdot 10^6$ cells$\cdot l^{-1}$. Observations showed (Tab. 1) that juveniles seem to be sensitive to G. cf. aureolum when density attains a threshold of $1.5 \cdot 10^6$ cells$\cdot l^{-1}$ (20 days tests). These preliminary tests performed to compare G. cf. aureolum's effects on scallops with two strains, have demonstrated the variability of G. cf. aureolum toxicity. A nine month culture of G. cf. aureolum (Tinduff strain) has been used in the experiments. During this time the specific toxicity of G. cf. aureolum has probably decreased considerably as shown by the results presented in table 1. This argument can also be applied to the Plymouth strain which shows a rather low toxicity. Another explanation could be a specific low toxicity of this strain.

Observations conducted in the Tinduff nursery (9) showed that the scallops post-larvae mortality was probably caused by starvation. Nevertheless, the experiments conducted in vitro in April 1988 demonstrated that in the case of post-larvae and juveniles, G. cf. aureolum densities of $5 \cdot 10^5$ cells$\cdot l^{-1}$ and $1.5 \cdot 10^6$ cells$\cdot l^{-1}$ do affect the filtration rate of forage algae without altering the ingestion process, and G. cf. aureolum was ingested together with forage algae. During these in vitro experiments, growth processes were inhibited. Nevertheless, when the juveniles return to their natural environmental conditions (Fig. 2B), they grew up again but independently of experiments they have undergone. However, the average growth was lowest for the juveniles exposed to the tests than for the juveniles grown in situ (Fig. 2B).

The environmental conditions for the tests, in batch cultures, may have important consequences on the *G.* cf. *aureolum* toxicity. Nevertheless, it was interesting to note that, after the recovery, experimental juveniles showed a disturbance growth ring as opposed to the control juveniles. *G.* cf. *aureolum* has provoked a stress ring without important shell damage.

Densities of *G.* cf. *aureolum* (cell. l^{-1})	Dates of tests	Times of tests (days)	Post-larvae death rate		Juveniles death rate	
			control samples	exp. samples	control samples	exp. samples
Plymouth strain						
1,600,000	Nov. 1987	2	0	0		
2,200,000	May 1987	2	4	6		
6,200,000	Nov. 1987	2	0	10		
Tinduff strain						
500,000	April 1988	2	0	0		
		4	8	12		
		6			3	4
		10			5	5
		20			9	10
800,000	Nov. 1987	2	0	15		
		4	5	30		
1,500,000	April 1988	2	0	15		
		4	5	46		
		6			0	4
		10			0	6
		20			4	18

TABLE I. Lethal effects of two strains of *G.* cf. *aureolum* on post-larvae and juveniles *P. maximus* (rate per cent).

In spite of few data on the toxicity of *G.* cf. *aureolum* to scallops, these observations in ths Bay of Brest, and the experiments conducted *in vitro* have demonstrated a toxic and inhibitory effect of *G.* cf. *aureolum* on *P. maximus*, even if the cultivated strain showed a loss of toxicity. The effects varied according to *G.* cf. *aureolum* density. In the Bay of Brest, the harmful effect may appear in the presence of low densities of *G.* cf. *aureolum.* This suggests that specific environmental conditions are responsible for the degree of toxicity of *G.* cf. *aureolum*, and may induce the production of either high or low toxicity according to the occurrence of specific environmental parameters in the water column.

ACKNOWLEDGEMENTS

The authors wish to thank D. Buestel, A. Guénolé and A. Gérard for providing unpublished *in situ* data concerning *P. maximus*, and M.P. Crassous for technical support. We are also grateful to Dr. Y. Monbet for his valuable comments. This work was supported by the IFREMER Red Tides Program.

REFERENSES

1. E.M. Hulburt, Biol, Bull. Mar. Biol. Lab. Woods Hole *112* (2):196-219 (1957).
2. T. Braarud and H. Heimdal, Nytt Mag, Bot. *17* (2):91-97 (1970).
3. E. Dahl, D.S. Danielssen and B. Bohle, Flödevigen, Rapport serie 4:1-15 (1982).
4. M.M. Helm, B.T. Herper, B.E. Spencer, P.R. Walne, J. Mar. Biol. Assoc. UK 54 (4):857-869 (1974).
5. K. Tangen, Sarsia *63* (2):123-133 (1977).
6. F. Partensky et A.Sournia, Cryptogam. Algol., *7*:251-275 (1986).
7. C. Belin, J.P. Berthomé, P. Lassus IFREMER-Nantes (FR), unpuhlished data.
8. R.R. Guillard and J.H. Ryther, Can. J. Microbiol. *8*, 229-239 (1962).
9. A. Gérard, C.L.P.M. Le Tinduff (FR), personal communication.

TROUT MORTALITY ASSOCIATED TO DISTEPHANUS SPECULUM

E. ERARD-LE DENN and M. RYCKAERT
IFREMER - Centre de Brest - DERO/EL - B.P. 70 - 29263 Plouzané - France

ABSTRACT

Numerous fishes mortalities associated with algal blooms have been observed in recent years and attributed to dinoflagellates, but there are harmful blooms species among taxa, such as diatoms, chrysophytes, and silicoflagellates.

In France, in a sea trout farm of Douarnenez, a sudden and important fish mortality occurred during the night from April 12-13th 1987. The following morning, water was turbid and caged fishes (*Salmo gairdneri* Richardson) showed symptoms of distress, coming to the surface with signs of asphyxia. Death appeared rapidly. The mortality corresponded to 15 tons of dead fish, that meant 70 % of sea-trout live-stock.

The pathological examination showed no evidence of bacterial or parasitic presence. Trout examination revealed gill clogging by mucus in which many phytoplankton cells : *Distephanus speculum* (Ehrenberg) were visible. Gill histological sample presented oedema and important hyperplasia, and liver histological samples showed a recent degeneration with nuclear pyknosis.

When caged fishes died, water had been yellowish discoloured for one day. Microscopic observation revealed the presence of the skeleton-form of the silicoflagellate *D. speculum* (70 % of dominance cells). On April 13th, *D. speculum* cell densities varied between 1.3 x 10^6 to 1.3 x 10^5 cells/litre (morning and afternoon sampling).

Oxygen depletion due to the respiratory activities of silicoflagellates in darkness and/or the decay of the bloom could be suggested as a possible explanation of fishes mortalities. During the night dissolved oxygen could have been insufficient for the survival of sea-trout bred with normal densities between 15 to 20 kg/m^3 in these very calm weather conditions during the low tide period.

Nevertheless from our observations, it appears that siliceous skeleton of *D. speculum* irritated gills, and its high densities caused mucus secretions which lowered gill exchange potential. Evidence of damage to fishes during red tide of *D. speculum* with a skeleton has been reported in the Kattegat and Belt Sea area (G. Aertebjerg and J. Borum (1984)).

Such fishkills have been reported in southwestern Denmark in 1983, but were due to a naked stage of *D. speculum* (O. Moestrup and J. Larsen, this volume).

Toxic Marine Phytoplankton
Edna Graneli et al., Editors

CYSTS OF PROROCENTRUM MARINUM (DINOPHYCEAE) IN FLOATING DETRITUS AT TWIN CAYS, BELIZE MANGROVE HABITATS

MARIA A. FAUST
Smithsonian Institution, Museum Support Center, 4201 Silver Hill Road, Suitland, MD 20560, USA

ABSTRACT

Excystment process in the life cycle of the benthic dinoflagellate Prorocentrum marinum (Cienkowski) Dodge & Bibby, 1973 from floating detritus at protected mangrove areas was investigated. Hypnocysts were found attached to detritus. Mature hypnocysts were 70 µm in diameter, spherical, brown, with smooth, organic triple layered wall. The archeopyle of hypnocysts was round. Excystment probably began when the protoplast emerged through the hypnocyst's archeopyle which completely opens. The protoplast was enclosed in a double layered division cyst. Cell division in the division cysts resulted in two daughter cells, and tetrads, each of which resembled a normal vegetative cell. This is the first report of the presence of hypnocysts of P. marinum in plankton with a circular archeopyle.

INTRODUCTION

Resting cysts are sources of seed populations in the development of toxic dinoflagellate blooms (1,14), and provide long-term survival under hostile environments, species dispersal, reproduction, and preservation of genetic variation (9). The extent to which a cyst can fulfill any particular one of these functions depends upon the biology and environmental regulation of life cycle of the given dinoflagellate species. So far, the life cycle of a limited number of freshwater and temperate marine dinoflagellates has been determined in cultures and some in natural populations (12).

Reports on Prorocentrum cysts are limited. Early reports indicate two types of Prorocentrum cysts. One type, from marine samples, is a brown, spherical resting cyst of P. micans (2,4), and also described as aberrant forms found inside old valves in old cultures (7,8). The second type is a thin cyst of Exuviella marina in which the development of two daughter cells were reported (11,15), enlarging in length and width inside the cyst (8). More recently, the sexual life cycle of P. micans was described in actively growing cultures (3), and the existence of a hypnozygote of P. lima was suggested in old cultures (14).

During field experiments conducted in Belize, an unexpected excystment event was observed after a period of heavy rains and temperature increase in Hidden Lake, a mangrove habitat. Patches of detritus floated to the surface of warm, highly-oxygenated water exposed to high level of sunlight. This environment triggered excystment of resting cysts of P. marinum and formation of vegetative cell populations. The observations described here do not allow a detailed description of the entire excystment process, but enable me to recognize most stages of excystment

Toxic Marine Phytoplankton
Edna Graneli et al., Editors

occurring in a natural population. The described stages were all observed simultaneously in samples taken from floating detritus. We believe that the mangrove environment plays an important role in the excystment process.

STUDY AREA

Field studies reported here were conducted at Twin Cays, Belize (16° 48' N, 88° 05' W), a section of the barrier reef of the Northern Hemisphere extending 220 km from the Mexican border in the north to the Gulf of Honduras in the south (13). Twin Cays is an undisturbed, diverse, intertidal mangrove island divided into two parts by an S-shaped channel. It has various habitats such as lakes, ponds, and mud flats. *Prorocentrum* assemblages were found at protected sites at Hidden Lake and the Lair. Hidden Lake is a shallow (30-90 cm deep) body of water with bottom sediment comprised of organic detritus, mud, sand, peat and carbonate silt surrounded by red mangroves *(Rhizophorae)*. In contrast, the Lair is a 2-3 m deep lagoon with a bottom composed of detritus, carbonate silt, mud, and meadows of turtle grass *(Thalassia)*. Specimens were collected during February, 1987 and again from late January to early February, 1989. The water temperature ranged from 24-30 C, and the salinity level was from 32-34 °/oo. *Prorocentrum marinum* was present at both locations, but was more abundant at Hidden Lake. Other species were present in low numbers: *P. lima, P. mexicanum, P. concavum, P. emarginatum, Prorocentrum* sp., *Scripsiella subsalsa, Amphidinium herdmanii,* and *Thecadinium kofoidii*.

MATERIAL AND METHODS

Samples of floating detritus were collected from the water surface by a pitcher; sediment surface samples were collected by a hand-held pump. Samples in 1 liter plastic containers were returned to the laboratory within 30 min after collection, sieved through 150 μm and 90 μm mesh filters, and concentrated into 20 ml sample. A portion of each fraction was fixed with 4 % buffered gluteraldehyde (10). Cell and cyst dimensions were determined by measuring the length and width of 50 fixed cells using an eye-piece micrometer with differential interference contrast (DIC) optics of a Carl Zeiss light microscope at appropriate magnification. Specimens were also examined with a scanning electron microscope (SEM) (10). The nuclear DNA in fixed cells and cysts was stained with fluorochrome 4'6-diamidino-2-phenylindole (DAPI) (5). Specimens were depigmented by 90 % acetone to remove interfering chlorophyll autofluorescence.

RESULTS AND DISCUSSION

Observations: Resting cysts of *P. marinum* were found adhered to detrital and sediment particles in floating detritus. Blue-green algae, diatoms and bacteria were regularly seen attached to the cyst surface. In the sediment, the aquatic environment is anoxic, the light level is low and the water temperature is relatively cool: 20-22 °C. Resting cysts are abundant in this environment. Detritus floats to the surface in early spring by visible expanding gas bubbles. Cysts of *P. marinum* and other stages of the excystment process were isolated from the floating

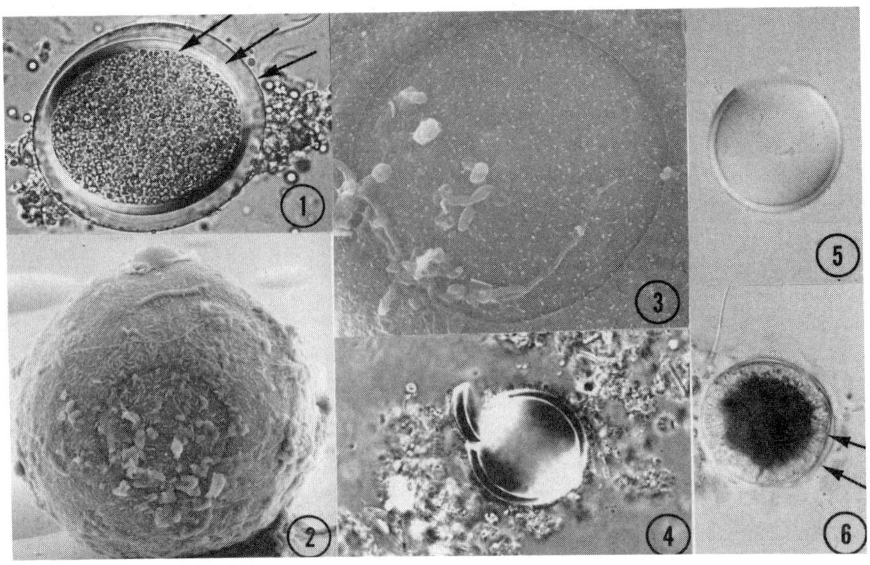

FIGS. 1-6. Prorocentrum marinum. FIG. 1. Hypnocyst with starch, and triple-layered wall (arrow), (DIC) 2950X. FIG. 2. Hypnocyst covered with papillae and a circular archeopyle shown, (SEM) 3000X. FIG.3. Archeopyle at higher magnification (SEM) 4600X. FIG. 4. Archeopyle partially detached from empty cyst wall (DIC) 2950X. FIG. 5. Archeopyle missing (DIC) 2730X. FIG. 6. Newly formed division cyst with double wall (arrows), (DIC) 2600X.

detritus, which on the water surface was exposed to warm temperature, high level of sun light, and oxygen, all factors needed for the excystment process (16). On three consecutive days (1/28, 1/31, and 2/6) excystment populations of P. marinum were as follows per ml of sample: hypnocysts 1253, 823, and 360; division cysts 697, 240, and 300; and free vegetative cells 2212, 2550, and 3859, respectively. The emerging vegetative cells of P. marinum provided confirmation of species identity.

Cyst Morphology: Resting cysts of P. marinum are oblong to spherical, diameter 60 to 70 µm, and considered to be hypnocysts (Fig. 1). They have a triple-layered cyst wall, 3-7 µm thick. Hypnocysts are round, brown, and composed of organic matter destroyed by acetolysis. The outer wall of the cysts is smooth, unornamented, and without paratabulation. This wall was often lost during sample preparation (Fig. 2). The middle wall is cytoplasm, nucleus (left), adjacent red body (arrow), and ovoid large bodies, (DIC), and Fig. 10. same cell stained with DAPI, covered with flexible papillae (Figs. 1, 2, 3). The inner wall has no distiguishable features. DAPI fluorescent dye could not

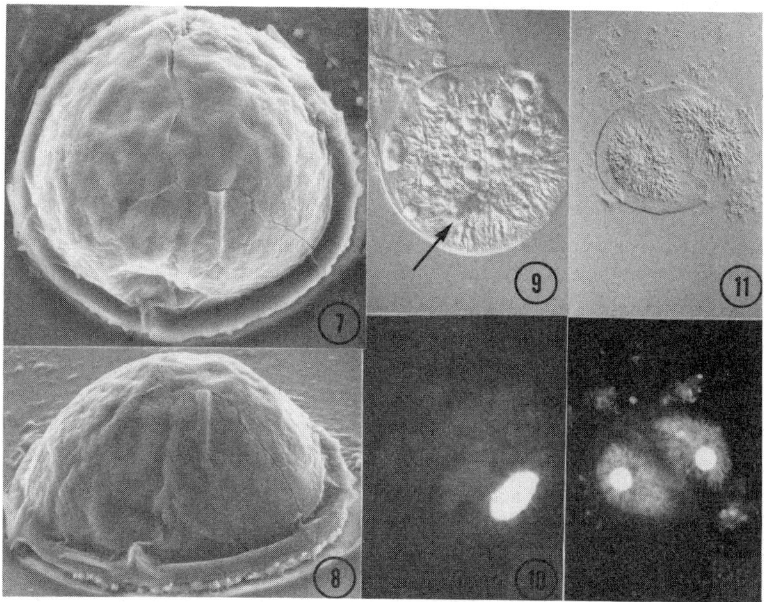

FIGS. 7-12. Prorocentrum marinum. FIGS.7-8. Division cyst with smooth surface, distinct rim shown top and side view, (SEM) 4300X. FIG. 9. Division cyst (depigmented) showing radial strands of cytoplasm, nucleus (left), adjacent red body (arrow), and ovoid large bodies, (DIC) and FIG. 10. same cell stained with DAPI, (DIC) 3500X. FIG. 11. Division cyst shown with two developing cells, (DIC), and FIG. 12. same cell's nuclei and developing plastids stained with DAPI (DIC) 3000X.

penetrate the cyst walls. The archeopyle is a circular opening (Fig. 3), and it opens completely to give a free operculum (Figs. 4, 5). Hypnocysts contain no plastids, and they are densely packed with storage granules in the form of starch, a red body, and a large nucleus.

Excystment: The excystment of the hypnocyst probably begins with the emergens of a protoplast, the reorganization of the cytoplasm, chloroplast development, the appearance of golden pigmentation, and the disappearance of starch granules. The protoplast is enclosed in a thin, transparent, colorless double-layered envelope, designated as the division cyst (Fig. 6). The division cyst has a smooth surface outer layer with a distinct rim at the periphery (Figs. 7, 8), and contains centrally aggregated cellular components (Fig. 6). The division cysts contain a large, ovoid nucleus with long, paired chromosomes, round to ovoid bodies of various sizes, and a red body, and cytoplasmic strands radiating inward from the inner membrane (Figs. 9, 10). Cell division takes place inside the division cyst. The division cysts are larger than hypnocysts (85-110 μm)

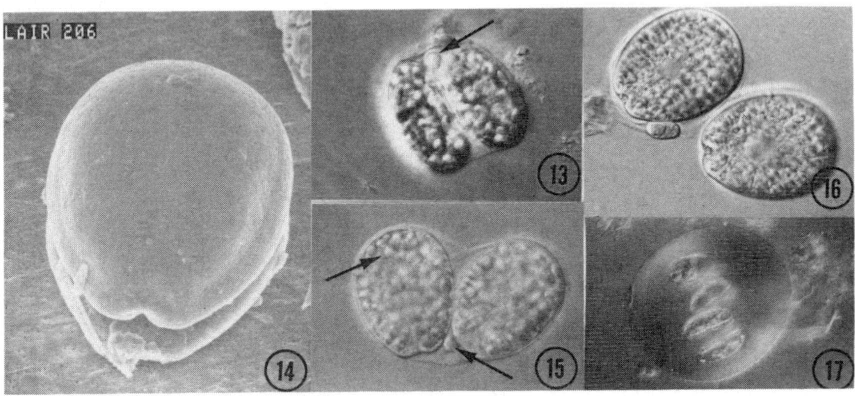

FIGS. 13-17. Prorocentrum marinum. FIG. 13. Two cells enclosed in membrane shown with red body (arrow), (DIC) 3000X. FIG. 14. Cell division not yet complete, (SEM) 2600X. FIG. 15. Cells fully developed in envelope filled with starch, and red body (arrow), (DIC) 3000X. FIG. 16. Cells emerging from envelope, red body dislodged, (DIC) 3000X. FIG. 17. Tetrads enclosed in envelope, (DIC) 2100X.

and they are non motile.

In the division cyst, indication of cell division begins when the inner membrane contracts inward at opposite sides, and is followed by development of two cells by longitudinal division (Figs. 11, 12). Each daughter cell has a large nucleus and developing plastids, and a red body situated at the anterior part of the cells, but no flagella was observed (Fig. 13). The process is concluded when the inner membrane ruptures, cells separate, the red body dislodges, and pigmentation becomes golden in color (Figs. 14, 15, 16). The emerging daughter cells look similar to vegetative cells with the exception that they appear to be filled with starch grains. On several occasions, tetrads were observed within division cysts (Fig. 17), probably involving two consecutive cell divisions. Cells were highly pigmented, golden in color, fully developed, and the red body was present. Division cysts at the two-or four-cell stage were only observed in floating detritus samples, and not in the sediment.

Conclusion: Dale in his review (9) noted that no resting cysts of Prorocentrum have been found in sediment often examined from areas where this genera is common in plankton. Apparently, Prorocentrum's organic-walled cysts do not survive in the environment. Here we report P. marinum hypnocysts from the Twin Cays mangrove ecosystem which is rich in Prorocentrum species, where we observed blooms in the past. The hypnocysts of P. marinum are different from those described for other dinoflagellates reviewed by Dale (9,) and Pfiester & Anderson (12). The archeopyle of P. marinum is round and is distinctly different from the intercalary archeopyles of Protoperidinium (9).

The field observation does not allow us to identify the hypnozygote nature of the hypnocysts we describe here. In cultures, Bhaud et al. (3) described the course of sexual reproduction of P. micans. They describe several steps as part of the excystment process. It appears from this description that the first meiotic division produces two daughter cells which appear to be followed by a second meiotic division producing tetrads. Our observations show that in P. marinum the two daughter cells are enclosed in a division cyst, and tetrads may be present occasionally. The frequency of observed tetrads was low. The existence of tetrads may indicate that P. marinum, similarly to P. micans, may undergo two consecutive meiotic divisions while still enclosed in the division cyst. The red accumulation body was part of the excystment process and was always present at the anterior end of the dividing cells. When the division cysts ruptured and the daughter cells separated, the red accumulation body dislodged. The morphological features of the archeopyle of hypnocysts is considered an important taxonomical character (9). Here a round archeopile is reported for P. marinum which is similar to those reported by Wall & Dale (17) for some quarternary calcerous dinoflagellate cysts.

ACKNOWLEDGEMENT

This paper is contribution 277 from the Caribbean Coral Reef Ecosystem Program at the Smithsonian Institution. This research was supported in part by grants from the Exxon Corporation.

REFERENCES

1 Anderson, D. M. and Wall, D. 1978. J. Phycol. 14:224-234.
2 Bergh, R. S. 1882. Der Organismus der Cilioflagellaten. Morphol. Jahrb. 7:259.
3 Bhaud, Y., Soyer-Gobbillard, M. O. and Salmon, J. M. 1988. J. Cell Biol. 89:197-206.
4 Breemen van Pieter, J. 1905. Tijdschr. Nederl. Dierk Ser. 2 9:180.
5 Coleman, A. W., Maguire, M. and Coleman, J. R. 1981. J. Histochem. Cytochem. 29:959-968.
6 Bursa, A. S. 1962. Grana Palinologica 3:54-66.
7 Braarud, T. and Rossavik, E. 1951. Avhandl. Norske Videnskaps-Acad. Oslo. Mat. Naturv. Kl. 1:3-18.
8 Bursa, A. 1959. Canad. J. Bot. 37:1-31.
9 Dale, B. 1983. In: G. A. Fryxell (ed) Survival Strategies of the Algae. Cambridge Univ. Press. Cambridge. pp 69-139.
10 Faust, M. A. 1974. J.Phycol. 10:315-322.
11 Lebour, M. V. 1925. The Dinoflagellates of Northern Seas. Plymouth Marine Biol. Assoc. pp 250.
12 Pfiester, L. A. and Anderson, D. M. 1987. In: F.J.R. Taylor (ed) The Biology of Dinoflagellates. Blackwell Sci. Publ. Oxford. Botanical Monog. 21:611-648.
13 Ruetzler, K. and Feller, C. Oceanus 30:16-24.
14 Steidinger, K. A. 1983. Progress Phycol. Res. 2:147-188.
15 Wood, E. F. J. 1954. Austral. J. Marine & Freshwater Res. 5:1-351.
16 Anderson, D. M. Taylor, C. D. and Armbrust, E. V. 1987. Limnol. Oceanogr. 32:340-351.
17 Wall, D. and Dale, B. 1978. J. Paleontology 42:1395-1408.

VERTICAL NUTRIENT TRANSPORT DURING PROLIFERATION OF *GYMNODINIUM CATENATUM* Graham IN RIA DE VIGO, NORTHWEST SPAIN

F.G. FIGUEIRAS AND F. FRAGA
Instituto de Investigaciones Marinas, Eduardo Cabello 6, CSIC, 36208 Vigo, Spain

ABSTRACT

At the end of summer 1986, a red tide of *Gymnodinium catenatum* in Ria de Vigo was sampled three times within a 4-week period. The first sampling on 4 September contained phytoplankton that consisted mainly of diatoms together with some small flagellates. This species composition coincided with a situation with weak upwelling. Fifteen days later, an advection of oceanic surface water caused downwelling in the mouth of the ria, and two distintict horizontal zones could be recognised on the basis of phytoplankton species composition and chlorophyll concentration. *G. catenatum* was relatively abundant for the first time and confined to the downwelling zone. On 3 October, with a clear sign of upwelling in the innermost part of the ria, *G. catenatum* was found in the nutrient poor surface waters throughout the rest of the ria and platform. The distributions of the conservative chemical parameters "NO", "PO" and "CO" at 3 October were in accordance with the possibility of vertical transport of nutrients and vertical segregation between photosynthesis and nutrient assimilation during a red tide of *G. catenatum*, showing nutrient uptake at depth without photosynthesis and photosynthesis at the surface where there were no nutrients.

INTRODUCTION

Spectacular red tides are often seen in the Rias Bajas of Galicia in late summer to early autumn and are sometimes toxic [1]. When this occurs, there is strong hydrographic stratification and nutrients are scarce in the upper layers but plentiful at depth. A steep nutricline located at about 10m separates the two distinct layers [2,3]. Under such circumstances, it is an advantage for red tide organisms to carry out daily vertical migration obtaining nutrients from deeper layers during the night and benefitting from light at the surface during the day [4,5,6,7,8]. For behavior of this kind to be useful for species forming red tides, it is necessary that the synthesis of structural and energetic components be separated in space and time, with carbohydrates formed at the surface during the day, and nutrient uptake and protein synthesis at depth at night at the expense of previously photosynthesized carbohydrates [9,10,11]. This paper shows that the distributions of the chemical parameters "NO", "PO" and "CO" are in accordance with behavior of this kind during a bloom of *Gymnodinium catenatum* in the Ria de Vigo.

The "NO" concept was introduced by Broecker in 1974 [12], defining a chemical constant for each water type resulting from the sum of oxygen and nitrate, the last multiplied by a factor *a* corresponding to the quantity of moles of oxygen consumed in the liberation of 1 mol of nitrate during the oxidation of organic matter, or conversely, moles of oxygen produced by photosynthesis during the assimilation of 1 mol of nitrate.

$$"NO" = O_2 + \underline{a}NO_3$$

Toxic Marine Phytoplankton
Edna Graneli et al., Editors

"PO" and "CO" can be defined in the same way. So these parameters are independent of synthesis or degradation of biomass if nutrient uptake and photosynthesis are proceeding at the same place.

MATERIAL AND METHODS

During a cruise (G-9) of the RV "Garcia del Cid", 8 stations located in the Ria de Vigo and adjacent platform (to 150m isobath) were visited on 4 and 21 September and 3 October, 1986 (Fig. 1).

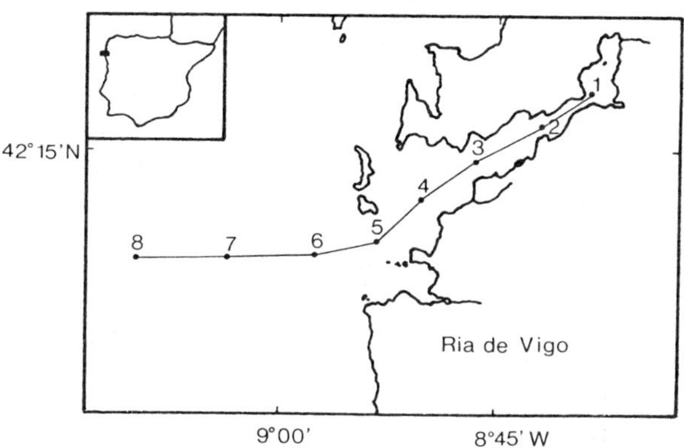

FIG. 1. Chart showing Ria de Vigo and station positions.

Water samples were collected from the surface to bottom at standard intervals (0,5,10,20,30,40,50,60,80,100,120,150m) using 1.7 and 5l Niskin water bottles with attached thermometers. Aliquots were separated to determine salinity, pH, alkalinity, oxygen, nitrate , nitrite , phosphate , chlorophyll and phytoplankton species composition. The raw data are given elsewhere [13]. "NO", "PO" and "CO" were calculated from these data using the following equations [14].

$$\text{"NO"} = O_2 + 10.3\ NO_3 + 9.8\ NO_2 + 8.3\ NH_4$$

$$\text{"PO"} = O_2 + 175\ PO_4 - 0.5\ NO_2 - 2\ NH_4$$

$$\text{"CO"} = O_2 + 1.36[tCO_2 - \tfrac{1}{2}(A + NO_3)] - 0.8\ NO_2 - 1.32\ NH_4$$

where tCO_2 is total CO_2 and A is alkalinity. These equations take into account the uptake and regeneration of nitrites and ammonium that can be important in coastal waters.

RESULTS AND DISCUSSION

Weak upwelling existed on 4 September, as shown by the distribution of density and nitrates (Fig. 2). The corresponding phytoplankton was composed of 75% diatoms (Fig. 3). Dominant species were *Leptocylindrus danicus* and *minimus*, small *Chaetoceros* spp., *Thalassiosira nana*, *Rhizosolenia alata gracillina* and *Skeletonema costatum*. The remaining 25% consisted of small flagellates; the whole flora can be considered as typical for these

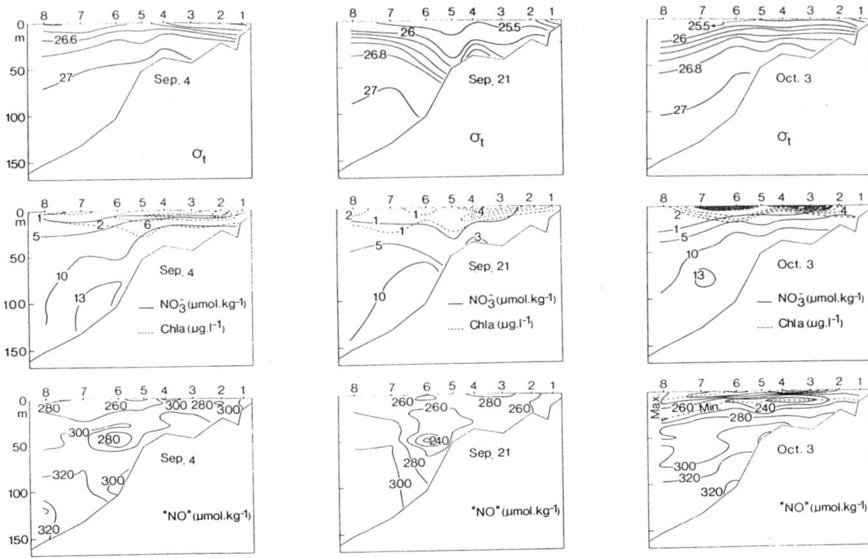

FIG. 2. Distribution of sigma-t, nitrates, chlorophyll and "NO" in Ria de Vigo on 4 and 21 September and 3 October, 1986.

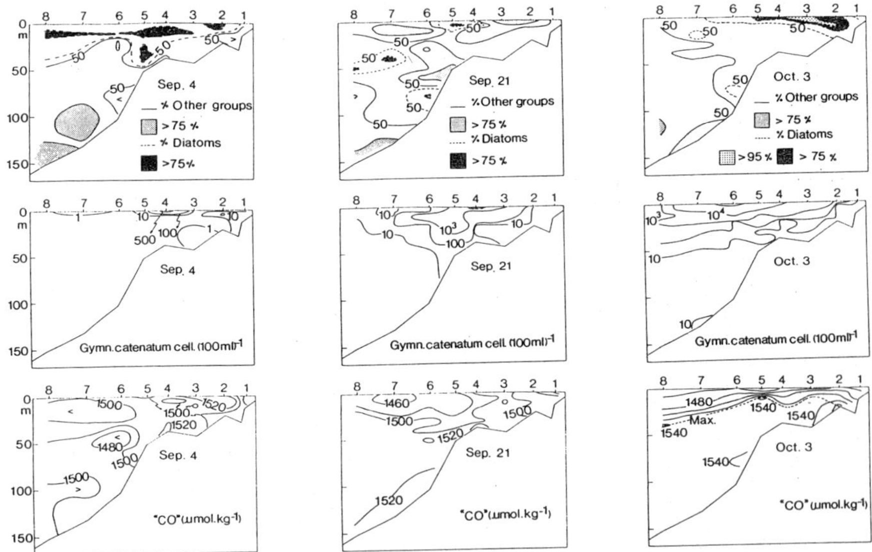

FIG. 3. Distribution of phytoplankton groups, concentration of Gymnodinium catenatum and "CO" in Ria de Vigo on 4 and 21 September and 3 October, 1986.

situations [15].

Seventeen days later, a change in the coastal circulation with a jet of warm water fluxing northward was unfavorable to upwelling [16] and warm oceanic surface water penetrated the area, leading to downwelling in the mouth of the ria and stopping the estuarine circulation (Fig. 2). Various small flagellates and Solenicola setigera were the dominant organisms in this influx. The rest of the ria (st. 1, 2, 3 and 4) showed a totally different phytoplankton community where Heterosigma akashiwo (Hada) Hada was the dominant species. The concentration of Gymnodinium catenatum, in contrast to the earlier situation, began to be perceptible. It was confined to the zone of downwelling at stations 4 and 5, as if trapped in a local circulation feature.

By 3 October, the circulation restarted, with slight upwelling (Fig. 2) which only reached the surface in the innermost part of the ria, and which provided nutrients at 10m in the rest of the ria, and at 20m at stations on the platform. Under this condition Skeletonema costatum, small centric species and various Chaetoceros spp. dominated the interior stations and comprised 95% of the total cell number (Fig. 3). The small flagellates were only important in the oceanic stations, but Gymnodinium catenatum was present throughout the ria, concentrated in the surface layer (Fig. 3). At this time, in contrast to the earlier sampling days, "NO" and "PO" (only "NO" is shown as the "PO" distributions are similar) maxima (300-340 µmol/kg) at the surface coincide with the distribution of Gymnodinium catenatum. "NO" minima at the interior stations at 10m and at the outer stations at a slightly grater depth coincide in both cases with the nutriclines.

"NO" distribution on 3 October (contrasting with the earlier sampling days) is only possible if phytoplankton photosynthesize carbohydrates at the surface layer where there is no nitrate, producing oxygen and consuming CO_2, and take up nutrients at depth without producing oxygen but releasing CO_2. Such behavior is only possible in species which migrate daily from the surface layer during the day to deeper water at night. Carbohydrates would be synthesized during the day, which would supply the energy necessary for basal metabolism and migration, which is a small fraction of the total energy requirements [17], and for subsequent acquisition of nutrients and synthesis of structural compounds.

The hypothesis of a vertical segregation between photosynthesis and nutrient assimilation is reinforced by the distribution of "CO" on 3 October (Fig. 3), where the deep maximum (1540 µmol/kg) corresponds to the "NO" minimum. Gymnodinium catenatum was the only species known to exhibit migratory behavior and sufficientely abundant to generate these distributions.

Species with this behavior must be favoured in environments like that described here, where a nutricline is established at such a depth that they can reach it by diurnal migration. Its concentration in the upper layer during the day can provoce shading, preventing the light reaching depths where nutrients are and so eliminating diatom competition.

ACKNOWLEDGEMENTS

We are grateful to all participants in the cruise Galicia-9 of the RV "Garcia del Cid". This work was supported by the grant 469 of the Comisión Asesora de Investigación Científica y Técnica.

REFERENCES

1. M.J. Cámpos, S. Fraga, J. Mariño and F.J. Sánchez, ICES, C.M. C.M./L:27.
2. F.G. Figueiras, F.X. Niell and C. Mouriño, Inv. Pesq. 50, 97-115 (1986).
3. F.G. Figueiras and F.X. Niell, Inv. Pesq. 51, 293-320 (1987).
4. R.W. Eppley, O. Holm-Hansen and J.D.H. Strickland, J. Phycol. 4, 333-340 (1968).
5. D. Kamykowski and S.J. Zentara, Limnol. Oceanogr. 22, 148-151 (1977).
6. S. Yamochi and T. Abe, Marine Biology 83, 255-261 (1984).
7. S.I. Heaney and R.W. Eppley, J. Plank. Res. 3, 331-344 (1981).
8. J.J. Cullen and S.G. Horrigan, Marine Biology 62, 81-89 (1981).
9. J.J. Cullen, M. Zhu, R.F. Davis and D.C. Pierson in: Toxic Dinoflagellates, D.M. Anderson, A.W. White and D.G. Baden eds. (Elsevier, Amsterdam 1985) pp. 189-194.
10. J.J. Cullen in: Migration: Mechanisms and Adaptative Significance, M.A. Rankin, ed. (Contr. Mar. Sci. 27 1985) pp. 135-152.
11. M. Watanabe, K. Kohata and M. Kunigi, J. Phycol. 24, 22-28 (1988).
12. W.S. Broecker. Earth and Planetary Sci. Lett. 23, 100-107 (1974).
13. F.G. Figueiras, R. Prego, F.F. Pérez, A.F. Rios and F. Fraga, Datos Informativos Inst. Inv. Mar. de Vigo 21, 1-129 (1987).
14. A.F. Rios and F. Fraga, Scienta Mar. in press (1989).
15. F.G. Figueiras and F.X. Niell, Inv. Pes. 51, 371-409 (1987).
16. F. Fraga and R. Prego in: Las purgas de mar como fenómeno natural. Las mareas rojas, F. Fraga and F.G. Figueiras, eds. (Sem. Est. Galeg. in press).
17. J.A. Raven and K. Richardson, New Phytol. 98, 259-276 (1984).

GYMNODINIUM CATENATUM BLOOM FORMATION IN THE SPANISH RIAS

SANTIAGO FRAGA, BEATRIZ REGUERA AND ISABEL BRAVO
Instituto Español de Oceanografia.
Apdo. 1552, 36280 Vigo, Spain

ABSTRACT

At least two mechanisms are responsible for Gymnodinium catenatum blooms in the rias of NW Spain. Coastal upwelling is the most relevant hydrographic feature in this area, so bloom formation and development largely depend on it. Bloom data from 1985 to 1988 related to meteorological conditions suggest a strong relation between upwelling relaxation and bloom development, at the end of the summer and beginning of autumn. During summer, warm nutrient depleted surface water was found offshore, while near the coast upwelling kept coastal waters cold and rich in nutrients and diatom populations predominated. When upwelling relaxes, offshore water moved towards the coast and caused an increase in temperature near the coast, and a change from diatom to flagellate populations. Once an initial population of a bloom species had been concentrated by advection of surface water into areas of convergence, short periods of upwelling appear to have provided injections of nutrients which produced conspicuous blooms, as was the case in 1986. If these upwelling pulses do not occur, high concentrations of algae are not recorded, but if the bloom species are toxic, they will still be sufficient to produce shellfish toxicity. When oceanographic conditions are different from those just described, a different bloom mechanism may be involved, as was the case in August 1988. On that ocassion, a bloom of G. catenatum may have been related to the presence of a cyst population. These assumptions have been used successfully in "red tide" forecasts in recent years to give warnings of shellfish toxicity to health officials and mariculturists.

INTRODUCTION

Since 1976, toxic outbreaks of PSP caused by the chain-forming dinoflagellate Gymnodinium catenatum Graham have been the cause of numerous closures of shellfish harvesting in Galicia (NW Spain), especially the farmed blue mussel with an estimated annual production of more than 200.000 Tm. The importance of this resource required prediction of the possible occurrence of toxic algae, to prevent damage to public health and large economic losses.

The western coastal waters of the Iberian Peninsula experience wind-driven upwelling (1,2). This is part of the general upwelling system at the Eastern North Atlantic (3) and is the cause of high primary production in the area. On the coast of Galicia, upwelling occurs from June to October and this seasonality together with that of insolation determines phytoplankton productivity and succession. The correlation between wind stress and coastal upwelling intensity in this area is well documented. Blanton et al. (4) showed good correlation between an upwelling index, based on Ekman transport derived from surface winds (5), and sea temperatures in the Rias. The same correlation was found by Fraga et al. (6) who described blooms of G. catenatum and A. affine in the Ria de Vigo in 1985. In this case, when the prevailing north winds that induced upwelling during summer changed to south winds and caused a relaxation of upwelling, the offshore surface water that had been warmed through the summer moved towards the coast and created

Toxic Marine Phytoplankton
Edna Graneli et al., Editors

downwelling areas. This warm water carried established populations of dinoflagellates, some of them typical warm water species, that were concentrated in the rias. These conditions are favourable for the survival of efficient swimmers such as chain forming dinoflagellates which can overcome downwelling (7). Then, the concentration of the toxic G. catenatum became high enough to produce toxicity in shellfish and an embargo on the marketing of molluscs was declared. (FIG 1). This mechanism of bloom formation was repeated in subsequent years with some variations that are the subject of this paper.

The region of intense mussel farming is a group of rias which usually show positive estuarine circulation. Surface waters flow out of the rias while at depth there is an input of water that in winter is mixed surface water and during summer is nutrient rich upwelled water. Part of the biomass produced in the rias is exported and subsequently sinks and remineralizes, enriching the bottom water even more (1). During summer, the commonest red tide species in the rias is the ciliate Mesodinium rubrum (8) a typical red tide species from upwelling regions (9,10). However, when upwelling relaxes at the end of the summer, dinoflagellate blooms are a common phenomenon in the Rias Bajas.

RESULTS AND DISCUSSION

Fig. 1a shows the evolution of water temperature at a station near the mouth of the Ria de Vigo during the autumn of 1985. The sharp increase in temperature during the first week of November was due to the entry of offshore water driven by southerly winds. The bloom of G. catenatum clearly coincides with this event (Fig. 1b), (6).

In September 1986, as in November 1985, wind induced upwelling relaxed in a similar way. The progressive increase in water temperature inside the Ria (Fig. 2a) was caused by the entrance of warm water from offshore. That warm offshore surface water brought an initial population of G. catenatum to the Rias where it was concentrated following the same mechanism as that already described for the same area during autumn 1985. Once an initial bloom had developed, (Fig. 2b) a weak upwelling event injected cold water along the bottom of the Rias that did not reach the surface. This phenomenum caused the water of the Rias to be thermally stratified, and very rich in nutrients in the bottom layers (11). Although this upwelling event was the cause of a decrease in cell concentrations, it was not enough to completely disperse the bloom. Upwelling of bottom water ceased with the onset of calms. But although the upwelled water flowed to the deeper layers of the shelf, due to its density, there was enough time for a big bloom to develop due to the high initial population of G. catenatum. Once the bloom developed, it showed some variations due to water movement, and declined slowly.

In the autumns of 1987 and 1988, (Figs. 3, 4) G. catenatum blooms responded to the same mechanism: when upwelling relaxed, the input of cold water from the shelf stopped, and the water temperature inside the Rias, increased due to the entrance of warm offshore surface water, and concentrated good swimmers, such as chain-forming dinoflagellates, in areas of convergence.

The close relationship between winds, sea temperature and dinoflagellate blooms has been used successfully to give warnings to health authorities when the danger of a toxic dinoflagellate bloom looks high. As can be seen in Figs. 1 to 4 the quarantine period instigated by toxicity levels higher than 80 µg of saxitoxin equivalent per 100 g of shellfish meat lags the temperature patterns, and these in turn lag the weather patterns.

Nevertheless, during summer 1988, a bloom of G. catenatum was recorded (Fig. 4) in very different conditions from those described for the autumn blooms. In this case, bottom temperatures were relatively low and corresponded to typical summer upwelling conditions, while the upper layers were warm with some periods of cooling when upwelling was stronger. The

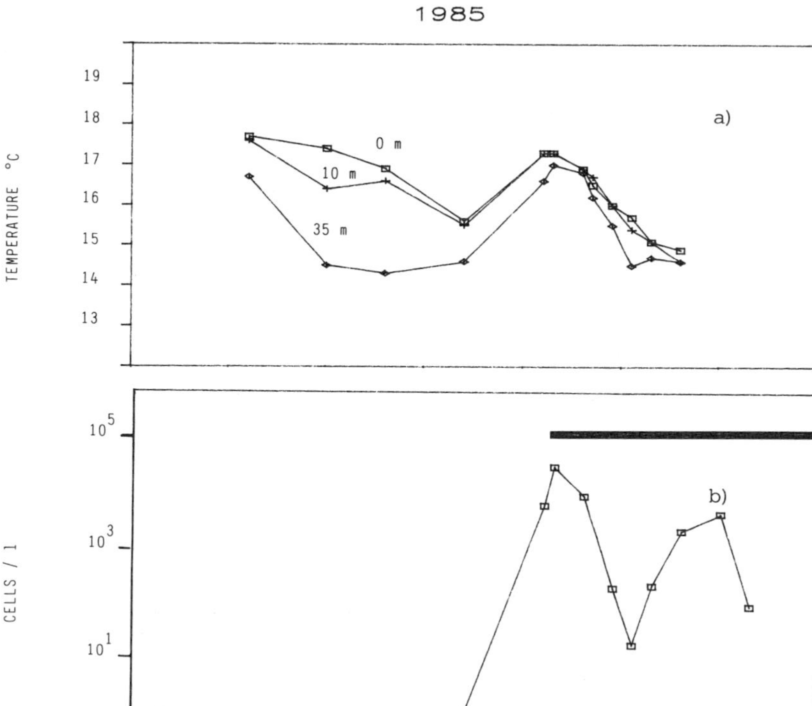

FIG. 1. a) Temperature evolution at a station near mouth of Ria de Vigo. b) G. catenatum cell concentration averaged from discrete samples at 0m, 5m, 10m and 20m. Bar shows shellfish quarantine period.

distribution of the cells in the water column was also different from the autumn bloom. During the autumn blooms, the maximum cell concentrations are in the top 5 meters, and the fluorescence maximum is usually at about 1m, and coloured patches sometimes appear at the surface. But in summer 1988 the maximum cell concentrations were mainly between 10 and 15m when surface waters were dominated by diatoms. We do not have light measurements, but the reason for the different behavior patterns could be the higher insolation of summer compared with autumn, such that G. catenatum cells have enough light at greater depths. It was also observed that the summer bloom lasted longer (80 days), and did not show large variations in cell density, while autumn blooms are usually more ephemeral with sharper increases and decreases in numbers. In the autumn blooms, the environmental conditions were much more variable than during the summer, and the increases and decreases in numbers depended mainly on processes such as advection and dilution, while the summer bloom might be more related to in situ growth and reseeding by benthic cysts (12,13).

From the scarce literature concerning this species, it looks as though it may dominate in different environments through different mechanisms of bloom formation. Hallegraeff et al. (14) stated that the the growth patterns of G. catenatum in southern Tasmania are markedly different from those in Spain and Mexico. Recent reports of blooms of this dinoflagellate at Senzaki

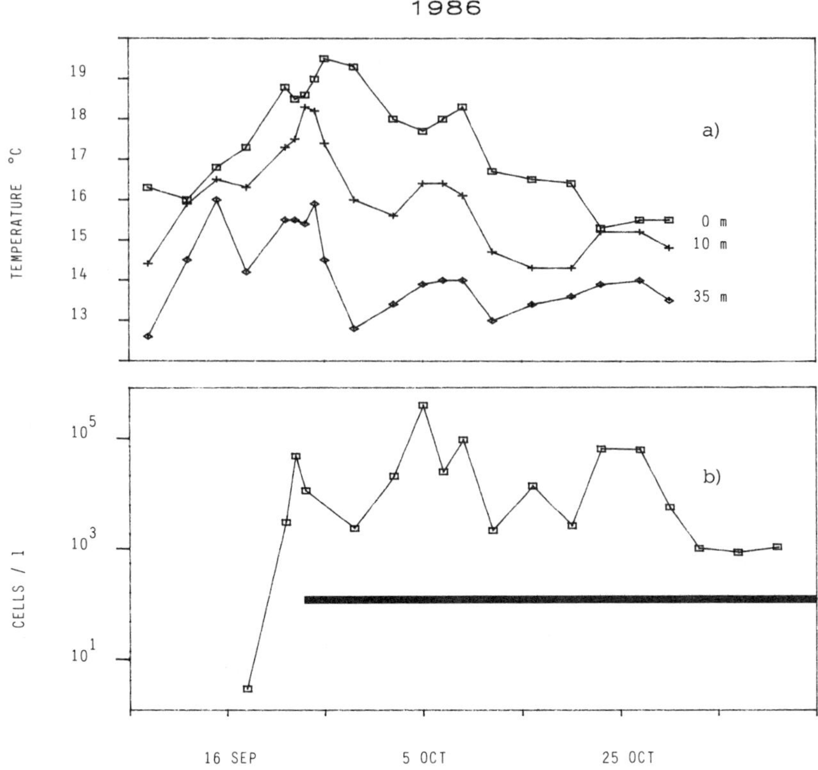

FIG. 2. a) Sea temperature at the same station as in Fig. 1. b) G. catenatum cell concentration at surface. Bar shows shellfish quarantine period.

Bay (Japan) (15), at a euhaline lagoon in Italy (16), and in southern Spain (17) are examples of the ability of G. catenatum to develop blooms in different environments. The comparative study of these environments and the behavior of G. catenatum blooms, will give a better understanding of the different mechanisms.

ACKNOWLEDGEMENTS

This paper was supported by funds of the Instituto Español de Oceanografia and by the USA-Spain Joint Committee for the Scientific and Technological Cooperation, grant No. CCA-8411089. The help of the crew of the R/W Jose Maria Navaz is gratefully acknowledged.

REFERENCES

1. F. Fraga in: Coastal Upwelling. F.A. Richards ed. (American Geophysical Union, Washington, 1981). pp. 176-182.
2. A.F.G. Fiuza in: Coastal Upwelling. E. Suess and J. Thiede eds. (Plenun Publishing corporation, 1983).
3. W.S. Wooster, A. Bakun, and D.R. McLain. J.Mar.Res., 34(2), 131-141.

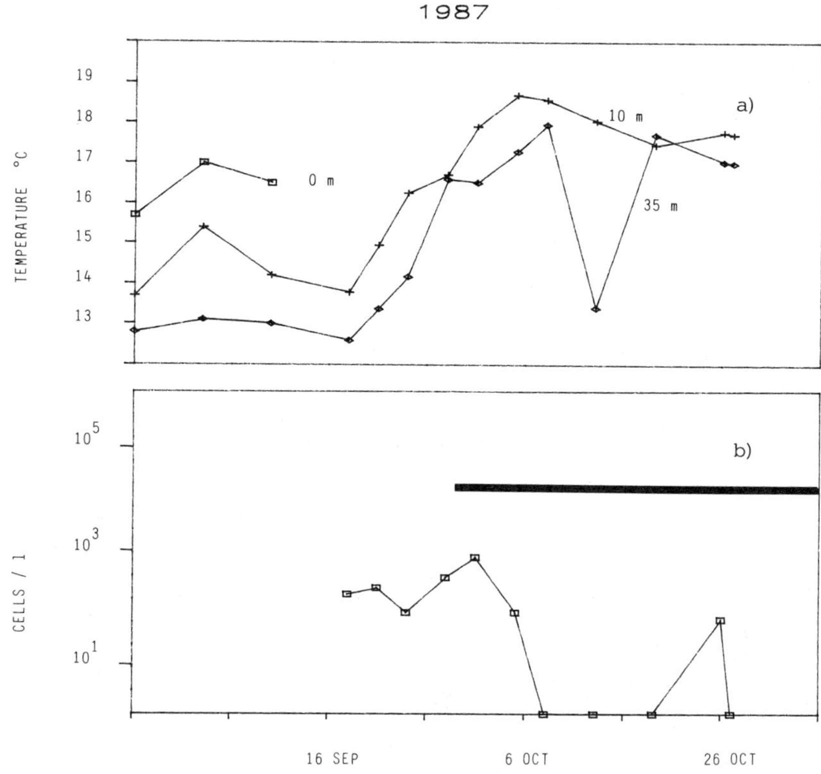

FIG. 3. a) Sea temperature at the same station as in Fig. 1. b) G. catenatum cell concentration integrated from 0m to 15m. Bar shows shellfish quarantine period.

(1976).

4. J.O. Blanton, L.P. Atkinson, F. Fernandez de Castillejo and A. Lavin Montero. Rapp. P.-v. Reun. Cons. int. Explor. Mer, 183, 79-90 (1984).
5. A. Bakun. NOAA Tech.Rep., NMFS SSRF-671, U.S. Dept. of Commerce, 103 pp. (1973).
6. S. Fraga, D.M. Anderson, I. Bravo, B. Reguera, K.A. Steidinger, C.M. Yentsch. Estuar. Coast. Shelf Sci. 27,349-361 (1988).
7. S. Fraga, S.M. Gallager and D.M. Anderson. in: Red Tides: Biology, Environmental Science, and Toxicology. T. Okaichi, D.M. Anderson and T. Nemoto eds. (Elsevier Science Publishing, 1989) pp. 281-284.
8. S. Fraga in: Las purgas de mar como fenomeno natural; Las mareas rojas. F. Fraga and F.G. Figueiras eds. Seminario de Estudos Galegos. (1989).
9. D. Blasco. in: Toxic Dinoflagellate blooms, V.R. LoCicero, ed. (Mass. Sci. and Technol. Foundn., Wakefield, 1975). pp. 113-119.
10. R. Margalef, M. Estrada and D. Blasco. in: Toxic Dinoflagellate Blooms, D.L. Taylor and H.H. Seliger eds. (Elsevier North-Holland, New York 1979). pp. 89-94.
11. R. Prego, F.F. Perez, A.F. Rios, F. Fraga and F.G. Figueiras. Datos Informativos Inst. Inv. Marinas 23, 106 pp. (1988).
12. I. Bravo. Inv. Pesq. 50(3), 313-321 (1986).
13. D.M. Anderson, D. Jacobson, I. Bravo and J.H Wrenn. J.Phycol. 24, 255-

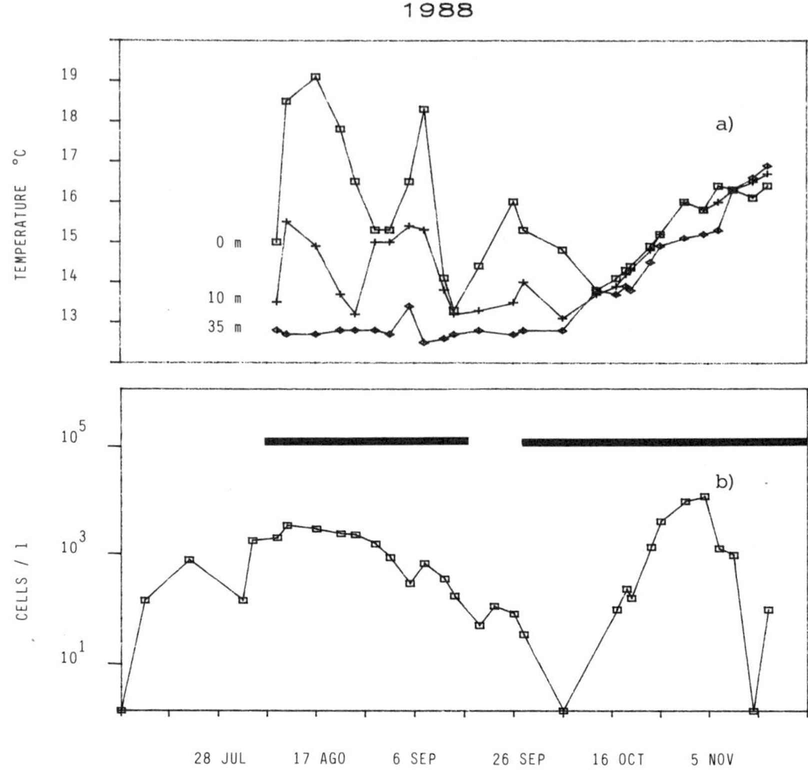

FIG. 4. a) Sea temperature evolution in 1988 at the same station as in Fig. 1. b) G. catenatum cell concentration integrated from 0m to 15m. Bar shows shellfish quarantine period.

262 (1988).

14. G.M. Hallegraeff, S.O. Stanley, C.J. Bolch and S.I. Blackburn. in: Red Tides: Biology, Environmental Science, and Toxicology. T. Okaichi, D.M. Anderson and T. Nemoto eds. (Elsevier Science Publishing. 1989). pp. 77-80.
15. T. Ikeda, S. Matsuno, S. Sato, T. Ogata, M. Kodama, Y. Fukuyo and H. Takayama. in: Red Tides: Biology, Environmental Science, and Toxicology. T. Okaichi, D.M. Anderson and T. Nemoto eds. (Elsevier Science Publishing. 1989). pp. 411-414.
16. C.G. Carrada, R. Casotti and V. Saggiomo. Rapp. Comm. int. Mer Medit., 31,2 (1988).
17. I. Bravo, B. Reguera, A. Martinez and S. Fraga. (This Conference).

HUMIC COMPOUNDS AND GROWTH RESPONSE OF PHYTOPLANKTON DOMINATED BY DINOFLAGELLATES (LATE SPRING BLOOM): OBSERVATION IN COASTAL WATERS OF THE SOUTHERN BALTIC

DANUTA GĘDZIOROWSKA,* AND MARCIN PLIŃSKI**
* Institute of Oceanology, Polish Academy of Sciences, Powst. Warszawy 55, 81-967 Sopot, Poland; **Centre of Marine Biology, Polish Academy of Sciences, Czołgistów 46, 81-378 Gdynia, Poland.

ABSTRACT

Exceptional phytoplankton blooms, dominated by dinoflagellates, have been observed recently in the Bay of Gdańsk (southern Baltic Sea). This work presents the results of studies on the influence of humic type substances, isolated from sea water, upon growth rate and the species composition of the natural phytoplankton samples collected in the Bay of Gdańsk during the phytoplankton bloom in May - June 1988. The results indicate that the humic compounds isolated from the waters of the Bay of Gdańsk enhanced the chlorophyll *a* and carotenoid concentrations, as well as the number of cells in the cultures. On the basis of the results we may conclude that the increased concentrations of humic compounds in coastal sea waters can play an important role in phytoplankton life as a selective and growth promoting factor. Besides, it has been stated that the humic compounds stimulated phytoplankton growth much more effectively in the presence of the major nutrients i.e. nitrates and phosphates.

INTRODUCTION

In light of the dinoflagellate blooms occured in the North Sea and the Danish Belt we intensified our attention on the exceptional phytoplankton bloom phenomena in the Polish coastal zones of the Baltic Sea, especially the Bay of Gdańsk. The Bay of Gdańsk area is very much endangered by different inorganic and organic compounds, including dissolved humic type substances (DHS) delivered mainly by the Vistula River.

During the last decade, in the Bay of Gdańsk a cyclical appearance of phytoplankton blooms, has been observed namely:
an early spring diatom bloom; a summer bloom strongly dominated by blue-green algae (mainly by *Aphanizomenon flos-aquae* and *Nodularia spumigena*); an autumn bloom, with

Toxic Marine Phytoplankton
Edna Graneli et al., Editors

distinct dominance of the centric diatom *Coscinodiscus granii*. Recently, we have also observed the appearance of a strange (to some extent) and a new phytoplankton bloom which occured in May - June, and which was dominated by the dinoflagellate *Heterocapsa triquetra*. It has been noted, however, that in the waters of the southern Baltic, including the Bay of Gdańsk, the phytoplankton bloom species are presently not toxic to fish.

Previous results showed that DHS isolated directly from the sea and fresh waters enhanced the growth rate of the marine algal monocultures and phytoplankton communities. The stimulating effects of DHS both at the natural concentrations found in sea and river waters (and higher concentrations) have been observed [1]. It has been noted that DHS affected the volutine accumulation and increased the chloroplast number in the algal cells [1]. It also influenced the alkaline phosphatase activity [1,2,3], modified the elementary composition of algae cells [1,2], as well as enhanced the uptake of iron ions [1]. Ortner et al. [4] showed in their experiments that isolated natural marine fulvics and marine fulvics synthesized in the laboratory affected the bioavailability of trace metals to marine phytoplankton.

The aim of our studies presented here was to determine the influence of dissolved humic compounds isolated from the Bay of Gdańsk waters on the natural phytoplankton communities, collected from the same area, during the May - June 1988 phytoplankton bloom. Studies were performed in the presence of enriched levels of nitrate and phosphate. The work is a continuation of the studies carried out to evaluate interactions between major nutrients taken up by algae in the presence of humic compounds in the growth medium.

MATERIALS AND METHODS

Origin of the studied materials

Sea water samples with indigenous phytoplankton communities for algal assay experiments and surface sea water for DHS isolation were collected from the Bay of Gdańsk (Sopot pier area), May - June 1988. The natural concentration of DHS during the studied period was 2.30 mg/dm^3. DHS concentrations have also been detected in the two estuaries entering the Bay of Gdańsk waters, i. e. Vistula Estuary (6.30 mg/dm^3) and Puck Bay (4.48 mg/dm^3). The initial concentrations of nitrate (NO_3) and phosphate (P) were 0.28 μmol/dm^3 and 1.92 μmol/dm^3, respectively.

Isolation of DHS

An extraction scheme using Amberlite XAD-2 according to [5,6] was applied for the concentration of humic substances:
1) Sea water samples were filtered through GF/F glass fibre filter and the pH was lowered afterwards to 2.0 with

concentrated HCl.

2) Acidified sample was passed through column of XAD-2 at an average flow 0.2 "bed volume"/min. Aquatic humic compounds were adsorbed to resin (XAD-2 resin Soxhlet extraction with acetone, prior to use).

3) Elution of DHS from XAD-2 resin was carried out in reverse direction using 1 molar aqueous ammonia. Before elution, the charged resin was rinsed with pure water, or 0.01 molar HCl, to remove salts. NH_4OH was evaporated and humic extract was rinsed with pure water.

4) Elementary analysis of DHS was performed using a Carlo Erba C,H,N analyser. The isolated DHS (from the local Sopot pier water) contained 41% C and 5.8% N, giving a C/N (atomic) ratio of 7:1.

5) For algal assay experiments, DHS was provided in aqueous phase.

Algal assay experiments

Sea water with indigenous phytoplankton communities was filtered through 100 μm nylon net to remove large zooplankters. The phytoplankton was cultivated in 3 l glass cylinders containing 2 l of culture medium (local sea water with salinity about 8‰), enriched with DHS (10 mg/dm^3) and basic nutrients as $NaNO_3$ (60 μmol/dm^3), K_2HPO_4 (1.5 μmol/dm^3).

Algae were incubated under controlled laboratory conditions at 17°C on a 16:8 hrs light:dark cycle with a photon flux density (PFD) of 130 μE m^{-2}s^{-1} from OSRAM cool-white fluorescent tubes. PFD was measured using the phytophotometer FF-01 "SOMOPAN".

After 8 days of algal cultivation, the chlorophyll *a* [7] and total carotenoid [8] concentrations were determined. Cell numbers were counted in a Palmer Maloney chamber calibrated against the Utermöhl method [9].

RESULTS AND DISCUSSION

The influence of DHS and major nutrients on the growth of the natural phytoplankton communities is shown in Figure 1. Maximal chlorophyll *a*, total carotenoid and cell number concentrations were detected in the samples enriched with DHS together with nitrate (N) and phosphate (P). The initial chlorophyll *a* and carotenoid levels in the study area were 2.6 and 1.5 mg/m^3, respectively. After 8 days of cultivation, the amounts of chlorophyll *a* and carotenoids increased to 51.7 and 13.6 mg/m^3 in the treatments containing P+N+DHS. In every case, the enrichment of culture medium with DHS (DHS alone, N+DHS, P+DHS, P+N+DHS) enhanced the cell number, chlorophyll *a* and total carotenoid concentrations more than in the control samples and in samples enriched with nitrate (N) and phosphate (P) alone. In experiments performed earlier [12], we showed that the addition of DHS alone to the natural phytoplankton cultures caused almost the same increase of chlorophyll *a* and cell number as in the samples enriched with nitrate alone. The latter has been observed for the three periods, i. e. autumn 1984, spring and summer 1985. According to Prakash and Rashid [10], the growth level of the algae in the presence of humic acids is dependent of major nutrients. Rashid [11] has pointed out that information concerning the interaction of humic compounts with major nutrients is limited with regard to the influence on phytoplankton responses in sea water.

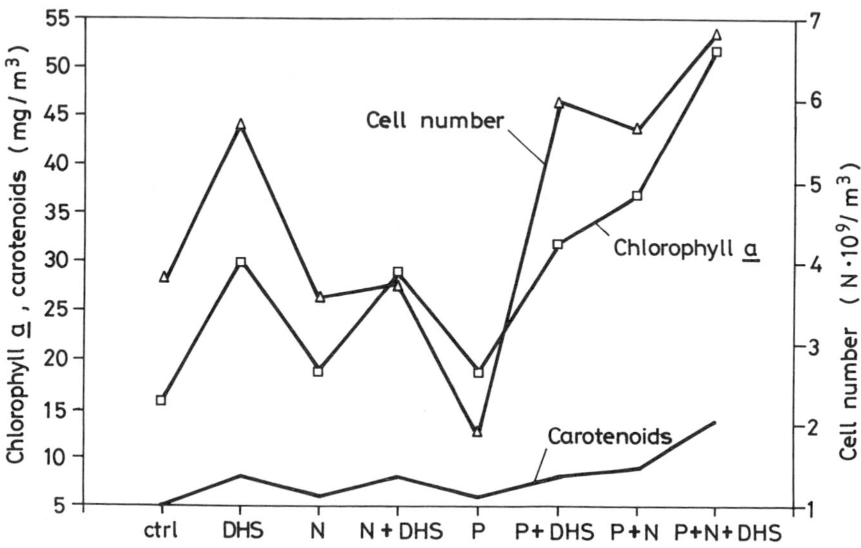

FIG. 1. Influence of DHS and nutrients (N,P) on the growth of natural phytoplankton communities in the Bay of Gdańsk waters. Concentrations used - DHS: 10mg/dm^3, $NaNO_3$: 60µmol/dm^3, K_2HPO_4: 1.5µmol/dm^3.

Figure 2 illustrates the responses of individual phytoplankton species belonging to different taxonomic groups in the phytoplankton community after 8 days of cultivation. The data indicate highest growth stimulation characterized the autotrophic flagellates, represented mainly by *Eutreptia viridis*. Lower responses characterized the diatoms, dominated by *Sceletonema costatum*. These results confirm the previous findings [12]. The green algae were dominated by *Scenedesmus quadricauda* and *Ankistrodesmus falcatus*. Within these three groups of phytoplankton, we observed increased cell numbers in the samples enriched with DHS, N+DHS and P+DHS. It is interesting to note the limited response of the dinoflagellates (after 8 days of cultivation), which were represented by *Heterocapsa triquetra* and *Diplopsalis sp*. Their behaviour was contrary in comparison to that of the other three groups of algae, i.e. autotrophic flagellates, diatoms and green algae. It is necessary to recall that dinoflagellates were the dominat group in the study area, i.e. at the start of the experiments. In the case of dinoflagellates, the highest cell numbers were detected in the control samples and in the samples enriched with nitrate and phosphate. Few blue-green algae were noted.

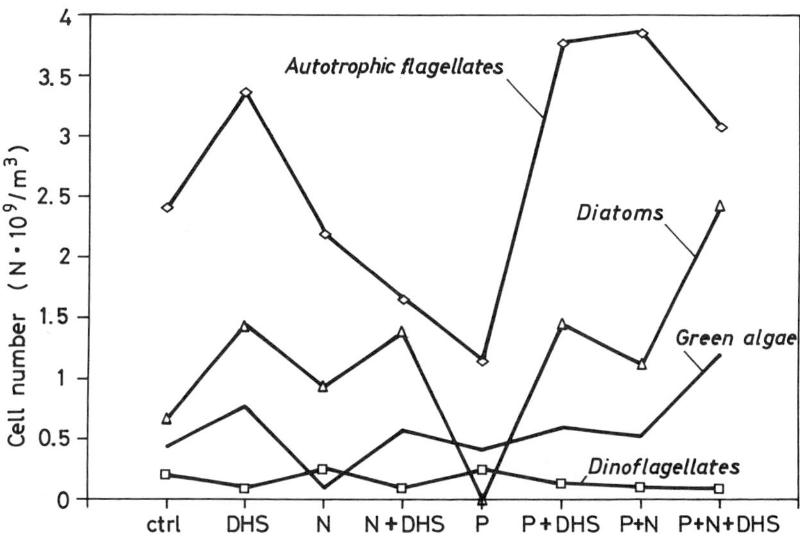

FIG. 2. Share of the phytoplankton fractions after 8 days of cultivation. DHS and nutrients concentrations cf. FIG. 1.

The environmental observations in coastal waters and the data obtained in the laboratory experiments showed that humic type substances, together with major nutrients, can stimulate phytoplankton blooming. According to Rashid [11], humic type substances and trace elements appear to be responsible for the biological conditioning of coastal waters in addition to other chemical and physical environmental factors. These biological conditions can initiate directly, or indirectly, a red tide.

In conclusion, it can be said that our experiments during the May-June phytoplankton bloom support results obtained earlier [1,2,10,11,12]. On the other hand, they indicate the necessity to undertake further research in order to elucidate the mechanisms of the observed interactions between the humic type substances, major nutrients and phytoplankton organisms. These interactions have great relevance in shallow sea waters, where organic and inorganic effluents can be observed to affect the existing biological life.

ACKNOWLEDGEMENTS

We wish to thank Mrs. Olimpia Bławat for her technical assistance. This work was financially supported by Polish Academy of Sciences, CPBP 03.10 project.

REFERENCES

1. D. Gędziorowska, Ph.D.Thesis, Gdańsk University (1986).
2. E. Granéli, L. Edler, D. Gędziorowska, and U. Nyman in: Toxic Dinoflagellates, D.M. Anderson, A.W. White, and D.G. Baden, eds.(Elsevier, New York, Amsterdam, Oxford 1985) pp. 201-206.
3. A.J. Stewart, and R.G. Wetzel, Freshwater Biol. 12, 369-380 (1982).
4. P.B. Ortner, C. Kreader, and G.R. Harvey, Nature 301, 57-59 (1983).
5. R.F.C. Mantoura, and J.P. Riley, Anal. Chim. Acta 76, 97-106 (1975).
6. D.H. Stuermer, and G.R. Harvey, Deep-Sea Research 24, 303-309 (1977).
7. S.W. Jeffrey, and G.F. Humphrey, Biochem.Physiol. Pflanzen (BPP) 167, 191-194 (1975).
8. J.D. Strickland, and T.R. Parsons, Fish. Res. Board Can. Bull. 125, 117-127 (1965).
9. L. Edler, Recommendations in methods for marine biological studies in the Baltic Sea. BMB Public. 5, 6-23 (1979).
10. A. Prakash, and M.A. Rashid, Limnol.Oceanogr. 13, 598-606 (1968).
11. M.A. Rashid, Geochemistry of Marine Humic Compounds, (Springer-Verlag 1985) pp. 248-272.
12. D. Gędziorowska, and M. Pliński, Kieler Meeresforsch., Sonderh. 6, 256-264 (1988).

A THEORETICAL CASE OF COMPETITION BASED ON THE ECTOCRINE PRODUCTION BY GYRODINIUM CF. AUREOLUM

P. GENTIEN and G. ARZUL
IFREMER - Centre de Brest - DERO/EL B.P. 70 - 29263 Plouzané - France

ABSTRACT

Gyrodinium cf. aureolum produces ectocrines which inhibit diatom growth. A simple theoretical model taking into account this ectocrine production was developed in order to simulate inhibitory exclusion. It is shown that final dominance of the ectocrine producer depends on two threshold values : the inoculum concentration at initial time and the specific toxicity production rate. Oversimplification of the system precludes its direct utilization.

INTRODUCTION

Since the late sixties, summer outbreaks of Gyrodinium aureolum have been implicated in massive kills of marine fauna and in particular in the Brest Bay (1). These blooms are nearly monospecific (2, 3) for unknown reasons. It has been suggested (4) that monospecific blooms are developed in coastal zones by lack of predation, inhibitory exclusion, or competitive advantage through physiological rates. Release of toxic substances by dinoflagellates is well documented (5). In this paper, we examine the inhibition of diatoms by sea water in which G. aureolum has been growing in the field or in culture. As the inhibitory exclusion seems to be a founded hypothesis in the case of G. aureolum, a theoretical model was developed in order to test the implications of such a process in the competition between an ectocrine producer (hereafter named E.P.) and an alga such as a diatom in a very simplified environment.

MATERIALS AND METHODS

Algae

A unialgal strain of G. cf. aureolum (6) was isolated in July 1987 from aquaculture tanks naturally contaminated by G. cf. aureolum. This strain was kindly provided by Dr F. Partensky (Roscoff Marine Station). All sea water used as growth medium for maintenance was purified on C18 and Florisil adsorbents in order to remove respectively lipophilic compounds and G. aureolum ectocrines. Cultures in F/2 medium (7) were performed at 18°C under dark and light (fluorescent lamps) cycles (12 : 12) at 50 μmol$\cdot m^{-2} \cdot s^{-1}$. Chaetoceros gracile was grown in the same conditions.

Bioassays

The growth rate of C. gracile was measured in F/2 medium using different sea waters. G. aureolum growth media (either natural or laboratory) were gravity filtered through glass fiber filter (Whatman GF/F).
Blanks were performed with either sea water without G. aureolum or with the water to be tested passed through a Florisil cartridge. Bioassays were conducted under the same conditions as the algal cultures. Algal biomass was estimated by fluorescence (arbitrary units). This technique was found satisfactory for cells in exponential growth phase.

RESULTS

A rapid change (2 days) was observed in Brest Bay in July 1987; from a Chaetoceros sp. dominated population to a population dominated by G. cf. aureolum. G. cf. aureolum, reached a 95 % dominance with more than $3 \cdot 10^5$ cells $\cdot l^{-1}$ and was only accompanied by Prorocentrum micans. This situation lasted a little more than one week.

Seawater containing between 3-4.3·10dinoflagellate cells per liter wasinoculated with Chaetoceros gracile after gravity filtration and enrichment G. cf. aureolum water delayed the growt of this diatom when compared to blanksmade up with water collected outside of the bay. Replicate samples of Brest Bay waters were treated on adsorption cartridges (Florisil: MgO, SiO_2 (16 : 84)) before enrichment.These bioassays did not show any diatom growth inhibition. A typical bioassay result is presented in Fig. 1. A group of

Toxic Marine Phytoplankton
Edna Graneli et al., Editors

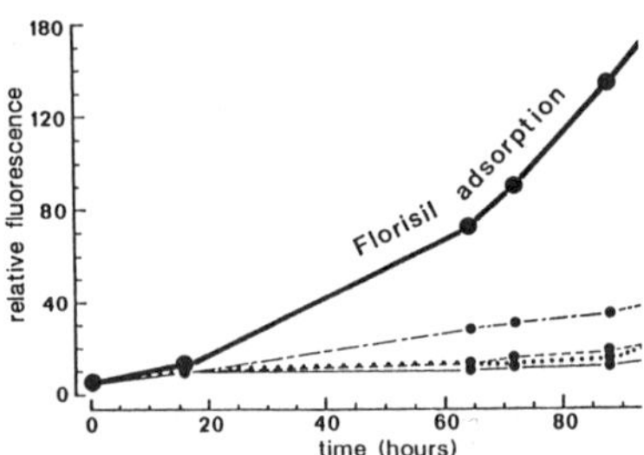

FIG. 1. Typical response curve of a C. gracile experiment. Fluorescence is expressed in arbitrary units. Broken lines represents growth with water containing between 3 and 4.3 $\cdot 10^5$ G. aureolum cells$\cdot l^{-1}$.

compounds present in the water inhibited diatom growth and could be removed on a polar adsorbent. A strain of G. cf. aureolum isolated during this July 1987 event was cultured in order to identify the role of this dinoflagellate in the diatom inhibition. As the active compounds were retained on Florisil, dinoflagellate growth medium wasprepared with sea water treated with this adsorbent. The dinoflagellate culture was then collected while exponentially growing ($1.6 \cdot 10^4$ cells$\cdot l^{-1}$) and C. gracile tests performed as described above. Blanks were prepared with growth medium after Florisil treatment. The diatom inhibition was still observed, strongly suggesting that G. cf. aureolum exuded ectocrines detrimental to the growth of C. gracile. The toxic principle has been isolated on HPLC silica column and contains at least two compounds which present hemolytic activity. Detailed procedures of the chemical analysis will be published elsewhere.

These bioassays conducted with overenriched media do not allow any extrapolation to the ecological significance of the ectocrine production. However, field samplings, in 1988, showed that samples with low densities of G. cf. aureolum ($1.6 \cdot 10^5$ cells $\cdot l^{-1}$) were still active against C. gracile. These lower densities give some relevance to a possible ecological significance of the ectocrine production. Further complications arise because high variabilities in the specific ectocrine rate of production occurred during the culture of G. cf. aureolum in the laboratory. Environmental conditions regulating this rate of production are unknown at present.

Unfortunately, we were unable to test this ectocrine effect on diatoms in a bialgal culture with nutrient concentrations representing a coastal ecosystem because, at that time, the dinoflagellate in culture had a very low ectocrine production rate.

In order to understand how an ectocrine may give an advantage through competitive exclusion of another species, a simple theoretical model was developed. It takes into account the maximum growth rate, the halfsaturation constant for nitrate and a constant antagonistic effect and it is supposed that all other factors are optimum. Basic equations are the following for diatom (Diat), ectocrine producer (EP), nitrate (N), detritus (Det) :

$$\frac{d(Diat)}{dt} = \left(\frac{\mu 1 \times N}{K1 + N} - \delta - \frac{Q}{V} - \gamma \,.\, EP\right) Diat$$

$$\frac{d(EP)}{dt} = \left(\frac{\mu 2 \times N}{K2 + N} - \delta - \frac{Q}{V}\right) EP$$

$$\frac{d(N)}{dt} = - \frac{\mu 1 \times N \times Diat}{K1 + N} - \frac{\mu 2 \times N \times EP}{K2 + N} + r \times Det + \frac{Q}{V}\,(N_0 - N)$$

$$\frac{d(Det)}{dt} = \delta \times Diat + \delta \times EP + \gamma \times Diat \times EP - r \times Det - \frac{Q}{V} \times Det$$

with:

$\mu 1$ = diatom maximum growth rate (2.5 d^{-1})
$K1$ = diatom half saturation constant (1.1 µmol $NO_3 \cdot l^{-1}$)
$\mu 2$ = EP maximum growth rate (1.3 day^{-1}) as estimated from our cultures
$K2$ = EP half saturation constant (3.5 - 8.5 µmol $NO_3 \cdot l^{-1}$)
δ = natural mortality rate ($5 \cdot 10^{-2}$ day)
Q/V = flushing rate (0.2 - 0.6 day^{-1})
r = remineralization rate (10^{-2} day^{-1}).
γ = toxicity factor per EP cell (variable 10^{-3} to 10^{-7} $day^{-1} \cdot cell^{-1}$)
N_0 = inflow nutrient concentration (1 - 20 µmol $NO_3 \cdot l^{-1}$).

Nitrogen to cell conversions factors are $6 \cdot 10^{-7}$ and $4 \cdot 10^{-6}$ for diatom and E.P. respectively (8).

For a given flushing rate and a constant nutrient inflow concentration, the nature (diatom or ectocrine producer) of the stable equilibrium depends only on the number of the E.P. inoculum (Figs. 2 and 3).

A plot of the time necessary to obtain 80 % dominance by the E.P.versus the number of inoculum cells is presented below for different nutrient inflow concentrations (Fig. 4). Each of these curves presents an asymptotic limit, showing that under a certain inoculum concentration, the ectocrine producer cannot overcome the other population. This effect can also be seen on Fig. 5 which represents the threshold surface for final E.P. dominance.

Above the surface, E.P. can achieve final dominance. This graph demonstrates clearly that the limit inoculum for E.P.dominance decreases continuously with the residence time and increases with the nutrient inflow concentrations. If the E.P. half-saturation constant for nitrate is lowered from 8.5 to 3.5, the same behaviour can be reproduced with a lowering of the inoculum limits. All these simulations were run with a constant toxicity factor. Running the model with variable toxicity factor 10^{-3} to 10^{-7} . day^{-1} $cell^{-1}$ and keeping all other parameters constant, leads to the results represented in Fig. 6 for two flushing rates (0.2 and 0.6 day^{-1}). The two curves are close to hyperbolic which means that, above a certain threshold, a specific toxicity rate increase will only change slightly the outcome of competition.

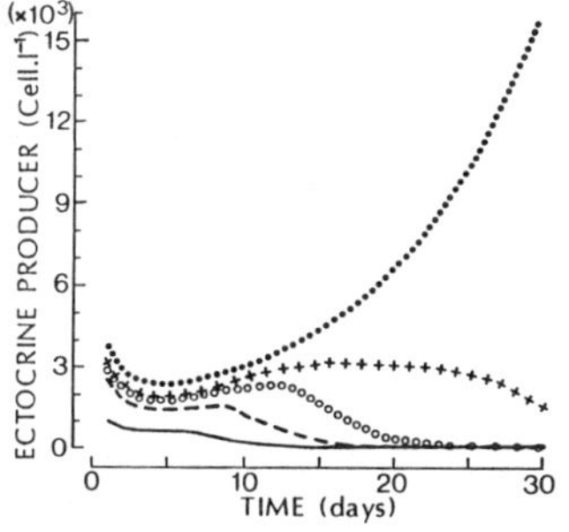

FIG. 2. Diatom population evolution different producer inoculum ($Q/V = 0.4$; $N_0=6$; $\gamma = 5 \cdot 10^{-4}$). (—— = 1000, --- = 3000, ∘∘∘ = 3200 +++ = 3250, ••• = 3500 $cells{\cdot}ml^{-1}$ of E.P.(ectocrine producer) initially.

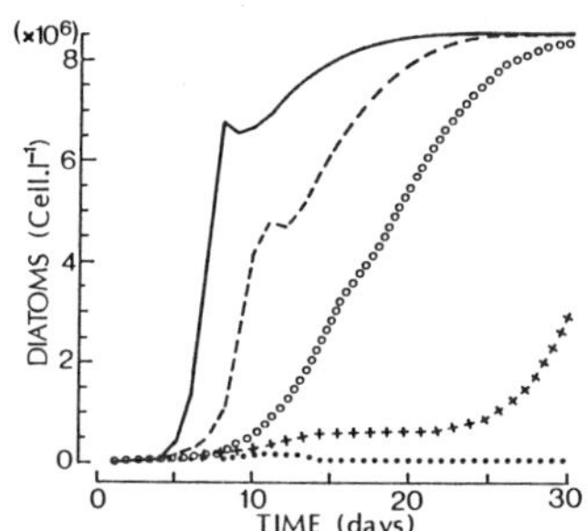

FIG. 3. Ectocrine producer theoretical case for evolution (same parameters as for ectocrine Fig. 2).

The introduction in the set of equations of a very simple toxicity formulation leads to highly nonlinear behaviour of the system which exhibits at least two thresholds for the inoculum and for the specific toxicity. The direct relationship shown by this simulation between the nutrient inflow concentration and the risk of dominance of the ectocrine producer does not necessarily imply that proliferations of this type of algae are directly in relation to eutrophication as this model is much too simplified. Before such an assessment can be made, it is at least necessary to determine ectocrine <u>in situ</u> during a bloom in order to identify the threshold values if they exist.

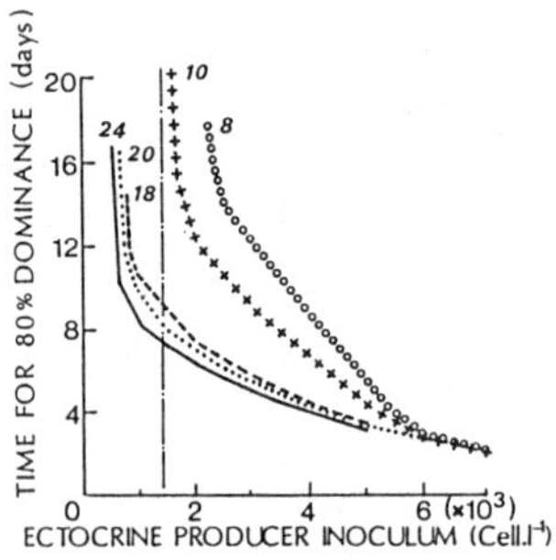

FIG. 4. Plot of the time necessary to obtain 80 % dominance of E.P. versus inoculum concentration for different nutrient inflow concentrations (8 - 24 µmol $NO_3 \cdot l^{-1}$ ($Q/V = 0.4$; $\gamma = 5 \cdot 10^{-4}$)).

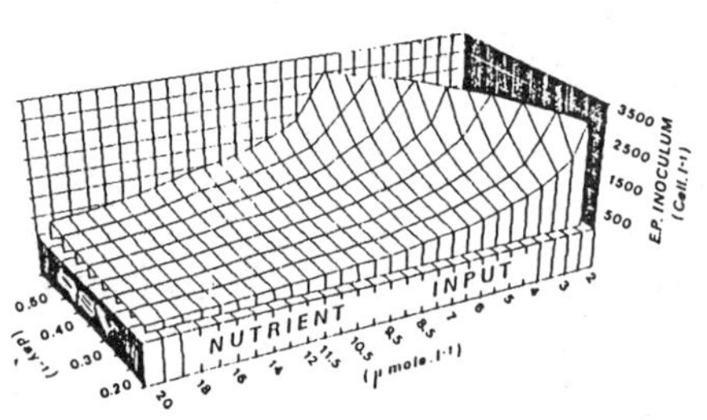

FIG. 5. Limit inoculum surface for different flushing rates (Q/V) and nutrient inflow concentrations.

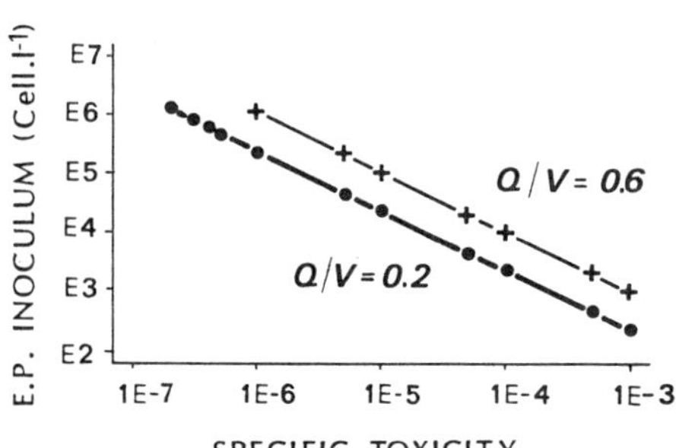

FIG. 6. Inoculum limit versus toxicity factor (flushing rates 0.2, 0.6 ; $N_0 = 10$).

ACKNOWLEDGEMENTS

This work has been supported by the IFREMER (Institut Français de Recherche pour l'Exploitationde la Mer) Red Tides Program. We wish to thank M. LUNVEN, M.P. CRASSOUS and A. YOUENOU for their technical assistance.

REFERENCES

1. E. Erard-Le Denn, M. Morlaix and J.C. Dao (This volume).
2. T. Braarud and B.R. Heimdal, Nytt. Mag. Bot. 17, 91-97 (1970).
3. P.H. Holligan and D.S. Harbour, J. Mar. Biol. Ass. U.K. 57 (4), 1075-1093 (1977).
4. K. Steidinger and K. Haddad, Bioscience, 31, 814-819 (1980).
5. S.Y. Maestrini and D.J. Bonin, Can. Bull. Fish. Aquat. Sci. 210, 323-338 (1981).
6. F. Partensky and A. Sournia, Cryptog. Algol, 7, 251-275 (1986).
7. R.R.L. Guillard and J.H. Ryther, Can. J. Microbiol. 8, 229-239 (1962).
8. J. Moal, V. Martin-Jezequel, R.P. Harris, J.F. Samain, S.A. Poulet, Oceanol. Acta 10 (3), 339-346 (1987).

ANNUAL CYCLE OF MOTILE CELLS OF GYMNODINIUM NAGASAKIENSE AND ECOLOGICAL FEATURES DURING THE PERIOD OF RED TIDE DEVELOPMENT

TSUNEO HONJO,* SHIGEYA YAMAMOTO,** OSAMU NAKAMURA,*** and MINEO YAMAGUCHI*
*Nansei Regional Fisheries Research Laboratory, Ohno-cho, Saeki-gun, Hiroshima 739-04 JAPAN; **National Research Institute of Aquaculture, Nansei-cho, Watarai-gun, Mie 516-01 JAPAN; ***Nansei Mariculture Center, Nansei-cho, Watarai-gun, Mie 516-02 JAPAN

ABSTRACT

A study of the annual cycle of motile cells of *Gymnodinium nagasakiense* and of ecological features pertinent to their development into red tides was conducted from 1984 to 1988 in Gokasho Bay in central Japan. Motile cells were observed throughout the year. Summer blooms originated each year from stocks of overwintering motile cells; the size of these stocks appeared to be dependent on the mean temperature of the water column in the winter. Population growth in the spring was slow, about 0.4 divisions/day. The depth distribution of cells during the daytime was related to cell abundance. Downward migration at night was observed; the rate was estimated to be 1.3 m/hour. The bottom waters contained more ammonia and phosphate than did the surface waters. From these observations it was concluded (1) that the overwintering cells of *G. nagasakiense* initiate growth in the spring in the middle layer where environmental conditions are stable, and (2) that spring growth develops to summer red tide through uptake of nutrients in the bottom layer associated with downward nocturnal migrations.

INTRODUCTION

The dinophycean *Gymnodinium nagasakiense* Takayama et Adachi [1] is one of several species of flagellates causing prodigious red tides in coastal waters of Japan. A red tide of this organism was first recorded by Iizuka and Irie [2] at Omura Bay in 1965. Since then several detailed investigations have been carried out [3,4]. Iizuka [5] proposed two factors as being important for *G. nagasakiense* red tides - rainfall and anoxic bottom water. The importance of the latter was confirmed subsequently through *in vitro* experiments; the organism is resistant to sodium sulfide [6], and enhanced growth results from anoxic water [7].

However, questions remained concerning the annual cycle of motile cells of *G. nagasakiense* and ecological conditions during development of their red tides. Therefore, we investigated these matters in Gokasho Bay following an extensive mortality of fish in net pens during a red tide in June and July 1984, in which losses exceeded US$ 2.7 million.

METHODS

Motile cells of *G. nagasakiense* were monitored in surface samples collected between 0900 and 1100 hours at Station A (8 m total depth) in Gokasho Bay (Fig. 1) from 1984 through 1987. Daily water temperatures at four depths (0, 2, 5, and 7 m) during the same hours were recorded for Station A from May 1984 through May 1988 and the daily mean temperature of the water column was calculated from those data. Observations on diurnal

Published 1990 by Elsevier Science Publishing Co., Inc.
Toxic Marine Phytoplankton
Edna Graneli et al., Editors

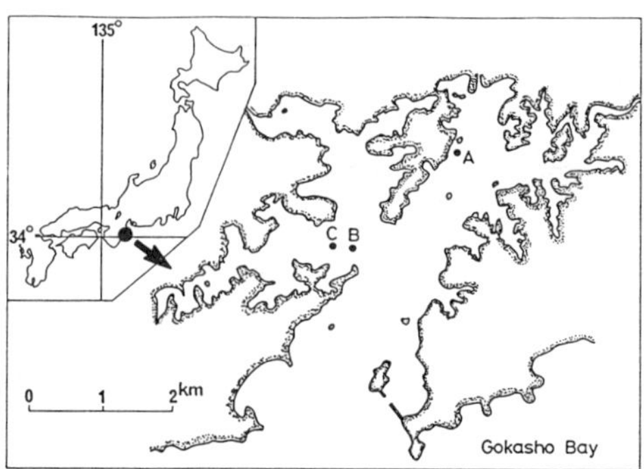

FIG. 1. Sampling stations and depths in Gokasho Bay.

vertical migration were made at Station A on June 28, 1986 and May 12, 1988. During the spring of 1987, weekly observations of daytime vertical distribution profiles were made at Station B (18 m depth) and monthly oceanographic observations were made at Station C (18 m depth). Monitoring of motile cells in surface samples was also conducted at Station B from November 1984 through May 1985.

Live cells of G. nagasakiense were counted microscopically. Cell counts were made using a Thoma hemocytometer when densities exceeded 10^4 cells/ml. When cell densities were in the range of 1 to 10^4 cells/ml, cells were counted in a specially devised chamber. When densities were less than 1 cell/ml, cells were concentrated 100-3,000 times at atmospheric pressure on glass-fiber filters (Whatman GF/C; effective membrane area 16.6 cm^2). An ultrafiltration apparatus (Amicon, model 2000; effective membrane area 144 cm^2) was used for pretreatment at atmospheric pressure for samples exceeding 0.5 l.

Growth rates were calculated as:

$$\text{divisions/day} = \ln(C_1/C_0)[1/\{(t_1-t_0)\cdot\ln 2\}]$$

where C_0 and C_1 are cell numbers at times t_0 and t_1. For the calculation of growth rates in the exponential phase, C_0 (at t_0) and C_1 (at t_1) were the points selected from the regression line.

Nutrient concentrations were measured according to the methods of Strickland and Parsons [8]. Water samples were filtered through 0.45-μm Millipore membrane filters before analysis.

RESULTS AND DISCUSSION

The abundance of G. nagasakiense motile cells in the surface water at Station A when plotted against the mean temperature of the water column over the four-year period (Fig. 2) showed bimodal patterns as well as clear decreases during the periods of minimum and maximum water temperature. Stock sizes in the winter were related to water temperature. In 1985 and 1987 the stock sizes were between 10^{-2} and 10^{-1} cells/ml in months when mean temperatures were high, but the stock size was much less (<10^{-4} cells/ml) in 1986 when low temperatures occurred in January and February. The critical temperature seemed to be about 12 °C.

Growth occurred both in spring and autumn. Spring growth originated from the motile cell stocks that overwintered; autumn growth arose from cells that survived the high-temperature period. Spring growth was initiated

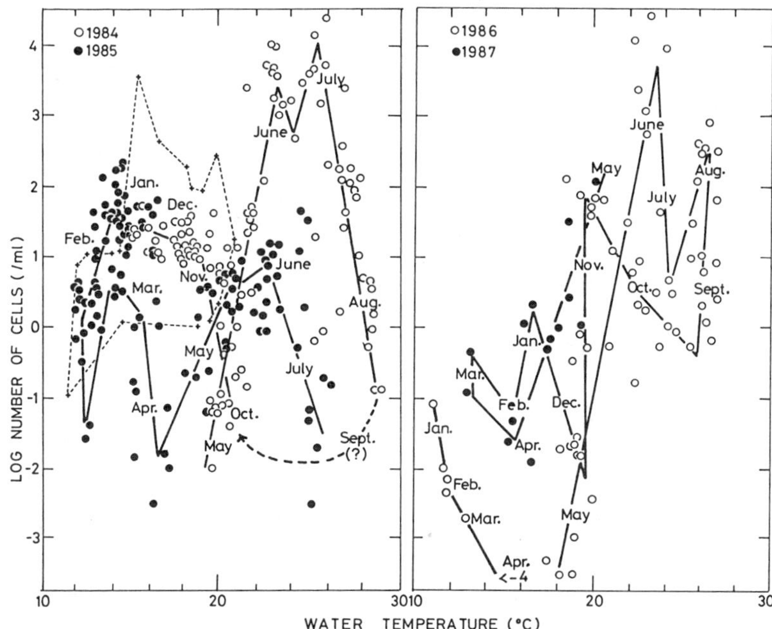

FIG. 2. Annual cycle of motile cells of G. nagasakiense at Station A versus the mean temperature of the water column (solid line). Dotted-fine line (+) shows part of the cycle of motile cells at Station B.

when cell densities were low (10^{-4} to 10^{-2} cells/ml); during April and early May the rise in temperature above about 16 °C resulted in the initiation of growth culminating in summer red tides from mid-May through early July. Autumn growth occurred when the water temperature declined from about 22 °C to 16 °C; a localized red tide was observed at Station B from December 1984 to January 1985. Although overwintering stock sizes appear to be dependent on temperature, growth of G. nagasakiense seems to be independent of rising or falling temperatures and seems to occur over a wide temperature range.

It is clear from Fig. 2 that the motile form of G. nagasakiense overwinters in Gokasho Bay. Nakata and Iizuka [9] reported concentrations of 20-60 motile cells/l in Omura Bay in January and February 1983, and Terada et al. [10] observed about 10^{3} cells/ml in a harbor of the Seto Inland Sea where water temperatures fell below 10 °C. Thus it seems that the seed population for the summer red tide of G. nagasakiense may well be the overwintering population of motile cells. Clarification of related matters, such as similarities with the life cycle of G. breve in Florida waters [11, 12], must await future studies.

The maximum growth rate for G. nagasakiense has been estimated as 1.0 division/day from both in situ and in vitro studies [13, 14]. It is evident from Fig. 3 that spring growth rates in this study, ranging from 0.32 to 0.47 divisions/day, were considerably lower than this. These low rates result in a long interval between the low cell densities observed at the time of growth initiation and those required for red tides. Developing populations most likely encounter unfavorable conditions, such as mixing by

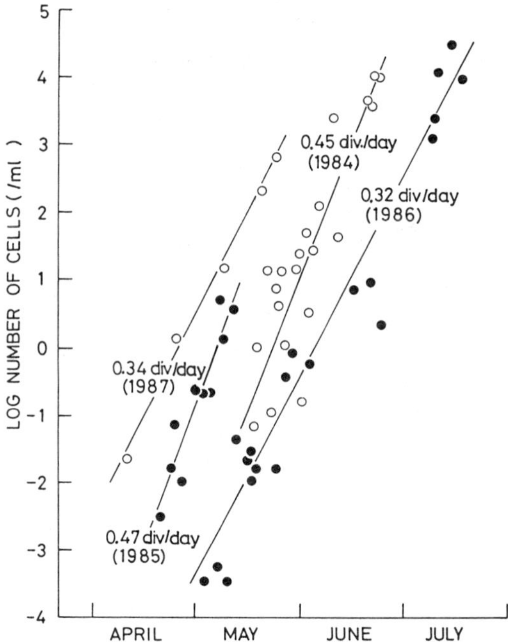

FIG. 3. In situ growth rates during the periods of development of summer red tides at Station A in 1984 through 1987.

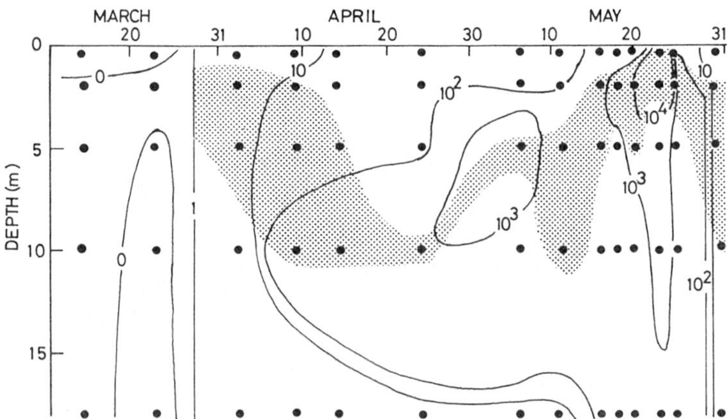

FIG. 4. Changes in the vertical distribution of G. nagasakiense in the daytime at Station B. Dotted area indicates the layer of maximum cell density, containing more than 80% of the cells in the water column.

TABLE I. Diurnal vertical migration and percentage of dividing cells in the water column at Station A

[May, 1988]

Date	12						13						
Time	1300	1500	1700	1900	2100	2300	0100	0300	0500	0700	0900	1100	1300
	Cell Number (cells/ml)*												
Depth													
0 m	110	75	70	230	550	210	2	34	380	1300	0	4	48
2 m	50	180	255	350	105	20	18	76	210	500	3	28	330
4 m	485	270	690	450	70	325	180	575	475	925	180	470	210
6 m	140	350	1225	850	200	1075	400	525	310	450	120	180	5
8 m	30	300	450	490	230	1075	1138	1725	510	300	50	10	10
Dividing cells(%)	0	0	0.07	0.7	1.4	5.9	13.2	5.8	2.1	0.2	0	0	0

[June, 1986]

Date	28							29				
Time	1615	1800	1900	2030	2140	2240	2330	0040	0140	0300	0500	0830
	Cell Number (cells/ml)*											
Depth												
0 m	120	42	940	410	1060	1	0	1	2	11	0	16
2 m	4140	1030	530	500	420	27	2	14	10	7	7	566
5 m	1290	560	2030	1140	1470	1000	933	1933	216	633	1366	1300
8 m	1030	60	14	200	130	4033	3983	3133	1950	1983	1450	100
Dividing cells(%)	0	0	0.35	4.0	6.5	1.4	3.4	4.4	1.4	2.3	0.6	0

*Total cell numbers in the water column appear to differ from time to time. This is probably related to the movement of the water past the sampling station.

TABLE II. Mean concentrations (μg-atoms/l) of nutrients in the surface layers (0, 2, 5 m depths) and bottom layers (10 m, 15 m depths), and dissolved oxygen (D.O., mgO_2/l) of the bottom water at Station C in 1987

	23 March		24 April		19 May	
	Upper	Bottom	Upper	Bottom	Upper	Bottom
Ammonia	2.8	4.4	6.2	20.6	2.4	5.3
Nitrite	0.2	0.4	tr	0.2	tr	tr
Nitrate	5.7	3.7	0.3	0.6	1.1	0.6
Phosphate	0.3	3.5	0.7	2.4	0.6	0.6
Silicate	17	13	6	8	61	37
D.O.		6.2		4.5		3.3

wind and reduced salinity from rainfall. Despite these suboptimal conditions, red tides frequently develop in Gokasho Bay. An upward shift in vertical distribution was noted at Station B as the density of the organisms increased; most cells occurred at 5 to 10 m at densities of 10^2 to 10^3 cells/ml but at 0 to 2 m at densities of 10^3 to 10^4 cells/ml (Fig. 4). The ability to maintain themselves in the middle layer at low cell densities may be important in avoiding the frequently unfavorable conditions encountered at or near the surface.

Downward migrations were clearly observed on two occasions at population levels of 10^2 to 10^3 cells/ml (Table I). Dividing cells occurred during the evening and early morning, with the highest proportion occurring around midnight. From the data of June 28, 1986 the rate of downward migration was estimated to be 1.3 m/hour. This suggests that the population can migrate 8 m downward in about six hours and that most cells at 10 m in the daytime could reach the bottom mud at 18 m at Station B by midnight. Bottom waters in the bay contained more ammonia and phosphate than did surface waters (Table II). In addition, growth-promoting substances are present in the interstitial water of the mud [7] and in the water in and around fish cages [15]. Thus the capability for vertical migration appears to be another important factor in the development of these red tides.

Based on the present data, it is concluded that growth of *G. nagasakiense* is initiated during the spring from overwintering cells in the middle layers under stable environmental conditions and that summer red tides develop in association with nutrient uptake from rich bottom layers as a result of downward nocturnal migration.

REFERENCES

1. H. Takayama and Adachi, Bull. Plankton Soc. Japan, 31, 7-14 (1984).
2. S. Iizuka and H. Irie, Bull. Fac. Fish. Nagasaki Univ., 21, 67-101 (1966).
3. S. Iizuka and H. Irie, Bull. Fac. Fish. Nagasaki Univ., 27, 19-37 (1969).
4. S. Iizuka and H. Irie, Bull. Plankton Soc. Japan, 16, 99-115 (1969).
5. S. Iizuka, Bull. Plankton Soc. Japan, 19, 22-33 (1972).
6. S. Iizuka and T. Nakashima, Bull. Plankton Soc. Japan, 22, 27-32 (1975).
7. K. Hirayama and K. Numaguchi, Bull. Plankton Soc. Japan, 19, 13-21 (1972).
8. J.D.H. Strickland and T.R. Parsons, Bull. Fish. Res. Bd. Can. 167, (1968).
9. K. Nakata and S. Iizuka, Bull. Plankton Soc. Japan, 34, 199-201 (1987).
10. K. Terada, H. Ikeuchi and H. Takayama, Bull. Plankton Soc. Japan, 34, 201-203 (1987).
11. K.A. Steidinger and E.A. Joyce Jr., Fla. Dep. Nat. Resour. Mar. Res. Lab. Educ. Ser., 17, 1-26 (1973).
12. L.M. Walker, Bioscience, 32, 809-810 (1982).
13. S. Iizuka, in: D.L. Taylor and H.H. Seliger, eds. Toxic Dinoflagellate Blooms, Elsevier, N. Y., 111-114 (1979).
14. S. Iizuka and K. Mine, Bull. Plankton Soc. Japan, 30, 139-146 (1983).
15. A. Nishimura, Bull. Plankton Soc. Japan, 29, 1-7 (1982).

NOXIOUS PHYTOPLANKTON BLOOMS IN WESTERN WASHINGTON WATERS. A REVIEW

Rita A. Horner,* James R. Postel,** and John E. Rensel***
*4211 N.E. 88th St., Seattle, WA 98115; **School of Oceanography, ***School of Fisheries, University of Washington, Seattle, WA 98195

ABSTRACT

This paper reviews studies on two kinds of noxious phytoplankton blooms that occur in Puget Sound, Washington. The first causes paralytic shellfish poisoning (PSP) and poses a serious problem for shellfish harvesting. The second primarily involves diatoms and causes mortality of finfish, particularly Atlantic salmon, reared in commercial net-pens.

PARALYTIC SHELLFISH POISONING (PSP)

In Washington (Fig. 1), only three deaths have been attributed to PSP since the 1930s. These occurred in 1942 and regulations were established shortly thereafter. However, shellfish toxicity is likely to occur each year along the open ocean coast and appears to be caused by blooms originating in the ocean, not in the bays [1]. Some beaches are affected annually, while others are affected less frequently. As a result, ocean beaches are closed to all shellfish harvest, except razor clams, from 1 April to 31 October each year. Along the Strait of Juan de Fuca and in inland waters of Puget Sound (Fig. 2), blooms originate in situ and the frequency of shellfish toxicity and beach closure varies [1].

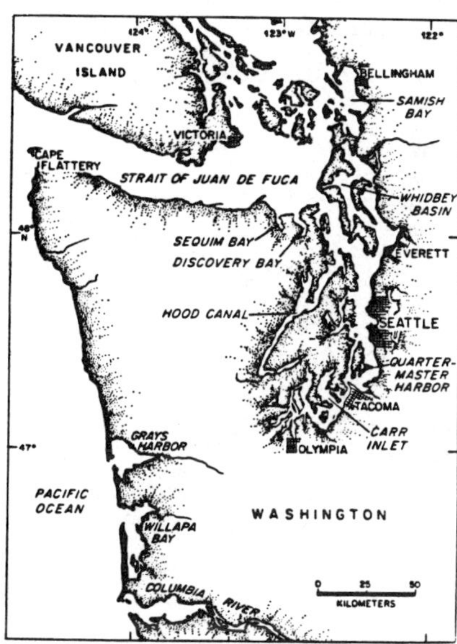

FIG. 1. Washington coastline with areas where closures have been required because of high toxin levels.

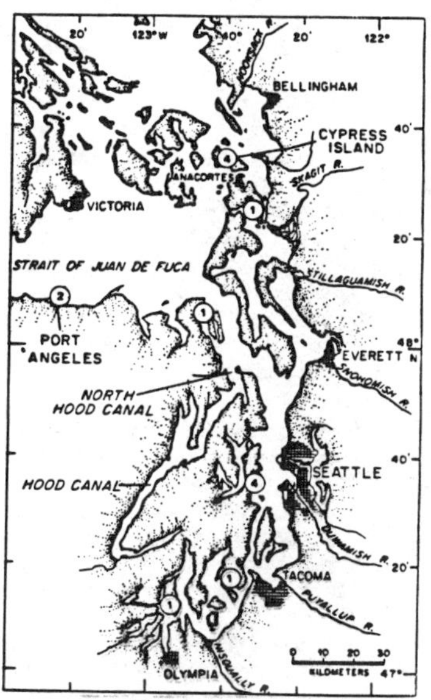

FIG. 2. Puget Sound with locations of fish farms. Numbers indicate the number of farms at each location as of spring, 1989.

Toxic Marine Phytoplankton
Edna Graneli et al., Editors

Monitoring of areas used for sports harvesting began in northern Puget Sound in 1958, and high toxin levels required closure for the first time in 1971 [2]. A commercial shellfish operation in Samish Bay, also monitored since 1958, was closed for the first time in 1974 [2]. This was followed in 1978 by a massive bloom of _Alexandrium catenellum_ in the Whidbey Basin with toxicity levels in bay mussels, _Mytilus edulis_ (L.), over 30,000 µg toxin/100 g mussel meat. Ten people reported PSP symptoms after eating recreationally harvested bay mussels and pink scallops (_Chlamys hastata_ Sowerby), but no deaths occurred.

Since then, toxicity has apparently spread southward with motile cells and cysts of _A. catenellum_ and low levels of toxin in shellfish being found in the south Sound at least since 1981 (Nishitani unpubl. data). It is not known if this is an actual spread of toxicity because the south Sound had not been regularly monitored until recently [2]. The first closures in southern Puget Sound occurred in the fall of 1988 when five people became ill from eating commercially harvested oysters. Now the only area in western Washington waters where closures have not been required is Hood Canal.

Local research

Early work showed that _A. catenellum_ was abundant in Puget Sound prior to the outbreak of toxicity and its abundance was related to water temperature, decline of diatom abundance, and amount of sunlight. The toxin content per cell varied inversely with growth rate and declined following the end of the log growth phase [1]. More recently, data from Quartermaster Harbor suggested that _A. catenellum_ is a vertical migrator [3]. Cells are near the surface during the day and move downward at night. Strong density gradients or wind- or tide-driven turbulence may influence distribution. This migratory pattern might be important for both exposure of _A. catenellum_ to different environmental conditions such as temperature, salinity, and nutrients, and exposure of shellfish to _A. catenellum_ because shellfish toxicity varies with the length of exposure [3]. This could have major implications for shellfish aquaculture practices depending on tide height and whether the shellfish are growing in raft culture or in shore beds [3].

A. catenellum grows slowly with a maximum of one doubling per day in cultures and with about 50% of the cells dividing per day in nature. However, as few as 10 _A. catenellum_ cells per milliliter can cause PSP levels higher than the closure level of 80 µg toxin/100 g shellfish meat [3, 4]. Population increases occur after stratification begins in the spring and the water temperature rises to 14 °C, conditions that develop faster in bays than in adjacent channels. High toxin levels in mussels are generally associated with these conditions [3]. Limiting concentrations of nitrogen or phosphorus may also influence the onset and duration of blooms [2].

A. catenellum populations in Quartermaster Harbor were reduced by removal of cells by tidal washout, encystment and settling out of cells, and by consumption of motile cells by zooplankton, benthic invertebrates, and a dinoflagellate parasite, _Amoebophrya ceratii_ (Koeppen) Cachon. The possibility of using the parasite as a biological control was investigated and rejected after finding that it also attacks other species that may be dominant in the phytoplankton [5, 6].

The degree of coincidence between El Nino/Southern Oscillation (ENSO) events in 1942 and 1957-1984 and exceptional PSP outbreaks in Washington and British Columbia has been examined [7]. Exceptional outbreaks occurred in both Washington and British Columbia in seven of nine ENSO events suggesting that conditions for the growth of _Alexandrium_ may have been increased by such widespread, long-term events. Shifts in the seasonality of PSP outbreaks were marked by earlier growth in spring and later growth in fall.

In addition, relatively high levels of toxin occurred in oysters and razor clams along the open coast of Washington during strong ENSO events. In four of 14 years without ENSO events, PSP outbreaks occurred when sea surface temperatures were abnormally high. Thus, perhaps weather conditions and sea surface temperatures can be used to help predict periods of Alexandrium blooms and the probability of increased PSP in shellfish [7].

The effects of PSP toxins on finfish have also been studied [8]. During some blooms, about 95% of the phytoplankton crop available to zooplankton may be A. catenellum. Copepods grown on A. catenellum were then fed to Pacific herring and chum and pink salmon that might be naturally exposed to A. catenellum in the marine environment and to young coho salmon still in the freshwater stage that would not be exposed to PSP toxins under natural conditions. Only pink salmon showed no effects at the toxin levels tested, while the coho salmon died at very low toxin concentrations [8].

Management and monitoring

The shellfish monitoring program by the Department of Social and Health Services and the county health departments has greatly reduced PSP risks to human health. There have been only five confirmed illnesses from commercial shellfish operations, currently valued at about $20 million, since the monitoring program began in 1957. Recreational harvesting has been severely restricted since the mid 1970s because of the more frequent occurrences and high levels of toxicity. Despite a large increase in the number of areas with high toxicity, the only reported PSP illnesses from sports-harvested shellfish were 10 cases at the beginning of the unexpected bloom in 1978 in a previously safe area, 3 cases from clams in 1979, and 2 cases from scallops in 1985 collected in areas where harvesting was closed [2].

Some level of toxicity exceeding closure levels occurs every summer in nearly all parts of Puget Sound. As of mid-1989, only Hood Canal has not had closures. Bays will probably be more affected than the main channels [3]. Because warm water and stratification are important to bloom development, periods of warm weather in the spring or fall and periods of high river runoff could cause increases in toxicity levels. Human activities may alter nutrient availability in some restricted waters, so the potential effect of these activities must be considered. In addition, the possibility now exists that finfish and other wildlife resources might be affected by PSP, so stomach contents should be analyzed for PSP toxins during unexplained massive kills of fish, birds, and mammals [4].

NET PEN SALMON MORTALITY

In Washington State there are 14 net-pen farms that raise Atlantic (Salmo salar) or Pacific (Oncorhynchus spp.) salmon (Fig. 2). Current production is estimated to be about 9000 tons in 1989. In 1987, phytoplankton blooms were involved in the mortality of at least 250,000 Atlantic salmon of all age classes with monetary losses over $500,000. Fish growers consider phytoplankton-related problems to be their number one research need because they threaten current production with chronic and acute losses and compound the risk and expense related to new site development.

A number of species from several algal phyla have been implicated in fish kills in western Washington waters, but the Chaetoceros convolutus Castracane/C. concavicornis Mangin diatom complex is apparently a major problem. It has been involved in the mortality of all ages of pen-reared Atlantic salmon. It is thought that cells of these organisms cause mechanical injury or clogging, possibly leading to mucus production and suffocation when spines break off and enter the gill tissue, but the actual mechanism is not known [9, 10]. Whole cells of Chaetoceros species and other diatoms have also been seen on and in gill tissue [11, 12].

Experimental work on the effects of C. convolutus on Pacific salmon in British Columbia [9] and Alaska [12] showed that about 7 x 10^5 cells per liter or 10^4 to 10^5 chains per liter killed juveniles and fingerlings, but these cell concentrations are orders of magnitude greater than the number of cells of this species usually found in Puget Sound [9, 11]. Thus, there is a need for chronic exposure studies that simulate conditions in the field. Atlantic salmon appear to be more susceptible to Chaetoceros than Pacific salmon. In the last eight years, approximately 40% of the Atlantic salmon smolts pen-reared at a National Marine Fisheries Laboratory died between late May and July (C. Mahnken and R. Waknitz, pers. comm.) while few losses occurred in adjacent pens containing coho or chinook salmon.

Local research

Preliminary studies have been done at two existing net-pen locations (Cypress Island and Port Angeles harbor), in an area where net-pen rearing has been proposed (central Hood Canal), and near a place where net-pen production was terminated, in part due to phytoplankton blooms (northern Hood Canal) (Fig. 2). These sites have diverse hydrographic characteristics ranging from stratified to well-mixed conditions. Water depths at all sites are generally <100 m, usually being 25 to 40 m. Fish kills occurred at the two existing net-pen sites.

At Cypress Island, sampling in October, 1988, was done late in a bloom and after severe fish losses occurred. At that time, Chaetoceros species comprised between 30 and 40% of the population in the well-mixed water column with C. Phaeoceros being <25% of the total population.

During a fish kill in Port Angeles harbor in late May-early June, 1988, samples were collected at the farm site and at a reference site about one mile offshore in the Strait of Juan de Fuca (Table 1). Chaetoceros Phaeoceros species, usually thought to be the culprits during these events, comprised <0.5% of the phytoplankton population in the harbor, while Chaetoceros Hyalochaete species were far more abundant at >60% (Table 1). Moribund and active fish did not have phytoplankton cells or lesions within

TABLE I. Comparison of hydrographic and phytoplankton data from Port Angeles harbor and nearby Strait of Juan de Fuca during a Chaetoceros bloom, 27 May 1988 (modified from [11]).

	Port Angeles Harbor			Strait of Juan de Fuca		
Secchi disk (m)	3.5			8.0		
Depth sampled (m)	2	10	25	2	10	25
Water temperature (°C)	10.8	10.1	9.5	10.7	9.9	9.1
Salinity (o/oo)	30.5	30.9	31.6	30.1	30.1	30.9
Dissolved oxygen (mg/l)	10.0	8.8	6.4	8.2	7.6	5.3
Chlorophyll a (μg/l)	13.0	9.2	2.8	1.1	2.6	0.6
Nitrogen (μg-at/l)	7.5	13.4	21.6	21.2	20.0	27.1
Total cells/liter (x10^6)	3.1	1.9	0.9	0.6	1.8	0.5
PERCENT OF TOTAL CELLS						
Chaetoceros (Phaeoceros)	0.4	0.2	0.0	0.0	0.0	0.0
Chaetoceros (Hyalochaete)	75.0	68.0	63.0	29.0	71.0	61.0
All other diatoms	12.0	23.0	22.0	41.0	18.0	20.0
All dinoflagellates	2.0	0.5	1.0	2.0	0.9	2.0
All flagellates	11.0	8.5	14.0	28.0	10.0	17.0

the gill tissue, but some *Chaetoceros* and other diatom cells were found between gill filaments (R. Elston pers. comm.).

In the stratified water of central Hood Canal, surface waters are seasonally nutrient-depleted. A few cells of *Chaetoceros Phaeoceros* were present in June, 1987, *Chaetoceros Hyalochaete* were relatively abundant in August, but small flagellates generally dominated the phytoplankton population [11]. At the same time, northern Hood Canal waters were well-mixed. *Chaetoceros Phaeoceros* comprised <1% of the population in late June and was not present in July or August. *Chaetoceros Hyalochaete* species comprised 20% of the population in June and July, but only 4% in August. All diatoms were generally more abundant in June and July with small, unidentified flagellates dominating in August [11].

Thus, knowledge of hydrographic conditions alone is not sufficient to predict the composition of phytoplankton populations in closely spaced areas of Puget Sound. Species suspected of causing fish mortality are not always the numerically dominant taxa and, in fact, species from several algal phyla may be involved.

Management and monitoring

Mitigative means to protect net-pen fish from phytoplankton blooms are either passive or active. Passive measures avoid stressing fish and include such things as not feeding or handling them. This reduces the activity of the fish and decreases their metabolic rate. Active measures try to isolate the fish from the phytoplankton and/or introduce phytoplankton-free water. One method involves the installation of skirts made of phytoplankton-opaque material around the pens. This must be accompanied by pumping relatively phytoplankton-free water from below the euphotic zone into the pens, or pumping may be done without the installation of skirts. However, most net-pens in Puget Sound are located in water only about 25 m deep and this may not be deep enough to be out of the euphotic zone in summer. Only one farm site currently has an adequate upwelling system for some of its pens. A second active measure currently being evaluated is to provide filtered seawater, but only for smolts when they are first introduced into salt water. This would allow young fish time to become acclimated to salt water without the added challenge of potentially noxious phytoplankton.

We are gradually learning which noxious species are present in Puget Sound along with their abundance and temporal and spatial distributions. We are not sure if *Chaetoceros convolutus* and *C. concavicornis* are the only diatom species that cause fish deaths locally, or if other species of *Chaetoceros* are involved as appeared to be the case at Port Angeles. We don't know how environmental parameters influence the magnitude and timing of blooms of *Chaetoceros* in Puget Sound. We don't know if diatoms contribute to existing physiological imbalances or if they act in some other way to cause fish mortality. However, fish farmers are assisting us in building a data base that will lead to a better knowledge of phytoplankton blooms in Puget Sound and their effects on finfish.

SUMMARY

Two kinds of noxious phytoplankton blooms occur in western Washington waters. The first, involving dinoflagellates, causes paralytic shellfish poisoning (PSP) and poses serious problems with regard to shellfish harvesting. There has been a drastic increase in the number of open ocean and inland areas requiring closure since 1975 with closures required in southern Puget Sound for the first time in 1988. The second kind of noxious bloom involves diatoms. Blooms of *Chaetoceros* have been implicated in the deaths of pen-reared Atlantic salmon. The actual causes of fish kills are not known and different mechanisms may exist during different events. The problem is receiving increased attention because of expansion of the net-pen fish farming industry and is compounded because the seasonal, geographic, and vertical distributions of apparent problem species are not known.

ACKNOWLEDGEMENTS

The PSP studies were conducted by L. Nishitani and G. Erickson in Dr. K. K. Chew's laboratory in the School of Fisheries at the University of Washington. They were funded by the U.S. Environmental Protection Agency, Washington Department of Social and Health Services, and the Washington Sea Grant Program. The finfish mortality project has been done primarily by J. Rensel, J. Postel, and R. Horner with funding and logistics assistance from local fish growers, including Sea Farm Washington Inc., Scan Am Fish Farms, International Marine Farms, Inc., and Swecker Salmon Farms, Inc.

REFERENCES

1. Norris, L. and K.K. Chew, in: Proceedings of the First International Conference on Toxic Dinoflagellate Blooms, V.R. LoCicero, ed. (Science and Technology Foundation, Wakefield, MA 1975) pp. 143-152.

2. Nishitani, L. and K.K. Chew, J. Shellfish Res. 7, 653-669 (1988).

3. Nishitani, L. and K.K. Chew, Aquaculture 39, 317-329 (1984).

4. Nishitani, L., G. Erickson, and K.K. Chew in: Proceedings First Annual Meeting on Puget Sound Research, Vol. 2. (Puget Sound Water Quality Authority, Seattle, WA 1988) pp. 392-399.

5. Nishitani, L., R. Hood, J. Wakeman, and K.K. Chew, in: Seafood Toxins, ACS Symposium Series, No. 262, E.P. Ragelis, ed. (American Chemical Society, Washington, D.C. 1984) pp. 139-149.

6. Nishitani, L., G. Erickson, and K.K. Chew, in: Toxic Dinoflagellates, D.M. Anderson, A.W. White, and D.G. Baden, eds. (Elsevier, New York 1985) pp. 225-230.

7. Erickson, G. and L. Nishitani, in: El Nino North, W.S. Wooster and D.L. Fluharty, eds. (Washington Sea Grant Program, Seattle, WA 1985) pp. 283-290.

8. Erickson, G. M.S. Thesis (University of Washington, Seattle, WA 1988) 453 pp.

9. Bell, G.R., W. Griffioen, and O. Kennedy, in: Proceedings of the Northwest Fish Culture Conference, 25th Anniversary. (Seattle, WA 1974) pp. 58-60.

10. Brett, J.R., W. Griffioen, and A. Solmie, Department of Fisheries and the Environment Canada, Fisheries and Marine Service Tech. Rep. No. 845, 1974. 10 pp.

11. Rensel, J.E., R. Horner, and J.R. Postel, Northwest Environmental Journal 5, 53-69 (1989).

12. Farrington, C.W. M.S. Thesis (University of Alaska - Southeast, Juneau, AK 1988) 80 pp.

THE CULMINATION OF THE CHRYSOCHROMULINA POLYLEPIS (Manton & Parke) BLOOM ALONG THE WESTERN COAST OF NORWAY

TORBJØRN M. JOHNSEN AND EVY R. LØMSLAND.
Department of Marine Biology, University of Bergen
N - 5065 Blomsterdalen, Norway.

ABSTRACT

The Chrysochromulina polylepis bloom in the Skagerrak-Kattegat area in May 1988 spred to the southern part of the west coast of Norway before it culminated. During 34 hours the standing stock of C. polylepis in the front area was reduced by 93% most probably due to prolonged nutrient starvation. Low concentrations of chlorophyll a in the algal cells and an increasing ratio between phaeopigments and chlorophyll a support this assumption. Generally, the total number of other planktonic algae declined heavily when the concentration of C. polylepis exceeded 10^6 cells/l. The cryptophyceans were the most affected algal class being reduced by 98%. Among the dinoflagellates, the ceratians showed high mortality. The Prasinophyceae reacted similarly to the cryptophyceans, but the effect was not so dramatic. No evidence of negative effects of C. polylepis on Chrysophyceae and Bacillariophyceae was found. Choanoflagellates and ciliates increased slowly during the bloom, and reached their maxima when the bloom culminated.

INTRODUCTION

The unexpected bloom of Chrysochromulina polylepis in May 1988 was a disaster to fish mariculture along the southern coast of Norway. The bloom was transported by The Norwegian Coastal Current from Kattegat where it was most probably initiated. It reached the west coast of Norway, but culminated at the island Karmøy (59^0 15´ N, 05^0 07.8´ E) before it reached the county Hordaland with the highest density of mariculture units in Norway. The culmination occurred quite suddenly.

Data presented are from a cruise with the research wessel "Håkon Mosby" during the period 25. May to 6. June. The purpose of the cruise was to locate the algal front and follow its movement to predict the position from one day to the next in order to make an algal forecast.

The purpose of this paper is to discuss the sudden collapse of the Chrysochromulina polylepis bloom and the effect of this alga on coexisting species.

MATERIALS AND METHODS

The investigated area was the southern part of western Norway. Four regions of this area were chosen for further examination (Fig. 1). Region A had high concentrations of C. polylepis from the start and the highest measured in the research area. In region B, only sporadic observations of C. polylepis were made when the sampling started and in C the initial concentrations where under the lethal dose for caged fish [1]. The

Toxic Marine Phytoplankton
Edna Graneli et al., Editors

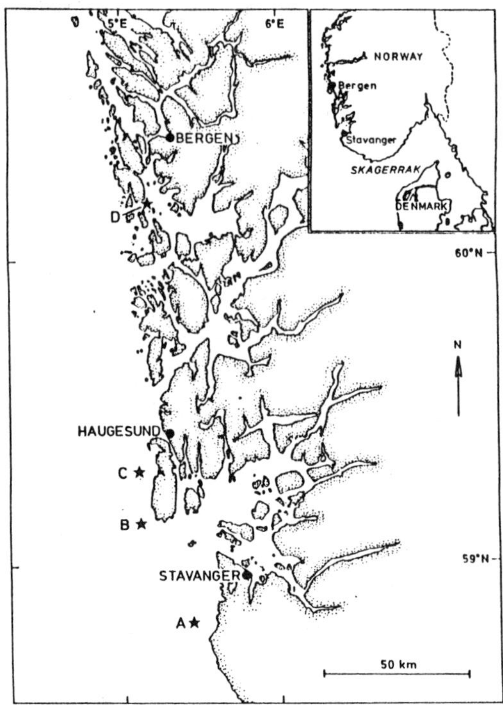

FIG. 1. Map of the investigation area.

development of C. polylepis and the phytoplankton community, as a whole, were followed closely in these regions until the culmination of the C. polylepis bloom. Ciliates and choanoflagellates were also enumerated. Region D was not affected by the Chrysochromulina bloom and is used as a reference. Here there was a bloom of the prymnesiophycean Emiliania huxleyi which is a yearly phenomenon in these parts.

In region A, B and C, possible biomass maxima were detected by in situ fluorescence profiling (Q-fluorometer). Samples were collected from selected depths for estimation of chlorophyll a and phaeopigments, nutrients and phytoplankton. Salinity and temperature measurements were made by CTD-profiling. For region D sampling was done by a small boat allowing only Salinoterm measurements of salinity and temperature. Samples were taken at fixed depths for estimation of nutrients and phytoplankton.

C. polylepis cell counts were made on board on live samples in a Fuchs-Rosenthal counting slide and by use of a flowcytometer. Phytoplankon samples were also preserved for later examination in two ways, acidified Lugol´s solution and formaldehyd neutralized with hexamethylenetetramine. The formaldehyd samples were used to quantify calcareous species like coccolithophorids. Cells were counted in 10 ml aliquots using the Utermöhl technique, although the ceratians occasionally had to be counted in 50 ml aliquots.

Fluorometric estimation of chlorophyll a after homogenisation and extraction in 90% aceton was done by a Turner filter fluorometer [2,3].

Nitrate, phosphate and silicate were analysed after the cruise by an autoanalyser. The samples were preserved with chloroform and kept refrigerated until analysis.

RESULTS AND DISCUSSION

Hydrography

During the first days in regions A, B and C there was a mixed water coloumn above 25-30 m where distinct halo- and thermoclines were situated. The surface temperatures were 9-10^{0}C. The salinity in surface water was 29‰ in region A and 30-31‰ in B and C. Later, the upper coloumn generally became more stratified and the surface temperatures increased by 1.5^{0}C and the salinity decreased by 2-3‰. The difference in salinity between region A and regions B and C can be explained by the location of A nearer Skagerrak In Skagerrak the salinity was 25-28‰ at the time of the bloom [4]. The reduction in salinity during the investigation period, and as the algal front moved north, may be attributed to a more pronounced influence of this water. Also in region D the surface temperature increased during the time of investigation from 9 to 11^{0}C. The general increase in temperature was attributed to the high insolation at the time. Region D was distinguished from the other regions by rather stable surface salinities of 32-33‰ during the investigation period.

Nutrients

The concentrations of all the nutrients were low in general for the whole period. However, a tendency toward higher concentrations in surface water near shore was registrated in region A. This may be attributed to the high landbased agricultural activity in the area. Here nutrient values ranged from 0.1-0.6 µg-at N/l, 0.04-0.3 µg-at P/l and 0.4-1.2 µg-at Si/l. Corresponding values for regions B and C were 0.03-0.3, 0.03-0.1 and 0.2-0.7 µg-at/l. The variation was due mainly to regional differences and not to development over time. Region D had the lowest values ≤0.1, ≤0.1 and 0.2-0.6 µg-at/l, respectively for nitrate, phosphate and silicate.

Biomass

In region A the maximum concentration of *C. polylepis* was 10^7 cells/l with a corresponding chlorophyll a quantity of 2 µg/l. Equivalent values for region B and C were 50% lower. Supposing that only *C. polylepis* contributed to the biomass, this would mean 0.2 pg chlorophyll a/cell. Since *C. polylepis* was not in monoculture, the actual cell content is even lower. Reported values for the alga in culture are 0.50-0.66 pg chlorophyll a/ cell [4]. The low concentration just before the culmination of the bloom may be an indication of reduced cell content of chlorophyll a because of nitrate exhaustion [5]. This result indicates that *C. polylepis* has a need for considerably higher N concentration than *Emiliania huxleyi*, blooming in region D at the same time and where nitrate levels were even lower. Low half-saturation constant for nitrate is reported for *E. huxleyi* [6].

In the frontal areas the ratio phaeopigment/chlorophyll a changed considerably during the period of investigation (Table 1), increasing from 0.3 to more than 1. An increasing ratio of phaeopigment/chlorophyll a combined with a low level of chlorophyll a/cell indicates that the physiological state of *C. polylepis* was constantly deteriorating during the

Table I. Development of the phaeopigment/chlorophyll a-ratio in the algal front areas during the investigation period.

Date	26.May	27.May	28.May	30.May	1.June
Phaeop./ Chl.a	0.3	0.4	0.8	0.8-1.0	1.0-1.3

investigation period. This lead to an inevitable collapse due to lack of nutrients. The break down came suddenly, and the standing stock was reduced by 93% during 34 hours. Influence of hydrographical processes on this collapse seems to be a factor of minor importance [7].

Development of coexisting species

The general picture is that the total number of other algae declined heavily when the concentration of C. polylepis exceeded 10^6 cells/l (Fig. 2). However, the responses of the different classes were not the same. The

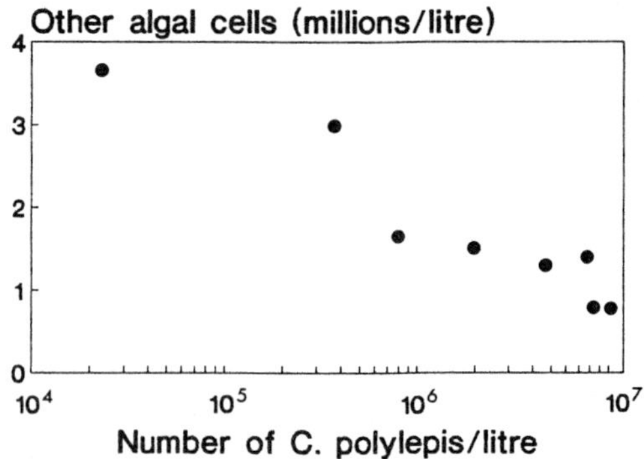

FIG. 2. Number of C. polylepis versus number of other identified algae. Unidentified monads and flagellates are left out.

most affected algal class seemed to be the Cryptophyceae. After the appearance of C. polylepis, the number of cryptophyceans declined to only 2% of the initial concentration of $1\text{-}2{*}10^6$ cells/l (Fig.3). In region D which was not affected by C. polylepis, the cryptophyceans were constantly above 10^6 cells/l. A very similar but less powerfull effect was registrated on the class Prasinophyceae. However, the Dinophyceae increased vigorously as C. polylepis culminated (Fig. 3). This was due to a great increase in Entomosigma peridinioides and small thecate dinoflagellates. A corresponding bloom of these dinoflagellates was not observed in region D. The ceratians, however, showed a quite opposite reaction. Both live and dead cells were seen. When the concentration of C. polylepis passed $2{*}10^6$ cells/l the mortality was more than 80% (Fig.4). The species concerned were C. furca, C. fusus, C. longipes, C. tripos. Dead ceratians were also reported from Skagerrak [4]. The same effect was not noted for other thecate dinoflagellates. High numbers of prymnesiophyceans characterized all four investigated regions (Fig.3). In regions A, B and C C. polylepis was the dominating species, but in region D E.huxleyi was dominant and increased during the investigation. This coccolithophorid was also observed in the other three regions, but the number did not exceed $2{*}10^5$ cells/l. After the culmination of C. polylepis the numbers of E. huxleyi were even lower. This in spite of a general higher nutrient level in these regions. On the basis

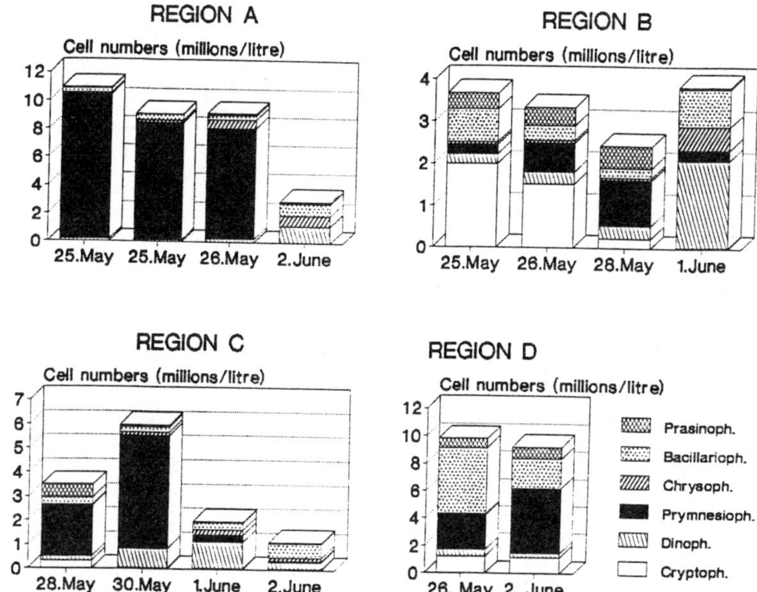

FIG. 3. Development of the different algal classes during the investigation period.

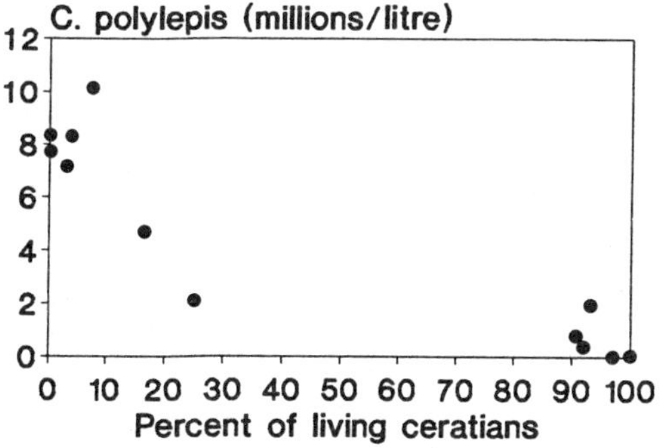

FIG. 4. Percent living ceratians versus number of C. polylepis.

of the existing data there is no evidence of a negative effect on the classes Chrysophyceae and Bacillariophyceae. In region A, B and C principally, the number of choanoflagellates increased slowly in the presence of C. polylepis, and reached their maxima after the culmination of C. polylepis. In region D, however, the development was quite the opposite, showing a decrease in choanoflagellates during the time of investigation. The general development of the ciliates were rather similar to that of the choanoflagellates. In region A, B and C the naked ciliates were quite prominent, while in region D the tintinnids were more prominent.

ACKNOWLEDGEMENT

We are particularly grateful to the Institute of Marine Research, Directorate of Fisheries, Bergen, who carried out a large number of nutrient analysis and to A. Aadnesen for doing the pigment analyses. We thank B.R. Heimdal for comments on the manuscript, E. Holm for help with the figures and D. Evensen for correcting the English. This work was supported by grants from the Directorate for Nature Management.

REFERENCES

1. H. Leivestad and B. Seringstad, Norwegian Institute of Marine Research, Rep. BKO 8803, 29 pp. (1988).
2. C.S. Yentsch and D.W. Mentzel, Deep-Sea Res. 10, 221-231 (1963).
3. O. Holm-Hansen, C.J. Lorenzen, R.W. Holmes and J.D.H. Strickland, J. Cons. perm. int. Explor. Mer. 30, 3-15 (1965).
4. E. Dahl, O. Lindahl, E. Paasche and J. Throndsen in: A novel phytoplankton bloom. Causes and impacts of recurrent brown tides, E.M. Cosper et al., eds. (Springer Lecture Notes on Coastal and Estuarine Studies 1989) in press.
5. N.J. Antia, C.D. McAllister, T.R. Parsons, K. Stephens and J.D.H. Strickland, Limnol. Oceanogr. 8, 166-183 (1963).
6. R.W. Eppley, J.N. Rogers and J.J. McCarthy, Limnol. Oceanogr. 14, 912-920 (1969).
7. I. Dundas, O.M. Johannessen, G. Berge and B.R. Heimdal, Oceanography 2, 9-14 (1989).

A RANDOM WALK MODEL EXAMINING HOW PHYTOPLANKTON DISTRIBUTE IN THE UPPER MIXED LAYER

DANIEL KAMYKOWSKI
Department of Marine, Earth and Atmospheric Sciences, Box 8208, North Carolina State University, Raleigh, N. C. 27695, USA

ABSTRACT

A random walk model was formulated to represent recent ideas on the character of vertical water motion in the upper mixed layer. The water column was divided into three layers with order of magnitude decreases in vertical eddy diffusivity coefficients in each successive depth stratum. The upper layer also possessed differential upwelling and downwelling to mimic the spatial and energetic character of Langmuir circulation. The direction and magnitude of vertical water motion in all three layers changed each time step (0.1 hours) according to the random selections. Populations of 100 individuals each representing one of 6 phytoplankton motility types (positively, neutrally or negatively buoyant cells, flagellates or two ciliates) were seeded at the surface at the beginning of a run. Each individual within a population behaved similarly. The model provides a composite of the individual trajectories due to stochastic physical (turbulence and advection) and deterministic biological vectors for the population of each phytoplankton type. A comparison of the different phytoplankton types provides insight into the mechanisms that determine multi-species phytoplankton distributions, including nuisance blooms, in the upper mixed layer.

INTRODUCTION

The vertical trajectories of phytoplankton in the upper mixed layer determine phytoplankton growth conditions. The contributing vectors due to inherent particle motion and vertical water motion, however, are poorly known. In the former case, estimates of inherent particle motion by non-motile phytoplankton rely either on Stokes law for which cell shape and density can be difficult to specify [1] or on direct measurements that require careful interpretation [2] especially for field populations. Motile phytoplankton, on the other hand, adjust their speed [3] and direction [4,5] in response to their internal state and a variety of environmental factors. Both non-motile and motile phytoplankton appear able to change their capabilities within a few hours [6], and different members of a population may possess different capabilities at the same time [7]. In the latter case, estimates of vertical water motion must include the integrated influences of all aspects of turbulence [8] and advection [9]. Routine estimates of integrated vertical water motion in the upper mixed layer will require continued technological advances [9].

One innovative study [10] examined the net effect of inherent particle motion and vertical water motion on the ensemble distribution of fish eggs. Vertical profiles of fish eggs characterized by empirically determined ascent velocities were

Toxic Marine Phytoplankton
Edna Graneli et al., Editors

related to the effect of wind mixing and supported the computation of mean vertical eddy diffusivity coefficients (VEDCs). The resulting coefficients were higher than usual [11] because they integrated the effects of turbulence and advection.

The present paper adapts this fish egg approach to a more realistic water column. Ensembles of different types of phytoplankton with different deterministic motility capabilities are exposed to stochastic vertical water motion in the same water column for 24 hours to determine how the abilities of the organisms influence their simultaneous distributions.

METHODS

An IMSL subroutine for random number generation was used to generate two sets of 100 groups of 240 steps between 0 and 1. One set was used to multiply the chosen maximum possible vertical water speed every 0.1 hours to determine the instantaneous rate; the second set was used to determine the simultaneous instantaneous vertical water direction. These sets of random numbers were read into a biophysical model with a three layer ocean. The top layer between 0 and 15 m depth penetrated half the mixed layer [9] with the highest VEDC. It had downwelling velocities that were twice as fast as the upwelling velocities but that occurred only half as often to simulate Langmuir circulation [9]. The downwelling and upwelling velocities were uniform in the middle layer between 15 and 30 m depth and in the bottom layer below 30 m depth; it decreased by tenfold both between the top and the middle layers and between the middle and the bottom layers. Water temperature was 20° C throughout the water column, incident radiation followed a sine function through a 12 hour light period that peaked at 1500 uEin/m^2/s and water column extinction was 0.085 m^{-1}.

The VEDCs required to yield realistic results were determined by comparing the output of the biophysical model with the published fish egg results [10]. Fish eggs were assigned an ascent velocity of 1.8 mm/s. The biophysical model required VEDCs that were 8 times higher (Fig. 1) than those (K_s) in the simpler

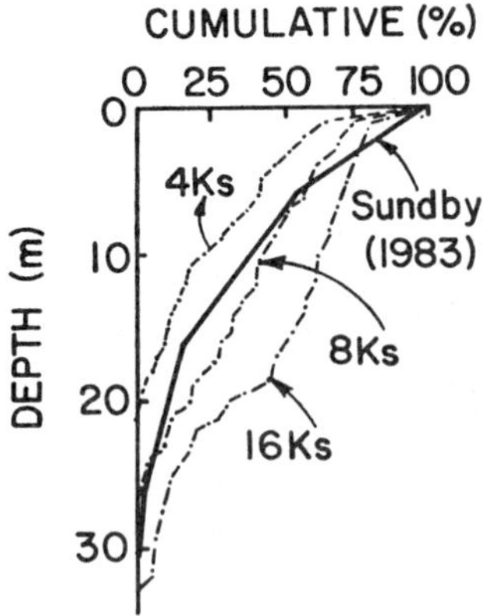

FIG.1. Calibration of the model (broken lines) against the fish egg study (solid line).

water column previously assumed [10]. The modeled VEDCs (K_z) for the phytoplankton comparisons were 16, 64, 144, 256, 400, and 576 cm^2/s. Instantaneous displacement (z(t)) at a given time (t) can be calculated from $z(t)=(2K_zt)^{1/2}$ for comparison with the biological velocities [11]. These VEDCs are equivalent to values below the residual vertical water motion observed in the fish egg study under calm conditions but provide a wide range of environmental forcing for present purposes. Phytoplankton motility was simulated using Stokes law [1] for non-motile phytoplankton (buoyancy rates: positive 50 um/s; neutral 0 um/s; negative -50 um/s) and a previously published motility model [3] for both dinoflagellates (~ 400 um/s) and ciliates (2000 or 5000 um/s) [12] that included the effects of gravity, temperature and light. The non-motile phytoplankton did not change direction with time of day; the motile phytoplankton, however, underwent imposed diel vertical migrations with daylight ascents (beginning at sunrise) and night descents (beginning at sunset). Inherent particle motion was combined with the vertical water motion through vector addition.

RESULTS

Fig. 2 examines the time course of fish egg distribution at a modeled VEDC of 710 cm^2/s, a value that approximates the residual vertical water motion in the fish egg study just after the wind

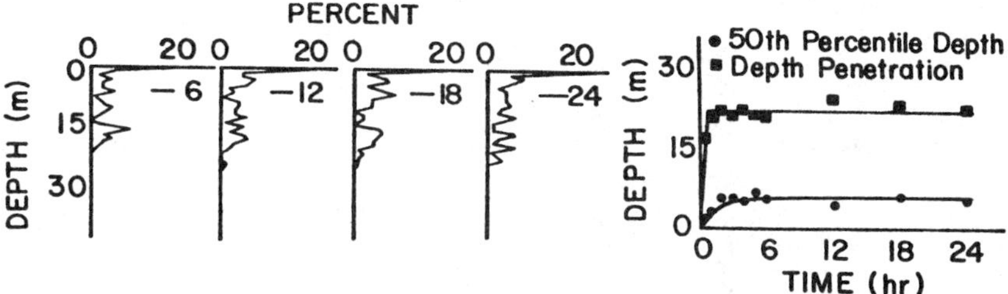

FIG.2. Time course of the model (12: sunrise) to reach equilibrium expressed in vertical profiles (left) with 50th percentiles (bars) and in an asymptotic plot of depth against time (right).

ceases. The left figure demonstrates that although the population started at the surface at the beginning of the run (time 0), all of the plotted profiles appear very similar. The right figure demonstrates that the depth penetration attains equilibrium after 1 hour; the 50th percentile requires about 3 hours. A single profile after 24 hours will therefore be used to compare the effects of water motion on the various phytoplankton motility types.

If the water column was static at all depths for 24 hours, the negatively buoyant cells would all end up at 4.3 m depth and the remaining phytoplankton motility types would all be near the surface. Fig. 3 successively compares the vertical profiles and the depth of the 50th percentiles of the different phytoplankton

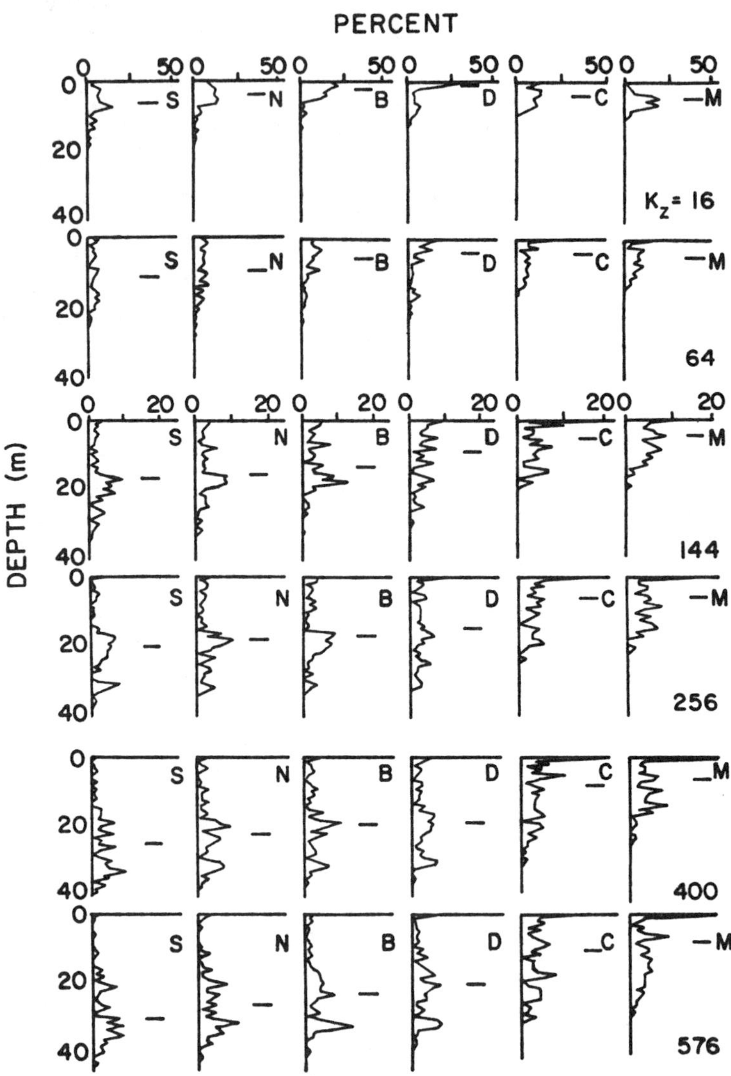

FIG. 3. Depth profiles (m) of the different phytoplankton types (S-negatively buoyant, N-neutrally buoyant, B-positively buoyant D-flagellate, C-slow ciliate and M-fast ciliate) exposed to increasing vertical water motion energy (K_Z).

types (in order of increasing capabilities from left to right) after 24 hours under increasingly energetic conditions of water motion (K_z). Even the weakest water motion of 16 cm^2/s is sufficient to influence the distribution of the cells. The depth penetration decreases from left to right, but the 50th percentile is shallowest for the flagellates which exhibit a surface peak. All of the other phytoplankton types exhibit subsurface peaks. At 64 cm^2/s, the depth penetration is deeper for all phytoplankton types. The 50th percentiles have also descended, but the flagellates remain higher in the water column than the ciliates. Distinct surface peaks are apparent for the flagellate and the slow ciliate. The other phytoplankton types are distributed more uniformly through a larger water column. At 144 cm^2/s (note the scale change), the depth penetration continues to increase and the effect of the middle layer becomes evident for the three non-motile phytoplankton types which show distinct subsurface peaks around 20 m depth. The three motile phytoplankton types still exhibit surface peaks; these are especially well developed for the ciliates. The 50th percentiles are now arranged in ascending order from left to right. At 256 cm^2/s, the effect of the bottom layer becomes evident for the negatively buoyant population. The other phytoplankton types exhibit some deepening, but the patterns are not very different from the previous case. The ciliates are now much higher in the water column than the other phytoplankton types. At 400 and 576 cm^2/s, the first four panels from the left appear quite similar. The rate of depth penetration has declined and peaks occur in the middle and bottom layers. The ciliates, however, remain much higher in the water column and continue to exhibit surface peaks.

Fig. 4 summarizes how the rate of water motion influences depth penetration (left) and the depth of the 50th percentile (right).

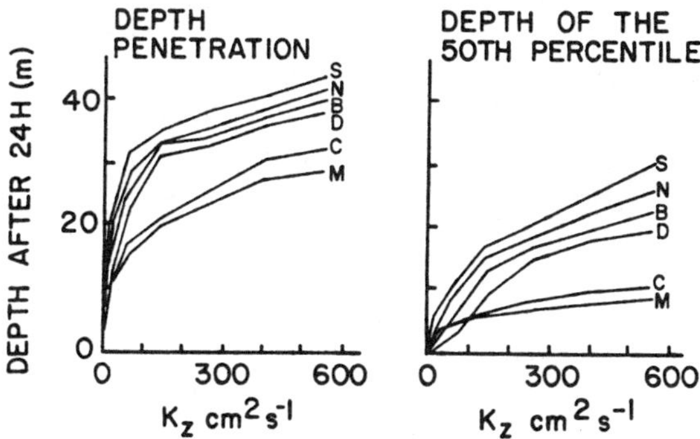

FIG.4. Summary of how increasing water motion influences depth penetration and the depth of the 50th percentile.

Both relationships are hyperbolic with the greatest changes occurring while the cells are exposed to the upper layer. As the intensity of the water motion increases, the less capable cells quickly penetrate below the upper layer. The ciliates, however, maintain an ability to occur in the upper water column.

DISCUSSION

The model combines recent views of water motion in the upper mixed layer with a range of phytoplankton capabilities to examine how these factors simultaneously influence cell distributions. As the energy of water motion increases, the range of the water column over which the organisms are distributed increases. The depth penetration and the 50th percentiles of the distributions of the different phytoplankton motility types, however, clearly exhibit the influence of the different phytoplankton capabilities. The various profiles under a given set of conditions can be combined to yield instantaneous profiles of phytoplankton community structure. Although the overall distributions appear relatively stable, the associations among individual cells are transient as a result of the complex trajectories from the random walk model.

ACKNOWLEDGEMENTS

Research supported by NSF grants 82-00159, 85-00589 and 88-18292, ONR contract N0014-85-C-3095, ONR grant 00014-89-J-1584 and a NCSU ICRAC grant. I thank G.J. Kirkpatrick, H. Yamazaki and E. Woloszyn for helpful discussions.

REFERENCES

1. A.E. Walsby and C.S. Reynolds in: The Physiology and Ecology of Phytoplankton I. Morris, ed. (University of California Press 1980) pp. 923-968.
2. P. Bienfang, E. Laws and W. Johnson, Experimental Marine Biology and Ecology 30, 283-300 (1977).
3. D. Kamykowski, S.A. McCollum and G.J. Kirkpatrick, Limnology and Oceanography 33, 66-78 (1988).
4. J.J. Cullen and S.G. Horrigan, Marine Biology 62, 81-89 (1981).
5. J.J. Cullen, M. Zhu, R.F. Davis and D.C. Pierson in: Toxic Dinoflagellates D.M. Anderson, A.W. White and D.G. Baden eds. (Elsevier, Amsterdam 1985) pp. 189-194.
6 S. Puiseux-Dao, Canadian Bulletin of Fish and Aquatic Sciences 210, 130-149 (1981).
7 S.W. Chisholm, Canadian Bulletin of Fish and Aquatic Sciences 210, 150-181 (1981).
8. J.C.J. Nihoul, Marine Turbulence (Elsevier, Amsterdam 1980).
9. R.A. Weller and J.F. Price, Deep Sea Research 35, 711-747 (1988).
10 S. Sundby, Deep Sea Research 30, 645-661 (1983).
11 K.L. Denman and A.E. Gargett, Limnology and Oceanography 28, 801-815 (1983).
12. T. Lindholm, Advances in Aquatic Microbiology 3, 1-48 (1985).

ON THE DEVELOPMENT OF THE *Chrysochromulina polylepis* BLOOM IN THE SKAGERRAK IN MAY - JUNE 1988

LINDAHL, O.,* and DAHL, E.,**
*Kristineberg Marine Biological station, S-450 34 Fiskebäckskil, Sweden; **Flødevigen Biological Station, N-4817 His, Norway

ABSTRACT.

In May and June 1988 a toxic and harmful bloom of *Chrysochromulina polylepis* occurred in the surface waters of the Skagerrak coasts of Sweden and Norway. The bloom first developed along the coasts of Sweden and Norway and later spread into offshore areas of the Skagerrak. The bloom of *C. polylepis* was unexpectedly toxic, and great harm was done to fish kept in farms as well as to a wide selection of organisms in the natural habitat. There are no previous reports that this species have occurred in massive blooms nor that it may become highly toxic.

INTRODUCTION.

A harmful bloom of the Prymnesiophycean flagellate *Chrysochromulina polylepis* developed in May 1988 in the Skagerrak area (Fig. 1). About twenty species of *Chrysochromulina* have been found in the Skagerrak and Kattegat area and it seems that the genus is relatively common in the area with a preponderance of occurrence in the period April to August [2, 5, 10, 11]. *Chrysochromulina* spp. has been recorded regularly during the 1980's in the Kattegat (T. Wilén, Environ. Prot. Bd., Uppsala, Sweden), although never before as an almost monospecific and harmful bloom. It should be emphasized that this group of species is easily overlooked by non-specialists examining phytoplankton samples and that electron microscopy is needed for identifying *Chrysochromulina* to species level.

The harmful bloom of *C. polylepis* initiated a large number of studies. The main purpose of this paper is to provide a brief description of the development, propagation and effects of the bloom in the Skagerrak area.

ENVIRONMENTAL AND NUTRIENT CONDITIONS BEFORE AND DURING THE BLOOM.

The winter and spring of 1988 in northern Europe were mild and very rainy. The precipitation during this period was approximately 2.5 times larger than the long-term mean value [1, 14]. This was the most conspicuous anomaly among the environmental conditions and led to increased nitrogen concentration in the waters of the sea before the spring bloom in March [1, 14].

During the development of the *C. polylepis* bloom in early May the concentrations of nitrate and phosphate in the Kattegat and Skagerrak surface waters were generally low (NO_3: 0.5 - 1.5 µM and PO_4: 0.05 µM) [1, 3, 14]. However, in April prior to the bloom high concentrations of nitrate (18 µM) were

Toxic Marine Phytoplankton
Edna Graneli et al., Editors

measured at some stations along the west coast of Denmark and in the Kattegat at or just below the pycnocline [1]. In the surface water the nitrate/phosphate, as atomic N/P ratio, ranged from about 10 along the Swedish coast [1] to 30 or more along the Norwegian coast [3] during the development of the C. polylepis bloom.

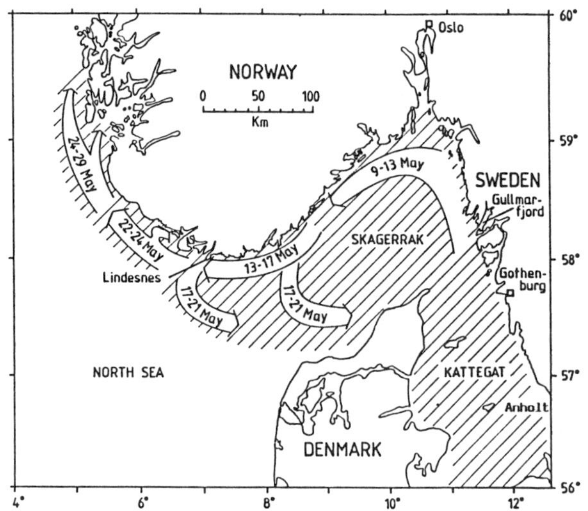

FIG. 1: The Skagerrak and Kattegat areas. The arrows indicate dates and spreading of the bloom and the shaded area shows where the bloom of C. polylepis occurred.

The nutrient concentrations in the deep water below the pycnocline were high during spring 1988 (NO_3: 14 μM, PO_4: 0.6 μM) with an unusually high N/P ratio (on average 24), indicating a surplus of nitrate to phosphate compared to the Redfield N/P ratio of 16 [1]. The nitrate-rich deep water occurring in the area during spring 1988 was identified as a former winter surface water orginating from either the central North Sea [14] or from the German Bight [1].

In the eastern part of the Kattegat and Skagerrak the harmful bloom developed in surface water with salinities of 17-23 o/oo [3, 14], above a pycnocline at 15 m depth which was more shallow and pronounced than normal [1]. The strong stratification was due to a large outflow of brackish water from the Baltic Sea in April and May combined with a rapid warming up (from 6 to 12 °C) of the surface layer due to calm and sunny weather during the two first weeks of May.

THE SPREADING OF THE BLOOM.

The first signs of a harmful bloom were recorded on 5 May as anomalous behavior of encaged fish, followed by the death of fish on 9 May. This occurred at a fish farm in the Gullmar fjord on the west coast of Sweden [14]. The following days harmful concentrations of C. polylepis were advected northwards and westwards along the coasts of the Skagerrak by coastal currents in the area [3, 14]. This was traced as severe mortalities at fish farms hit by the bloom. The plankton samples from the early stage of the fish kill contained up to $2 \cdot 10^{5}$ cells·l^{-1} of the dinoflagellate *Gyrodinium aureolum*, and it was first believed that this specie was the cause to the observed effects [14]. However, *G. aureolum* disappeared and by 17 May the causative organism was identified as *Chrysochromulina polylepis* [3].

By the middle of May the bloom had reached Lindesnes, the southernmost point of Norway. At that time a shift in the wind direction from easterly to westerly retarded the westward advection of the bloom and instead caused a spreading to the central parts of the Skagerrak [3]. After a few days the wind again changed to easterly and forced harmful concentrations of the algae westwards and closer to the Norwegian coast. At the end of May the bloom culminated on the southwest coast of Norway and then started to decline due to nutrient deficiency [9] without penetrating into the fjords. In central Skagerrak the bloom ended in the beginning of June [4].

THE DEVELOPMENT AND DYNAMICS OF THE BLOOM.

The first water sample containing *C. polylepis*, at a concentration of 18 400 cells·l^{-1}, was taken on 4 May 1988 near Anholt in the Kattegat (pers.comm. H. Kaas., Danish Lab. of Marine Pollution, Copenhagen, Denmark). The bloom was not visible as a discolouration of the water as it spread along the Swedish and Norwegian coasts, although the highest concentration of *C. polylepis* was found in the 0 - 5 m depth interval. However, the advection of the bloom was easily traced by its effects on a wide selection of marine organisms, both in cultivation and in the natural habitat. The concentration of cells during this first period of the bloom (5 - 15 May) was in the range $3 - 7 \cdot 10^{6}$ cells·l^{-1} [3, 14]. The chlorophyll *a* concentration was in the range 3 - 5 µg·l^{-1} [3, 14]. The toxicity of the bloom was substantial as testified by the fish kills and kills of other organisms, which occurred during this period [3, 14].

From 15 to 22 May the water along the Swedish coast became yellow-greenish and the plankton flora was almost a monoculture of *C. polylepis* [14]. The horizontal and vertical cell concentrations were patchy and variable with values up to $10 - 20 \cdot 10^{6}$ cells·l^{-1} [14]. The maximal concentration of the population at this time was found at 5 - 10 m depth and displayed a tendency to accumulate in strata [14]. Chlorophyll *a* concentrations up to 10 µg·l^{-1} were found during this period in samples taken in the above-mentioned strata [14]. In the laboratory the toxic effect during this period was conspicuous when organisms were exposed to sea water containing the algae. It was found, for example, that a *C. polylepis* concentration of $0.2 \cdot 10^{6}$ cells·l^{-1} was lethal to the copepods *Acartia* and *Daphnia* (LC50, 48 h) and that rainbow trout died after 1.5 h exposure, while cod were alive but affected (inert and swimming upside down) after 4.5 h exposure to a concentration of $12 \cdot 10^{6}$ cells·l^{-1} [14].

By measuring the primary productivity *in situ* by the ^{14}C-technique at 10 depths the daily primary production was calculated to 2.6 gC·m^{-2}·d^{-1} on 19 May in the Gullmar fjord [13]. This was a large, but not extreme, daily production compared to earlier records of primary production in the area (Lindahl, unpubl. data) [18]. The largest productivity over depth was found between 2 and 4 m and hardly any ^{14}C-uptake was measured below 8 m depth (Secchi depth = 5 m) [13]. This was in contrast to the primary productivity profile of C. polylepis found in early June in the Kattegat where the bloom formed a subsurface population and was actively photosynthesizing under the in situ light conditions at the pycnocline at about 15 m depth [17]. The standing stock of C. polylepis during the period 15 to 24 May was estimated to about 9.8 $\cdot 10^{10}$ cells$\cdot m^{-2}$ (= 3.1 $gC \cdot m^{-2}$) by integrating cell concentrations over depth [13], but the distribution was patchy [14]. From these results the turn over of C. polylepis was found to be about 1.2 days [13]. This is a rather high, but not exceptional, growth rate for a phototrophic alga at the ambient temperature of 12 °C [4], and consistent with preliminary data on the growth rate of the algae in cultures [4].

During the last stage of the C. polylepis bloom (23 - 30 May), cell concentrations between 40 and 80 $\cdot 10^{6}$ cells$\cdot l^{-1}$ could be found in thin layers close to the pycnocline at 15 - 20 m depth [4, 15]. The growth rate of C. polylepis had decreased considerably compared to the previous period (daily primary production 0.5 $gC \cdot m^{-2} \cdot d^{-1}$) and the generation time was calculated to be about 6 days [13]. The toxic effect of C. polylepis also seemed to have decreased in comparison to the previous period of the bloom since for example cod survived without any visible problems for 44 h in an algal concentration of 6 $\cdot 10^{6}$ cells$\cdot l^{-1}$ [14].

During the bloom the primary production (gross production = 27 $gC \cdot m^{-2}$), the biomass of the C. polylepis population, and the chlorophyll a concentration [13] were not exceptional compared to the phytoplankton blooms which have occurred during the last decade in the Kattegat and Skagerrak area [7]. The net production (new production ≈ 9.4 $gC \cdot m^{-2}$) of the bloom was estimated from the sedimentation of particulate organic carbon (POC) during the bloom. This material was caught by a cylindrical trap, with an aspect ratio of 6, suspended at 20 m depth in the Gullmar fjord and the carbon content was measured by a CHN-analyzer [13]. The macro-nutrient consumption (N ≈ 74 $mmol \cdot m^{-2}$, P ≈ 7 $mmol \cdot m^{-2}$) of the bloom [14] was calculated from the estimated net production [13] and the carbon, nitrogen and phosphate content of C. polylepis [4]. From these data it was suggested that the net production of the bloom may have been supplied by entrainment of nutrients (caused by the coastal currents) from the deep water during the bloom period [1, 14]. Thus, the amounts of nutrients available in the water prior to the bloom were sufficient to develop and sustain the produced biomass of C. polylepis. The high N/P ratio in the sea during the development of the bloom may have been of special importance because phosphate-limitation of C. polylepis in cultures seems to increase the toxicity of the algae [4, 12].

EFFECTS ON FARMED FISH.

Farmed trout and salmon died in order of decreasing size (personal observations and fish farmers reports [3, 14]). In Sweden it was observed during an early stage of the bloom that trout kept in water of low salinity (< 14 o/oo) survived even when large concentrations of the algae (> 20 $\cdot 10^{6}$ cells$\cdot l^{-1}$) were present [14]. These findings were confirmed by further studies in Norway [16]. The effect of _C. polylepis_ on trout and cod was investigated at different salinities and cell concentrations. It was found that osmoregulatory failure caused the fish mortality and furthermore, that the toxin of _C. polylepis_ interfered with gill membrane integrity [16].

About 120 fish farms were successfully moved into brackish water or into fjords where the bloom was either not harmful or not present. Nevertheless, about 800 metric tonnes of rainbow trout, atlantic salmon and cod were lost along the Swedish and Norwegian coast.

EFFECTS IN THE NATURAL HABITAT.

A number of observations of effects of the toxic bloom of _C. polylepis_ on the marine flora and fauna were made [6, 15]. These reports can be summarized as follows: many starfish (_Asterias_), molluscs (_Buccinum_, _Littorina_), crabs (_Carcinus_), sea urchins (_Psammechinus_) and fish (e.g. Rock cook (_Centrolabrus exoletus_ and _C. rupestris_), Ballan wrasse (_Labrus bergylta_) and Tadpole-fish (_Raniceps raninus_)) living in the algal belt (0 - 10 m depth) were killed. From underwater photographs it was apparent that other animals living in the algal zone suffered from the bloom but eventually survived. Sea-anemones (_Metridium senile_) for example, were retracted and loosing mucus [14]. In many areas divers found on average 1 - 5 dead fish per m^2 below rocky slopes, but in some localities much larger numbers of dead fish were found, especially in cracks and gorges. Furthermore, it was found that eggs of blue mussels (_Mytilus edulis_) and ascidians (_Ciona intestinalis_) would neither fertilize nor develop in water containing _C. polylepis_ [8].

Dead or dying red seaweed (_Delesseria sanguinea_) was also observed in many places. Laboratory experiments showed that _D. sanguinea_ changed colour in water contaning _C. polylepis_ from red to orange and then to green indicating pigment break-down [15]. The bloom seemed also to have affected phytoplankton species as empty shells of _Ceratium_ spp. were found offshore in the Skagerrak where _C. polylepis_ was present while at the same time "normal" cells of _Ceratium_ were found nearby "outside" the bloom [4]. Extremely few zooplankton were seen in the plankton samples taken during the period when _C. polylepis_ was dominating the plankton flora (authors observations). In the laboratory feeding experiments with a tintinnid (_Favella ehrenbergii_) and the copepods (_Acartia_ sp, _Centropages hamatus_ and _Temora longicornis_) showed that _C. polylepis_, depending on cell concentration, caused effects from reduced production of eggs [14] and reduced grazing [14, 19] to the death of the grazers [14, 19].

CONCLUDING REMARKS.

During the last decade there have been several exceptional and harmful blooms in the Skagerrak and Kattegat area [7,18]. These blooms may be an indication of eutrophication of the area (both through local runoff and through transport of water from adjacent sea areas) altering the nutrient balance of the system and thus the ecological balance and competition between phytoplankton species. However, the bloom of _C. polylepis_ was an unusual event [1, 3], and there have been many speculations

as to its cause. It is unlikely that it can be related to one single factor. Rather it was an unusual combination of physical, chemical and biological factors which contributed to create an ideal situation for growth and toxin production of *C. polylepis*. Factors causing the bloom seem to have been:
a. the preceeding mild winter with heavy precipitation and land run-off, and thereby increased nitrate-concentrations and N/P ratios of the coastal waters of northern Europe,
b. the bright, calm and warm weather, and a strong and shallow pycnocline during the development of the bloom and
c. the toxic properties of *C. polylepis* which may have killed co-occurring phytoplankton species and reduced the grazing of potential grazers.

ACKNOWLEDGEMENTS.

Special thanks to an unknown referee and the editors for valuable comments to the manuscript.

REFERENCES.

1. D.L. Aksnes, J. Aure, G.K. Furnes, H.R. Skjoldal and R. Sætre, Report BSC 89/1, Bergen Scientific Centre, Bergen, Norway (1988).
2. T. Christensen, C. Koch and H.A. Thomsen, Univ. Copenhagen, Universitetsbogladen, Copenhagen, 64 pp. (1986).
3. E. Dahl, Vann 3B: 512-524 (in Norwegian) (1988).
4. E. Dahl, O. Lindahl, E. Paasche and J. Throndsen, In E.M. Cosper *et al*. (eds.): A novel phytoplankton bloom. Causes and impacts of recurrent brown tides. Springer Lecture Notes on Coastal and Estuarine Studies (1989).
5. G. Espeland and J. Trondsen, Sarsia 71: 201-226 (1986).
6. J. Gjøsæter, Vann 3B: 524-535 (in Norwegian) (1988).
7. E. Granéli, P. Carlsson, P. Olsson, B. Sundström, W. Granéli and O. Lindahl, In E.M. Cosper *et al*. (eds.): A novel phytoplankton bloom. Causes and impacts of recurrent brown tides. Springer Lecture Notes on Coastal and Estuarine Studies (1989).
8. A. Granmo, J. Havenhand, K. Magnusson and I. Svane, J.Exp.Mar.Biol.Ecol. 124: 65-71 (1988).
9. T.M. Johnsen and E.R. Lømsland, this volume.
10. B.S.C. Leadbeater, Sarsia 49: 65-80 (1972).
11. B.S.C. Leadbeater, Sarsia 49: 107-124 (1972).
12. B. Edvardsen, F. Moy and E. Paasche, this volume.
13. O. Lindahl, in O. Skulberg (ed.): Toksinproduserende alger, symposium proceedings. Norwegian Inst. for Water Res. (NIVA), Oslo, Norway (1989).
14. O. Lindahl and R. Rosenberg, Report SNV PM 3602, Swedish Environmental Protection Board, Stockholm, Sweden, (in Swedish with English summary) (1989).
15. R. Rosenberg, O. Lindahl and H. Blanck, Ambio 17: 289-290 (1988).
16. H. Leivestad and B. Serigstad, Inst. Mar. Res. Bergen: Rep. BKO 8803 (1988).
17. H. Kaas, F.B. Pedersen, F. Møhlenberg and K. Richardson, this volume.
18. O. Lindahl and L. Hernroth, Mar.Ecol.Prog.Ser. 10: 119-126 (1983).
19. P. Carlsson, E. Granéli and P.J. Hansen, this volume.

MODELLING DINOPHYSIS BLOOMS : A FIRST APPROACH

A. MENESGUEN *, P. LASSUS **, F. DE CREMOUX ** and L. BOUTIBONNES **
Institut Français de Recherche pour l'Exploitation de la Mer (IFREMER)
* Centre de Brest - B.P. 70 - 29263 Plouzané ; ** Centre de Nantes - B.P. 109 - 44037 Nantes Cedex 01

ABSTRACT

This first attempt at numerical modelling of *Dinophysis* blooms is based on a 5-year survey in Vilaine Bay (southern Brittany, France) and on general ecological parameters for dinoflagellates reported in the literature. The purpose of the model was to determine from field data the role of classical processes involved in coastal oceanography rather than to develop a predictive tool. The biological submodel is based on 3 state variables : cell numbers and cell nitrogen and phosphorus quotas ; the driving variables are temperature, light, inorganic nitrogen and phosphorus and total chlorophyll content of water. The biological equations are of the classical Droop type. The physical submodel is a simple discretization of the water column into 4 boxes. Mixing among the boxes is due to vertical eddy diffusion which is time-variable according to sea-state. Daily alternative vertical migration of *Dinophysis* was also simulated. The results would seem to confirm the major role of vertical turbulence in inhibiting *Dinophysis* blooms. In the absence of vertical migration, the model indicates that there cannot be significant population growth due to the severe light limitation caused by water turbidity. The simulations argue for summer nitrogen limitation but do not account for the very acute bursts in cell numbers.

INTRODUCTION

It is only since 1983 that *Dinophysis* spp. have been considered responsible for diarrhetic poisoning of mussel consumers along the southern Brittany and Normandy coasts of France [1]. However, gastrointestinal disorders had been previously reported, especially in Vilaine Bay (southern Brittany) in 1978 and 1981. Consequently, forecasting *Dinophysis* blooms has been included as a future goal of the national monitoring program set up in 1984 to protect the shellfish industry and consumer health. Numerical modelling would appear to be the appropriate tool for forecasting, but it can only prove reliable if all major processes are taken into account and if relevant ecophysiological parameters are included in the model. The latter requirement has been just partially met due to a lack of durable pure *Dinophysis* sp. culture and the fact that only rough estimations relevant to other dinoflagellate cultures are available. However, existing *in situ* surveys can be used in conjunction with an ecological model to test the relative importance of the major physical and biological processes currently involved in dinoflagellates blooms. This paper concerns such an attempt but does not pretend to provide definitive answers. Preliminary results need to be validated on other sites, and the ecophysiological parameters updated once experiments on a *Dinophysis* culture are feasible.

Published 1990 by Elsevier Science Publishing Co., Inc.
Toxic Marine Phytoplankton
Edna Graneli et al., Editors

THE AVAILABLE DATA

Vilaine Bay (fig. 1) has been chosen as the pilot area for the monitoring of the hydro- biological context of Dinophysis sp. blooms. Each year since 1984, numerous sampling stations have been surveyed about every two weeks from April til September [2]. According to multivariate statistical analysis [3] of current hydrobiological data (temperature, salinity, dissolved inorganic nitrogen, dissolved inorganic phosphorus, silicates, chlorophyll, turbidity), the stations have been grouped into three main clusters, relative respectively to the estuarine, coastal and oceanic subregions of Vilaine Bay (fig. 1).

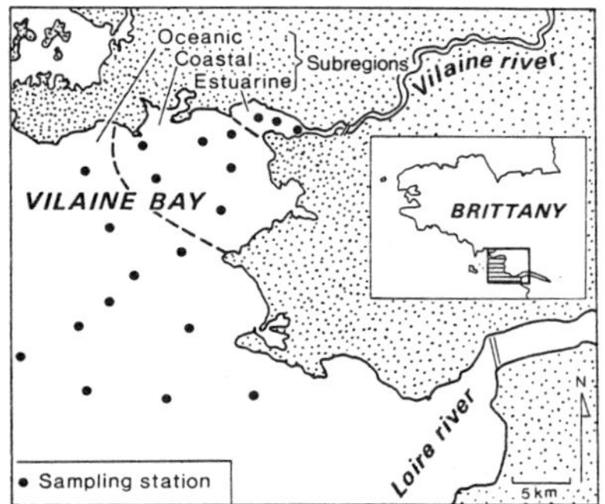

FIG. 1. Location of Vilaine Bay and the three subregions.

Due to the incompleteness of the data base in recent years, consideration here will be limited to 1984, a year in which Dinophysis counts were only performed at the - 1 m level. Figures 2 and 3 show the mean changes of two of the four environmental factors retained as driving variables for the model : temperature, dissolved inorganic nitrogen and dissolved inorganic phosphorus, chlorophyll a. For the oceanic subregion, the original sampling depths (- 3 m and - 5 m) were pooled, as were the depths - 20 m and - 30 m, in accordance with the spatial discretization used in the model (see below).

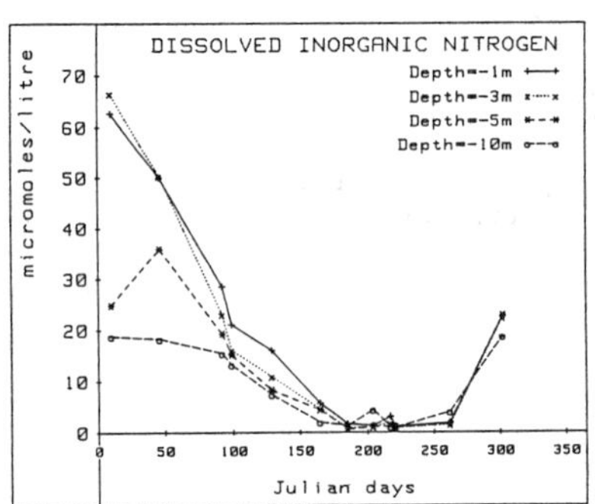

FIG. 2. Dissolved inorganic nitrogen in the coastal subregion.

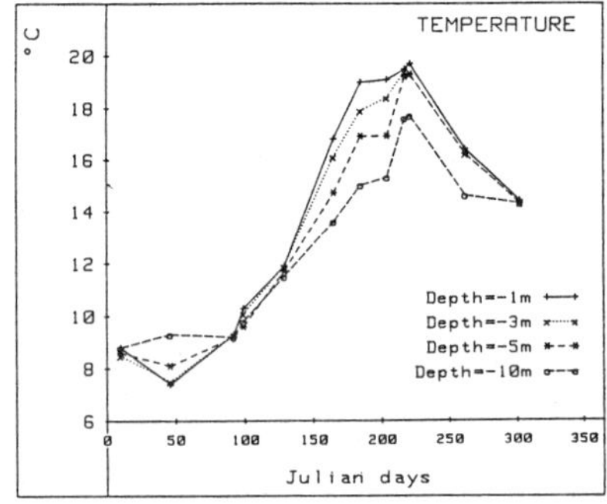

FIG. 3. Temperature in the coastal subregion.

During June, very strong stratification occured at - 6 m, caused by massive freshwater inflow from the Vilaine river in late May (haline component), which was followed by very fine weather during three weeks, so that a strong thermal stratification developed and reinforced the haline stratification (fig. 3). The persistence of such a stratified regime is related to the particularly weak local tidal residual drift [4] ; correlatively, wind-induced mixing becomes the main process involved in the disappearance of stratification. A very rough estimation of this physical driving variable is given by the daily sea-state (fig. 4). The final, indispensable driving variable involved is the photosynthetically active radiations (P.A.R.) : figure 5 shows an estimation of this variable from insolation measured at the nearest weather station (Nantes), following the computational procedure described by Brock [5].

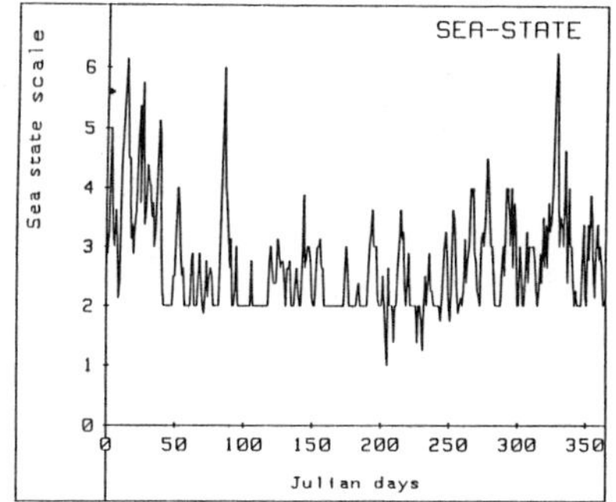

FIG. 4. Sea-state in Vilaine Bay.

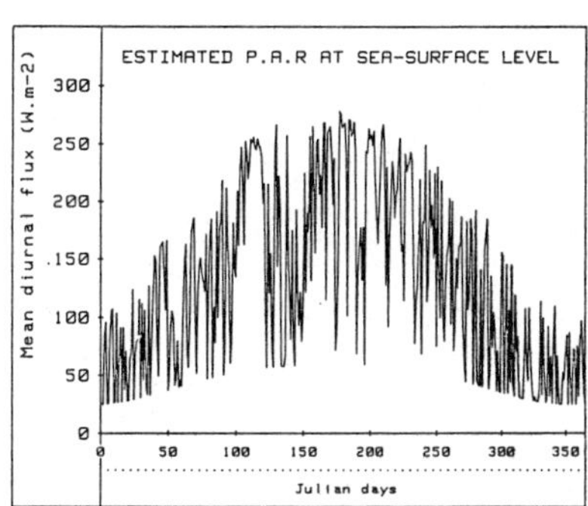

FIG. 5. Solar incident radiation in Vilaine Bay.

THE NUMERICAL MODEL

The role of advection in the seeding of *Dinophysis* populations in coastal embayments like Vilaine Bay, when practically no residual drift occurs because of absence of wind (*i.e.*, during major periods of *Dinophysis* blooming), was considered negligible with respect to vertical processes. Therefore the hydrodynamic submodel is reduced to a local, vertical pile of four boxes exchanging water by means of eddy diffusion alone. The eddy diffusion coefficient was empirically linked to the sea-state in an exponential form, and the bottoms of the boxes were set respectively at - 2 m, - 4 m, - 6 m and - 14 m in the coastal subregion and - 2 m, - 6 m, - 14 m and - 30 m in the oceanic subregion.

The biological submodel was constructed according to a principle of economy : only those processes were simulated which seemed necessary, and mechanisms considered to be of minor importance were either excluded or not explicitly formulated. The validity of this choice will be discussed below in terms of the goodness of fit of the model to data. Owing to the photosynthetic capabilities of *Dinophysis* [6], light limitation was simulated (Steele's formula) and shading taken into account (Riley's formula). The demonstrated capability of *Dinophysis* to migrate in the 10 m-deep subsurface layer ([6] & [7]) was also retained as a meaningful feature of *Dinophysis* ecology and simulated as a daily, purely sinusoïdal vertical movement between the first three boxes.

The puzzling discrepancy between observed blooming of Dinoflagellates in nutrient impoverished waters and the relative low absorption capacities of these species would also seem to be a critical feature. To monitor the role of nutrient limitation, a Droop model [8] with cell nitrogen and phosphorus quotas was adopted.

Finally, the sharp decrease in *Dinophysis* abundance as soon as the sea-state was equal to or greater than 3 probably resulted from the harmful effect of mechanical stress on cells. This link between turbulence and mortality of Dinoflagellates has often been mentioned but has not yet been quantitatively explored ; an empirical exponential relationship was set in the model.

Table I gives the differential equations governing the three state variables, and table 2 the parameter values. The scarcity (or variability) of published values for some parameters led to considerable inaccuracy in *a priori* values and required a longer and less reliable calibration stage.

TABLE I. Differential equations governing the model.

Number of *Dinophysis* cells in the j th box :

$$dX1j / dt = (\mu - mo \cdot \exp(\beta . s)) \cdot X1j + w/hj \cdot (X1i - X1j) + Ko \cdot \exp(\gamma . s) \cdot [(X1j{-}1 - X1j) / \Delta j,j{-}1 + (X1j{+}1 - X1j) / \Delta j,j{+}1] / Vj$$

Internal nitrogen stock in the j th box :

$$dX2j / dt = VNmax \cdot Nj / (KN + Nj) \cdot X1j - (\mu + mo \cdot \exp(\beta . s)) \cdot X2j + w/hj \cdot (X2i - X2j) + Ko \cdot \exp(\gamma . s) \cdot [(X2j{-}1 - X2j) / \Delta j,j{-}1 + (X2j{+}1 - X2j) / \Delta j,j{+}1] / Vj$$

Internal phosphorus stock in the j th box :

$$dX3j / dt = VPmax \cdot Pj / (Kp + Pj) \cdot X3j - (\mu + mo \cdot \exp(\beta . s)) \cdot X3j + w/hj \cdot (X3i - X2j) + Ko \cdot \exp(\gamma . s) \cdot [(X3j{-}1 - X3j) / \Delta j,j{-}1 + (X3j{+}1 - X3j) / \Delta j,j{+}1] / Vj$$

where : θ = temperature s = sea-state I = P.A.R. at the surface

vertical swimming velocity $w = wmax \cdot \sin(\pi . t/12)$ t in hours

box index $i = j + sgn(w)$ (boxes being counted from surface to bottom)

box thickness : hj box volume : Vj

distance between boxes j and j+1 : $\Delta j,j{+}1 = (hj + hj{+}1) / 2$

growth rate : $\mu = \mu 20 \cdot \exp(\alpha(\theta j - 20)) \cdot \min(Ilimj, Nlimj, Plimj)$

nitrogen limitation : $N\ limj = (X2j / X1j - qNO) / (X2j / X1j)$

phosphorus limitation : $P\ limj = (X3j / X1j - qPO) / (X3j / X1j)$

light limitation :

$$I\ limj = 1/hj \cdot \int_o^{hj} \int_o^{24} (Iz,t / Iopt) \cdot \exp(1 - Iz,t / Iopt) \cdot dt \cdot dz$$

available P.A.R. at depth Z :

$$Iz,t = It \cdot \exp(-(ko + 0.0088 \cdot Chl + 0.054 \cdot Chl^{2/3}) \cdot Z)$$

RESULTS AND DISCUSSION

Acceptable agreement between data and simulation can be achevied according to parameter values mentioned in Table II (see fig. 6a & b). However, the sharp increase in cell numbers at the beginning of July cannot be accounted for solely by the processes simulated by the model : for instance, either a concentration mechanism or a chemical stimulation of growth could be put forward.

TABLE II. Parameter values.

Parameter	Unit	Value	Comments
$\mu 20$	day-1	1.3	max growth rate at 20°C for Prorocentrum micans [8] Gymnodinium splendens [9].
α	°C-1	0.07	temperature coefficient giving a Q10 $\simeq$ 2.
qNO	pM.cell-1	.3	nitrogen minimum cell quota for Gymnodinium sp. [10].
qPO	pM.cell-1	.01	phosphorus minimum cell quota for Gymnodinium sp. [10].
VNmax	pM.cell-1.day-1	.35	
VPmax	pM.cell-1.day-1	.04	
KN	µM.l-1	8.5	
KP	µM.l-1	0.4	
Iopt	W.m-2	120	
wmax	m.day-1	15	after [7]
ko	m-1	.6	for coastal subregion } in situ measurements
		.325	for oceanic subregion } in situ measurements
mo	day-1	0.002	death rate in still water (s=0).
β		1.5	specific augmentation rate of death due to sea-state.
Ko	m2.s-1	10-5	diffusion coefficient in still water.
γ		1	specific augmentation rate of dispersion due to sea-state.

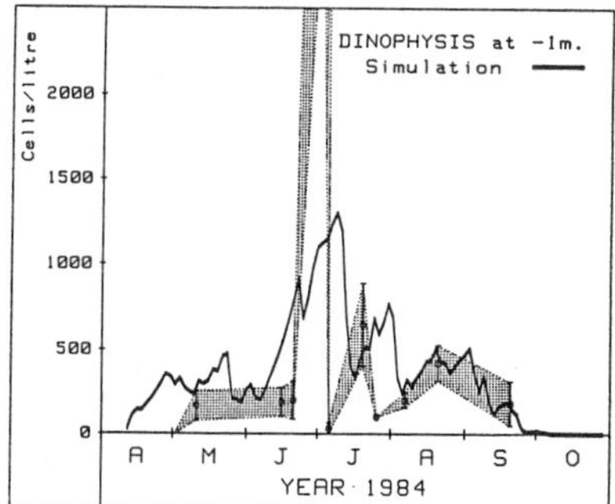

FIG. 6a. Simulated vs observed cell counts in the coastal subregion.

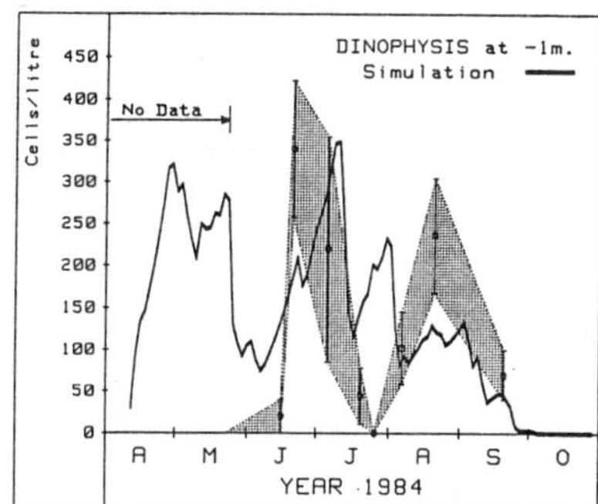

FIG. 6b. Simulated vs observed cell counts in the oceanic subregion

Despite its limitations, in part due to the lack of ecophysiological data, this model can be used to test the relevance of strong ecological mechanisms. Figure 7 shows the improbability of obtaining high abundances in these turbid waters when vertical migration is not allowed. On the contrary, figure 8 shows that, in summer and with migration, population should burst out when only light-limited. The nutrient limitation would seem to bring here an additional limitation strong enough to prevent such very high summer abundance. Moreover, trials with physiological parameters exaggerating the phosphorus limitation did not provide as realistic simulation as those obtained with prevailing nitrogen limitation. The exhaustion of phosphorus in the water occurs earlier (end of April) than that of nitrogen (end of June), especially in the coastal subregion. However, gradual replenishment of water phosphorus stock, and hence, of internal algal buffer, occurs during summer, whereas nitrogen concentrations remain low until late september. Such a nitrogen limitation of summer Dinoflagellates populations is not unplausible, since many red tides have been reported in eutrophicated waters (see [11]), where this nitrogen limitation often does not exist anymore. Some Gyrodinium aureolum blooms have been reported to be nitrogen-controlled ([12] & [13]). Measurements of nitrogen and phosphorus cell-quotas on in situ populations of Dinophysis are now needed to confirm or refute this hypothesis.

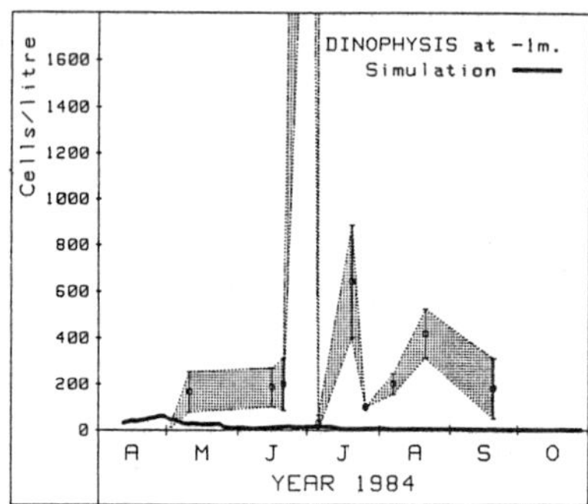

FIG. 7. Simulation in the coastal subregion without daily migration.

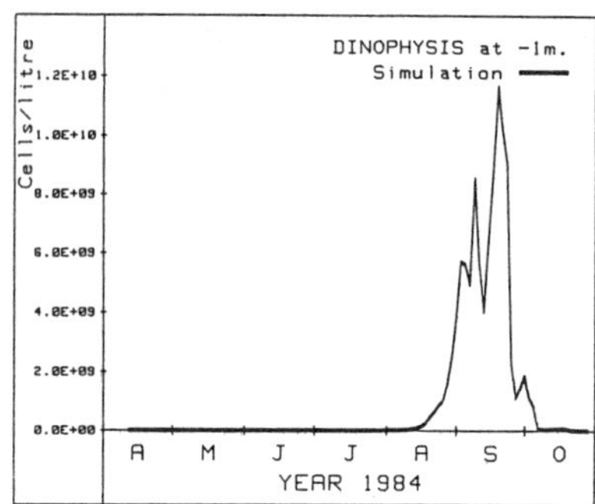

FIG. 8. Simulation in the coastal subregion without any nutrient limitation.

REFERENCES

1. P. Lassus, M. Bardouil, I. Truquet, P. Truquet, C. Le Baut, and M.J. Pierre, in : Toxic Dinoflagellates, D.M. Anderson, A.W. White and D.G. Baden, eds. (Elsevier, North Holland, 1985), pp. 159-164
2. P. Maggi, L. Soulard, I. Truquet, and J. Chauvin, in : Toxic blooms on French coasts : four-year studies, P. Lassus ed. (IFREMER Report No. DERO-88.03-MR, 1988), pp. 150-172.
3. G. Mer, Traitement mathématique des données hydrologiques acquises en Baie de Vilaine en 1984 et 1985 (IFREMER Report No. DERO-86.08-MR), 72 p.
4. J.C. Salomon, P. Lazure, Etude par modèle mathématique de quelques aspects de la circulation marine entre Quiberon et Noirmoutier. (IFREMER Report No. DERO-88.26-EL, 1988), 104 p.
5. T.D. Brock, Ecol. Modelling 14, pp. 1-19 (1981).
6. M. Durand-Clément, J.C. Clément, A. Moreau, N. Jeanne, S. Puiseux-Dao, Mar. Biol. 97, 37-44 (1988).
7. P. Lassus, F. Proniewski, C. Pigeon, L. Veret, L. Le Déan, M. Bardouil, and P. Truquet, Aquat. Liv. Resources (to be published).
8. J.M. Kain, G.E. Fogg, J. Mar. Biol. Ass. U.K., 39, 33-50 (1960).
9. H. Kayser, Mar. Biol. 52, 357-369 (1979).
10. J. Grim, Int. Rev. Hydrobiol. 39, 139-315 (1939).
11. F.J.R. Taylor, ed. The biology of Dinoflagellates. Botanical Monographs. Vol. 21 (1987). Blackwell Scientific Publ., 785 p.
12. P. Le Corre, J.L. Birrien, and P. Morin, in : Toxic blooms on French coasts : four-year studies, P. Lassus ed. (IFREMER Report No. DERO-88.03-MR, 1988) pp. 127-149.
13. G.K. Dixon and P.M. Holligan, J. Plank. Res. 11 (1), 105-118 (1989).

ROLE OF TEMPERATURE, SALINITY, AND LIGHT ON THE SEASONALITY OF *PROROCENTRUM LIMA* (Ehrenberg) DODGE

STEVE L. MORTON AND DEAN R. NORRIS
Florida Institute of Technology
Department of Oceanography and Ocean Engineering
Melbourne, FL 32901 USA

ABSTRACT

In laboratory temperature experiments of the toxic dinoflagellate, *Prorocentrum lima* (Ehrenberg) Dodge, temperatures >31° and <21°C limited division rate, however growth was possible from 16° to 33°C. Optimum growth occurred at 30°/oo salinity, division rates at 20 and 40°/oo were 75% and 30% of maximum respectively. In light experiments, fastest growth rates were achieved under blue wavelengths (450 nm) of light. *P. lima* could not grow at light intensities below 120 µW/cm² and was inhibited at light intensities above 4,400 µW/cm². Using optimum conditions of light, salinity, and temperature, maximum growth rates of >0.6 divisions/day could be sustained.

INTRODUCTION

Effects of temperature, salinity, light intensity, and light quality on the growth of the ciguatera related dinoflagellate *Prorocentrum lima* is not understood. *P. lima* grew best at water temperature of 27°C in the Florida Keys [1] and during the summer months on Papeete reef [2]. However, a slight correlation with water temperature was discovered in the Virgin Islands [3]. Maximum populations of *P. lima* commonly occur at depths of <3m [1,3] where light intensities are high. However, *P. lima* was found to be inhibited in shallow, sand bottom environments [4].
Recently *P. lima* was reported attached to sub-surface drift algae where light intensities are >10% of full sun light [5]. In regard to salinity, *P. lima* has been collected from mangrove roots, where salinities commonly reach over 40°/oo [1]. Production of okadaic acid by *P. lima* is found in the tropics but reduced or absent in the temperate seas [6].

Due to the differences in the abiotic factors reported for maximum biomass of *P. lima*, we experimentally examined the effects of temperature, salinity, light intensity, and light quality on growth rate of a clone from the Florida Keys. Finally, the optimum values of these parameters were applied to batch cultures to maximize cell yield.

METHODS

P. lima cells were isolated from *Heterosiphonia gibbesii* (Harvey) Falkenberg collected at Knight Key, Florida in November 1985 by J. Bomber. Clonal cultures, PL100A are kept in a modified K medium at 27°C on a 14:10 light:dark cycle [7]. The K medium was modified by omitting Tris, copper, and silica. Natural aged sea water was used for all stock cultures and

Toxic Marine Phytoplankton
Edna Graneli et al., Editors

experiments. Sea water was filtered through a 0.22 µm filter prior to autoclaving and enrichment. For all experiments, light was supplied by Vitalite fluorescent tubes (Dura-Lite).

All experiments were conducted in 25mm diameter tubes with approximately 25 ml of medium. Growth rates were assessed by vortexing the cultures on a Scientific Products Co. Mixer and then inserting the tube into a Hach Model 2100A Turbidimeter. Cell density was calibrated using a hemacytometer and tested for correlation to NTU's by the Pearson test [8]. Successive counts were used to calculate divisions/day [9]. Only values from log phase were used to calculate division rates.

Temperature Experiments

A temperature-gradient block [10] was constructed of an aluminum block set upon wooden legs. Light was supplied to the cultures through the bottom of the apparatus. Heat was supplied by a block heater on one end, cooling by cold water was supplied by a water cooler and run through copper tubing coiled at the opposite end of the block. This system produced a temperature gradient from 16° to 45°C. The inoculum for each treatment was acclimated at a rate of 5°C per 24 hours.

Salinity Experiments

Salinity experiments and light experiments were conducted in a water bath at 27±0.5°C. Salinity of sea water was increased by evaporation or reduced with double-distilled, carbon-filtered water. The inoculum for each treatment was acclimated at a rate of 5°/₀₀ per 24 hours.

Light Experiments

Light intensity received by the culture tube was determined using a Kahl Underwater Irradiometer Model 288WA310 set upon the tubes themselves. Irradiance values were recorded as µW/cm^2. Light intensity optimum of *P. lima* was determined by using a light intensity range of 120 to 4,400 µW/cm^2. Growth-rate comparison of *P. lima* was also conducted using irradiance approximate to that in three different water types: oceanic at the 1% isolume (425 nm peak intensity), shallow oceanic (450 nm) and coastal water (495 nm). These water types were simulated using plastic filters in conjuction with the light source. The exact peak wavelength of light transmitted by the plastic filters was measured using an IBM U.V. Visable Spectrometer Model 9420.

The effects of light quality on the growth of *P. lima* was compared by calculating gamma slopes [11]. *P. lima* was grown at four light limiting intensities with a minimum of 16 replicates. The resulting slopes of each plot of divisions/day vs. irradiance for each light quality was calculated using linear regression.

RESULTS

Temperature and Salinity

Ecological studies of the Florida Keys [1] very closely agree with laboratory data (Figure 1) in which peak growth (0.30

divisions/day) occurred at 26°C. Cultures at temperatures <19°C yielded no growth and could not be revived by gradually increasing the temperature. The coefficient of variation among replicates in the estimation of growth rates in the temperature experiments was 10.2%.

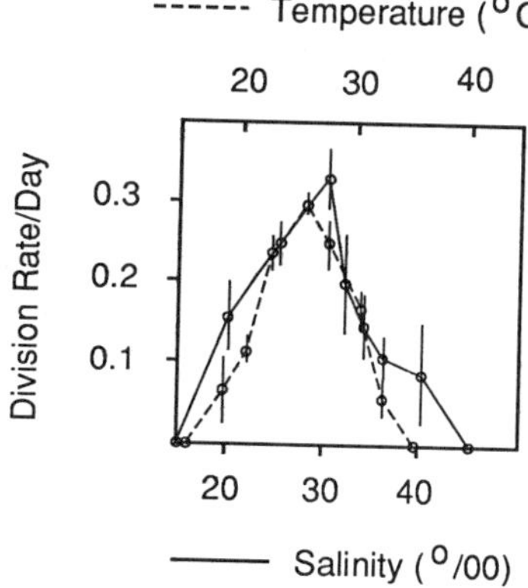

FIG. 1: Divisions of P. lima vs. temperature and salinity. Each point is the mean of four replicates and the bars are one standard deviation.

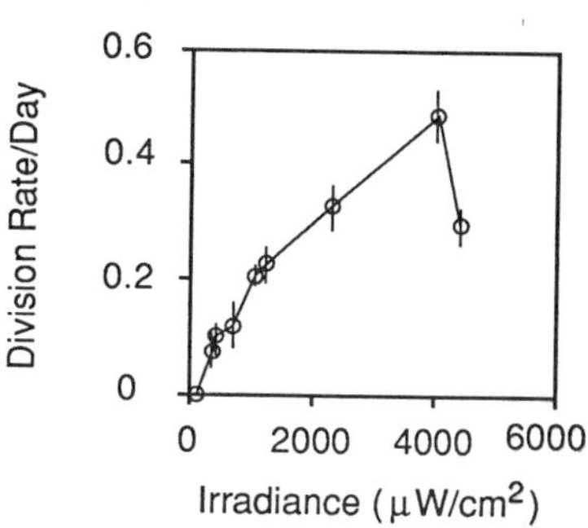

FIG. 2: Divisions of P. lima vs. irradiance. Each point is the mean of four replicates and the bars are one standard deviation.

Salinity optimum experiments are compiled in Figure 1 which indicates P. lima grew best at 30°/oo. The mean value was significantly different between growth rates from 20°/oo to 34°/oo. However, no significant difference was established between 34°/oo and 40°/oo. Compared to 30°/oo, division rates at 20°/oo and 40°/oo were 75% and 30% of maximum respectively. The coefficient of variation among replicates in the salinity-tolerance experiments was 9.5%.

Light Intensity and Light Quality

Light intensity experiments (Figure 2) reveal the maximum growth rate of 0.47 divisions/day acheived with an intensity of 4,000 $\mu W/cm^2$. This light intensity corresponds to aproximately 10% of full sunlight for mid-low light intensities on a clear, sunny day at noon [12]. Light intensities greater than 4,000 $\mu W/cm^2$ inhibited growth. Growth was limited (<0.2 divisions/day) at light intensities less than 680 $\mu W/cm^2$.
P. lima grew best under blue light (Figure 3) with a gamma slope of 0.58. Green light had a gamma slope of 0.24 and blue-violet light had a gamma slope of 0.38. If each linearly regressed slope was extended to the y axis, both the 450 nm and the 495 nm appear to grow without light. This may be a factor of the linearly regressed lines not taking in account the sudden decline in growth rate below 680 $\mu W/cm^2$.

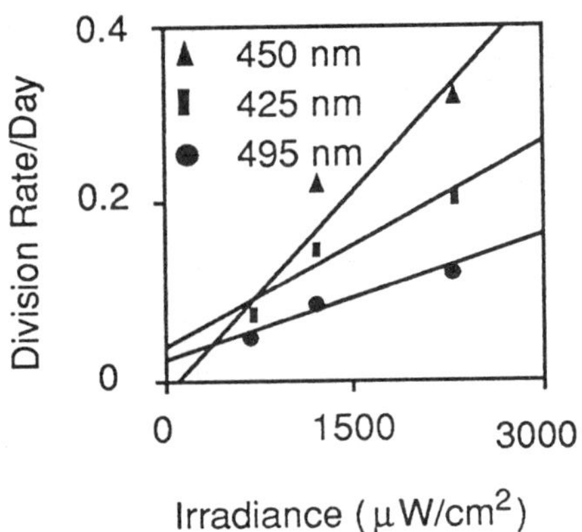

FIG. 3: Division of P. lima vs. irradiance. Each point is the mean of eight replicates.

DISCUSSION

Our temperature optimum of 26°C agrees with previous ecological studies [1,3]. However, this result disagrees with the findings from the Virgin Islands [3], where P. lima was correlated to water temperature.

Salinity experiments show that P. lima is inhibited by high salinities, growth does not occur above 40°/₀₀. This is surprising since P. lima is found in shallow near shore areas where salinities above 40°/₀₀ are common. The inhibition of growth at low salinities agrees with findings from ecological studies from the Gambier Islands [4] and may explain the positive correlation between precipitation and P. lima [3]. Faust (pers. comm.) found natural populations of P. lima at ca. 38°/₀₀ and 37° C at Twin Cays, Belize in May 1989.

P. lima was expected to have a high light intensity optimum because of the general preference to shallow water where light intensity exceeds 10% of full sunlight. Recently, P. lima has been observed as an epiphyte on drift algae [5] where it could experience high light intensities. However, P. lima was found to grow optimally at 10% of full sunlight which is average for dinoflagellates [12]. Results of light quality studies suggest P. lima grows best in waters where blue wavelengths dominate. This has also been shown for Gambierdiscus toxicus [13] and for P. maria-labouriae [14].

P. lima appears to lack any sustainable growth tolerence to periodic adverse abiotic tropical conditions (high temperature, salinity, and light). Bomber et al.[13] came to the same conclusion for G. toxicus. Okadaic acid produced by P. lima may act as an intracellular modulator to enhance or inhibit photoresponsiveness [15] or may act an allelopathic compound to other microalgae. The non-toxin production of P. lima in temperate climates may be a result of the absence of one or more of these stressful abiotic conditions. Growth over many generations of a toxic clone at low temperatures and low irradiance may lead to a lack of toxic production.

The greatest growth rate of P. lima in combined optimum conditions yielded a maximum growth rate of 0.62 divisions/day. This growth rate is greater than that reported for G. toxicus [13]. Since P. lima is found in greater numbers than G. toxicus in the Florida Keys, the species may have significant impact on ciguatera.

ACKNOWLEDGEMENTS

The authors would like to give special thanks to Dr. J.W. Bomber for help and advice on this project. Thanks also to M.S. Morton for operation of the Visable Spectrometer. This research was supported by Florida Institute of Technology, Department of Oceanography and Ocean Engineering.

REFERENCES

1. J.W. Bomber M.S. thesis, Florida Institute of Technology, Melbourne, FL, 104 pp (1985).
2. R. Bagnis, J. Bennett, C. Prieur, and A.A. Legrand in: Toxic Dinoflagellates D.M. Anderson, A.W. White, and D.G. Baden, eds. (Elsevier, Amsterdam 1985) pp. 177-182.
3. R.D. Carlson and D.R. Tindall in: Toxic Dinoflagellates D.M. Anderson, A.W. White, and D.G. Baden, eds. (Elsevier, Amsterdam 1985) pp. 171-176.
4. T. Yasumoto, Y. Oshima, A. Inoue, T. Ochi, R. Adachi, and Y. Fukuyo, Report to the Ministry of Education, 1980.
5. J.W. Bomber, S.L. Morton, J.A. Babinchak, D.R. Norris, and J.G. Morton, Bull Mar Sci 43, 204-214 (1985)
6. M. Kat in: Toxic Dinoflagellates D.M. Anderson, A.W. White, and D.G. Baden, eds. (Elsevier, Amsterdam 1985) pp. 73-77.
7. M.D. Keller and R.R.L. Guillard in: Toxic Dinoflagellates D.M. Anderson, A.W. White, and D.G. Baden, eds. (Elsevier, Amsterdam 1985) pp. 113-116.
8. R. Sokal and F.J. Rohlf, Biometry (W.H. Freeman & Co., California 1981) 859 pp.
9. R.R.L. Guillard in: Handbook of Phycological Methods J.R. Stein ed. (Cambridge University Press, Cambridge 1973) pp. 289-212.
10. W.F. Blankley and R.A. Lewin, Limnol. Oceanogr. 21, 457-462 (1976).
11. H.E. Glover, M.D. Keller, and R.W. Spinrad, J. Exp. Mar. Biol. Ecol. 105, 137-159 (1987).
12. R.R.L. Guillard and M.D. Keller in: Dinoflagellates D. Spector ed. (Academic Press, New York 1984) pp. 391-442.
13. J.W. Bomber, R.R.L. Guillard, and W.G. Nelson, J. Exp. Mar. Biol. Ecol. 115, pp. 53-65 (1988).
14. M.A. Faust, J.C. Sager, and B.W. Meeson, J. Phycol. 18 pp. 349-356 (1982).
15. K.A. Steidinger and D.G. Baden in: Dinoflagellates D. Spector ed. (Academic Press, New York 1984) pp. 201-261.

THE PERIODICITY OF GYMNODINIUM BREVE (DAVIS) IN SARONICOS GULF, AEGEAN SEA

K. PAGOU* AND L. IGNATIADES**
*National Centre For Marine Research, Aghios Kosmas, 166 04, Hellinikon, Greece; **N.R.C. "Democritos", Aghia Paraskevi, 153 10, Attiki, Greece.

ABSTRACT

The abundance and annual periodicity pattern of the unarmored dinoflagellate *Gymnodinium breve* (Davis) was studied in an inshore eutrophic station, fertilized continuously by domestic sewage. The station was located in Saronicos Gulf Aegean Sea and samples were collected monthly from 1m depth, during a seven year period (1977 - 1983) as well as in 1987. Inspection of the raw data showed that the *G. breve* population fluctuated rather regularly during the seven year period. Thus, a spectral analysis was applied on the log transformed data and the frequency spectrum was estimated using the Fast Fourier Transform procedure. The results showed that the spectrum was characterized by a large peak at the frequency of one cycle every twelve months. This cyclical pattern could be attributed to the seasonal nature of the data and not to random events. The same conclusion was reached by calculating the autocorrelation of *G. breve* and the cross - correlation between *G. breve* and temperature.

INTRODUCTION

The unarmoured dinoflagellate *Gymnodinium breve* (Davis) is the species commonly responsible for the "red tide" phenomena, fish kills and health implications (1). Previous studies have mainly been based on culture work and relate either to the physiology (2,3) or to the isolation and chemical characterization of toxins of this species (4). Our information on the ecological factors affecting the growth of *G. breve* in the natural environment are scanty (5,6).

In this work an attempt has been made to investigate the annual periodicity of *G. breve* in an inshore eutrophic environment. Interpretation of a series of 7-year data was made by spectral analysis.

MATERIALS AND METHODS

Data were collected monthly from January 1977 to December 1983, as well as during January - December 1987, from the Saronicos Gulf, Aegean Sea. The sampling station was located in an eutrophic environment near the Athens sewage outfall (7). Samples were drawn from 1m depth by a van Dorn bottle and analyzed for phosphates, nitrates and ammonia (8). Temperature and salinity were also recorded. Data collections might be limited to monthly sampling at lm depth but the length of sampling period (8 years) and the number of samples (96) are large enough to include the major pulses and interactions and assure statistically significant interpretations. Phytoplankton samples were preserved with Lugol's solution and counted in an inverted microscope. Species identification of *Gymnodinium breve* Davis, 1948 = *Ptychodiscus brevis* (Steidinger, 1979), using electron microscopy was not performed.

The data on cell abundances of *G. breve* were log transformed and analysed as follows:

Spectral analysis

This analysis is based on representing the series as a sum of sinusoids at the Fourier frequencies. The Fast Fourier Transform (FFT) procedure was used to compute power spectral estimates and squared amplitudes of the sinusoids were plotted (9).

Auto and cross correlation analysis

These procedures have been described elsewhere (7).

Trend removal

Smoothing of data by the moving average technique (9) did not improve them,

Published 1990 by Elsevier Science Publishing Co., Inc.
Toxic Marine Phytoplankton
Edna Graneli et al., Editors

therefore time series analysis was performed on unsmoothed data.

RESULTS AND DISCUSSION

The ranges and means of certain hydrographic parameters and G. breve concentrations during the period 1977 - 1983 are given in Table 1. Temperature fluctuated regularly with time (7) having well defined maxima (24.8 - 27.4°C) in summer and minima (12.0 - 13.9°C) in winter. The narrow range of salinity (36.30-38.99°/₀₀) indicates no considerable variation of this parameter with time.The levels of nutrient concentratins indicate the eutrophic character of thesamplingstation.

TABLE I. Range, mean and standard deviation of selected hydrographic parameters and Gymnodinium breve concentrations in 1m depth of Saronicos Gulf, during the period 1977 - 1983.

	Temperature °C	Salinity ‰	$P\text{-}PO_4$ µg-at/l	$N\text{-}NO_3$ µg-at/l	$N\text{-}NH_3$ µg-at/l	G.breve cells/l
Range	12.0-27.4	36.3-38.9	0.10-12.03	0.13-20.39	0.14-20.39	$2.0X10^2$-$2.7X10^7$
Mean	19.43	37.94	1.95	3.44	5.30	$6.7X10^5$
S.D.	4.67	0.53	1.91	3.67	4.48	$3.4X10^6$

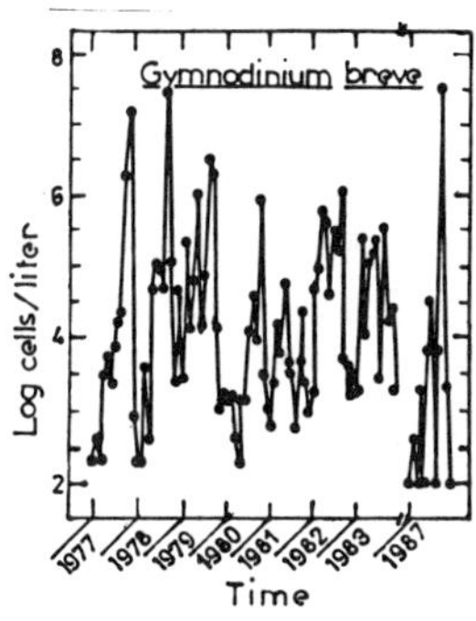

FIG.1. Raw data of cell concentrations of the species Gymnodinium breve during the period 1977-1983 and 1987.

The seasonal variation in abundance (raw data) of G. breve (Fig.1) showed certain maxima (mainly in autumn) and minima, but the regularity of fluctuation with time is not clear. Thus an attempt has been made to interpret these data by autocorrelation and spectral analysis.

The autocorrelogram (Fig.2) indicated a significant oscillation of the order of 12 months. This indication was further examined by spectral analysis. This method confirmed that the dominant frequency of the variance spectrum was 1cycle/12 months (Fig.3).

An attempt to interpret the annual fluctuations of G. breve was further made by computing the cross-correlation coefficients between the abundance of this species and certain physical (temperature, salinity) and chemical ($P\text{-}PO_4$, $N\text{-}NO_3$ and $N\text{-}NH_3$) parameters. Oscillations for most combinations were less apparent and out of phase (Fig.4B, C, D, E). Significantly positive cross-correlation coefficients were recorded only for the combination abundance - temperature (Fig.4A), indicating a strong effect of temperature upon G. breve growth.

All blooms of G. breve during the seven year experimental period (1977 - 1983) as well as during 1987, occurred during the stratification period, over a temperature range of 22.0 - 27.0°C.

The observed significant association of G. breve seasonal fluctuations with temperature might be explained in terms of the physiological response of this species as well as the state of the physical environment. Temperature may affect several physiological processes of G. breve such as growth (10) and photosynthetic capacity (2) and furthermore may regulate the water column stability, which is a major structuring factor in the upper layer of the euphotic zone (11).

Blooms of G. breve have been observed in Florida waters over a temperature range of 15.0 - 33.0°C (5) whereas the optimum temperature in culture was 22.0°C (10).

Also, it should be pointed out that fish kills were observed only twice (November 1977 and October 1987).

The obtained data support the view that temperature has a priority among the factors affecting the growth of Gymnodinium breve in the natural environment.

SUMMARY

The abundance and annual periodicity pattern of the unarmoured dinoflagellate Gymnodinium breve (Davis) was studied in an eutrophic polluted environment (Saronicos Gulf, Aegean Sea) during 1977-1983 and 1987. Spectral analysis of the data, as well as auto and cross correlation analysis confirmed the existence of a 12 month cyclic variation of Gymnodinium breve populations, regulated mainly by temperature.

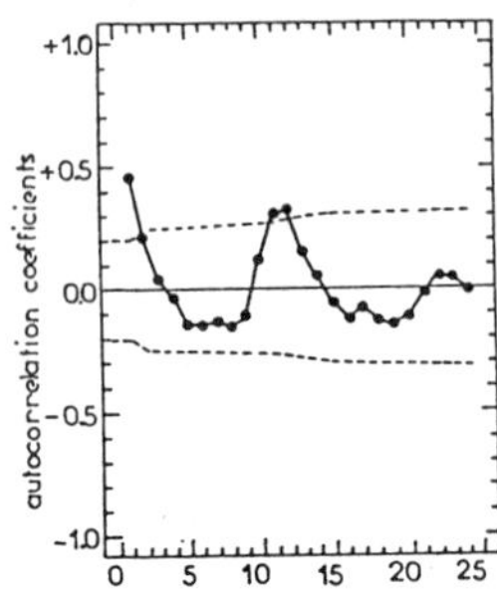

FIG.2. Autocorrelation for abundance of the species Gymnodinium breve for data of the period 1977-1983. Dashed lines indicate the confidence interval for a probability of 0.05.

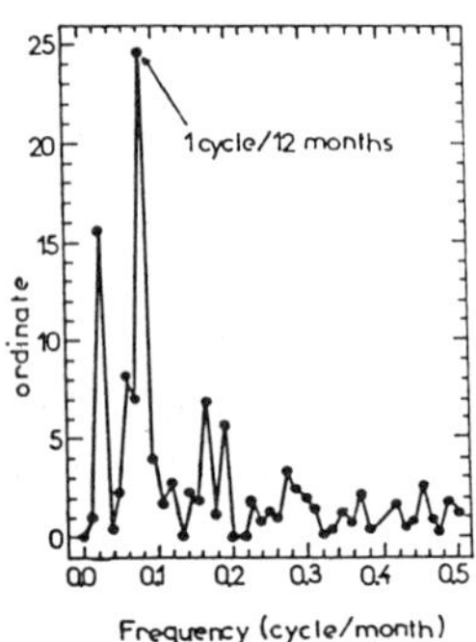

FIG.3. Variance spectrum for abundance of the species Gymnodinium breve for data of the period 1977 - 1983.

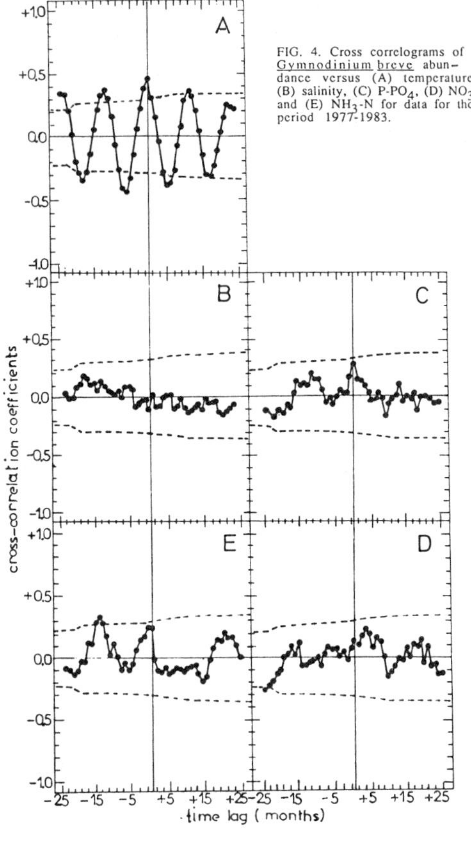

FIG. 4. Cross correlograms of Gymnodinium breve abundance versus (A) temperature, (B) salinity, (C) P-PO_4, (D) NO_3-N and (E) NH_3-N for data for the period 1977-1983.

REFERENCES

1. W.H. Hemmert in: Proc. First Intern. Conf. Toxic Dinofl. Blooms, V.R. LoCicero, ed., (Mass. Sci. Tech. Found., Wakefield, Mass. 1975) pp. 489-497.
2. D.V. Aldrich, Science 137, 988-990 (1962).
3. K.A. Steidinger, W,E. Truby, and C.J. Dawes, J. Phycol. 14, 72-79 (1978).
4. J. Golik, J.C. James, K. Nakanishi, and Y.Y. Lin, Tetrahedron Lett. 23, 2535-2539 (1982).
5. H.D. Baldbridge in: Proc. First Intern. Conf. Toxic Dinofl. Blooms, V.R. LoCicero, ed., (Mass. Sci. Tech. Found., Wakefield, Mass. 1975) pp. 69-79.
8. G.A. Rounsefell and A. Dragovich, Bull. Mar. Sci. 16, 404-422 (1988).
7. K. Pagou and L. Ignatiades, Biol. Oceanogr. 5, 229-241 (1988).
8. J.D. Strickland and T.R. Parsons, Bull. Fish. Res. Bd Can., 127 (1968).
9. L. Legendre and P. Legendre, Numerical Ecology (Elsevier, New York 1983).
10. D.L. Eng-Wilmot, W.S. Hitchcock, and D.F. Martin, Mar. Biol. 41, 71-77 (1977).
11. T. Wyatt and J. Horwood, Nature 244, 238-240 (1973).

MASS MORTALITY OF FISHES CAUSED BY DINOFLAGELLATE BLOOM IN GWADAR BAY, SOUTHWESTERN PAKISTAN

MOHAMMAD M. RABBANI*, ATIQ-UR-REHMAN* AND CLARENCE E. HARMS**
* National Institute of Oceanography, 37-K/6 P.E.C.H.S., Karachi Pakistan; ** Department of Biology, Westminster College, New Wilmington, Pennsylvania 16172, U.S.A.

ABSTRACT

This paper describes the natural event of mass mortality of fishes and its association with a dinoflagellate bloom. *Prorocentrum minimum* in the waters of Gwadar Bay, Pakistan (25° 07'N, 62° 23'E) during November 1987 caused brown waters due to its massive development. The bloom extended over 7 km^2 and lasted for 3 days. This is the first record of a massive bloom with this species in this part of the Arabian Sea. The cell number ranged from 8 to 45 x 10^6 per Liter, while all other species comprised only 4 to 6% of the total numbers. The massive development of bloom caused death to a unusual quantity of fishes: *Congresox* sp. (pike conger) was the most abundant and was estimated to comprise at least 60% of the mortality; *Pomadasys maculatum* (saddle grunt) and *Terapon puta* (small scaled terapon) were the next most common species to be severely affected. Many factors appear to have been responsible for the development of the bloom at that particular location and time. The relatively high temperatures of the surface water in the Bay (27 to 29° C) associated with high light intensity and moderate winds for most of the duration of the bloom were initiative factors to the development of the bloom. Its maintenance was supported by the permanent availability of the phosphate and nitrogen compounds in the mixed layer of the water column.

INTRODUCTION

Monospecific blooms of marine planktonic algae which cause red tides may be accompanied by death of fishes, invertebrates and sea birds or they may show no signs of discolouration nor fish and invertebrate kills [1].

Dinoflagellate and diatom blooms causing red tides are common in coastal and estuarine waters around the world, particularly in temperate and subtropical regions. The majority of red tides in the Indian Ocean are basically harmless events (caused by the blue-green alga, *Trichodesmium*; the dinoflagellates, *Noctiluca scintillans* and *Gonyaulax polygramma*; and the prymmesiophyte, *Phaeocytis*). Only in exceptional cases have such plankton blooms caused fish kills by the generation of anoxic conditions in waters. Subramanian and Purushothaman reported a bloom of *Hemidiscus hardmanianus* (Bacillariophyceae) in the coastal and estuarine waters of southern India which was associated with heavy mortality of fishes and invertebrates [2]. Such examples are rare in this region or they might not have been documented in the past due to lack of scientific facilities or expertise.

East Gwadar Bay, a semienclosed shallow coastal body of water, is highly productive and situated at the western coast of Pakistan. Due to its productive nature, there is significant fisheries activity in the bay. But in the last few years, the mortality among fish in association with red tide blooms was observerd by local residents. In November, 1987 the situation

Toxic Marine Phytoplankton
Edna Graneli et al., Editors

created panic among the fishermen and was reported in national newspapers. This attracted the attention of scientists. The National Institute of Oceanography (NIO) sent a team to Gwadar with the aim of investigating the primary causes for the red tide and its possible impact on fish mortality.

Red tides due to toxic dinoflagellates are well known from Europe, North and South America, and Japan. But no such scientific report has so far been made in the coastal areas of Pakistan. The present investigation is the first study of its kind from this region.

MATERIALS AND METHODS

There was intensive sampling of the area in which the marine life was severely affected. Samples were taken on a daily basis using Nansen bottles on board the R/B Kazi for a week until the bloom ceased.

Between 11 and 19 November, 1987, four sampling stations (Fig. 1) were established in the East Gwadar Bay waters on the western coast of Pakistan (25° 07'N, 62° 23'E), 360 km west northwest of Karachi. At each station, samples were collected from the surface at high tide for the determination of phytoplakton species composition, chlorophyll-a, temperature, salinity, dissolved oxygen and nutrient salts. Temperature was observed with simple thermometers and salinity was determined using an inductive salinometer. Subsamples for the various biological and chemical analyses were drawn immediately and processed using standard techniques. Wind velocity as well as light intensity and air temperatures were obtained from routine data of the Hydraulic Research Station, National Insititute of Oceanography, Gwadar, Pakistan.

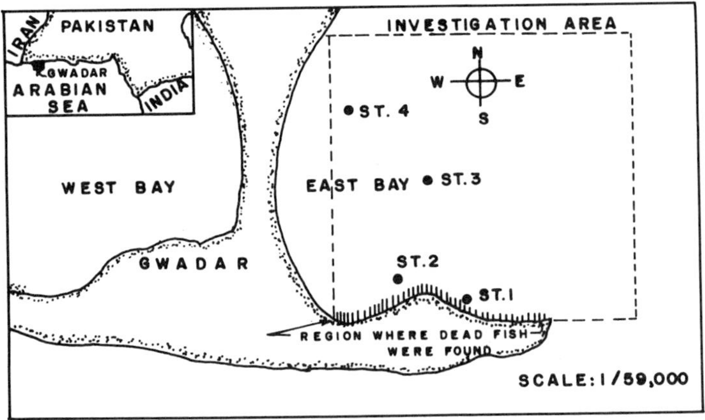

FIG. 1. Location of investigation area in East Gwadar Bay, Pakistan showing sampling stations and region where dead fish were found.

The water samples, for determining the number of phytoplankton cells per liter, were preserved in Lugol's iodine and analyzed by the Utermöhl technique [3]. In addition, plankton samples were collected with a small, 20 micrometer mesh net for further detailed studies. These samples were perserved in 4% formalin. Samples were studied under the inverted light microscope.

Chlorophyll-a was determined photometrically, using a Schimadzu UV-260 spectrophotometer, after concentration of samples on glass fiber filters and extraction in 90% acetone [4]. Determinations of dissolved inorganic phosphate, nitrite, nitrate and silicate were carried out on filtered sea water samples. The analytical techniques used to determine nutrient concentrations in water were those described in Strickland and Parsons [5]. Dissolved oxygen was measured according to the Winkler Method [5].

Dead fishes were collected along the shores of the bay, counted and identified on the spot. They were not preserved for further study.

RESULTS

The record of the National Insitute of Oceanography, Pakistan indicate that a few dead fishes are always found on the beach during the Northwest Monsoon (area marked on the map, Fig 1) but the quantity of dead fishes accumulating here (at the rate of 150 to 300 per day for a total of 592) over a three day period was highly unusual. The quantity of fishes collected is noted only for the east Gwadar beach and an unknown number of fishes would have been transported toward the open sea. Unfortunately the total number in the fish kill is not known. Most of the dead fishes collected (59.97%) were *Congresox* sp. (Pike conger). There were eight other species collected during the period 14 to 16 November. These are listed, with their numbers, in Table I. No dead fishes appeared on the beach after 16 November.

TABLE I. Dead fishes collected from East Gwadar beach during phytoplankton bloom, 14-16 November, 1987.

Scientific Name	English Name	Total No.	Per cent
Congresox sp.	Pike conger	355	59.97
Pomadasys maculatum	Saddle grunt	65	10.98
Terapon puta	Small scaled terapon	56	9.46
Platycephalus sp.	Flathead	25	4.22
Scoliodon sp.	Milk shark	25	4.22
Protonibea diacanthus	Spotted croaker	23	3.89
Arius sp.	Sea catfish	23	3.89
Nibea maculata	Blotched croaker	18	3.00
Otolithes ruber	Long-toothed salmon	2	0.34
TOTALS		592	100.%

The occurrence of the dinoflagellate, *Prorocentrum minimum*, in the waters of East Gwadar Bay on 11 November was initially traced from routine samples by a team of NIO Scientists engaged in a hydrographic survey of the bay. It was observed that the bloom extended over seven square kilometers (see Fig. 1). The colour of the sea over that area was yellow; then it became reddish-brown but never the clear blue as in the West Bay of Gwadar. Because of the obvious color, the NIO Scientists decided to immediately set up four monitoring stations in the East Bay area (Fig. 1).

The results of the investigation are conveyed in Tables II, III and IV as well as Figs. 2 and 3. These give details about the environmental conditions and biological parameters which were observed during 11 to 19 November, 1987.

Daily changes in cell density of *Prorocentrum minimum* at Station II are shown in Figure 2. The cells of this species were detected for the first time on 11 November when the concentration of cells was estimated to be about 19 x 10^6 cells per Liter; but on the next day the number decreased slightly. Thereafter numbers increased rapidly to a maximum in the period 12 to 14 November. This represented the formation of the real bloom. On 14 November, Prorocentrum minimum was in the stationary phase and the density was 45 x 10^6 cells per Liter. After a decline (9.5 x 10^6 cells per Liter) on 16 November, cells again began to increase and formed a one day bloom on 17 November. Thereafter the number decreased rapidly to a minimum (1.9 x 10^6 cells per Liter) on 19 November, until the ending of observations.

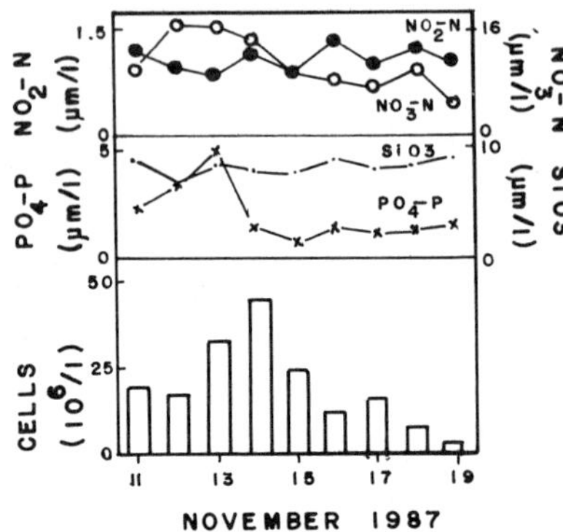

FIG. 2. Nutrients and abundance of Prorocentrum minimum at Station II during observation period.

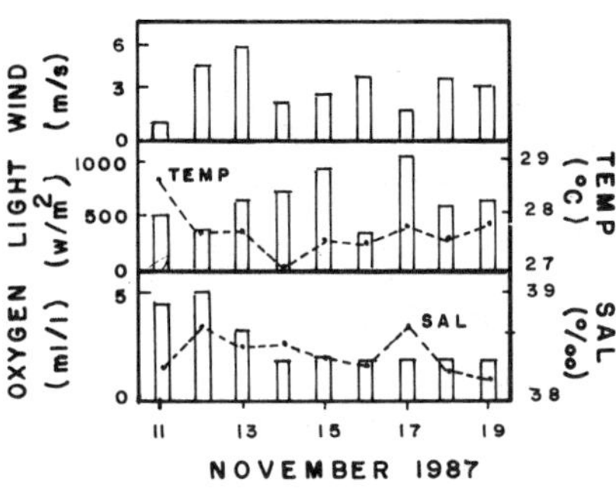

FIG. 3. Hydrological and metero-logical data at Station II during observation period.

The species composition of phytoplankton in various locations during the bloom period is summarized in Table II. It can be seen that the phytoplankton composition at all four stations is fairly constant. Among the species detected, Prorocentrum minimum, Prorocentrum micans, Peridinium sp., Gymnodinium sp ., Rhizosolenia sp. and Coscinodiscus sp. were always present. But Prorocentrum minimum was the largest component of the plankton at all four stations. The cell numbers ranged from 8.4 to 44.9 x 10^6 cells per Liter; all the other species comprised only 2 to 6% of the total in the investigated area.

TABLE II. Species composition of dominant phytoplankton in surface water at Stations I to IV on 14 November, 1987. Cell density is given in 10^6/Liter

Species name	St. I	St. II	St. III	St. IV
Prorocentrum minimum	15.72	44.89	18.17	8.40
Prorocentrum micans	0.36	1.82	0.46	0.08
Peridinium sp.	0.03	0.60	0.31	0.02
Gymnodinium sp.	0.01	0.03	0.01	0.02
Rhizosolenia sp.	0.06	0.06	0.09	0.01
Coscinodiscus sp.	0.06	0.08	0.01	0.02

The aerial distribution of Chlorophyll-a biomass in the surface water is shown in Table III. The maximum concentration (87.46 micrograms per Liter) during bloom period occurred at Station II whereas values as low as 13.41 micrograms per Liter were observed at Station IV during the same day (14 November).

Nutrients measured on 14 November at the four stations are presented on Table III. This shows high concentrations of nitrite and nitrate at Station II while silicate was found less at Station II. The phosphate concentration was more or less similar at all stations.

TABLE III. Chlorophyll-a (micrograms per Liter and Nutrient salts micro moles per Liter) in surface water at Stations I to IV on 14 November, 1987.

Station No.	Chlorophyll-a	Nitrite	Nitrate	Phosphate	Silicate
I	28.09	0.70	1.83	2.08	16.60
II	87.46	1.15	13.60	1.98	7.13
III	33.87	0.60	4.70	2.05	10.80
IV	13.41	0.85	0.80	2.15	10.25

Fig. 2 indicates that concentrations of nitrate and phosphate in the initial phase of bloom are higher than in post-bloom; whereas nitrite and silicate seem more or less similar during the entire observation period at Station II.

TABLE IV. Air temperature, water temperature, salinity and oxygen in surface waters at Stations I to IV on 14 November, 1987.

Station No.	Air Temp. (°C)	Water Temp. (°C)	Salinity (‰)	Oxygen (ml/L)
I	27.8	27.5	38.0	5.15
II	27.2	27.0	38.54	1.95
III	28.0	28.5	38.17	2.16
IV	28.0	28.8	38.15	4.42

The temporal fluctuations of oxygen concentration in water at Station II is shown in Fig. 3. The oxygen concentration fell to its lowest level during post phase of bloom. However, in the initial phase the concentration was about six times higher than those in post-bloom.

On the days of this study the mean wind speed was 3.09 m/s and maximum light intensity was 1020 W/m^2 (shown in Fig. 3). Temperature and salinity measurements were found to be in the range of 27-29°C and 38-39 ‰, respectively (Fig. 3). Additional information on environmental parameters is given in Table IV.

DISCUSSION

The bloom of *Prorocentrum minimum* in Gwadar Bay, particularly at Station II, might be associated with tidal mixing due to its particular location (see Station II in Fig. 1) and time (post summer monsoonal period which is characterized by weak northeast winds) [6]. However, the relatively high temperatures, high light intensity and moderate winds for most of the duration of the bloom were initiative factors to the bloom. This agrees with the observations by Kimor et.al. [7] in the Kiel Fjord.

A common feature of different hydrographic conditions for occurrences of dinoflagellate bloom supposes an appropriate combination between the stability of the water column and the continuous flow of nutrients [8]. In our case the bloom was supported by nitrogen and phosphate compounds which were supposedly released from bottom sediments (Fig. 2 and Table III). The results also indicate that high cell abundance of Prorocentrum minimum was only at near-shore Station II, where higher nutrient level was observed (Tables II and III).

The characteristic of this organism and its biogeography are two of the most important factors conditioning the appearance of this bloom in East Gwadar Bay. In general, this species is described as euryhaline and eurythermal [9], and other information about its biogeography is given in detail by Graneli [10]. There is, no doubt, a rapidly growing literature on this species, particularly in connection with its ecology and outbreaks of red tides; but its implication in red tides that kill fishes and invertebrates is inadequately reported. Tangen [9] found that Prorocentrum minimum together with other dinoflagellates, caused brown water through their extensive growth in Oslo Fjord during August to September, 1979. The same species was also described as potentially toxic to humans through shellfish poisoning [9]. Kimor et.al. [7], reported that Prorocentrum minimum appeared for the first time on the coast of Kiel Fjord. It is significant to note that the bloom of this species in the water of East Gwadar Bay was toxic because fish mortality was recorded from the area of East Gwadar Bay only during the bloom (Table II and Fig. 2). In fact, by the end of 16 November (post phase of bloom) a low cell density ($<16 \times 10^6$ Liter) was observed at the same time when the oxygen concentration was <2 ml/Liter (Fig. 3). Kimor et.al. [7] found no evidence of mortality among marine organisms during the bloom of the same species when its maximum density was 31.2×10^6 cell/ Liter in the water of Kiel Fjord.

Prorocentrum minimum is a potentially toxic dinoflagellate, as reported by several investigators. The numerical abundance (45×10^6 cell/ Liter) observed here certainly describes a real toxic bloom. All the other species comprised only 6% of the phytoplankton (Table II and Fig. 2). It is, therefore, suggested that a toxic species only has effect on marine organisms when its density reaches a threshold sufficient to react critically on marine life. As we found in our case on 14 November, 1987, mass mortality of fishes was caused by massive abundance of Prorocentrum minimum in the East Gwadar Bay, Pakistan.

REFERENCES

1. Steidinger, K.S. and K. Haddad, Bio. Science, 31, 814-819 (1981).
2. Subramanian, A. and A. Purushothaman, Limnol. Oceanogr. 30(4), 910-911 (1985).
3. Utermoehl, H. Lunndogee 9, 1-38 (1958).
4. SCOR-UNESCO, Monographs on Oceanog. Method. 1, 69 p. (1966).
5. Strickland, J.D. and T.R. Parsons, Bull. Fish. Res. Bd. Canada, 167, 208 p. (1968).
6. Quraishee, G.S., M. Ali and A. Lakhani, Hydraulic Research Station at Gwadar, NIO-TR-17/88, 105 pp (1988).
7. Kimor, B., A.G. Moigis, V. Dohms and C. Steienen, Mar. Ecol. Prog. Ser. 27, 209-215 (1985).
8. Tangen, K. Blyttia, 38, 145-158 (1980).
9. Tangen, K. Sarsia, 68, 1-7 (1983).
10. Graneli, E. Natural Swedish Research Board, 3293, 133 pp (1987).

DINOPHYSIS SPP TOXICITY AND RELATION TO ACCOMPANYING SPECIES

M. A. de M. SAMPAYO*, P. ALVITO**, S. FRANCA** and I. SOUSA**
*Instituto Nacional de Investigação das Pescas (INIP), Av. Brasilia, 1400 Lisboa; ** Instituto Nacional de Saúde (INSA), 1699 Lisboa Codex

ABSTRACT

In Portuguese coastal waters there is evidence that *Dinophysis acuta* Ehr. and *D. sacculus* Stein are the main responsible species for DSP-ocurrence in bivalves. However the relation between cell numbers and toxicity detection is dependent on the relative abundance of the non toxic accompanying species. In the present work an account is given of the found relation between toxicity detection, variable concentrations of the toxic species, and quantities of the non toxic accompanying species.

INTRODUCTION

The capacity to remove particles over a broad range of sizes, from large volumes of water is highly developed in many species of bivalve molluscs of commercial importance [1, 2] The particles removed from the environment are concentrated to levels many times those found in the water; the concentration factor may be in the range 10 to 1000 or more. Some of these captured particles are destined to be partially or completely degraded, thus making their components available to the nutritional pool of the bivalve. Certain compounds, for example phytotoxins may be selectively retained in some of the tissues, like those involved in diarrhetic shellfish poisoning (DSP) which are retained mainly in the hepatopancreas [3].

Filter feeding bivalve molluscs that have ingested toxic phytoplankton are known to cause shellfish poisoning risking consumers health. As a public health provision in the chain of activities involved in production, harvesting and marketing of bivalve molluscs for human consumption a monitoring programme has been established in Portugal since 1986. The programme includes screening of water samples from shellfish producing areas for toxic species, and mice bioassays for PSP and DSP toxins [4].

Although no human DSP intoxication, have been reported, positive DSP results have been found [5], mainly in human bivalve molluscs from areas where *Dinophysis acuta* and/or *D. sacculus* occur.

However, the cell numbers of the toxic species needed to contaminate shellfish is highly variable, and show a positive relation with the relative abundance of the non toxic accompanying species.

MATERIALS AND METHODS

Water samples from shellfish producing areas, were collected for phyto-plankton analyses from the surface or as an integrated 5 m water column, depending on the depth of the sampled area. Both fresh seawater and samples preserved in 4% formaldehyde-acetic acid solution were examined within 24 hours. Identification and counting of the species were carried out using a standard Zeiss photomicroscope. A Palmer-Maloney slide was used for quantification after concentration using a centrifuge [6].

For DSP detection in bivalve molluscs Yasumoto's method (1978) was used according to the practice in France [5, 7, 8].

Temperature and salinity range during the studied period were respectively 16-22°C and 26-35.5 ‰.

The presented data are from 1988.

Toxic Marine Phytoplankton
Edna Graneli et al., Editors

RESULTS AND DISCUSSION

In tables 1-6 the weekly data for D. acuta, D. sacculus, D. spp, total non toxic accompanying species, and toxicity degree in bivalve molluscs, from the first occurrence of the toxic species until the last detection of DSP are shown per different regions.

In most cases there is at least an interval of 2-3 weeks between the first occurrence of the toxic species and the detection of DSP in shellfish (Tables I-III). The length of the time lag is related to the concentration of the non toxic phytoplankton species co-occurring with Dinophysis-species. Between the last occurrence of toxic species and the disappearance of DSP from bivalve molluscs there is as well an interval of 2-3 weeks (Tables I-IV) also depending on the relative abundance of the non toxic phytoplankton.

At two regions, Albufeira lagoon in the Setubal area, and Algarve coastal lagoons, the potentially toxic species were detected without any contamination of shellfish with DSP (Tables V-VI). Albufeira lagoon is a highly productive lagoon with almost permanant phytoplankton blooms. It is the only site in Portugal where mussels are grown on rafts with very good results. Although the toxic species D. acuta and D. sacculus occurred in three consecutive weeks (40-42) in this lagoon in relatively high numbers, there toxicity was not detected. Non toxic accompanying species occurred in bloom proportions (Table V). The toxic species were found more sporadically in Algarve coastal lagoons (Table VI). They were recorded during the weeks 29, 31 and 35 and only for one of the sampling stations were found two times with a week interval. At this station the non toxic phytoplankton was more abundant compared to the other stations.

The toxicity detected in the shellfish was more persistent at the Aveiro region (coastal zone and lagoon) and at Figueira da Foz (Mondego estuary) (Tables I and II). Toxicity was related to the occurrence of D. acuta in combination with low numbers for the non toxic accompanying species. Toxicity disappeared two and three weeks, respectively after the last occurrence of the toxic species.

At the Obidos lagoon (Table III) toxicity was more intermittent, corresponding to longer period, (41 to 44th week), with low phytoplankton abundance.

At Setubal coastal zone (Table IV) the picture is not so clear as in the above regions. The potentially toxic species occurred in the area during the weeks 27, 28 and 30 without any DSP being detected in shellfish. During week35th the first positive toxicity result was found, although this occurred two weeks after a period of low concentration of non toxic species.

Our results suggest that toxin accumulation by shellfish through ingestion of toxic Dinophysys species is influenced by the relative abundance of non toxic species co-occurring with the toxic ones. The time needed to clear the bivalves from the toxins is related to the abundance of the non toxic phytoplankton species occurring in the affected area after the disappearance of the toxic species.

All facts are most likely connected with the filter feeding activity of bivalves with sufficient food supply of non toxic phytoplankton to meet the physiological needs of the bivalves, less toxins are accumulated and/or more toxins are metabolized/excreted [2].

The presented evidence pointing towards an influence from the relative proportion between toxic and non-toxic phytoplankton, and not just the absolute amount of toxic species, is of importance for shellfish fisheries management during DSP toxicity episodes.

TABLE 1

AVEIRO REGION - COASTAL ZONE AND LAGOON (1988)

SAMP. WEEK	SAMP. SITE	DINOPHYSIS (CELLS/L) ACUTA	SACCUL.	SPP	OTHER SPP C.*10^6/L	TOX. DEGREE
** 17	C.NOVA	-	100	200	0.17	-
** 25	C.NOVA	100	-	100	0.11	-
	S.JACINTO	200	-	800	0.13	-
28	C.NOVA	-	-	500	0.05	+++
29	C.NOVA	1000	-	1700	0.12	+++
30	C.NOVA	1000	-	1000	0.10	+++
	S.JACINTO	-	-	1000	0.10	+++
31	C.NOVA	9500	-	10800	0.16	+
	S.JACINTO	3000	-	3100	0.14	+++
32	C.NOVA	5000	-	5350	0.13	+
	S.JACINTO	1000	-	1200	0.06	+++
33	C.NOVA	4000	-	5350	0.06	+++
	S.JACINTO	1500	-	1700	0.04	+++
34	C.NOVA	8000	-	8200	0.16	+++
	S.JACINTO	500	-	500	0.02	+++
35	C.NOVA	200	-	200	0.61	+++
	S.JACINTO	300	-	300	0.05	?
36	C.NOVA	200	-	200	0.24	+++
	S.JACINTO	?	?	?	?	+++
37	C.NOVA	6500	-	7300	0.19	+++
	S.JACINTO	5000	-	7300	0.19	+++
38	C.NOVA	8500	-	10000	0.09	+++
	S.JACINTO	4500	-	4600	0.06	+++
39	C.NOVA	600	-	600	0.06	+++
	S.JACINTO	1000	-	1100	0.05	+++
40	C.NOVA	25500	-	27000	0.18	+++
	S.JACINTO	100	-	100	0.02	+++
41	C.NOVA	1000	-	2000	0.09	+++
	S.JACINTO	500	-	500	0.07	+++
42	C.NOVA	100	-	600	0.21	+++
	S.JACINTO	-	-	-	0.09	+++
43	C.NOVA	24000	-	24700	0.12	+++
	S.JACINTO	1000	-	1000	0.02	+++
44	C.NOVA	1500	-	1600	0.04	+++
	S.JACINTO	2500	-	2500	0.09	++
45	C.NOVA	-	-	200	0.09	+++
	S.JACINTO	-	-	-	0.05	+++
46	C.NOVA	(+)	-	(+)	0.04	+++
	S.JACINTO	-	-	-	0.01	+++
47	C.NOVA	-	-	100	0.05	?
	S.JACINTO	200	-	700	0.04	++
48	C.NOVA	-	-	-	0.06	++
	S.JACINTO	-	-	-	0.09	+
49	C.NOVA	-	-	-	0.10	-
	S.JACINTO	-	-	-	0.07	++

** weeks 18-24 and 26-27 without Dinophysis spp.
(+) present

TABLE 2

FIGUEIRA DA FOZ - MONDEGO ESTUARY (1988)

SAMP. WEEK	DINOPHYSIS(CEL./L) ACUTA	SPP	OTHER SPP $C.*10^6/L$	TOX. DEGREE
29	100	100	0.03	-
30	100	100	0.05	-
31	-	-	2.06	+++
32	600	1100	0.05	+++
33	6500	6700	0.01	+++
34	8000	8200	0.23	+++
35	100	100	1.10	+++
36	500	500	0.02	+++
37	8000	8400	0.34	+++
38	800	800	0.14	+++
39	3000	3500	0.07	++
40	12500	13000	0.15	+++
41	?	?	?	+++
42	2500	2500	0.22	+++
43	100	100	0.27	+++
44	500	500	0.05	+++
45	-	-	0.03	+++
46	-	-	0.14	++
47	-	-	0.93	++

TABLE 3

OBIDOS LAGOON (1988)

SAMP. WEEK	DINOPHYSIS (CELLS/L) ACUTA	SACCUL.	SPP	OTHER SPP $C.*10^6/L$	TOX. DEGREE
**25	-	100	100	17.70	-
29	500	-	500	1.60	-
30	-	100	100	8.50	-
31	3600	7000	10600	7.00	+++
32	4000	600	5400	0.70	++
33	200	100	300	19.00	-
**34	200	-	400	0.58	-
36	500	-	-	7.00	-
37	500	-	500	0.80	-
38	1000	-	1100	1.54	-
39	?	?	?	?	++
40	1000	-	1100	5.14	-
41	100	-	100	0.07	+++
42	500	-	1700	0.06	+++
43	-	-	-	0.06	+++
44	-	-	-	0.90	++

** weeks 26 to 28 and 35 without <u>Dinophysis</u> <u>Spp</u> in the plankton

TABLE 4

SETUBAL REGION - COASTAL ZONE (1988)

SAMP. WEEK	DINOPHYSIS (CELLS/L) ACUTA	SACCUL.	SPP	OTHER SPP C.*10^6/L	TOX. DEGREE
27	500	-	1100	0.10	-
**28	100	-	100	0.09	-
30	500	300	1100	0.11	-
33	-	-	(1)2000	0.07	-
35	-	-	-	0.02	+++
36	-	-	300	0.10	+++
37	100	-	1600	0.05	+++
38	-	-	200	0.07	-
39	100	-	(2) 300	0.04	+++
40	-	-	-	0.05	++
41	-	-	-	0.10	+
42	-	-	-	0.10	++

** weeks 29, 31, 32, 34 without Dinophysis Spp in the plankton

(1) 600 cells/l D. acuminata

(2) 100 cells/l D. acuminata

TABLE 5

SETUBAL REGION, ALBUFEIRA LAGOON (1988)

SAMP. WEEK	DINOPHYSIS (CELLS/L) ACUTA	SACCUL.	SPP	OTHER SPP C.*10^6/L	TOX. DEGREE
33	(+)	-	100	0.34	-
40	700	-	800	0.39	-
41	100	100	400	1.60	-
42	-	2000	2000	14.40	-
46	-	500	500	5.50	-
53	-	500	-	1.60	-

(+) present

TABLE 6

ALGARVE COASTAL LAGOONS (1988)

SAMP. WEEK	DINOPHYSIS (CELLS/L) ACUTA	SACCUL.	SPP	OTHER SPP C.*10^6/L	TOX. DEGREE
29	100	-	100	0.21	-
	300	-	300	0.30	-
	300	-	300	0.31	-
	(3)1000	-	2000	0.23	-
31	(3) 500	200	1950	0.43	-
	100	-	800	0.10	-
35	500	-	1000	0.21	-

(3) same sampling station

SUMMARY

1. In Portuguese coastal waters DSP has been detected in bivalves since 1987 although no human health problems have been reported.
2. DSP detection has followed roughly two weeks after the appearance in the phytoplankton of Dinophysis acuta and/or D. sacculus.
3. The time needed for shellfish to become toxic depends not only on the presence of the toxic species but mainly on the relative abundance of the non toxic accompanying species.
4. The clearance of bivalves from toxins is also dependent on the abundance of the non toxic phytoplankton area after the toxic species have disappeared.
5. In highly productive regions with high densities of non toxic phytoplankton species the presence of potentially toxic species in numbers which would have affected shellfish in less productive areas, do not affect the bivalves.
6. These findings are related to the filter feeding activity of bivalves in order to collect food supply sufficient for the physiological needs:
   a) If the non toxic accompanying species are scarce bivalves have to filtrate larger volumes of water, which has a concentrating effect on the toxic species;
   b) If the non toxic phytoplankton is abundant bivalves need to filtrate smaller volumes of water, diminishing the accumulation of toxic algal cells.
7. Our findings have practical implications for shellfish fisheries management during DSP toxicity episodes.

ACKNOWLEDGEMENTS

Partial support for the research by JNICT (Portugal), European Science Foundation and European Medical Research Council are greatly appreciated. Thanks are due to Calouste Gulbenkian-Foundation (Portugal) and to the local Committee of the Fourth International Conference on Toxic Marine Phytoplankon for support to attend the meeting.

REFERENCES

1. S. Galassi and W. J. Canzonier, Atti Soc. Ital. Sci. Nat. e Museo Stor. Nat., Milano 118, 198-206 (1977).
2. J. E. Winter, Aquaculture 13, 1-33 (1978).
3. M. Murata, M. Shimatani, H. Sugitani, Y. Oshima and T. Yasumoto, Bull. Jap. Soc. Sci, Fish. 48, 549-552 (1982).
4. S. Franca and J. F. Almeida in: Red Tides, Okaichi, Anderson and Nemoto, eds. (Elsevier, New York 1989) pp 93-96.
5. P. Alvito, I. Sousa, S. Franca and M. A. M. Sampayo in: 4th Int. Conf.on Toxic Marine Phytoplankton, Lund, Sweden, June 26-30 (1989).
6. A. Sournia, ed. Phytoplankton manual, Unesco, Page Brothers Ltd, U, K.(1978).
7. T. Yasumoto, Y. Oshima and M. Yamaguchi, Bull, Jap. Soc. Sci, Fish. 43, 207-211 (1978).
8. C. Marcaillou-Le Baut, D. Lucas and L. Le Dean in: Toxic Dinoflagellates, Anderson, White and Baden eds. (Elsevier, Amsterdam 1985) pp 485-488.

DINOFLAGELLATE-MICROZOOPLANKTON INTERACTIONS IN CHESAPEAKE BAY

KEVIN G. SELLNER AND DAVID C. BROWNLEE
The Academy of Natural Sciences, Benedict Estuarine Research Laboratory, Benedict, MD, USA, 20612

ABSTRACT

Large cross-bay differences in phytoplankton species and biomass were observed in mesohaline Chesapeake Bay in summers of 1986-1988. Eastern shore and central channel regions were typified by flagellates and small diatoms with low phytoplankton biomass. In contrast, phytoplankton densities increased along the western shore of the Bay primarily in response to dinoflagellate blooms. Summer chlorophyll levels in the blooms were significantly higher (up to 535 μg L^{-1}) than further east while phytoplankton biomass reached 2097 μgC L^{-1}. Over the three years, dinoflagellate carbon ranged from 219-1703 μgC L^{-1}. Using radioisotope labeling techniques, grazing rates on bloom and non-bloom phytoplankton assemblages and associated bacteria were determined for individual rotifer and tintinnid populations in 1986. Overall, *in situ* microzooplankton grazing pressure would remove little of ambient dinoflagellate standing stocks (21-53%) leaving the majority of dinoflagellate carbon for other heterotrophs in the system. It is therefore likely that dinoflagellate blooms frequent in mesohaline regions of the Chesapeake Bay from January through October provide a large and labile carbon reservoir for the "microbial loop" and indirectly high water column oxygen demand that may lead to water column hypoxic/anoxic conditions typically observed in summer in stratified portions of the Bay.

INTRODUCTION

Recent observations of high phytoplankton biomass and productivity in summer mesohaline Chesapeake Bay and the relationships of these phytoplankton accumulations (blooms) with nutrient recycling and oxygen demand in bottom waters of the stratified estuary [1] coupled with concerns over the possible increasing extent (15 fold) and duration of hypoxic/anoxic conditions in the Bay [2] resulted in a multi-year, multi-disciplinary program in the region to investigate relationships between stratification, local production and hypoxia/anoxia in mesohaline Chesapeake Bay. An intensive field program was initiated to: (1) determine the frequency and extent of phytoplankton blooms in the region; (2) determine whether pycnocline tilting could account for elevated production and oxygen demand in the region; (3) estimate the importance of bloom production in water column food webs as opposed to sedimentation of the bloom, decomposition at depth and further oxygen demand in oxygen-poor bottom waters of the region.

METHODOLOGY

Mesohaline Chesapeake Bay is characterized by a shallow (<10 m) broad (7 km) western shore region, a deep main channel (26 m) and a relatively shallow and narrow (2.6 km) littoral zone on the eastern shore. From spring, 1986-fall, 1988, cross-bay distributions of

Toxic Marine Phytoplankton
Edna Graneli et al., Editors

chlorophyll in the region (76°21′-30′W, 38°29′-30′N) were obtained from horizontal and vertical profiles of in vivo fluorescence subsequently calibrated with chlorophyll a concentrations determined spectrophotometrically on 90% acetone extracted grab samples [3]. In addition, vertical profiles of temperature, salinity, dissolved oxygen (DO) and plankton species composition, density and biomass were determined in 3-5 stations in each transect. In 1986, water samples were subsequently collected from chlorophyll-rich bloom regions (western shore; 76°30′W, 38°30′N) and non-bloom (central channel; 76°24′W, 38°29′N) areas and using $^{14}C-NaCO_3$ and ^3H-methyl thymidine labeling techniques, grazing rates of individual microzooplankton populations on bloom and non-bloom phytoplankton assemblages and associated bacterioplankton were estimated [4,5]. Total grazing demand by the entire microzooplankton community was estimated from the product of daily clearance rates, microzooplankton densities and phytoplankton biomass. The ratio of total grazing demand to available phytoplankton carbon represented the fraction of ambient biomass consumed each day.

RESULTS AND DISCUSSION

Lateral Heterogeneity in Phytoplankton

Strong vertical gradients in salinity, DO and chlorophyll were observed along the transect from the eastern-western shore of Chesapeake Bay over the three summer field seasons: as expected, chlorophyll concentrations and DO were highest in surface waters and declined with depth while salinity distributions showed an opposite pattern. More interesting, however, was the marked horizontal heterogeneity in chlorophyll concentrations, phytoplankton species composition and biomass along the transects. Throughout the study period, chlorophyll concentrations in surface waters were elevated along the western side of the Bay relative to pigment concentrations encountered in mid-channel or eastern shoreline regions (Fig. 1); maximum chlorophyll concentrations in blooms were 535 $\mu g\ L^{-1}$ (9 August 1988). Elevated chlorophyll levels generally resulted from high dinoflagellate densities and biomass. Dinoflagellates, principally *Gyrodinium*, *Gymnodinium* and *Ceratium*, were observed as the biomass-dominant taxa in western shore samples in 9 of 10 collections, replaced by flagellates and diatoms further offshore in mid-channel waters and eastern shore regions. Total phytoplankton biomass along the western shore of the Bay in August 1986 and 1987 ranged between 799-2097 $\mu gC\ L^{-1}$ with dinoflagellate biomass ranging from 374-1703 $\mu gC\ L^{-1}$. In contrast, mean dinoflagellate biomass never exceeded 388 $\mu gC\ L^{-1}$ in the surface mixed layer while diatom biomass ranged from 80-1288 $\mu gC\ L^{-1}$ in the main channel of the Bay.

Although high salinities and low DO concentrations typical of bottom waters in the main channel were observed in western shore, dinoflagellate-rich regions on some occasions, high dinoflagellate densities on the western shore were noted over a wide range of salinities and oxygen concentrations. These observations imply that cross-bay displacements of the pycnocline and intrusion of nutrient-rich, high salinity bottom waters into shallow western shoreline areas is only one mechanism that might lead to phytoplankton blooms in the region; other mechanisms must favor dinoflagellate growth and/or accumulation in the region including wind- or tidal current-induced resuspension of settled populations or cysts [6], introduction of external populations through passage of internal waves [7-9] and associated plankton [10], seaward transport of lower salinity waters and plankton along the western shore due to the earth's rotation and the development and passage of fronts [11,12].

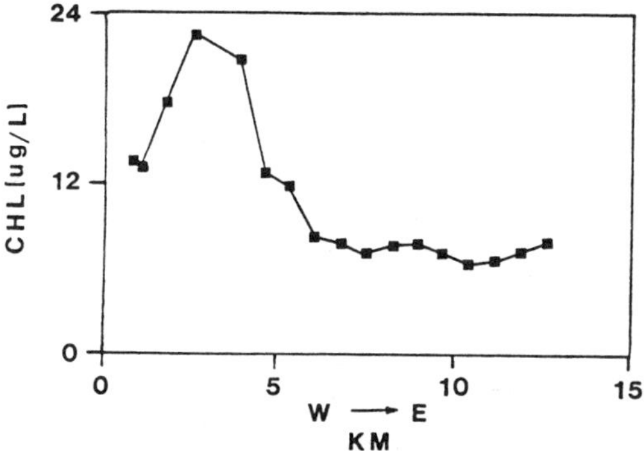

FIG. 1. Cross-bay distribution of chlorophyll, August, 1988.

Dinoflagellate Blooms and Microzooplankton Grazing

In 1986, microzooplankton densities followed a similar pattern to that observed for dinoflagellate biomass across mesohaline Chesapeake Bay: microzooplankton densities were highest in western shore regions of the Bay and declined to similar and lower levels in the main channel and eastern shore areas. For example, dinoflagellates (Gymnodinium, Ceratium and the heterotrophic species Polykrikos) dominated western shore phytoplankton biomass on 27 August along a transect at 38°30'N with cell densities reaching 10^6 L^{-1}. Dinoflagellate densities in central channel and eastern shore surface mixed layers were 0.2 and 0.1 x 10^6 L^{-1}, respectively. Microzooplankton densities, in turn, exceeded 9500 L^{-1} on the western shore with tintinnids (4880-5320 L^{-1}), oligotrichs (3280-4120 L^{-1}), other ciliates (<480 L^{-1}) and rotifers and copepod nauplii (~194-282 L^{-1}) smaller contributors. Microzooplankton densities in stations to the east ranged from 2901-6452 individuals L^{-1} and decreasing contributions for all groups.

Grazing rates for three microzooplankton populations were estimated for dinoflagellate-rich (western shore) and poor (central channel) phytoplankton assemblages and bacterioplankton along the same transect. Dinoflagellate biomass along the western shore and from the central channel approximated 325 and 90 μgC L^{-1}, respectively. The ciliate Favella sp., identified as a major grazer on dinoflagellates in other coastal systems [13], was found at densities of 213 and 104 individuals $^{-1}$ L^{-1}, respectively. As an herbivore, the tintinnid cleared 2.06 and 1.77 μl (individual·h)$^{-1}$ when feeding in high and low dinoflagellate biomass assemblages, respectively. Favella herbivory would remove, respectively, 11.8 and 3.9 μgC (L·d)$^{-1}$, very minor fractions (<2%) of available phytoplankton biomass in dinoflagellate-rich and poor assemblages suggesting that Favella control of dinoflagellates and other phytoplankton in the region would be minimal. Bacteria were removed at even lower rates from bloom and non-bloom plankton communities; clearance rates on ^{3}H-labeled bacteria were 0.07-0.09 μl (Favella·h)$^{-1}$.

An oligotrich, Laboea sp., was observed at densities of 2600 and 200 L^{-1} in dinoflagellate-rich and poor regions, respectively. This ciliate was typified by the highest clearance rates measured in the study, incorporating ^{14}C-label at rates of 31.2 and 2.9 μl (individual·h)$^{-1}$ from the two suites of phytoplankton. The high clearance rate for the small oligotrich is somewhat suspect perhaps indicating some mixotrophy in this population [14]. Applying the lower rate to inshore and main channel Laboea populations, the oligotrich would have removed 167 and 12 μgC $(L\cdot d)^{-1}$ from dinoflagellate-rich and poor assemblages, respectively, or 18% and 1% of total phytoplankton biomass in each phytoplankton assemblage daily. As in Favella, bacterivory in Laboea was low. Clearance rates on bacteria ranged from 3.6-6.5 μl $(\text{oligotrich}\cdot h)^{-1}$, removing <0.7 μgC $(L\cdot d)^{-1}$.

In contrast to the two ciliates, Synchaeta stylata preferred bacteria to phytoplankton prey. Grazing rates for this rotifer were 4-10 times higher on ^{3}H-labeled bacteria {2.1-4.5 μl $(\text{individual}\cdot h)^{-1}$} than on ^{14}C-labeled phytoplankton {0.4-0.7 μl $(\text{rotifer}\cdot h)^{-1}$} resulting in bacterial carbon forming 61% and 34% of total carbon ingested by the rotifer in dinoflagellate-rich and poor assemblages. S. stylata herbivory removed insignificant quantities of phytoplankton carbon in the regions typified by high and low dinoflagellate biomass. Total rotifer demand amounted to 0.33 and 0.03 μgC $(L\cdot d)^{-1}$, far less than 1% of available phytoplankton biomass in either of the two phytoplankton assemblages.

Estimates of total microzooplankton herbivory were obtained from the products of "average" clearance rates of the dominant taxa encountered in May (not reported here) and August and ambient microzooplankton densities assuming no net growth in either plankton component. For August, 1986, total microzooplankton would consume 21-53% (highest demand in oligotrich-dominated communities) of the available phytoplankton carbon. Using this range of grazing pressures in bloom and non-bloom regions, total microzooplankton demand in shallow, dinoflagellate-rich western shore areas versus demand in dinoflagellate-poor central channel regions has been estimated (Table 1) indicating high algal biomass remains ungrazed (2.2-7 gC m^{-2}) in mesohaline Chesapeake Bay, potentially supporting aperiodic menhaden schools, or more likely, pico-to-microheterotrophic metabolism in the system (see below).

TABLE 1. Average daily post-grazing bloom and non-bloom phytoplankton carbon in the surface mixed layer (6 m), summer mesohaline Chesapeake Bay.

PERIOD	BLOOM (gC m^{-2})	NON-BLOOM (gC m^{-2})	BLOOM/NON-BLOOM
AUG 1986	4.4-7.0	1.8-2.9	2.4
AUG 1987	3.7-6.0	1.8-3.0	2.1
JUL 1987	3.3-5.3	1.1-1.7	3.0
AUG 1988	2.2-3.5	1.3-2.1	1.7

These general results suggest that dinoflagellate blooms in mesohaline Chesapeake Bay might support carbon and nutrient flow and oxygen demand as depicted in Figure 2. There are indications that the disappearance of dinoflagellate blooms from the region is not accompanied by massive sedimentation events as noted for spring diatom blooms. Boynton et al. [15] have deployed sediment traps at mid-depth in the main channel of the study area over the last four years. Species composition of trap-collected particles is dinoflagellate-poor even though dinoflagellate blooms dominate the water column only several kilometers to the west (see above). One of several mechanisms supporting the absence of the ungrazed dinoflagellates in the traps is a loss of cellular integrity of the bloom dinoflagellates in the surface mixed layer, as suggested for Ptychodiscus brevis in Florida waters [16], with an accompanying release of cytoplasmic materials, ideal substrates for bacterioplankton in supra-pycnocline depths. With bacterial densities approximating 10^7 cells ml^{-1} in summer surface waters [17,18] and strong correlations between bacterial densities and water column respiration [17], bacterial utilization of post-bloom DOM pools could result in even higher bacterial densities and associated oxygen demand or elevated predation of bacteria by larger numbers of heterotrophic flagellates and rotifers like Synchaeta stylata, either scenario leading to higher oxygen consumption in the "microbial loop" in the upper water column.

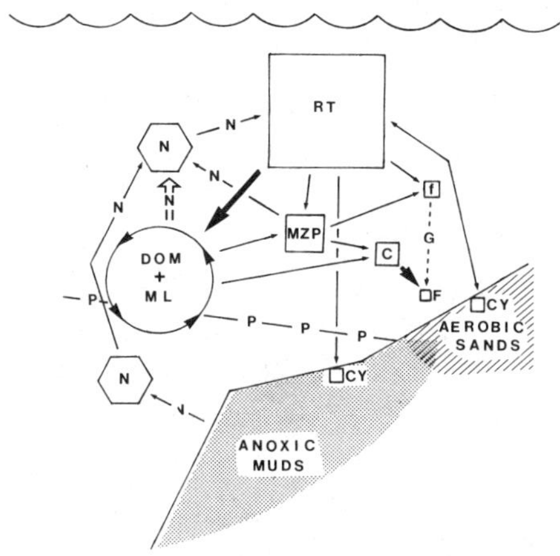

FIG.2. Conceptual model of carbon and nutrient cycling in dinoflagellate bloom in mesohaline Chesapeake Bay. Solid arrows represent movement of energy or biomass between compartments with dominant carbon flow from dinoflagellate blooms (RT = red tide) to and within the DOM pool-microbial loop (ML) and from copepods (C) to fish (F). Nutrient cycling is denoted with arrows labeled with N; nutrient supply is via diffusion through the pycnocline, "tilting" of the pycnocline, destratification and regeneration within the microbial loop. Dotted arrow labeled with G indicates (G)rowth from herbivorous fish (f; e.g. young menhaden) to adults. Other symbols are: P-P-P = pycnocline; MZP = microzooplankton; CY = dinoflagellate cysts.

Data presented above suggest that dinoflagellate blooms are frequent events in mesohaline Chesapeake Bay, occurring at time scales more rapid than might be explained by cross-bay displacements of the pycnocline. Dinoflagellate biomass accumulates in shallow depths along the western shore of Chesapeake Bay in excess of microzooplankton grazing pressure and as measured previously, macrozooplankton demand as well [19]. However, due to the absence of substantial dinoflagellate biomass signals in sediment traps moored in the region, it appears dinoflagellate-associated pools of C, N and P might be remineralized in the surface mixed layer perhaps maintaining large required nutrient pools in lighted euphotic depths for continued phytoplankton production. A repeating summer cycle of dinoflagellate growth, decline and remineralization above the pycnocline removes some of the dependence of continued high surface phytoplankton production on stochastic destratification events or slow upward cross-pycnocline diffusion of nutrients.

ACKNOWLEDGEMENTS

The authors would like to acknowledge the support from NOAA, MD Sea Grant College as well as the assistance provided by K.R. Braun, S.G. Brownlee, M.H. Bundy, L. Chaput, D.C. DeMattia, J.R. Dolan, S. Hedrick, R.V. Lacouture, M.M. Marsh, B.B. Wagoner and G. Ziegler.

REFERENCES

1. T.C. Malone, W.M. Kemp, H.W. Ducklow, W.R. Boynton, J.H. Tuttle, and R.B. Jonas, Mar. Ecol. Prog. Ser. 32, 149-160 (1986).
2. G.B. Mackiernan, D.A. Flemer, W. Nehlsen, and V.K. Tippie in: Chesapeake Bay: A Framework for Action, U.S. Environmental Protection Agency (Philadelphia 1983) pp. 15-35.
3. J.D.H. Strickland, and T.R. Parsons, A Practical Handbook of Seawater Analysis (Fish. Res. Bd. Canada, Ottawa 1972).
4. M.R. Roman, and P.A. Rublee, Mar. Biol. 65, 303-309 (1981).
5. E.J. Lessard, and E. Swift, Mar. Biol. 87, 289-299 (1985).
6. L.M. Walker in: Marine Plankton Life Cycle Strategies, K.A. Steidinger, and L.M. Walker, eds. (CRC Press, Inc., Boca Raton, FL 1984) pp. 19-34.
7. A. Brandt, C.C. Sarabun, and D.C. Dubbel, EOS 66, 1269 (1985).
8. D.C. Dubbel, A. Brandt, and C.C. Sarabun, EOS 66, 1269 (1985).
9. C.C. Sarabun, C.J. Vogt, and A. Brandt, EOS 66, 1269 (1985).
10. D. Kamykowski, Mar. Biol. 50, 289-303 (1979).
11. R.D. Pingree, P.R. Pugh, P.M. Holligan, and G.R. Foster, Nature 258, 672-677 (1975).
12. H.H. Seliger, K.R. McKinley, W.H. Biggley, R.B. Rivkin, and K.R.H. Aspden, Mar. Biol. 61, 119-131 (1981).
13. D. Stoecker, R.R.L. Guillard, and R.M. Kavee, Biol. Bull. 160, 136-145 (1981).
14. D.K. Stoecker, A. Taniguchi, and A.E. Michaels, Mar. Ecol. Prog. Ser. 50, 241-254 (1989).
15. W.R. Boynton, W.M. Kemp, J. Garber, J.M. Barnes, L.L. Robertson, and J.L. Watts, Chesapeake Bay Water Quality Monitoring Program, Ecosystems Processes Component (University of Maryland, Solomons, MD 1988).
16. R.H. Pierce, M.S. Henry, and L.S. Proffitt in: Toxic Marine Phytoplankton (this volume) (Elsevier, NY 1989).
17. J.H. Tuttle, R.B. Jonas, and T.C. Malone in: Contaminant Problems & Management of Living Chesapeake Bay Resources, S.K. Majumdar, L.W. Hall, Jr., and H.M. Austin, eds. (Penn. Acad. Sci., Philadelphia 1987) pp. 442-472.
18. H.W. Ducklow, E.A. Peele, S.M. Hill, and H.L. Quinby in: Understanding the Estuary: Advances in Chesapeake Bay Research, M.P. Lynch, and E.C. Krome, eds. (U.S. EPA and CRC, Inc., Solomons, MD 1988) pp. 511-523.
19. K.G. Sellner, and M.M. Olson in: Toxic Dinoflagellates, D.M. Anderson, A.W. White, and D.G. Baden, eds. (Elsevier, NY 1985) pp. 245-250.

TOXIC BLOOMS OF THE DOMOIC ACID CONTAINING DIATOM *NITZSCHIA PUNGENS* IN THE CARDIGAN RIVER, PRINCE EDWARD ISLAND, IN 1988

JOHN C. SMITH[1], ROLAND CORMIER[1], JEAN WORMS[1], C.J. BIRD[2], M.A. QUILLIAM[2], ROGER POCKLINGTON[3], RANDALL ANGUS[4] and LOUIS HANIC[5]
[1]Gulf Fisheries Centre, P.O. Box 5030, Moncton, N.B., E1C 9B6, Canada; [2]Atlantic Research Laboratory, National Research Council of Canada, 1411 Oxford Street, Halifax, N.S., B3J 2S7, Canada; [3]Department of Fisheries and Oceans, Science Directorate, Bedford Institute of Oceanography, P.O. Box 1006, Dartmouth, N.S., B2Y 4A2, Canada; [4]Miminegash Research Station, Miminegash, P.E.I., C0B 1S0, Canada; [5]Biology Department, University of Prince Edward Island, Charlottetown, P.E.I., Canada.

ABSTRACT

As in 1987, blooms of *Nitzschia pungens* again appeared in the Cardigan Bay region of eastern Prince Edward Island, Canada in the fall of 1988. These blooms contained the toxin domoic acid and were associated with domoic acid toxicity in cultivated mussels. The time course of the development of the *N. pungens* blooms and their relationship to the uptake and depuration of domoic acid in mussels are described. Monitoring the phytoplankton population provided an early warning of the impending domoic acid buildup of mussels. The concurrent changes in physical, chemical and meteorological variables suggest that the blooms were initially limited by the supply of nitrogen and declined because of low winter temperatures and irradiance. The initial supply of nitrogen was correlated with rainfall patterns and wind events, but the precise source and mechanism of nitrogen supply are yet to be determined.

INTRODUCTION

In late 1987, many people became seriously ill after consuming mussels cultured in eastern Prince Edward Island. There were over 100 confirmed cases with either gastrointestinal or gastrointestinal and neurological symptoms. Although most of the victims apparently recovered fully, about 12 were left with severe memory loss and from three to five elderly persons died. The chronology of these events [1] and the epidemiology of the poisoning [2] have been described elsewhere.

The symptoms of the poisoning were not those of either paralytic or diarrhetic shellfish poisoning and it was quickly shown that the new toxin was domoic acid [3]. There is very strong evidence that, in this case, the producer of the domoic acid was the diatom *Nitzschia pungens* Grunow f. *multiseries* Hasle [4], although it has not yet been possible to culture this organism axenically.

In 1987, mussels contaminated with domoic acid were found in a restricted area of eastern Prince Edward Island, with the majority of the highly toxic animals coming from two sites located at the mouths of the Seal and Mitchell Rivers in the Cardigan River system (Fig. 1). This posed a number of important questions concerning the possible uniqueness of the Cardigan area, the mechanisms underlying the bloom of 1987 and the probability of this event being repeated in the Cardigan River or

Toxic Marine Phytoplankton
Edna Graneli et al., Editors

elsewhere in 1988. To address these problems, it was necessary to study 1) the conditions under which *N. pungens* forms blooms and produces domoic acid, and 2) the relationship between the toxic blooms and the uptake and depuration of domoic acid by shellfish. The results of these investigations are reported here together with an evaluation of the possibility of monitoring phytoplankton populations as an early warning system for imminent domoic acid toxicity events.

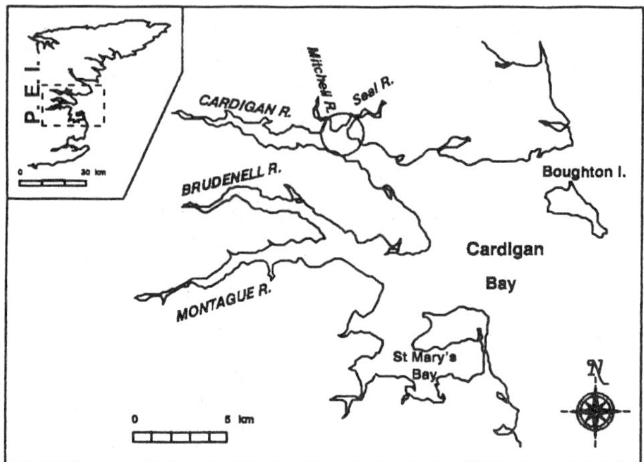

FIG. 1. Map of the Cardigan Bay showing the main sampling area (circled).

MATERIALS AND METHODS

Sample collection

Samples were collected near the mouth of the Seal River (Fig. 1) from one to three times per week from the beginning of October to the end of December. Mussels were collected from commercially grown suspended cultures and were retained alive until they could be processed (less than two days). Phytoplankton samples were collected by pumping with a plastic hand bilge pump, generally from a depth of 4 m. To concentrate such samples for domoic acid analysis, 100 L of water was passed through either a 28 or 20 µm mesh size Nitex net. The retentate in the net was made up to 500 mL with seawater and transported to the laboratory on ice for domoic acid analysis and cell counts; samples obtained in this manner are referred to as netplankton samples. Unfiltered (unconcentrated) seawater samples were retained for taxonomic analysis, cell counts, chlorophyll and nutrient determinations. Cell numbers per unit volume of sample were determined by counting all the cells in 20 µL aliquots.

Chlorophyll and Nutrient Analysis

Water samples for chlorophyll analysis were concentrated by filtration of 100 or 250 mL onto a 25 mm diameter GF/F glass fibre filter. For chlorophyll size fractionation, 500 mL of water was filtered through a 47 mm diameter, 1 µm diameter pore size Nuclepore screen. This retentate is the >1 µm or large fraction. An aliquot of the filtrate (400 mL) was then passed through a 47 mm diameter, 0.2 µm pore size Nuclepore screen and this retentate is the <1 µm or picoplankton fraction. Filters were stored dry, in the dark, at −20°C until analysis. Chlorophyll was extracted with 90% acetone and measured fluorometrically.

Samples for nutrients were filtered (0.2 µm Nuclepore) and stored frozen (−20°C) in polyethylene bottles until analysis. Nitrate and phosphate levels were measured with a Technicon Autoanalyzer.

Domoic Acid Analysis

Domoic acid was extracted from mussel tissue by the procedure of the Association of Official Analytical Chemists for paralytic shellfish poisons [5]. Domoic acid was extracted from phytoplankton by sonicating the cells in seawater and filtering the resulting solution through a 0.22 µm Millipore filter to remove debris. Domoic acid concentration in mussel extracts was determined by reverse phase HPLC using ultraviolet detection [6,7]. The high sensitivity 9-fluorenyl-methoxycarbonyl chloride (FMOC) precolumn derivitization method followed by reversed phase HPLC with fluorescence detection [8] was used to determine the domoic acid concentration in phytoplankton samples.

Meteorological Data

Wind velocity and direction and rainfall data were obtained from the Environment Canada Atmospheric Environment Service.

RESULTS AND DISCUSSION

The concentrations of domoic acid in the netplankton fraction and in mussels during the *N. pungens* bloom in the Cardigan River in 1988 are shown in Figure 2. At the peak of the bloom, ≈99% of the netplankton visible in

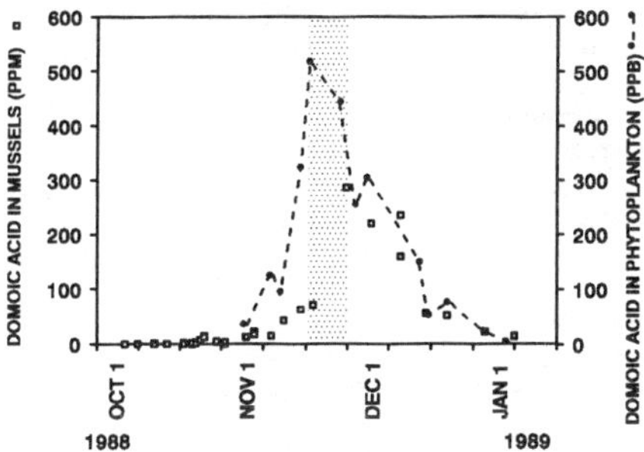

FIG. 2. The concentration of domoic acid and the time lag (shaded area) between its peak concentration in phytoplankton and mussels respectively.

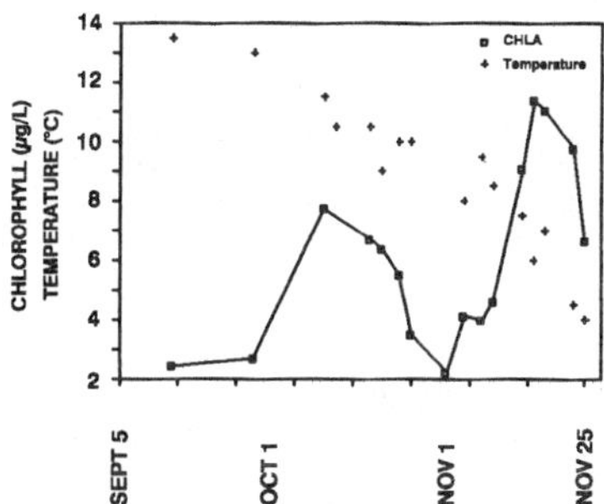

FIG. 3. Chlorophyll concentration and temperature (1m) during the ***Nitzschia pungens*** bloom in 1988.

the light microscope was *N. pungens*. However, in chlorophyll size fractionation experiments (6-8 Dec. 1988) in the Cardigan River, 9.32 ± 1.89 per cent (n=25) of the chlorophyll was in the picoplankton fraction. Also, domoic acid in the picoplankton was determined as the difference between the amount retained by a 1 µm pore diameter screen and a GF/F filter. This difference amounted to 26.24 ± 2.20 per cent (n=6) of the total retained by the GF/F filter. Although the fractionation procedures were somewhat different for chlorophyll and domoic acid, it is evident that significant amounts of both substances passed a 1 µm screen and that the proportion of domoic acid found in the picoplankton fraction was nearly three times that for chlorophyll. It is clear that the domoic acid content of the retentate of a 20 µm net represents only a portion of the particulate domoic acid present in the water column and the values shown in Figure 2 are minimal estimates of the availability of domoic acid to shellfish. These findings also suggest the possibility of production of domoic acid by picoplanktonic photoautotrophs or heterotrophs, or the excretion of domoic acid by *N. pungens* and its subsequent uptake and metabolism or recycling in the food web by various heterotrophic microorganisms. Also, excreted, dissolved domoic acid could conceivably be adsorbed by living or non-living suspended particles.

The peak level of domoic acid in mussels lags behind that for the netplankton by about 9 days. This period is represented by the shaded area in Figure 2. However, the depuration of domoic acid from mussels shows no such lag and is coincidental with the decline in the domoic acid concentration in the net plankton. Depuration occurs at a time when the environmental temperature is falling rapidly (Fig. 3), and takes about two months to complete. Depuration of domoic acid and the possible acceleration of this rather slow process are prime concerns of the mussel growers in eastern P.E.I.

The lag between the appearance of domoic acid in the phytoplankton and its uptake by mussels means that monitoring domoic acid in phytoplankton could give a sufficiently early warning of an impending health hazard. Although an early warning of this magnitude would avoid illness in consumers, the interval is too short to allow an orderly suspension of harvesting activities or the enactment of other measures to prevent undue economic losses to the industry. A phytoplankton monitoring program was carried out on a weekly basis in 1988 to determine if it was possible to obtain earlier indications of potential toxic blooms by identifying and enumerating known and putative toxic algal species. *Nitzschia pungens* began to appear in phytoplankton samples as early as the middle of August and became quite prominent by the end of September. There was about a two week interval before the appearance of domoic acid in the netplankton during which it was apparent that there was a potential for a toxic bloom. Thus, in the case of *N. pungens*, at least, a phytoplankton monitoring program proved very efficient as an early warning system.

The change in *N. pungens* cell numbers during the bloom in the Cardigan River is shown in Figure 4. There were actually two blooms, the first minor one peaking in mid October at a little over 200,000 cells/L, which then declined until 1 Nov., and a second larger bloom which peaked in the middle of November at about 1.2 million cells/L and declined thereafter, finally disappearing in early January. The bimodal nature of the blooms is more clearly seen in the changes in chlorophyll concentration (Fig. 3) but is difficult to visualize in the plot of domoic acid in mussels (Fig. 2) because of the vertical scaling of the figure. There was also a change in the *N. pungens* cellular domoic acid content during the bloom period (Fig. 4), indicating that qualitative changes were occurring in the cells.

It is important to attempt to understand the factors which resulted in the rise and decline of these blooms. Prior to the period under consideration here, there was a prolonged, intense bloom of *Skeletonema* (more than 7 million cells/L) which reduced the nitrate concentration in the water column below the limit of detection. Between 15 Sept. and 29 Sept.

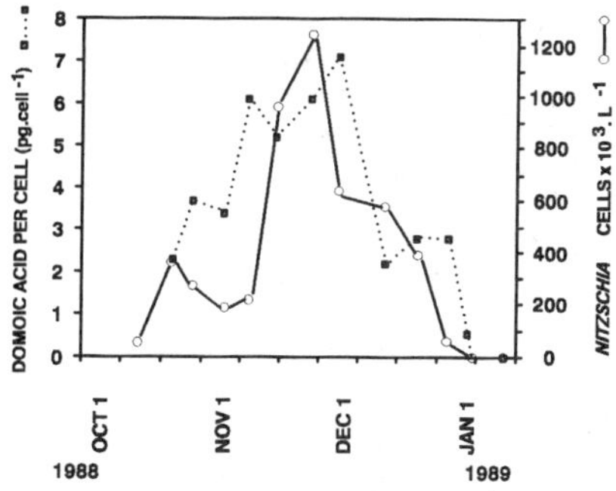

FIG. 4. ***Nitzschia pungens*** cells per unit volume of sea water and domoic acid per cell during the bloom of 1988.

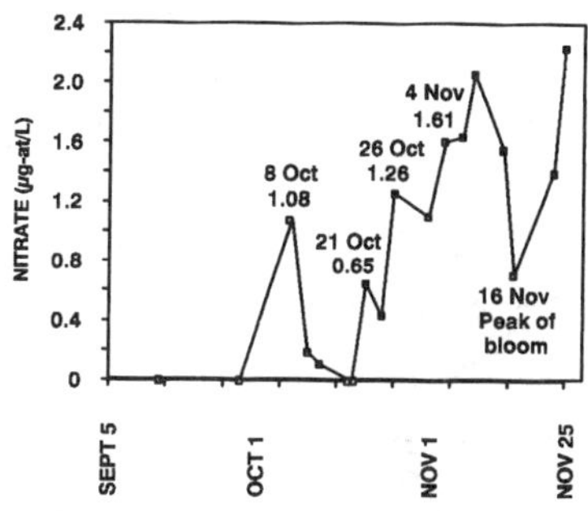

FIG. 5. Nitrate concentration (1m). Pulses of nitrates are indicated by dates.

no nitrate was detectable while chlorophyll remained at ≈2.5 µg/L (Fig. 3); presumably this is the level which could be maintained by nitrogen regeneration and ammonia production by animals such as the mussels. By 8 Oct., chlorophyll had increased to ≈4 µg/L and nitrate (Fig. 5) was 1.08 µg-at/L. Nitrate decreased to 0.19 µg-at/L and chlorophyll peaked at 7.8 µg/L by 11 Oct. Thereafter, chlorophyll gradually decreased to 2.3 µg/L (1 Nov.); nitrate could not be detected by 19 Oct. The first *N. pungens* bloom and its decline thus seem to be correlated with the availability of nitrate. Nitrate again appeared in the water column by 21 Oct. and by 1 Nov., the second *N. pungens* bloom was about to start. There had also been a further increment of nitrate (26 Oct.) and this was also observed on 4 and 9 Nov. The second bloom reached its peak on 16 Nov. (11.4 µg/L) when nitrate had been declining since 9 Nov. (from 2.06 to 0.71 µg-at/L). Nitrate began to increase after 16 Nov. and reached levels in excess of 5 µg-at/L (not shown in Fig. 5). These elevated nitrate levels did not affect the decline of the second bloom which eventually disappeared in early January. In this instance it is reasonable to attribute the disappearence of the bloom to low environmental temperatures (Fig. 3) or low solar irradiance, both of which were declining rapidly at the time.

Although nitrate levels seem to be involved in the progress of the *N. pungens* blooms in the Cardigan River, it is less clear what mechanisms produce the periodic pulses of nitrate. An inspection of meteorological data for the region suggests that rain or wind velocity or both are involved. Thus, the nitrate pulse of 8 Oct. was preceded by 3 days of rain (30.4 mm) and strong winds (45 km/h), the pulse of 21 Oct. was preceded by 56.4 mm of rain and that of 26 Oct. by 39.8 mm, both events being accompanied by sustained brisk winds. This pattern holds true for all the increments of nitrate observed throughout October and November 1988. As well, October was an unusually rainy month both in 1987 and 1988 and in October 1987 there were also several periods of strong winds associated with the rainfall. These associations suggest a number of hypotheses. One is that since the Cardigan River is surrounded by agricultural land, the source of the nitrate is terrestrial, the nitrate pulses being associated with freshwater runoff following rains. Another possible hypothesis is that, even though the water column was not stratified throughout this period (and hence there was no subpycnocline reservoir of nitrate), benthic nitrate could nevertheless be mixed into the water column by very strong winds resuspending sediments. Very strong winds were observed to stir up sediment into the surface layers on only one occasion (2 Nov.), however, so the sediments may be an unlikely source of the nitrate pulses. A further possibility is that storm winds could upwell nutrient rich offshore waters into Cardigan Bay; however, there did not appear to be any relationship between the direction of the wind and the timing of the pulses of nitrate, nor were there any indications of intrusions of cold bottom water, so this possibility appears less likely. Further study is required to determine the source of the pulses of nitrate, but, if this should prove to be terrestrial, then it is possible that the magnitude of the toxic *N. pungens* blooms might be diminished by small alterations of agricultural practices.

REFERENCES

1. R.F. Addison and J.E. Stewart, Aquaculture 77, 263–269 (1989).
2. T.M. Perl, J.C. Hockin, L. Bedard, T. Kosatsky, E.C.D. Todd, and R.S. Remis, Presented at: Symposium on Domoic Acid Toxicity, Ottawa, Ont., 10 April 1989.
3. J.L.C. Wright, R.K. Boyd, A.S.W. de Freitas, M. Falk, R.A. Foxall, W.D. Jamieson, M.V. Laycock, A.W. McCulloch, A.G. McInnes, P. Odense, V.P. Pathak, M.A. Quilliam, M.A. Ragan, P.G. Sim, P. Thibault, J.A. Walter, M. Gilgan, D.J.A. Richard and D. Dewar, Can. J. Chem. 67, 481–490 (1989).
4. S.S. Bates, C.J. Bird, A.S.W. de Freitas, R. Foxall, M. Gilgan, L.A. Hanic, G.R. Johnson, A.W. McCulloch, P. Odense, R. Pocklington, M.A. Quilliam, P.G. Sim, J.C. Smith, D.V. Subba Rao, E.C.D. Todd, J.S. Walter and J.L.C. Wright, Can. J. Fish. Aquat. Sci. 46 in press (1989).
5. J.F. Lawrence, C.F. Charbonneau, C. Ménard, M.A. Quilliam and P.G. Sim, J. Chromatography 462, 349–356 (1989).
6. M.A. Quilliam, P.G. Sim, A.W. McCulloch and A.G. McInnes, Atlantic Research Laboratory Tech. Rep. 55, NRCC 29015, p 17, (1988).
7. M.A. Quilliam, P.G. Sim, A.W. McCulloch and A.G. McInnes, Int. J. Environ. Anal. Chem., in press, (1989).
8. R. Pocklington, J.E. Milley, S.S. Bates, C.J. Bird, A.S.W. de Freitas, M.A. Quilliam, Int. J. Environ. Anal. Chem., in press (1989).

VARIABLE TOXICITY OF THE BROWN TIDE ORGANISM, AUREOCOCCUS ANOPHAGEFFERENS, IN RELATION TO ENVIRONMENTAL CONDITIONS FOR GROWTH

Gregory Tracey[1], Richard Steele[2] and Lorraine Wright[3]
[1]Science Applications International Corporation, c/o U.S. Environmental Protection Agency, Narragansett, Rhode Island 02882; [2]U.S. Environmental Protection Agency, Environmental Research Laboratory -Narragansett, Narragansett, Rhode Island 02882; [3]Marine Ecosystems Research Laboratory, Graduate School of Oceanography, University of Rhode Island, Narragansett, RI 02882.

ABSTRACT

Variation in toxicity of the brown tide organism, *Aureococcus anophagefferens*, was investigated in relation to environmental conditions for its growth. In a series of laboratory culture experiments, light (60 μE m^{-2} s^{-1}), temperature (20°C) and nutrient regimes (1/8 strength Provasoli's Enriched Seawater ("P/8")) for *Aureococcus* growth were modified, and toxicity assayed by measurement of feeding response in the blue mussel, *Mytilus edulis*. *Aureococcus* toxicity in late-exponential phase growth was 1) reduced in low light (30 μE m^{-2} s^{-1}) relative to high light (60–120 μE m^{-2} s^{-1}), 2) enhanced at 24°C relative to 20°C, and enhanced at high nutrient concentration (P/2 - P/8 media) relative to lower nutrient concentration (P/16 media). Results indicate that these environmental parameters in addition to cell density may be important factors modifying algal toxicity in the field.

INTRODUCTION

Many factors are involved in determining the quality of food available to suspension feeding bivalves such as the blue mussel, *Mytilus edulis*. Among these are seston concentration, algal concentration, and the size, shape and ingestability of food particles [1-3]. The importance of algal toxicity in bivalve nutrition was demonstrated vividly during the summer of 1985 in Narragansett Bay, RI, where reduced feeding, reproductive failure and massive mortalities in *M. edulis* populations occurred during an extremely dense bloom of the chrysophycean alga, *Aureococcus anophagefferens* [4,5]. Mortality in mussel populations during this "brown tide" are attributed to an algal-induced reduction in feeding, causing starvation [5]. Similar brown tide-related mortalities were also observed in bay scallop (*Argopecten irradians*) populations from the Peconic Bay system [6,7]. Specific cell toxins in *Aureococcus* have not been identified, although morphological and experimental evidence suggests a cell surface component is involved [8,9].

The purpose of this study was to examine the potential for variable toxicity of *Aureococcus* in relation to environmental conditions for its growth. Previous field studies with dinoflagellates, including *Alexandrium* (formerly *Protogonyaulax*) *tamarensis* [10], have documented significant variation in cellular toxin content. Laboratory studies using single clones have found toxin content to depend upon environmental conditions for growth, including water temperature and light intensity [11]. The data presented here will demonstrate that both cellular toxin content and cell density of *Aureococcus* are important factors determining the impact of brown tide blooms on the blue mussel and possibly other filter feeding species.

Toxic Marine Phytoplankton
Edna Graneli et al., Editors

MATERIALS AND METHODS

Algal culture. A clone of Aureococcus anophagefferens was obtained from E. Cosper (SUNY-Stonybrook) and maintained routinely at 20°C, 60 $\mu E\ m^{-2}\ s^{-1}$ on modified P/8 media (Table 1; [12,13]). The alga was acclimated to various test conditions for 3-5 days prior to a 5-7 day grow-out for mussel feeding experiments. Cultures were harvested during their late exponential phase of growth. In addition to Aureococcus, mussels were fed the chrysophyte, Isochrysis galbana (clone T-iso), a commonly used food source for bivalves [14]. In all experiments, algae were grown with a 16 h light:8h dark cycle using cool-white fluorescent lamps. Separate batch cultures were tested for each replicate within treatments.

TABLE I. Formula for modified Provasoli's Enriched Seawater [12,13].

Compound	Amount
$NaNO_3$	0.66 mM
Na_2 Glycerophosphate	25.0 μM
EDTA	26.9 μM
Fe (as $FeCl_3$)	1.8 μM
Zn (as $ZnSO_4$)	0.8 μM
Mn (as $MnSO_4$)	7.3 μM
Co (as $CoSO_4$)	0.17 μM
Bo (as H_3BO_3)	185.0 μM
Vitamins:	
Thiamine HCl	20.0 μg/L
B_{12}	1.6 μg/L
Biotin	0.8 μg/L

Toxicity assessment. Differences in mussel clearance (feeding) rate responses assessed whether various conditions for Aureococcus growth altered its toxicity. Mussels were collected from a subtidal population in lower Narragansett Bay (7° 24.0' W by 4° 29.4' N), sized (3.0 $\pm$ 0.2 cm length) and acclimated to laboratory conditions (15°C; unfiltered Narragansett Bay seawater) for > 2 wk prior to experiments.

During feeding experiments, mussels received a mixture of Aureococcus and Isochrysis. Isochrysis, a commonly used algal food for bivalves [14], was added to diets such that effects of Aureococcus on mussel feeding could be measured directly. Isochrysis is of sufficient size (5-7 μm diameter) to allow its clearance to be measured independently of Aureococcus within a mixed diet. In this size range, particle retention efficiency is near 100% [15], and thus feeding responses in absence of Aureococcus-induced effects are expected to be optimal.

Mussels were placed individually into 1 L beakers (V) containing experimental diets at 15°C for clearance rate determination. Cell concentration and size-frequency distribution of particles were measured using an electronic particle counter equipped with a 50μm aperture and calibrated with known-sized polystyrene spheres (Coulter Electronics, Hialeah, FL). Concentrations of particles (C) greater than 4 μm were measured at 30 min intervals (T) for 2 hours. Clearance rates (CR, ml min^{-1}) were determined using a variation of Coughlan's method [16];

$$CR = (\log C_1 - \log C_2)/(T_2 - T_1) \times V,$$

where the change in particle concentration vs. time (= d(log C)/dT) was determined by linear regression [17]. Data were inspected graphically to include only the linear portion of each curve (i.e. constant clearance rate). Clearance rates of 3 mussels were measured and averaged ($\pm$ 1 S.D.) in all experiments. Statistical differences between treatments were tested by one-way analysis of variance (P = 0.05).

Isochrysis was added to treatments at 4-6 x 10^4 cells ml^{-1}, a concentration at which clearance rates of mussels are independent and maximal yet pseudofeces production is minimal [1]. In light experiments, *Aureococcus* concentrations were 3.6 $\pm$ 0.1, 4.9 $\pm$ 0.1, 5.0 $\pm$ 0.1 and 4.4 $\pm$ 0.05 x 10^5 cell ml^{-1} in the 30, 60, 90, and 120 $\mu E\ m^{-2}\ s^{-1}$ treatments, respectively. At test temperatures of 16, 20 and 24°C, *Aureococcus* concentrations were 8.9 $\pm$ 0.2, 10.2 $\pm$ 0.04 and 9.2 $\pm$ 0.04 x 10^5 cell ml^{-1}, respectively. *Aureococcus* concentrations tested in P/2, P/4, P/8 and P/16 nutrients were 6.7 $\pm$ 0.2, 7.9 $\pm$ 0.2, 8.7 $\pm$ 0.1 and 8.4 $\pm$ 0.1 x 10^5 cell ml^{-1}, respectively.

RESULTS

Aureococcus was cultured under different light, temperature and nutrient enrichment conditions to determine whether these factors would alter cell toxicity. *Aureococcus* toxicity was similar over a range of 60-120 $\mu E\ m^{-2}\ s^{-1}$ light intensity, reducing mussel clearance rates by about 35% relative to the control (Fig. 1A). *Aureococcus* toxicity was significantly less (P= 0.05) when cultured at 30 $\mu E\ m^{-2}\ s^{-1}$ relative to higher light intensities, with clearance rates reduced by about 20% relative to the control.

Effects of culture temperature on *Aureococcus* toxicity were examined. Toxicity was significantly greater at 24°C than at the 20°C culture temperature (Fig. 1B). At 24°C, mussel clearance of food algae was reduced by 35% relative to the control, whereas *Aureococcus* cultured at 20°C did not cause significant clearance rate reductions. *Aureococcus* cultured at 16°C was also toxic, but caused clearance rate reductions of only 23%.

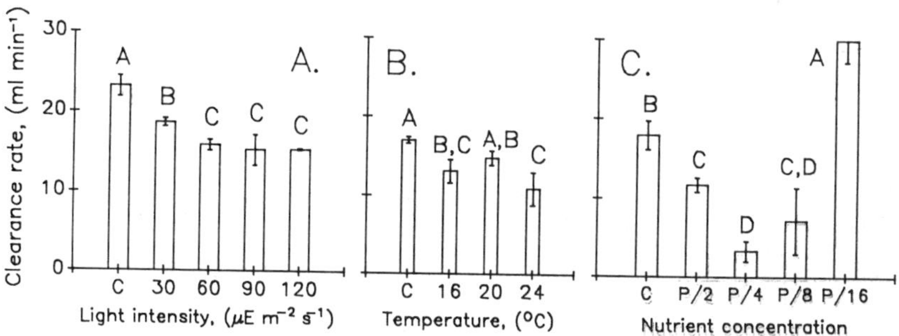

FIG. 1. *Mytilus edulis.* Mussel clearance rates (mean $\pm$ 1 S.D.) of *Isochrysis galbana* mixed with *Aureococcus anophagefferens* grown under different A) light, B) temperature, and C) nutrient conditions. C = Control treatment (no *Aureococcus* added). P = Provasoli's Enriched Seawater media (Table 1.). Bars with different letter annotations indicate significant differences among treatments.

Aureococcus was also cultured over a range of nutrient concentrations to determine whether this would affect toxicity. Mussel clearance rates were reduced 35, 72, and 60% relative to the control at P/2, P/4 and P/8 nutrient concentrations, respectively (Fig. 1C). Among the *Aureococcus* treatments, algae grown in P/4 media was significantly more toxic than other nutrient media. In marked contrast, *Aureococcus* grown in P/16 nutrients was non-toxic and increased mussel clearance rates by 76% relative to the control.

DISCUSSION

Previous experiments with *Aureococcus* show that conditions of 90-120 $\mu E\ m^{-2}\ s^{-1}$ light intensity, 20-22°C and P/8 media provide good conditions for maintaining cultures of this species relative to other combinations tried thus far [13,18]. Results of the present study suggest that cultures grown under reduced light (30 $\mu E\ m^{-2}\ s^{-1}$) were less toxic (Fig. 1A), possibly related to light limitation of growth rate. *Aureococcus* grown at 24°C was more toxic than at 20°C, possibly related to faster growth rate at the higher temperature (Fig. 1B). Nutrient conditions may have also altered *Aureococcus* toxicity. Toxicity was highest at an intermediate nutrient concentration (P/4), whereas very high (P/2) and low (P/16) nutrients reduced toxicity (Fig. 1C). These nutrient extremes may represent growth inhibition and nutrient limitation conditions for the P/2 and P/16 treatments, respectively. Thus the data suggest that *Aureococcus* cell toxicity may be possibly linked to its growth rate. These results are preliminary, since complete cell concentration-toxicity relationships for each culture condition using algae from different growth phases are required to evaluate growth rate effects on *Aureococcus* toxin production.

The environmental significance of variable *Aureococcus* toxicity was demonstrated in a previous study [19]. Samples of bloom waters containing *Aureococcus* obtained from south shore Long Island embayments during 1988 were assayed for toxicity by the mussel clearance rate bioassay. Peconic Bay-derived *Aureococcus* appeared toxic (i.e. reduced clearance rates) at about 1-4 x 10^4 cell ml^{-1}. In comparison, *Aureococcus* obtained from Great South Bay was toxic at about 5-10 x 10^4 cell ml^{-1}, and Narragansett Bay-derived *Aureococcus* was toxic when cell concentrations exceeded 25-50 x 10^4 cell ml^{-1} [5]. Observed bloom densities, although numerically remarkable [20], do not necessarilly cause feeding reduction effects; another similar-sized picoplankter, *Synechococcus*, fed at concentrations up to 10^6 cells ml^{-1} does not suppress mussel clearance rates [8]. Thus, similar reductions in mussel feeding, and accordingly growth [19] were observed despite a 50-fold variation in *Aureococcus* density.

The apparent toxicity of *Aureococcus* in the field has not been adequately simulated in the laboratory. *Aureococcus* in culture is non-toxic below 40-70 x 10^4 cell ml^{-1} [19,21]. We expect that *Aureococcus* toxicity in laboratory cultures will be enhanced as the specific growth requirements for this species are better defined. Nevertheless, results of the present study corroborate field observations of variable toxicity in *Aureococcus* [19], suggesting that one mechanism for variation in toxicity may be ambient environmental conditions. Thus, an adequate assessment of potential impacts on shellfish populations due to *Aureococcus* blooms as well as factors controlling bloom dynamics (e.g. grazing pressure) necessitate knowledge of both density and toxicity of this algal species.

ACKNOWLEDGEMENTS

Contribution No. 1082 of the U.S. Environmental Protection Agency, Environmental Research Laboratory-Narragansett (ERLN). This research was funded in part under EPA Contract no. 68-03-3529 to Science Applications International Corporation, Allen D. Beck, Project Officer. The authors are indebted to D. Phelps, S. Schimmel, W. Nelson and J. Prager for support of the research and critical review of the manuscript. The contents of the manuscript do not necessarily reflect views or policies nor does mention of trade names or commercial products constitute endorsement or recommendation for use by the U.S. Environmental Protection Agency.

REFERENCES

1. Foster-Smith, R.L., J. exp. mar. Biol. Ecol. 17, 1-22 (1975).
2. Kiorboe, T. Mohlenberg, F. and O. Nohr, Ophelia 19, 193-205 (1980).
3. Bricelj, V.M., Bass, A.E. and G.R. Lopez, Mar. Ecol. Prog. Ser. 17, 57-63 (1984).
4. Tracey, G.A., Trans. Amer. Geophys. Union 66(51), 1303 (1985).
5. Tracey, G.A., Mar. Ecol. Prog. Ser. 50, 73-81 (1988).
6. Cosper, E.M., Dennison, W.C., Carpenter, E.J., Bricelj, V.M., Mitchell, J.G., Kuenstner, S.H., Estuaries 10, 284-290 (1987).
7. Bricelj, V.M, J. Epp and R.E. Malouf, Mar. Ecol. Prog. Ser. 36, 123-137 (1987).
8. Tracey, G.A., Hargraves, P.E., Johnson, P.W. and J. McN. Sieburth, J. Shellfish Res. 7, 671-675 (1988).
9. Draper, C., L. Gainey, S. Shumway and L. Shapiro, In: Abstracts 4th Int. Conf. Toxic Mar. Phtyo., Univ. of Lund, Sweden, 26-30 June (1989) p. 53.
10. Cembella, A.D., J.-C. Therriault and P. Beland, J. Shell. Res. 7, 611-622 (1988).
11. Ogata, T., T. Ishimaru and M. Kodama, Mar. Biol. 95, 217-220 (1987).
12. McLachlan, J., In: J.R. Stein (ed.) Handbook of Phycological Methods. Cambridge Univ. Press, Lond. (1973) pp. 25-51.
13. Steele, R.L., Tracey, G.A., Wright, L.C. and G.B. Thursby, In: Novel Phytoplankton Blooms: Causes and Impacts of Recurrent Brown Tides and Other Unusual Blooms. Lecture Notes on Coastal and Estuarine Studies, E.M. Cosper, E.J. Carpenter and V.M. Bricelj, Eds., Springer-Verlag, Berlin (1989).
14. Ewart, J.W. and Epifanio, C.E., Aquaculture 22, 297-300 (1981).
15. Silvester, N.R. and M.A. Sleigh, J. Mar. Biol. Assoc. U.K. 64, 859-879 (1984).
16. Coughlan, J., Mar. Biol. 2, 356-358 (1969).
17. Snedecor, D.W. and W.G. Cochran, In: Statistical Methods. Iowa Univ. Press, Ames. (1980) 507 pp.
18. Cosper, E.M., E.J. Carpenter and M. Cottrell, In: Novel Phytoplankton Blooms: Causes and Impacts of Recurrent Brown Tides and Other Unusual Blooms. Lecture Notes on Coastal and Estuarine Studies, E.M. Cosper, E.J. Carpenter and V.M. Bricelj, Eds., Springer-Verlag, Berlin (1989).
19. Tracey, G. A., J. Gatzke, R. L. Steele, L. Wright and D. K. Phelps, In: Novel Phytoplankton Blooms: Causes and Impacts of Recurrent Brown Tides and Other Unusual Blooms. Lecture Notes on Coastal and Estuarine Studies, E.M. Cosper, E.J. Carpenter and V.M. Bricelj, Eds., Springer-Verlag, Berlin (1989).
20. Sieburth, J. McN., Johnson, P.W. and P.E. Hargraves, J. Phycol. 24, 416-425 (1988).
21. Kuenstner, S.H., M.S. thesis, State University of New York at Stonybrook, 84 pp. (1988).

SPATIAL DISTRIBUTION OF RESTING CYSTS OF *ALEXANDRIUM SPP.* IN SEDIMENTS OF THE LOWER ST. LAWRENCE ESTUARY AND THE GASPÉ COAST (EASTERN CANADA)

J. TURGEON*, A.D. CEMBELLA**, J.-C. THERRIAULT** AND P. BELAND***
*Dept. of Oceanography, Université du Québec at Rimouski, 300 des Ursulines, Rimouski, Québec, Canada G5L 3A1; **Maurice Lamontagne Institute, Dept. of Fisheries and Oceans, 850 route de la Mer, Mont-Joli, Quebec, Canada G5H 3Z4; ***St. Lawrence National Institute of Ecotoxicology, 310 des Ursulines, Rimouski, Quebec, Canada G5L 3A1

ABSTRACT

The spatial distribution of resting cysts (hypnozygotes) of the toxic dinoflagellate genus *Alexandrium* was studied in the lower St. Lawrence estuary, along the Gaspé coast and in Gaspé Bay and Harbor. The distribution revealed a definite tendency for cyst accumulation along the shores of the estuary and gulf, at depths < 100 m. On the north shore of the lower estuary, higher cyst concentrations also tended to be associated with sediments dominated by fine sand. Gaspé Bay and Harbor were apparently unfavorable environments for the accumulation of benthic dinoflagellate cysts. A strong correlation was found between cyst abundance and PSP toxicity levels in shoreline molluscs. In the St. Lawrence ecosystem, the dominant hydrodynamic feature of the surface waters is the nearshore-flowing Gaspé current, which originates from the freshwater plume formed by the combined outflows of the Manicouagan and Aux-Outardes rivers (M-O). Our results suggest that the spatial distribution of benthic *Alexandrium* cysts is mostly restricted to the zone of influence of this current.

INTRODUCTION

The estuary and gulf of St. Lawrence have long been recognized as regions significantly affected by paralytic shellfish poisoning (PSP) [1]. Several authors have shown that the resting cysts of certain *Alexandrium* species (referred to as *Protogonyaulax tamarensis*, *Gonyaulax tamarensis* or *G. excavata* in previous literature) are toxic [2-6]. It has even been proposed that these cysts may have been the direct cause of toxicity in molluscs on the Atlantic coast [7]. Benthic dinoflagellate cysts represent relatively stationary populations compared to their pelagic counterparts. Therefore, a more complete historical record of resident blooms should, in principle, be obtained by studying the abundance and distribution of benthic cysts, rather than pelagic cells. Coupled with a better understanding of dominant hydrodynamic and granulometric processes, the information on benthic cysts should, therefore, increase considerably our capacity to provide explanations for the development and persistence of toxic blooms.

The objectives of the present study were: 1) to determine the spatio-temporal distribution of *Alexandrium* cysts in coastal sediments of the lower St. Lawrence estuary and the Gaspé coast, as well as the Bay and Harbor of Gaspé (Fig. 1); 2) to determine whether or not there exists a relationship between this distribution and the type of sediment present; and 3) to examine the possible correlation between the concentration of these benthic cysts and PSP toxin levels in molluscs at adjacent stations.

MATERIALS AND METHODS

Sediment sample collection and analyses have been previously described by Cembella et al.[8]. In summary, the study area encompassed the lower St. Lawrence estuary and Gaspé coast, including Gaspé Bay and Harbor (Fig.1).

Toxic Marine Phytoplankton
Edna Graneli et al., Editors

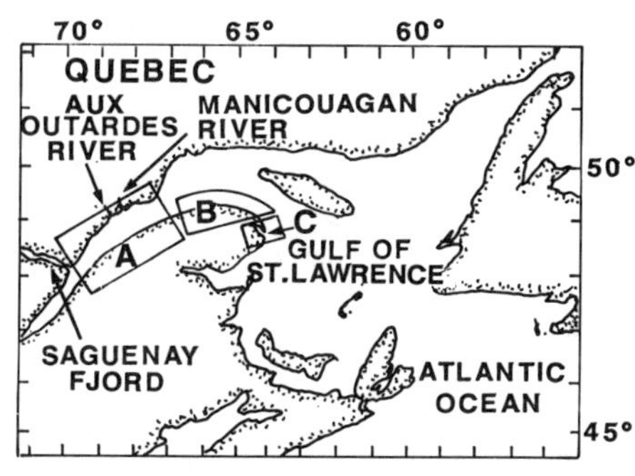

FIG. 1. Map showing the sampling regions: (A) Lower St. Lawrence Estuary; (B) Gaspé coast; (C) Gaspé Bay and Harbor.

Sampling was carried out over three years (1986-88), during which a total of 209 stations were visited. The sediment samples containing the benthic dinoflagellate cysts were obtained using a Shipek sampler. The sampling program was carried out during the summer, in order to verify the relative abundance of benthic cysts when blooms of motile cells in the water column were maximal. The same stations were revisited in October, several weeks after the disappearance of *Alexandrium* from surface waters, to examine the cyst distribution following the termination of the summer blooms. A final sampling was conducted in May, before the initiation of pelagic blooms of excysted populations, to confirm the persistence in the distribution of the benthic cyst populations throughout the winter. The particle size characterization of the sediment was accomplished using a recognized granulometric method [9]. The mollusc PSP toxin data were obtained from mouse bioassay results provided by J.E. Reid (Health Protection Branch, Dept. of Health and Welfare, Ottawa).

RESULTS

Figure 2A shows that in May and October, the highest concentrations of *Alexandrium* cysts (>500 cysts cm^{-3}) were found along the coastal shelf of both the north and south shores of the St. Lawrence estuary, at depths <100 m. The Laurentian Channel, with depths >300m, was generally impoverished in benthic cysts (<250 cysts cm^{-3}), except within the zone under the direct influence of the M-O plume, where the number of cysts was markedly elevated (>500 cysts cm^{-3}). The cysts were generally more abundant at the downstream end of the lower estuary along both shores. Figure 2B also reveals a tendency for cysts to accumulate near the shoreline at shallow depths, rather than further offshore. There was a decreased number of cysts (<250 cm^{-3}) in the downstream region of the Gaspé peninsula. Only low concentrations of cysts were present in Gaspé Bay and Harbor. However, higher numbers of cysts (>500 cysts cm^{-3}) were found at the mouth of the bay (Fig. 2C). In summer, the cyst concentrations were generally low in the lower estuary (Fig. 2A). Only one station near the downstream end, on the north shore, showed a substantial concentration of cysts (>400 cysts cm^{-3}).

Figure 3A reveals a strong dominance of silty sand (see legend of Fig. 3A for nomenclature) along almost the entire north shore of the lower estuary, with which highly variable cyst concentrations (13-1,500 cysts cm^{-3}) are associated. The sediment of the Laurentian Channel is composed primarily of clayey silt, except within the plume-dominated region, where the surface sediment is largely composed of silty clay. It is within this latter zone that the highest cyst concentrations were found. The sediments of the south shore exhibit a gradient in grain size, increasing in the downstream direction, with which variable cyst concentrations were associated. The sediments on the Gaspé coast are largely dominated by gravelly sand, alternating with sandy gravel on the coastal shelf (Fig. 3B). Once again, the cyst concentrations observed in this type of sediment were highly variable. In general, at depths >100 m, sediments were dominated by silty

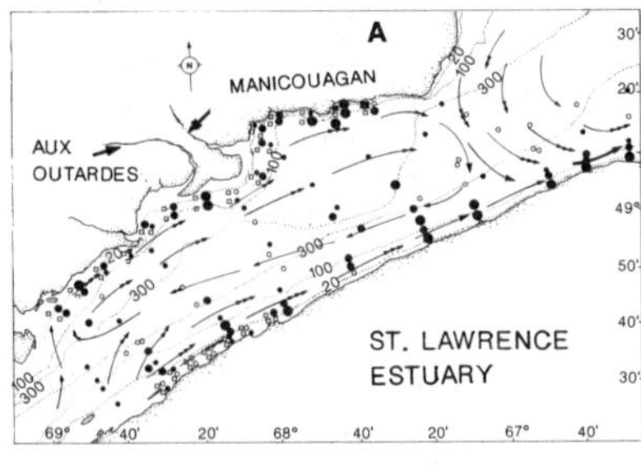

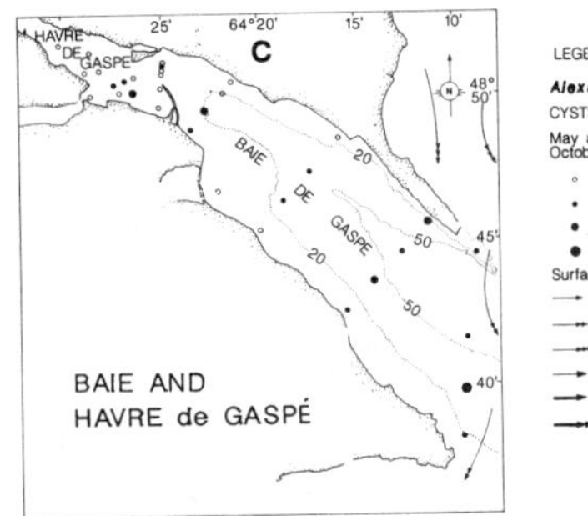

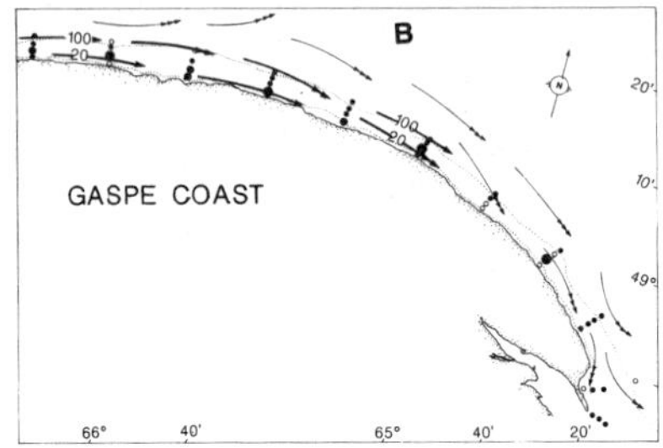

FIG. 2. Alexandrium cyst distribution in: (A) Lower St. Lawrence Estuary; (B) Gaspé coast and (C) Baie and Havre de Gaspé. Generalized pattern of residual surface current adapted from El-Sabh [17].

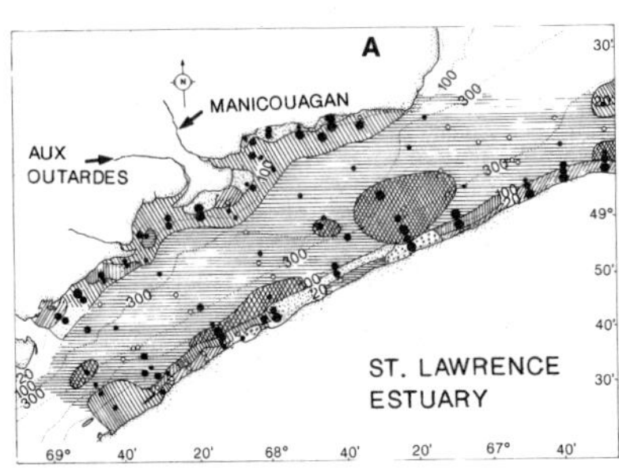

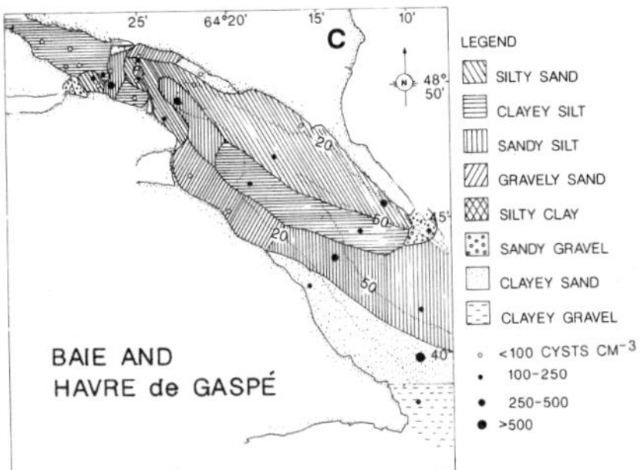

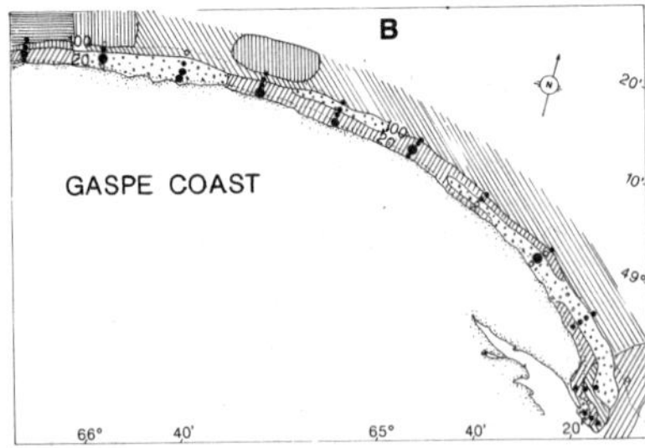

FIG. 3. Granulometric features and distribution of resting cysts of Alexandrium in: (A) Lower St. Lawrence Estuary; (B) Gaspé coast and (C) Baie and Havre de Gaspé. The granulometric nomenclature for grain size is given in % of the dominant size class [silty sand (57-95%), clayey silt (64-97%), sandy silt (56-83%), gravely sand (69-98%), silty clay (55-63%), sandy gravel (63-93%), clayey sand (54-83%) and clayey gravel (58-75%)].

sand, containing only low cyst numbers. In Gaspé Harbor, the predominance of clayey silt was accompanied by relatively few cysts (Fig. 3C). Gaspé Bay was characterized by a large variability in grain size, sometimes associated with higher cyst concentrations. The increased number of cysts at the mouth of the Bay was particularly noteworthy (Fig. 3C).

A correlation analysis (Pearson) did not show a significant relationship between the distribution of benthic Alexandrium cysts and the sediment type, when all three geographical regions were combined. However, a detailed examination of Fig. 3 indicates a clear tendency for cysts to accumulate within sediments dominated by fine sand along the north shore and along the Gaspé coast, including the mouth of Gaspé Bay.

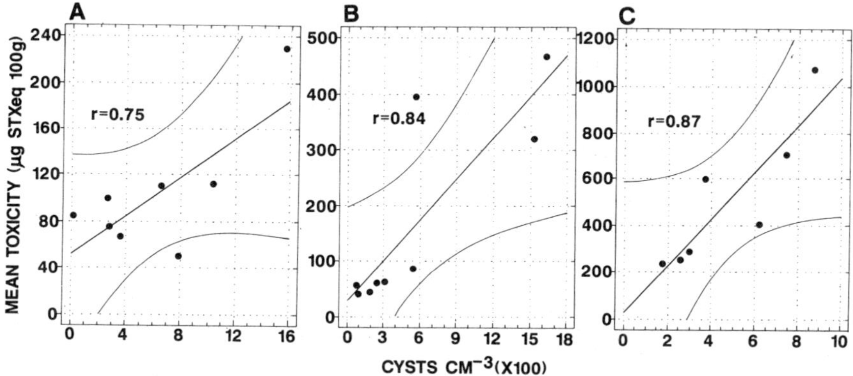

FIG. 4, Fitted linear correlation (Pearson) and 99% confidence interval of mean annual paralytic shellfish toxicity in shoreline molluscs (Mya arenaria and Mytilus edulis), and Alexandrium cyst concentrations in surface sediment of the north (A) and south (B) coast of the Lower St. Lawrence estuary and the Gaspé coast (C).

A significant relationship was found between the distribution of hypnozygotic cysts of Alexandrium and PSP levels in molluscs at shoreline stations for the north coast (Fig. 4A) and the south shore (Fig. 4B) of the lower estuary, as well as along the Gaspé coast (Fig. 4C). The mean annual toxicity levels for two mollusc species (Mya arenaria and Mytilus edulis) were significantly correlated ($p \geq 0.01$) with cyst abundance at neighboring stations along the north ($r = 0.75$, $n = 8$) and south ($r = 0.84$, $n = 9$) shores, and in the Gaspé region ($r = 0.87$, $n = 7$).

DISCUSSION

In terms of cyst morphology, the various species of Alexandrium Balech are virtually indistinguishable. Nevertheless, Alexandrium cells which arise from benthic cyst populations in the St. Lawrence ecosystem, usually possess the angular theca and prominent ventral pore characteristic of A. excavatum (Braarud) Balech & Tangen. The presence of hypnozygotes provisionally classified as A. excavatum (= Protogonyaulax tamarensis) in sediments of the St. Lawrence ecosystem presupposes either an exogenous transport or a direct sedimentation from the overlying water column. Clearly, these mechanisms are not mutually exclusive. The direct sedimentation hypothesis implies a distribution which mirrors that of blooms in the upper water column, this being largely a function of the hydrodynamics of surface water masses. It was previously observed [10,11] that the spatial distribution of Alexandrium blooms in surface waters of the lower estuary in August was closely associated with the extent of the M-O plume. Since it is known that this plume serves as a major driving force to generate the Gaspé current [11,12], Therriault et al. [11] have hypothesized that motile Alexandrium cells should be abundant in surface waters along the entire extent of this current. This was verified indirectly by examining the spatial distribution of PSP toxicity in the estuary and Gulf of St. Lawrence.

Given the above evidence, it is reasonable to hypothesize that the depositional zones of Alexandrium hypnozygotes in the sediment will closely follow the geographical extent of the Gaspé current. This pattern is clearly evident from Fig. 2. Therefore, the high cyst concentrations present in spring along the north shore within the zone influenced by M-O plume (Fig. 2A) can be considered as a local benthic reservoir for the initiation of Alexandrium blooms. The motile bloom populations of Alexandrium would then be transported across the estuary by the M-O plume, to join the Gaspé current on the south shore. The sedimentation pattern of the cysts therefore tracks the movement of the surface water masses. As proof of this, it is significant that large numbers of cysts in the Laurentian Channel were only found in the region which is directly influenced by the M-O plume. We know, however, that the Gaspé current is a strong nearshore jet current which flows close to the Gaspé peninsula (see Fig. 2A: surface current data of El-Sabh). This current is separated from offshore waters (>100 m) by a well defined density front [13]. The sedimentation of dinoflagellate cysts along the south shore is thus confined to this narrow longitudinal band, which serves as a sediment trap for cysts. Therefore, the main contraint on cyst distribution appears to be the hydrographic barrier between the Gaspe current and deeper offshore waters. The presence of high cyst concentrations at the entrance of Gaspé Bay indicates that cyst sedimentation is also significant in this region. The low numbers of cysts found within Gaspé Bay and Harbor suggest that Alexandrium blooms, and the consequent shellfish toxicity [14] have an external origin, from water entrained into the bay from the Gaspé current.

The presence of a similar spatial distribution of benthic cysts before and after a pelagic bloom of Alexandrium indicates that once deposited on the bottom, the cyst populations are relatively stationary. Accordingly, the hypothesis of exogenous winter transport of cysts can be eliminated. Moreover, the absence of cysts during the summer (Fig. 2A) demonstrates that excystment within the region is the ultimate source of these regional blooms. Since the physico-chemical conditions on the near bottom region vary only slightly on an annual basis [15], there must exist a physiological mechanism or a biological clock, as proposed by Anderson and Keafer [16], which provokes excystment.

In the St. Lawrence ecosystem, where hydrodynamics play a large role, the highest cyst concentrations along the north shore of the lower estuary were associated with sediments dominated by fine sand (Fig. 3A). This tendency suggests that within this plume-dominated region, the presence of such fine grained sediment may be useful as an indicator of favorable depositional zones for Alexandrium cysts. However, the presence of large numbers of cysts (>500 cm^3) in at least four of the eight sediment types present along the shore of the estuary and the Gaspé coast indicates that such an index cannot be universally applied, particularly in the most hydrodynamically active zones.

A significant relationship was observed between the distribution of Alexandrium hypnozygotic cysts and PSP (Fig. 4). This confirms the results of Therriault et al. [11], which considered the abundance of motile cells in surface waters, rather than cysts. Nevertheless, the relationship with benthic cysts does not necessarily imply that cysts are directly responsible for PSP contamination. The motile vegetative cells alone may account for the toxicity. However, the presence of high cyst concentrations certainly indicates the potential for rapid production of large quantities of toxic motile cells in the water column.

In conclusion, this study clearly demonstrates that the Gaspé current is a most important hydrodynamic feature in the St. Lawrence ecosystem, one which exerts a strong controlling influence on the spatial distribution of toxic blooms of Alexandrium.

ACKNOWLEDGEMENTS

The authors wish to thank Richard Larocque and the crew of the M.V. Grèbe for their assistance in the collection of benthic cyst samples.

REFERENCES

1. A. Prakash, J.C. Medcof and A.D. Tennant, Bull. Fish. Res. Bd Can. 177, 88 pp. (1971).
2. B. Dale, C.M. Yentsch and J.W. Hurst, Science 201, 1223-1224 (1978).
3. C.M. Yentsch, C.M. Lewis and C.S. Yentsch, BioScience 30, 251-254 (1980).
4. J.W. Hurst and C.M. Yentsch, Can. J. Fish. Aquat. Sci. 38, 152-156 (1981).
5. A.W. White and C.M. Lewis, Can. J. Fish. Aquat. Sci. 39, 1185-1194 (1982).
6. A.D. Cembella, C. Destombe and J. Turgeon, this volume (1989).
7. C.M. Yentsch and C.F. Mague in: Toxic Dinoflagellate Blooms, D.L. Taylor and H.H. Seliger, eds. (Elsevier-North Holland, New York 1979) pp. 127-130.
8. A.D. Cembella, J. Turgeon, J.-C. Therriault and P. Béland, J. Shellfish Research 7, 597-609 (1988).
9. A. Rivière, Méthodes granulométriques: techniques et interprétation. Masson, Paris 1977).
10. J.-C. Therriault and M. Levasseur, Naturaliste Can. 112, 77-96 (1985).
11. J.-C. Therriault, J. Painchaud and M. Levasseur in: Toxic Dinoflagellates, D.M. Anderson, A.W. White and D.G. Baden, eds. (Elsevier, New York 1985) pp. 141-146.
12. J. Lacroix, M.I. El-Sabh, A. Condal and J.M.M. Dubois in: Télédétection et Gestion des Ressources- L'aspect opérationnel, Bernier, Lessard et Gagnon, eds. (L'association Québécoise de Télédétection, Québec 1985) pp. 435-444.
13. C.L. Tang, J. Geophys. Res. 85, 2787-2796 (1980).
14. M. Desbiens, F. Coulombe, J. Gaudreault, A.D. Cembella and R. Larocque, this volume (1989).
15. N. Silverberg, H.M. Edenborn and N. Belzile in: Marine and Estuarine Geochemistry, A.C. Sigleo and A. Hattori, eds. (Lewis, Chelsea 1985) pp. 69-80.
16. D.M. Anderson and B.A. Keafer, Nature 325, 316-317 (1987).
17. M. I. El-Sabh, Naturaliste Can. 106, 53-73 (1979).

NITROGEN AND PHOSPHATE ACCUMULATION BY *CHATTONELLA ANTIQUA* DURING DIEL VERTICAL MIGRATION IN A STRATIFIED MICROCOSM

MASATAKA WATANABE,* KUNIO KOHATA,* AND MASAYUKI KUNUGI*
Laboratory of Marine Environment,
National Institute for Environmental Studies,
Onogawa 16-2, Tsukuba, Ibaraki 305, Japan

ABSTRACT

The ecological advantage of diel vertical migration on the nutrition and accumulation of *C. antiqua* was examined by using a large axenic culture tank, in which vertical stratified conditions of salinity, temperature and nutrient, analogous to conditions observed in the Seto Inland Sea, were simulated. It was found that *C. antiqua* was capable of migrating through a very sharp salinity and temperature gradient. At night this species migrated to the deep nutrient-rich water and took up nutrients. During the daytime this species migrated to the nutrient depleted surface water and utilized the accumulated nutrient for photosynthesis. It was observed that nitrogen uptake was synchronized with phosphate uptake. ^{31}P-NMR spectroscopy of this species revealed that *C. antiqua* has no capability of synthesizing polyphosphate, which was considered as phosphate reservoir. It is demonstrated from the experiments using a large scale culture tank that under stable stratification diel vertical migration is essential for *Chattonella* to develop a population capable of competing against coastal diatoms.

INTRODUCTION

Prior to 1960, red tides seldom occurred in the Seto Inland Sea (av. depth : 30m), but the number of outbreaks has increased in parallel with eutrophication. *Chattonella antiqua* (Hada) Ono, a spindle shaped Raphidophyte (size 50 - 130 μm) is the flagellate most commonly responsible for the red tides, which caused severe damage to aquacultured yellowtail in the summers of the 1970s and 1980s. It can colour the sea surface a dark brown over areas of 100 - 1000 km^2. Cell concentrations typically reach 200 - 3000 cells·ml^{-1}. Extensive field observations of a *Chattonella* red tide throughout an entire outbreak period in the summer of 1987 [1] revealed the following oceanographic characteristics not observed in years in which there was no red tide occurence: A nutrient cline was observed at a shallow depth (5-10m), with no nutrients in the surface water and high nutrient concentrations in deeper water. In 1987, the presence of a shallow phosphorus cline (at a depth of 5-7m) was most characteristic compared with years without an occurrence [1].

Our hypothesis to explain the outbreak of *Chattonella* red tide under such environmental conditions was that this species can migrate across the stratified layer and take up nutrients during the night in deep water. Although the ecological advantages of vertical migration for nutrient assimilation have been discussed many times [2-5], these have not been validated clearly for *Chattonella* due to the difficulty in large scale culture experiment. A large axenic culture tank [5] was used to permit simulation of the diel vertical migration of *C. antiqua* under stable stratification, thus providing conditions analogous to those observed in the Seto Inland Sea [1], and the nocturnal uptake of nutrient was examined in

Toxic Marine Phytoplankton
Edna Graneli et al., Editors

detail.

EXPERIMENTAL

The clonal axenic culture Chattonella antiqua (Hada) Ono (strain NIES-1, Microbial Culture Collection of the National Institute for Environmental Studies) was inoculated into the NIES tank (2 m high by 1 m diameter, working volume of 1 m^3) to give an initial cell concentration of 28 $cells \cdot ml^{-1}$. Initial concentration of P_i (NaH_2PO_4) in the f/2 medium was ajusted to 1.0 μM and salinity was 31.0‰. Temperature was kept at 25°C and irradiance was 530 $\mu E \cdot m^{-2} \cdot s^{-1}$ (average at the surface) under a 12:12 h LD regime. The specific growth constant under fully mixed conditions was μ = 0.72 (ln unit of $increase \cdot day^{-1}$). P_i in the medium was completely depleted on the 6th day after inoculation. At 0900 h on the 7th day, the cell concentration reached 1300 $cells \cdot ml^{-1}$ and the P cell quota was 0.91 $pmol \cdot cell^{-1}$, a level similar to the minimum P cell quota (1.0 $pmol \cdot cell^{-1}$). C. antiqua accumulated at the surface due to vertical migration when aeration was stopped at 0930 h on the 7th day. At 1300 h, 100 L of the new medium, which was enriched with P_i (18.5 μM) and of higher salinity (33.5‰), was introduced into the bottom of the tank. Then the migration experiment was conducted. Initial distributions of temperatures and salinities in the tank were 25°C, 31.0‰ in the surface water and 24°C, 33.5‰ in the deep water, respectively (FIG. 1a). Initial P_i concentrations were 0μM in the surface water and 18.5μM in the deep water (FIG. 1b). Large differences in salinity (Δs = 2.5‰) prevented vertical mixing and a stable stratification of P_i was maintained throughout the migration experiment. Samplings were repeated alternately at 1300 h (in the surface) and at 2300 h (in the bottom) for the measurements of phosphate, nitrogen and ^{31}P-NMR. The methods for PCA extraction, ^{31}P-NMR measurement, phosphate and nitrogen measurements were the same as before [6].

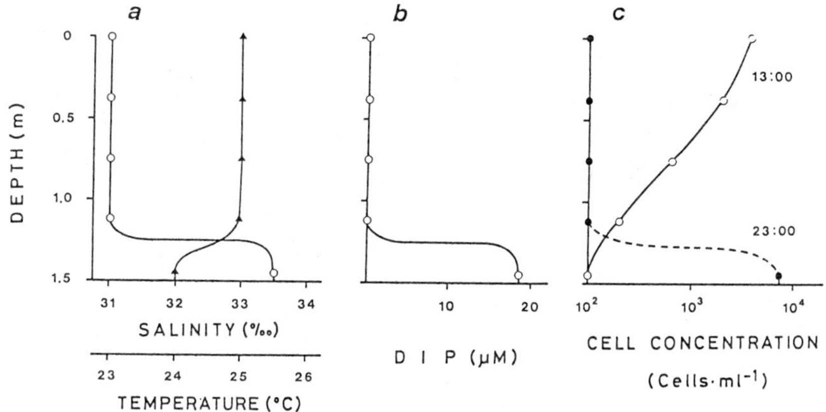

FIG. 1(a),(b). Vertical distributions of salinity(o), temperature(▲) and P_i(o) imposed at the beginning of the migration experiment in the tank. (c) Vertical distributions of C. antiqua cell concentration, (o) : measured at 1300 h of the 1st day of the migration experiment, (●) : measured at 2300 h on the same day.

RESULTS AND DISCUSSION

Initially, most of the C. antiqua cells accumulated on the surface (3700 $cells \cdot ml^{-1}$ at the surface, 1300 h, FIG. 1c) and the P cell quota at

Table 1 Variations in cell quota of P intermediates in *C. antiqua*

	Sampling location	Sugar P (%)	Pi (%)	Phospho -diester (%)	P cell quota (pmol/cell)	N cell quota (pmol/cell)
(A) 1300h First day P-starved	surface	47.6	23.6	23.4	0.91	26.3
(B) 2300h First day	bottom	17.0	46.7	13.1	1.71	28.2
(C) 1300h Second day	surface	16.0	39.7	15.7	0.97	21.1
(D) 2300h Second day	bottom	14.5	51.0	17.8	3.51	38.0
(E) 1300h Third day	surface	18.6	38.5	19.2	1.22	24.1

Percentages of each P intermediate in PCA extract of the cells were obtained from the integration of peak areas under the ^{31}P-NMR signals.

the surface was 0.91 pmol$\cdot$cell^{-1} in the P starved condition (Table 1). *C. antiqua* started to move downward to deep water before the lights went off and formed a dense layer above the bottom (7200 cells$\cdot$ml^{-1} at the bottom, 2300 h, FIG. 1c). The P-starved cells reached the phosphate-rich deep water and took up phosphate rapidly. The P cell quota at the bottom increased to 1.71 pmol$\cdot$cell^{-1} at 2300 h of the first day of the migration experiment (Table 1). The timings of the migratory ascent and cell division of this species overlapped. The cells started to move upward to the water surface before the lights went on and accumulated as a surface layer (1300 h, FIG. 1c). At this time the P cell quota at the surface decreased to 0.97 pmol$\cdot$cell^{-1} due to cell division (Table 1). The P cell quota at the bottom increased again to 3.51 pmol$\cdot$cell^{-1} at 2300 h on the 2nd day of the migration experiment (Table 1). This pattern of migration and P_i uptake at the bottom during the night was repeated throughout the experiment. In this experiment nitrogen was sufficient in the medium and not limiting. However, the N cell quota showed synchronized behavior with the P cell quota and the N cell quota increased at the bottom during the night (Table 1). These data suggest that P_i and N uptakes were synchronized with each other.

^{31}P-NMR spectra of metabolites associated with vertical migration and phosphate uptake at the bottom by *C. antiqua* cells are shown in FIG. 2. Strangely, *C. antiqua* does not have the capability to synthesize polyphosphate, although many algae accumulate and store large quantities of phosphate as polyphosphate under unlimited phosphate conditions [6-10]. When the P-starved cells (FIG. 2a)migrated to the phosphate-rich deep water (2300 h, 1st day), the increase in the P cell quota was due solely to the increase in P_i, with no other intermediates synthesized (FIG. 2b). When the cells migrated to the photic surface water where the ambient P_i

concentration was 0 (1300 h, 2nd day), accumulated P_i in the cells must have been utilized directly for photophosphorylation (FIG. 2c).

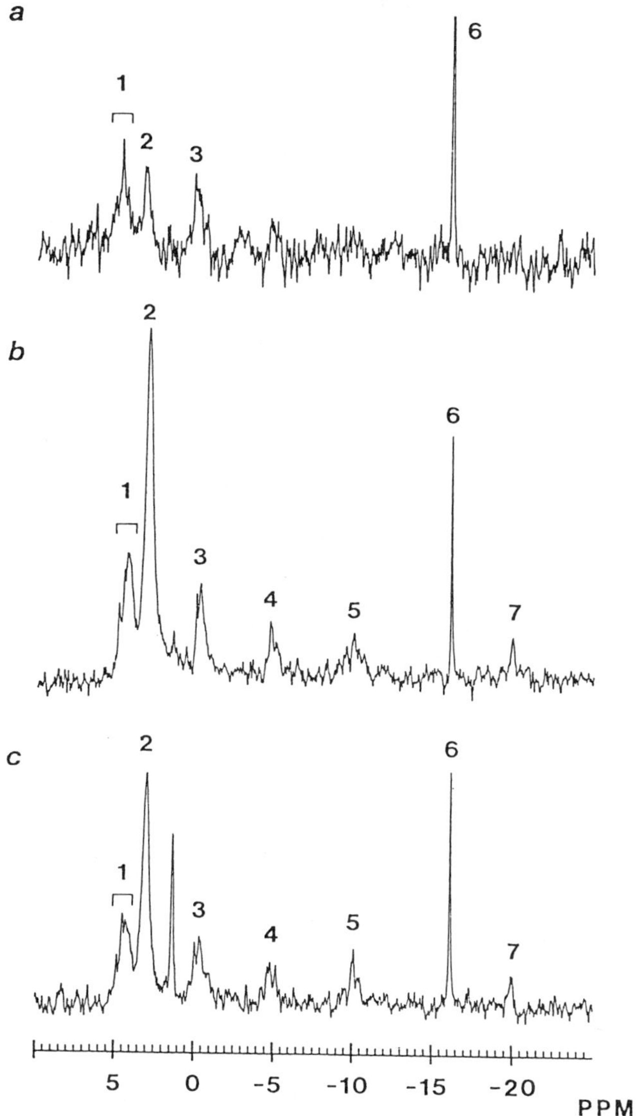

FIG. 2. ^{31}P-NMR spectra of PCA extracts from C. antiqua cells. (a) Cells accumulated at the surface at 1300 h on the 1st day of the migration experiment. Cells were in the P-starved condition; (b) Cells accumulated at the bottom at 2300 h on the 1st day of the migration experiment; (c) Cells accumulated at the surface at 1300 h on the 2nd day of the migration experiment. The identities of the peaks are : 1 = sugar phosphate, 2 = P_i, 3 = phosphodiester, 4 = NTPγ, 5 = NTPα, 6 = external reference, 7 = NTPβ.

The laboratory experiment proved that 1) C. antiqua is capable of migrating through a very sharp salinity gradient(Δs=2.5‰); 2) following phosphorus starvation, uptake into the inorganic phosphate pool occurs in the dark in deep water, and 3) there are subsequent migrations to nutrient-depleted, photic surface waters, where the internal phosphate pools are used for photophosphorylation. It was observed that N uptake was synchronized with P uptake and therefore nocturnal uptake of N and P in deep water resulting from diel vertical migration provides a strong ecological advantage for C. antiqua under conditions with phosphorus and nitrogen clines.

ACKNOWLEDGEMENTS

The authors wish to thank Mr. T. Kimura and Ms. Ohkawa for their assistance.

REFERENCES

1. Y. Nakamura, T. Umemori, and M. Watanabe, J. Oceanogr. Soc. Japan (in press).
2. R.W. Eppley, O. Holm-Hansen, and J.D.H. Strickland, J. Phycol. 4, 333-340 (1968).
3. R.W. Holmes, P.M. Williams, and R.W. Eppley. Limnol. Oceanogr., 12, 503-512 (1967).
4. S.I. Heaney, and R.W. Eppley, J. Plankton Res. 3, 331-344 (1981).
5. S. Yamochi, In Red Tides. [Eds] T. Okaichi, D. Anderson and T. Nemoto. Elsevier. pp.253-255 (1989).
6. M. Watanabe, K. Kohata, and M. Kunugi, J. Phycol. 24, 22-28 (1988).
7. G.Y. Rhee, J. Phycol. 9, 495-506 (1973).
8. L. Sicko-Goad, and T.E. Jensen, Am. J. Bot. 63, 183-188 (1976).
9. M.E. Callow, and L.V. Evans, Br. Phycol. J. 14, 327-337 (1979).
10. A. Elgavish, and G.A. Elgavish, J. Phycol. 16, 368-374 (1980).

DO ALGAL BLOOMS PLAY HOMEOSTATIC ROLES?

T. WYATT
Instituto de Investigaciones Marinas del CSIC Eduardo Cabello, 6, 36208 Vigo, Spain

ABSTRACT

It has been suggested that some coccolithophorids may act through dimethylsulphide secretion as part of a "putative planetary thermostat", and that saxitoxin production by dinoflagellates could perform a chemostatic rôle vis a vis nitrogen. A more general hypothesis can be proposed in which anomalies of the Redfield ratio, or its equivalent for other nutrient species, lead to growth of phytoplankton species with biochemical machinery able to reduce the anomalies. Then one axis of Margalef's mandala is accessible to biological control.

Chlorophyll power spectra often fail to fit models based on the interaction between eddy diffusion and phytoplankton growth rates, even at quite high wavenumbers. That being the case, the value of conventional models of blooms based on mass action may be more limited than is often assumed. There may be biological means by which a progressive alteration of the internal wave field is accomplished. It can be argued that in some species certain morphological features (e.g. cell shape, chain formation) and biochemical activities (specifically polymer secretion) have been evolved to regulate local shear patterns. Species lacking such abilities might then colonize the environments created by organisms which precede them during succession. Then the course of turbulent decay which forms the second axis of the mandala would also be under increasing biological control as phytoplankton stock increased.

INTRODUCTION

Margalef et al. (21) presented a mandala (Fig. 1) in which phytoplankton succession was plotted in a space defined by turbulence (A) and growth rate (r) as a function of nutrient concentration. Rotation generates two new axes, $(rA)^{1/2}$ representing production potential, and $(r/A)^{1/2}$ with the dimensions of wavenumber as in the classical model of Kierstead and Slobodkin (18). The positions of some species in this space are given by Margalef (20). The functional response of r to nutrients is defined in Michaelis-Menten terms. Red tides typically lie to the left of the r-K axis of this space, in the quadrant where nutrients are high and turbulence low.

In a treatment of the problem of chlorophyll distributions, Denman and Platt, (6) compared the time to transfer kinetic energy from eddies of length scale d to eddies of 0.5 d with r^{-1}. They distinguished regimes $T \ll r^{-1}$ and $T \gg r^{-1}$ separated by a reciprocal wavenumber k_c with a value of about 1 km, below which spatial pattern is dominated by r. These wavenumbers should be considered as upper bounds since it is the realized rather than the intrinsic growth rate which determines their values (26). The reduction of the grazing coefficient during red tides (31) can be expected to lead to somewhat higher values of k. In further developments of this model, it has been found that many observed chlorophyll power spectra fail to fit it (26, 12, 13), and that the processes which determine the shape of these spectra "cannot be entirely physical, unless we can rationalize a progressive change in the turbulence or internal wave field concomitant with the development of the spring bloom" (26, p. 81).

This paper comments on the axes of Margalef et al.'s mandala, and suggests that there may be some biological control of mixing processes.

THE NUTRIENT AXIS

There is circumstantial evidence that the increase in the extent, frequency, and persistence of red tides in coastal zones is related to eutrophication (2, 27). This view still needs to be treated cautiously, for several reasons:

i) the increase in bloom frequency which began in some areas in the 1960s was coincident in time with other trends in marine populations which are rather clearly related to climatic phenomena;

Published 1990 by Elsevier Science Publishing Co., Inc.
Toxic Marine Phytoplankton
Edna Graneli et al., Editors

ii) some well explored and heavily polluted areas have not witnessed any obvious changes in bloom frequency;

iii) some relatively pristine areas seem to have experienced an increase in bloom frequency.

It is likely that both climate change and eutrophication are implicated. we should not forget that the atmosphere is being enriched too!

Strong support for the eutrophication link comes from Japanese experience in the Seto Inland Sea where legislation regulating the quality of discharged wastes has led to a reduction in the frequency and toxicity of outbreaks. One hypothesis proposed to account for this relationship invokes a geophysiological response, and suggests that saxitoxins are synthesized in order to remove excess nitrogen from the water column (32). There is experimental support for this view. Gonyaulax tamarensis responds to nitrogen enrichment in culture by producing up to four times more toxin than when grown in balanced media (4).

The inactivation of nitrogen in relatively refractory saxitoxins may be a mechanism for adjusting intracellular nutrient ratios. Species without such an ability are assumed to be at a competitive disadvantage when these blooms occur. The subsequent burial of the toxins in coastal sediments would be the next step of a geophysiologial response. It may be worthwhile to examine other kinds of enrichment from the same viewpoint. Are there for example outbreaks of certain phytoplankton species which are provoked by excess of phosphorus? The phosphorus would not necessarily be incorporated into the cell, but could be excreted into the atmosphere as phosphine or in some reduced gaseous form (7). There is little doubt that the necessary anoxic microzones exist in the planktonic environment (24), and they might be rather abundant in extreme bloom conditions. But a phosphorus containing proteolipid seems to be the toxic agent in the flagellate Prymnesium parvum, and perhaps too in the related Chrysochromulina polylepis which devastated Scandinavian coastal waters in May and June last year.

If the "putative planetary thermostat" of Charlson et al. (5) is an example of a broader phenomenon, then other geophysical and geochemical regulatory systems dependent on microorganisms should eventually be identified. Charlson et al. link dimethylsulphide (DMS) production (and the subsequent abundance of cloud condensation nuclei) to algae in the tropical oligotrophic ocean, perhaps coccolithophorids. In one member of this group, Emiliania huxleyi, it is possible to use the degree of unsaturation of long-chain ketones which coccolithophores synthesize to establish palaeotemperatures. This species and more oceanic species like Coccolithus pelagicus can also be identified in bloom conditions with satellite sensors (11). It is therefore in principle possible to use correlations between palaeoclimate and coccolith abundance in the sediments to construct predictive models relevant for example to the greenhouse effect. One might predict that outbreaks of these species will become more frequent as global warming occurs under the influence of greenhouse gases. Phaeocystis, Eutreptiella and some dinoflagellates are also capable of DMS production (e.g. 15).

In more general terms, it may be said that red tides are not responses to eutrophication per se, but to deviations from the normal nutrient ratios (such as the Redfield ratio) which drive succession along the r-K axis of the mandala. Thus toxic blooms of Gonyaulax tamarensis and Gymnodinium catenatum may be responses to nitrogen enrichment; Phaeocystis may respond to phosphorus enrichment. Blooms of Eutreptiella and Prorocentrum triestenum are stimulated by sulphides in pulp wastes (23). We should not be surprised that flagellates tend to replace diatoms, since it is usually nitrogen and phosphate amongst the classical nutrients, and not silicate, that are increased in eutrophic situations.

THE TURBULENCE AXIS

Many red tides and other massive algal blooms are characterized by high production rates of extracellular polysaccharides of high molecular weight. Boney (3) called mucilage "the ubiquitous algal attribute"; there are some astonishing accounts of its more bizarre manifestations. We know little about the functions of these products, but judging by the extent of the investment -more than 50% of carbon fixation sometimes- they must be important. Speculation may be premature as the basic phenomelogy is not yet established but it is hard to resist suggesting a few of the possibilities. I refer here only to the high molecular weight polymers.

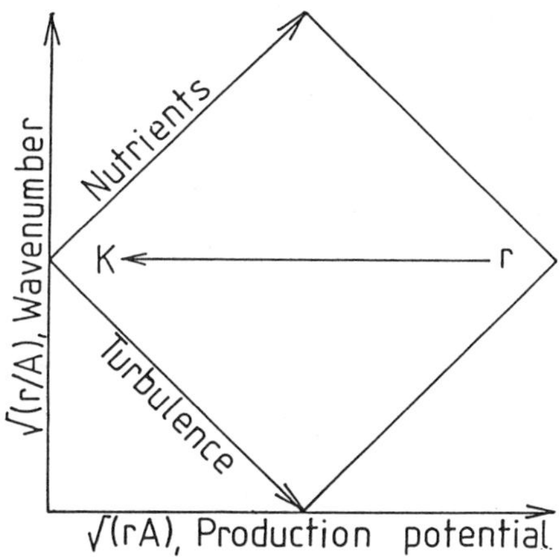

FIG. 1. Margalef's phytoplankton mandala

i) They may act as ion-exchange resins. This suggestion has been made both for slime producing bacteria and mammalian connective tissues (10). Cell concentrations encountered during dense phytoplankton blooms may not be very different from those in some connective tissues. Such a property has been proposed for the mucilage of Phaeocystis colonies (19).

ii) They may be useful to remain suspended. Some species of (Thalassiosira form colonies by secreting cobweb-like structures (9). Other diatoms (and desmids) possess processes through which mucus is secreted, e.g. Odontella sinensis, in which the cells appear to be suspended (25). The suggestion has frequently been made that oxygen bubbles produced by photosynthesis and trapped in such awnings assist flotation (8, 22), but the bubbles are not necessary if long threads are locked in the turbulent structure of the water (cf. aerial spiders).

iii) With different architecture, mucus threads can also promote sinking by causing entanglement or aggregation (1).

iv) In bacteria, and in the egg masses of the squid Illex illecebrosus (which are a metre or more in diameter), external mucilage may interfere with phagocytosis and predation respectively. The same is likely to be true of algal colonies (30). In parenthesis, we may add that mucilage can also provide food (28).

v) Exudates (also those of low molecular weight) could provide information to nearby cells (which will often be of the same clone) and thus have a social function, related for example to the timing of encystment or the initiation of the sexual phase.

vi) It is well known that even very low concentrations of such substances (« 1%) can cause remarkable alterations in the viscosity of the solvent phase (14). Several studies have shown that this is so in a variety of algal species in culture, including some well known red tide agents (14, 15, 16, 17).

Similar rheological changes can be achieved in particle suspensions in the absence of dissolved polymers. The aspect ratios, chain lengths, and the flexibility of chains are all likely to be important parameters in this context. This subject has received little attention in phytoplankton blooms, but studies of suspensions of inert particles indicate that the densities reached by phytoplankton in coastal waters at least are sufficient to generate these effects.

If polymer secretion and form selection in phytoplankton are responsible for the lack of fit to which Platt draws attention, then they can be viewed as rheological tools. As a bloom developed, turbulence would be suppressed at some wavenumber (greater than that predicted by physical mechanisms alone. Then (the well known succession of phytoplankton species composition may demonstrate an increasingly effective structuring of the physical environment, rather than an inevitable response to the decay of turbulence. That being so, the rheological changes wrought by phytoplankton could be viewed as dynamic analogues of the trails and burrows of terrestrial ecosystems and the simple principle of mass action would early in succession cease to operate; put more simply, plankton is not soup. Arguing from a different stance, the same conclusion is reached by Smetacek and Pollehne (29).

Could we then think of outbreaks of certain species of phytoplankton as responses to exceptional patterns of turbulence? If such an argument could be maintained, a short leap would bring us to a putative planetary turbidostat, to parallel the thermostat, and any real or imagined chemostats such as those for nitrogen and phosphorus already suggested. The thought that "veils of sticky slime" (29, p. 415) are secreted by phytoplankton to exploit local patterns of shear, or even to create more agreeable ones, tempts comparison with the technology of kites and with the trailing warps of yachtsmen in heavy weather (different Reynolds regimes notwithstanding). In these conditions, the restrictions imposed by Stokes'law, which has provoked such an abundant literature, are evaded.

These remarks may be condensed as follows. Exceptional algal blooms, like other kinds of outbreaks, do not sit easily within the framework of theoretical ecology. We may progress by seeking exceptional causes. It is suggested here that such blooms are often responses to extreme circumstances, generated either climatically or anthropogenically, and that their impact may be homeostatic.

REFERENCES

1. A.L. Alldredge and C.C. Gotschalk, Deep-Sea Res. 36,159-171 (1989).
2. D.M. Anderson in: Red tides, biology environmental science, and toxicology. T. Okaichi, D. M. Anderson and T. Nemoto, eds. (Elsevier, 1988) pp. 11-16.
3. A.D. Boney, Br. Phycol. J. 16, 115-132 (1981).
4. G.L. Boyer et al., Mar. Biol. 96, 123-128 (1987).
5. R.E. Charlson et al., Nature 326, 655-661(1987) .
6. K. Denman and T. Platt, Mem. Soc. Roy. Liége, 6e serie, 7, 19-30 (1975).
7. I. Dévai et al., Nature 333, 343-345 (1988).
8. F.E. Fritsch, The structure and reproduction of the algae, vol. I (Cambridge 1948).
9. G.A. Fryxell et al., Br. phycol. J., 19, 141-156 (1984).
10. A.G. Gristina, Science 237, 1588-1593 (1987).
11. P.M. Holligan et al., Nature 304, 339-342 (1983).
12. J.W. Horwood, J. mar. biol. Ass. U.K., 58, 487-502 (1978).
13. J.W. Horwood, J. Cons. int. Explor. Mer, 39, 261-270 (1981).
14. J.W. Hoyt, Trends in Biotechnol., 3,17-21 (1985).
15. K. Ishida, Mem. Coll. Agric. Kyoto Univ., 94,7-52 (1968).
16. I.R. Jenkinson, Nature 323, 435-37 (1986).
17. P.R. Kenis, Nature, 217, 940-942 (1968).
18. H. Kierstead and L.B. Slobodkin, J. Mar. Res., 12,141-147 (1953).
19. C. Lancelot and S. Mathot, Mar. Biol., 86, 227-232 (1985).
20. R. Margalef, Oceanol. Acta I, 493-509 (1978).
21. R. Margalef et al. in: Toxic Dinoflagellate Blooms, D.L. Taylor and H.H. Seliger, eds. (Elsevier, Amsterdam 1979).
22. M.J. Mc.Conville and A Wetherbee, J. Phycol. 19, 31-439 (1983).
23. T. Okaichi and I. Yagyu, Bull. Plankton Soc. Japan, 16,126 (1969).
24. H.W. Paerl and R.G. Carlton, Nature, 332, 260-262 (1988).
25. J.D. Pickett-Heaps et al., J. Phycol., 22, 334-339 (1986).
26. T. Platt in: Spatial pattern in plankton communities, J.H. Steele, ed. (Plenum 1978) pp. 73-84.
27. A. Prakash in: Proc. Ist. Intern. Conf. on Toxic dinoflagellate blooms, V.R. Lo Cicero, ed. (Mass. Sc. Technol. Foundation) pp. 1-6.
28. E.B. Sherr, Nature, 335, 348-351 (1988).
29. V. Smetacek and F.Pollenhe, Ophelia 26, 401-420 (1986).
30. T. Wyatt in: The ecology of the seas, D.H. Cushing and J.J. Walsh, eds. (Blackwells 1976) pp. 59-77.
31. T. Wyatt and J. Horwood, Nature, 24, 238-240 (1973).
32. T. Wyatt and B. Reguera in: Red tides, biology, environmental science, and toxicology, T. Okaichi, D.M. Anderson and T. Nemoto, eds. (Elsevier 1988) pp. 33-36.

MODELLING THE SPREAD OF RED TIDES - A COMMENT

T. WYATT
Inst. de Invest. Marinas. Vigo, Spain

ABSTRACT

An epidemiological approach to the spread of red tides is proposed, which can be linked to hydrodynamic and physiological models through mixing coefficients and nutrient kinetics respectively. The so-called SEIR models distinguish susceptible (S), exposed (E), infective (I), and recovered [R) sites. The approach is viewed in terms of increasing nitrogen inputs. It is suggested that preventive rather than therapeutic management strategies would bear consideration.

AN EPIDEMIOLOGICAL MODEL

The strategics behavior of red tide species might usefully be explored using epidemic models of the form

$$dt/dt = axy - by$$

Here X is the number of unoccupied sites which could support blooms, y is the number of potential seed populations (of vegetative cells or of cysts), a is the colonization rate and b is the death rate.

There is a value of x >/b/a which gives positive values of dy/dt, hence allows net colonization to take place. An artificial increase in x - due for example to eutrophication - will change the equilibrium distribution "without any change in the climate or in the plants' biology" (1).

This approach is given a novel twist by Watson and Lovelock (2), who imagine two species of daisies with different albedos, which coexist in the face of increasing solar radiation. Here we substitute an increasing nitrogen: phosphorus ratio for radiation, and observe that a species able to more effectively manipulate the ratio, perhaps by synthesizing saxitoxins (3), will outcompete species which cannot- see upper part of figure l.

A statistical analysis of the response of algal communities along pollution gradients (phosphate, ammonium, BOD, ...) in freshwater
is given by Fricke and Steubing (4).

A provocative aspect of the model is its abllity to maintain temperature, or in the present case the N:P ratio, over long time periods despite a continual increase in the external forcing, see lower part of figure 1.

Here we distinguish two kinds of sites, susceptible [S) and infective (I) in epidemiological terms. The SEIR models also distinguish exposed (E) and recovered (R) sites. The former (E) are a subclass of S, and one can imagine that noxious blooms have failed to appear in some areas because infection has not occurred. Recovered sites would be those in which a bloom has already taken place in response to some threshold value of the rising N:P ratio, and the ratio as a result has once more been reduced below the threshold. The threshold will clearly he partly, controlled by the rates of nitrogen input and dispersion, thus providing a link between epidemiological and hydrodynamic models.

Published 1990 by Elsevier Science Publishing Co., Inc.
Toxic Marine Phytoplankton
Edna Graneli et al., Editors

DISCUSSION

There must however be a sink for the nitrogen. Intermittent or permanents deoxygenation in the lower water column accompanied by sapropel formation provides a potential sink. Nitrogen (and phosphorus) are normally preferentially removed from sediments (relative to carbon), but sorption, humification, and anoxia can combine to sequester organic nitrogen in refractory forms which are subsequently buried. The resistant walls of cysts must aid this process .

Regions which combine density stratification and weak circulation below the thermocline are specially prone to deoxygenation processes. The New York Bight is such a region in summer. Areas influenced by the Baltic outflow are similarly at risk following unusual conditions such as the winter of 1987-88. The enriched waters of the Southern Bight of the North Sea (but with a reduced N: P ratio) are perhaps protected by an energetic tidal regime which ensures year round mixing. The fascinating patterns of toxicity in British Columbia described by Gaines and Taylor (5) might be related to changes in susceptibility dependent on relatively subtle oceanographic changes, controlled ultimately by ENSO events. Small variations around critical dispersion rates are a possibility at the local scale (6).

These remarks raise the curtain on the familiar scenarios of anoxia, mortalities and so on, on the well known consequences of massive algal blooms. The synthesis of toxins is not essential for most of them. So far, with the exception of the Inland Sea of Japan, management strategies have principally aimed to prevent or alleviate the potentially undesirable consequences of red tides to public health and aquaculture. These are therapeutic strategies. The radical manager may prefer to follow the Japanese example, and interrupt the whole play before the curtain rises, instead of following protocol and instigating yet more monitoring. Preventative strategies are acknowledged to be more cost-effective in epidemiology. They may be too in coastal zone management.

REFERENCES

1. R.N. Carter and S. D. Prince. Nature 293, 644-455 (1981).
2. A.J. Watson and J.E. Lovelock, Tellus 35B, 284-289 (1983).
3. T. Wyatt and B. Reguera in: Red tides: biology, environmental science, and toxicology T. Okaichi, D.M. Anderson, T. Nemoto eds. (Elsevier, N. York 1989) pp 33-36.
4. G. Fricke and L. Steubing. Archiv. für Hydrobiol. 101, 361-372 (1984).
5. G. Gaines and F.J.R. Taylor in: Toxic Dinoflagellates D.M. Anderson, A.W. White, D.G. Baden, eds. (Elsevier , N. York1985) pp 439-444.
6. T. Wyatt and J. Horwood, Nature 244, 238-240 (1973).

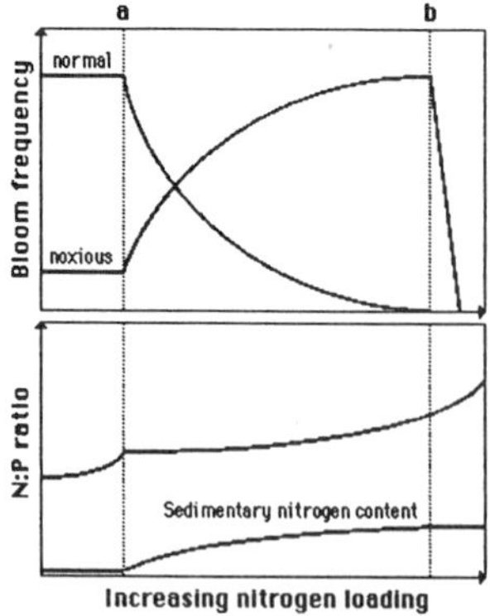

FIG. 1. At (a), a treshold is reached which alters the competitive balance between <<normal>> and <<noxius>> species in the latter's favour. At (b) the sediments can no longer absorb nitrogen, and the system collapses.

V PHYSIOLOGY AND BIOCHEMISTRY

UPTAKE AND TISSUE DISTRIBUTION OF PSP TOXINS IN BUTTER CLAMS

Mark K. Beitler and John Liston
Institute for Food Science & Technology, HF-10
Seattle, WA 98195, USA

ABSTRACT

In two feeding experiments, 48-50 nontoxic butter clams (*Saxidomus giganteus*) were fed clonal *Alexandrium catenella* containing gonyautoxin (GTX) I-IV and neosaxitoxin (NEO) but no detectable saxitoxin (STX) for several days. Nine tissues were dissected periodically over 0-83d and analyzed for PSP toxins using HPLC and mouse bioassay (MBA) methods. Toxins rapidly concentrated in the visceral mass (VM) with toxin profiles similar to that of the alga, and gradually decreased afterwards. GTX I-IV rapidly appeared in the mantle (M)+ kidney (K)+ pallial muscle (PM) during feeding while the gill (G) contained low concentrations of GTX II+III. STX was detected mainly in the siphon (S) after a delay. STX was detected in the VM, pericardial gland (PG), mantle muscle (MM) and in the M+K+PM. HPLC and MBA results were comparable. The origin of STX is still being investigated.

INTRODUCTION

Butter clams retain primarily STX mostly in the S and can remain toxic for years [1]. Two explanations have been proposed for the origin of STX in this clam. First, the S could accumulate STX from a low initial concentration of STX from ingested *Alexandrium* spp. [2,3]. Second, the high concentration of STX in the S may result from a conversion of GTX to STX [3,4]. The transformation of GTX I-IV and NEO to STX has been associated with a scallop and *Pseudomonas* sp. and *Vibrio* sp. isolated from the viscera of marine invertebrates [4,5].

The purpose of this investigation was to determine whether nontoxic butter clams fed *A. catenella* containing GTX I-IV and NEO but no STX would accumulate STX in the S. This process was examined by analyzing toxin profiles of various butter clam tissues sampled throughout the experiments.

MATERIALS & METHODS

Feeding Experiments

Two experiments were conducted using 6-14 yr old, nontoxic butter clams collected from Seabeck, WA, a site historically free of *A. catenella* blooms. Twenty-four and 25 shellfish were placed into each of two, identical 20 gal aquaria containing 50L filtered, aerated sea water and held for 5d for acclimation. Large clams were used to obtain sufficient volumes of tissue for toxin extraction.

Exp. 1: Clonal *A. catenella* (GC 84-40) was grown in 2-3L f/2 media in 2.8L Fernbach flasks or 3 and 4L Erlenmeyer flasks at 14 C under continuous illumination at a 55 $\mu E\ m^{-2}\ s^{-1}$ light intensity for 34d until the cells reached late exponential to early stationary growth. Cultures containing 5.59×10^3 and 1.80×10^4 cells/ml were pumped at 5 ml/min into each aquarium for 24 and 29.5h, respectively, starting on day 0. Clams were held without

Toxic Marine Phytoplankton
Edna Graneli et al., Editors

additional food in the aquaria up to 58d after the initiation of the feeding. Sea water was changed at approximately 6d intervals [6].

Exp. 2: The algal culture was grown in 9-18L f/2 media in 9.5L bottles and 20L carboys at the same temperature and light intensity as in exp. 1. About twice as many algal cells were fed to the clams for a longer period than in exp. 1 to more closely simulate a natural bloom and to improve the detection of toxins in tissues. For 21 and 19h, culture containing 8.86×10^3 and 8.92×10^3 cells/ml, respectively, were pumped at 5.8 ml/min into each tank beginning on day 0. Because the pump malfunctioned on day 2, 11.0L, 7.4L, and 7.5L of culture containing 1.30×10^4, 1.39×10^4, and 9.45×10^3 cells/ml, respectively, were added to each aquarium on days 2, 7, and 9, respectively, with sea water replacements at 2-5d intervals. The clams were held in the tanks without additional food up to 83d following the start of the feeding. The water was replaced at 3-9d intervals.

Toxin Analyses

Clams exp. 1: 2 clams were randomly sampled periodically from each tank on days 1-58 after the initiation of the feeding. The S, VM, M+K+PM, MM, PG, G, adductor muscle, foot, and heart were dissected per clam and separately homogenized in 0.10 N HOAc.

Clams exp. 2: 2-3 surviving clams were sampled per tank on days 2-27 and 83 following the start of the feeding. Tissues were dissected, paired to increase toxin concentrations, and homogenized in 0.10 N HOAc.

Tissues were centrifuged and filtered following the method of Boczar *et al.* [7].

Alga exp. 1: 2.46×10^5- 5.00×10^5 cells were collected from *A. catenella* cultures fed and centrifuged at 2,100 x G for 15 min.

Alga exp. 2: 8.86×10^6 - 1.39×10^7 cells from each of the 5 volumes of culture fed were taken and filtered through 20 µm Nitex.

Toxins were extracted from the cells and filtered following the method of Boczar *et al.* [7]. Algal culture samples were enumerated with an inverted microscope and a Coulter Counter.

Tissue and cell extracts were analyzed 2-3 times on different days by the HPLC method of Sullivan and Wekell [8]. Tissue extracts were tested for toxicity by the MBA at the Washington State PSP Laboratory, Seattle, WA [9]. Toxicities of the cultures fed in exp. 1-2 were calculated using appropriate conversion factors from Boyer *et al.* [10].

RESULTS

Table I illustrates the amount of toxin and the number of *A. catenella* cells consumed by the clams and the toxicities of the cultures fed in exp. 1-2. The algal cells contained GTX I-IV and NEO, mostly GTX I+IV, but no STX. The cultures used in exp. 2 were more toxic than those in exp. 1.

TABLE I. The total µmol of PSP toxins and number of *A. catenella* cells ingested by nontoxic butter clams in two experiments and mean toxicities of the cultures fed.

	µmol					Toxicity
Exp.	GTX I+IV[a]	GTX II+III[a]	NEO[b]	Total	No. of Cells (10^8)	µMU/cell[c]
1	2.36(72)[d]	0.60(18)	0.33(10)	3.29	4.00	14
2	14.51(83)	1.43 (8)	1.65 (9)	17.59	7.00	43

[a] GTX = gonyautoxin [b] NEO = neosaxitoxin [c] MU = mouse unit [d] mol%

Butter clams in exp. 1-2 absorbed and retained PSP toxins (FIG. 1-2). The shellfish were initially free of detectable PSP toxins based on HPLC and MBA results. Foreign components in tissue extracts prevented the quantitation of NEO with any certainty. GTX I-IV rapidly increased in the VM during feeding and gradually decreased afterwards. STX was detected late in the VM. STX primarily accumulated in the S after a lag period following the initiation of the feedings and persisted for 4 or more weeks, attaining the highest level after 83d. In exp. 1, negligible concentrations of toxins were detected in the 7 other tissues. In exp. 2, toxins were detected in the PG after a lag period; GTX II+III increased in concentration with time and GTX I+IV persisted through day 83 when STX was first detected.

The M+K+PM showed increased concentrations of GTX I-IV during feeding and retained GTX I+IV for 27d and GTX II+III for 83d. STX was noted in the M+K+PM on the same days as its appearance in the S. The G contained GTX II+III at low concentrations. In exp. 2, STX was detected at 0.48-0.63 nmol/g tissue in two MM samples on days 22 and 28. Toxins were not detected in the 3 other tissues.

MBA and HPLC values ranged from <37-113 and <37-159 µg STX equivalents/100g tissue, respectively. Out of 35 samples which had >37 µg STX equivalents/100g tissue based on calculations from HPLC results, 33 had detectable toxicities when analyzed by the MBA.

In exp. 1, shellfish were exposed to mean algal cell densities ranging from 46-265 cells/ml over 2d. Mean cell densities in the aquaria in exp. 2 ranged from 276-3,208 cells/ml. Algal cells were detected in the aquaria for a total of 4 and 11d in exp. 1 and 2, respectively. Fecal samples collected after day 14 in exp. 2 did not contain intact algal cells nor detectable toxins by HPLC analyses. During the fasting periods, the clams did not pump water as actively as when they were feeding. No mortalities occurred in exp. 1 but 16 shellfish died in exp. 2.

The clams in exp. 1-2 did not experience weight loss. In exp. 1, the total wet tissue weight of the clams was 48.127 $\pm$ 10.406g and 43.021 $\pm$ 3.912g on days 1 and 58, respectively. On days 2 and 83 in exp. 2, the total wet tissue weight of shellfish was 28.961 $\pm$ 8.448g and 24.477 $\pm$ 3.483g, respectively.

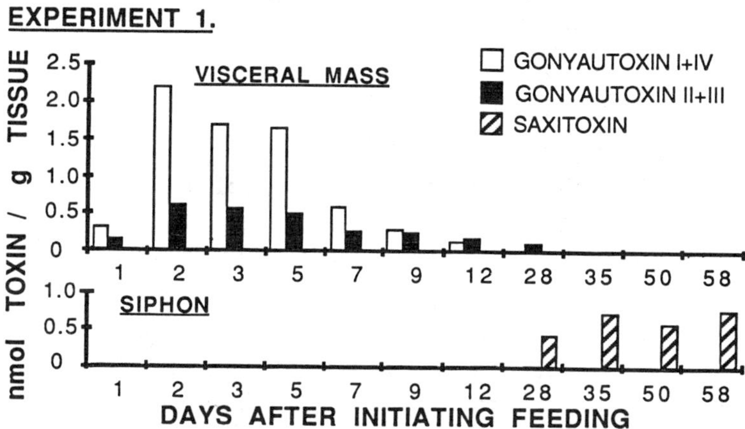

FIG. 1. Paralytic shellfish toxin composition of tissues of nontoxic butter clams (*Saxidomus giganteus*) fed *Alexandrium catenella* for 4d. Each bar represents the mean of 2-3 analyses of 4 tissues from 4 clams. Note: the time scale was disproportionally drawn to conserve space.

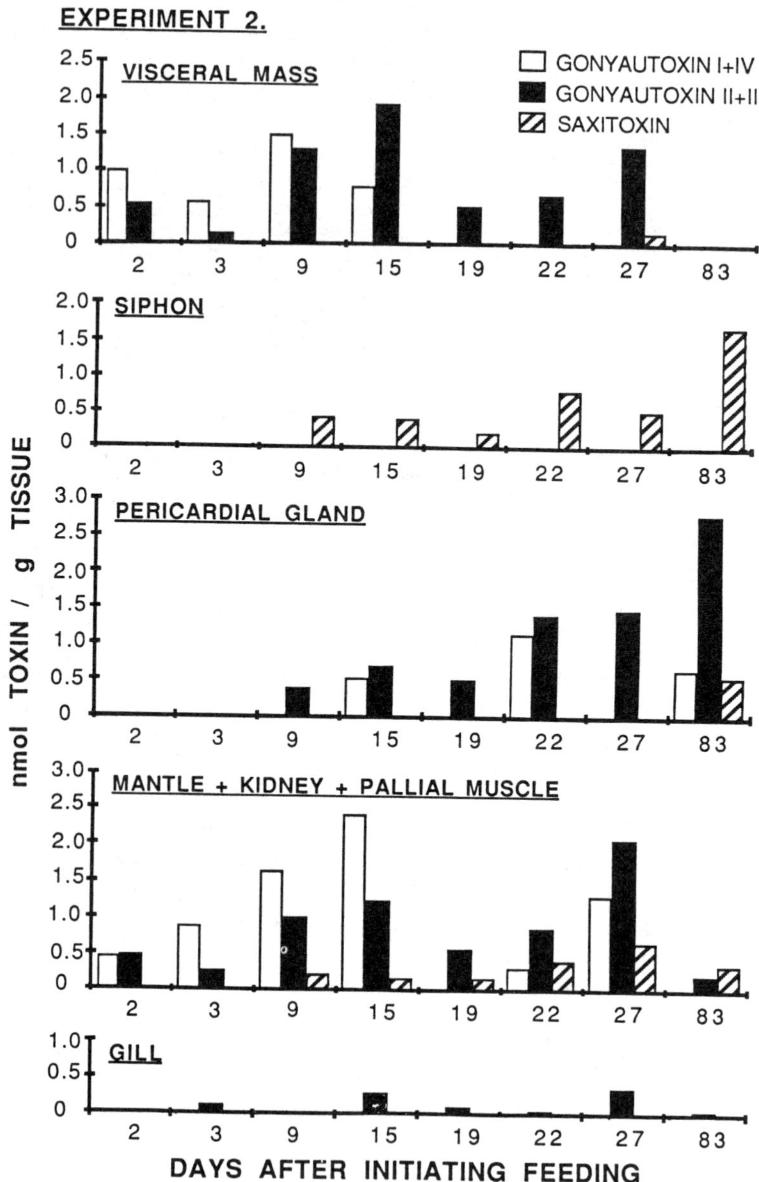

FIG. 2. Paralytic shellfish toxin composition of tissues of nontoxic butter clams (*Saxidomus giganteus*) fed *Alexandrium catenella* for 11d. Each bar represents the mean of 2-3 analyses for 2-3 pairs of pooled tissues from 4-6 clams. Note: the time scale was disproportionally drawn to conserve space.

DISCUSSION

The results of this study confirm that toxins are widely distributed through most of the tissues of butter clams after feeding on toxigenic *A. catenella* and that STX is the dominant toxin found in the S. It is noteworthy that STX levels increase in the S for 4 or more weeks after a lag period. By feeding the shellfish more algal cells and pooling matching tissues, toxins were detected in the PG, M+K+PM, and G. Results from field studies suggest the PM contributed to the STX detected in the M+K+PM. In molluscs, the PG filters compounds from the blood and the K absorbs and excretes solutes [11]. The concentration of GTX II+III in the PG suggests it may abstract toxins from the blood, but it is unknown whether the K reabsorbs GTX I-IV over time.

The findings in this investigation are comparable to those reported by Sullivan [6] from a limited study in which 9 butter clams were fed *A. catenella* containing C2 and B1 for 2d. As described by the authors of this manuscript and Sullivan [6], STX mainly accumulated in the S after a period of time during which the concentration of toxins in the VM rapidly increased and then decreased to low levels. It is interesting that STX was detected in the S of butter clams following the ingestion of *A. catenella* containing either carbamate or N-sulfocarbamoyl toxins but no detectable STX.

Shifts in the pattern of occurrence of toxins other than STX do not clearly point to a biological transformation of GTX or NEO to STX though this is theoretically possible [4,5]. STX was not detected in the algal cultures and this, together with the slow build up of STX in the S, suggests that some other transformation is occurring. It is not likely that this is simply a concentration of STX from less than detectable levels in the culture because of the final total amount calculated. Separate experiments using homogenates of butter clam tissues have not demonstrated a conversion of GTX or NEO to STX, as reported for scallops [4]. Bacterial conversion [5] seems unlikely because STX was not detected in the VM during and soon after feeding. Analysis of *A. catenella* culture filtrates are underway to test the possibility that any STX produced by bacterial action was absorbed from this source [12]. This is unlikely because STX should have been detected in the viscera or gills.

A cryptic origin of STX arising from substances produced by the *A. catenella* or the clam [13] cannot be ruled out. A precursor produced by the dinoflagellate could conceivably be acted upon by enzymes in the S or adjacent tissues to release STX. Alternatively, transformation through a cryptic form within the clam could be the source. The S is highly specific in retention of STX and rejection of most other toxins; this is supported by field research. Such observations raise interesting questions concerning the biological significance of the STX concentration for the butter clam.

CONCLUSIONS

STX accumulates in the S of nontoxic butter clams feeding on toxigenic *A. catenella* lacking detectable STX. Some type of synthesis or biotransformation of GTX I-IV and/or NEO to STX in the S or closely associated tissue occurs *in vivo*. STX in the S does not obviously appear to be synthesized in another tissue and transported to the S.

ACKNOWLEDGMENTS

The authors thank G. Skow and L. Nishitani and Drs. B. Rasco, J. Sullivan, and B. Boczar for technical assistance and/or advice. We appreciate the generous donations of PSP toxin standards from Dr. S. Hall and butter clams from J.E. Emel and E. Mears. This research was supported by grants from the Washington Sea Grant Program.

REFERENCES

1. D.B. Quayle, Fish. Res. Bd. Can. Bull. 168 (1968).
2. E.J. Schantz and H.W. Magnusson, J. Protozool. 11, 239-242 (1964).
3. Y. Shimizu, W.E. Fallon, J.C. Wekell, D. Gerber Jr., and E.J. Gaughlitz Jr., J. Agric. Food Chem. 26, 878-881 (1978).
4. Y. Shimizu and M. Yoshioka, Science 212, 547-549 (1981).
5. Y. Kotaki, Y. Oshima, and T. Yasumoto, Bull. Jpn. Soc. Sci. Fish. 56, 1009-1013 (1985).
6. J.J. Sullivan, Ph.D. thesis, University of Washington (1982).
7. B.A. Boczar, M.K. Beitler, J. Liston, J.J. Sullivan, and R.A. Cattolico, Plant Physiol. 88, 1285-1290 (1988).
8. J.J. Sullivan and M.M. Wekell in: Seafood Quality Determination Developments in Food Science, D.E. Kramer and J. Liston, eds. (Elsevier, New York 1987) pp. 357-371.
9. "Official Methods Analysis," Assoc. Off. Anal. Chem. (14th ed.), Arlington, VA (1984).
10. G.L. Boyer, J.J. Sullivan, R.J. Andersen, F.J.R. Taylor, P.J. Harrison, and A.D. Cembella, Mar. Biol. 93, 361-369 (1986).
11. E.B. Andrews in: The Mollusca, Vol. 11, E.R. Truman and M.R. Clarke, eds. (Academic Press, Inc., San Diego 1988) pp.381-448.
12. A. Andrasi in: Toxic Dinoflagellates, D.M. Anderson, A.W. White, and D.G. Baden, eds. (Elsevier, New York 1985) pp. 401-406.
13. M. Kodama, T. Ogata, Y. Takahashi, T. Niwa, and F. Matsuura, J. Biochem. 92, 105-109 (1982).

PIGMENT COMPOSITION AND LOW-LIGHT RESPONSE OF FOURTEEN CLONES OF *GAMBIERDISCUS TOXICUS*

J.W. Bomber[1], D.R. Tindall[1], C.W. Venable[2] and D.M. Miller[2]. Department of Botany[1] and Department of Physiology[2], Southern Illinois University, Carbondale, Illinois 62901.

ABSTRACT

The pigment composition of the ciguatera-causing dinoflagellate *Gambierdiscus toxicus* is disputed. The present study examined the pigment composition of 14 acclimated clones of *G. toxicus* from both Atlantic and Pacific regions by TLC, HPLC, Scanning Spectrophotometry and ^1H NMR spectrometry. There is significant variability in the pigment concentration and low light response of these clones indicating the existence of different ecotypes. Highly toxic clones produced less chlorophyll than less toxic clones (r = -0.85), suggesting that a relationships exists between toxigenesis and photosynthesis.

INTRODUCTION

Monospecific clonal lines of marine diatoms exhibit differences in cellular pigment content, even when grown in the same environment [1]. This disparity may arise from genetic divergence into races which are selected according to their indigenous light regimes. Similar studies with dinoflagellates are incipient although their toxinology and enzymology have been explored systematically [2, 3]. Considering that dinophycean polycyclic ether toxins are potentially photosynthetic modulators [4], pigment/toxin studies are needed to better understand the roles of these toxins. These efforts have both systematic and biochemical value. The present study explores the pigment composition and low-light response of 14 clones of the ciguatera-causing dinoflagellate *Gambierdiscus toxicus*. Particular attention is given to the chlorophyll c component as the presence of chlorophyll c_1 in *G. toxicus* is a matter of current debate [5, 6]. The data is compared to clonal potency data generated in related work [2] in order to hopefully further illuminate knowledge of the genetics and toxigenesis of this important alga.

MATERIALS AND METHODS

Acclimation. Fourteen clones of *Gambierdiscus toxicus* (Table 1) were acclimated [7,8] for one year in 500 ml and 1 L flasks in K medium [9] at 28° C and 1,800 lux (vita-lite bulbs). The cultures were subsequently shifted to 900 lux and acclimated for up to 200 additional days.

Harvesting, Extraction, Pigment Analyses. The 500 ml cultures were harvested with a 38 µm mesh screen in mid-log phase. Preliminary studies indicated that pigment decomposition occurred rapidly after this period and made results ambiguous. The cells were frozen at -20° C, lyophilized until dry, extracted with 100% de-gassed acetone with sonication in an ice bath and then filtered through a 0.2 µm polyethylene filter. Cells and extracts were processed in darkened fume hoods and stored under nitrogen at -20° C. More than 90% of the samples were analyzed within two weeks of harvest.

The crude extracts were assayed for chlorophylls as per Jensen [11] on a Perkin Elmer, Lambda 3-B scanning spectrophotometer. Chlorophyll c_1 was not quantified at this stage. All clones were subsequently analyzed by analytical thin layer chromatography (silica gel H, Supelco, Inc.), co-chromatogrammed with known chlorophyll c_1-containing algae *Isochrysis*

Toxic Marine Phytoplankton
Edna Graneli et al., Editors

galbana and *Amphora costata*. ß-carotene constituted < 2% of the total pigment content and for this reason it was difficult to quantify spectrophotometrically. Consequently, the concentration of ß-carotene was not used in calculations of chlorophyll/carotenoid ratios. All other thin layer eluates were examined spectrophotometrically [11]. In addition, the eluates were analyzed via reverse phase High Performance Liquid Chromatography (HPLC) by isocratic elution with methanol as the mobile phase on a Waters C_8 column. The eluates were monitored at 210 and 440 nm. Extracts of clone 175 were also purified via cellulose (medium grade, Sigma Chemical Co.), silica gel H (Sigma), silicic acid (SIL-LC, Sigma) and polyethylene (Polysciences, Inc.) column chromatography [12] in order to detect chlorophyll c_1. The relatively lipid-free chlorophyll c mixture was precipitated in petroleum ether [12]. The precipitate was next examined by ^1H NMR spectrometry in deuterated methanol on a Varian model XR 500 spectrometer.

TABLE I. Clones of *Gambierdiscus toxicus* used in the study.

Clone Number	Location	°N Latitude	Isolator
175	Martinique, Caribbean	14	J. Babinchak
350	Virgin Gorda, V.I.	18	D. Tindall
177	Hawaii	20	N. Withers
196	Marathon Key, FL	24	J. Bomber
199	ibid		
200	ibid		
300	ibid		
158	Drift Algae, Gulf Stream	25	J. Bomber
163	Gingerbreads, Bahamas	25	J. Bomber
170	ibid		
171	ibid		
169	Drift Algae, Gulf Stream	26	J. Bomber
172	Great Isaacs Light, Bahamas	26	J. Bomber
135	Bermuda	32	L. Brand

Statistical Analyses. We intended a-priori to conduct paired Analyses of Variance (ANOVA's) on the data. However, because of an observed interaction between clones and light intensities (see results), we could not conduct the tests without violating the rules of paired ANOVA's [10]. Consequently, we arbitrarily separated low latitude (< 21°N) and high latitude (>21°N) clones into what appeared to be two distinct pigment content groups (low and high) and compared them via the t-method [10]. Association between chlorophyll a content and acclimated clonal potencies were tested by the Pearson correlation test [10].

RESULTS AND DISCUSSION

G. toxicus required an unusually long acclimation period of up to one year for some isolates although others acclimated within 16 weeks. Cultures of *Prorocentrum concavum* grown simultaneously with *G. toxicus* during this work acclimated in 2/3 less time. This supports the hypothesis set forth by Bomber et al. [13] that *G. toxicus* is accustomed to a relatively stable chemico-physical environment and is not very adaptive to changes in these parameters. Consequently, in order to make comparisons of quantitative traits, complete acclimation was paramount to hypothesis testing. The cultures made changes during acclimation including an initial decrease in the chlorophyll a/c ratio when moved to low light. However, the cultures returned to their previous a/c ratio when completely acclimated. Figure 1 shows the chlorophyll a/c response of clone 172 to low light and this was typical of most high latitude (> 21° N) clones. These data suggest that the pigment composition of a clone should be monitored via a time-course study in order to be certain of its true biochemical response to a change in conditions. Had we not waited until three successive reproduction rates were the same and terminated the study earlier, we may have been misled to state that the chlorophyll a/c ratio plummets when grown in low light and is "probably" maintained at that level. In fact, the decrease in the ratio may be an initial response to stress and when given time the clone can readjust its pigmentation to its "normal" ratios.

The clones examined produced slightly more pigment at the lower light intensity (Table 2). This is consistent with work on other dinoflagellates at low light intensities [14]. However, in the case of *Gambierdiscus* it is difficult to generalize. First, we did not monitor changes in cell size between the different light intensities and such changes could have affected the results. Secondly, early in this project it became clear that pigment content between the two light intensities could not be compared by the paired ANOVA's [10] planned a-priori. This conclusion was based on the amount of "scatter" detected by a principal components analysis (Statview 512, BrainPower, Inc.) of the pigment data (Fig. 2). Two factors were extracted from the coded (observation - mean/standard deviation) data. The "scatter' indicated that a light x clone interaction exists and therefore clonal lines could not be utilized as replicates. This scatter also prevented us from testing for differences among clones by using the data from the two light intensities as replicates. Indeed, paired Analyses of Variance must assume no interaction [10].

TABLE II. Comparisons between the two light intensities in mean pigment content (pg per cell), pigment ratios and reproduction rates of all clones.

Character	900 lux	1,800 lux
Chlorophyll a	616	580
Chlorophyll c_2	203	177
Peridinin	189	176
Dinoxanthin	25	18
Diadinoxanthin	21	16
Total pigment	1054	967
Chlorophylls/carotenoids	3.5	3.6
Chlorophyll a/c_2	3.0	3.3
Reproduction rate (division per day)	0.07	0.15

The clonal "scatter" can be explained in part by the differences between low and high latitude clones. Clones 175, 177 and 350 had difficulty acclimating to low light and died during post-acclimation transfers. These clones also had significantly different and lower pigment content than the high latitude clones (Table 3). A paired t-test [10] comparison indicates that these groups have significantly different pigment concentration even when the data are normalized to account for differences in cell size among clones (chlorophyll a, ts = 3.957; chlorophyll c_2 ts = 2.709 where $t_{.05\,[27]} = 2.052$). The poor growth in low light of the low-latitude group is also supported by reproduction rate/light intensity plots in Bomber et al. [2] which show that the Hawaiian clone (177) and the two Caribbean clones (350 and 175) cannot grow in low light. The slopes also overlap, indicating that the clones may be adapted to specific light regimes and consequently experience different selective forces [15]. An ANOVA [2] indicated that there is in fact significant variation among the reproduction rates of these clones (significant at 0.0001 level). Consequently, the work of Bomber et al. [2] and the present work suggest that the clones examined are genetically distinct races. The data are consistent with the environments from which the clones were isolated. The low-latitude clones would experience higher light intensities than the high latitude clones due to their proximity to the equater. Thus, the low latitude clones would require less pigment.

Their lower pigment content was in fact observed in this study. We intend to examine additional clones to test these hypotheses further.

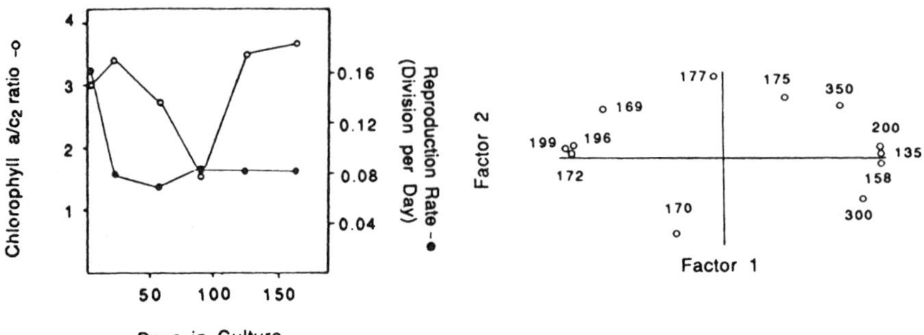

FIG. 1. Changes in the chlorophyll a/c_2 ratio for clone 172 when acclimated from 1,800 lux (first data point) to 900 lux (last 5 data points).

FIG. 2. Principal components analysis of the pigment data from twelve clones.

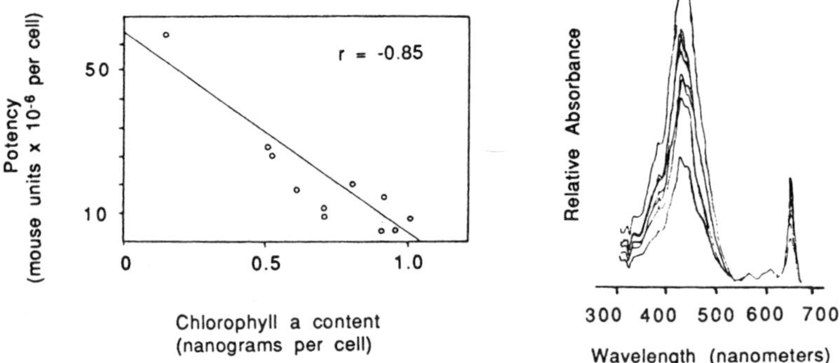

FIG.3. Acclimated Chlorophyll a content (1,800 lux) of eleven clones vs. their acclimated potencies (1,800 lux). The potency data is derived from Bomber et al. [2].

FIG. 4. Composite absorbance spectra of eight clones of *G. toxicus*.

The chlorophyll a content was plotted against the toxin content (mouse units per mg dried cells, Fig. 3) of these clones generated by mouse bioassay from acclimated cultures in related research [2]. There is a significant negative correlation between the variables. All three low latitude clones are consistently more toxic than high latitude clones. This leads us to speculate that the trade off for reduced pigment synthesis may be increased toxin synthesis. Without question more research is necessary to test this hypothesis, especially considering that some Caribbean clones are reportedly non-toxic [16]. However, this observation should encourage research into the possible mode of action of maitotoxin as a photosynthetic modulator.

TABLE III. A comparison between low and high latitude clones. Pigments are in picograms per cell and represent the mean from both light intensities (n=8). The potency data is from Bomber et al. [2].

Character	< 21° N	> 21° N
Chlorophyll a	381 ± 136	662 ± 83
Chlorophyll c_2	122 ± 48	238 ± 67
Peridinin	101 ± 28	228 ± 48
Reproduction rate (1,800 lux)	0.19 ± 0.03	0.13 ± 0.02
Cell size (1,800 lux, transdiameter)	71.6 ± 6.4	82.8 ± 2.5
Potency (1,800 lux, mouse units per cell)	36.3 ± 16	10 ± 6

The pigments found in the fourteen clones were given in Table 2. The clones contained chlorophylls a and c_2, peridinin and two xanthophylls as determined by TLC. These identifications were made from characteristic absorbtion spectra and their maxima. However, the designations of dinoxanthin and diadinoxanthin should be considered tentative without further analysis. There were no qualitative differences in pigment composition detected among the twelve clones. The similarity is evidenced by homogeneity of the absorbance spectra shown for 8 clones in Fig. 4. We could not detect chlorophyll c_1 in *G. toxicus* by TLC, although we could for *Isochrysis* and *Amphora*. We next examined the chlorophyll c component via HPLC and again observed only a single component (Fig. 5). However, this peak has a broad base, indicating that two compounds may be present. When we scaled-up to column chromatography of crude extracts on cellulose followed with lipid removal on silicic acid, two bands of chlorophyll c occurred. Nevertheless, these two bands overlapped considerably and absorbance spectra were still unconvincing. Finally, when the chlorophyll c component was precipitated in petroleum ether and then separated on polyethylene [12], a clearer picture emerged (Fig. 6). Four types of spectra were consistently seen in separate bands eluted from polyethylene (Fig. 6). These spectra were similar in several clones and are identical to those of Jeffrey [12] for chlorophyll c_1 and c_2 and their alteration products phaeophytin c_1 and c_2. The final piece of evidence came from the ^1H NMR spectrum of the precipitated chlorophyll c which shows a signal at 3.7 ppm, indicative of the ethyl group at C_4 in ring II in chlorophyll c_1, whereas chlorophyll c_2 has a vinyl group in this position.

Indelicato and Watson [1] reported that *G. toxicus* does not contain chlorophyll c_1. In contrast, our report is in accord with Durand and Berkaloff [2] who state that *G. toxicus* contains chlorophyll c_1. There are three possibilities for the differences in our results. First, Indelicato and Watson [1] based their determination primarily on results from silica gel thin layer chromatography which may be insufficient. Jeffrey [12] is probably correct in stating that polyethylene must be used in order to detect chlorophyll c_1 with certainty. However, an HPLC method was recently reported for separation of chlorophylls c_1 and c_2 using gradient elution [17]. Secondly, perhaps there are clones of *G. toxicus* which do not produce chlorophyll c_1. Finally, it is possible that *G. toxicus* produces chlorophyll c_1 only intermittently. Given the homogenous absorbtion spectra among the fourteen clones that we examined and the difficulties encountered in elucidating the presence of the two chlorophyll c's, the first explanation is most likely. It is possible that several other dinoflagellates may contain both chlorophylls c_1 and c_2. Until such time as other species can be re-tested we cannot say that the presence of chlorophyll c_1 in *G. toxicus* warrants it special taxonomic consideration.

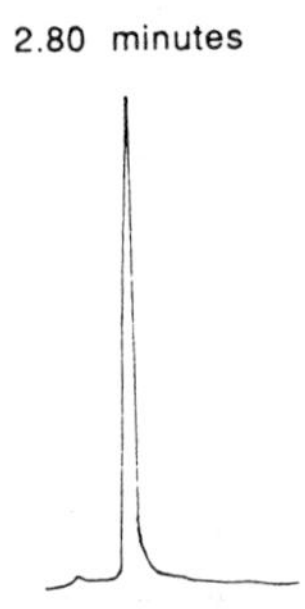

FIG. 5. HPLC chromatogram (440 nm) of Chlorophyll c components eluted from TLC.

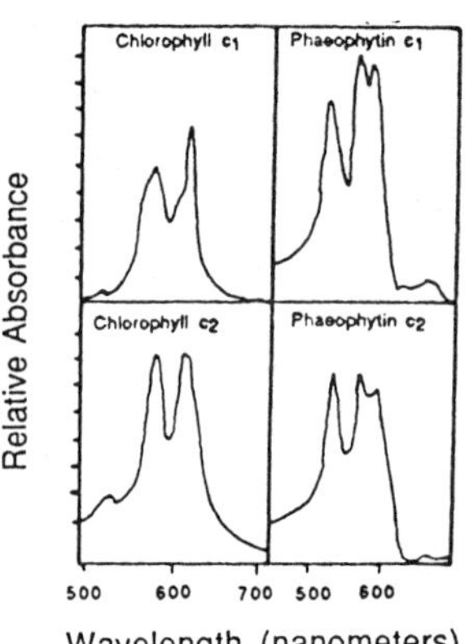

FIG. 6. Spectra of chlorophyll c's and their alteration products purified by polyethylene chromatography.

ACKNOWLEDGEMENTS

This work was supported by the U.S. Army Institute of Infectious Diseases (DAMD17-87-C-7002). We also thank an anonymous reviewer, Patricia Tindall, Deanna Hasenstab and Melody Pierce for their contributions.

REFERENCES

1. J.C. Gallagher and R.S. Alberte, J. Exp. Mar. Biol. Ecol. 94, 233-250 (1985).
2. J.W. Bomber, D.R. Tindall,.and D.M. Miller, J. Phycol. (in press).
3. A.D. Cembella and F.J.R. Taylor in: Toxic Dinoflagellates, D.M. Anderson, A.W. White and D.G. Baden, eds. (Elsevier, Amsterdam 1985) pp. 55-60.
4. K.A. Steidinger and D.G. Baden in: Dinoflagellates, D. Spector, ed. (Academic Press, New York 1984) pp. 201-261.
5. S.R. Indelicato and D.W. Watson, Mar. Fish. Rev. 48, 44-47 (1986).
6. M. Durand and C. Berkaloff, Phycologia 24, 217-223 (1985).
7. J.W. Bomber, S.L. Morton, J.A. Babinchak, D.R. Norris, and J.G. Morton, Bull Mar. Sci. 43, 204-214 (1988).
8. L.E. Brand, R.R.L. Guillard, and L.S. Murphy, J. Plankton Res. 3, 193-201 (1981).
9. M.D. Keller, R.C. Selvin, W. Claus and R.R.L. Guillard, J. Phycol. 23, 633-638 (1987).
10. R.R. Sokal and F.J. Rohlf, Biometry (W.H. Freeman, New York 1981).
11. A. Jensen in: Handbook of Phycological Methods: Physiological and Biochemical Methods, J.A. Hellebust and J.S.Cragie eds. (Cambridge University Press 1978) pp. 59-70.
12. S.W. Jeffrey, Biochim. et Biophys. Acta, 279, 15-33 (1972).
13. J.W. Bomber, R.R.L. Guillard, and W.G. Nelson, J. Exp. Mar. Biol. Ecol. 115, 53-65 (1988).
14. M.A. Faust, J.C. Sager, and B.W. Meeson, J. Phycol. 18, 349-356 (1982).
15. L.E. Brand, L.S. Murphy, R.R.L. Guillard, and H.-t Lee, Mar. Biol. 62, 103-110 (1981).
16. T.R. Tosteson, D.L. Ballantine, C.G. Tosteson, A.T. Bardales, H.D. Durst, and T.B. Higerd, Mar. Fish. Rev. 48, 57-59 (1986).
17. M.W. Fawley, Plant Physiol. 86, 76-78 (1988).

UPTAKE OF *Alexandrium fundyense* BY *Mytilus edulis* AND *Mercenaria mercenaria* UNDER CONTROLLED CONDITIONS

V.M. Bricelj*, J.H. Lee*, A.D. Cembella** and D.M. Anderson***
* Marine Sciences Research Center, State University of New York, Stony Brook, N.Y. 11794, USA; ** Maurice Lamontagne Institute, Dept. of Fisheries and Oceans, 850 route de la mer, P.O. Box 1000, Mont-Joli, Quebec, Canada G5H 3Z4; *** Biology Dept., Woods Hole Oceanographic Institution, Woods Hole, MA 02543, USA.

ABSTRACT

We report preliminary results of an experimental study on the kinetics of PSP toxin uptake in bivalve molluscs. *Mytilus edulis* were exposed to *A. fundyense* (66 pg STX eq $cell^{-1}$) at a mean density of 256 cells ml^{-1} over 17 days. Mussels ingested *A. fundyense* at a constant rate of 0.8×10^6 cells g^{-1} tissue wet weight day^{-1}. Toxin composition of different tissue pools was examined by HPLC. Mussels attained saturation toxin levels of ca. 4.5×10^4 µg STX eq/100g within 10-13 days of initial exposure. The viscera, which constitute only 30% of total tissue weight, contributed 96% of total toxicity. *Mercenaria mercenaria*, reported to remain non-toxic during red tides, ingested ca. 3.4×10^5 *A. fundyense* cells g^{-1} day^{-1} when exposed to toxic cells supplemented with a known good algal food source. Toxin analysis on this species is in progress.

INTRODUCTION

Bivalve molluscs accumulate paralytic shellfish toxins by suspension feeding on toxic dinoflagellates. Among these, *Alexandrium fundyense* is the principal species responsible for paralytic shellfish poisoning (PSP) in the North Atlantic. Considerable information is available on toxin dynamics in bivalves under field conditions [1,2]. Rates of toxin uptake and loss are species-specific [2], and related to the number of dinoflagellate cells in the water column [3]. Data derived from monitoring programs, however, are based on bulk toxin analysis by mouse bioassay and provide no information on the behavior of individual toxins. These vary considerably in specific toxicity and can undergo bioconversions in shellfish [4,5]. Few studies [3,5] have attempted to model the kinetics of PSP toxins in shellfish under controlled conditions. Such studies are necessary in order to develop improved predictive capabilities on the fate of PSP toxins in nature.

Marked differences in toxin accumulation among bivalves species were correlated with the degree of sensitivity of their nerves to PSP toxins [6]. *Mytilus edulis*, commonly used as a sentinel organism to monitor PSP, is resistant to the toxins. It also accumulates toxins more rapidly and attains higher toxic levels than other bivalves. *Mercenaria mercenaria*, a commercially exploited bivalve on the East coast of the U.S., is anomalous in that it is also resistant to PSP toxins [6],

Toxic Marine Phytoplankton
Edna Graneli et al., Editors

but is not reported to accumulate toxins during red tides [7].

The main objectives of the present study were: 1) to quantify the relationship between ingestion of A. fundyense and the accumulation of individual PSP toxins in mussel tissue pools under controlled conditions, and 2) to determine whether M. mercenaria is capable of ingesting toxic cells and thus becoming a vector of PSP. Rates of toxin uptake by hard clams, and detoxification rates in the two species will be reported in a subsequent paper.

METHODS

Mussels collected from Long Island, N.Y. populations with no prior PSP exposure history were acclimated to the experimental temperature, 15-17°C, for 3 wks prior to experiments. A. fundyense (isolate GtCA29 from the Gulf of Maine) was grown in batch culture in K medium [8] on a 16:8h LD cycle, with an irradiance of ca. 550 µE m^{-2} s^{-1}, at 16°C.

Thirty-six mussels (43.8 mm mean length; mean wet tissue weight = 2.30 g)) were held in a recirculating tank containing A. fundyense in 0.22 µm filtered seawater (salinity = ca. 28 $^{o}/_{oo}$). Mixing of the suspension was gentle enough to prevent disruption of algal cells and bivalve feces. Algae were continuously metered into the tank from a concentrated stock with a peristaltic pump. The stock was replenished twice a day from cultures in late log phase grown in 20 l carboys. The volume of the tank was completely replaced every 24-48 h. Mussels were thus exposed to an approximately constant cell density, averaging 256 cells ml^{-1}, over 17 days. Preliminary experiments showed that maximal ingestion rate of toxic cells occurs at 100 to 300 cells ml^{-1}.

Mussels (n=4) were removed periodically for toxin analysis and replaced with new individuals (not previously exposed to PSP), which were sampled at the end of the experiment, thus providing a duplicate time series (Series II). Four tissue pools were dissected: foot, viscera including the digestive gland, muscle including adductors and pedal retractors, and mantle plus gills. Weighed tissues were frozen in liquid nitrogen and stored at -70°C. Toxins from algal cells and from lyophilized tissues were extracted with 0.03 N acetic acid, and analyzed by HPLC at the Maurice Lamontagne Institute, Quebec, Canada, following previously described methods [9]. Toxicity was calculated from published conversion factors of Mouse Units (MU)/µmol of toxin [9], assuming 0.18 µg saxitoxin equivalents (STX eq)/MU [10].

Clams (n=171; 25.8 mm mean length), obtained from a hatchery source, were exposed to a mixed suspension of GtCA29 and the diatom Thalassiosira weissflogii to stimulate their feeding activity. Initially, the two algal species were mixed in 50:50 volume equivalent proportions. From day 8 of the experiment onwards, the proportion of GtCA29 was increased to 70% of the total volume (155 cells GtCA29 ml^{-1}).

RESULTS

Cumulative weight-specific ingestion rate of GtCA29 by M. edulis over the experimental period increased linearly at a constant rate of 0.81 x 10^6 cells day^{-1} g wet tissue $weight^{-1}$ (1.88 x 10^6 cells day^{-1} $mussel^{-1}$) (Fig.1). Thus the cumulative amount of toxin ingested by mussels also increased linearly with time. Mean toxicity of A. fundyense cultures was equal to 65.7 pg STX eq $cell^{-1}$ (SD = 34.6).

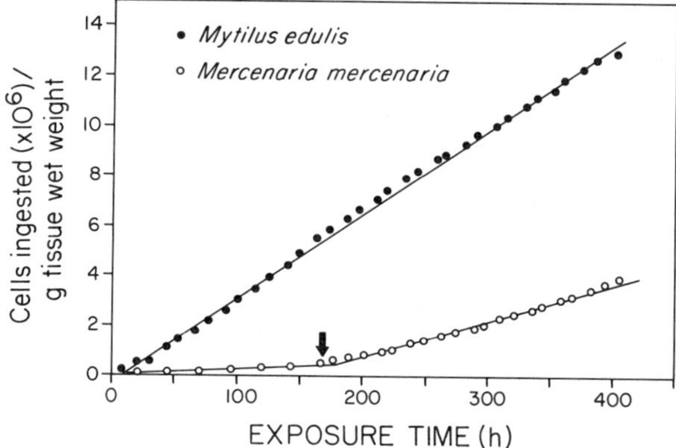

Fig.1. Cumulative ingestion rate of A. fundyense (GtCA29) by mussels and hard clams over the 17 d experimental period.

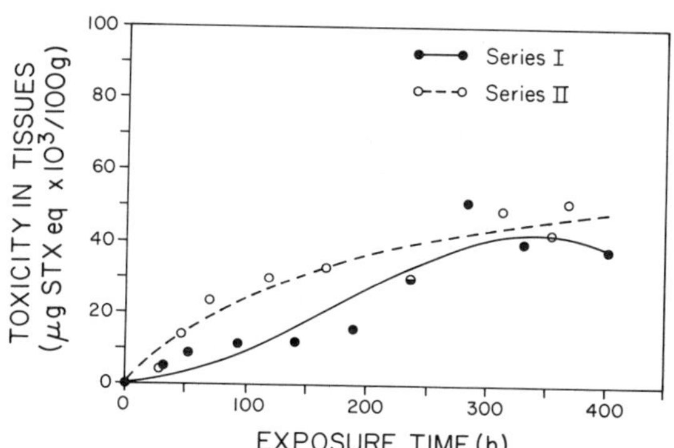

Fig.2. Toxin levels in Mytilus edulis tissues during experimental intoxication with A. fundyense (see text).

The toxin content of total tissues attained saturation levels at 12-13 days (Fig.2). For Series II mussels a Michaelis-Menten equation provided a good description of toxin uptake kinetics:

$$\text{Toxicity } (\mu g \text{ STX eq}/100g) = [71{,}678(t - t_o)]/[196 + (t - t_o)]$$

where t_o = 1.14 and t = time in hours. Series I mussels spawned during the first day of the experiment and showed somewhat lower initial toxin accumulation, which was best described by a polynomial function:

$$\text{Toxicity } (\mu g \text{ STX eq}/100g) = 5.3t + 1.03t^2 - 0.002t^3 \quad (r^2 = 0.95)$$

Maximum toxin body burden, calculated by averaging the last three data points in Fig.2, was 4.3×10^4 µg STX eq/100g after

11.8 days of exposure for Series I and 4.7 x 10^4 μg STX eq/100g after 13 days for Series II. Thus, when toxin saturation was achieved, mussels had incorporated 78% of the toxin ingested.

Approximately 96% of the total toxin body burden accumulated in the viscera, although this tissue represents only 30% of the total tissue wet weight. Toxin accumulation in viscera and mantle/gill is shown in Fig.3. A pattern of toxin uptake similar to that of the mantle was obtained for the muscle and foot. Maximum toxin levels averaged 41.9, 36.4 and 31.9 μg STX eq/g in the mantle, muscle and foot respectively. Toxin levels in mantle and muscle declined somewhat after reaching a maximum, although mussels ingested toxic cells at a constant rate throughout the experiment.

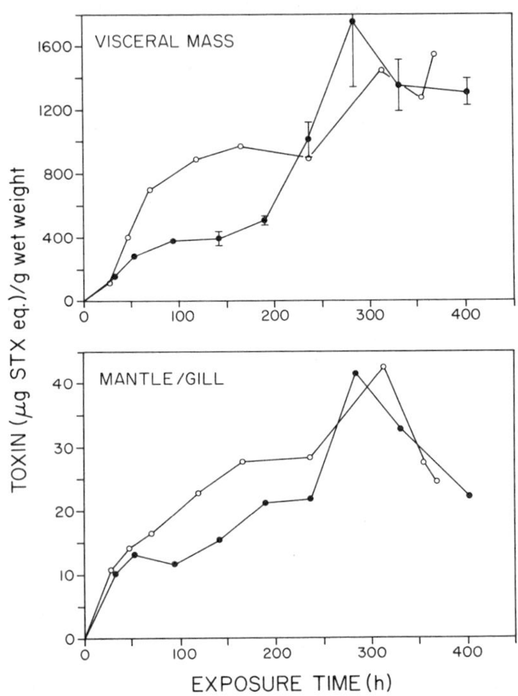

Fig.3. Toxin accumulation in visceral mass and mantle/gill of <u>Mytilus</u> <u>edulis</u>. (Series I and II as in Fig.2.)

The toxin composition of each tissue pool, averaged over the experimental period, is compared to that of ingested cells in Fig.4. Mantle, muscle and foot showed significant enrichment in STX and the gonyautoxins GTX_{1+4}, and a decrease in neosaxitoxin (neoSTX) and GTX_{2+3}. The viscera showed intermediate levels of GTX_{2+3} and STX.

Ingestion rate of GtCA29 by clams was initially low, at 0.68 x 10^5 cells day^{-1} g^{-1}, and increased to 3.44 x 10^5 cells day^{-1} g^{-1} after the sixth day. The inflection point marked by the arrow in Fig.1 indicates the point when the relative contribution of <u>A.</u> <u>fundyense</u> increased from 50 to 70% of the total algal volume. On two occasions, early and late in the experiment, we exposed clams to a diet composed solely of

GtCA29. Clams closed their shells and did not resume pumping

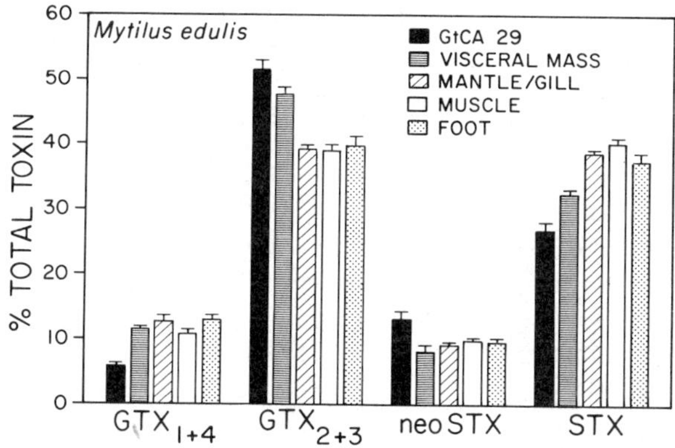

Fig.4. Percent composition of individual toxins in A. fundyense (isolate GtCA29) and mussel tissues (Series II). (The N-sulfocarbamoyl toxins C_1/C_2, not shown in the graph, accounted for 2.7% of total toxicity in this dinoflagellate).

until a low density of T. weissflogii was present in the water column. Clams ingested T. weissflogii and GtCA29 in the same proportion as offered in the suspension, thus exhibiting no ingestion selectivity. Ingestion of dinoflagellates was confirmed microscopically by the presence of numerous intact cells in fecal strands.

DISCUSSION

Toxicity of A. fundyense isolates from the N.W. Atlantic coast ranges from undetectable to ca. 52 pg STX eq $cell^{-1}$, and can vary 2 to 4-fold with culture conditions [10]. Thus the isolate used in our study is one of relatively high toxicity. The experimental concentration used (256 cells ml^{-1}) was selected to maximize ingestion of toxic cells and to simulate Alexandrium bloom densities, typically in the range of 10^5 cells l^{-1} [3]. Thus peak toxicities attained by mussels in this study (4.5×10^4 µg STX eq/100g) are expected to reflect near maximal intoxication rates. They are indeed comparable to those reported during massive red tides. For example, M. edulis attained up to 1.0×10^4 µg STX eq/100g during the 1972 New England red tide [6], and 5.0×10^4 µg STX eq/100g in the Argentine Sea in 1980 [11]. Our results further suggest that under optimal conditions, mussels exposed to a highly toxic strain such as GtCA29, could exceed the regulatory level (80 µg STX eq/100 g) in less than 1 h. This underscores the difficulty in providing adequate warning through routine monitoring programs conducted on a weekly basis.

In M. edulis the muscles, mantle/gill and foot accumulated similar toxin levels when normalized on a per unit weight basis. In contrast, the surf clam Spisula solidissima [12] and the scallop Placopecten magellanicus [2] do not accumulate significant toxin levels in the adductor muscle. Prior studies have shown that the toxin profile in bivalve tissues may differ from that of the cells upon which they feed [5]. Mya arenaria tissues showed an increase in STX and reduction in neoSTX relative to A. fundyense cells [13]. In vitro studies [4] demonstrated that PSP toxins undergo reductive conversions in scallop locomotor tissues, particularly in the adductor muscle, resulting in an increase in STX, and decrease in neoSTX and GTX 1, 2 and 3. Thus the increase in STX, and reduction in neoSTX and GTX_{2+3} observed in our study for M. edulis is consistent with toxin conversions reported in other bivalve species. Sullivan [5] found that M. edulis accumulated toxins in similar proportions to its dinoflagellate diet, suggesting that metabolic conversions were probably not significant in this species. The P. catenella isolate used in his study was rich in N-sulfocarbamoyl toxins, while carbamate toxins are dominant in our GtCA29.

We have demonstrated that hard clams are capable of ingesting A. fundyense cells, although at a slower rate than mussels. They only do so, however, when a non-toxic algal food is present in the suspension. Rapid increase in A. fundyense cell densities and its dominance during red tides may thus preclude hard clams from accumulating significant levels of PSP toxins in nature. Toxin analyses now in progress will indicate the degree of toxin retention by this species.

ACKNOWLEDGEMENTS

We thank J. Christie and R. Larocque for their technical assistance. This work is the result of research sponsored by the NOAA Office of Sea Grant, U.S. Department of Commerce, under Grant #NA86AADSG045 to the New York Sea Grant Institute.

REFERENCES

1. J.W. Hurst and E.S. Gilfillan, Tenth Nat. Shellfish Sanitation Workshop, E.S. Wilt, ed. (U.S. Dept. Health, Education and Welfare 1977) pp. 152-161.
2. S.E. Shumway, S. Sherman-Caswell, and J.W. Hurst, J. Shellfish Res. 7, 643-652 (1988).
3. A. Prakash, J. Fish. Res. Bd. Can. 20, 983-996 (1963).
4. Y. Shimizu and M. Yoshioka, Science 212, 547-549 (1981).
5. J.J. Sullivan, Ph.D. dissertation, Univ. of Washington, WA, 259 pp (1982).
6. B.M. Twarog and H. Yamaguchi in: Proc. First Int. Conf. on Toxic Dinoflagellates Blooms, V.R. LoCicero, ed. (Mass. Science and Technology, MA 1974) pp. 382-393.
7. S.E. Shumway, in prep.
8. M.D. Keller, R.C. Selvin, W. Claus, and R.R.L. Guillard, J.Phycol. 23, 633-638 (1987).
9. G.L. Boyer, J.J. Sullivan, R.J. Anderson, F.J.R. Taylor, P.J. Harrison, and A.D. Cembella, Mar. Biol. 93, 361-369 (1986).
10. L. Maranda, D.M. Anderson, and Y. Shimizu, Est. Coastal Shelf Science 21, 401-410 (1985).
11. J.I. Carreto, R.M. Negri, H.R. Benavides, and R. Akselman in: Toxic Dinoflagellates, D.M. Anderson, A.W. White and D.G. Baden, eds. (Elsevier, New York 1985) pp. 147-152.
12. W.J. Blogoslawski and M.E. Stewart, Mar. Biol. 45, 261-264 (1978).
13. Y. Oshima, L.J. Buckley, M. Alam, and Y. Shimizu, Comp. Biochem. Physiol. 57C, 31-34 (1977).

FLUENCE AND WAVELENGTH DEPENDENCE OF MYCOSPORINE-LIKE AMINO ACID SYNTHESIS IN THE DINOFLAGELLATE ALEXANDRIUM EXCAVATUM

CARRETO, J.I., V.A. LUTZ, S.G. DE MARCO and M.O. CARIGNAN
Instituto Nacional de Investigación y Desarrollo Pesquero. C.C. 175, Playa Grande s/n. 7600, Mar del Plata, Argentina.

ABSTRACT

The effect of light quality on the synthesis rate of mycosporine-like amino acid pigments in the dinoflagellate *Alexandrium excavatum* was studied spectrophotometrically, by incubating "low"-light adapted cells (20 $\mu E \cdot m^{-2} \cdot s^{-1}$) at several higher irradiances (PAR) with different spectral composition ("white", blue, green and red light). Synthesis of these pigments depends on fluence rate of radiation, but is dramatically affected by chromatic conditions. Tungsten light is less effective than sunlight. In the visible region (400-700 nm), only blue light (400-500 nm) induces the synthesis of aminomycosporines, although in the studied irradiances (up , to 250 $\mu E \cdot m^{-2} \cdot s^{-1}$) light saturation is not obtained. Green (500-600 nm) or red (600-700 nm) light are ineffective up to 500 $\mu E \cdot m^{-2} \cdot s^{-1}$ (PAR). Conversely, a small decrease of the initial amounts of aminomycosporines was found. These results persist provided that the photosynthetically usable radiation (PUR) is made equal. In the UV region, the synthesis of these compounds is strongly activated by UV-A (320-400 nm) radiation. Only small amounts of UV-A are needed. Our data suggest that besides changes in light intensity, spectral changes are probably the main information sources used by *Alexandrium excavatum* during their vertical migration.

INTRODUCTION

Some red tide dinoflagellates grown at high light intensities produce substances that strongly absorb in the near UV region of the spectrum (1-6). Although the chemical nature of these substances is scarcely known, some of their properties (spectral absorption, polarity, etc.) indicate their close, structural relationship with mycosporine-like amino acids (5). Mycosporines were first found in sporulating mycelia (7). Various mycosporine-like amino acids have been isolated from different plants and marine animals (8, 9,10). In the dinoflagellate *Noctiluca miliaris*, Okaichi (pers. comm.) isolated several mycosporine-like amino acids. Our first results show that the substances that cause the absorption in the near UV region in *A. excavatum* are a complex mixture of mycosporine-like amino acids (ll). Although their function is scarcely known, their UV photoprotective role and sporogenetic activity has been demonstrated for fungal mycosporines (12,13). Their role as protective filters has also been postulated for cyanobacteria (2,14), corals (9), brown and red algae (14,15) and dinoflagellates (2,4,5). It has recently been demonstrated that the synthesis of these substances is regulated by the light intensity present during growth. Changes in irradiance from "low" to "high" light intensity induce a rapid response, which has been interpreted as an adaptive mechanism that confers upon these organisms a competitive advantage at high light intensity and short wavelengths (4). However, our results have shown that besides light intensity, the synthesis of these substances is highly dependent on the spectral composition of the light source. In this paper, the effect of light quality on the synthesis rate of mycosporine-like amino acids is analyzed in the dinoflagellate *A. excavatum*, by incubating "low"-light adapted cells at various higher irradiances, with different spectral composition.

EXPERIMENTAL

Cultures of *A. excavatum* (clone AM l) were grown in f/2 medium without silicon addition. All experiments were started from cultures in logarithmic growth phase, which were adapted to a 12:12 light-dark cycle at 20 $\mu E \cdot m^{-2} \cdot s^{-1}$ provided by fluorescent tubes (l5 daylight, sylvania), and incubated at l8°C in a culture chamber. Cultures were distributed into vials (20 ml), and then transferred to the new light conditions for 6 h.

Published 1990 by Elsevier Science Publishing Co., Inc.
Toxic Marine Phytoplankton
Edna Graneli et al., Editors

Incubations were carried out in an acrylic thermoregulated bath. Both vials and bath walls present high light transmission in the UV-A region (Fig. 1). Illumination was provided through a frontal window by a Leitz Prado slide projector, equipped with a 500 tungsten lamp (OSRAM,). The different light spectral ranges (Fig. 1) were obtained by interposing appropriate filters: blue (blue Corning Glass + 2A Kodak Gelatin), green (Kodak Gelatin) and red (Corning Glass). In one experiment, UV-A fluorescent light was provided by a 8 lamp (BLB NIS) with maximum emission at 360 nm, through a lateral window. In the case of sunlight incubations, the different light intensities were obtained by using metal sieves as neutral filters. Since not all experiments were carried out simultaneously, a vial was incubated as reference at 200 $\mu E \cdot m^{-2} \cdot s^{-1}$ in the culture chamber. In other experiments, "low"-light (20 $\mu E \cdot m^{-2} \cdot s^{-1}$) adapted cultures were transferred to a higher light (150 $\mu E \cdot m^{-2} \cdot s^{-1}$) in the culture chamber, with and without DCMU (3-(3,4-dichlorophenyl) -1,1-dimethylurea) (Final concentration = 10^{-5} M). Light intensities (PAR) were measured with a LI-185 Lambda Licor quantameter. UV-A radiation was measured with a RK-5200 Power Radiometer. Values of PUR under blue, green and red light regimes were derived from the absorption spectrum a (λ) of the cells by normalizing this spectrum with respect to its maximum $a_{max}(\lambda$ = 438 nm)(16). "In vivo" absorption spectra were recorded as previously described (4). Mycosporine-like amino acids were separated by HPLC (8). A prepacked KNAUER column (Hypersil-ODS, 5 µm, 4.6 mm x 25 cm) was used with 0.2 %. v/v acetic acid as mobile phase. Compounds were identified from retention time predetermined with reference mycosporine-like amino acids, isolated from red algae (17). Since palythene and usujirene cannot be recovered from the charcoal column (8), in order to supplement mycosporine-like amino acids identification, TLC chromatography was carried out on Silica Gel G plates (ethanol-water; 70:30 v/v). These compounds were characterized by chemical reactions (10) and UV absorption maxima. Relative amounts of uv-absorbing substances were estimated by spectrophotometry of methanolic extracts. In all cases, a Shimadzu UV 210 A dual beam spectrophotometer was used.

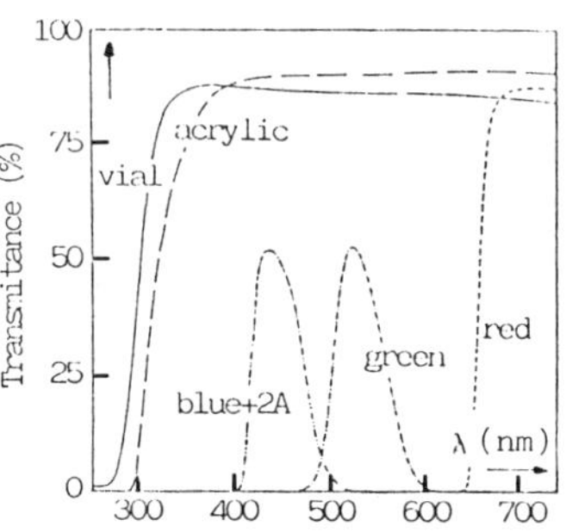

FIG. 1. Transmission characteristics of different materials and filters used in the experiments.

RESULTS

The result of transferring cultures of A. excavatum from "low" (20 $\mu E \cdot m^{-2} \cdot s^{-1}$) to high "white" light intensities is revealed in the methanolic extracts by a rapid absorption increase in the UV region of the spectrum (Fig. 2). This increase is produced by the synthesis of a complex mixture of mycosporine-like amino acids (Fig. 3 and 4), in which compounds with F maximum absorption at 334 nm (porphyra-334 and shinorine) and at 360 nm (palythene and in usujirene) predominate (see also Fig. 2).

Since each component has a different extinction coefficient, the employed method only allows an estimation of the relative changes of these substances. Thus, it is appropriate to study the response of A. excavatum at different regimes of light intensity and spectral composition. Synthesis of these substances is dependent on the light intensity of transference, as well as on the spectral composition of the light source (Fig. 5). Exposure to sunlight results in a much more efficient synthesis than exposure to tungsten light. In the experiment with sunlight, synthesis rate of the mycosporine-like amino acid substances reach light saturation values at around 150 $\mu E \cdot m^{-2} \cdot s^{-1}$. At these light intensities, no changes are evident in the experiment with tungsten light, and the synthesis rate seems to reach saturation values at around 650 $\mu E \cdot m^{-2} \cdot s^{-1}$. Sunlight produces a marked photoinhibition of the synthesis above these light intensities. The shape of the synthesis curve of mycosporine-like amino acids versus light intensity, as well as the inhibitory effect produced by addition of DCMU (Table I), show that the synthesis of these substances is closely coupled to photosynthesis. Thus, the effect observed with light source of different spectral compositions may be the result of a marked difference in the PUR:PAR ratio of both light sources. Alternatively, it can be hypothesized that synthesis of mycosporine-like amino acids, although photosynthesis-dependent, is only induced by light of a certain wavelength range. This hypothesis was tested for the visible region of the spectrum, in the experiments carried out with blue (400-500 nm), green (500-600 nm) and red (640-700 nm) light (Fig. 6). The obtained results show that only blue light is effective, and that green or red light produce a

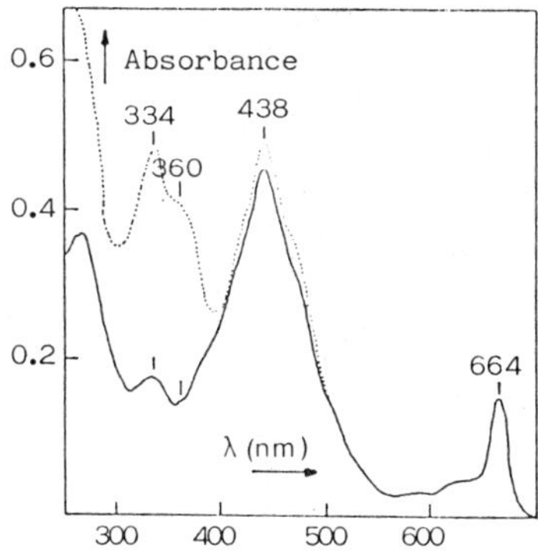

FIG. 2. Changes in the absorption spectra of ethanolic extracts of A. excavatum upon transference from "low" (20 $\mu E \cdot m^{-2} \cdot s^{-1}$) (---) to "high" (200 $\mu E \cdot m^{-2} \cdot s^{-1}$) (......) sunlight.

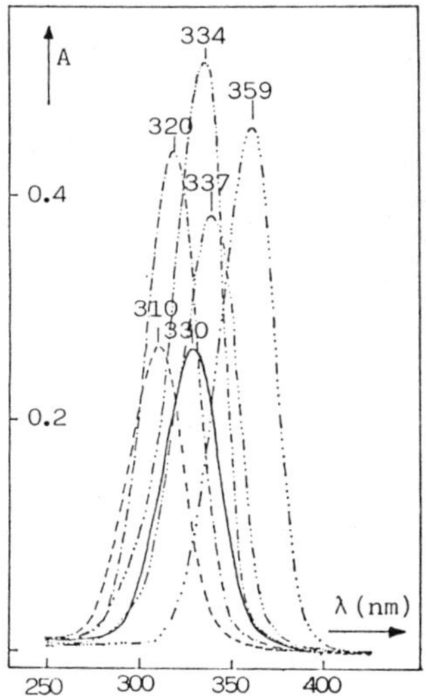

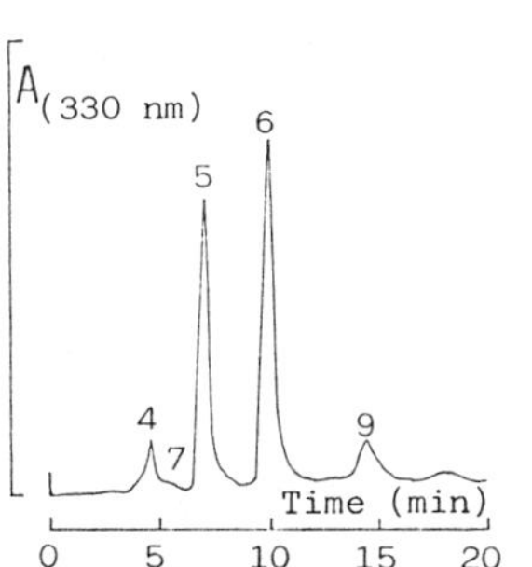

FIG. 4. HPLC separation of mycosporine-like amino acids from A. excavatum: (4) palythine; (7) asterina-330; (5) shinorine; (6) porphyra 334 and (9) unknown (λ max = 337 nm).

FIG. 3. UV-light absorption spectra of several mycosporine-like amino acids isolated from A. excavatum. Solvent: distilled water.

TABLE I. Absorbance change with and without DCMU at 334 nm and 360 nm of A. excavatum "high" light (HL, 150 $\mu E \cdot m^{-2} \cdot s^{-1}$) cells with respect to those exposed to "low" light (20 $\mu E \cdot m^{-2} \cdot s^{-1}$).

	HL	HL+DCMU
Absorbance at 334 nm (%)	+41.3	-34.8
Absorbance at 360 nm (%)	+77.8	-19.4

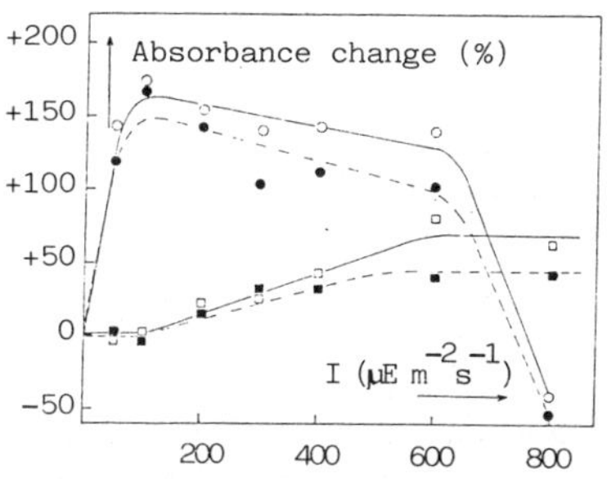

FIG. 5. Response of A. excavatum to increasing light levels of sunlight (o ●) and tungsten light (□■). Absorbance changes at 334 nm (●■) and 360 nm (o□).

small opposite effect up to 500 μE·m^{-2}·s^{-1} PAR. Since the PAR:PUR ratio of the spectrally integrated values is 1.08, 1.92 and 2.17 for blue, green and red light, respectively, the response obtained is normalized for an equal PUR range (Figs. 6a and 6b). Then, blue light produces a specific effect on the synthesis of these substances. However, sunlight is more efficient than blue light, since the slope of the synthesis rate versus light intensity curve for the former is higher than for the latter. Furthermore, in the experiment with blue light, the maximum light intensity assayed (250 μE·m^{-2}·s^{-1}) is not sufficient to saturate the synthesis of mycosporine-like amino acids, while with sunlight saturation is reached at only 150 μE·m^{-2}·s^{-1} . Preliminary experiments (results not shown) with filtered and unfiltered sunlight using UV absorbing materials indicate that, in addition to blue light, the synthesis of these compounds is activated by UV-A radiation (320-400 nm). These results are confirmed by irradiating cultures of A. excavatum adapted to low "white" light intensities, with higher green light intensities, supplemented with constant UV-A radiation (0.6 μE·m^{-2}·s^{-1}). The results obtained confirm the catalytic effect of UV-A radiation on the synthesis of mycosporine-like amino acids (Fig. 7). However, the absorption spectra of the methanolic extracts indicate that the composition of the iminomycosporine mixture is very different from that observed at other light conditions. The metabolism of these compounds is outside the scope of this work, since it requires the individual analysis of the components in the mixture.

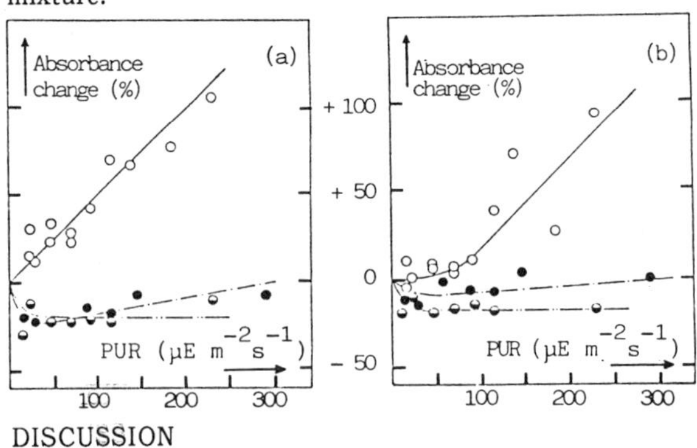

FIG. 6. Response of A. excavatum to increasing levels of blue (o), green (•) and red (◐) light. Absorbance changes at 334 nm (a) and at 360 nm (b).

FIG. 7. Response of A. excavatum to increasing light levels of green + constantUV-A radiation. Absorbance change at 334 (o) and 360 nm(o).

DISCUSSION

Many organisms show both negative and positive phototaxis, but in dinoflagellates only positive phototaxis been reported (18). The migration patterns of various red tide dinoflagellates have shown that under abundant nutrient conditions, cells accumulate at relatively high light intensities and short wavelengths (4,5,6). Under these conditions, cells may be exposed to photochemical damage, especially that produced by UV radiation (19). Our previous results with A. excavatum and Prorocentrum micans support the idea of an UV photoprotective function for the mycosporine-like amino acids present in these organisms (4,11). Vernet et al., (5) studied the "in vivo" absorption and photosynthetic action spectra of natural populations of P.micans and Gonyaulax polyedra and arrived at the same conclusions. Synthesis of these substances in A. excavatum is dependent on light intensity, as well as on the spectral light composition. The dependence on light intensity reflects the requirements of photosynthetic activity.This is demonstrated by the fact that DCMU inhibits the synthesis of these substances. The dependence on photosynthesis may be related to the synthesis of mycosporine-like amino acid precursors, although biogenesis of these substances in dinoflagellates is unknown. However, in the deuteromycete Trichotecium roseum, it has been demonstrated that the basic unit of these compounds - a β -cyclohexenoneis synthetized during the first steps of the shikimate-pathway. The incorporation of amino acids and/or amino-alcohols seems to be a later step (13). The fact that the synthesis of these compounds depends on light quality indicates the existence of unidentified specific photoreceptor/s, with high activity in the UV-A range, and, a much lower activity in the blue region. The UV-synthesis induction of these substances has also been reported for fungi (12), corals (9) and brown algae (15), thus suggesting a common biosynthetic mechanism. The effect of blue light is more difficult to explain, since it is only observed at relatively high light intensities, which suggests that the photoreceptor/s has little activity at these wavelengths. The action of blue light in determining the nature of photosynthetic products is, nevertheless, known. Generally, blue light favours the synthesis of amino acids and proteins over that of carbohydrates (20). It is reasonable to suppose that there exists a close relationship between the phototactic response of a photosynthetic organism and the selection of a habitat with optimal ambient light for the population growth. The positive phototaxis exhibited in dinoflagellates requires a photosynthetic system with great plasticity, able to support exposure to a wide range of light intensities and qualities. Thus, the same signal could activate the phototactic mechanism and the photoprotective response. It has previously been noted that the action spectra of various dinoflagellates show activity in the UV and in the blue region (18). Although the nature of photoreceptors is unknown, they seem to include caroteno-proteins, with absorption characteristics which resemble those of the photocollector component peridinin-chlorophyll a-protein (21). Light-dependent differences between the UV absorption spectra of A. excavatum, as well as those observed in cultures of P. micans (4) and in a natural bloom (5) indicate the existence of a regulation mechanism for the composition of mycosporine-like amino acids, which is still poorly understood. The response time for the synthesis and interconversion of these substances in dinoflagellates is consistent with the rapid changes of light intensity and spectral composition during their vertical migration. However, the absorption spectra of the dinoflagellate Gyrodinium cf. aureolum at different depths in the euphotic zone during a bloom did not show significant differences up to 10 % of the incident light intensity (6).

REFERENCES

1. F.T. Haxo in: Comparative Biochemistry of Photoreactive Systems, M.B. Allen, ed. (Academic Press 1980) pp. 339-360.
2. C.S. Yentsch and C.M. Yentsch in: The Role of Solar Ultraviolet Radiation in Marine Ecosystems, J. Calkins ed. (Plenun Press 1982) pp. 691-706.
3. M. Balch and F.T. Haxo, J. Plankton Res. 6, 515-525 (1984).
4. J.I. Carreto, S.G. De Marco and V.A. Lutz in: Red Tides: Biology, Environmental Science and Toxicology, T. Okaichi, T. Nemoto and D.M. Anderson, eds. (Elsevier 1989) pp. 333-336.
5. M. Vernet, A. Neori and F.T. Haxo, Mar. Biol. (in press).
6. R.M. Negri, J.I. Carreto, H.R. Benavides, V.A. Lutz and R. Akselman. In this volume.
7. J. Favre-Bonvin, N. Arpin and C. Brevard, Can. J. Chem. 54, 1105-1113 (1976).
8. H. Nakamura, J. Kobayashi and Y. Hirata, J. Chromatogr. 250, 113-118 (1982) and references cited therein.
9. W.C. Dunlap and B.E. Chalker, Coral Reefs 5, 155-159 (1986) and references cited therein.
10. I.Sekikawa, C. Kubota, T. Iraoki and I. Tsujino, Jap. J. Phycol. 34, 185-188(1986).
11. J.I. Carreto, M.O. Carignan, D. Daleo and S.G. De Marco, Abstract. Fourth Int. Conf. on Toxic Marine Phytoplankton. 26-30 June, 1989. Lund , Sweden .
12. H. Young and V.J. Patterson, Phytochemistry 21, 1075-1077 (1982).
13. J. Favre-Bonvin. Structure et Biosynthèse des Mycosporines. Thesis. Université Claude Bernard.-Lyon I (1986) 124 pp.
14. P.M. Sivalingam, T. Ikawa, Y. Yokohama and K. Nizisawa, Bot. Mar. 17, 2329 (1974).
15. W.F. Wood, Mar. Biol. 96, 143-150 (1987).
16. J. Gostan, C. Lechuga-Deveze and L. Lazzara, J. Phycol. 22, 63-71 (1986).
17. I. Tsujino, K. Yabe, I. Sekikawa and N. Hamanaka, Tetrahedron Lett. 16, 1401-1402(1978).
18. N. Ekelund and D-P. Häder, Plant Cell Physiol. 29, 1109-1114 (1988).
19. B. Bühlmann, P. Bossard and U. Uehlinger, J. Plakton Res. 9, 935-943 (1987).
20. I. Morris in: Physiological Bases of Phytoplankton Ecology, T. Platt, ed.(Can., Bull. Fish. Aquat. Sci.) pp. 83-102.
21. B.B. Prezelin in: The Biology of Dinoflagellates, F.J.R. Taylor, ed. (Oxford, Bradwell) pp. 174-223 (1987).

AN ENVIRONMENT-SYNCHRONIZED INTERNAL CLOCK CONTROLLING THE ANNUAL CYCLE OF DINOFLAGELLATES

E. COSTAS, M.NAVARRO, AND V. LOPEZ-RODAS
Facultad de Veterinaria (Genética). Universidad Complutense.
28040 Madrid. Spain.

ABSTRACT

The occurrence of marine dinoflagellates has been related to environmental conditions. Nevertheless an environmentally-modulated internal clock could be a pacemaker for an annual cycle. Circannual rhythms under laboratory conditions rended as constant as possible has been detected for growth rates and cyst formation in different species, showing genetic variability and differentiation. Since these rhythm periods exceeds the life span of dinoflagellates, it could suggest the existence of a temporal organization at a supragenerational level, whose molecular mechanisms still remain elusive, but that could have an important role in dinoflagellate ecology. When temporal structure of the circannual rhythms in laboratory cultures, and the yearly cycle in nature were compared using COSINOR rhythmometric techniques, a statistically significant agreement between both rhythms was evident, suggesting the influence of internal clocks on dinoflagellate occurrence.

INTRODUCTION

Many studies have been conducted on phytoplankton blooms in relation to oceanographic conditions. In spite of this we are still far from being able to predict bloom appearance. An alternative approach is to study biological and stochastical factors. An endogenous clock synchronized by environmental conditions could be responsible for the circannual biological response, Circannual rhytms regulated by internal clocks have been reported in a wide array of organisms from single cells to man (1, 2).

Rhythmical phenomena are traditionally explained as the result of a complicated interaction between an internal clock mechanism and several environmental factors that act as synchronizers. The environmental factors ("synchronizers"), such as dark-light cycle, photoperiod, etc., act to keep biological cycles in phase with periodic fluctuations in the environment. In the absence of such synchronizers, the cycles continue but can begin to drift out of phase ("free-running") (3, 4). The general theory proposed by Edmunds and Adams (5) implies that generation time of the organisms exceeds the rhythm period.

CIRCANNUAL RHYTHMS IN DINOPHYCEAE

Recently, circannual rhythms have also been observed in dinoflagellates growing under constant laboratory conditions, These rhythms have been found in cyst formation and growth-rates.

Fukuyo et al. (6) could not germinate dinoflagellate cysts during the fall months. Anderson & Keafer (7) reported an endogenous circannual clock mechanism controlling the germination of *Gonyaulax tamarensis* cysts. Furthermore, differences in the germination rhythms between shallow and deep water populations of *Gonyaulax tamarensis* cysts were detected. Cyst formation seems to be regulated by a circannual endogenous rhythm as well.

Toxic Marine Phytoplankton
Edna Graneli et al., Editors

Costas & Varela (S) recently demonstrated the existence of a circannual rhythm in cyst formation in _Scrippsiella trochoidea_ and found genetic variability among clones.

Yentsch & Mague (9) provided evidence suggesting an annual cycle in maximal growth rates and cell yield of _Gonyaulax excavata_. Costas & Varela (10) demonstrated the existence of an endogenous circannual rhythm under constant laboratory conditions for reproduction rates of _Gonyaulax excavata_ and _Prorocentrum micans_, detecting intra- and interspecific differences. _Scrippsiella trochoidea_ also shows a circannual rhythm in reproduction rate (8). Reproduction rate and cyst formation rhythms have been studied simultaneously only for _Scrippsiella trochoidea_. The temporal concordance between the rhythms is surprising; maximum reproduction rate coincides with minimum cyst formation, while minimum reproduction rate coincides with maximum cyst formation (Fig. 1). This suggests the existence of a single clock mechanism that controls both circannual rhythms simultaneously .

Despite this concordance, however the cyst and reproduction rate rhythms apparently have different explanations. The cyst germination rhythn can be explained on a chronophysiological level. On the contrary, the reproduction rate rhythm seems to be anomalous because the rhythm period exceeds the generation time, thus contradicting Edmunds and Adams (5) chronobiological general theory that there should not be rhythms whose period exceed the generation time. The explanation for this phenomenon presents a challenge. The hypothesis of communication among organisims with different information, as a cause of rhythms of anomalous periods (11, 12) cannot explain this because clonal cultures were used. Neither is it an amplification of the circadian clock since many generations are involved. The above information suggest that there could exist a supragenerational chronobiological organization, of great ecological and evolutionary relevance, that acts to produce rhythms that involve many generations.

Although reproduction rate and cyst formation could not be related directly to the annual cycle of abundance of the dinoflagellate, when we compare the rhythms under constant laboratory conditions with in nature, we find a statistically significant concordance for the acrophase of both rhythms (Fig. 2). This concordance suggests a temporal coupling of the different circannual rhythms, often dinoflagellate regulated probably by a single clock mechanism that could control partly the circannual response. Although these are only a few observations and no generalizations can be made at present, these results seem to suggest that an internal clock mechanism could play a role in the seasonal appearance of dinoflagellates.

A CHRONOBIOLOGICAL MODEL OF ANNUAL CYCLES

We tentatively suggest a general model that integrates the effects of internal clocks with the environmental and stochastical factors in the annual cycle of dinoflagellates giving a general view of the process, even though at present we cannot estimate the relative importance of each single factor (Fig.3).

The internal clock, whose molecular mechanism is still unknown, could be set in time by some kind of environmental factor -probably photoperiod- that acts as a synchronizer. To be set in time the clock needs only a little input of information. In this way the clock seems adaptative, while acting like a temporal memory. Furthermore, for the annual rhythms to take place, it is necessary that none of the model routes are omitted.

However, the appearance and abundance cycle of dinoflagellates is a complicated phenomenon in which many factors take part. The interacting components can be so many and so varied that it may be impossible to predict blooms.

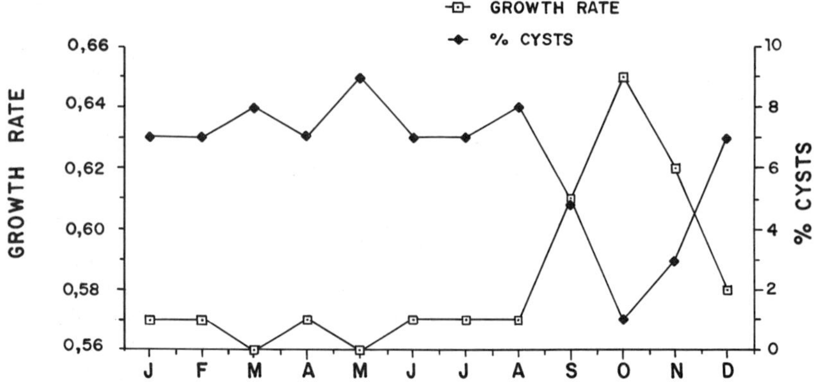

FIG. 1. Annual rhythm in growth rate and cyst formation in S. trochoidea under constant laboratory conditions.

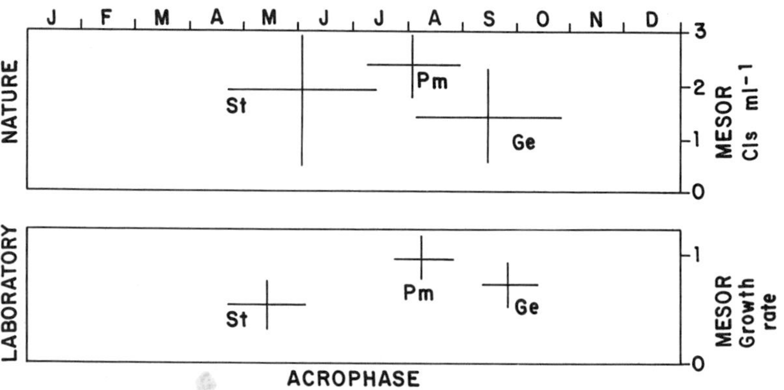

FIG. 2. MESOR-amplitude-acrophase map (13) of the growth rate under constant laboratory conditions and the annual cycle of abundance in nature (Galician shores N.W. Spain), showing a statistically significant concordance for the acrophase of both rhythm. Pm: P. micans, St: S. trochoidea, Ge: G. excavata.

ACKNOWLEDGEMENTS

To profs. F. Halberg and V. Goyanes. This paper was supported by Spanish F.C.A.E.C.C.

REFERENCES

1. E. T. Pengelley, in: Annual Biological Rhythms American Association for Advancement of Science Symposium (Academic Press inc, New York 1974) pp.523.

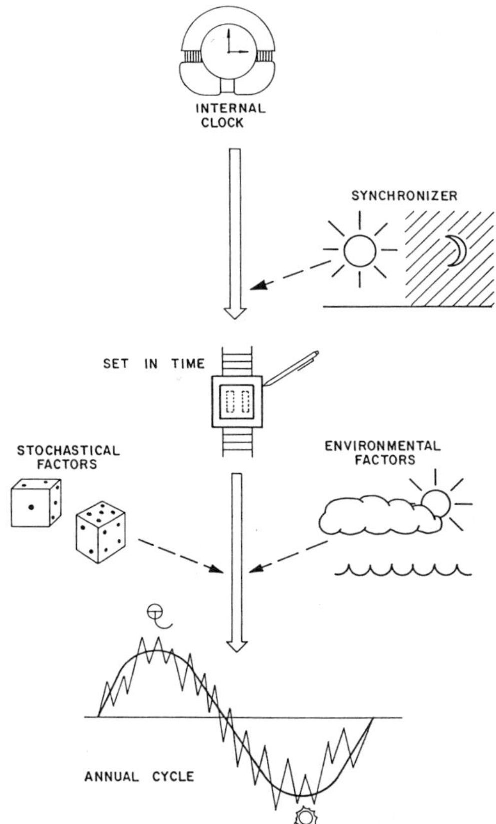

FIG. 3. A chronobiological model of the annual cycle of dinoflagellates .

2. B. Sweeney, in: Rhythmic Phenonena in Plants. Academic Press, New York 1969, pp. 87-91.
3. F. Halberg, M. B. Visscher, and J. J, Bittner, Amer. J. Physiol. 179, 229-235(1954).
4. J, Aschoff, Naturwissenschaften 41, 49-56 (1954).
5. L. N. Edmunds, and K. L, Adams, Science 211, 1002-1013 (1981).
6. Y. Fukuyo, M. M. Wantanabe, and M. Wantanabe, in: The Coastal Marine Environment. (NationaI Institute Japan for Environmental Studies 1982) pp. 43-52.
7. D.M.Anderson, and B.A.Keafer, Nature 325, 616-617 (1987).
8. E. Costas and M. Varela, Chronobiologia (in press).
9. C.M. Yentsch and F. C. Mague, Int. J. Chronobiol. 7, 77-84 (1980).
10. E. Costas and M. Varela, Chronobiologia 15, 1-4 (1988).
11. F. Halberg, W. Hastings, G. Cornelissen and H. Broda, Chronobiologia 12, 185 (1985).
12. G. Cornelissen, H. Broda, and F. Halberg, Cell Biophysics 8, 69-85 (1986).
13. W. Nelson, Y. Tong, K. Lee, and F. Halberg, Chronobiologia 6, 305-362 (1979).

HEMOLYTIC ACTIVITY IN EXTRACTS OF *CHRYSOCHROMULINA POLYLEPIS* GROWN AT DIFFERENT LEVELS OF SELENITE AND PHOSPHATE

B. EDVARDSEN, F. MOY, AND E. PAASCHE
Section of Marine Botany, Department of Biology, University of Oslo,
P.O. Box 1069 Blindern, N-0316 Oslo 3, Norway.

ABSTRACT

We examined the hemolytic activity of *Chrysochromulina polylepis* cultures grown to the stationary phase with phosphorus, selenium, or neither of these as the yield-limiting nutrient. An extract from a phosphate-limited culture of *C. polylepis* caused 100% lysis of cod erythrocytes at a test concentration corresponding to 2.2 10^6 cells ml^{-1}. Hemolytic activity was stronger than controls for concentrations equivalent to 0.14 10^6 cells ml^{-1}. In extracts from cultures grown with excess phosphate, with or without selenium limitation, hemolytic activity was clearly stronger than controls only at test concentrations corresponding to 2.2 10^6 cells ml^{-1} or higher.

The results indicate that *C. polylepis* produces toxin(s) with hemolytic activity when starved of phosphate. The toxicity of the bloom in 1988 may have been due to phosphorus limitation, possibly in combination with other factors.

INTRODUCTION

In May and June 1988 a bloom of the Prymnesiophycean flagellate *Chrysochromulina polylepis* occurred in Kattegat and Skagerrak, off the coasts of Denmark, Sweden and Norway. The bloom covered an area of approximately 60 000 km^2 [1], and concentrations of up to 80 000 cells ml^{-1} were recorded [2]. The algal bloom caused the death of a wide range of naturally occurring marine organisms, as well as of salmonid fish in fish farms.

In the original description of *C. polylepis* by Manton and Parke [3], based on cultures started from material collected off southern England, the authors state that this species is not toxic to fish. Jebram [4], using the same isolate, found that an actively growing culture was non-toxic to a bryozoan. The manifest toxicity of this alga during the bloom in 1988 may have been due to strain differences or to conditions during the bloom event being different from those obtained in earlier laboratory experiments. In order to clarify possible strain differences and to better understand the influence of environmental parameters on the toxicity of the alga, we set out to examine strains of *C. polylepis* isolated from the bloom.

Toxicity of the related *Prymnesium parvum* is promoted by phosphorus deficiency [5]. Toxins from *P. parvum* exhibit various biological effects, including a hemolytic activity which has been studied extensively [5,6]. We have investigated the hemolytic activity of extracts prepared from one of our *Chrysochromulina polylepis* cultures grown at different phosphorus levels. *C. polylepis* has an easily demonstrable requirement for selenite (L. Edler and R.R.L. Guillard, personal communication). We therefore included measurements on selenium-limited cultures in our experiment.

Toxic Marine Phytoplankton
Edna Graneli et al., Editors

MATERIALS AND METHODS

Cultures

A non-axenic strain of *C. polylepis* was used. It originated from a serial dilution series carried out with water collected in inner Skagerrak, near the entrance to the Oslofjord, May 29, 1988. The stock cultures were checked by electron microscopy to ensure that all the scale types characteristic of *C. polylepis* [3] were present.

Cultures were grown in six 1-litre Erlenmeyer flasks with 600 ml filtered seawater diluted to about 25 $^{o}/oo$ salinity and enriched with vitamins as in 1/2-strength IMR medium and nitrate and chelated trace metals as in 1/4-strength IMR medium [7]. Four flasks received a spike of 12.5 µM phosphate and two out of these, 0.01 µM selenite. The medium was sterilized by gentle autoclaving (4 min at 110°C). Inocula were chosen to give starting cell concentrations of 4000 cells ml^{-1} in flasks with P and Se added (termed +P+Se) or with P only (+P-Se), and 7000 cells ml^{-1} in flasks with no P or Se added (-P-Se). The cultures were grown at 12°C under daylight fluorescent light with a quantum flux of 40 µM m^{-2} s^{-1} and a 12h:12h L:D cycle. Cell concentrations were determined in an electronic particle counter. The cultures were harvested and frozen (-20°C) after they had entered the stationary growth phase.

Preparation of fish erythrocytes

Blood was obtained by heart puncture from cod, *Gadus morhua*, anaesthesized with MS-222. Small amounts of heparin were added. The blood was centrifuged at 2800 *g* for 5 min and the pellet was subsequently mixed with cold buffer (8°C) with the composition: 150 mM NaCl, 3.20 mM KCl, 1.25 mM $MgSO_4$ and 12.2 mM Tris base. The Tris buffer was adjusted to pH 7.4 at 10°C with 1 N HCl. The crude erythrocyte suspension was washed three times in the buffer. The blood cells were concentrated by centrifugation and then resuspended in new buffer. $CaCl_2$ was added to the final buffer, to a concentration of 3.75 mM.

Extraction of algal cultures

Methanol (300 ml) was added to frozen cultures of *C. polylepis* (600 ml). When thawed the methanol-water mixtures were filtered through glass fibre filters (Whatman GF/C), and the filters washed with methanol (2 x 50 ml). The filtered solutions were then extracted five times with chloroform (1:10 $CHCl_3$:MeOH,H_2O). The pooled chloroform fractions were evaporated to dryness with a rotavapor. The extracts were weighed and redissolved in small amounts of ethanol. Tris buffer (pH 7.4) containing Ca^{2+} was then added to give final concentrations equivalent to 5-15 10^6 cells ml^{-1} extract. The extracts were sonicated for 5 min to attain homogeneity.

Hemolysis assay

A 1.4 ml portion of erythrocyte stock suspension was mixed with 1.6 ml of extract. Dilutions of the solution to be tested corresponded to algal concentrations of 0.07-2.2 10^6 cells ml^{-1} for P-starved cultures and 0.28-8.8 10^6 cells ml^{-1} for cultures with added P and Se. The erythrocyte suspension was diluted to 10% compared to the original blood. Erythrocytes and extracts were incubated at 8°C for 4 h. Following centrifugation for 5 min at 2800 *g*, absorbance of the supernatant was read at 540 nm in a spectrophotometer. Hemolysis is expressed as per cent of total hemolysis. Total hemolysis was attained using a high concentration of the extract from culture -P-Se. Spontaneous hemolysis was determined by adding buffer to the 10% erythrocyte suspension, and lysis due to ethanol by adding ethanol at the same concentrations as used in the test. Hemolysis was also determined after 19 h to investigate possible delayed effects.

RESULTS

Growth of *Chrysochromulina polylepis*

Growth curves of cultures with or without added P and Se are shown in Fig. 1. With limiting amounts of phosphorous (-P-Se), growth ceased after 10 days reaching a maximum concentration of 30 000 cells ml^{-1}. Chemical

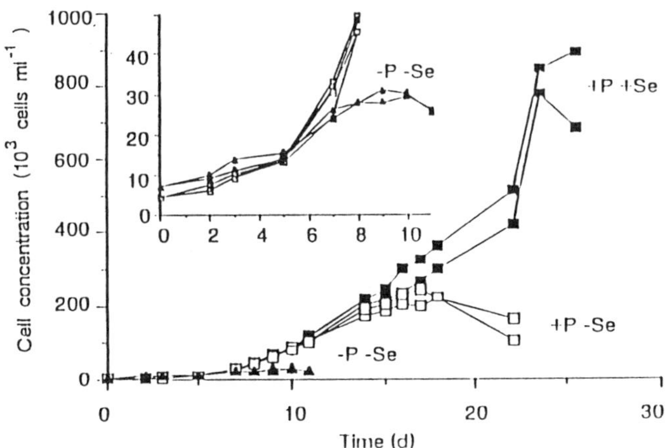

FIG. 1. Growth curves of *C. polylepis* cultures with or without added phosphate (P) and selenite (Se).

analysis confirmed that growth ceased here because phosphate had become exhausted from the medium. Cultures with phosphate in excess but with limiting amounts of selenite (from the seawater) (+P-Se) attained 140 000 cells ml^{-1}, while those with additions of both phosphate and selenite (+P+Se) eventually reached up to 890 000 cells ml^{-1} after 25 days. Although no selenite analyses were performed, these data show that the cell yield in flasks +P-Se was limited by a lack of selenium.

Hemolytic activity in algal extracts

After 4 h of incubation strong hemolytic activity was observed for the extract from the phosphate-limited culture of *C. polylepis* (Fig. 2). For this culture, test concentrations corresponding to 2.2 10^6 cells ml^{-1} caused 100% lysis of cod erythrocytes, and hemolytic activity was stronger

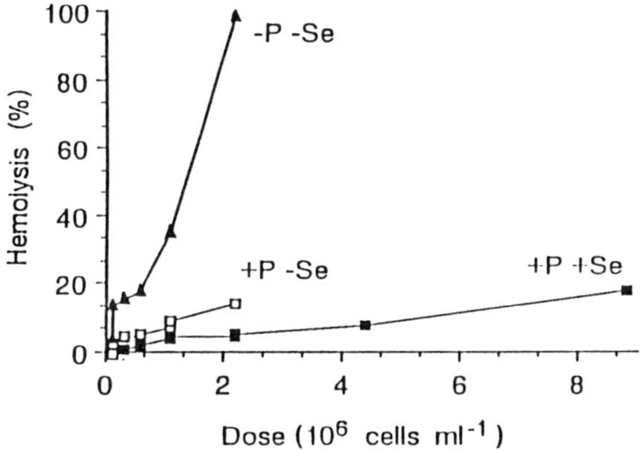

FIG. 2. Hemolytic activity (% of total lysis) in extracts of *C. polylepis* grown with or without added phosphate (P) and selenite (Se). Incubation time was 4 h.

than controls for concentrations equivalent to 0.14 10^6 cells ml^{-1}. Extracts from all cultures grown with phosphate in excess displayed lower hemolytic activity, and the activity was clearly stronger than controls only at test concentrations corresponding to 2.2 10^6 cells ml^{-1} or higher. After incubation for 19 h, hemolytic activity was observed for all extracts of *C. polylepis* (Fig. 3). Hemolysis was independent of ethanol concentration and spontaneous hemolysis was 2.0-5.2 % of the total after 4 h and 1.8-8.4 % after 19 h.

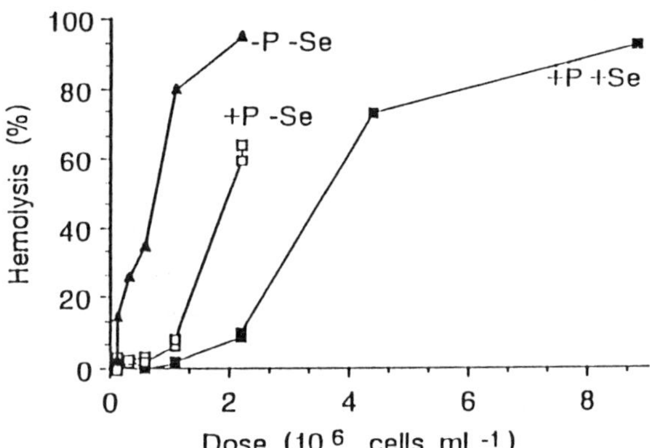

FIG. 3. Hemolytic activity (% of total lysis) in extracts of *C. polylepis* grown with or without added phosphate (P) and selenite (Se). Incubation time was 19 h.

DISCUSSION

The hemolytic activity of extracts prepared from a strain of *C. polylepis* from Skagerrak increases markedly in response to phosphorus limitation. Our results confirm observations by Aune (personal communication quoted by [1]) who found, in a preliminary test, a hepatocytotoxic effect of extracts from phosphorus-starved cultures grown by us. Some hemolytic activity can also be seen in cultures limited by selenium in the 4-h test. Further experiments are needed to decide whether the hemolytic activity after 19 h is a delayed effect.

Phylogenetic considerations suggest that the hemolytic substance(s) in *C. polylepis* may be related to those in *Prymnesium parvum*, in which they have been chemically characterized [8]. We do not yet know how far the hemolytic effect in *C. polylepis* might be accompanied by a toxicity towards fish or other organisms, nor whether such a toxicity would be under nutritional control. However, nutrient analyses conducted in Skagerrak showed nitrate-N:phosphate-P ratios well above the Redfield ratio of 16:1 during an early phase of the bloom [9]. Thus phosphorus may have been in short supply relative to nitrogen. On this basis, Dahl et al. [1] suggested that phosphorus deficiency may have been the reason for the patent toxicity of the *Chrysochromulina* infested waters. This hypothesis will be strengthened once it can be demonstrated that phosphorus-limited *C. polylepis* cells are deleterious to representative species of marine animals and plants.

Stimulation of toxin production by a lack of phosphorus may not be restricted to Prymnesiophytes among the marine plankton algae. Boyer et al. [10] found an increased toxicity of the dinoflagellate *Alexandrium tamarense* (= *Gonyaulax excavata)* when starved of phosphorus. This may not be entirely relevant, as the dinoflagellate neurotoxins are chemically unrelated to the membrane-type toxins produced by *Prymnesium* and *Chrysochromulina*. Further, it is not clear what ecological advantages the dinoflagellates might gain from their toxins. In *Chrysochromulina*, Estep and MacIntyre [11] have argued that the toxin will enable the cells to "tax" other plankton algae for nutrients. In such a context the link between toxicity on phosphorus metabolism takes on an interesting ecological aspect.

Toxicity in dinoflagellates may be under the control of a number of variables such as intraspecific (interclonal) genetic makeup [12] and the composition of the bacterial flora in the growth medium [13]. We are at the moment growing several clones of *C. polylepis* isolated from two different sites on the Norwegian Skagerrak coast. Very preliminary results indicate that there is some interclonal variation in regard to toxicity that may or may not be correlated with morphological differences, e.g., in regard to the presence of atypical spiked body scales.

We have obtained bacteria-free cultures of some of our clones. This put us in a position to find out whether the presence of bacteria is in any way connected with toxicity.

It seems possible at present that a number of variables besides phosphorus deficiency may have affected the toxicity of *C. polylepis* during the bloom in Skagerrak in 1988.

REFERENCES

1. E. Dahl, O. Lindahl, E. Paasche, and J. Throndsen in: Novel Phytoplankton Blooms: Causes and Impacts of Recurrent Brown Tides and other Unusual Blooms, E.M. Cosper, E.J. Carpenter, and M. Bricelj, eds. (Springer Verlag, Berlin 1989).
2. R. Rosenberg, O. Lindahl, and H. Blanck, Ambio 17, 289-290 (1988).
3. I. Manton and M. Parke, J. mar. biol. Ass. U.K 42, 565-578 (1962).
4. D. Jebram, Zool. Jb. Syst. 107, 368-390 (1980).
5. M. Shilo in: Microbial toxins, Vol. 7, S. Kadis, A. Ciegler, and S.J. Ajl, eds. (Academic Press, New York 1971) pp. 67-103.
6. J. Yariv and S. Hestrin, J. gen. Microbiol. 24, 165-175 (1961).
7. R.W. Eppley, R.W. Holmes, and J.D.H. Strickland, J. exp. mar. Biol. Ecol. 1, 191-208 (1967).
8. H. Kozakai, Y. Oshima, and T. Yasumoto, Agric. Biol. Chem. 46, 233-236 (1982).
9. E. Dahl, Vann 23(3B), 512-523 (1988). [In Norwegian.]
10. G.L. Boyer, J.J. Sullivan, R.J. Andersen, P.J. Harrison, and F.J.R. Taylor in: Toxic dinoflagellates, D.M. Anderson, A.W. White, and D.G. Baden eds. (Elsevier/North Holland, New York 1985) pp. 281-286.
11. K.W. Estep and F. MacIntyre, Mar. Ecol. Prog. Ser. Submitted.
12. P.J. Hansen, Mar. Ecol. Prog. Ser. 53, 105-116 (1989).
13. T.R. Tosteson, D.L. Ballantine, C.G. Tosteson, V. Hensley, and A.T. Bardales, Appl. environ. Microbiol. 55, 137-141 (1989).

BREVETOXIN BINDING IN THREE PHYLOGENETICALLY DIVERSE VERTEBRATES

Richard A. Edwards*, Adam M. Stuart** and Daniel G. Baden*.
*University of Miami, Rosenstiel School of Marine and Atmospheric Science, Division of Biology and Living Resources, 4600 Rickenbacker Causeway, Miami, FL 33149 USA; **University of Miami School of Medicine.

ABSTRACT

Equilibrium binding of neurotoxins isolated from the Florida red tide dinoflagellate Ptychodiscus brevis to rat (Indian River strain), turtle (Pseudemys scripta) and fish (Tilapia sp.) brain synaptosomes is described. Binding studies of tritiated brevetoxin PbTx-3 were undertaken and specific binding was measured using a rapid centrifugation technique. Specific binding was saturable and accounted for 80-90% of total binding at label concentrations near the K_d. Rosenthal analysis of specific binding yielded K_d values of 1.5-6.1nM and B_{max} values of 1.5-6.8pmoles/mg and suggested a single class of binding sites. Tritiated PbTx-3 was displaced from its specific binding site by unlabeled PbTx-1, -2 and -3. Relative displacing ability correlates to in vivo ichthyotoxicity and to toxin hydrophobicity. It is concluded that factors affecting brevetoxin binding in excitable tissue are similar in these three diverse vertebrates and that toxicity in these tissues would be approximately the same.

INTRODUCTION

Neurotoxins (herein referred to as brevetoxins) produced by the Florida red tide dinoflagellate, Ptychodiscus brevis, exert their effects primarily by altering membrane properties of excitable cells [1]. As a result of toxin binding to site 5 associated with the voltage-sensitive sodium channel (VSSC), a portion of the membrane channels are activated at normal resting potential resulting in membrane depolarization [2]. The affinity of the toxin for its binding site and the number of binding sites (VSSC) present greatly affect the rapidity and extent of toxin binding and depolarization. Previous studies have also suggested that the physical properties of the region surrounding the toxin binding site may play a critical role in the abilities of these brevetoxins to bind to their active site [3].

While it is known that these toxins affect a variety of organisms, it is not known to what extent brevetoxin binding affinity is correlated between species. Studies were undertaken to examine the binding of these toxins to neural tissue of three phylogenetically-diverse vertebrates in order to determine interspecific similarities with regard to toxin binding affinity, number of binding sites and binding site microenvironment.

MATERIALS AND METHODS

Brevetoxins PbTx-1, Pbtx-2 and PbTx-3 [1] were prepared from laboratory cultures of P. brevis by a combination of solvent-solvent extraction, thin-layer and flash chromatography, and high performance liquid chromatography [3]. Tritium-labeled PbTx-3 was prepared by chemical reduction of PbTx-2 using sodium borotritiide [4] yielding a probe with a specific activity of 10-15 Ci/mmole. This tritium label is covalently linked and is non-exchangeable [1].

Published 1990 by Elsevier Science Publishing Co., Inc.
Toxic Marine Phytoplankton
Edna Graneli et al., Editors

Synaptosomes were prepared using brain tissue from rats (225-250g), turtles (Pseudemys scripta;500-750g) and fish (Tilapia sp.;10-20cm). The brains were removed and homogenized at 4°C with 10 strokes in a Potter-Elvehjem tube containing a 0.32M sucrose (0.7M;fish) buffer. The homogenate was centrifuged at approximately 2,000xg for 10 minutes, and the supernate was collected. The rat and turtle tissue was layered over 1.2M (rat) or 1.0M (turtle) sucrose and centrifuged at 105,000xg for 30 minutes. The interface layer was collected, suspended above 0.8M (rat) or 0.6M (turtle) sucrose and centrifuged at 130,000xg for 35 minutes. The resulting synaptosomal pellet was resuspended in a synaptosomal binding medium [1]. The fish tissue supernate was simply centrifuged at 33,000xg for 60 minutes. The resulting crude synaptosomal pellet was resuspended in synaptosomal binding medium which also contained 370mM sucrose to maintain osmolarity equal to the 0.7M sucrose used in vesicle isolation [5]. Further attempts to purify the fish synaptosome preparation were unsuccessful [5]. As a result, the crude synaptosomal preparation was used. While this most likely resulted in an increase in nonspecific binding, the values were only slightly higher than those for the other two preparations.

Binding assays were performed using 1ml total volume of binding medium consisting of 50mM HEPES (pH 7.4), 130mM choline chloride, 5.5mM glucose, 0.8mM magnesium sulfate, 5.4mM potassium chloride, 1mg/ml BSA, 1mM EGTA, 370mM sucrose (fish only) and 0.01% Emulphor EL-620 emulsifier. This last addition prevents toxin micelle formation in aqueous solution [1]. Synaptosome suspension (0.05-0.1mg protein) in 0.1ml binding medium was added to a mixture of ^{3}HPbTx-3 and other effectors in 0.9ml binding medium in 1.5ml polypropylene microfuge tubes. For ^{3}HPbTx-3 binding assays the label concentration ranged from 0.78-100nM. During competitive displacement assays the competitor brevetoxin concentration ranged from 0.1-100nM, while ^{3}HPbTx-3 concentration remained constant at 12.5nM. After incubation for 60 minutes at 4°C (rat and turtle)[1] or at 23°C (fish)[5], tubes were centrifuged at 15,000xg for 3 minutes. The supernate was sampled for free toxin and the remaining supernate was aspirated. Pellets were rapidly washed with several drops of ice-cold wash medium consisting of 5mM HEPES (pH 7.4), 163mM choline chloride, 1.8mM calcium chloride, 0.8mM magnesium sulfate, 370mM sucrose (fish only), and 1mg/ml BSA. Pellets were transferred to liquid scintillation vials and bound radioactivity was determined. Nonspecific binding was determined in parallel tubes containing a saturating concentration (0.01mM) of unlabeled toxin, and was subtracted from total binding to yield specific binding. Protein was determined as described by Bradford [6].

RESULTS AND DISCUSSION

Radiolabeled brevetoxin binds to synaptosomes from all three organisms with high affinity and specificity. Specific binding was saturable and at a label concentration corresponding to the K_d, accounted for approximately 80-90% of total binding in all cases. The fish preparations were generally lowest, the turtle was somewhat higher and the rat demonstrated the highest percent specific binding. Rosenthal analysis of specific binding data suggests a single class of non-interacting binding sites. Linear regression yielded a dissociation constant (K_d) of 2.9nM, 1.5nM and 6.1nM and a maximum binding capacity (B_{max}) of 6.8, 2.25, and 1.5 pmoles toxin/mg protein in rat, turtle and fish, respectively (Table I). Differences in K_d may be the result of slight variations in the hydrophobic nature of the binding site region [3]. The differences in B_{max} may be due to varying numbers of binding sites per synaptosome or to different quantities of synapse-rich brain regions in the three species. However, it is equally likely that these differences simply reflect the relative purities of the synaptosomal preparations.

TABLE I. Equilibrium dissociation constants (Kd) and maximum binding capacities (Bmax) of ^{3}HPbTx-3 binding to rat, turtle and fish synaptosomes.

	Kd (nM)	Bmax (nM)
RAT	2.9	6.8
TURTLE	1.5	2.25
FISH	6.1	1.5

Unlabeled PbTx-3 displaced ^{3}HPbTx-3 in a dose-dependent manner, as did PbTx-1 and PbTx-2 (Table II). IC_{50} values at ^{3}HPbTx-3 concentrations

TABLE II. Brevetoxin competitive displacement and icthyotoxicity.

	IC_{50} (nM)[a]			Icthyotoxicity[b]
	Rat	Turtle	Fish	(nM)
PbTx-1	3.76	4.95	30	4.4
PbTx-2	15.0	12.0	70	21.8
PbTx-3	11.4	10.98	110	10.9

a) inhibits 50% of specific binding b) from Baden et al. [3]

corresponding to the K_d were 3.76, 15.0 and 11.4nM in rat, 4.95, 12.0, and 10.98nM in turtle and 30, 70 and 110nM in fish for the toxins PbTx-1, -2 and -3, respectively. While the displacement of labeled brevetoxin requires significantly more competitor in the fish assay, the relative potencies of the brevetoxins is similar to those observed in the other tissues. PbTx-1 is the most potent brevetoxin [7,8], is the most hydrophobic [8] and is also the most potent on a molar basis at displacing ^{3}HPbTx-3 from the specific binding site. Although the displacement ability of PbTx-2 and -3 are not statistically different, they are certainly less efficacious than is PbTx-1. This correlates with previous work [3] which demonstrated differences in displacement ability of the two molecular backbone types present in the brevetoxins. The difference in the relative displacement abilities of the brevetoxins is most likely due to their relative hydrophobicity [3].

Previous LD_{50} studies using Gambusia affinis [3] have demonstrated that the brevetoxins exert their toxicity in the nanomolar concentration range (Table II). PbTx-1 is the most potent (4.4nM) and PbTx-2 and -3 less so (10-20nM). Thus the binding of brevetoxin observed in these experiments appears to represent physiologically relevant toxin binding, since the K_d values are approximately the same as concentrations that result in acute toxicity.

Several criteria of specific toxin-receptor binding [9] have been met in this study: saturation of specific binding, displacement of specific binding by an unlabeled toxin of the same structure, a finite (small) number of binding sites (B_{max} in the pmole/mg range), high affinity binding (K_d in the nM range), and a demonstration of binding equilibrium at concentrations approximating in vivo LD_{50} concentrations. Thus it appears that in all three tissues the observed specific binding represented a specific ligand-receptor interaction. In spite of the phylogenetic diversity of these three organisms, the mechanism of brevetoxin binding appears to be remarkably similar. This suggests that the brevetoxin binding site associated with the voltage-sensitive sodium channel, and the region directly surrounding it, have been conformationally conserved over evolutionary time. The displacement data also suggest that those factors contributing to the hydrophobic nature of this region of the VSSC have also been preserved. It appears probable that these toxins would exhibit equally toxic effects in excitable tissue from these three organisms. It remains to be seen what effect biological barriers have on the uptake of brevetoxins and on their ability to express their toxicity in rat and turtle neural tissue in vivo.

REFERENCES

1. M.A. Poli, T.J. Mende, and D.G. Baden. Mol. Pharmacol. 30 129-135 (1986).
2. J.M.C. Huang, C.H. Wu and D.G. Baden. J. Pharmacol. Exp. Therapeut. 229 615-621 (1984).
3. D.G. Baden, T.J. Mende, A.M. Szmant, V.L. Trainer, R.A. Edwards and L.E. Roszell. Toxicon 22 97-103 (1987).
4. D.G. Baden, T.J. Mende, J. Walling, and D.R. Schultz. Toxicon 22, 783-789 (1984).
5. A.M. Stuart and D.G. Baden. Aquatic toxicology 13 271-280 (1988).
6. M.M. Bradford. Anal. Biochem. 72 248-252 (1976).
7. W.A. Catterall and M. Risk. Mol. Pharmacol. 19 345-348 (1981).
8. Y. Schimizu, H.N. Chou, H. Bando, G. van Duyne and J.C. Clardy. J. AM. Chem. Soc. 108 514-515 (1986).
9. D.R. Burt in: Neurotransmitter Receptor Binding, H.I. Yamamura, S.J.Enna, and M.J.Kuhar, eds. (Raven Press, New York 1985) pp. 41-60.

ACKNOWLEDGEMENTS

This work was supported in part by the U.S. Army Medical Research and Development Command, Contract Numbers DAMD17-85-C-5171, DAMD17-87-C-7001 and DAMD17-88-C-8148.

With regard to our research using animals, the investigator adhered to the Guide for the Care and Use of Laboratory Animals, prepared by the Committee on Care and Use of Laboratory Animals of the Institute of Laboratory Animal Resources, National Research Council (NIH Pub. # 86-23, revised 1985).

Opinions, interpretations, conclusions and recommendations are those of the authors and are not necessarily endorsed by the U.S. Army.

THE BIOCHEMICAL PROCESSES DURING CYST FORMATION IN ALEXANDRIUM CATENELLA

THAITHAWORN LIRDWITAYAPRASIT*, SHACHIO NISHIO**, SHIGERU MONTANI* AND TOMOTOSSHI OKAICHI*
*Faculty of Agriculture, Kagawa University Miki-cho, Kita-gun, Kagawa-ken, 761-07 Japan, **Shikoku Women's University, Ojin-cho, Tokushima City, Tokushima-ken, 771-11 Japan

ABSTRACT

The changes of the biochemical components in the life cycle of Alexandrium catenella were examined. Two mating types (Pc1 and Pc2) of this species were isolated from Kii Channel and Kagoshima Bay and used for induced cyst formation. Two weeks after inoculating the two mating types into an encystment medium the cyst yield was about 2 % of the maximum vegetative cell number (Pc1 + Pc2) and the planozygote yield was about 40%.

The cellular ATP contents of type Pc1 and Pc2 were 4.7 pg cell^{-1} and 5.4 pg cell^{-1}, respectively. An analysis of PSP toxins by reversed phase ion-pairing HPLC showed that GTX 4 was dominant in both Pc1 and Pc2 whereas PX 4 (C4) was dominant in Pc1 + Pc2 and cysts. The total toxin concentrations in the cysts were notably higher than those in Pc1, Pc2 and PC1 + PC2.

Amino acids were also analyzed. Glutamic acid and arginine contents changed during cyst formation, but there were minimal differences in the levels of other amino acids.

INTRODUCTION

Recently the life cycle of Alexandrium catenella (Whedon et Kofoid) Balech has been docutmented by Yoshimatsu [1] and the important roles of the resting stage in Protogonyaulax spp. have been pointed out [2]. A lot of biochemical work has been concentrated on PSP toxins profile and their toxicity. A high toxicity in the cyst was reported by Dale et al. [3], who stated that the resting cysts were more toxic than motile cells. In Japan, Oshima et al.[4] also found that the toxicity of natural cysts of Protogonyaulax spp. isolated from Ofunato Bay was as much as 5 to 10 times that of vegetative cells germinated from the same cysts. To obtain more physiological and biochemical information on the life cycle of A.catenella, the cellular ATP, PSP and amino acids contents during cyst formation of A.catenella were examined.

MATERIALS AND METHODS

Materials

The two mating types of A.catenella used in this study were obtained from Kii Channel (Wakayama Pref. 1986, Pc1 mating type +) and Kagoshima Bay (Kagoshima Pref. 1988, Pc2 mating type -). Both were provided by T. Yoshimatsu (Akashiwo Res. Inst. Kagawa Pref.). Axenic cultures of the two types were cultured separately in ESM medium [5], at 21 ± 1 °C and were illuminated by cool white fluorescent tubes (ca. 4,000 lux) on a 14:10 L:D cycle. At the middle of the log growth phase, 7 ml of Pc1 and Pc2 were combined in 300 ml Erlenmeyer flasks containing 150 ml of modified ESM - encystment medium (2,8 μM HPO_4^{2-}) and grown under the same conditions as mentioned above. All the glass culture vessels used in this study were coated with Surfa Sil [6].

Toxic Marine Phytoplankton
Edna Graneli et al., Editors

At the early, middle and post log phase stages of growth, cells of each mating type were collected by flitration on ignited GF/C glass fiber filters. Cells of Pc1 + Pc2, swimming above the cysts at stationary phase, were also collected by pasteur pipette from several flasks to get enough samples for chemical analysis. Cysts at the bottom of flasks were loosened with a rubber scraper and finally cysts were separated from the remains of swimming cells by Percoll-sorbital solution [7]. All samples on ignited GF/C glass fiber filters were stored at -20 °C prior to analysis.

Analysis of PSP toxins

Each sample was homogenized in 80% EtOH containing 1% ACOH with a glass homogenizer for 10 min. and then centrifuged. The supernatant was washed 3-5 times with $CHCl_3$ and the toxin layer was concentrated under vacuo evaporation and freeze dried. The dried toxins were then dissolved in 1 ml of distilled water and analyzed with a reversed phase HPLC system (Shimadzu LC-6A) – a slight modification of the method described by Nagashima et al.[8].

Analysis of ATP

The extraction of cellular ATP was done by the direct pouring of 1 ml of the culture sample into 10 ml of boiling TRIS buffer, to preclude filtering artefacts. Extracts were assayed in a SAI integration photometer, SAI technology, Model 2000 [9].

Analysis of amino acids

Amino acids for each sample were extracted in boiling Nano pure water for 15 min. and filtered on ignited GF/C filters. The hot water soluble, combined amino acids were hydrolyzed with 12 N HCl (1:1 vol.) at 105 °C for 22 hrs. The residues on GF/C filters were hydrolyzed with 6 N HCl at the same conditions. Hydrolysed samples were analyzed by the ninhydrin method with an amino acid analyzer (Hitachi, L-8500).

RESULTS

Growth and cyst production

In the encystment medium, the two mating types grew exponentially at a rate of 0.35 divisions day^{-1} (Fig. 1). The first cyst appeared five days after inoculation and a maximum cyst yield was reached within 2 weeks. The final cyst yield was about 2% and the planozygote yield was about 40% of the maximum vegetative cell number. Most of the planozygotes did not develop into hypnozygotes (resting cysts) even after a week.

ATP

Cellular ATP of Pc1 and Pc2 in the early log phase was 4.7 and 5.4 pg $cell^{-1}$, respectively. ATP contents in both mating types slightly decreased throughout the growth process and after cyst formation it was found to be 4.8 pg $cyst^{-1}$ (Fig. 2).

PSP toxins

The PSP toxin composition of cells and cysts are summarized in Table 1. GTX 5 and STX were not detected in all samples. GTX 4 was dominant in both Pc1 and Pc2 followed by GTX 6. GTX 2,3 and PX 1 (epi GTX8) were detected in Pc2, but only in very small amounts. The total toxin concentration in Pc1 exceeded that in Pc2 by a factor of 1.5.

PX 4 (C4) was the dominant toxic component in both PC1 + PC2 and in cysts, followed by GTX 4. GTX 2,3 and PX1 could not be detected in Pc1 + Pc2 cells but were found in cysts in small amounts. The total toxin concentration of cysts exceeded that of Pc1, Pc2 and Pc1 + Pc2 by a factor of 4, 6 and 1.3, respectively. The GTX 1 concentration in cysts exceeded that in Pc1 , Pc2 and Pc1 + Pc2 by a factor of 24, 20 and 6.7, respectively. Neo STX could not be detected in cysts.

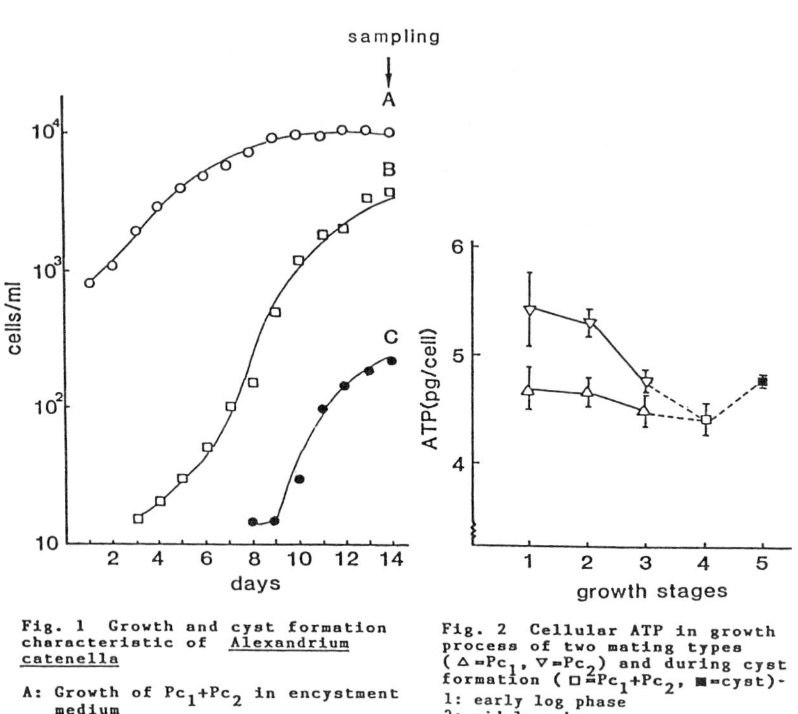

Fig. 1 Growth and cyst formation characteristic of Alexandrium catenella

A: Growth of Pc_1+Pc_2 in encystment medium
B: Planozygote production from A
C: Cyst production from B (first cyst appeared on day 5 in 5 cells ml^{-1})

Fig. 2 Cellular ATP in growth process of two mating types (△ =Pc_1, ▽ =Pc_2) and during cyst formation (□ =Pc_1+Pc_2, ■ =cyst)-
1: early log phase
2: mid log phase
3: before stationary phase
4: Pc_1+Pc_2 stationary phase including planozygote
5: cyst

Amino acids

Glutamic acid was found to be the dominant component in hot water extracts of cells at middle log growth phase of both mating types (Fig. 3A). The molar percent composition of amino acids in the two mating types showed almost the same pattern except for glutamic acid in Pc2, which was notably higher than in Pc1, and for arginine and proline in Pc1 which were higher than in Pc2. Cystine and methionine could not be detected in Pc1 and Pc2.

Arginine in cysts and in Pc1 + Pc2 were higher than in Pc1 and Pc2. Cystine, methionine, tyrosine and proline could not be detected in the resting stage.

The amino acid composition in the residue of cells and cysts showed almost the same pattern except that proline in cysts was very high (Fig. 3B).

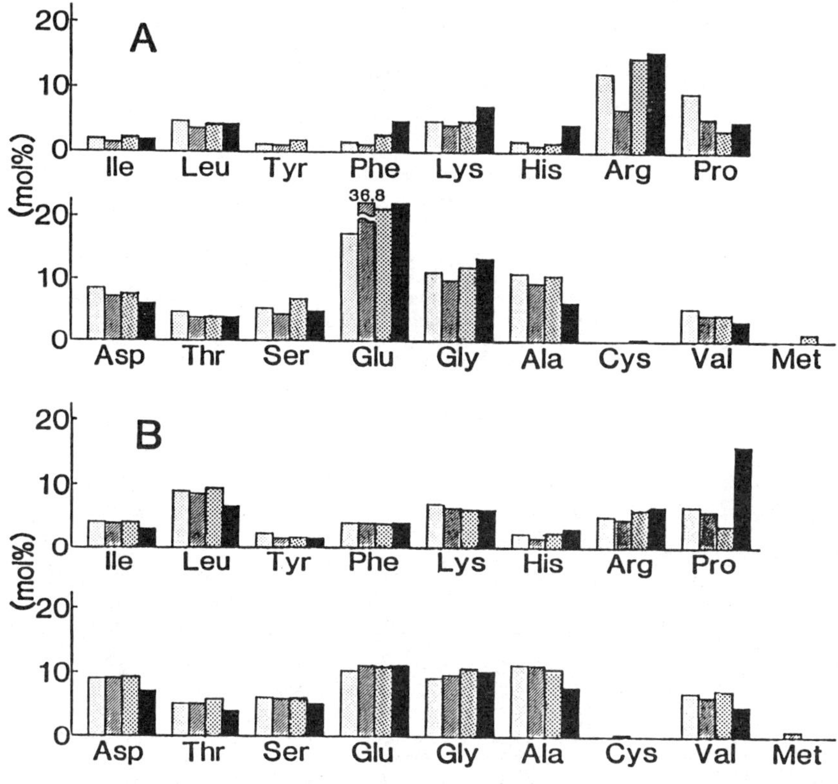

Fig. 3 The composition of amino acids in hot water extracts (A) and residues of cells and cysts (B)

- Pc_1 mid log phase
- Pc_2 mid log phase
- Pc_1+Pc_2 stationary phase
- cyst

Table 1. Cellular PSP toxins composition of cells and cysts of <u>Alexandrium</u> <u>catenella</u> (f mole/cell or cyst)

	GTX 1	GTX 2	GTX 3	GTX 4	GTX 6	PX 1 (epi-GTX8)	PX 2 (GTX8)	PX 3 (C3)	PX 4 (C4)	neo-STX	Total
Pc_1*	0.2	-	-	6.9 (38)	3.5 (19)	-	2.7 (12)	0.6	3.9 (21)	0.5	18.3 (100)
Pc_2*	0.2	0.02	trace	6.6 (57)	2.6 (22)	0.02	0.8	0.2	0.9 (8)	0.1	11.5 (100)
Pc_1+Pc_2**	0.7	-	-	13.4 (25)	5.4 (10)	-	3.5 (6)	2.9	26.3 (48)	2.5	54.7 (100)
Cyst	4.6	0.1	0.3	14.6 (20)	11.3 (15)	0.1	0.4	12.1 (16)	30.4 (41)	-	73.9 (100)

* Cells before inoculation into encystment medium
** Cells in encystment medium
GTX 5 and STX were not detected

DISCUSSION

The facts that a few planozygotes appeared on day 3, and the first cyst appeared within 5 days after inoculation into the encystment medium suggest that the two mating types could fuse at least 1 or 2 days after inoculation. The planozygotes required at least 2 days as a maturation period before encystment. However, the cyst production was rather low compared with that of <u>P</u>. <u>tamarensis</u> reported by Anderson and Lindquist [10]. Most of the planozygotes did not develop into hypnozygotes in this study. The same observation was also reported for planozygotes of <u>Gymnodinium</u> <u>uncatenum</u> [11] and it was pointed out that several factors might have inhibited cyst formation after sexual induction, and that all seemed to be associated with batch culture conditions.

The concentrations of the major toxin components (the GTX group) were almost the same for Pc1 and Pc2. However, concentrations of the minor toxin components (the PX group) differed markedly between the two (Table 1). These results suggest that the toxin profile could probably be used to characterize the mating types of this species. This is also supported by the fact that GTX 2, 3 and PX 1 could be detected both in Pc2 and in cysts but not in Pc1. These differences might be a result of the genetic combination of the two different mating types, but such a conclusion needs support from further studies. Resting cysts were more toxic than motile cells, especiailly with respect to GTX 1, which is very highly toxic and was found in high concentrations in cysts in comparison with motile cells (Table 1).

The decrease of ATP in Pc1 and Pc2 observed during growth might have been an effect of low concentrations of phosphorus in the medium, as reported by Sakshaug and Holm-Hansen [12]. After cyst formation, the cellular ATP content increased to around an intermediate level between Pc1 and Pc2, possibly reflecting the fusion of Pc1 and Pc2 gametes.

The changes in amino acid composition were probably dependent on the physiological state at each stage of the life cycle. Arginine concentrations in hot water extracts of cysts were higher than that of the other samples. Interestingly, arginine has been reported as one of the precursors of PSP [13]. An increase in arginine concentration during cyst formation in <u>Scrippsiella</u> <u>trochoidea</u> was also observed by the present authors (unblished).

ACKNOWLEDGMENTS

The authors are greatly indebted to Mr. S.Yoshimatsu for providing *A.catenella* and to Miss A. Nakai for her assistance.

REFERENCES

1. S.Yoshimatsu, Bull. Plahkton. Soc. Japan. 28, 131-139 (1981)
2. D.M.Anderson in: Sea food Toxins, E,P.Ragelis, ed. (American Chemical Society, Washington, 1984,) pp,125-138
3. B.Dale, C.M.Yentsch and J.W.Hurst, Science 201, 1223-1224 (1978)
4. Y.Oshima, H.T.Singh, Y.Fukuyo and T.Yasumoto, Bull. Japan. Soc. Sci. Fish. 48, 1303-1305 (1982)
5. T.Okaichi and Y.Imatomi in: Toxic Dinoflagellate Blooms, F.J.R.Taylor and H.H.Seliger eds. (Elsevier, North Holland, 1979) pp.385-388
6. D.M.Anderson, D.M.Kulis and B.J.Binder, J.Phycol. 20, 418-425 (1984)
7. C.A.Price, E.M.Readon and R.R.L.Guillard, Limnol. Oceanogr. 23, 548-553 (1987)
8. Y.Nagashima, J.Maruyama, T.Noguchi and K.Hashimoto, Nippon Suisan Gakkaishi 53, 819-823 (1987)
9. O.Holm-Hansen and C.R.Booth, Limnol. Oceanogr. 11, 510-519 (1966)
10. D.M.Anderson and N.L.Lindquist, J. exp. mar. Biol. Ecol. 86, 1-13 (1985)
11. D.M.Anderson, D.W.Coats and M.A.Tyler, J. Phycol, 21, 200-206 (1985)
12. E.Sakshaug and O.Holm-Hansen, J. exp. mar. Biol. Ecol. 29, 1-34(1977)
13. Y.Shimizu, S.Gupta, M.Norte, A.Hori, A.Genenah and M,Kobayashi in: Toxic Dinoflagellates, D.M.Anderson, A.W.White and D,G.Barden, eds. (Elsevier, New York, 1985) pp.271-274

INVESTIGATION OF THE SOURCE OF DOMOIC ACID IN MUSSELS

LUCIE MARANDA, RONGHUA WANG, KAZUO MASUDA AND YUZURU SHIMIZU
Department of Pharmacognosy and Environmental Health Sciences, College of Pharmacy, The University of Rhode Island, Kingston, RI, U. S. A. 02881

ABSTRACT

In the fall of 1987, blue mussels, *Mytilus edulis* L. from Prince Edward Island, Canada, caused serious human illnesses and, in a few cases, death. Domoic acid, a known neurotoxic amino acid, was identified in the shellfish and the source of this water-soluble toxin was sought in the dietary plankton from the area. Domoic acid was produced by a 7-day-old culture of *Amphora coffaeiformis* Ag. (*ca.* 0.2 pg/cell), though not in a consistent yield. Minimal production of domoic acid (less than 0.02 pg/cell) was also seen in cultures of the dominant plankter, *Nitzschia pungens*. The particulate material (> 10 μm size fraction, mostly phytoplankton) harvested in the fall of 1988 from the water column contained 0.1 mg of domoic acid per g (wet weight), whereas 60 mg of pure crystals was isolated from 1 l of the filtrate (< 10 μm fraction). Physical constants including optical rotation and spectroscopic evidence indicate that the isolated compound is the same optical isomer as that reported earlier from *Chondria armata* Okamura.

INTRODUCTION

In November 1987, blue mussels (*Mytilus edulis* L.) from eastern Prince Edward Island (Canada) became toxic, causing many serious and, in a few cases, fatal human poisonings [1]. The cultured shellfish contained an excitatory neurotoxin, which was identified as domoic acid (Figure 1). The compound is a small water-soluble amino acid (M. W. 312) [2,3] and, until then, unknown to affect human health. It was thus paramount to determine the origin of this natural toxin and investigate its acquisition by the mussels and possibly by other edible shellfish.

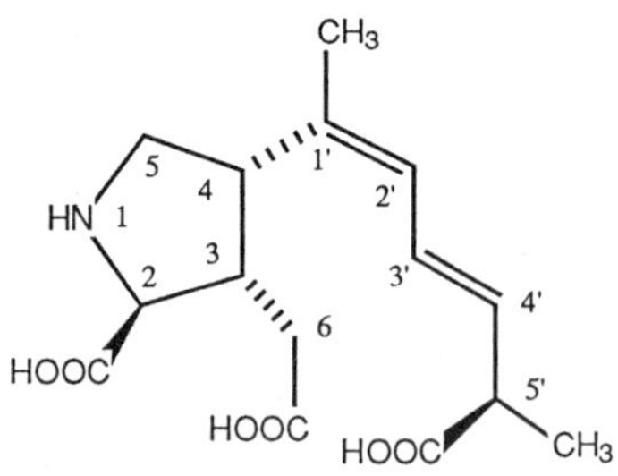

FIG. 1. Structure of domoic acid.

Domoic acid had previously been reported from some red algae of the genus *Chondria* and *Alsidium* [4,5]. Of those two genera, *Chondria baileyana* is documented to be of a rare occurrence in the affected area [2], making it an unlikely culprit of the outbreak. Since mussels feed mostly on live and detrital particulate

Published 1990 by Elsevier Science Publishing Co., Inc.
Toxic Marine Phytoplankton
Edna Graneli et al., Editors

material present in the water column, phytoplankton from the area was first investigated as a possible source of domoic acid.

Here, evidence is presented for the occurrence of domoic acid in extracts of particulate material (mostly phytoplankton > 10 µm) from the water column, in "liquid" recovered after filtration of this material (seawater and material < 10 µm) and in the pennate diatom *Amphora coffaeiformis* Ag., isolated at the time of the outbreak.

MATERIAL AND METHODS

Phytoplankton was collected from a fixed site on Seal River, off Cardigan Bay (P. E. I.) in early December 1988 with 10 and 70 µm mesh nets set against the tidal flow. The particulate material accumulated in the net as a thick slurry which was rinsed with seawater and "squeezed" to the consistency of a paste. Two samples were thus obtained: the "squeezed" particulate material (SPM) and the drip "liquid" (DL). These samples were analysed according to scheme 1 (Figure 2).

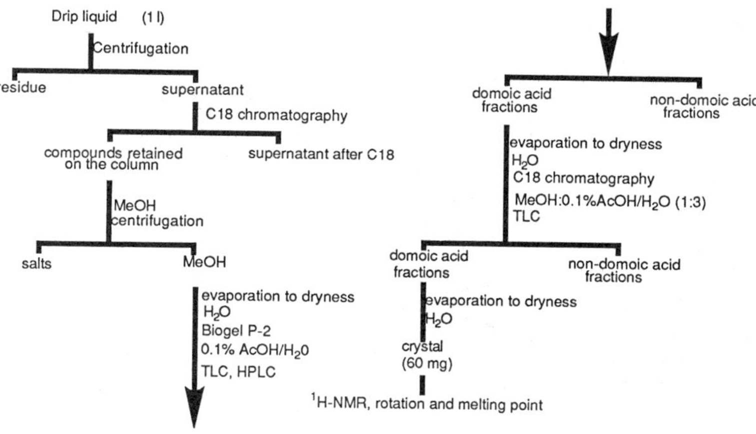

FIG. 2. Analysis scheme 1 for the isolation of domoic acid in the filtrate (< 10 µm) of the "squeezed" particulate material.

Following the 1987 outbreak, phytoplankton and bacterial cultures were started from the pallial fluid of toxic mussels. Following inoculation of the water into different nutrient-enrichment media (f/2-t [6], MET-44 [7], f/2-t enriched with peptone and f/2-t enriched with trypticase, yeast extract and glucose), single phytoplankton cells were isolated from the resulting flora and some bacterial cultures were plated for subsequent isolations. During the 1988 fall-winter bloom, single cells were isolated from unfiltered water samples into f/20-t.

Phytoplankton in batch cultures were grown at 5°C under a 10 h light: 14 h dark cycle or at 15°C under a 13: 11 cycle, in 500-ml Erlenmeyer, 3-l Fernbach flasks or 20-l carboys, in medium f/4-t. Cell number was determined on live samples using a Palmer-Maloney slide or a haemacytometer. Samples were processed according to scheme 2 for regular screening (Figure 3) and to scheme 3 for follow-up assays (Figure 4), involving large-volume cultures.

Identification of domoic acid was made by comparison with a standard sample (Sigma Chemical Co., 90% pure) using high-performance liquid and thin-layer chromatography, UV spectrum and ^{1}H-NMR (D_2O) analyses. HPLC of the samples was achieved via one or both of two solvent systems, 10% CH_3CN/H_2O + 0.1% TFA (system 1), 20% $MeOH/H_2O$ + 0.1% AcOH (system 2), on a C18 analytical column (5

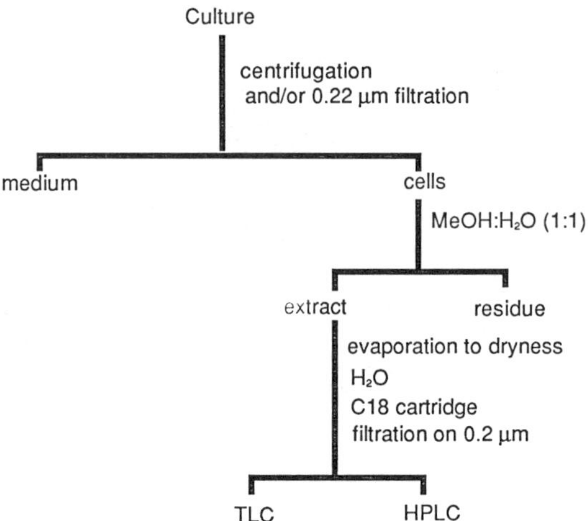

FIG. 3. Analysis scheme 2 for the detection of domoic acid in cultured algae and bacteria.

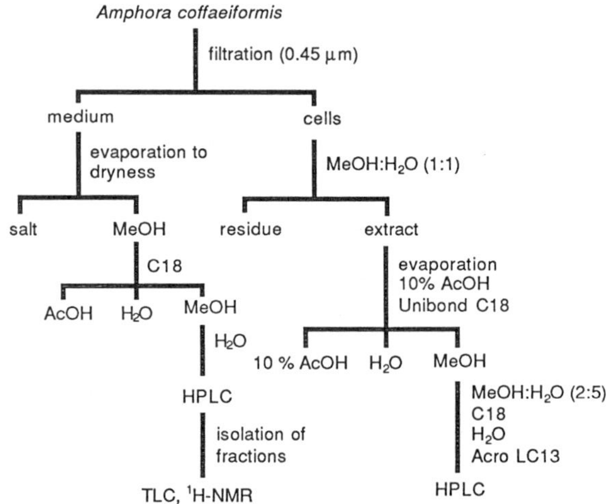

FIG. 4. Analysis scheme 3 for the detection of domoic acid in a large culture (8 l) of *Amphora coffaeiformis*.

μm), and domoic acid was detected at 242 nm. Detection limits were in the 5 to 10-ng range. TLC (Whatman HP-KF) was developed with nBuOH:AcOH:H_2O (7:2:3), and domoic acid was detected by UV and ninhydrin. Optical rotation of domoic acid in water was determined with an Autopol III polarimeter (Rudolph Research). Melting point measurement was made on a Fisher-Johns melting point apparatus.

RESULTS

Of the various bacterial cultures isolated and of the mixed and unialgal cultures (diatoms, blue-green and green algae) obtained, twelve and twenty were tested respectively for domoic acid production. The main species of the diatom bloom, *Nitzschia pungens* f. *multiseries*, was not successfully established from the 1987 outbreak specimen, whereas 16 clones were isolated and maintained from the 1988 bloom.

Only one cell extract among the samples analysed gave a recognizable peak in HPLC, which had the retention time in both systems matching that of the standard. It was from a 7-day-old culture (9 x 10^7 cells total) of a pennate diatom (clone BPT-11), which was later identified as *Amphora coffaeiformis* Ag. Upon further work on a larger sample, only traces of domoic acid could be detected in the cell extract, whereas the extract from the medium yielded the expected HPLC peaks. Fractions suspected to contain domoic acid were purified by HPLC to a single component and examined by TLC and ^{1}H-NMR. TLC of standard domoic acid and the isolated sample gave an identical R_f value (0.47). Proton-NMR (6650 scans, 90° pulse) revealed the proton signal of two methyl groups (δ1.12d (C-5' -Me) and 1.70s (C-1' -Me). From HPLC data, it was calculated that *A. coffaeiformis* produces *ca.* 0.2 pg of domoic acid per cell.

Of the sixteen clones of *Nitzschia pungens* isolated from the 1988 fall-winter bloom, two have been analyzed by HPLC so far. Only trace amounts of domoic acid were detected in both samples (less than 0.02 pg/cell).

Domoic acid was identified by HPLC analysis in both components of the phytoplankton sample (SPM and DL) from the 1988 algal bloom dominated by *Nitzschia pungens* f. *multiseries*. From HPLC data, it was calculated that "squeezed" particulate material contained 0.1 mg of domoic acid per g (wet weight). From 1 l of the drip "liquid", 60 mg of pure crystals, needles from MeOH-H_2O, mp. 209-217° (decomp.), reported: 217° (decomp.) [4], was isolated. The optical rotation of the crystalline sample, $[\alpha]^{22}_D$ -117.5° (C = 0.228, H_2O), was close to the reported value, $[\alpha]^{12}_D$ -109.6°, for a specimen of *Chondria armata* [4], indicating that they are the same optical isomer. The ^{1}H-NMR spectra are also identical.

DISCUSSION

Given the natural occurrence of domoic acid [4,5], its presence in cultured mussels was hypothesized to come from the filtered food or seawater. *Nitzschia pungens* f. *multiseries* was the dominant phytoplankter at the time of the 1987 incident. Domoic acid was identified in the > 10 μm size fraction of the particulate material (mostly phytoplankton) harvested from the water column. A large amount of domoic acid was also found in the filtrate from the "squeezed" particulate material. Therefore it is not clear if domoic acid found in the >10 μm fraction was in the residual "liquid" or in the algal cells or in both.

Domoic acid was identified from a culture of *Amphora coffaeiformis* isolated from the outbreak area. Although the production does not appear to be constant, the maximum amount produced by *A. coffaeiformis* was within the range reported by Subba Rao *et al.* (1988) for *Nitzschia pungens* [8]. Results with our 1988 isolates indicate that the production of domoic acid by *N. pungens* in culture is minimal, if at all. It must be noted that neither our clone of *A. coffaeiformis* nor the clone of *N. pungens* used by Subba Rao *et al.* were axenic. Under these circumstances, the possible involvement of associated organisms (e.g. bacteria, fungi) in domoic acid production cannot be ruled out. The solution of this issue will require careful studies of domoic acid production by axenic and non-axenic clones under various growth conditions.

ACKNOWLEDGEMENTS

Dr. P.E. Hargraves (University of Rhode Island) kindly identified our isolate of

Amphora coffaeiformis Ag. Dr. L.A. Hanic (University of Prince Edward Island) provided live and frozen material from the fall-winter bloom of 1988. Ms. C. Koertig-Walker, Drs. N.K. Gulavita and A.V. Krishna Prasad provided much valuable help. This work was done under NIH grants, GM24425 and GM28754, awarded to Y. Shimizu.

REFERENCES

1. Canada Diseases Weekly Report, Health and Welfare Canada 13-49, 224-226 (1987).
2. J. L. C. Wright, R.K. Boyd, A.S.W. de Freitas, M. Falk, R.A. Foxall, W.D. Jamieson, M.V. Laycock, A.W.McCulloch, A.G. McInnes, P. Odense, V. Pathak, M.A. Quilliam, M.A. Ragan, P.G. Sim, P. Thibault, J.A. Walter, M. Gilgan, D.J.A. Richard and D. Dewar, Can. J. Chem. 67, 481-490 (1989).
3. Y. Shimizu, S. Gupta, K. Masuda, L. Maranda, C. K. Walker and R. Wang, Pure Appl. Chem. 61, 513-516 (1989).
4. K. Daigo, Yakugaku Zasshi 79, 353-356 (1959).
5. G. Impellizzeri, S. Mangiafico, G. Oriente, M. Piattelli, S. Sciuto, E. Fattorusso, S. Magno, C. Santacroce and D. Sica, Phytochem. 14, 1549-1557 (1975).
6. R.R.L. Guillard and J.H. Ryther, Can. J. Microbiol. 8, 229-239 (1962).
7. H.K. Schone and A. Schone, Bot. Mar. 25, 117-122 (1982).
8. D. V. Subba Rao, M. A. Quilliam and R. Pocklington, Can. J. Fish. Aquat. Sci. 45, 2076-2079 (1988).

NMR SPECTROSCOPY OF CHLOROPHYLL(S)-A ISOLATED FROM *GAMBIERDISCUS TOXICUS*

DONALD M. MILLER *, DONALD R. TINDALL** AND CHARLES W. VENABLE*, *Department of Physiology and **Department of Botany, Southern Illinois University, Carbondale, IL 62901, USA.

ABSTRACT

A Caribbean clone of *Gambierdiscus toxicus* (SIU #350) was grown in large-scale culture. A total of 18 fifteen liter carboys, totaling 226L were harvested to yield 81.58 grams fresh weight and 12.1 grams dry weight of cells. The cells were first extracted in methanol, and subsequently in methanol:hexane (1:1). The extracts and further purification products were examined at a concentration of 1 mg/ml using ^{1}H and ^{13}C NMR spectroscopy. Among the components examined were lipids and three forms of chlorophyll-a. The data accumulated has been extremely helpful in designing extraction procedures for large scale separation of pigments and toxins.

INTRODUCTION

The isolation of toxins on a large scale from *Gambierdiscus toxicus* is made more difficult by lipid and pigment components in the cell extracts. In an effort to understand the nature of the components interfering with purification, we have made extractions and examined them using UV-Visible and NMR spectroscopy. In the course of this study we have isolated three types of chlorophyll-a pigments.

METHODS AND MATERIALS

A Caribbean clone of *Gambierdiscus toxicus* (SIU #350) was grown in large-scale culture. A total of 18 fifteen liter carboys, totaling 226 liters were harvested to yield 81.58 grams fresh weight and 12.1 grams dry weight of cells after the method of Tindall [1]. Methanol extracts were subjected to brief pulses of a Insonater 1000 ultrasound generator (4 mm diameter probe) operated at 50 V for 30 sec. The slurry was centrifuged and the resultant one liter supernatant was evaporated to one half and collected. An equal volume of hexane was then added to the methanol extract, the material shaken and the phases separated. This process was repeated until the hexane fraction was clear. The hexane extract was separated on a Waters liquid chromatograph using a Delta 3000 Preparative and Semi-preparative HPLC interlinked with SIM modules to an 840 Controller System and a 10x300 mm column (ODS, 10 μm). The three chlorophylls were eluted isocratically using methanol

Toxic Marine Phytoplankton
Edna Graneli et al., Editors

over a period of 30 min (flow rate of 20 ml•min-1) and monitored at 440 nm using a Waters 481 UV-Vis detector. NMR spectra were performed using two state of the art NMR instruments presently available for our use: specifically a 300 MHz Varian VXR-300 and a 500 MHz Varian VXR-500 multinuclear spectrometer system. Both instruments operate in the pulse Fourier Transform mode and are equipped with a liquid helium VXR superconducting magnet and acquisition hardware.

RESULTS AND DISCUSSION

When separated on the preparative HPLC three distinct green pigments were apparent as separate peaks (Table 1).

TABLE I. Characteristics of Spinach Chlorophyll-a and Three Fractions from *Gambierdiscus toxicus* Isolated by Preparative HPLC.

Fraction	Chl-a	#13	#14	#15
Color	Grass Grn	Grass-Grn	Blue-Grn	Grass-Grn
Elution Time	35.9	35.92	38.59	40.50
Pigment (mg)	1.0	8.5	2.5	2.1
Lipid (mg)	0	26.5	2.5	18.6
Total (mg)	1.0	35.0	5.0	20.7

Spectroscopic data of Fractions 13, 14, and 15 from the hexane extract of methanolic crude cell suspensions, indicate that there are three different chlorophyll-a (chl-a) type compounds (Fig. 1).

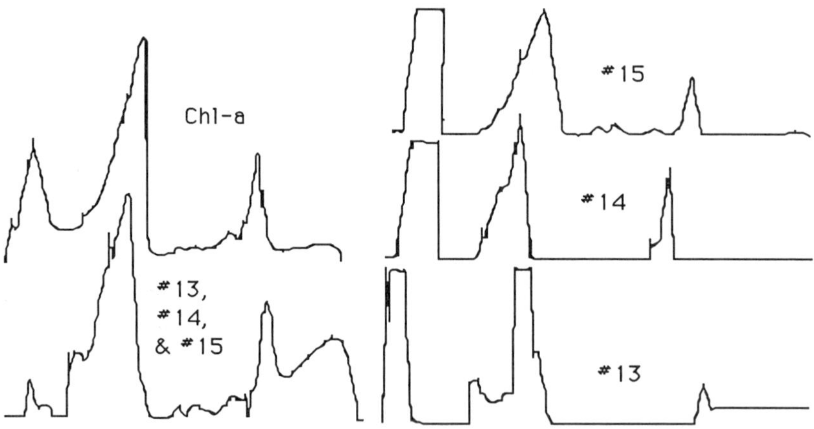

FIG. 1. UV-Visible Scans of Spinach Chlorophyll-a and Three Fractions from *Gambierdiscus toxicus* Isolated by Preparative HPLC.

All three fractions contain varying amounts of unsaturated lipid, predominantly linolenic acid. The distribution of chlorophylls and lipid was determined by direct weighing and integration of ^{1}H NMR signals.

TABLE II. UV-Visible Scan Data for Spinach Chlorophyll-a and Three Fractions from *Gambierdiscus toxicus* Isolated by Preparative HPLC.

Fraction	Chl-a	#13	#14	#15
	672	673.6	661	681
	609.5	610.		617
	538.3	537		546
	507.5	508		515
	415.1	416.5	425	417

In the UV-Vis spectrum of Fraction 13 (in methanol), predominate peaks due to chl-a are seen at 673.6, 416.5, and 316.7. These correlate very closely to the major absorption of pure chl-a at 673, 416 and 316 (Fig. 1, Table 2).

TABLE III. Table of ^{1}H NMR Shift Resonances for Spinach Chlorophyll-a and Three Fractions from *Gambierdiscus toxicus* Isolated by Preparative HPLC.

	Chlor a	#13	#14	#15
β	9.51 s	9.60 s	9.60 s	9.65 s
α	9.38 s	9.46 s	9.38 s	9.65 s
δ	8.54 s	8.61 s	8.38 s	8.0 bs
2	8.00 dd	8.0 dd	8.0 dd	7.95 dd
2'	6.28 d	6.29 d	6.18 d	6.30
2''	6.16 d	6.18 d	6.0 d	6.17
10	6.22 s	5.5 s	-	5.15 s
p2	5.30 t	5.2 t	4.98 t	5.08 t
8	4.76 q	5.0 q	4.20 q	4.82 q
P,A	4.45 q	4.51 t	4.08 m	4.3 m
P,B	4.40 q	4.45 t	4.08 m	4.22 m
7	4.19 d	4.27 dd	4.20 d	4.56 d
10AMe	3.86 s	3.70 s	3.68 s	3.72 s
4CH2	3.69 m	3.70 m	3.70 m	3.68 m
5Me	3.67 s	3.60 s	3.52 s	3.61 s
1Me	3.38 s	3.40 s	3.31 s	3.58 s
3Me	3.21 s	3.22 s	3.28 s	3.26 s
7A(2)	2.6,2.4 m	2.6,2.45m	4.48 m	2.38-2.30 m
7B(2)	2.3, 2.15 m	2.25, 2.22 m	2.26 m	2.2, 2.06
P-4	1.82 q	1.98 q	1.84 q	--
8Me	1.78 d	1.57 d	1.43 d	1.41 d
4Me	1.68 t	1.68 t	1.69 t	1.70 t
P-3Me	1.52 s	1.58 s	1.48 s	1.44 s
P5/15CH2	1.3-1.0 m	1.3-1.0 m	1.3-1.0 m	1.3-1.0 m
P7Me	0.80 d	0.82 d	0.84 d	0.84 d
P11Me	0.80 d	0.82 d	0.84 d	0.84 d
P15	0.75 d	0.78 d	0.81 d	0.83 d
P16 Me	0.77 d	0.76 d	0.78 d	0.81 d

While both ^{1}H NMR and UV absorption spectroscopy indicate the presence of the porphyrin ring structure of chl-a for all three compounds,

Fraction 13 is definitely chl-a as denoted in the literature [2], the most abundant of the three green pigments. All of the proton resonances of chl-a are visible even in the presence of the fatty acid contaminant which is present in a 3:1 ratio over the pigment. The only differences from the spectrum of pure chl-a (from spinach) as run on our system is (1) the upfield shift of the H-10 singlet from 6.22 to 5.5 ppm and (2) the 8 Me doublet upfield shift from 1.78 to 1.57 ppm. These somewhat significant shifts must undoubtedly be due to lipid interaction at rings IV and V near the phytyl ester bond. Well defined lipid resonances in Fraction #13 are seen at 5.3, 2.8, 2.3, 2.05, 1.68 and 0.95 ppm.

Fraction 14 collected from the hexane extract appeared as a deep blue green. Examination of the ^{1}H spectrum revealed that it contained the porphyrin ring structure of chl-a. Low field ring methine singlets at 9.6, 9.4 and 8.4 ppm; vinyl protons at 8.0 ppm dd (16, 11 Hz); 2 vinyl terminal methylene protons 6.2 ppm d (16z) and 6.0 ppm d (11 Hz); and four ring methyl singlets at the 1, 3, 5 and 10a positions adsorbing between 3.0 and 4.0 characterize the chl-a porphyrin ring structure. The H-10 singlet is noticeably absent from this chlorophyll's ^{1}H spectrum, but may possibly be shifted under the 5.3 lipid resonance. In Fraction #13, this signal was shifted upfield from 6.1 to 5.5 ppm. The resonances in the 4.0-5.0 ppm region also vary from those of chl-a. The 4.76 q, H-8 is shifted to 4.20 while the P-1 CH_2 is shifted from 4.40 to 4.08. As these signals involve ring IV of the porphyrin structure, these chemical shift differences may suggest an alternative structure of this proton of the chl-a porphyrin ring. Signals of the phytyl chain are easily seen: P-2, t at 4.98 ppm, P-1 CH_2 at 4.08 ppm, P-3 CH_3 at 1.48 ppm, chain methyl doublets P-7, P-11 at 0.84 ppm and terminal chain methyl doublets P15, and P16 at 0.81 and 0.78 ppm.

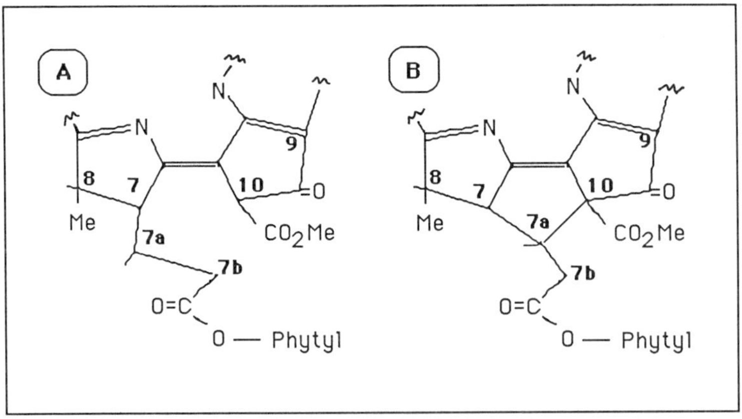

FIG. 2. Suggested structure of chlorophyll extracted in Fraction #14 from *Gambierdiscus toxicus* Isolated by Preparative HPLC.

It is proposed that the chlorophyll corresponding to Fraction #14 is a different type of chlorophyll with the C-7a carbon bonded to C-10 (see Figure 2) and is supported by the following in the NMR spectra:

a. H-10 is completely missing,
b. A new singlet proton appears at 4.48 (7a), and
c. The P-1 protons are shifted to higher field due to closer proximity of the C-10 methyl ester and phytyl ester group.

As Fraction #14 contains only 2.5 mg lipid 2.5 mg pigment, the deviations of observable resonances from chl-a due to lipid interaction would appear to be almost nil. Fraction #13 contained 26.5 mg lipid and 8.5 mg of pigment and all but one of the resonances were within ±0.2 ppm of those of pure chl-a obtained commercially. In the UV-Vis absorption spectra, major pigment peaks seen at 660 and 425 nm were somewhat shifted from the comparable major absorption peaks of chl-a peaks of 473 and 415 nm. This perhaps indicates some modification of the standard chl-a peaks of 473 and 415 nm.

Fraction #15 as isolated from the preparative HPLC column was a bright grass green, very similar in appearance to Fraction #13 which identifies it as a chl-a. This fraction was the least pure of the three chlorphylls having about 2 mg of pigment and 18 mg of lipid. Deviations from the ^{1}H NMR spectrum of chl-a[1] are observed in ring I of the porphyrin structure. The proton spectrum reveals that (1) the δ methine singlet is shifted from 8.4 to 8.0 ppm. (2) the two terminal methylene protons from the 2 vinyl group are reversed. The 11.0 Hz cis proton is shifted to lower field, 6.3 ppm and the 16.0 Hz trans proton is at higher field 6.18 ppm. Just the reverse was observed in Fractions #13 and #14. (3) The 1-methyl ring singlet was shifted to 3.58 ppm. This 0.2 ppm shift represents the greatest chemical shift for any of the ring methyl singlets of the three chlorophyll types. These chemical shifts may possibly be related to a different orientation of the 2-vinyl group on the macrocyclic ring. Also to be noticed is the absence of the H-10 singlet normally found near 6.0 ppm. Once again, it may be shifted into the region of the large 5.3 peak arising from lipid. Of note in this spectrum is the possible appearance of a broad singlet at 6.0 ppm. This might possibly arise from an OH group at position 10 as has been previously been observed. Signals of the phytyl chain are visible but more submerged in the lipid peaks due to the relatively large amount of lipid present in Fraction #15.

CONCLUSIONS

The UV-Visible and proton NMR data shift changes may possibly indicate a conformation change or an overall structural change in the chl-a molecule. Our investigation of Fraction #14 appears to enhance this interpretation. The absence of the H-10, 6.1 ppm singlet first led us to speculate the possibly of an OH or other substituent at C-10. However examination of the ^{13}C NMR spectra of Fraction #14 revealed a signal at 65.6 ppm which closely matches the 66.2 ppm signal assignment of C-10 in Chl-a extracted from clover leaves [2]. This indicates that ring V of the porphyrin ring structure is intact. We confirmed the complete absence of the H-10 singlet from a spectrum obtained from the sample purified to a 3:1 pigment to lipid ratio.

We also speculated that possibly a double bond existed in Fraction #14 between C-7a and C-7b. This would undoubtedly alter the proton NMR pattern as well as the visible UV-Vis absorption of chl-a. Repeated decoupling throughout the 5.35 ppm multiplet did not significantly alter the 4.45 ppm (6.0 Hz, d). As no other olefinic protons in Fraction #14 were apparent other than the 2 vinyl groups and 5.3 ppm resonance of the lipid protons, this was the only place olefinic protons of a possible 7a-7b double bond could possibly be. In the ^{13}C spectrum of Fraction #14 signals at 30.0 and 31.95 closely match the assignments of c-7a, 30.9 and C-7b 30.1 reported for Chl-a (15). However, our ^{13}C data is tentative due to the limited amount of sample, 2.5 mg, a very minimal amount for ^{13}C work. Obviously a more highly purified amount of sample is needed to further investigate the differences in these altered forms of chl-a. (Support by USAMRIID Contract DAMD17-87-C-7002)

REFERENCES

1. D.R. Tindall, R. Dickey, R. Carlson, and G. Morey-Gaines in: Seafood Toxins, ACS Symposium Series, 262, E. Ragelis, ed. (1984) pp.225-240

2 S. Lotjonen and P. H. Hynninen, Org. Magnetic Res. 21(12): 757-767 (1983)

PRODUCTION OF PARALYTIC SHELLFISH TOXINS BY BACTERIA ISOLATED FROM TOXIC DINOFLAGELLATES

TAKEHIKO OGATA*, MASAAKI KODAMA*, KEIICHI KOMARU*, SETSUKO SAKAMOTO*, SHIGERU SATO* AND USIO SIMIDU**
*Laboratory of Marine Biological Chemistry, School of Fisheries Sciences, Kitasato University, Iwate 022-01, Japan; **Ocean Research Institute, University of Tokyo, Nakano, Tokyo 164, Japan

ABSTRACT

Bacteria isolated from toxic strains of dinoflagellates were found to possess sodium channel blocking activity. Previously, we reported that *Moraxella* sp. isolated from an Ofunato strain of *Protogonyaulax tamarensis* produces paralytic shellfish toxins (PSP toxins). However, bacteria isolated from other strains of *P. tamarensis* or other species of dinoflagellates were different from *Moraxella* sp., suggesting some relationship between strains or species of toxin-producing bacteria and toxic dinoflagellates. Production of PSP toxins was confirmed by HPLC-fluorometric analysis in *Bacillus* sp. isolated from *Gymnodinium catenatum* and PTB-5 from a Harimanada strain of *P. tamarensis*, respectively.

INTRODUCTION

We have earlier reported that *Moraxella* sp. isolated from an Ofunato strain of *Protogonyaulax tamarensis* produces paralytic shellfish toxins (1, 2). This bacterium was isolated from an apparently axenic culture of *P. tamarensis* grown in a medium containing antibiotics, which suggests that it originated from inside the cells of *P. tamarensis*. Our findings not only show that bacteria produce paralytic shellfish toxins (PSP toxins), but also suggest that bacteria are associated with the toxin production in dinoflagellates.

We report here on the isolation of different species of bacteria which produce PSP toxins in different *P. tamarensis* strains and in other dinoflagellate species.

MATERIALS AND METHODS

Strains of *P. tamarensis* were obtained from various areas: PT-1 and 2 from Ofunato Bay; PT-3 and 4 from Okkirai Bay; PT-5 from Harimanada; PT-6 from Kushimoto coastal water; and PT-7 and 8 from the Gulf of Thailand. Strains of *P. catenella* were also collected: PC-1 and 2 from Ofunato Bay; and PC-3 from Seto Inland Sea. *P. cohorticula* strains (PCO-1 and 2) were collected from the Gulf of Thailand. A strain of *Gymnodinium catenatum* (GC-1) was isolated from Senzaki Bay, Yamaguchi Prefecture. As a non toxic control, 2 strains of *Fragilidium* sp. (FR-1 and 2) and *Alexandrium* sp. (AL-1) were isolated from Suruga Bay and Tolo Harbour, Hong Kong, respectively. The dinoflagellates were cultured in T1 medium (3) under 4000 lux (16hr:8hr, LD cycle) at 15°C (PT-1 to 4), 20 °C (PT-5 and 6, PC-1 to 3, GC-1, FR-1 and 2), and 25 °C (PT-7 and 8, PCO-1 and 2, AL-1), respectively. The cells were harvested at the late exponential growth stage and assayed for toxicity by i.p. mouse bioassay using male mice (ddY strain). With this technique, 1 mouse unit (MU) is defined as the amount of toxin needed to kill a 20 g mouse in 15 min (4).

Isolation of bacteria from dinoflagellates was carried out as described previously (1). Dinoflagellate cells grown in the medium containing

Published 1990 by Elsevier Science Publishing Co., Inc.
Toxic Marine Phytoplankton
Edna Graneli et al., Editors

antibiotics (1) were harvested, washed thoroughly with sterilized seawater, and homogenized. The homogenates were incoulated to agar plates (Marine Agar, Difco). Bacterial clones were isolated from colonies appearing on the plates after 5 days incubation at 20 °C. The bacterial isolates thus obtained were grown as standing cultures in 200 ml of a liquid medium (Marine Broth, Difco) for 10 days at 15 °C. Cells were harvested by centrifugation and extracted with 80% ethanol containing 1% acetic acid. The extracts were evaporated to remove ethanol, defatted with dichloromethane and made up to 10 ml with distilled water. A 200 µl portion of each extract was assayed for sodium channel blocking activity by the mouse neuroblast cell culture method according to Kogure et al. (5). Portions of harvested cells were treated with formalin, resususpended in saline, and then reacted with anti-PSP toxins-producing Moraxella sp. serum which was prepared by immunization of rabbit with Moraxella sp. cells. Cross reactivity was judged from aggregation of cells under microscopic observation.

Among the bacterial clones obtained, those from G. catenatum and the Harimanada strain of P. tamarensis were inoculated into 4 L of autoclaved seawater, left standing for 14 days at 15 °C, and then harvested by filtration through membrane filter with an opening size of 0.45 µm. Harvested cells were homogenized together with filter membranes in 80% ethanol containing 1 % acetic acid, sonicated for 10 min and centrifuged. After removal of ethanol by evaporation, the supernatant extracts were defatted with dichloromethane and treated with a SEP-PAK C18 cartridge. Toxin components in the extracts were analyzed by a HPLC-fluorometric analyzer for PSP toxins (6).

RESULTS

Toxic species of dinoflagellates showed various levels of toxicity from non-detectable to highly toxic (Table I). In the case of P. tamarensis, strains from the northern part of Japan (PT-1,2,3,4) showed high or mild toxicity, while those from southern waters (PT-5 and 6) were weakly toxic or nontoxic. Strains from the Gulf of Thailand were nontoxic as reported before (7). The toxicity of P. catenella strains also varied. PC-1 and 2 from Ofunato Bay showed significant toxicity, whereas PC-3 from Seto Inland Sea was nontoxic. Both P. cohorticula and G. catenatum were highly toxic. Fragilidium sp. and Alexandrium sp. were nontoxic. All the strains grew normally in the medium containing antibiotics, indicating that the antibiotics used did not affect the growth, as reported before for a P. tamarensis strain (1).

Bacteria were isolated from all the toxic strains of dinoflagellates except from a weakly toxic strain of P. catenella (PC-1). Several species of bacteria were obtained from some of the strains. However, no bacteria were obtained from nontoxic dinoflagellate strains or species, with the exceptions of PT-7 and AL-1. The bacterial isolates grew well in the medium used (Marine Broth, Difco). Within 2 days, all the cultures showed maximum turbidity. In the cross-reactivity test with anti-Moraxella sp. serum, anti-serum reacted only with bacterial isolates from P. tamarensis and P. catenella collected from Ofunato Bay and Okkirai Bay. It did not react with other isolates, not even with that of the Harimanada strain of P. tamarensis (Table I). Sodium channel blocking activity appeared in the bacteria from toxic strains of dinoflagellates, when they had been cultured in Marine Broth for 10 days. In contrast, no activity was detected in the bacteria from nontoxic strains or species of dinoflagellates under the same conditions. Toxic extracts thus obtained were not suitable for HPLC-fluorometric analysis, because relatively high peaks of impurities interfered with those of the toxins.

The isolates from G. catenatum and the Harimanada strain of P. tamarensis did not react with anti-Moraxella sp. serum, which suggests that they belong to different species than Moraxella sp. In order to confirm the production of PSP toxins, further culture experiments were carried out on these species. Although little is known about the conditions for toxin production of bacteria, toxin production of Moraxella sp. seems to increase when grown under starved conditions (8). Thus, the culture conditions for Moraxella sp. were applied to these isolates. After these isolates had been inoculated in seawater and left for 2 weeks at 15 °C, the cultures became slightly turbid, implying a significant bacterial growth. The extracts of cells from both cultures, collected by filtration trough membrane filters, showed sodium channel blocking activity in the neuroblast cell culture assay. Fig. 1 shows HPLC chromatograms of toxins in the extracts of both isolates. Bacterial strain GCB-2 from G. catenatum showed peaks of gonyautoxin (GTX) 1, 2, 3 and 4 and saxitoxin (STX) along with some unidentified peaks. PTB-6 from the Harimanada strain of P. tamarensis showed peaks of GTX 1, 2, 3 and 4 but STX and neoSTX were not detected. The results indicate that these strains of bacteria produce PSP toxins.

The bacterial strain GCB-2 was a Gram-positive, rod type bacterium which made spores. These characteristics indicate that it belongs to the genus, Bacillus, according to the criteria proposed by Shimidu (9).

TABLE I. Characteristics of bacterial isolates, and their host dinoflagellates.

Dinoflagellates			Bacteria		
Species	Strain	Toxicity ($MU/10^4$cells)	Isolate	Cross reactivity with anti-serum[a)]	Sodium channel blocking activity[b)]
P. tamarensis	PT-1	0.750	PTB-1	+	+
	PT-2	0.325	PTB-2	+	+
			PTB-3	-	-
	PT-3	1.340	PTB-4	+	+
	PT-4	1.920	PTB-5	+	+
	PT-5	0.020	PTB-6	-	+
	PT-6	-[c)]	-[d)]		
	PT-7	-	PTB-7	-	-
	PT-8	-	-		
P. catenella	PC-1	0.050	-		
	PC-2	0.200	PCB-1	+	±
	PC-3	-	-		
P. cohorticula	PCO-1	0.617	PCOB-1	±	+
			PCOB-2	-	+
	PCO-2	1.333	PCOB-3	-	+
			PCOB-4	-	-
			PCOB-5	-	-
G. catenatum	GC-1	1.935	GCB-1	-	-
			GCB-2	-	+
Fragilidium sp.	FR-1	-	-		
	FR-2	-	-		
Alexandrium sp.	AL-1	-	ALB-1	±	-

a) Reactivity was examined with anti-PSP producing Moraxella sp. serum.
b) Activity was assayed by mouse neuroblast cell culture method [5].
c) No toxicity was detected.
d) No bacterium was found.

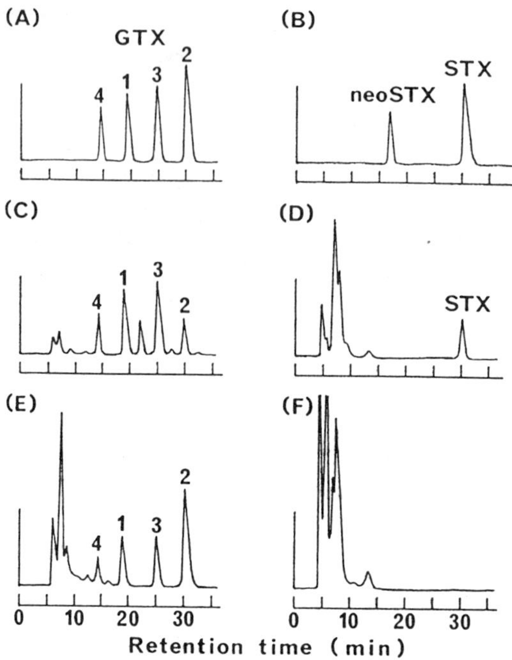

FIG. 1. Identification of bacterial toxins by HPLC-fluorometric analysis using Develosil C8 according to Oshima et al. (6). Mobile phase in (A), (C), and (E): 2 mM 1-heptanesulfonic acid/ 10 mM phosphoric acid (pH 7.2); mobile phase in (B), (D), and (F): 2 mM heptanesulfonic acid/ 10 mM phosphoric acid/ 10% acetonitrile (pH 7.2).
(A): standard gonyautoxins (GTXs), (B): standard saxitoxin (STX) and neoSTX, (C) and (D): extracts of bacterial strain (GCB-2) from _G. catenatum_, (E) and (F): extract of bacterial strain (PTB-6) from the Harimanada strain of _P. tamarensis_.

DISCUSSION

Bacteria were isolated from homogenates of dinoflagellate cells grown in a medium containing antibiotics. The bacteria derived mostly from toxic strains, but some isolates were from nontoxic strains or species of dinoflagellates. However, bacteria possessing sodium channel blocking activity were obtained only from toxic strains of dinoflagellates. We have earlier shown the that a PSP toxins-producing bacterium, _Moraxella_ sp., was isolated from a Ofunato strain of _P. tamarensis_ (1). The present results suggest that toxic dinoflagellates are always contaminated by toxin-producing bacteria.

In order to examine whether _Moraxella_ sp. is the only species contaminating toxic dinoflagellates, anti-_Moraxella_ sp. serum was prepared and its cross reactivity was tested on bacterial isolates from toxic dinoflagellates. Interestingly, the antiserum did not react with all of the isolates, but only with those from strains of _P. tamarensis_ and _P. catenella_ collected from Ofunato Bay and the neighbouring Okkirai Bay. Isolates of *P. tamarensis* collected at Harimanada, the southern part of Japan, did not react with the antiserum. These findings show the occurrence of other toxin-producing bacteria than *Moraxella* sp. The present data also suggest that the relationship between toxin-producing bacteria and dinoflagellates is variable, even within the same species of dinoflagellates.

The amount of toxins produced by *Moraxella* sp. varied even when grown under the same conditions, indicating the presence of unknown factor(s) regulating toxin production (1). However, toxin productivity of *Moraxella* sp. increased when *Moraxella* sp. was grown under starved condition. Thus, the growth condition for *Moraxella* sp. that showed the highest toxin productivity was applied to bacterial isolates from *G. catenatum* and the Harimanada strain of *P. tamarensis*, which differ from *Moraxella* sp., in order to identify the toxin components they produce. When grown in natural seawater, the cultures became slightly turbid after 14 days, indicating significant bacterial growth. The amount of toxins obtained from cells in 1 L of the seawater cultures was similar to that obtained from much denser Marine Broth cultures, showing that the toxic potency of cells grown in seawater is much higher than that for cells grown in an artificial culture medium. The crude extracts of the cells thus obtained were analyzable by HPLC-fluorometry without purification. Both strains were shown to produce PSP toxins.

Isolates from *G. catenatum* were assigned to *Bacillus* sp. according to various characteristics.

The results show that PSP toxin producing bacteria are widely distributed and belong to several species.

ACKNOWLEDGEMENTS

The authors express their sincere thanks to Dr. Oshima, Tohoku University and Red Tide Institute of Kagawa Prefecture for supplying strains of *P. tamarensis* (PT-5 and 6) and *P. catenella* (PC-3). This work was supported in part by grants from Ministry of Agriculture, Fisheries and Forestry of Japen and Kaiun Mishima Foundation.

REFERENCES

1. M. Kodama, T. Ogata and S. Sato, Agric. Biol. Chem. 52, 1075-1077 (1988).
2. M. Kodama, T. Ogata and S. Sato in: Red Tides. Biology, Environmental Science, and Toxicolocy, T. Okaichi, D.M. Anderson, and T. Nemoto, eds. (Elsevier, New York 1989) pp. 363-366.
3. T. Ogata, T. Ishimaru and M. Kodama, Mar. Biol. 95, 217-220 (1987).
4. S. Williams in: Official Methods of Analysis, S. Williams, ed. (Association of Official Analytical Chemists, Arlington 1984) pp. 344-345.
5. K. Kogure, M.L. Tamplin, U. Shimidu and R.R. Colwell, Toxicon 26, 191-197 (1988).
6. Y. Oshima, M. Hasegawa, T. Yasumoto, G. Hallegraeff and S. Blackburn, Toxicon 25, 1105-1111 (1987).
7. M. Kodama, T. Ogata, Y. Fukuyo, T. Ishimaru, S. Wisessang, K. Saitanu, V. Panichyakarn and T. Piyakarnchana, Toxicon 26, 707-712 (1988).
8. M. Kodama, T. Ogata, S. Sakamoto, K. Komaru, S. Sato, and T. Honda, submitted to Toxicon.
9. U. Shimidu in: Kaiyo Biseibutsu Kenkyuho (Methods for investigation of marine microorganisms), H. Kadota and N. Taga, eds. (Japan Scientific Societies Press, Tokyo 1985) pp. 228-233.

RESPONSE OF GYMNODINIUM CATENATUM TO INCREASING LEVELS OF NITRATE: GROWTH PATTERNS AND TOXICITY

B.REGUERA, + AND Y.OSHIMA ++
+ Instituto Español de Oceanografía, Centro Oceanográfico de Vigo. Aptdo. 1552. 36280 Vigo. Spain. ++ Faculty of Agriculture, Tohoku University. Tsutsumidori, Sendai 980. Japan.

ABSTRACT

Gymnodinium catenatum, responsible for PSP outbreaks in NW Spain, was cultured in K media with increasing levels of nitrate. Growth rates were calculated from measurements of in vivo fluorescence and cell counts. Toxicity was determined by HPLC. Toxin concentrations per cell reached a peak with nitrate concentrations of 110-220 µM. Much of the increase in toxin concentration was due to a higher production rate of GTX 6, showing that the toxin profile is not a conservative property in this species.

INTRODUCTION

G.catenatum has for some years been the main organism responsible for toxic red tides in the Rías Bajas of Galicia in NW Spain, Portugal, Tasmania, Mexico, and more recently in Japan. Ecophysiological studies of this species are however still scarce. It is known that in some dinoflagellates growth rates and toxin production rates are influenced by physiological and environmental factors. For example, toxin production in Protogonyaulax spp. varies significantly in response to variations in temperature, salinity, light intensity, cell size and the stage of the culture (1, 2). Ogata et al (3) studied the effect of temperature and light intensity on growth rate and toxin production in a Japanese strain of G.catenatum. They found that photosynthesis was essential for toxin production, and that quantities of toxin produced increased as growth rate decreased when this was controlled by temperature.

The effect of nitrogenous compounds on growth and toxin production is particularly interesting in view of the high nitrogen content of saxitoxins. Boyer et al (4, 5) found that toxin production per cell in P.tamarensis decreased in nitrogen limited cultures, while phosphorus limited cultures showed a dramatic increase in toxin production at the end of the stationary phase. They also found that despite changes in the quantities of toxins produced, the relative proportions of the different toxic components remained constant. These results suggest that toxic cells require a particular N:P ratio for growth, and that excess nitrogen may result in further toxin production. Wyatt and Reguera (6) suggested that toxin production is a mechanism of detoxification in nitrogen-enriched environments. This could account for the apparently increasing occurrence of PSP episodes in regions with increasing pollution by domestic and agricultural runoff.

In this paper we study the effect of increasing concentrations of nitrate on growth rate and toxin production in a strain of G.catenatum from Ría de Vigo in NW Spain. In earlier experiments (7), it was found that G.catenatum grew better on ammonium than on nitrate, but that they suffer from ammonia intoxication at concentrations higher than 50 µM. For this reason , nitrate was chosen as the nitrogen source. But in field populations, other sources such as ammonium or organic nitrogen might be more important.

Published 1990 by Elsevier Science Publishing Co., Inc.
Toxic Marine Phytoplankton
Edna Graneli et al., Editors

MATERIALS AND METHODS

G.catenatum (GC 21 V) was isolated from Ría de Vigo by I.Bravo in 1986, and has since been maintained in K medium (8). Two sets of experiments were carried out under the same conditions, in October 88 and January 89. In the first, there were two replicates for each treatment, and in the second there were four. The main difference between the two series was that in the first the inoculum used was from a culture in mid-exponential phase, while in the second the inoculum was from a culture in late stationary phase. In the first set, all cultures remained undisturbed until the moment of harvesting. They were grown in 25 mm diameter fluorometer tubes at 18-20° C (14:10 L:D cycle, 16.6 watts/m^2) using local seawater from Ría de Vigo enriched with K medium from which nitrate and ammonium chloride had been excluded. Nitrate was then added at concentrations of 55 µM, 110 µM, 220 µM, 440 µM, and 880 µM which represent 1/16, 1/8, 1/4, 1/2 and 1/1 respectively of the nitrate concentration normally used in the preparation of K medium. One pair of cultures was also grown without added nitrogen. In all cases, a third transfer grown in the experiments conditions was used for the definitive experiment. In vivo fluorescence (using a Turner Designs fluorometer) and cell numbers (in a Sedgwick-Rafter counting chamber) were estimated every other day at the same time of day. Cells for toxin analysis were harvested at the end of the exponential phase (9th-10th day), centrifuged at 1000 rpm during 15 mn, and the drained pellets deep frozen inmediately. They were air freighted to Japan in dry ice. HPLC toxin analysis was performed as described by Oshima et al. (9) using three different conditions for the gonyautoxins, saxitoxins, and C1-C4 toxins respectively.

RESULTS

Cells from cultures without added nitrogen grew well in the first transfer in both sets of experiments, showing growth rates similar to the cultures with added nitrates. In the first set, cells with no added nitrogen grew well during the second transfer too, but after the third transfer the cells were very pale and sedimented to the bottom of the tubes. In the second set, these characteristics were already apparent in the second transfer, and in the third transfer did not grow at all. It was not therefore possible to obtain enough cells for pellet preparation.

Exponential growth rates calculated from in vivo fluorescence were very similar in all cultures with added nitrogen and ranged between 0.27 and 0.30. Growth rates calculated from cell numbers during the exponential phase were also similar in all cultures, and ranged between 0.35 and 0.43. But there were differences in the duration of the exponential phase, and hence in the final yield (Table 1).

Nine different toxins were identified: C1, C2, C3, C4, GT5, GTX6, dcGTX2, dcSTX and neoSTX. About 90% of the amounts of toxin were sulphocarbamates, i.e., with low potency (Table 2). Similar compositions of toxins have been found in G.catenatum from Tasmania (9), Japan (10) and Spain (11). The highest proportions of neoSTX and decarbamoyl toxins were found in the N limited cultures.

Notable differences were found in toxin production, not only in the amounts produced, but also in the relative proportions of individual toxins (Fig. 1). Total toxin production was highest at nitrate concentrations of 110 and 220 µm. The most marked changes in total toxin content per cell were due to GTX6. Nevertheless, the quantities of toxins produced per unit volume of culture (cell counts x toxin/cell) are similar for all concentrations of nitrate from 110 µM upwards -about 500 pmol/ml.

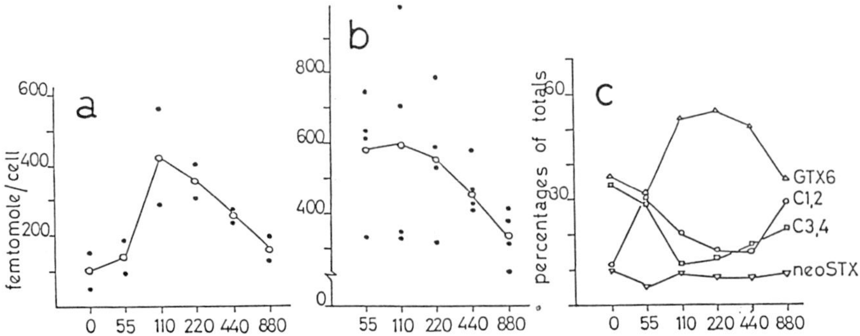

FIG. 1. a) Total toxin content of G.catenatum at different nitrate concentrations: close circles are replicate cultures, open circles the means of each pair (October experiments). b) Total toxin content as in a) (January experiments). c) Percentages of GTX6, C1+C2, C3+C4 and neoSTX (October experiments).

DISCUSSION

Some dinoflagellates can store nitrogen in intracellular pools for use during times of limitation (12, 13). This also appears to be true of G.catenatum. The growth of cultures without added nitrogen cannot be explained on the basis of nitrogen in the medium since the cultures collapsed at the third transfer using the same water source.

The pale colour and low fluorescence of N-limited cells have been noted in other species (5, 7). Motility is also lost in these conditions (14).

The correspondence between the increase in fluorescence and cell multiplication is poor, and is not due to differences in cell sizes. Thus fluorescence is not a good indicator of growth. It may indicate different degrees of cell stress.

Mucus production is characteristic of G.catenatum cultures, and the amounts increases with time. In the late exponential phase, the chains adhere to each other to form mucilaginous filaments when the tubes are stirred for fluorescence measurements.

Except in those cultures with 55 µM nitrogen or less, the amounts of toxin found in these experiments are very high. The decrease in the toxin content per cell with the higher levels of nitrate suggests that oversaturation or even intoxication occurs at these concentrations. Toxin concentrations in field populations in Ría de Vigo (in 1986) were in the range 60-120 fmol toxin/cell, and in cells cultured in K medium between 110 and 190 fmol/cell (11). Of course, as in all batch cultures experiments, the concentrations of nitrate used here exceeds those likely to be encountered in nature, even under conditions of extreme eutrophication. We think that the K medium used for routine cultures of dinoflagellates is excessively rich in nitrate, and its concentration could be halved without reducing the yield. In our experiments, some cultures reached values of 700 to 900 fmol/cell. These results are in contrast to those found in P.excavata, where natural populations contain more toxins than cultures (15).

The higher values of toxins in cultures may be due to the very high concentration of nitrogen compounds in K medium, which has an N:P ratio of 93:1, very much higher than the typical Redfield ratio of 15-18:1. In the present experiments, only nitrate was provided as a nitrogen source. Culture media for G.catenatum generally employ 1 to 50 µM of ammonium chloride. The response of G.catenatum to nitrate is described elsewhere (7).

The change in toxin composition with different culture conditions in this species differs from results found with P.tamarensis (16). In G.catenatum, the toxin composition appears not to be a conservative property of a strain, and the toxin profiles are clearly not suitable as potential taxonomic markers.

TABLE I. Concentrations of G.catenatum cells at begining (No) and end (Nf) of exponential phase, calculated growth rates and final yield (Y). All figures are means of four replicates.

Nitrate (µM)	No (cells/ml)	Nf (cells/ml)	k (div/day)	Y (cell/ml)
55	31	521	0.40	521
110	40	791	0.43	971
220	49	806	0.40	834
440	48	1158	0.35	1158
880	58	1551	0.37	1551

REFERENCES

1. D.M. Anderson, J.J. Sullivan, and B. Reguera, Toxicon (in press).
2. M. Kodama, Y. Fukuyo, T. Ogata, T. Igarashi, H. Kamiya, and F. Matsuura, Bull. Jap. Soc. Sci. Fish. 48, 567-571 (1982).
3. T. Ogata, M. Kodama, and T. Ishimaru in: Red Tides: Biology, Environmental Science and Toxicology, T. Okaichi, D.M. Anderson, and T. Nemoto, eds. (Elsevier, New York 1989) pp. 423-426.
4. G.L. Boyer, J.J. Sullivan, R.J. Andersen, P.J. Harrison, and F.J.R. Taylor, in: Toxic Dinoflagellates, D.M. Anderson, A.W. White, and D.G. Baden, eds. (Elsevier, New York 1985) pp. 281-287.
5. G.L. Boyer, J.J. Sullivan, R.J. Andersen, P.J. Harrison, and F.J.R. Taylor, Mar. Biol. 96, 123-128 (1987).
6. T. Wyatt, and B. Reguera in: Red Tides: Biology, Environmental Science and Toxicology, T. Okaichi, D.M. Anderson, and T. Nemoto, eds. (Elsevier, New York 1989) pp. 33-36.
7. B. Reguera, C.M. Yentsch, and D.M. Anderson, J. Plank. Res (Submitted).
8. M.D. Keller, and R.R.L. Guillard, in: Toxic Dinoflagellates, D.M. Anderson, A.W. White and D.G. Baden, eds. (Elsevier, New York 1985) pp. 113-116.
9. Y. Oshima, K. Sugino and T. Yasumoto in: Mycotoxins and Phycotoxins 1988, S. Natori, K. Hashimoto and Y. Ueno, eds. (Elsevier, New York in press).
10. T. Ikeda, S. Matsuno, S. Sato, T. Ogata, M. Kodama, Y. Fukuyo, and H. Takayama, in: Red Tides: Biology, Environmental Science and Toxicoly, T. Nemoto, eds. (Elsevier, New York 1989) pp. 411-414.
11. D.M. Anderson, J.J. Sullivan, and B. Reguera, Toxicon (in press).
12. A.D. Cembella, N.J. Antia and P.J. Harrison, Crit. Rev. Microbiol. 10, 317-339 (1984).
13. Q. Dortch, J.R. Clayton Jr., S.S. Thoresen, and S.I. Ahmed, Mar. Biol. 81,237-250 (1984).
14. J.J. Cullen, and S.G. Horrigan, Mar. Biol.(Berl.) 62, 81-87 (1985).
15. A.W. White, Toxicon 24: 605-610 (1986).

ISOZYME AND CROSS ANALYSIS OF MATING POPULATIONS IN THE ALEXANDRIUM CATENELLA/ TAMARENSE SPECIES COMPLEX

YOSHIHIKO SAKO, CHANG HOON KIM, HIDEYUKI NINOMIYA, MASAO ADACHI, AND YUZABURO ISHIDA
Laboratory of Microbiology, Department of Fisheries, Faculty of Agriculture, Kyoto University, Kyoto 606, Japan

ABSTRACT

Isozyme patterns and cross experiments of the toxic dinoflagellates *Alexandrium catenella/tamarense* species complex were examined to distinguish mating populations and also genetic variability of these organisms isolated from two different Japanese Bays. Most isolates had heterothallic sexual reproduction showing mating type + or -. All isolates from Tanabe Bay were identified as *A. catenella* according to morphological features. On the other hand, four strains from Ofunato Bay were designated *A. catenella* while other strains were identified as *A. tamarense.* All isolates were subjected to polyacrylamide gel electrophoresis and banding patterns were examined for malate dehydrogenase (MDH), malic enzyme (ME), NAD(P) reductase (NAD(P)R), xanthine dehydrogenase (XDH), octanol dehydrogenase (ODH), and tetrazolium oxidase (TO). As a result, all isolates were grouped into two different populations. A high degree of similarity among Tanabe isolates was observed. Four isolates from Ofunato Bay showed a pattern similar to that of the Tanabe type. Few differences were found for some other isolates from Ofunato Bay. The results of isozyme analysis corresponded with those of the morphological features. However, a cross experiment showed that both species fused and formed zygotes with each other.

INTRODUCTION

The taxonomy of *Alexandrium* and other dinoflagellates based on morphology remains controversial and confusing (1,2). The main reason is changeability of morphological features *in vivo* and also *in vitro* with environmental conditions. Recently morphological intermediates between *A. catenella* and *A. tamarense* were found in coastal waters of the North East Pacific and then biochemical discrimination by enzyme electrophoresis was tried on this group (3). However, the isozyme patterns of *A. tamarense/catenella* species complex revealed a high degree of genetic polymorphism and were not well correlated with the morphological features.

In a previous study we clarified the heterothallic sexual reproduction in *A. catenella* (4) and examined the isozyme patterns of several dinoflagellates (5), showing that isozyme analysis was a very useful technique to distinguish microalgae at the species or subspecies level, and examine their genetic variability in clonal axenic culture.

In order to distinguish mating populations and genetic variability of the *A. catenella/tamarense* species complex, we first isolated cysts from two completely different coastal regions. In Tanabe Bay located in southern Japan where the temperature is high we found *A. catenella*. In Ofunato Bay located in northern Japan we observed both species. From both bays we isolated cells which had germinated from cysts, examined isozyme patterns and made cross experiments with biochemical and biological approaches.

MATERIALS AND METHODS

Organisms

Cysts of the genus *Alexandrium* were isolated from the sediments in Tanabe Bay (southern Japan, 1987) and Ofunato Bay (northern Japan, 1984, 1988), and after

Toxic Marine Phytoplankton
Edna Graneli et al., Editors

germination 24 clonal axenic cultures were established (Table I). All strains were isolated when germinated cells were dividing into two or more vegetative cells. Morphological taxonomy followed the criteria of Fukuyo (2) .

Cross experiments

Mating types (+,-) were determined with the reference strains of *A. catenella* TN7(-) and TN22(+) which were isolated from Tanabe Bay. All isolates were grown to a late exponential phase and mixed with each other. Fusing cells and zygote formation were observed under an inverted microscope.

Culture and harvest

All isolates were grown in SWIIm medium. Standard cultures were maintained at 20°C (Tanabe strains) or I5°C (Ofunato strains) with a 14:10 h light:dark cycle at 5,000 lux. Cells were harvested at the late exponential phase by centrifugation, washed with buffer A (20 mM Tris-HCI, lmM EDTA, 5 mM 2-mercaptoethanol, pH 7.5) and stored at -90°C.

TABLE I. Strains of *Alexandrium.*

Strains	Abbreviation	Mating type
A. catenella TN4 (Tanabe Bay, 1987)	TNX4	+
A. catenella TN7 (Tanabe Bay, 1987)	TNY7	-
A. catenella TN8 (Tanabe Bay, 1987)	TNY8	-
A. catenella TN11 (Tanabe Bay, 1987)	TNY11	-
A. catenella TN12 (Tanabe Bay, 1987)	TNX12	+
A. catenella TN22 (Tanabe Bay, 1987)	TNX22	+
A. tamarense OF031 (Ofunato Bay, 1988)	OFY031	-
A. tamarense OF034 (Ofunato Bay, 1988)	OFX034	+
A. tamarense OF041 (Ofunato Bay, 1988)	OFY041	-
A. tamarense OF045 (Ofunato Bay, 1988)	OFX045	+
A. tamarense OF8401 (Ofunato Bay, 1984)	OFX051	+
A. tamarense OF8404 (Ofunato Bay, 1984)	OFY054	-
A. catenella 5C4-1 (Ofunato Bay, 1988)	OFY071	-
A. catenella 5C4-2 (Ofunato Bay, 1988)	OFX072	+
A. catenella D3-1 (Ofunato Bay, 1988)	OFY101	-
A. catenella D3-2 (Ofunato Bay, 1988)	OFX102	+
A. tamarense CC1-1 (Ofunato Bay, 1984)	OFX151	+
A. tamarense CC1-2 (Ofunato Bay, 1984)	OFY152	-
A. tamarense BC6-1 (Ofunato Bay, 1984)	OFY161	-
A. tamarense BC6-2 (Ofunato Bay, 1984)	OFX162	+
A. tamarense AC2-1 (Ofunato Bay, 1984)	OFX181	+
A. tamarense AC2-4 (Ofunato Bay, 1984)	OFY184	-
A. tamarense AC4-1 (Ofunato Bay, 1984)	OFX191	+
A. tamarense AC4-3 (Ofunato Bay, 1984)	OFY193	-

Enzyme extraction and electrophoresis

Cells were suspended in buffer A and disrupted by ultrasonication at 0°C. The sonicates were centrifuged at 15,000 g for 30 min and the supernatant was applied to the gels. Electrophoresis was carried out with 7,5% polyacrylamide gels as previously described (5). Gels were stained for the following enzymes : NAD dependent malate dehydrogenase (MDH), malic enzyme (ME), NAD(P) reductase (NAD(P)R), xanthine

STRAIN NO.		TNY 7	TNX 22	OFY 101	OFX 102	OFX 151	OFY 152
MATING TYPE		−	+	−	+	+	−
TNY 7	−	⧅	Z	⧅	Z	Z	⧅
TNX 22	+	Z	⧅	Z	⧅	⧅	Z
OFY 101	−	⧅	Z	⧅	Z	⧅	⧅
OFX 102	+	Z	⧅	Z	⧅	⧅	⧅
OFX 151	+	Z	⧅	⧅	⧅	⧅	Z
OFY 152	−	⧅	Z	⧅	⧅	Z	⧅

FIG. 1. Typical cross experiments of *Alexandrium*.
Z, Zygotes were observed.
⧅, Zygotes were not observed.

dehydrogenase (XDH), octanol dehydrogenase (ODH), tetrazolium oxidase (TO). The migration distance for each band was converted into a relative Rf value by reference to a standard strain.

RESULTS AND DISCUSSION

Table I shows a list of all strains examined. All isolates (6 strains) from Tanabe Bay were identified as *A. catenella* according to morphological features. Four strains from Ofunato Bay were morphologically identified as *A. catenella* while the other strains were designated as *A. tamarense*. In this Table all isolates show the mating type, but some isolates didn't fuse with both mating types.

Sexual reproduction was heterothallic in both species. Figure 1 shows the results of typical cross experiments. Tanabe isolates fused at a high frequency among the population of *A. catenella* in Tanabe Bay and also fused with *A. catenella* from Ofunato Bay. However, zygotes were also produced between *A. catenella* (Tanabe) and *A. tamarense* (Ofunato) at a low frequency.

Each isozyme band was scored as a phenotype and zymogram band patterns were compared pair-wise by use of the similarity coefficient of Jaccard (6). Linkage dendrograms were summarized by using unweighted pair-group arithmetic averages. Based on the coincidence of isozyme patterns of all isolates, multiple molecular forms of all enzymes were identified. In all enzymes, however, there were many common bands.

The isozyme patterns of MDH and TO showed two groups. The first group is the Tanabe isolates and 4 isolates (OFY071, OFX072, OFY1O1, OFX102) from Ofunato Bay. The second group is the remaining Ofunato isolates. Isozyme patterns of both enzymes were nearly the same in all Tanabe isolates and in 4 isolates from Ofunato Bay, regardless of the large geographical separation (>1,000 km). In other Ofunato isolates, the isozyme patterns of these enzymes were also very similar. As a result, similarity values for MDH (Fig. 2) and TO were very high in each group.

Figure 3 gives a summarized dendrogram based on 6 enzymes of *Alexandrium*. All isolates could be grouped into two different populations. A high degree of similarity among Tanabe isolates was observed. On the other hand, four isolates (belonging to *A. catenella* morphologically) from Ofunato Bay showed a pattern similar to that of the Tanabe type, but were different from the other 14 isolates from Ofunato Bay. These results of isozyme analysis corresponded with those of the morphological features. Results are further supported by preliminary assays of monoclonal antibodies obtained for several isolates. However, in a cross experiment both species, identified according to

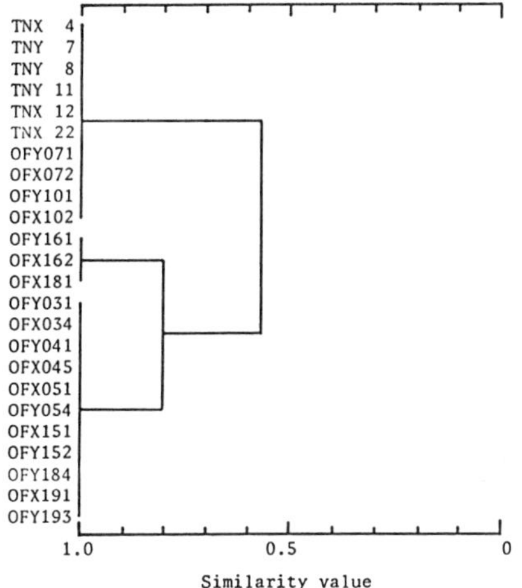

FIG. 2. Dendogram based on similarity values in the isozyme patterns of MDH.

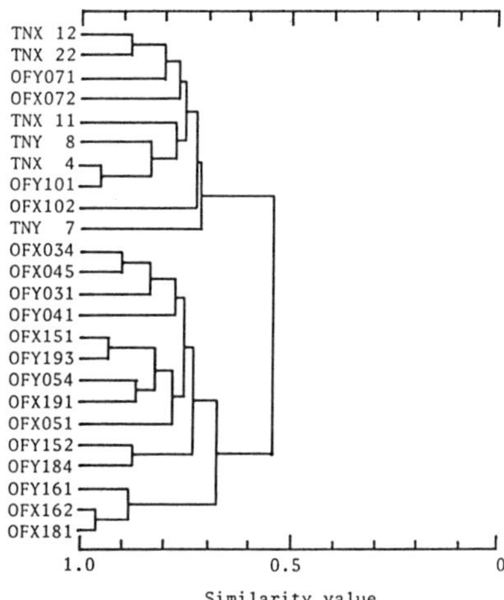

FIG. 3. Summarized dendogram based on the similarity values in six electrophoretic isozyme patterns (MDH, TO, ME, ODH, XDH, NADR).

morphological features fused with each other and formed a zygote .

In our study, a high degree of similarity was observed between each population, compared to a low similarity among the *A. tamarense /catenella* species complex in northeastern Canada (3). We probably obtained a high degree of similarity by use of axenic and clonal cultures.

ACKNOWLEDGEMENT

This work was supported in part by a Grant-in-Aid for Scientific Research (NOS. 01760152 and 62760152) from the Ministry of Education, Science and Culture of Japan, and by the Nippon Life Insurance Foundation.

REFERENCES

1. F.J,R. Taylor in: Toxic Dinoflagellates, D.M. Anderson, A.W, White and D. G. Baden, eds. (Elsevier, New York 1985) pp. 11-26.
2. Y. Fukuyo, K. Yoshida and H. Inoue in: Toxic Dinoflagellates, 0.M. Anderson, A.W. White and 0.G. Baden, eds, (Elsevier, New York 1985) pp. 27-32.
3. A.0. Cembella and F.J.R. Taylor, Biochem. System. Ecol. 14, 311-323 (1986).
4. Y. Sako and Y. Ishida in: Microbial Ecology 14 (Gakkai Schuppan Center 1986) pp. 99-113.
5. Y. Sako, A. Uchida and Y. Ishida in: Red Tides, T. Okaichi, D.M. Anderson and T. Nemoto, eds (Elsevier, New York 1989) pp. 325-328.
6. P.H.H. Sneath and R.R. Sokal in: Numerical Taxonomy, (W,H. Freeman, San Francisco 1973) pp. 635.

PHOSPHORUS DYNAMICS IN PTYCHODISCUS BREVIS: CELL PHOSPHORUS, UPTAKE AND GROWTH REQUIREMENTS

GABRIEL A. VARGO, DEBORAH HOWARD-SHAMBLOTT
Department of Marine Science, University of South Florida
140 Seventh Avenue South, St. Petersburg, Florida 33701, U.S.A.

ABSTRACT

Ptychodiscus brevis populations reach bloom proportions in the oligotrophic waters of the West Florida Shelf where inorganic phosphorus levels are 0.1 to 0.2 μg at l^{-1}. Aspects of this species P requirements were examined in semi-continuous cultures to determine growth requirements and cellular storage capacity.

Cell phosphorus decreased from 0.77 pg at $cell^{-1}$ to 0.33 pg at $cell^{-1}$ while external P decreased from 3.7 to 0.07 μg at l^{-1}. At minimum cell P, the hot water extractable fraction accounted for 40% of total cell P but increased to 85% of total cell P within 3 hrs. after the addition of 10 μM glycerophosphate. Growth rate was related to external P concentration with a calculated K_s of 0.18 μg at l^{-1}.

Uptake rates of P-deficient cells were almost two-fold greater than P-sufficient cells: 0.73 pg at P $cell^{-1}$ hr^{-1} and 0.4 pg at P $cell^{-1}$ hr^{-1} respectively. These rates yield cellular P turnover rates of 4.4 and 19 hrs. for P deficient and sufficient cells respectively. Populations at bloom levels (10^5 cells $^{-1}$) used 0.15 to 0.18 μg at P $doubling^{-1}$. With P standing stocks of 0.1 to 0.2 μg at P l^{-1} in oligotrophic waters, this species uptake rate, its half-saturation constant for growth and its long generation times allow it to maintain its growth rate in regions where the standing stock of P is low.

INTRODUCTION

Populations of the toxic, red-tide dinoflagellate, *Ptychodiscus brevis* regularly attain bloom proportions along the West Florida Shelf [1]. Although the major impact of these blooms occurs in shallow coastal waters, they are initiated 18 to 74 km offshore [1] and populations persist in oligotrophic Gulf of Mexico waters for months [2]. Transport to the east coast of Florida [3,4] via the Loop Current and more recently to North Carolina [5] through the Florida Current-Gulf Stream system, all highly oligotrophic water masses, suggests that *P. brevis* can maintain itself under low nutrient conditions.

Wilson [6] reported that 9.0 μg P l^{-1} (0.3 μM) would support growth of *P. brevis* whereas at 6.0 μg P l^{-1} (0.2μM) cells persisted but would not grow. Such phosphorus concentrations are characteristics of West Florida Shelf coastal waters [7]. There is essentially no information available on nutrient-growth interactions for this species. Therefore, we present some additional data on the phosphorus requirements of *P. brevis*, and the results of two preliminary uptake rate determinations.

METHODS

Cellular phosphorus concentration and growth-phosphorus relationships were determined using populations (Wilson's clone obtained from Dr. Karen Steidinger, FMRI) maintained in semi-continuous cultures at μE m^{-2} s^{-1} (12 hr. photoperiod) from combination of natural light and warm white/daylight fluorescent tubes at a temperature of approximately 25° C. The medium used was a modification of B-5 with 2X EDTA concentrations. The final medium composition is equivalent to f/10 without silicate and phosphate. Cultures were diluted at a rate of 16% to 33% per week depending upon the growth rate. Cell counts were done at 4X and 10X in a Palmer-Maloney counting

Toxic Marine Phytoplankton
Edna Graneli et al., Editors

chamber, soluble reactive phosphorus by the method of Strickland and Parsons [8] modified for 10 ml samples, total particulate phosphorus by the method of Solorzano and Sharp [9], hot-water extractable phosphorus according to Fitzgerald and Nelson [10] and perchloric acid extractable (PA) phosphorus after Rivkin and Swift [11]. The values presented for all phosphorus samples are the mean of replicate determinations.

Growth as a function of phosphorus concentration was calculated as doublings day^{-1} from changes in cell concentration over selected time periods in culture No. 1. SRP is the average phosphate concentration measured for the same time span. The K_s value was calculated as the X-intercept of a linear regression of S/K vs. S. The phosphorus required for a population doubling and the quantity of cells produced per unit phosphorus were calculated from the same culture over the same time periods.

Changes in cellular phosphorus fractions were measured in semi-continuous culture No. 2 after cells showed characteristics of P-deficiency (decreasing total cell phosphorus and alkaline phosphatase activity).

Uptake rates of P-deficient populations were calculated from data obtained during earlier measurements [12]. P additions were made to replicate flasks with SRP measured at the beginning and end of the experiment. Uptake rates of P-sufficient populations were measured on populations recently transferred to P-replete modified B-5 medium. SRP was measured in replicate subsamples taken at 5 to 15 min. intervals over a 2-hr. time course. However, uptake was calculated from the difference between the initial and the 2-hr. concentration.

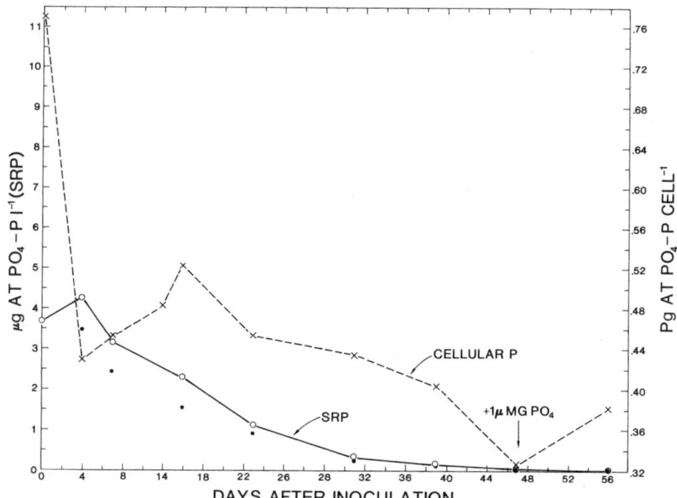

FIG. 1. The time-course changes in cellular phosphorus (dashed line) and soluble reactive phosphorus (SRP, solid line) in semi-continuous culture 1. Cell P in pg at P cell^{-1}, SRP in μg at PO_4-P^{-1}. The solid circles indicate SRP concentration after dilution.

RESULTS

Cellular phosphorus decreased rapidly within 4 days after addition of P-free medium (Fig. 1) and may represent cell release due to a 5‰ difference in salinity between the parent culture and the first dilution of the semi-continuous culture. The increase in SRP over the same time period is greater than can be accounted for by cellular release. Cell-P increased through day 14 to 67% of the initial concentration followed by a steady decline to the minimum cell concentration of 0.32 pg at cell^{-1} on day 47 (Fig. 1). SRP levels were reduced below 0.5μM after day 31. The addition of 1 μM glycerophosphate just prior to sampling on day 47 yielded a slight, but significant, increase in cell-P. Six days after this addition cell-P was 0.92 pg at cell^{-1}, however the populations had declined from 1200 cells ml^{-1} to 430 cell ml^{-1}. Many burst cells were seen and bacterial populations had increased. Therefore the cell-P data for this date were not included in the tabular data.

Table 1: Representative cellular particulate phosphorus fractions for *Ptychodiscus brevis* in semicontinuous culture No. 2 started 13 Jan. 1987. Glucose-6-phosphate, at 10μ M final concentration, was added just prior to the sample taken on day 22. PA is cold, perchloric acid extraction.

		pg at P cell^{-1}				
Date	Days after inoculation	Total (1) particulate	Hotwater (2) soluble	PA insoluble	PA soluble	% 2/1
1/13	0	0.31	--	--	--	--
1/30	17	0.32	0.036	--	--	11.5
2/3	21	0.27	0.12	--	--	44.2
2/4	22[a]	0.36	0.15	0.32	0.042	42.9
	22[b]	0.49	0.42	0.35	0.14	85.7
2/6	24	0.67	0.60	0.44	0.16	90.3

a) sampled immediately after the addition of 10 μg at ℓ$^{-1}$ G-6-P
b) 3 1/2 hrs. after the G-6-P addition

The population used in culture No. 2 had minimum cellular phosphorus concentrations prior to the addition of 10 μM glucose-6-phosphate (Table 1). Total cell-P increased by a factor of 1.4 within 2 hrs. after the addition and 1.8 times after 2 days. Both hot-water extractable and ice cold perchloric acid extractable phosphorus increased 3-fold within 2 hrs. and by a factor of 4 after 2 days; the hot-water soluble fraction accounted for 85-90% of the total cellular phosphate concentration whereas the perchloric acid soluble fraction accounted for 12% and 24% of the total before and 2 days after the addition, respectively (Table 1).

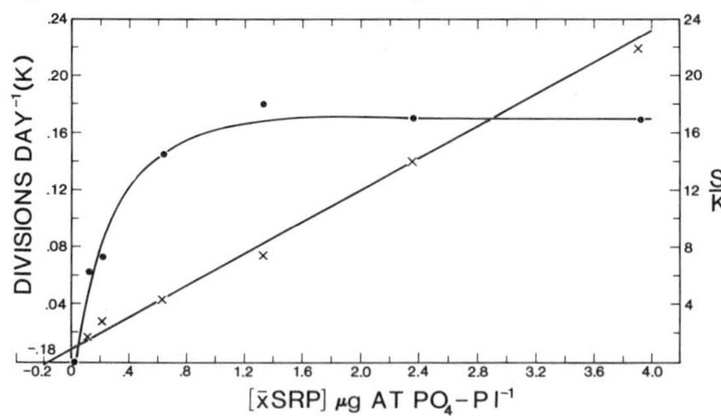

FIG. 2. Growth rate (divisions day^{-1}) as a function of the average SRP concentration in the medium during the time period used to calculate growth rates. The hyperbolic curve was fitted by eye. The S/K vs. S line was drawn to points calculated from a liner regression of the data and has an $r^2 = 0.994$.

Growth in culture No. 1 was related to the external SRP concentration (Fig. 2). The regression of S/K vs. S had a correlation coefficient (r^2) of 0.994 and the X-intercept yielded a K_s value of 0.18 μg at $P\text{-}PO_4\ l^{-1}$.

Estimates of the amount of phosphorus required for a population doubling and the number of cells supported per unit phosphorus were calculated from cell concentrations and the change in phosphate levels in culture No. 1. The culture was not diluted over the time intervals used for these estimates. Growth rates were low, with generation times that varied from 6 to 10 days (Table 2). Phosphorus requirements for a population doubling were similar over two time periods however the number of cells supported per unit P varied by more than a factor of 4. Calculations for days 39 to 47 indicate that populations were using P most efficiently. This time period coincided with the reduction of cellular P levels to their minimum value (see Fig. 1).

Table 2: Phosphorus-growth requirements for P. brevis populations in semicontinuous culture No. 1 started 17 Oct. 1986 and run through 23 Dec., 1986.

Days after inoculation	Δt (days)	k doublings d^{-1}	Δ PO_4 μg at d^{-1}	μg at P $doubling^{-1}$	cells produced per μg at P ($X10^{-6}$)
7 - 14	7	0.169	0.030	0.18	4.51
21 - 31	10	.145	0.087	0.60	1.70
39 - 47	8	.098	0.015	0.15	9.08

Phosphate uptake rates were 2 to 4- fold greater in P-deficient cells than P-sufficient cells (Table 3). A 2- fold increase in the uptake rate of P-deficient cells also occurred with a doubling of added phosphorus. The values listed in Table 3 must be considered conservative estimates since phosphate concentrations were only measured at the beginning and end of the experiment. Preliminary time course studies of uptake using P-sufficient populations showed undetectable changes in P conentration over the first 1 to 1 1/21 hrs. of sampling.

Table 3: Phosphorus uptake rates in P-deficient and P-sufficient P. brevis cultures. Rates are based on replicate flasks with phosphate determined at the beginning and end of the experiment.

P-status	P-additonal μg at ℓ^{-1}	ΔP	pg at P $cell^{-1}$ hr^{-1}
Deficient	+ 5	4.82	0.074
		4.72	0.072
	+10	7.1	0.190
		7.05	0.190
Sufficient	+ 5	0.12	0.043
		0.13	0.046

DISCUSSION

Persistence of dinoflagellate blooms, especially blooms which occur in oligotrophic waters such as the coastal Gulf of Mexico [12], should be related to the ability of the species to take up, store, and efficiently utilize nutrients for growth and maintenance. Total cellular phosphorus concentrations for P. brevis are in the mid-range of reports for other species [e.g., 13,14]. Therefore, given its' cell size, P. brevis does not possess unusual cellular phosphorus levels that might account for the ability of offshore populations to maintain themselves for extended periods in oligotrophic conditions.

Rapid uptake of phosphorus by P-depleted cells is common to many species. Storage pools have been characterized by a variety of extraction methods: Hot-water soluble [10], cold TCA extractable fraction [15,16]; and cold Perchloric Acid [17]. When P-depleted, P. brevis cells had the majority of total cellular phosphorus in the PA insoluble pool which is similar to that reported by Elgavish et al. [15] for several species from Lake Kinneret. However, after the addition of an organic phosphorus source, P. brevis rapidly increased both the hot-water soluble and PA soluble pools with the hot-water soluble pool containing a greater percentage of the total (85% compared to 24%, respectively). Although there is a question about the role of the hot-water soluble fraction as an indicator of nutritional status [16] and as an effective intracellular storage pool [15], P. brevis apparently has the ability to partition phosphorus, when available in excess, into two potential storage pools. The relationship of these two pools to growth for this species remains to be determined.

The K_s value of 0.18 μg at P 1^{-1} for growth is similar to that reported by Rivkin and Swift [14] for Pyrocystis (0.15) and for Ceratium furca (0.15) by Qasim et al. [18]. Given the low inorganic P concentrations in coastal waters P. brevis appears to be adapted for growth in this region. It is of interest to note that the K_s value is equivalent to about 5.6 μg P 1^{-1}; a value which is close to the 6 μg 1^{-1} determined by Wilson [6] as the minimum concentration required to maintain populations.

Uptake rates for P. brevis are more than an order of magnitude faster than rates reported for other dinoflagellates [19] based on autoradiographic methods but 2-3 orders of magnitude lower than uptake rates for Pyrocystis noctiluca [11], a larger cell. Cellular P turnover rates of 4.4 and 19 hrs. for P deficient and sufficient cells can be calculated from uptake rates. Such turnover rates are short relative to the generation time for cells in this experiment and when combined with the time required to fill intracellular storage pools, suggests that this species can rapidly acquire P.

Cell yield is one measure of the efficiency with which P. brevis uses available phosphorus. Although the values reported in Table 2 are variable, they are similar (within a factor of 2-5) to calculations reported by Wilson [6] for natural blooms of P. brevis. With P standing stocks of 0.1 μM P in nearshore waters, population levels of 10^5 cells per liter could develop without P limitation. Since the outwelling of P from Tampa Bay is sufficient to turnover the standing stock of P in 24 hrs. within 10 km of the mouth, bloom development would not be limited by phosphorus. In offshore, oligotrophic waters, this species uptake rate, half-saturation constant for growth and its long generation times contribute toward its ability to maintain populations in waters with low phosphorus concentrations.

REFERENCES

1. K.A. Steidinger, and K. Haddad. BioScience 31, 814 (1981).
2. G.A. Vargo, K.L. Carder, W. Gregg, E. Shanley, C. Heil, K.A. Steidinger, and K. Haddad. Limnol. Oceanogr. 32, 762 (1987).
3. E.B. Murphy, K. Steidinger, B.S. Roberts, J. Williams, and J.W. Jolley, Jr. Limnol. Oceanogr. 20, 481 (1975).
4. B.S. Roberts in: Toxic Dinoflagellate Blooms, Proc. 2nd Int. Conf. (Elsevier/North Holland 1979) pp. 1990202.
5. P.A. Tester, P.K. Fowler, and J.T. Turner in: Proc. Symp. on Novel Phytoplankton Blooms: Causes and Impacts of Recurrent Brown Tides and

Other Unusual Blooms. Stony Brook, NY (in press).
6. W.B. Wilson, Fla. Bd. Conserv. Mar. Lab., Prof. Paper Ser. (7), 1 (1966).
7. K.A. Fanning, K.L. Carder, and P.R. Betzer, Deep Sea Res. 29, 953 (1982).
8. J.D.H. Strickland, and T.R. Parsons, A practical handbook of seawater analysis, 2nd ed. Bull. Fish. Res. Bd. Can. 167 (1972).
9. L. Solorzano, and J.H. Sharp, Limnol. Oceanogr. 25, 754 (1980).
10. G.P. Fitzgerald, and T.C. Nelson, J. Phycol. 2, 32 (1966).
11. R.B. Rivkin, and E. Swift, J. Phycol. 18, 113 (1982).
12. G.A. Vargo, and E. Shanley, Mar. Ecol. PSZNI 6, 251 (1985).
13. E. Sakshaug, E. Graneli, M. Elbrachter and H. Kayser, J. Exp. Mar. Biol. Ecol. 77, 241 (1984).
14. R.B. Rivkin, and E. Swift, Mar. Biol. 88, 198 (1985).
15. A. Elgavish, M. Halmann, and T. Berman, Phycologia 21, 47 (1987).
16. D. Wynn, and T. Berman, J. Phycol. 16, 40 (1980).
17. M. Watanabe, K. Kohata, and M. Kunugi, J. Phycol. 23, 54 (1987).
18. S.A. Qasim, P.M.A. Bhattathirio, and V.P. Devassy, Mar. Biol. 21, 299 (1973).
19. E.S. Friebele, D.L. Correll, and M.A. Faust, Mar. Biol. 45, 39 (1978).

ACKNOWLEDGEMENTS

This study was partially supported by NSF grant INT 84-19963. Travel to the 4th International Toxic Phytoplankton Conf., Lund, Sweden was supported by the Division of Sponsored Research, University of South Florida, International Travel Program.

VI TOXIN CHEMISTRY

TOXIN COMPOSITION OF ALTERNATIVE LIFE HISTORY STAGES OF ALEXANDRIUM, AS DETERMINED BY HIGH-PERFORMANCE LIQUID CHROMATOGRAPHY.

ALLAN D. CEMBELLA*, CHRISTOPHE DESTOMBE* and JOANNE TURGEON**
*Maurice Lamontagne Institute, Dept. of Fisheries and Oceans, 850 route de la Mer, Mont-Joli, Quebec, Canada G5H 3Z4; **Dept. of Oceanography, Universite du Québec at Rimouski, 300 des Ursulines, Rimouski, Quebec, Canada G5L 3A1.

ABSTRACT

The toxicity and toxin composition of alternative stages in the life history of Alexandrium excavatum were determined by HPLC. The toxin spectrum of natural populations of motile cells from the water column and benthic hypnozygotes was compared with that of fractions of early stationary phase cultures of cells isolated from the lower St. Lawrence estuary. In general, the hypnozygotes contained less toxin on a molar basis than motile cells of natural populations from the same region. Hypnozygotes were relatively richer in two carbamate derivatives, saxitoxin and gonyautoxin 2, than other life history stages. Early stationary phase cultures consisted of a mixture of small and large vegetative cells, putative gametes, pellicular cysts and a few planozygotes. Analysis of the toxins present in fractions enriched in these alternative stages revealed that the toxin spectrum was relatively homogeneous. However, on a per unit cell volume basis, there were significant quantitative differences among certain fractions in the total amount of toxin present.

INTRODUCTION

Approximately a dozen paralytic shellfish toxins are known to be produced among various species of the genus Alexandrium (also known as members of the Protogonyaulax tamarensis/catenella species complex), a group of toxic red-tide dinoflagellates [1-3]. High-performance liquid chromatography (HPLC), based upon the alkaline oxidation of paralytic shellfish toxin components to fluorescent derivatives, is a powerful technique for the analysis of these dinoflagellate toxins [4,5]. The HPLC method has been frequently applied to the analysis of toxins from motile haploid vegetative cells in culture [5-7], often after they have been subjected to a variety of environmental manipulations [8-11]. More rarely, the toxin spectrum has been determined directly from motile Alexandrium cells collected from natural bloom populations in the water column [12].

In marine ecosystems, the resting cysts (hypnozygotes) formed at the termination of Alexandrium blooms have been proposed as a direct vector of toxin transfer to shellfish, particularly in offshore coastal waters [13,14]. Nevertheless, there remain discrepancies in the literature regarding the relative toxicity of resting cysts and motile vegetative cells from the same region [13-15].

In the present study, the toxin composition of hypnozygotes and vegetative cells of natural populations from the lower St. Lawrence estuary (Quebec) was compared with that of various life-history stages in an early stationary phase culture. This work was undertaken to determine if hypnozygotes were similar in toxicity and toxin composition to vegetative cells in the overlying water column, and whether or not the toxin spectra among various fractions of a cultured isolate were homogeneous.

Toxic Marine Phytoplankton
Edna Graneli et al., Editors

MATERIALS AND METHODS

Net samples of natural phytoplankton assemblages dominated by Alexandrium excavatum were size-fractionated, concentrated, and enumerated according to previously published methods [12,16], during a summer bloom in the lower St. Lawrence estuary. Hypnozygote samples were collected by benthic dredge from an offshore station during early spring. These rich cyst samples (>1100 cysts cm-3 of sediment) were further concentrated by sieving and centrifugation [17]. Toxin extractions were performed directly on the cyst-rich sediment fractions, as well as on hundreds of Alexandrium cysts isolated and concentrated by micropipette.

A unialgal strain of A. excavatum (Pr 210b), isolated from the St. Lawrence estuary, was cultured in 20 l glass carboys in f/2 nutrient enrichment, under environmental conditions as previously described [16]. Cells were harvested in early stationary growth phase, as determined by following the in vivo chlorophyll a fluorescence to a maximum value [7]. A novel method of obtaining alternative life history stages from cultures, by exploiting differences in motility and phototaxis was employed to yield four operationally defined fractions enriched in vegetative cells, putative gametes or pellicular cysts [18]. These fractions were further subdivided by sieving through a Nitex screen of 20 µm nominal pore size. Nuclei of cells from each of these fractions were stained with the fluorochrome DAPI, and quantitative DNA analysis was performed by epifluorescence microphotometry [6]. The discrimination of presumptive gametes from normal vegetative cells and pellicular cysts was confirmed by subsequent cell fusion and transfer experiments [18], and by observations under phase-contrast microscopy. Cell volume measurements were obtained by micrometric determinations using the optical microscope.

Toxic cell extracts were prepared by sonication in 0.03 N acetic acid. The HPLC analysis of the toxic fractions was performed according to established methods [5,7]. The concentrations of the 21-N-sulfocarbamoyl derivatives (C_1;C_2) were determined by hot acid hydrolysis in 0.1 N HCl to their respective gonyautoxin analogues [5,7], and also by isocratic ion-pair chromatographic separation [10].

RESULTS

Hypnozygotes from field populations in early spring were substantially larger than motile vegetative cells of Alexandrium collected during the summer bloom (Fig. 1). In fact, the vegetative cells from natural blooms were comparable in volume to the small cells from culture fractions FR 1, 3, 5 and 7, which passed through the 20 µm nominal mesh-diameter sieve. The small cells in these fractions from cultures were generally distinct in size from the respective fractions (FR 2, 4, 6 and 8) retained on the sieve.

Fractions FR 1 and 2 consisted entirely of non-motile ovoid cells with weak pigmentation and without thecae. On the basis of these criteria, they were judged to represent two size classes of pellicular cysts. Fractions FR 3 and 4 were dominated by pale cells of low motility with intact thecae, which were either neutrally or negatively phototactic. In general morphology, they resembled normal vegetative cells. The cells in both fractions FR 5 and 6 were neutrally or slightly positively phototactic, however, they were readily discriminated on the basis of size and pigmentation. The small cells in FR 5 were regular in shape, but approximately half were characterized by pale pigmentation. In contrast, all of the large cells in FR 6 were deeply pigmented and many exhibited the lumpy or irregular morphology typical of planozygotes or mitotically-arrested vegetative cells. In neither of these latter fractions were any duplets or fusing cells observed.

Fractions FR 7 and 8 represented different size-fractions which were

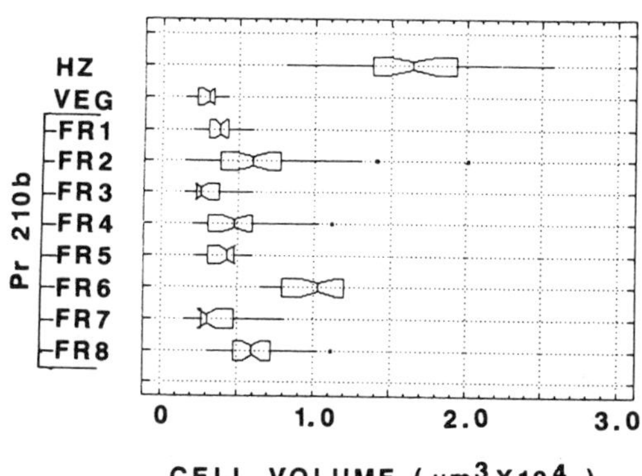

FIG. 1. Notched box-and-whisker plot of the distribution of cell volumes for a natural population of Alexandrium hypnozygotes (HZ) and motile vegetative cells (VEG), and isolate Pr 210b fractions (FR 1 to 8) in early stationary phase. The box includes the two central quartiles, with the whiskers' indicating extreme values within 1.5X the interquartile range. The notch overlap corresponds to the 95% confidence interval for the median; the width of each box is proportional to $\sqrt{n}$, where n > 30.

overwhelmingly dominated by well-pigmented, highly motile, positively phototactic individuals, with the appearance of normal vegetative cells.

When transferred to nutrient-replete medium, the small cells in FR 7 grew rapidly in size and began normal mitotic divisions. In contrast, the weakly-pigmented small cells in FR 5, failed to grow and divide under these conditions. Sexual fusion was observed between pairs of these pale cells [18], thus confirming their identity as gametes.

TABLE 1. Quantity of DNA per nucleus for Pr 210b fractions in stationary growth phase.

Fraction	DNA content (pg $cell^{-1}$) $X \pm$ s.e. (n = 30)
FR 1	37.2 ± 0.9
FR 2	31.5 ± 2.5
FR 3	25.2 ± 0.7
FR 4	27.6 ± 1.4
FR 5	29.5 ± 1.1
FR 6	27.5 ± 1.4
FR 7	23.1 ± 1.0
FR 8	59.3 ± 1.8

With two notable exceptions, there was no clear discrimination between the quantity of DNA per nucleus and the various fractions (Table 1). There was a tendency for the pellicular cyst fractions (FR 1 and 2) to have slightly more DNA than other cell types, except for the vegetative cells which dominated in FR 8. Cells in the latter fraction were clearly in G_2 phase of the mitotic cycle, having completed DNA replication, as the amount of DNA was approximately double that found in vegetative cells from other fractions.

Compared to the vegetative cells from natural populations in the region, the benthic hypnozygotes exhibited a reduced toxin spectrum, with saxitoxin (STX) as the dominant toxin, rather than the N-21-sulfocarbamoyl component, C_2. Between the two gonyautoxins present, the α-isomer, GTX_3, was characteristically favored over the β-isomer, GTX_2, in the vegetative cells, whereas the reverse was true for the hypnozygotes. On a per unit cell volume basis, the hypnozygotes contained much less toxin than the vegetative cells (Fig. 2). However, this difference was less pronounced when the toxicity per cell was compared for hypnozygotes (5.0 pg STXeq $cell^{-1}$) versus vegetative cells (21.4 pg STXeq $cell^{-1}$), given the larger cell volume (Fig. 1) and higher specific toxicity of the toxin components [7] found in hypnozygotes.

The fractions obtained from cultured isolate Pr 210b in early stationary phase were less toxic than motile vegetative cells from the field, even when cell volume differences were considered (Fig. 2). The toxin spectrum of this isolate in early stationary was unusual, in that all fractions were largely devoid of gonyautoxins. With the exception of the

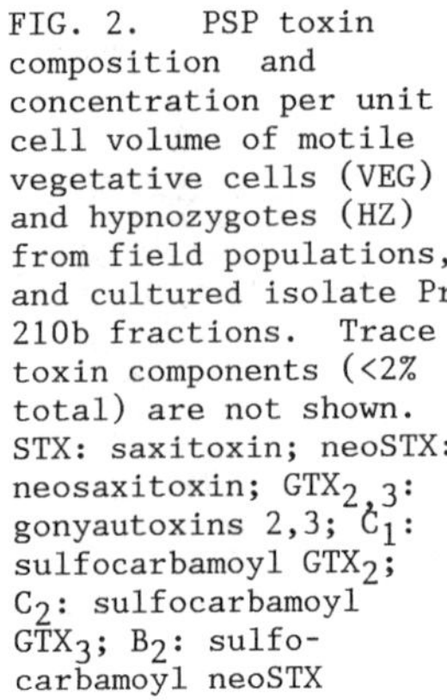

FIG. 2. PSP toxin composition and concentration per unit cell volume of motile vegetative cells (VEG) and hypnozygotes (HZ) from field populations, and cultured isolate Pr 210b fractions. Trace toxin components (<2% total) are not shown. STX: saxitoxin; neoSTX: neosaxitoxin; $GTX_{2,3}$: gonyautoxins 2,3; C_1: sulfocarbamoyl GTX_2; C_2: sulfocarbamoyl GTX_3; B_2: sulfocarbamoyl neoSTX

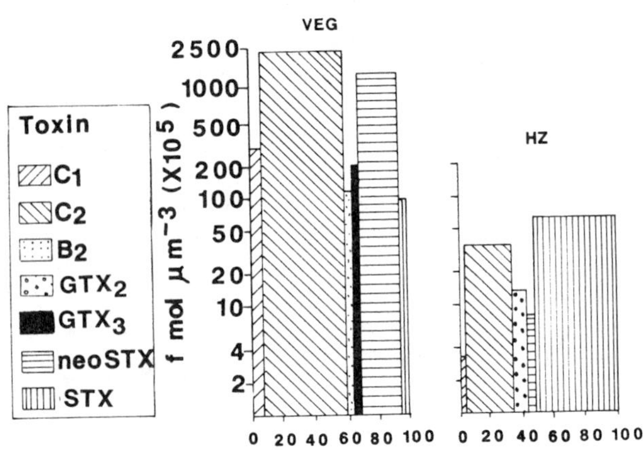

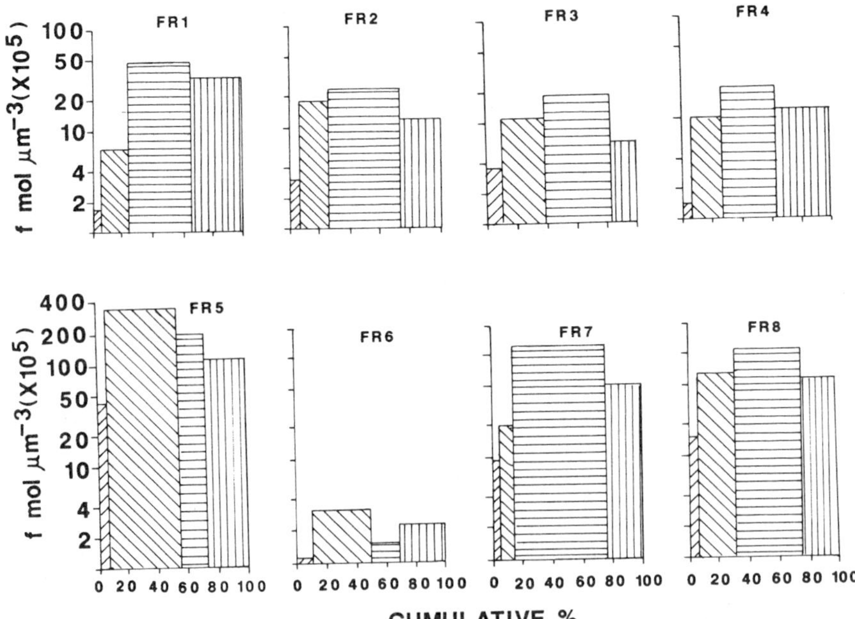

fractions FR 5 and 6, where C_2 was the dominant toxin, the toxin profile was similar among the fractions, with neoSTX as the most important component. Both the small and large vegetative cells from FR 7 and FR 8, respectively, contained more toxin per cell (and per unit cell volume) than pellicular cysts (FR 1 and FR 2) and the unhealthy-appearing weakly motile cells in FR 3 and FR 4. The small cells of FR 5 were even more toxic than the actively motile vegetative cells (FR 7 and FR 8). However, the large lumpy cells of low motility in FR 6 were much less toxic.

In general, there was no evident relationship between the total toxin concentration per unit cell volume and cell size or quantity of nuclear DNA per cell among the fractions (Table 1; Fig. 1 and 2).

DISCUSSION

Several significant conclusions may be drawn from the results of the present study. First, the toxicity and toxin profile may be substantially different among alternative life-history stages from natural populations, particularly between vegetative cells and dormant benthic hypnozygotes. Second, in early stationary phase batch cultures, a number of morphological, physiological, and life-history variants may coexist. Third, there can be both qualitative and quantitative differences in toxin content among these variants, although the toxin profile is typically rather conservative. Finally, under the present growth conditions, there are no marked differences in toxin concentration which can be related to a particular phase of the mitotic cycle, i.e. pre- or post-DNA replication. This observation is important, in view of the fact that nutrient limitation of growth has been shown to result in substantial changes in cell toxin concentration [8,11]. Under the current culture conditions, it was not definitively shown whether or not the onset of stationary growth phase was related to nutrient deficiency. However, growth limitation by the macronutrients N or P was unlikely, given the saturating levels of these components typically measured in early stationary phase batch cultures grown under identical conditions. In any case, regardless of the mechanism of growth-limitation, the induction of toxin heterogeneity among the various life-cycle stages simultaneously present in batch culture indicates a differential cellular response to the limiting factor.

Previous work on isolates and natural *Alexandrium* populations from the lower St. Lawrence estuary [12] demonstrated that the toxin profile obtained from the motile vegetative cells in culture was indeed typical of rapidly dividing cells in nature. The gonyautoxins, particularly GTX_2 and GTX_3, were commonly detected, although not usually abundant in both natural populations and isolates from the St. Lawrence region. The fact that the spectrum of toxins in cultured isolates generally reflected that found in natural blooms from this region, and that the toxin profile remained qualitatively consistent throughout the culture cycle, suggests that the absence of these toxins in isolate Pr 210b is peculiar to this isolate, and not a function of a short-term physiological response.

In previous studies, some authors [13,15,19] have claimed that resting cysts were at least an order of magnitude more toxic than motile cells. Others have considered cysts to be of approximately equal toxicity to vegetative cells [14], or even less toxic [20]. Although some discrepancies may arise through strain-specific differences and variation in cellular metabolic status, in certain studies [13-15,19] potentially misleading comparisons have been made between natural benthic cyst populations and vegetative cells derived from cultured isolates. In other studies, it was assumed that the toxicity of cultures produced from excysted cysts was typical of that of natural cyst populations [20]. It has been repeatedly established that cultured *Alexandrium* cells are typically less toxic than field samples from the same region [12,16], and therefore do not yield a representative toxicity value.

The relatively high proportions of STX and GTX_2 found in hypnozygotes from the St. Lawrence ecosystem were also observed in a similar HPLC analysis which compared hynozygotes from the Gulf of Maine with cultured vegetative cells [19]. The evidence from the toxin composition of hypnozygotes is consistent with their role as quiescent cells exhibiting low metabolic activity. Epimerization in *in vitro* toxin preparations also frequently results in a relative increase in the β-epimer, GTX_2, as was found in the hypnozygotes, whereas the α-epimer, GTX_3, typically predominates in fresh extracts from actively growing dinoflagellates [1].

Since STX ultimately serves as the precursor molecule for the synthesis of the other derivatives, it might also appear logical that the STX/neoSTX ratio should be higher in dormant cells, as was observed. Nevertheless, recent evidence from cultured isolates has actually indicated that the proportion of STX relative to neoSTX declined as the culture aged [10,11]. This elevated level of neoSTX was also seen in the vegetative cells from early stationary phase cultures in the present experiments.

In summary, it is evident that coexisting cells do not necessarily represent a homogeneous population, even when grown under apparently uniform conditions. Further studies on strain-specific toxin variations must therefore take this heterogeneity in life-history stages into account.

ACKNOWLEDGEMENTS

The authors thank J.J. Sullivan, formerly of the Dept. of Health and Human Services, FDA, Seattle, WA for reference toxins for HPLC calibration. The assistance of R. Larocque in the analysis of chromatographic data is also gratefully acknowledged.

REFERENCES

1. S. Hall and P.B. Reichardt in Seafood Toxins, ACS Symposium Series, No. 262, E. Ragelis, ed. (American Chemical Society, Washington 1984) pp. 113-123.
2. F.J.R. Taylor in Toxic Dinoflagellates, D.M. Anderson, A.W. White and D.G. Baden, eds. (Elsevier, New York 1985) pp. 11-26.
3. Y. Shimizu, in The Biology of Dinoflagellates, F.J.R. Taylor, ed., Blackwell, Oxford 1987) pp. 282-315.
4. J.J. Sullivan, J. Jonas-Davies, and L.L. Kentala in: Toxic Dinoflagellates, D.M. Anderson, A.W. White and D.G. Baden, eds., (Elsevier, New York 1985) pp. 275-280.
5. G.L. Boyer, J.J. Sullivan, R.J. Andersen, F.J.R. Taylor, P.J. Harrison, and A.D. Cembella, Mar. Biol. 93, 361-369 (1986).
6. A.D. Cembella and F.J.R. Taylor in: Toxic Dinoflagellates, D.M. Anderson, A.W. White and D.G. Baden, eds. (Elsevier, New York 1985) pp. 55-60.
7. A.D. Cembella, J.J. Sullivan, G.L. Boyer, F.J.R. Taylor, and R.J. Andersen, Biochem. System. Ecol. 15, 171-186 (1987).
8. G.L. Boyer, J.J. Sullivan, R.J. Andersen, P.J. Harrison and F.J.R. Taylor, Mar. Biol. 96, 123-128 (1987).
9. T. Ogata, T. Ishimaru and M. Kodama, Mar. Biol. 95, 217-220 (1987).
10. B.A. Boczar, M.K. Beitler, J. Liston, J.J. Sullivan and R.A. Cattolico, Plant Physiol. 88, 1285-1290 (1988).
11. D.M. Anderson, D.M. Kulis, J.J. Sullivan and S. Hall, submitted to Mar. Biol. (1989).
12. A.D. Cembella and J.-C. Therriault in Red Tides: Biology, Environmental Science and Toxicology, T. Okaichi, D.M. Anderson and T. Nemoto, eds. (Elsevier, New York 1989) pp. 81-84.
13. B. Dale, C.M. Yentsch, and J.W. Hurst, Science 201, 1223-1224 (1978).
14. A.W. White and C.M. Lewis, Can. J. Fish. Aquat. Sci. 39, 1185-1194 (1982).
15. Y. Oshima, H.T. Singh, Y. Fukuyo and T. Yasumoto, Bull. Japan. Soc. Sci. Fish. 48, 1303-1305 (1982).
16. A.D. Cembella, J.-C. Therriault, and P. Béland, J. Shellfish Res. 7, 611-621 (1988).
17. A.D. Cembella, J. Turgeon, J.-C. Therriault and P. Béland, J. Shellfish Res. 7, 597-609 (1988).
18. C. Destombe and A.D. Cembella, in prep., 1989.
19. J.W. Hurst, R. Selvin, J.J. Sullivan, C.M. Yentsch, and R.R.L. Guillard in: Toxic Dinoflagellates, D.M. Anderson, A.W. White and D.G. Baden, eds. (Elsevier, New York 1985) pp. 427-432.
20. Y. Onoue, T. Noguchi, J. Maruyama, Y. Ueda, K. Hashimoto and T. Ikeda, Bull. Japan. Soc. Sci. Fish. 47, 1347-1350 (1981).

SPECIFICITY AND CROSS-REACTIVITY OF AN ABSORPTION-INHIBITION ENZYME-LINKED IMMUNOASSAY FOR THE DETECTION OF PARALYTIC SHELLFISH TOXINS

ALLAN CEMBELLA*, YVES PARENT**, DANIELLE JONES** AND GILLES LAMOUREUX***
*Maurice Lamontagne Institute, Dept. of Fisheries and Oceans, 850 route de la Mer, Mont-Joli, Quebec, Canada G5H 3Z4; **Centre de Recherche en Sciences Appliquées en Alimentation (CRESALA) and ***Centre de Recherche en Immunologie, Institut Armand-Frappier (IAF), 531 Boulevard des Prairies, Laval, Quebec, Canada H7N 4Z9

ABSTRACT

A polyclonal anti-saxitoxin (STX) antibody was prepared by covalently coupling STX to poly-alanine lysine (Pal) and conducting a series of rabbit immunizations. In order to test the specificity of this antibody, the cross-reactivity was evaluated against a number of purified derivatives of STX, including neosaxitoxin, gonyautoxins 2 and 3, and a mixture of N-21-sulfocarbamoyl compounds. The STX-antibody possesses a high affinity for STX, while cross-reacting to various degrees with the other derivatives tested. No cross-reaction was evident with the sulfocarbamoyl derivatives, nor against the non-tetrapurine biotoxins assayed. The enzyme-immunoassay is at least four orders of magnitude more sensitive than the fluorescence-based HPLC method, when STX is the major toxin present. The relatively broad antigen specificity of this polyclonal STX-antibody demonstrates that when incorporated into an absorption-inhibition enzyme-linked immunoassay (ELISA), it may be used as a rapid diagnostic assay for paralytic shellfish toxins in contaminated shellfish.

INTRODUCTION

The toxins responsible for paralytic shellfish poisoning (PSP) comprise a group of low molecular weight water-soluble tetrapurine derivatives which are accumulated in marine fauna, particularly bivalves, through normal feeding activity. These potent neurotoxins are produced primarily by marine dinoflagellates, most notably by certain *Alexandrium* species (some of which are synonomous with members of the *Protogonyaulax tamarensis*/*catenella* species complex), as well as by *Pyrodinium bahamense* var. *compressa* and *Gymnodinium catenatum* [1].

Due to the relatively low sensitivity, lack of precision (±20%), and non-specificity of the conventional mouse bioassay [2], there is strong interest in alternative methods, especially immunological techniques, for PSP toxin monitoring. The first successful production of a polyclonal STX-antibody was achieved by Johnson et al. [3], although the method resulted in the formation of a relatively labile conjugate and poor antisera activity. In recent years, significant progress towards the development of immunological assays for the detection of PSP toxins, more specifically for STX, has been reported. Such techniques have included both radio-immunoassays (RIA) [4-6] and enzyme-linked immunosorbent (ELISA) assays [5,7]. In many cases, the cross-reactivity of these STX-antibodies towards related tetrapurine derivatives of STX (Fig. 1) was not examined. When the STX-antibody specificity was investigated, markedly less affinity for the N-1 hydroxy derivatives, such as neoSTX, was frequently observed [4,7]. Since both the causative dinoflagellates and contaminated shellfish usually contain a variety of N-1 hydroxy and sulfated derivatives [1,8-11], it is imperative that cross-reactivity be considered in immunological methods for PSP toxins. This study describes a novel combination of an absorption-inhibition and ELISA method, which displays cross-reactivity towards most of the major PSP toxins found in Atlantic shellfish.

Toxic Marine Phytoplankton
Edna Graneli et al., Editors

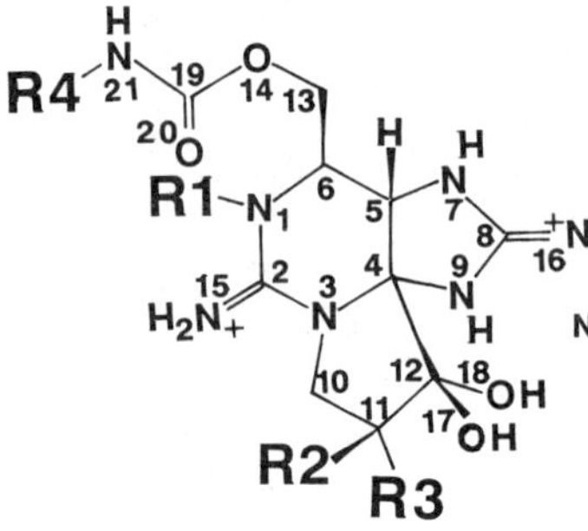

	R1	R2	R3	R4=H	$R4=SO_3^-$
SAXITOXIN GROUP	H	H	H	STX	B1
	H	H	OSO_3^-	GTX_2	C1
	H	OSO_3^-	H	GTX_3	C2
NEOSAXITOXIN GROUP	OH	H	H	NeoSTX	B2
	OH	H	OSO_3^-	GTX_1	C3
	OH	OSO_3^-	H	GTX_4	C4

FIG. 1. Structures of saxitoxin and various carbamate (R4 = H) and N-sulfocarbamoyl R4 = SO_3^-) derivatives. Saxitoxin = STX; neosaxitoxin = neoSTX; gonyautoxins 1,2,3,4 = $GTX_{1,2,3,4}$

MATERIALS AND METHODS

The STX-antigen was prepared by the covalent linkage of STX to poly-alanine-lysine (Pal), a synthetic carrier polypeptide, in the presence of glutaraldehyde. A series of immunizations into New Zealand rabbits was performed according to conventional methods [12,13]. A detailed description of the polyclonal antibody purification, and the novel features of the Institut Armand-Frappier (IAF)-ELISA technique for STX detection, has been given previously [14,15]. Briefly, the technique involves immobilized STX fixed to polystyrene batons, which is capable of binding free STX-antibody from the toxic sample-antibody incubation mixture. In the second step, the batons are incubated in horseradish-peroxidase-congugate, followed by a final development of colored reaction product in cuvettes containing the enzyme substrate solution. The optical density of the colored product is determined spectrophotometrically at 450 nm.

A number of STX analogues, specifically neoSTX, GTX_2 and GTX_3 (Fig. 1), were purified from a cultured isolate of _Alexandrium excavatum_ from the St. Lawrence estuary (Quebec), by molecular weight fractionation (10,000 M.W. cut-off), ion-exchange chromatography (Bio-Rex 70), and cellulose acetate electrophoresis (Gelman Sepraphore III). A mixture of N-21-sulfocarbamoyl (C_x = C_{1-4}) derivatives prepared from cultures of _Gymnodinium catenatum_ was also employed for these experiments. The pure toxins were diluted in 0.03 N acetic acid for analysis by reverse-phase ion-pair chromatography [9,10].

The IAF-ELISA technique was used to determine the binding affinity of various concentrations of STX and related analogues to the STX-antibody, when diluted to a concentration capable of binding 250 pg STX per ml assay (0.83 nM). Three other microbial toxins, domoic acid, okadaic acid and _Staphylococcus_ enterotoxin B, were also tested by this ELISA method.

In other experiments, the purified PSP toxins were reacted directly with the antibody in an overnight incubation, after which the unbound toxin residues were analyzed by HPLC. Similar tests were conducted with semi-purified and unpurified mixtures of toxins prepared from _Alexandrium_ cultures extracted in 0.03 N acetic acid [10], and contaminated mussels extracted according to the AOAC protocol [2]. In order to produce samples within the sensitivity range of the HPLC method (1 to 35 nM, depending upon the specific toxins present), the IgG stock solution containing the STX-antibody was concentrated by dialysis to yield a final concentration of 840 µg/ml. After incubation, the samples were ultrafiltered (Millipore Ultrafree membrane 10,000 M.W. cut-off) by centrifugation through membranes saturated with 0.2% BSA. In parallel experiments, the concentrated antibody was loaded onto affinity membranes (Nalgene activated U12), and the sample toxins were allowed to react with the antibody for several minutes. The antibody-bound toxin was determined as the difference in the HPLC toxin profile before and after incubation, after correction for non-specific retention of the toxins on the filters.

RESULTS

The STX-antibody, prepared in phosphate buffer as specified for the IAF-ELISA assay, exhibited a high affinity for purified STX at various concentrations. The spectrophotometric determination of STX-binding by this ELISA method indicated that 250 pg STX per assay were >85% adsorbed by the antibody within less than one minute incubation (Fig. 2). A similar rapid binding was shown towards the purified gonyautoxins in a short-term incubation, although the relative antibody binding affinity for GTX_3 (87%) was somewhat higher than for GTX_2 (74%) (Fig. 3). The antibody bound purified neoSTX at a level which was initially comparable to that for GTX_3. However, within a few minutes there appeared to be a progressive decrease in antigen adsorption, as revealed by the slight increase in optical density in the assay. The mixture of N-21 sulfocarbamoyl (C_x) derivatives yielded a negative binding response after five minutes, indicating virtually no antibody cross-reaction. There was no evidence of cross-reaction against the other marine phycotoxins tested, okadaic acid and domoic acid, nor against _Staphylococcus_ enterotoxin B.

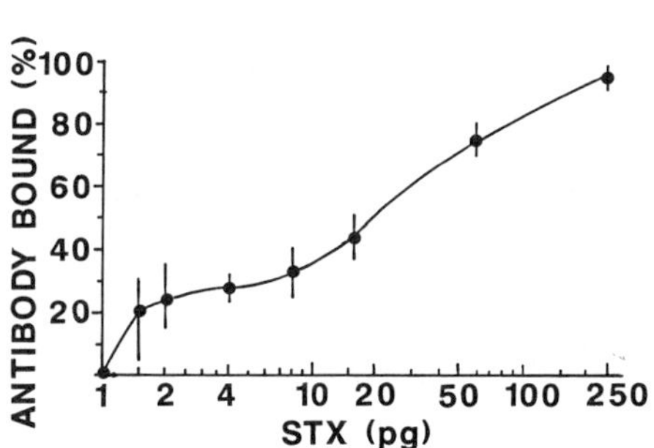

FIG. 2. Percent STX-antibody bound by various concentrations of purified STX in the IAF-ELISA assay.

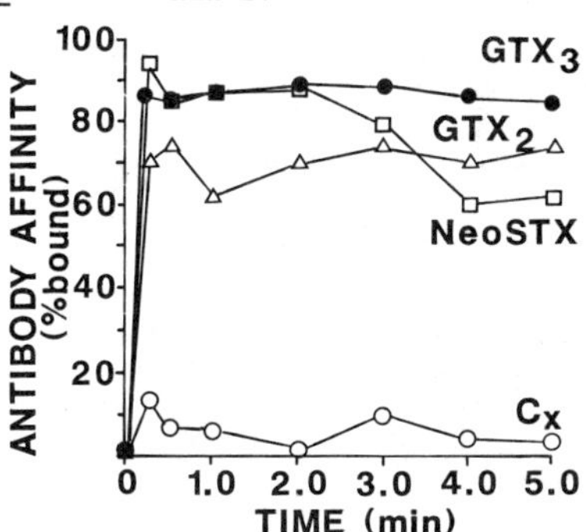

FIG. 3. STX-antibody binding to PSP toxins (250 pg) in a short time-series incubation.

The results obtained by the complexation-fractionation technique, followed by HPLC separation of the toxin components, were comparable to those achieved by the IAF-ELISA method, however, there were some significant differences. It was apparent that purified STX (10 ng) was irreversibly bound (>95%) to the concentrated antibody during an overnight incubation (Fig. 4). In both semi-purified extracts of PSP toxins from _Alexandrium_ cultures (Fig. 5) and crude mussel extracts containing PSP toxins (Fig. 6), STX appeared to be complexed with the antibody to a similar extent. However, the binding affinity for GTX_2 (50-60%) and GTX_3 (50-60%), was somewhat less than that obtained by the ELISA technique. The concentration of GTX_1 was too low for a reliable estimate of the relative binding affinity, although in most chromatograms there was a decrease in peak height after antibody treatment, which could not be explained by non-specific binding. Significantly, there was substantially less antibody affinity for neoSTX (15-20%) using the complexation-fractionation method, than was found by the IAF-ELISA technique. The former

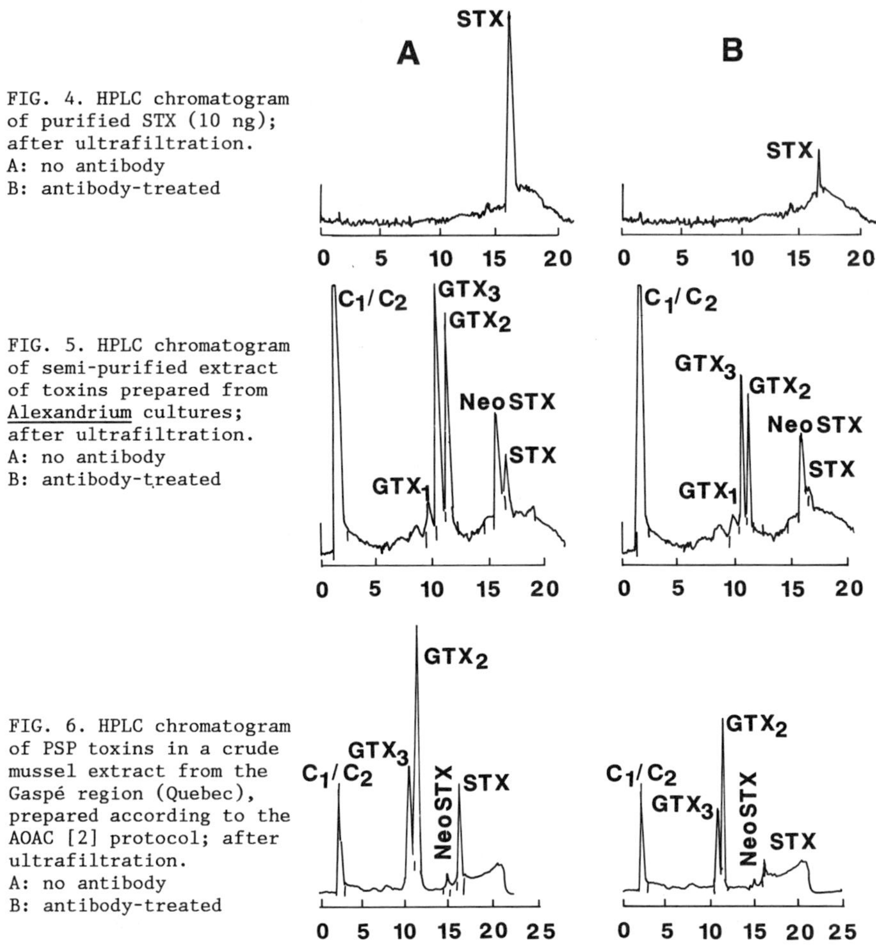

FIG. 4. HPLC chromatogram of purified STX (10 ng); after ultrafiltration.
A: no antibody
B: antibody-treated

FIG. 5. HPLC chromatogram of semi-purified extract of toxins prepared from *Alexandrium* cultures; after ultrafiltration.
A: no antibody
B: antibody-treated

FIG. 6. HPLC chromatogram of PSP toxins in a crude mussel extract from the Gaspé region (Quebec), prepared according to the AOAC [2] protocol; after ultrafiltration.
A: no antibody
B: antibody-treated

method confirmed the lack of antibody binding with two of the N-21 sulfocarbamoyl derivatives (C_1;C_2), which remained unaffected by overnight incubation with the concentrated antibody.

The length of the incubation period prior to ultrafiltration did not have a noticeable effect on binding affinity, except in the case of neoSTX, where a decrease in the concentration of this toxin in the filtrate was observed with progressively shorter term incubations, over several minutes.

The cross-reactivity, as determined using the affinity membranes, rather than ultrafiltration, was comparable for all of the toxin mixtures tested. However, when affinity membranes were used, a greater non-specific loss of toxin components to the filtrate through non-selective binding was evident.

DISCUSSION

Both the IAF-ELISA method and the complexation-fractionation-HPLC technique demonstrated the high affinity of the polyclonal antibody for purified STX. Furthermore, tests with impure toxin mixtures from dinoflagellates and molluscs confirmed that matrix-interference effects and non-specific antibody binding were, when not insignificant, at least correctable within the ELISA assay. Since this ELISA technique is based upon the optical density of the colored product resulting from a peroxidase-coupled reaction, there can be a significant departure from linearity as extreme toxin values are approached. This is the most likely explanation for the apparent high binding of the antibody with extremely low concentrations of STX (<10 pg).

The differences in cross-reactivity observed between the IAF-ELISA and the complexation-fractionation methods may be attributable to the much higher concentration of antibody employed in the latter technique. In the overnight incubation, more weakly bound toxins (neoSTX and GTX_2) would have a tendency to uncouple from the antibody, while re-establishing equilibrium. The substantially lower binding for neoSTX in impure toxin mixtures, than in pure form, also indicates that competitive interactions with other toxins may affect relative antibody affinity.

Although the nature of the antigen-antibody linkage has not been fully elucidated, the pattern of cross-reactivity suggests that binding occurs primarily through the guanidinium groups associated with the central nucleus, with both the N-1,2,3 and N-7,8,9 guanidinium rings serving as recognition sites for antibody coupling (Fig. 1). In this context, it would be informative to investigate the degree of cross-reactivity against decarbamoyl-STX, a natural degradation product of STX found in some shellfish, and against tetrodotoxin, another guanidinium neurotoxin with Na-channel activity. In any case, the differences observed in specific reactivity against particular PSP toxins indicate that steric hindrance and binding-site specificity do affect antibody affinity to some extent. The relatively high degree of cross-reactivity against GTX_2 and GTX_3 shows that the presence of the hydroxysulfate (OSO_3^-) group at the C-11 position (Fig. 1) does not unduly interfere with antibody affinity. However, the fact that more rapid and complete binding was achieved with GTX_3, than with GTX_2, indicates that differences in molecular configuration among isomers can exert an effect, with the α-isomer being slightly favored over the β-configuration for antibody binding. Since rapid antibody binding was followed by a decreasing affinity with time for neoSTX, some interference from the N-1 hydroxy group, perhaps weakening the linkage with the N-1,2,3 guanidinium ring, was evident. Although purified GTX_1 and GTX_4 were not available for these tests, it is likely that these members of the neoSTX family would react similarly.

The failure of the antibody to cross-react at more than trace levels with the N-21 sulfocarbamoyl toxin mixture suggests that the SO_3^- group at the N-21 position effectively blocks antigen attachment to the antibody (Fig. 1). Since these compounds are often the dominant toxins in PSP-producing dinoflagellates [8,10,11], this may at first seem to pose a major obstacle for the use of this immunological method as a quantitative screening test for PSP. However, these labile compounds are relatively much less abundant in contaminated shellfish (Fig. 6), due to enzymatic and pH conversion of toxins in the mollusc digestive system. Furthermore, the sulfocarbamoyl derivatives are substantially converted to their carbamate analogues [8] in the standard AOAC procedure for mouse bioassay samples [2], which involves boiling in 0.1 N hydrochloric acid.

One advantage of the IAF-ELISA procedure over the previously developed RIA method [4] is that the need for an expensive scintillation counter is eliminated. The IAF-ELISA method, with its high sensitivity (<10 pg per 1 ml assay), also exhibits a higher specificity for PSP toxins than at least one alternative technique of comparable sensitivity, the competitive displacement nerve receptor assay [16]. An additional advantage of the present immunological method is that STX-conjugation for antibody production was achieved without derivatization to saxitoxinol, as was required in a previous protocol [4]. By avoiding this derivatization, it should be possible to produce conjugates with some of the less labile STX analogues, for monoclonal antibody production.

In summary, the high sensitivity and relatively broad-spectrum cross-reactivity of the polyclonal antibody developed against STX demonstrates that it may be readily incorporated into a simple immunological procedure for the detection of several key PSP toxins.

ACKNOWLEDGEMENTS

The authors wish to thank Dr. J. Sullivan (formerly of the Public Health Service, FDA, Seattle, WA) for reference PSP toxins for HPLC calibration. Purified STX was obtained from the Public Health Service, FDA, Cincinnati, OH. A mixture of N-21-sulfocarbamoyl derivatives was graciously supplied by Dr. Y. Oshima (Tohoku University, Sendai, Japan). Catherine Poulin was primarily responsible for the production and testing of the STX-antibody serum at the Institut Armand-Frappier. This research was supported by unsolicited proposal UP-Q8-005 awarded to CRESALA-IAF by the Depts. of Fisheries and Oceans and Supply and Services (Canada).

REFERENCES

1. Y. Shimizu, in: The Biology of Dinoflagellates, F.J.R. Taylor, ed., (Blackwell, Oxford 1987) pp. 282-315.
2. Association of Official Analytical Chemists, Procedure 18.086-18.092 in: Official Methods of Analysis, 14th ed. (rev.), W. Horowitz, ed. (AOAC, Washington 1984).
3. H.M. Johnson, P. Allen-Frey, R. Angelotti, J.E. Cambell, and K.H. Lewis, Proc. Soc. Exp. Biol. Med. 117, 425-430.
4. Carlson, R.E., M.L. Lever, B.W. Lee, and P.E. Guire in: Seafood Toxins, ACS Symposium Series 262, E.P. Ragelis, ed. (Amer. Chem. Soc., Washington 1984) pp. 181-192.
5. S.R. Davio, J.F. Hewetson, and J.E. Beheler in: Toxic Dinoflagellates, D.M. Anderson, A.W. White and D.G. Baden, eds. (Elsevier, New York 1985) pp. 343-348.
6. J.W. Hurst, R. Selvin, J.J. Sullivan, C.M. Yentsch, and R.R.L. Guillard in: Toxic Dinoflagellates, D.M. Anderson, A.W. White and D.G. Baden, eds. (Elsevier, New York 1985) pp. 427-432.
7. F.S. Chu and T.S.L. Fan, J. Assoc. Off. Anal. Chem. 68, 13-16 (1985).
8. S. Hall, Ph.D. dissertation, University of Alaska, Fairbanks, Alaska, 196 pp. (1982).
9. J.J. Sullivan, J. Jonas-Davies, and L.L. Kentala in: Toxic Dinoflagellates, D.M. Anderson, A.W. White and D.G. Baden, eds., (Elsevier, New York 1985) pp. 275-280.
10. Cembella, A.D., J.J. Sullivan, G.L. Boyer, F.J.R. Taylor, and R.J. Andersen, Biochem. System. Ecol. 15, 171-186 (1987).
11. Cembella, A.D. and J.-C. Therriault in: Red Tides: Biology, Environmental Science and Toxicology, T. Okaichi, D.M. Anderson and T. Nemoto, eds. (Elsevier, New York 1989) pp. 81-84.
12. S. Avrameas, T. Ternynck, and J.L. Guesdon, Scan. J. Immunol. 8, 7-23 (1978).
13. R.N. Robbins and M.S. Bergdoll, J. Food Protection 47, 172-176 (1984).
14. C. Poulin, M.Sc. thesis, Inst. Armand-Frappier, Laval, Quebec (1988).
15. C. Poulin, R. Charbonneau and G. Lamoureux, 54th ACFAS Conference, May, (1986).
16. S.R. Davio and P.A. Fontelo, Anal. Biochem. 141, 199-204 (1984).

IDENTIFICATION OF THE CAUSATIVE ORGANISM OF A DSP-OUTBREAK ON THE SWEDISH WEST COAST

LARS EDLER* AND MATTS HAGELTORN**
*Department of Marine Ecology, University of Lund, Box 124, S-221 00 Lund, Sweden,
**Department of Food Hygiene, Faculty of Veterinary Medicine, Swedish University of Agricultural Sciences, Box 7009, S-750 07 Uppsala, Sweden

ABSTRACT

Diarrhetic Shellfish Poisoning is mostly connected with species of the genera *Dinophysis*. In Sweden, however, it has not been possible to correlate the presence and concentrations of *Dinophysis* species in the water before and during accumulation of DSP toxins in *Mytilus edulis*. In order to find the phytoplankton species responsible for the toxin production, the development of the okadaic acid concentrations during the annual DSP outbreak in September-October was studied. Concentrations of okadaic acid in the water reached a maximum of 125 ng L^{-1}, whereas mussels contained 0.4-17 µg okadaic acid g^{-1} hepatopancreas. Okadaic acid was mainly detected in the phytoplankton fraction 50-100 µm. Regression analysis of the okadaic acid concentration and the cell numbers of phytoplankton species suggests that *Dinophysis acuta* was responsible for the DSP outbreak on the Swedish west coast in the autumn 1987.

INTRODUCTION

Diarrhetic Shellfish Poisoning (DSP) has only been recognized since 1978 [1]. The likely overlook of the toxin production and its effects is probably explained by the similarity of the symptoms with gastroenteritis associated with the consumption of polluted shellfish [2]. The toxins responsible for DSP include okadaic acid and derivatives [3]. *Dinophysis fortii* and *Prorocentrum* spp. have shown to be the source of DSP in Japan [3, 4], whereas *D. acuminata, D. acuta, D. mitra, D. norvegica, D. sacculus* and *Prorocentrum lima* are considered to be the DSP toxin producing algal species in Europe [5, 6].

On the Swedish west coast DSP-outbreaks occur regularly since at least 1984 [7], but attempts to correlate the occurrence and concentration of *Dinophysis* species with the toxicity of the blue mussel *Mytilus edulis* have in general been unsuccessful. In order to identify the causative organism(s) for the DSP production in Sweden we studied phytoplankton concentrations and concentration of okadaic acid in phytoplankton and *Mytilus edulis* during September-October 1987.

MATERIALS AND METHODS

Cages with approximately 40 specimens of *Mytilus edulis* were kept at 1 and 10 m depth in the Gullmar fjord on the west coast of Sweden. Fifteen times during September-October two mussels from each depth were sampled and kept deep frozen until analysis of the okadaic acid content.

At the same time as the sampling of mussels 10 L of water was collected at 1 and 10 m depth. The water (3 - 5 L) was passed through a series of sieves retaining particles of 100, 50 and 15 µm respectively. The retained plankton were rinsed from the sieves with prefiltered sea water and the volume was made up to 50-150 ml. A subsample of 10 ml for

Toxic Marine Phytoplankton
Edna Graneli et al., Editors

phytoplankton enumeration was withdrawn, preserved with Lugol's solution and analyzed using the Utermöhl method. The 40 - 140 ml sample of concentrated phytoplankton was filtered through glassfiber filters (GF/C) and frozen until toxin analysis.

One digestive gland (hp) of known weight of a mussel was homogenized in aq. 80% MeOH in the ratio corresponding to 1 g/4 ml of glands to methanol solution. Following steps of the extraction procedure were performed according to Lee *et al.* [8]. Samples were tested in duplicate.

Filters with phytoplankton were placed in a 3 ml glass tube and 1.5 ml of 80% MeOH was added. The sample was sonicated and then extracted once with 1.5 ml of petroliumether. The lower phase was transferred to a 10 ml test tube and 0.6 ml of H_2O followed by 2.9 ml of chloroform was added. The mixture was shaken for 20 seconds on a Vortex and centrifuged. The lower phase was transferred to another test tube and the remaining phase was re-extracted with 2.9 ml of chloroform. The combined chloroform extract was finally evaporated.

An aliquot of 0.5 ml of the chloroform extract was dried under N_2 and the residue was esterified in 100 µl of 0.1% ADAM (9-Anthryldiazomethane) solution, synthesized according to Nakaya *et al.* [9] for one hour in dark at 25°C. After evaporation, the reaction product was transferred to a 100 µg BondElut® silica cartridge with 0.5 ml of hexane-chloroform (1:1) in two portions. The cartridge column was washed with 1 ml of the same solvent and then with 1 ml of chloroform. The esters of okadaic acid were eluted with 2 ml of chloroform - methanol (95:5) and evaporated under reduced pressure. The residue was dissolved in 0.1 ml of methanol and 10 µl of the solution was injected into the chromatograph. Algal extracts were derivatized and cleaned-up in the same way as the mussel extracts.

A SP 8700® chromatograph (Spectra-Physics) equipped with a LS-2 Filter Fluorimeter® (Perkin Elmer) and a pen recorder were used. The excitation and emission wavelengths were preset at 365 and 412 nm, respectively. Analyses were carried out at room temperature on 2 Chromsphe C 18® columns (100x3 mm, Chrompack) using MeCN-MeOH-H_2O (8:1:1) as the mobile phase.

RESULTS

Okadaic acid in low amounts was present in mussels from 1 m depth from the beginning of the study (Fig. 1). During the first part of September values ranged between 0.4 and 0.8 µg okadaic acid g^{-1} hepatopancreas. After September 17, okadaic acid increased almost exponentially to a concentration of 17 µg okadaic acid g^{-1} hepatopancreas on October 22 which was the last day of sampling.

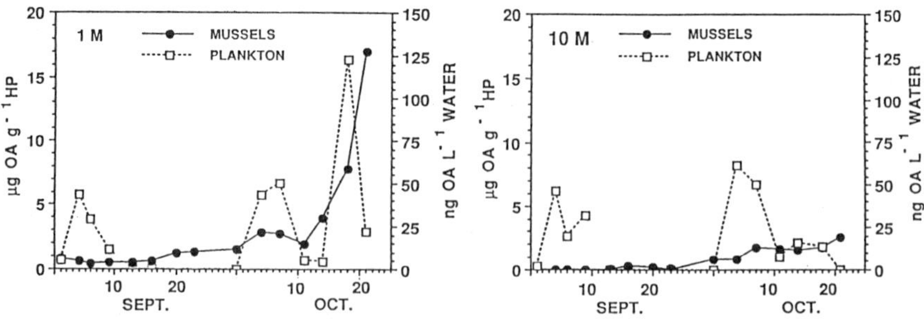

FIG. 1. Concentration of okadaic acid (OA) in mussels and phytoplankton at 1 m depth.

FIG. 2. Concentration of okadaic acid (OA) in mussels and phytoplankton at 10 m depth.

Mussels kept at 10 m depth were toxin free until September 17 (Fig. 2). From then on the okadaic acid concentration increased from 0.3 to 2.5 µg okadaic acid g^{-1} hepatopancreas.

The concentrations of okadaic acid, measured as the sum found in each fraction of phytoplankton showed a different development. At both 1 and 10 m depth there was an increase already in the beginning of September (Fig. 1 and 2). A maximum of 45-50 ng okadaic acid L^{-1} was measured on September 5. In the beginning of October another peak with concentrations in the range of 50-60 ng okadaic acid L^{-1} occurred. The values were slightly higher at 10 compared to 1 m depth. Towards the end of the study okadaic acid concentrations at 10 m depth decreased, whereas at 1 m depth an isolated peak of 122 ng okadaic acid L^{-1} was found on October 19.

The phytoplankton community during the time of study was dominated by dinoflagellates. Among other species only *Skeletonema costatum* and *Dictyocha speculum* occurred in high concentrations, but only during a short period in late September - early October. Among the dinoflagellates, *Ceratium furca* and *C. lineatum* showed highest concentrations, with maxima of 311 000 and 125 000 cells L^{-1}, respectively (Fig. 3). Their peaks, however, appeared only after the first okadaic acid peak in early September. *Prorocentrum minimum* was present in concentrations of 97 000 cells L^{-1} on October 1, i.e. just prior to the second increase of okadaic acid, but showed no other accordance with the okadaic acid concentrations of phytoplankton. *Prorocentrum micans* was present in high numbers both at the early September okadaic acid peak and the late October peak (Fig. 4). In September, however, its concentrations were still high when the okadaic acid in plankton had disappeared. *Dinophysis norvegica* (Fig. 4) developed its first peak shortly after, and the second peak just before, the increases of okadaic acid in phytoplankton.

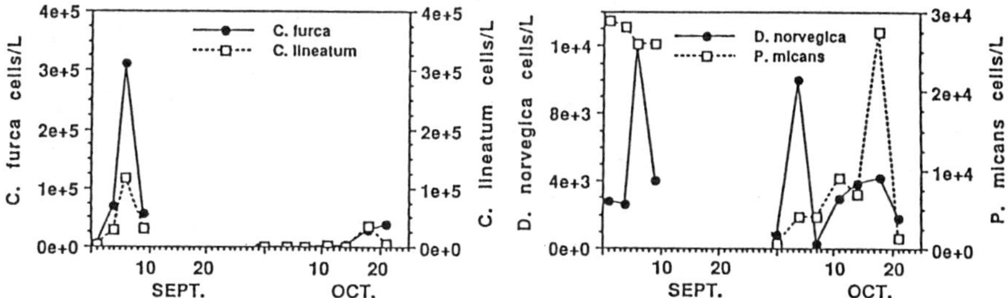

FIG. 3. Concentrations of *Ceratium furca* and *C. lineatum* at 1 m depth.

FIG. 4. Concentrations of *Dinophysis norvegica* and *Prorocentrum micans* at 1 m depth.

The abundance of two species, *Dinophysis acuminata* and *D. acuta* agreed well with the okadaic acid concentration of plankton with peak values of cell densities concomitant with the okadaic acid peaks (Fig 5).

In order to find the best correlation between cell concentrations of the different species and the okadaic acid content of the phytoplankton, the data from all fractions were used for regression analysis. The results, shown in Table I, reveal that, among the twelve dominating dinoflagellates, only two have a correlation coefficient (r) of more than 0.50; *Dinophysis acuminata* and *D. acuta* with 0.59 and 0.93, respectively.

On all occasions, except the first two sampling days, okadaic acid dominated in the

50-100 μm fraction (Fig 6), making up more than 65% of the total okadaic acid. In most cases more than 90% of the *D. acuta* cells appeared in this fraction, whereas *D. acuminata* dominated the 15-50 μm fraction which usually contained small amounts of okadaic acid.

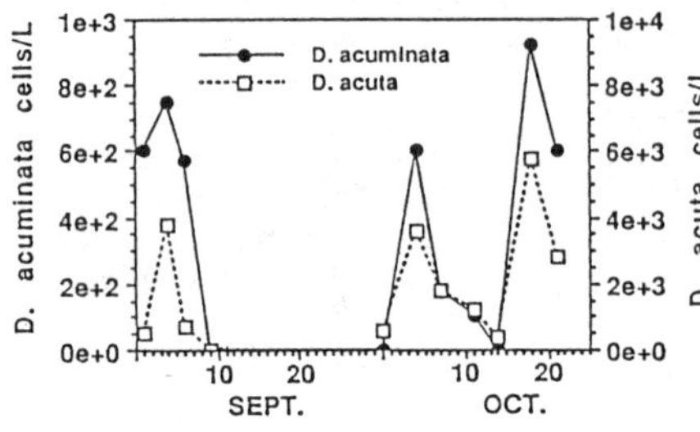

FIG. 5. Concentrations of *Dinophysis acuminata* and *D. acuta* at 1 m depth.

TABLE I. Correlation coefficients for the dominating dinoflagellates and the okadaic concentrations in phytoplankton.

SPECIES	r
Ceratium furca	0.17
C. fusus	0.18
C. lineatum	0.40
C. longipes	0.39
C. tripos	0.39
Dinophysis acuminata	0.59
D. acuta	0.93
D. dens	0.12
D. norvegica	0.46
D. rotundata	0.37
Prorocentrum micans	0.35
P. minimum	- 0.12

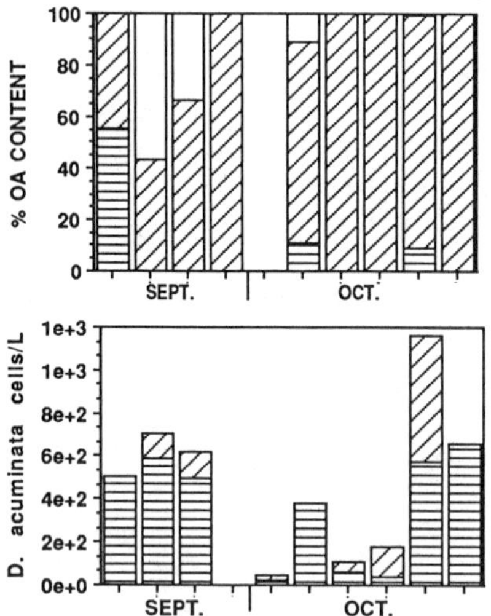

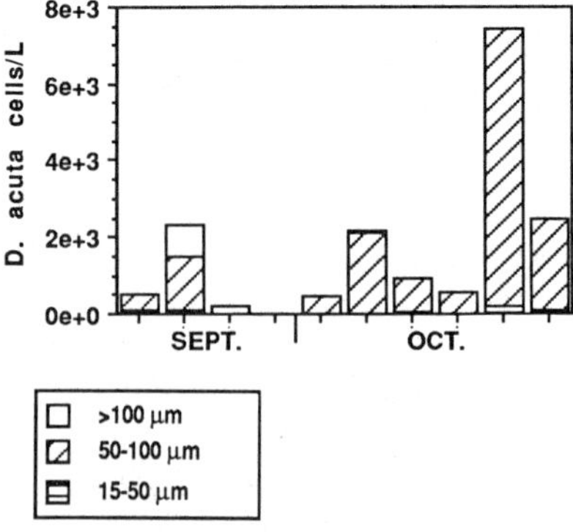

FIG. 6. Relative okadaic acid content and cell concentrations of *D. acuminata* and *D. acuta* in the different fractions.

DISCUSSION

Although all *Dinophysis* species observed during this study, except *D. dens*, have shown to produce okadaic acid [5,6], the close association between *Dinophysis acuta* and the okadaic acid concentration in the phytoplankton suggests this species to be the DSP transmitter in the Gullmar fjord in 1987. The time of appearance of *D. acuminata* correlated well with the okadaic acid in the phytoplankton, but most of the cells were found in the 15 - 50 µm fraction which showed very low or no okadaic acid concentration. In the case of *Dinophysis norvegica* the time of appearance did not correlate with the okadaic acid. Moreover, the cells of *D. norvegica* were found in about equal amounts in the 15-50 and 50-100 µm fractions We interpret this as *Dinophysis acuminata* and *D. norvegica* were not involved in the production of okadaic acid. It can, however, not be ruled out that also these species may be DSP transmitters along the Swedish west coast, as they have shown to produce DSP toxins in adjacent waters [2].

If our assumption that *Dinophysis acuta* was the only producer of okadaic acid is correct, it is possible to give a rough estimate of the okadaic acid concentration in each cell. Such a calculation results in a concentration of about 20 pg okadaic acid cell^{-1}. This means that it takes approximately 200 000 cells of *D. acuta* to reach 4 µg okadaic acid which is equal to one Mouse Unit. A similar estimate made on *Dinophysis fortii* [4], showed that it took about 13 000 cells to reach one Mouse Unit. Whether cell concentrations of okadaic acid are species specific and/or variable due to physiological stage, cell age or environmental factors remain to be studied.

ACKNOWLEDGEMENTS

This study was supported by funding from the Swedish Council for Forestry and Agricultural Research (SJFR), contracts 877472 and 823-85 L 30, and Ivar and Elsa Sandbergs Research Foundation.

REFERENCES

1 T. Yasumoto, Bull. Jap. Soc. Sci. Fish. 44, 1249-1255 (1978).
2. S. Shumway, ICES C.M. E:25, Mariculture Committee. pp. 52 (1989).
3. T. Yasumoto, M. Murata, Y. Oshima, M. Sano, G.K. Matsumoto and J. Clardy. Tetrahedron 41, 6, 1019-1025 (1985).
4. T. Yasumoto, Y. Oshima, W. Sugawara, Y. Fukuyo, H. Oguri, T. Igarashi and N. Fujita. Bull. Jap. Soc. Sci. Fish. 46, 1405-1411 (1980).
5. J.S. Lee, M. Murata and T. Yasumoto in: Mycotoxins and Phycotoxins '88, S. Natori, K. Hashimoto, Y. Ueno, eds. (Elsevier, Amsterdam 1989) pp. 327-334
6. P. Alvito, I. Sousa, S. Franca and M.A. De M. Sampayo. This volume..
7. M. Hageltorn in: Mycotoxins and Phycotoxins '88, S. Natori, K. Hashimoto, Y. Ueno, eds. (Elsevier, Amsterdam 1989) pp. 289-294
8. J.S. Lee , T. Yanagi, R. Kenma and T. Yasumoto T. Agric. Biol. Chem. 51(3), 877-881. (1987).
9. T. Nakaya, T.,Tomomoto and M. Imoto , Bull. Chem. Soc. Jap.. 40, 691-692 (1967).

IDENTIFICATION OF PARALYTIC SHELLFISH TOXINS IN MACKEREL FROM SOUTHWEST BAY OF FUNDY, CANADA.

K. HAYA, J. L. MARTIN, B. A. WAIWOOD, L. E. BURRIDGE, J. M. HUNGERFORD* and V. ZITKO
Department of Fisheries and Oceans, Biological Station, St. Andrews, N. B., E0G 2X0, Canada.
*Department of Health and Human Services, Food and Drug Administration, 22201 23rd Drive S.E., Bothell, WA 98021-4421, U. S. A.

ABSTRACT

During July 1988 a small bloom of *Alexandrium fundyense* occurred in southwest Bay of Fundy, New Brunswick, Canada. The highest concentration observed in a surface water sample was 7.5 x 10^3 cells/L on July 12. From July to the end of September, Atlantic mackerel (*Scomber scombrus*) were sampled from 9 locations. Concentrations of PSP toxins in liver extracts were measured by mouse bioassay and ranged from 40 to 209 µg saxitoxin (STX) equiv/100 g wet wt. PSP toxins (2-26 µg STX equiv/100 g wet wt.) were detected by HPLC analysis in 5 of 6 gastro-intestinal tract extracts. In all but one sample, saxitoxin was greater than 97% of the total PSP toxins content in liver. One liver and one GI tract extract contained only neosaxitoxin. B2, GTX II and GTX III were also detected in some of the extracts from both tissues. The data suggest that mackerel do not accumulate PSP toxins exclusively from *A. fundyense*.

INTRODUCTION

Blooms of marine algae can kill fish by production of toxins, by generation of anoxic water [1], and by disruption of gill function and osmoregulation through induction of ultrastructural changes in gill lamella [2] and physical mechanisms, for example, increased viscosity of seawater [2] . The first report of paralytic shellfish (PSP) toxins causing a fish kill was that for sand lance, *Ammodytes* sp., during a bloom of *Alexandrium fundyense* (= *Gonyaulax excavata* = *Protogonyaulax tamarensis*) [4]. The relationship between fish kills, PSP toxins and the marine food web have been reviewed [5]. Recently, the poisoning of bluenose dolphins on the Eastern seaboard of U.S.A. has been associated with blooms of *Ptychodiscus brevis* [6]. The deaths of humpback whales off Cape Cod, U.S.A. may have been caused by feeding on mackerel contaminated with PSP toxins [7].

During a bloom of *A. fundyense* fish could absorb a toxic dose of PSP toxins through the gills and by feeding on toxic cells and other organisms which have accumulated PSP toxins. A number of herbivorous zooplankters can accumulate PSP toxins and are probably the major biological vector through which PSP toxins reach pelagic fish [5] and larval fish [8]. In the present study, PSP toxins in mackerel gastro-intestinal (GI) tract and liver are identified, and the profiles of the PSP toxins suggest mackerel accumulate them by feeding on organisms other than *A. fundyense*.

Published 1990 by Elsevier Science Publishing Co., Inc.
Toxic Marine Phytoplankton
Edna Graneli et al., Editors

MATERIALS AND METHODS

Atlantic mackerel (*Scomber scombrus*) were sampled from herring weirs at 9 locations in southwest Bay of Fundy (Fig. 1). Based on their weight and length (211 ± 78 g; 29.0 ± 2.7 cm) the mackerel were less than 3 years of age [9]. The mackerel were shipped to the lab on ice and dissected within 8 hours. Concentrations of PSP toxins in the GI tract and liver dissected mackerel were determined by the standard AOAC mouse bioassay [10]. A pooled extract from ten mackerel was used for each assay. CD-1 female albino mice from Charles River, Montreal, Quebec, Canada were used. HPLC analysis of the standard AOAC extracts of liver and GI tract were by the method of Sullivan and Wekell [11]. Surface water samples were collected from 4 sites (Fig. 1) and analyzed for *A. fundyense* according to Martin and White [12].

RESULTS AND DISCUSSION

Analyses of surface water from southwest Bay of Fundy indicated that small blooms of A. fundyense occurred in May and July, 1988 (Fig. 2). Surface water samples are indicative of *A. fundyense* abundance as these algae are not diurnal and are concentrated at the surface (Martin, unpublished). The highest concentrations of *A. fundyense* in samples from Brandy Cove, Deadman's Harbour, Bliss Island and The Wolves, were 1.0 X 10^2, 4.8 X 10^3, 7.0 X 10^3 and 7.5 X 10^3 cells/L, respectively. Blooms of *A. fundyense* are an annual occurrence in the Bay of Fundy. For example, during *A. fundyense* blooms between 1980 - 1984 the concentrations were generally >1.0 X 10^5 cells/L and reached 1.8 X 10^7 in August 1980 [12]. During this period *A. fundyense* concentrations remained below 1 X 10^4 cells/L only in 1981.

The highest concentrations of *A. fundyense* were observed on July 12 in surface water from The Wolves and Bliss Harbour and on July 22 from Deadman's Harbour (Fig. 2). By mid-August cells counts had decreased to less than 20 cells/L. In mid-July mackerel began to make their annual summer migration into the Bay of Fundy and remained until the end of September (R. L. Stephenson, personal communication). Liver extracts of mackerel captured in southwest Bay of Fundy during this period were lethal to mice (Table 1). The symptoms observed in mice were typical of those caused by paralytic shellfish toxins. The PSP toxins concentrations in liver determined by mouse bioassay ranged from 40 to 209 µg STX equiv/100 g wet wt of liver (Table 1). Therefore, even during a "mini-bloom" of *A. fundyense* mackerel accumulated significant concentrations of PSP toxins.

PSP toxins concentrations determined by HPLC were higher than those determined by mouse bioassay (Table 1). In studies by others, the concentrations of PSP toxins in shellfish extracts determined by HPLC analysis have been both overestimated when they were above 30 µg STX equiv/100 g wet wt. [13] and underestimated above 80 µg STX equiv/100 g wet wt. [14] when compared to the results from mouse bioassay.

The majority of the mackerel were captured after the period of *A. fundyense* bloom. The liver extract with the highest PSP toxins concentration (209 µg STX equiv/100 g wet wt.) was that of mackerel caught in Deadman's Harbour on July 20, at the height of the bloom (Table 1). As the cell counts of *A. fundyense* in the surface water decreased the concentration of PSP toxins in the liver of mackerel from Deadman's Harbour decreased to 52-77 µg STX equiv/100 g wet wt. Significant concentrations remained in the liver up to one month after the disappearance of *A. fundyense*.

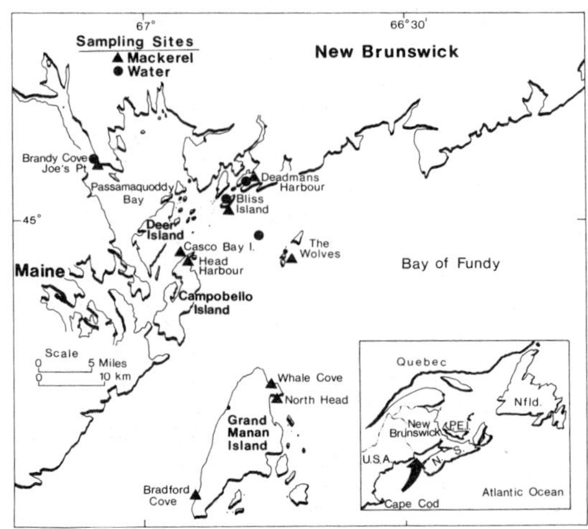

Figure 1. Surface water and Atlantic mackerel, *Scomber scombrus*, sampling sites in southwest Bay of Fundy, New Brunswick, Canada.

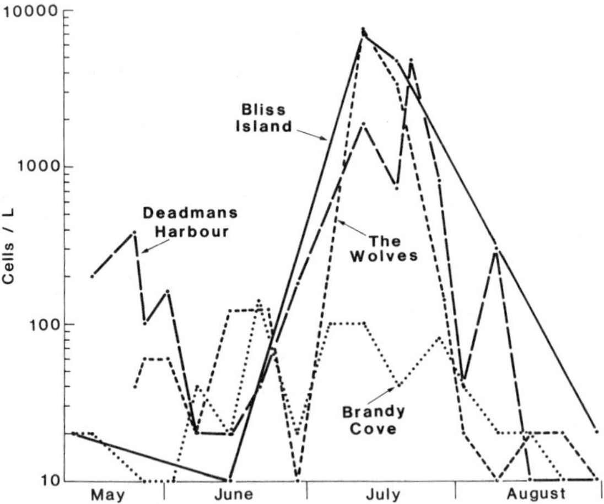

Figure 2. Concentration of *Alexandrium fundyense* in surface water samples from southwest Bay of Fundy, New Brunswick, Canada from May to August 1988.

TABLE 1. Concentration of paralytic shellfish toxins in liver and GI tract of mackerel (*Scomber scombrus*) from southwest Bay of Fundy determined by mouse bioassay and HPLC analysis.

		CONCENTRATION (μg STX EQUIV/100 g WET WT.)			
		LIVER		GI TRACT	
SAMPLE SITE	DATE (DAY/MONTH)	BIOASSAY	HPLC	BIOASSAY	HPLC
Deadman's Harbour	20/7	209	224	57	16
	3/8	77	170	-	6
	4/8	52	89	-	2
	9/8	76	-		
	15/8	65	173	-	-
Joe's Point	12/8	64	167	-	17
Bliss Island	15/8	174	-	26	
	29/8	95	-		
	15/9	122	-		
North Head	12/8	128	288	-	16
	8/9	104	-		
Bradford Cove	17/8	61	-		
Whale Cove	22/8	72	-		
	29/8	112	-		
Head Harbour	23/8	71	-		
	1/9	65	-		
Casco Bay	6/9	62	-		
The Wolves	28/9	40	-		

- = not detected; blank = not determined

This suggests that PSP toxins persist in mackerel food or that excretion rates of PSP toxins become very slow at concentrations below 80 μg equiv STX/100 g wet wt. Further evidence that PSP toxins are not readily excreted by mackerel is the report of the high concentrations found in mackerel removed from a dead humpback whale off of Cape Cod, U.S.A. in November 1987 [7]. These mackerel may have accumulated the PSP toxins during their summer migration to Canadian waters. No mackerel were captured after September 28 in southwest Bay of Fundy.

GI tracts from the mackerel that had the most toxic liver extracts by mouse bioassay had a PSP toxins concentration of 57 μg STX equiv/100 g wet wt (Table 1). This is considerably less than concentrations of 1414 and 64-245 μg STX equiv/100 g guts found in stomachs of herring killed during the 1976 and 1979 *A. fundyense* blooms, respectively, in southwest Bay of Fundy [15, 16]. PSP toxins were not detected in the GI tract from other mackerel by mouse bioassay. Since HPLC analysis is more sensitive than the mouse bioassay, PSP toxins concentrations of 2-26 μg STX equiv/100 g wet wt were measured in 5 of 6 mackerel GI tracts analyzed by this method. The presence of PSP toxins in the GI tract indicates that mackerel could accumulate them from food.

Profiles of the PSP toxins for the mackerel liver and GI tract analyzed by HPLC are given in Table 2. At least 97% of the PSP toxins found in the liver was saxitoxin except for liver extracts of mackerel from Joe's Point (0%) and Deadman's Harbour, August 15 (84%). In these two extracts neosaxitoxin, which was not found in the other liver extracts, was present in significant concentrations. The mackerel caught at Joe's Point was also different in that only neosaxitoxin is present in the liver. GTX II and GTX III were found in both liver and GI tract extracts from Deadman's Harbour mackerel, but in low concentration compared to that of saxitoxin in the liver. GTX II and GTX III were not detected in mackerel from other sites. Only B2 was detected in the GI tract of North Head mackerel. B2 was not detected in any of the other mackerel.

Table 2. Paralytic shellfish toxins profile in liver and GI tract of mackerel (*Scomber scombrus*) from southwest Bay of Fundy.

SAMPLE SITE	DATE (DAY/MONTH)	TISSUE	B2	GTX-III	GTX-II	NEO	STX
				CONCENTRATION (nmol/g WET WT)			
Deadman's Harbour	20/7	liver	-	0.08	0.10	-	6.16
	3/8	liver	-	-	-	-	4.66
	4/8	liver	-	0.02	0.03	-	2.40
	15/8	liver	-	-	-	0.76	3.98
Joe's Point	12/8	liver	-	-	-	4.60	-
North Head	12/8	liver	-	-	-	-	7.92
Deadman's Harbour	20/7	GI tract	-	0.42	-	-	0.10
	3/8	GI tract	-	0.19	0.04	-	-
	4/8	GI tract	-	-	0.09	-	-
	15/8	GI tract	-	-	-	-	-
Joe's Point	12/8	GI tract	-	-	-	-	0.47
Bliss Island	15/8	GI tract	-	-	-	0.71	-
North Head	12/8	GI tract	5.8	-	-	-	-

- = not detected

Mackerel feed both by filter feeding (gill rakers are used as filters) and individual selection of organisms [9]. A great variety of plankton, crustacean larvae, small squid, fish eggs and young fish are sources of food for mackerel. Profiles in the liver suggest that mackerel do not accumulate PSP toxins directly from ingestion of *A. fundyense*. The PSP toxins profile for *A. fundyense* in water samples from The Wolves on July 25, 1986 was: GTX IV, 39%; GTX III, 19%; neosaxitoxin, 19%; GTX I, 15%; saxitoxin 4%; GTX II, 3%; and C1/C2 and B1, <1%. GTX I and GTX IV, which make up 54% of PSP toxins concentration in *A. fundyense*, were not detected in mackerel GI tract and liver. Saxitoxin is essentially the only PSP toxin present in mackerel liver but is a minor component of the profile for *A. fundyense*. However, it is possible that GTX I and IV undergo rapid biotransformation in mackerel to saxitoxin. Mackerel feed mainly on young herring during the summer in southwest Bay of Fundy, and they are the most likely source for PSP toxins in mackerel. Herring in southwest Bay of Fundy accumulated lethal concentrations of PSP toxins during *A. fundyense* blooms in 1976 and 1979 [15, 16]. On the other hand, mackerel are preyed upon by other aquatic animals, for example dogfish, Atlantic cod, tuna, swordfish, porpoises and harbour seals [9], therefore PSP toxins may be transferred further up the food web.

ACKNOWLEDGMENT

We thank Sherwood Hall, FDA, Washington, D.C. for the PSP toxins used as primary standards, R. L. Stephenson for the mackerel, and Mr. F. Cunningham, B. Best, and S. Bellis for the preparation of the figures and manuscript.

REFERENCES

1. Ochi, T. In: Okaichi, T., Anderson, T. M., Nemoto, T., editors. Red Tides. Biology, Environmental Science and Toxicology. New York: Elsevier Science Publishing Co., Inc., 1989: 201-204.
2. Toyoshima, T., Shimada, M., Ozaki, H. S., Okaichi, T., Murakami, T. H. In: Okaichi, T., Anderson, T. M., Nemoto, T., editors. Red Tides. Biology, environmental sciences and toxicology. New York: Elsevier Science Publishing Co., Inc., 1989: 439-442.
3. Jenkinson, I. R. In: Okaichi, T., Anderson, T. M., Nemoto, T., editors. Red Tides. Biology, environmental science and toxicology. New York: Elsevier Science Publishing Co., Inc., 1989: 435-438.
4. Adams, J. A., Seaton, D. D., Buchanan, J. B., Longbottom, M. R. Nature, Lond. 220:24-25, 1968.
5. White, A. W. In: Ragelis, E. P., editor. Seafood Toxins. ASC Symposium Series, No.262. Washington, D. C.: American Chemical Society, 1984: 171-180.
6. Geraci, J., Toronto Globe and Mail. Thursday, Feb. 2, 1989.
7. Mackenzie, D. New Scientist. 117: 30, 1988.
8. White, A. W., Fukuhara, O., Anraku, M. In: Okaichi, T., Anderson, D. M., Nemoto, T., editors. Red Tides. Biology environmental science and toxicology. New York: Elsevier Science Publishing Co., Inc., 1989: 395-398.
9. Scott, W. B., Scott, M. G. Can. Bull. Fish. Aquat. Sci. 219: 452-455, 1988.
10. Association of Offical Analytical Chemists. Offical methods of analysis. 14th ed. Arlington, VA: AOAC, Inc, 1984: 344-345.
11. Sullivan, J. J., Wekell, M. M., and Kentala. J. Food Sci. 50: 26-29, 1985.
12. Martin, J. L., White, A. W. Can. J. Fish. Aquat. 45: 1968-1975, 1988.
13. Martin, J. L., White, A. W., Sullivan, J. J., This Book, 1989.
14. Sullivan, J. J., Simon, M. G., Iwaoka, W. T. Food Sci. 48:1312-1314, 1983.
15. White, A. W. Can. J. Fish. Aquat. Sci. 37: 2262-2265, 1980.
16. White, A. W. J. Fish. Res. Board Can. 34: 2421-2424, 1977.

OKADAIC ACID PRODUCES DRASTIC HISTOPATHOLOGIC CHANGES OF THE RAT INTESTINAL MUCOSA AND WITH CONCOMITANT HYPERSECRETION

S. LANGE[1], G.L. ANDERSSON[2], E. JENNISCHE[2], I. LÖNNROTH[3], X.P. LI[1] AND L. EDEBO[1]
[1]Dept. of Clinical Bacteriology, [2]Histology, and [3]Medical Microbiology, University of Göteborg, Guldhedsgatan 10, S-41346 Göteborg, Sweden

ABSTRACT

We have previously shown that okadaic acid (OA) rapidly induces secretion in ligated loops of the rat small intestine. The kinetics of the secretory response is extremely rapid, reaching peak values within two hours. The dose-response relationship is close to linear in the range 0.5-5.0 µg OA.

The present investigation demonstrate that OA produces rapid and characteristic changes of the mucosa morphology in ligated loops. After 90 minutes of OA challenge most of the enterocytes at the upper part of the villi are shed. The goblet cells, which are not affected by the toxin, are consequently accumulated. There is no sign of bleeding or inflammatory reaction, and no apparent barrier damage. The crypt regions appear unaffected.

INTRODUCTION

After consumption of mussels containing marine toxins, man react with intestinal disturbances such as diarrhoea, nausea, vomiting, abdominal pain and chills. Diarrhoea is in most cases the dominating symptom. We have found that the rat small intestine function as a most sensitive and reproducible organ for studies of the diarrhoeal effects caused by these toxins (1).

The present investigation was undertaken in order to study the histopathological effects caused by okadaic acid in the rat small intestine.

MATERIALS AND METHODS

Animals, operation procedure: Male Sprague-Dawley rats weighing 200 grams were used. The rats were anaesthetized with ether and one ligated loop, some 15 cm long, were prepared on the middle of jejunum and injected with 1 ml of different concentrations of OA dissolved in phosphatebuffered saline, pH 7.2. Controls were injected with buffer alone. Ether is used for anaesthesia, since all other anaesthetic influence the secretory response, and most probably also the histopathological reactivity of the small intestine.

Histological studies: Rats were fixed by transcardial perfusion with 4% buffered paraformaldehyde. The middle portion of the ligated loop was then cut into some 5 mm long pieces and postfixed in the same fixative.

a) Light microscopy. Small pieces of the intestinal loops were embedded into methacrylate plastic and 1 µm thick sections prepared. The sections were either stained with basic fuchsin/methylene blue/Azur III or with PAS/ hematoxylin using standard procedures.

b) Scanning electron microscopy (SEM). The fixed specimens were handled according to standard procedures.

Antisecretory factor (ASF). ASF (2) was prepared from porcine pituitary glands and injected intravenously via the penis vein in a volume of 2 ml 30 seconds before challenge with OA.

Toxic Marine Phytoplankton
Edna Graneli et al., Editors

RESULTS AND DISCUSSION

Light microscopic changes were observed within 15 min after injection of OA. Starting at the tips of the villi the enterocytes became swollen and were subsequently detached from the basement membrane and shed into the lumen. The goblet cells were not affected by the doses of toxin used. (1-5 µg OA)(Fig.1, 2).

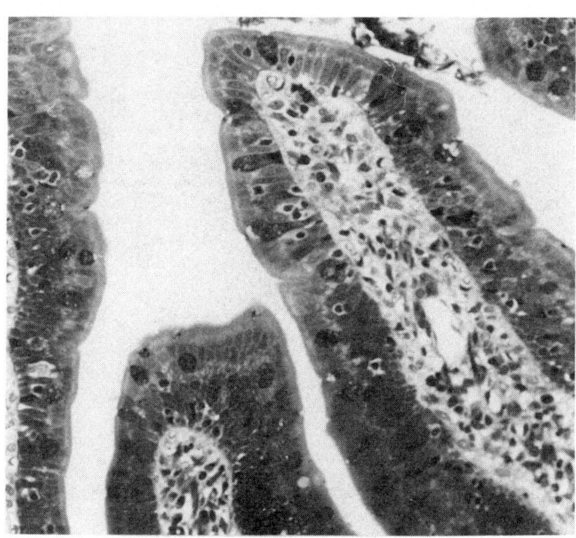

Fig.1. Control loop, challenge with PBS for 2 hrs. Normal morphology.

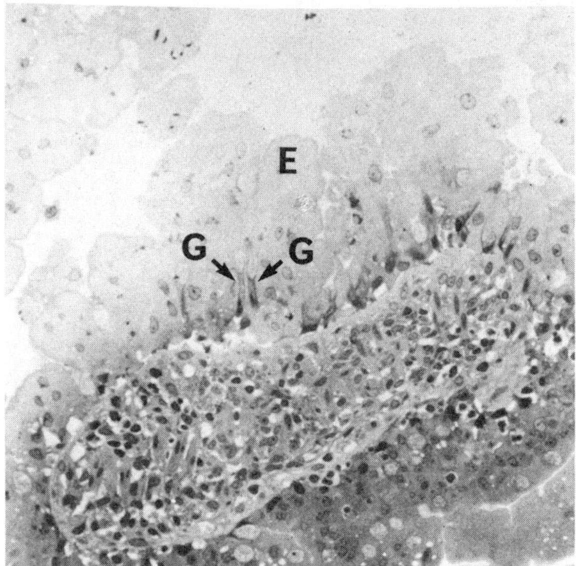

Fig.2. Challenge with 3 µg OA for 30 min. There is pronounced swelling of enterocytes (E) and some of them are detached from the basal membrane and shed into the lumen. The goblet cells (G) appear elongated, but remain attached to the basal membrane.

After 60-90 min most of the villus enterocytes were shed and large parts of the flattened villi were covered by goblet cells containing PAS-positive material. A low dose of OA ($<3\mu g$) affected only the top of the villi (Fig. 3A, 3B), while a higher dose (5 μg) involved reaction of the total villi, and collapse of the villi architecture.

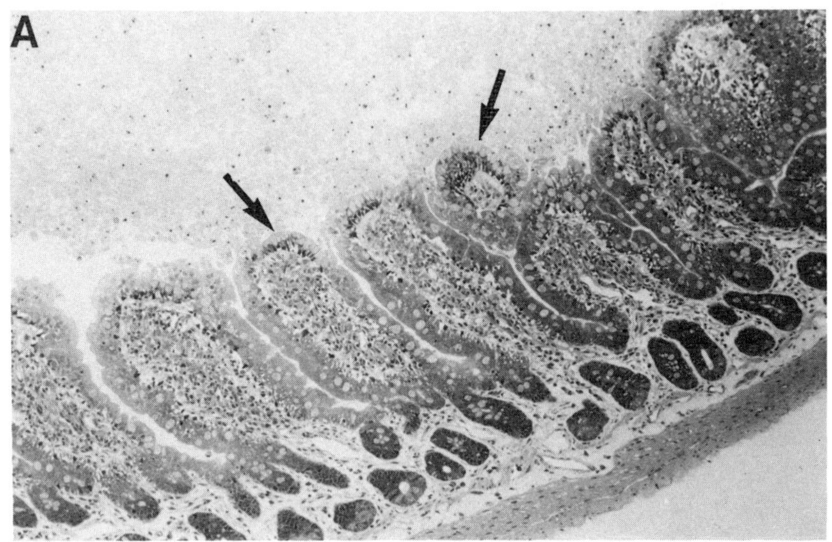

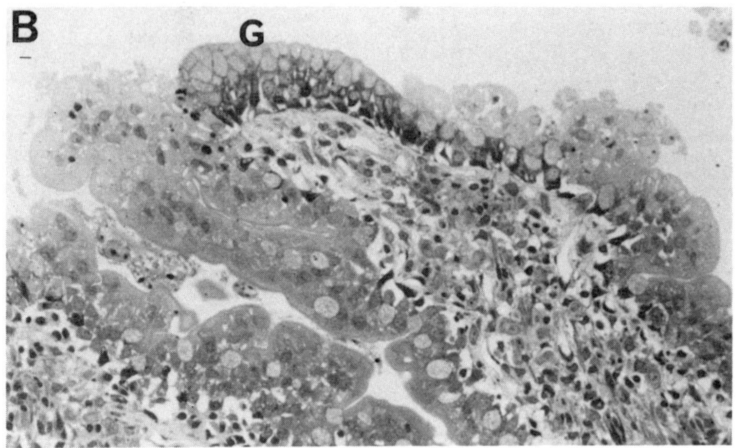

Fig. 3A. After 90 min most of the enterocytes at the upper parts of the villi are shed. The goblet cells, which are not affected by the toxin, are consequently accumulated (arrows, G). There is no obvious release of mucus from the goblet cells.

Fig. 3B. Higher magnification showing goblet cells (arrows, G).

By SEM, swelling and beginning detachment of single swollen cells could be observed within five minutes after injection of the toxin and after 30 min the mucosal surface was partly covered by shed epithelial cells and cellular debris.

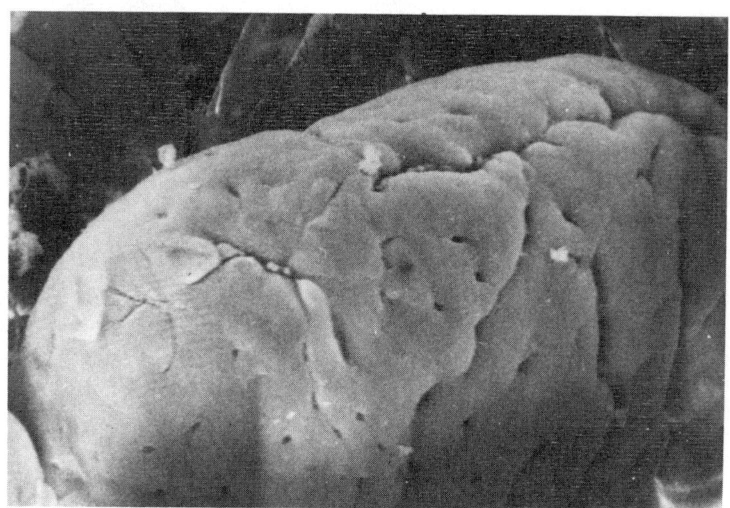

Fig.4. Control loop shows a part of a villus, note the smooth surface.

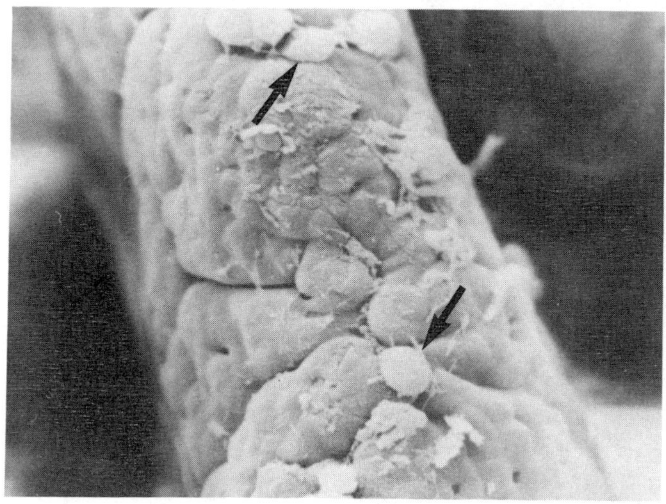

Fig.5. 5 min OA. Swollen enterocytes are beginning to detach from the basal membrane (arrows).

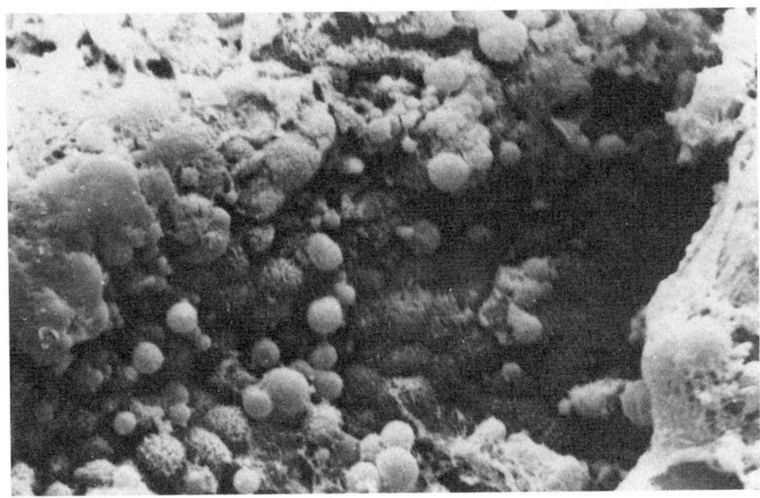

Fig.6. 30 min OA. A large number of swollen shed cells are covering the surface of the villus.

Administration of ASF before toxin challenge inhibited the secretion some 60 per cent, but did not inhibit the characteristic morphological changes. This suggests a specific, antisecretory action of ASF, not involving the morphological effects of OA.

Intravenous injection of OA also induced similar, but less extensive changes in loops as intraluminal administration suggesting that the enterocytes are specific target cells for OA.

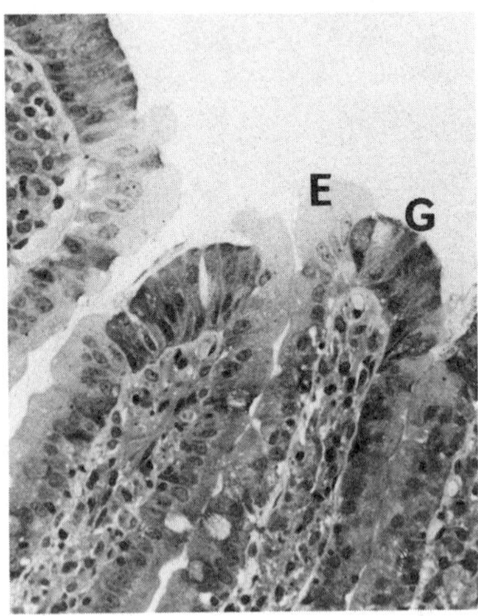

Fig.7. When OA is injected intravenously similar changes as those observed after intraluminal injection are seen, i.e. shedding of enterocytes (E) and accumulation of goblet cells (G).

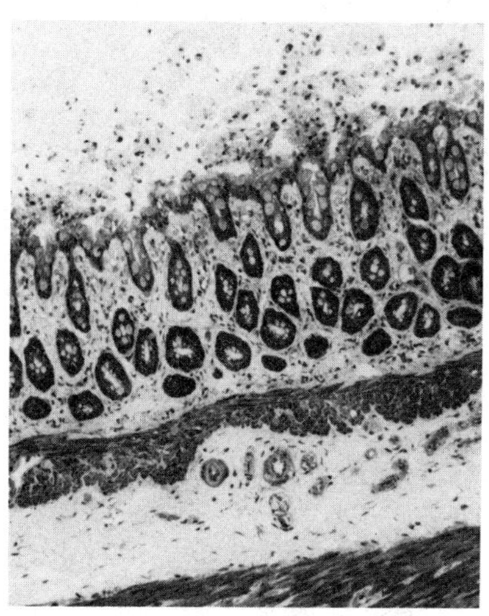

Fig.8. Colon after challenge with 5 µg OA for 2 hrs. Shedding of the upper epithelial lining.

CONCLUSION

We have demonstrated that OA induces rapid changes in the rat small and large intestine. These changes consist of selective shedding of the enterocytes, starting on top of the villi, resulting in accumulation of goblet cells. The goblet cells remain attached to the basal membrane and appear to form a covering lining of the luminal surface of the intestine. There is no sign of bleeding or inflammatory reaction, i.e. no apparant barrier damage. The crypt regions appear unaffected. When a high dose (5 µg OA) is used, a collapse of the villi architecture is obtained.

REFERENCES

1. Edebo, L., Lange, S., Li, X.P. and Allenmark, S.: Toxic mussels and okadaic acid induce rapid hypersecretion in the rat small intestine. Acta Path. Microbiol. Scand. 96: 1029-1035, 1988.

2. Lönnroth, I. and Lange, S.: Purification and characterization of the antisecretory factor - a protein in the central nervous system and in the gut which inhibits intestinal hypersecretion induced by cholera toxin. Biochimica et Biophysica Acta 883: 138-144, 1986.

OKADAIC ACID, APLYSIATOXIN AND TELEOCIDIN INDUCE RAPID CELLULAR CHANGES AND HYPERSECRETION IN THE RAT SMALL INTESTINE

S. LANGE[1], E. JENNISCHE[2], H. FUJIKI[3], AND L. EDEBO[1]
[1]Dept. of Clinical Bacteriology and [2]Histology, University of Göteborg, Guldhedsgatan 10, S-41346 Göteborg, Sweden and [3]Cancer Prevention Division, National Cancer Research Institute, 5-1-1 Tsukiji, Chuo-ku, Tokyo 104, Japan

ABSTRACT

The present results demonstrate that okadaic acid, aplysiatoxin and teleocidin seem to form a group of substances which rapidly induce activating effects on the rat intestinal mucosa when used in µg doses. These effects consist of shedding of enterocytes and accumulation of goblet cells on the tips of the villi. Simultaneously, hypersecretion is observed. The tumor promoting agent phorbol 12-teradeconate 13-acetate (TPA) did not induce reaction in doses less than 200 µg, neither did the chemotactic peptide N-formyl-L-methionyl-L-leucyl-L-phenylalanine.

INTRODUCTION

We have demonstrated that okadaic acid (OA) induces rapid cellular alterations in the rat intestinal mucosa with concomitant hypersecretion (accompanying communication, 1).It has also been shown that OA as well as a few other marine toxins as aplysiatoxin and teleocidin are cell activators and tumor promoting agents (2).

The aim of the present study was to evaluate <u>in vivo</u> the secretory response of the rat small intestine to aplysiatoxin and teleocidin. Simultaneously the morphological effects were observed and compared to those induced by OA. As reference, the induction of intestinal secretion was tested with the strong tumor promoting agent phorbol 12-tetradecanoate 13-acetate (TPA) and also with the chemotactic peptide N-formyl-L-methionyl-L-leucyl-L-phenylalanine.

MATERIALS AND METHODS

<u>Animals, operation procedure</u>: Male Sprague-Dawley rats weighing 200 grams were used. The rats were anaesthetized with ether and one ligated loop, some 15 cm long, were prepared on the middle of jejunum and injected with 1 ml of different concentrations of the agent to be tested. All agents were dissolved in phosphatebuffered saline, pH 7.2. Controls were injected with buffer alone. We use ether for anaesthesia, since all other anaesthetics influence the secretory response, and most probably also the histopathological reactivity of the small intestine.

<u>Histological studies</u>: Rats were fixed by transcardial perfusion with 4% buffered paraformaldehyde. The middle portion of the ligated loop was cut into some 5 mm long pieces and postfixed in the same fixative. Small pieces of the intestinal loops were then embedded into methacrylate plastic and 1 µm thick sections prepared. The sections were either stained with basic fuchsin/methylene blue/Azur III or with PAS/hematoxylin using standard procedures.

Toxic Marine Phytoplankton
Edna Graneli et al., Editors

RESULTS AND DISCUSSION

Hypersecretion. Aplysiatoxin, teleocidin and okadaic acid all induced strong and rapid secretory responses in the rat small intestine, and all toxins were found to be active in µg range (Fig.1). The results in fig.1 allow us to postulate that the secretory potency of the various toxins is almost equal.

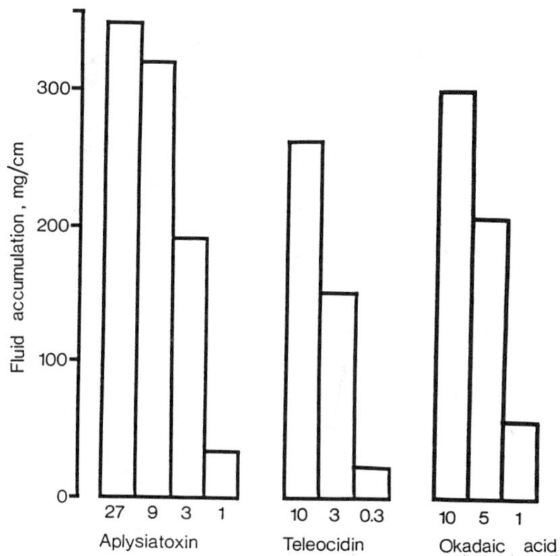

Fig.1. Diagram of the dose-response of aplysiatoxin, teleocidin and okadaic acid. The challenge time was 2 hrs, the number indicated under the staples represents doses in µg.

The time-response curve for aplysiatoxin was constructed with a 9 µg dose (Fig.2). The results demonstrate that already after one hour significant secretion was induced and the amount of liquid in the loop remained elevated for 12 hrs. As a reference TPA, which is a well-known and strong tumor promotor isolated from the oil of the seeds of *Croton tiglium* L., was tested for induction of intestinal secretion. Croton oil is more widely known as a drastic cathartic, a vesicant and a pustulant. However, no secretory effect was seen using doses less than 200 µg, but abundant secretion was obtained by 1 mg. The chemotactic peptide N-formyl-L-methionyl-L-leucyl-L-phenylalanine did not produce any secretory response at amounts of 1 mg and below. Thus, both TPA and the chemotactic peptide was far less potent than the toxins when tested as secretagogues.

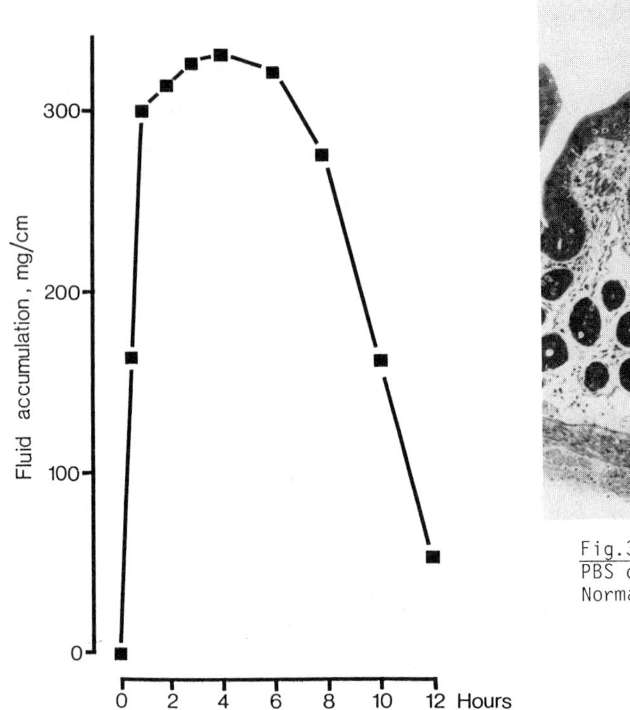

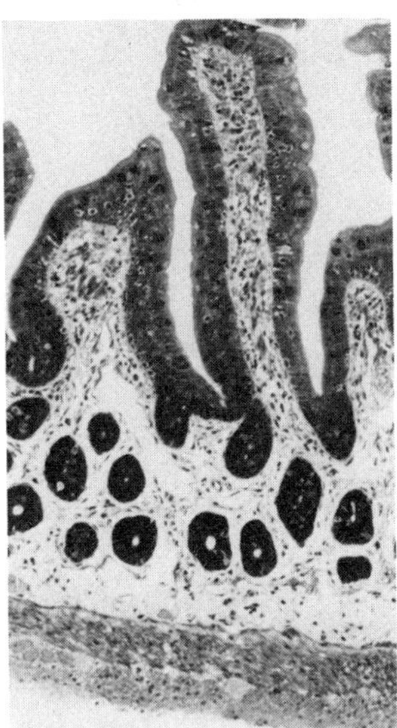

Fig.3. Control loop, PBS challenge for 2 hrs. Normal morphology.

Fig.2. Aplysiatoxin, time-response curve, challenge dose 9 µg.

Morphology. Tissue specimens were taken from the intestinal mucosa after 30 min and after 2, 3 and 6 hrs and stained for light microscopy. In loops injected with PBS, no morphological changes were observed. All toxins induced similar morphological changes in the small intestine. Changes induced by OA are described in the accompanying communication (1). Starting at the upper parts of the villi there was a selective shedding of enterocytes, resulting in an accumulation of goblet cells. The goblet cells appeared to cover the luminal aspects of the mucosa and there was no obvious barrier damage or inflammation. The deeper parts of the mucosa were unaffected. These changes were most obvious in loops injected with okadaic acid. The changes induced by aplysiatoxin were more discrete, and did not appear to progress between 30 min and 6 hrs.

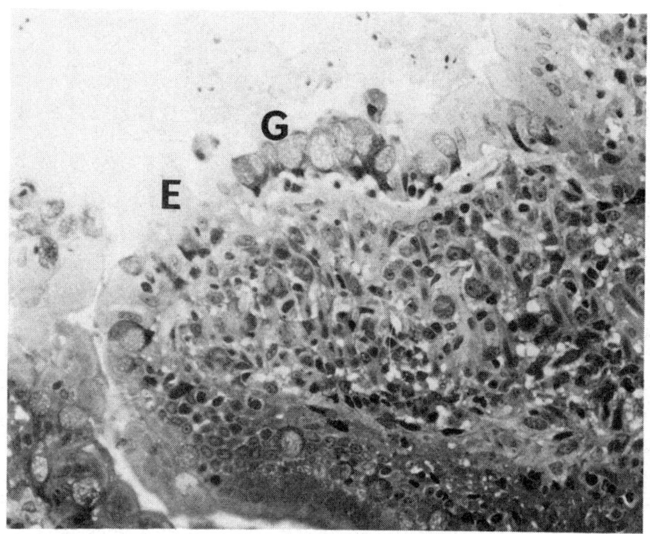

Fig.4. Aplysiatoxin, 3 µg challenge for 30 minutes. Note accumulation of goblet cells (G) and shedding of enterocytes (E).

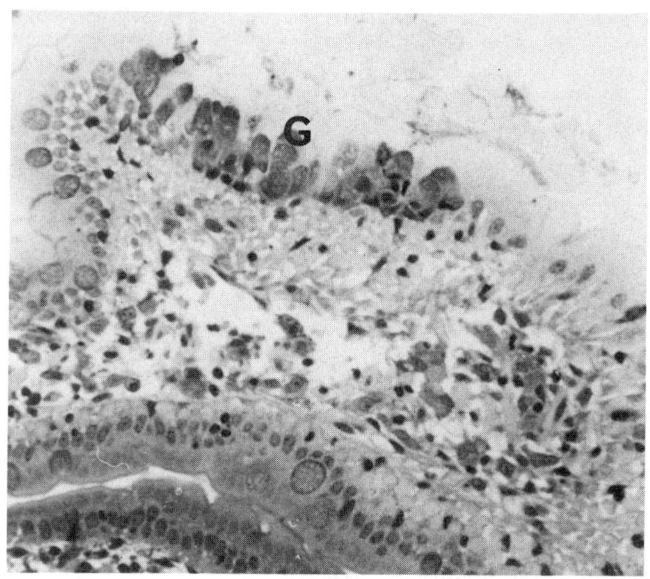

Fig.5. Aplysiatoxin, 3 µg challenge for 6 hrs. No progress of changes compared to 30 minutes challenge time.

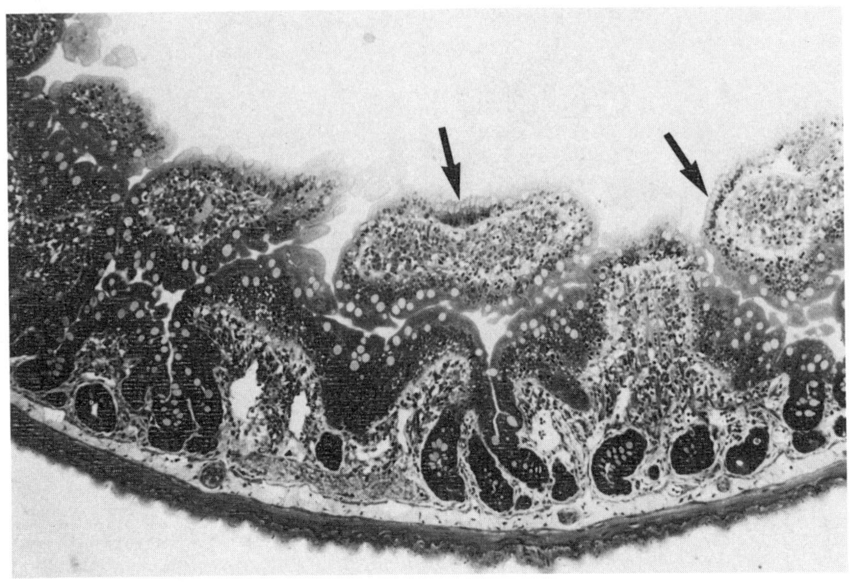

Fig.6A. Teleocidin, 3 μg challenge for 2 hrs. Note shedding of enterocytes and accumulation of goblet cells (arrows). The changes observed are discrete.

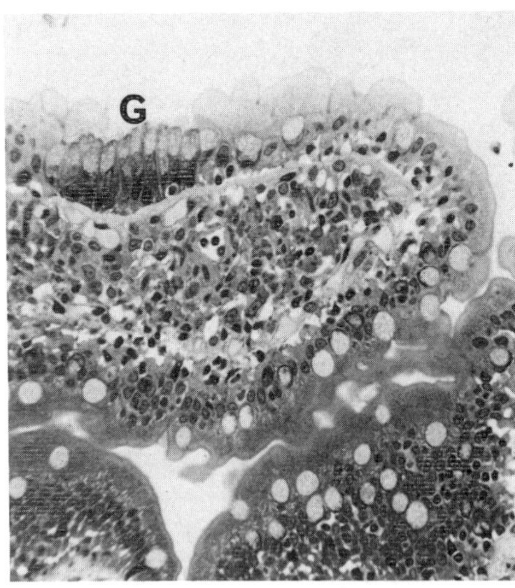

Fig.6B. Higher magnification of 6A, showing accumulation of goblet cells (G).

CONCLUSION

1. The marine toxins okadaic acid, aplysiatoxin and teleocidin seem to form a group of substances with rapid activating effects on several kinds of mammalian cells, and tumor promoting activity.

2. Histological changes in the intestinal mucosa show rapid swelling and consequtive shedding of enterocytes. The enterocytes on top of the villi are replaced by goblet cells. There is no barrier damage or inflammation, the deeper part of the mucosa are unaffected.

3. The effects observed in a number of experimental systems are similar to those seen with the classical cathartic and tumor-promoting agent TPA. The doses required are, however, considerably smaller when the marine toxins are used.

REFERENCES

1. Lange, S., Andersson, G.L., Jennische, E., Lönnroth, I., Li, X.P., and Edebo, L.: Okadaic acid produces drastic histopathologic changes of the rat intestinal mucosa and with concomitant hypersecretion. Present communication.

2. Suganuma, M., Fujiki, H., Suguri, H., Yoshizawa, S., Hirota, M., Nakayasu, M., Ojika, M., Wakamatsu, K., Yamada, K., and Sugmura, T.: Okadaic acid: An additional non-phorbol-12-tetradecanoate-13-acetate-type tumor promotor. Proc. Natl. Acad. Sci. USA, 85: 1768-1771, 1988.

ECOLOGICAL PARAMETERS ASSOCIATED WITH THE SEASONAL OCCURRENCE OF *ALEXANDRIUM* SPP. AND CONSEQUENT SHELLFISH TOXICITY IN THE LOWER ST. LAWRENCE ESTUARY (EASTERN CANADA)

RICHARD LAROCQUE AND ALLAN D. CEMBELLA
Maurice Lamontagne Institute, Dept. of Fisheries and Oceans, 850 route de la Mer, Mont-Joli, Quebec, Canada, G5H 3Z4.

ABSTRACT

Paralytic shellfish poisoning in the lower St. Lawrence Estuary (Quebec) is associated with seasonal blooms of the toxic dinoflagellate *Alexandrium*. The present detailed study along the south coast, confirms the role of water column stability and demonstrates the importance of the longshore Gaspé current in affecting the abundance and spatial distribution of bloom populations. The toxicity levels in shellfish suggest that there is a lag phase of approximately three weeks between the maximum abundance of *Alexandrium* and the resulting peak of shellfish toxicity. The patchy distribution of this dinoflagellate at stations adjacent to the south coast of the estuary may explain the small-scale variations in toxin levels observed in shoreline molluscs.

INTRODUCTION

The lower St. Lawrence estuary is subject to seasonal blooms of *Alexandrium* spp., the dinoflagellate responsible for paralytic shellfish poisoning (PSP) in this area [1]. Previous work along the north shore of the estuary has emphasized the importance of freshwater runoff and water column stability as the major factors controlling bloom dynamics [2, 3]. It is also known that the sediments adjacent to the Manicouagan peninsula, an area of the north shore strongly influenced by the freshwater input of a major river system, are an important reservoir of benthic *Alexandrium* cysts [2]. Until now, no studies have considered the population dynamics of this dinoflagellate near the south shore of the estuary, a region strongly dominated by the influence of the longshore Gaspé current which flows towards the Gulf of St. Lawrence. The molluscs on the Gaspé coast often become toxic due to PSP, with a general tendency for toxicity to increase eastward along the Gaspé peninsula. This present study was undertaken to identify the principal ecological factors associated with the seasonal blooms of *Alexandrium* in the near-shore area of Rimouski (Quebec). The relationship between bloom dynamics and the consequent shellfish toxicity was also examined.

MATERIAL AND METHODS

Weekly samples from the water column were collected at four nearshore stations located between Bic and Pointe Mitis (Fig. 1), between May and the end of August, 1987. Pump samples taken at five standard depths (surface, 2, 5, 11 and 20 m) were collected to determine the phytoplankton composition and, specifically, the concentration of *Alexandrium* spp., by optical microscopy. Pump samples were also filtered and used for inorganic nutrient (NO_3, PO_4) analysis [4]. To determine the stratification of the water column, temperature and salinity measurements were taken according to a standard method previously described [3]. The *in vivo* chlorophyll *a* profile was determined by fluorometry [2] and the light extinction coefficient (k) was also measured using a Secchi disk. Meteorological and tidal data were examined for possible relationships with

Toxic Marine Phytoplankton
Edna Graneli et al., Editors

bloom dynamics. Mouse bioassay results for PSP from previous years for three shoreline stations were obtained from J.E. Reid (Health Protection Branch, Dept. of Health and Welfare, Ottawa) and analyzed to determine inter-annual variations.

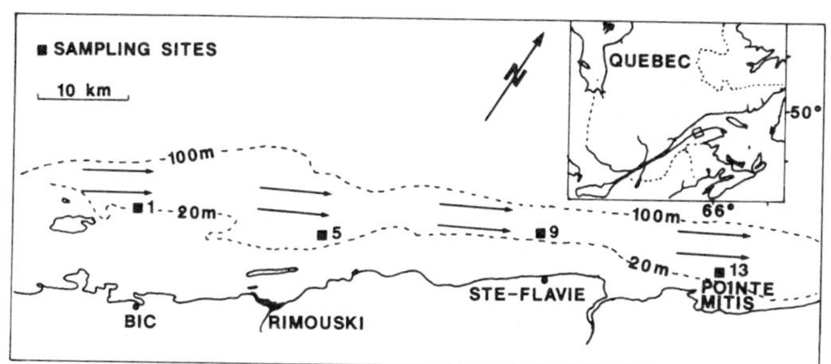

FIG. 1 Sampling stations along the south shore of the lower St. Lawrence estuary. Residual surface current direction according to El-Sabh [9].

RESULTS

Although there was some variation in overall thecal morphology, most Alexandrium cells exhibited the prominent ventral pore and angular theca characteristic of A. excavatum (Braarud) Balech & Tangen. A single bloom of Alexandrium occurred along the south shore of the estuary during the summer sampling period of 1987 (Fig. 2). The bloom appeared in August at all sampling sites and the maximum Alexandrium cell concentrations, which reached 4 x 10^5 cells l^{-1}, were observed at St. 1. During the rest of the summer, the number of motile cells of Alexandrium in the water column in nearshore areas was relatively low, ranging from 0 - 7 x 10^3 cells l^{-1}.

The highest A. excavatum cell concentrations were not consistently correlated with fluctuations in tidal amplitude throughout the sampling period (Fig. 2). Nevertheless, during the August bloom at St. 13, the sampling site closest to shore, peaks in Alexandrium cell numbers did coincide with periods when tidal amplitude was at a minimum. Furthermore, the decrease in the extinction coefficient, indicative of reduced turbidity in the water column, approximately followed the reduction in tidal amplitude. Finally, although there was no evident relationship between bloom dynamics and seasonal precipitation levels along the south shore, the peak of the bloom was associated with a period of reduced wind stress in the region (data from Transport Canada, Mont-Joli).

The vertical density profile (σ_t), obtained from converted salinity and temperature data at a representative sampling station (St. 13), indicated that within the sampling period, stratification of the water column was only briefly established in August (Fig. 3). This period of relatively high stratification coincided with the occurrence of a significant near-surface Alexandrium bloom, between 2 to 5 m. At this time, the peak of in vivo chlorophyll a corresponded closely with the vertical distribution of Alexandrium. The phytoplankton bloom population in late August was overwhelmingly dominated by A. excavatum, which at times approached 100% of the species composition in the size fraction >20 µm. In contrast, during the period immediately preceeding the bloom, the phytoplankton assemblage was largely dominated by large centric diatoms. This diatom-rich near-surface population was clearly reflected in the chlorophyll a peak at approximately 10 m during mid-July (Fig. 3).

A typical vertical profile of salinity, temperature, inorganic nutrient concentrations (NO_3, PO_4) and A. excavatum cell numbers taken at St. 5 off Rimouski (Fig. 1) throughout a diurnal tidal cycle during August is presented in Fig. 4. There was evidence of a shift in the vertical distribution of Alexandrium in the water column during the tidal cycle, even in the complete absence of physical stratification. This vertical displacement of Alexandrium followed the rising tide, from a near-surface maximum in late morning, to the surface in early afternoon. On the late afternoon falling tide, the peak of the bloom population shifted to the near-surface layer. Although there was some oscillation in the maximal vertical distribution of the two major macronutrients (N and P), peak abundance of Alexandrium tended to correspond with the consistent near-surface nutrient peak at between 2 to 5 m.

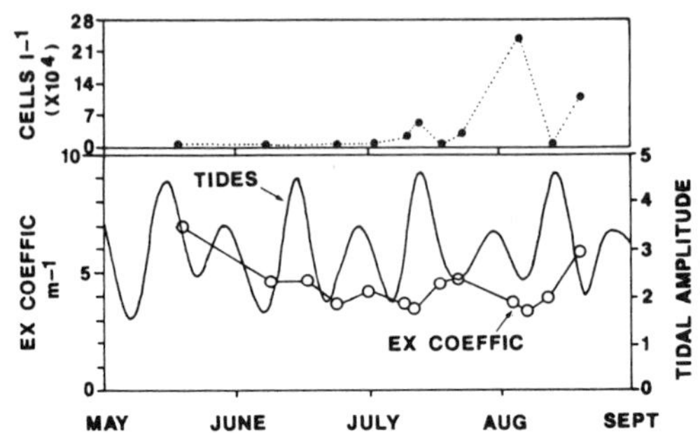

FIG. 2 Seasonal pattern of Alexandrium cell numbers, extinction coefficient (k) and tidal amplitude at St. 13.

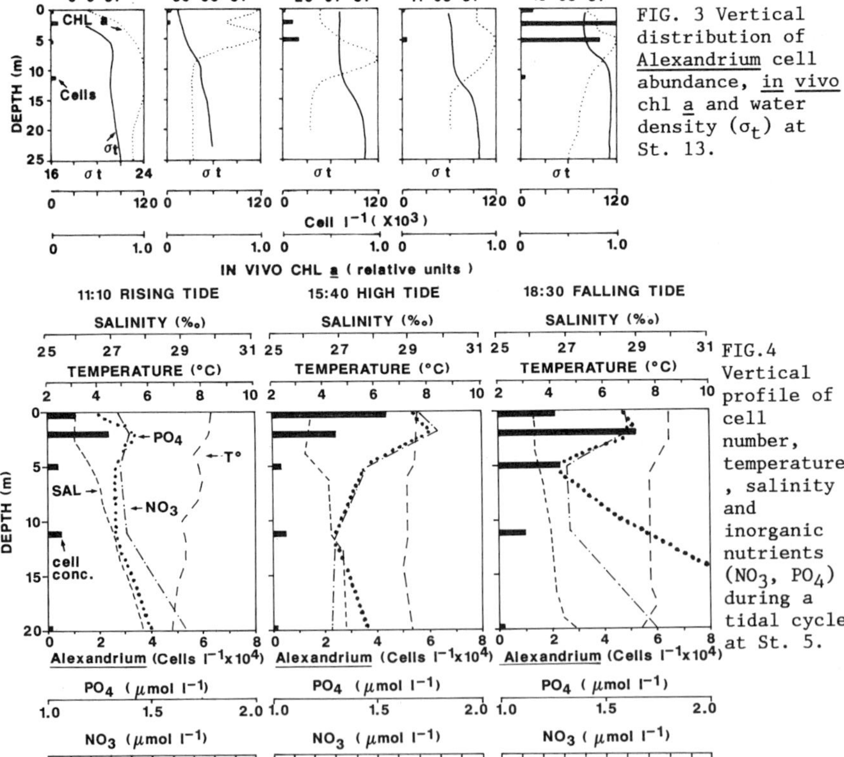

FIG. 3 Vertical distribution of Alexandrium cell abundance, in vivo chl a and water density (σ_t) at St. 13.

FIG.4 Vertical profile of cell number, temperature , salinity and inorganic nutrients (NO_3, PO_4) during a tidal cycle at St. 5.

The relationship between the temporal distribution of Alexandrium in the water column and the consequent increase in PSP toxicity in the soft-shell clam, Mya arenaria, at adjacent shoreline stations, is shown in Fig. 5. Of the three stations illustrated, St. 1 may be the least representative of the effects of bloom dynamics on shellfish toxicity, since it was located more than 15 km offshore. At the other two stations (St. 9 and 13), the maximum toxicity levels were reached at the end of August, approximately three weeks after the initial peak in Alexandrium cell numbers.

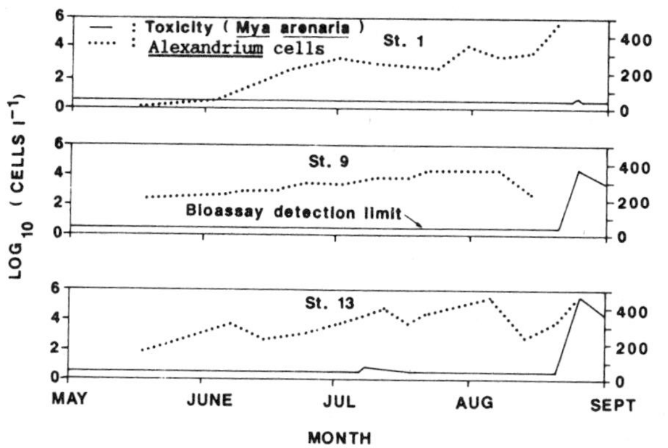

FIG. 5 Seasonal abundance of A. excavatum cells in the water column and PSP toxin levels in the soft-shell clam Mya arenaria at three adjacent stations along the south shore of the lower St. Lawrence estuary.

Recent historical data on shellfish toxicity in the Rimouski region confirm that PSP toxicity is an annual cyclical phenomenon (Fig. 6). However, in exceptional years, such as 1985, little or no PSP toxin was apparently accumulated in shoreline mollusc populations. As well as the substantial inter-annual variation in peak toxin levels, there was also a marked variation between two sampling sites located <20 km apart along the south shore. However, regardless of the fluctuations in toxicity from year to year, the maximum levels were characteristically attained in late August or early September, somewhat after the peaks in Alexandrium blooms have normally subsided.

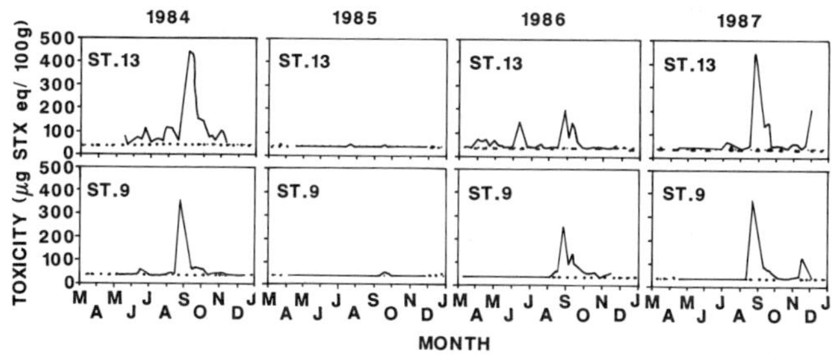

FIG. 6 Inter-annual variation in PSP toxin levels in Mya arenaria at two stations in the lower estuary.

DISCUSSION

The hypothesis that Alexandrium blooms are initiated through excystment from benthic cyst populations, which may be largely under the control of endogenous biological mechanisms [5] is particularly applicable in marine environments with relatively invariant bottom conditions, such as the lower St. Lawrence estuary. Accordingly, further attention should be focused on the ecological factors that permit a bloom population of motile cells to flourish and maintain itself. In the St. Lawrence ecosystem, it has been previously shown that water column stability arising from abundant freshwater run-off can contribute to the creation of favorable conditions for bloom development, especially along the north shore [3,4,6].

It has been recently demonstrated that the sediments in the Rimouski area are relatively impoverished in Alexandrium cysts [7]. It is therefore highly improbable that the observed blooms originate from local cyst populations. The surface circulation patterns across the estuary and along the south coast suggest that areas along the north shore, which are highly infested with Alexandrium cysts [2], could conceivably act as seed beds for blooms. The presence of trans-estuarine currents with the potential to transport blooms towards the south coast in surface waters has been recently confirmed [8].

Along the corresponding region of the south coast of the lower estuary, where the stabilizing input of fresh water from the shoreline is relatively minor and highly localized, the presence of the Gaspé current flowing toward the Gulf of St. Lawrence must be considered as the dominant factor affecting the macro-scale spatial distribution of phytoplankton. The surface circulation pattern determined by the Gaspé current in the Rimouski region (Fig. 1) has been repeatedly confirmed [9, 10] as a strong narrow current adjacent to the Gaspé coast. Most of the flow is concentrated in the surface layer above 20 m, where the majority of the cells in a bloom population of Alexandrium are typically found. In spring and early summer, the Gaspé current may reach velocities as high as 73 cm s^{-1}, however, usually the current strength decreases somewhat by August to approximately 30 cm s^{-1} [9]. This longshore jet of surface water combined with a weaker onshore current should have a tendency to concentrate offshore bloom populations towards the coast. In fact, this effect was observed during the bloom period, as Alexandrium cell numbers in surface waters tended to increase as the coast was approached from offshore.

The establishment of a distinct pycnocline may act as a physical barrier limiting vertical movements of phytoplankton. This tends to favor the accumulation of dinoflagellates near the surface, rather than diatoms, by virtue of the ability of flagellates to vertically migrate in the water column. Within the Rimouski study area, stratification of the water column and the development of a pycnocline did not occur during the early summer. However, by August, transient water stratification was achieved, as a result of the decreased velocity of the Gaspé current, less wind-driven vertical mixing and the progressive warming of the surface water. At this time, the subsurface phytoplankton biomass peak tended to reflect the vertical distribution of the Alexandrium bloom, rather than sinking diatom cells (Fig. 3).

It was not possible to directly correlate the occurrence of Alexandrium blooms with prevailing meteorological conditions, nor the limited freshwater run-off available along the south shore. Nevertheless, the accumulation of high cell numbers at St. 13 during a period when tidal amplitude, wind stress, and water column turbidity were at a minimum is undoubtedly significant. Such conditions should favor Alexandrium bloom development, by augmenting the penetration of light into the water column, and also by increasing the residence time of cells in the local area.

The maximum daily tidal amplitude can be considered as an indication of the importance of tidal currents in bloom dynamics in nearshore areas. In the present case, it is possible that the low velocity of the tidal currents during the August bloom period contributed to stabilizing the water column and helped to established the stratification. As evidence supporting this suggestion, it is significant that the extinction coefficient approximately followed the tidal cycle. Decreased turbidity in nearshore waters at times of reduced tidal amplitude implies that vertical mixing was diminished. The variations observed in inorganic nutrient profiles and in the vertical distribution of Alexandrium during the tidal cycle (Fig. 4) also confirm the importance of tidal mixing. Furthermore, the bloom was consistently associated with a near-surface maximum of inorganic nutrients, which would permit rapid growth rates.

A comparison of the shellfish toxicity data with the seasonal occurrence of substantial concentrations of motile Alexandrium cells in the water column yielded significant information on the temporal coupling of bloom dynamics and toxin accumulation. The apparent lag phase between the rise in Alexandrium cell numbers and toxin levels in shellfish may be the result of a combination of factors. One potential explanation for this phenomenon is that the distribution of Alexandrium blooms may be extremely patchy on a small-spatial scale (<5 km), such that phytoplankton sampling is unrepresentative of what is occurring in the nearby intertidal shellfish beds. The failure of shoreline molluscs to accumulate detectable toxin levels even when faced with offshore blooms of Alexandrium during August at St. 1 is a possible example of this effect (Fig. 5). Nevertheless, the results obtained at St. 9 and 13, showing a clear asynchrony between increasing cell numbers and toxin accumulation, likely yielded a good indication of the concentrations of toxic dinoflagellates to which the molluscs were subjected, since both stations were much closer to shore. In this case, the most probable explanation for the lag phase is related to the length of time required for the retention of significant levels of PSP toxins.

In summary, based on the evidence of the present study and corroborating data from previous work [3,4,6,8,10], the presence of the strong longshore Gaspé current is the main factor affecting Alexandrium bloom dynamics in the study area. The high velocity of this current, coupled with strong tidal currents in nearshore areas, does not allow for persistent stratification of the water column to occur before late summer. Given these conditions, the residence times of bloom populations, which may have been swept across from the north shore, are too short to permit the development of high cell numbers before August, even though the high nutrient levels appear to be sufficient to sustain both high growth rates and abundant primary production. The transient nature of these blooms results in an erratic accumulation pattern of PSP toxins in molluscs along the south shore. The failure of blooms to persist in the area results in PSP levels that are usually substantially less than those found along the north coast and further east along the south shore of the Lower St. Lawrence estuary.

REFERENCES

1. A. Prakash, J.C. Medcof and A.D. Tennant, Bull. Fish. Res. Bd. Can. 177, 88pp (1971).
2. A.D. Cembella, J. Turgeon, J.-C. Therriault and P. Béland, J. Shellfish. Res. 7, 597-609 (1988).
3. J.-C. Therriault, J. Painchaud and M. Levasseur, in: Toxic Dinoflagellates, D.M. Anderson, A.W. White and D.G. Baden, eds. (Elsevier, New York 1985) pp. 141-146.
4. J.-C. Therriault and M. Levasseur, Natur. Can. 112, 77-96 (1985). (1988).
5. D.M. Anderson and B.A. Keafer, Nature 325, 316-317 (1987).
6. A.D. Cembella and J.-C. Therriault, in: Red Tides: Biology, Environmental Science and Toxicology, T. Okaichi, D.M. Anderson and T. Nemoto, eds. (Elsevier, New York 1988) pp. 81-84.
7. J.Turgeon, A.D. Cembella and J.-C. Therriault, this volume (1989).
8. Y. Gratton, G. Mertz and J. Gagné, J. Geophys. Res. 93, 6947-6954 (1988).
9. M.I. El-Sabh, J. Fish. Res. Bd. Can. 33, 124-138 (1976).
10. M.I. El-Sabh and J. Benoit, Sc. Tech. Eau 17, 55-61 (1984).

CHROMATOGRAPHIC AND SPECTRAL EVIDENCE FOR THE PRESENCE OF MULTIPLE CIGUATERA TOXINS

Anne-Marie LEGRAND,* Phillippe CRUCHET,* Raymond BAGNIS,* Michio MURATA,** Yoshihiko ISHIBASHI** and Takeshi YASUMOTO**
*Institut Territorial de Recherches Medicales Louis Malarde, BP 30, Papeete, Tahiti, French Polynesia; **Faculty of Agriculture, Tohoku University, 1-1 Tsutsumidori-Amamiyamachi, Sendai 981, Japan

ABSTRACT

Ciguatera toxins were purified from moray eel viscera as well as from the causative dinoflagellate Gambierdiscus toxicus. Analyses by high performance liquid chromatography clearly indicated the presence of several minor toxins in moray eels in addition to ciguatoxin. One of the minor toxin was purified and suggested to have a molecular weight of 1056, which was smaller than that of ciguatoxin by three molecules of water. Four toxins detected in G. toxicus were less polar than moray eel ciguatoxin and were presumed to have less oxidized forms.

INTRODUCTION

Ciguatera is a term given to a disease caused by eating a variety of coral reef fishes and is most prevalent in the South Pacific area. The difficulty in dealing with ciguatera derives from the unpredictable nature of the fish toxicity and the implication of a large number of fish species. Even among the same species, fish toxicity may vary significantly from fish to fish and from place to place. For local fisheries economy, therefore, it is impractical to ban all the suspected fish species from the market because of their potential toxicity. In order to avoid risks of intoxication, however, it is necessary to have a rapid and sensitive method, such as an ELISA test to screen individual fish for toxicity before they are sent to the market. To develop such an assay, in turn, it is necessary to isolate the pure toxin and elucidate its chemical structure. A toxin named ciguatoxin was isolated from moray eel viscera by Scheuer's group in Hawaii [1-3] and has been presumed to be the principal toxin in ciguateric fishes, at least in carnivorous fishes. THe toxin was suggested to be a polyether compound on the basis of ^{1}H NMR data. However, its chemical structure has remained undetermined. Neither was it clear whether ciguatoxin was solely responsible for all the ciguatera syndromes or was accompanied by other toxins of related structures. Recently, three of the authors (MM, AML, TY) succeeded in determining the partial structure of ciguatoxin purified from moray eel viscera [4]. The presence of a primary alcohol at one terminal of the molecule suggested that we might selectively use this group to prepare a toxin-protein conjugate to immunize animals for an anti-ciguatoxin-antibody. The successful use of such an antibody may depend on the number and types of toxins present in fish. One of the aim of the present work is to survey and characterize the toxins in moray eel, which is one of the representative ciguateric fish at the top of the food chain. Another purpose of this work was to test the toxins (G.t-toxins) of the dinoflagellate Gambierdiscus toxicus, which had been presumed to be the origin of the fish toxin, [5, 6] to see whether the toxins are truly related to those in fish.

Toxic Marine Phytoplankton
Edna Graneli et al., Editors

MATERIALS AND METHODS

Materials

A total of 830 moray eels Gymnothorax(=Lycodontis) javanicus were collected from various archipelagoes of French Polynesia, and the viscera (125 kg including 43 kg of livers) were used for extraction of toxins. Samples of G. toxicus were collected in the Gambier Islands, French Polynesia, in 1979, by scratching the surface of dead corals, where the epiphytic dinoflagellates were proliferating on calcareous red algae, Jania sp. The organisms were roughly separated from coral and algal fragments by decantation and sieving [5, 6] and had been stored in a freezer at -20°C until used.

Extraction

The extraction of moray eel toxins was carried out following the procedures shown in Fig. 1. Further details of the procedures were described in a previous paper [7]. The dinoflagellate samples were extracted with acetone at room temperature. After removing the solvent by evaporation, the residue was suspended in water to be extracted first with diethyl ether and then with 1-butanol. After evaporation of the organic solvents, the residues were suspended in acetone and kept at low temperature to obtain G.t-toxins in the supernatant and maitotoxin in the precipitates. The toxins in the acetone solubles were treated first on Florisil columns (hexane-acetone 4:1, acetone-methanol 9:1, methanol) and then on a reversed phase silica column (Develosil ODS Lop) with solvents of varying concentration of aqueous methanol. Toxic fractions thus obtained were purified next by gel permeation chromatography on either a Sephadex LH-20 column or a Toyopearl HW-40 column. First gel permeation chromatography was carried out with methanol and the second with methanol-water (85:15). The whole procedures are shown in Fig. 2.

Moray eel viscera 125 kg
| Extracted wiht acetone
| Partitioned Et_2O/H_2O
| Defatted with $n-C_6H_{14}$

Toxic extract
| SiO_2 /$CHCl_3$-MeOH (97:3); (9:1)

$CHCl_3$-MeOH (9:1)
| Florisil
| /EtOAc; EtOAc-MeOH (9:1); (3:1)

EtOAc-MeOH (9:1)
| DEAE-cellulose
| /$CHCl_3$; $CHCl_3$-MeOH (1:1)

$CHCl_3$-MeOH (1:1)
| Sephadex LH-20 /$CHCl_3$-MeOH (6:4)

Toxic Fr.
| ODS cartidge /MeOH-H_2O (5:5); (6:4); (7:3); (8:2)

MeOH-H_2O (8:2)
| LiChroprep RP-18
| /MeOH-H_2O (8:2) twice
| /MeCN-H_2O (65:35) twice

CTX 350 µg (LD_{99} 330 ng/kg)

Fig.1. Isolation procedure of moray eel ciguatoxin

Acetone extract of wild G. toxicus
| Florisil

Acetone-MeOH (9:1)
| Develosil ODS Lop

90% MeOH & MeOH fr.
| Toyopearl HW-40
| /MeOH & 85% MeOH

Toxic fr.
| Develosil ODS-7
| /85% MeCN - 100% MeCN

Toxic fr.

Fig.2. Isolation of G.t-toxins

High Performance Liquid Chromatography (HPLC)

The last stages of the toxin purification were performed by HPLC. Ciguatoxin and a minor toxin-1 of the moray eels were chromatographed on a LiChrosorb RP-18 column (Brownlee, 10 μm, 4.6 x 250 mm) with methanol-water (8:2) and then on another LiChrosorb RP-18 column (5 um, 4.6 x 250 mm) with acetonitrile-water (65:35). Elution of less polar toxins in moray eels was carried out with methanol-water (9:1). Toxins of G. toxicus were chromatographed on a Develosil ODS-7 column (8 x 250 mm, Nomura Chemicals) by linear gradient starting from acetonitrile-water (85:15) to acetonitrile (0.5 ml/min for 100 min). Elution of toxins from the columns were monitored with a UV flowmonitor (215 nm) and by mouse bioassays as described previously [7].

Spectral Measurements

Fast atom bombardment (FAB) mass spectra of purified samples were measured on a JEOL JMS-DX303HF spectrometer under a positive mode. Glycerol, thioglycerol or 4-nitrobenzylalcohol was used as the matrix. An FTIR spectrum was taken on a Nicolet 7199 FT-IR spectrometer. ^{1}H NMR spectra were measured with a JEOL GSX-400 (400 MHz) spectrometer.

RESULTS AND DISCUSSION

From 125 kg of the moray eel viscera, 350 μg of the principal toxin was obtained in a pure form and was judged to be the same as ciguatoxin isolated by the Hawaii group, by comparison of ^{1}H NMR data [4, 7]. FTIR spectra of the toxin indicated absence of any carbonyl groups in the molecule. Intraperitoneal mouse lethality tests indicated ciguatoxin to have an LD_{50} of 0.33 μg/kg. Ciguatoxin accounted for about 90% of the mouse toxicity of the moray eel viscera. Three minor toxins had the polarity relatively close to that of ciguatoxin. One of such toxins, tentatively coded minor-1 was isolated in a pure form. The retention time of the minor-

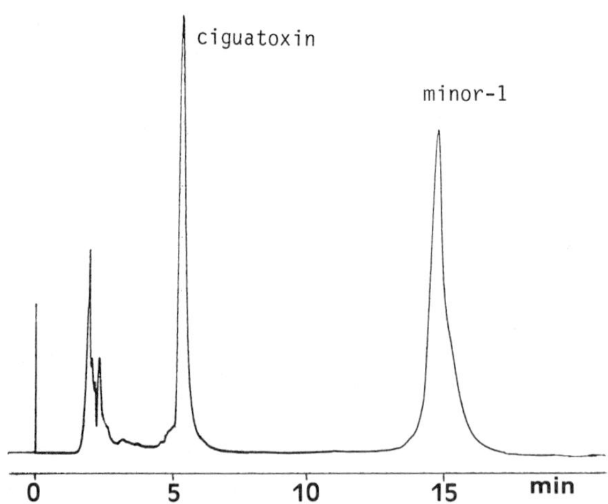

Fig. 3. HPLC chromatogram of ciguatoxin and toxin minor-1

Column: LiChrosorb RP-18-5 (ϕ 4.6 x 220 mm)
Solvent: 65% MeCN; Flow rate: 1.0 ml/min; Detection: 215 nm (UV)
*The peak heights do not reflect the relative ratio of the two toxins.

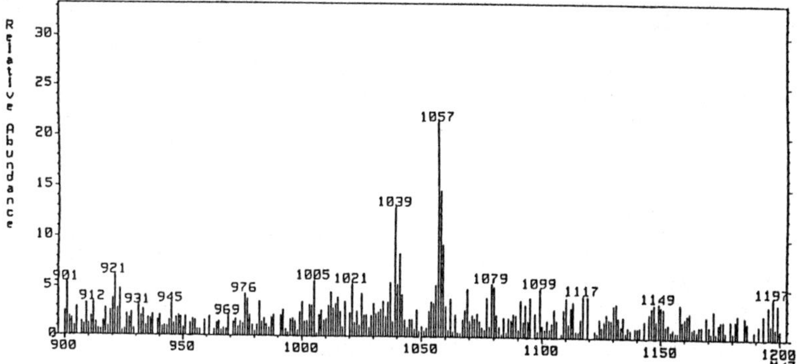

Fig. 4 FAB mass spectrum of the toxin minor-1.

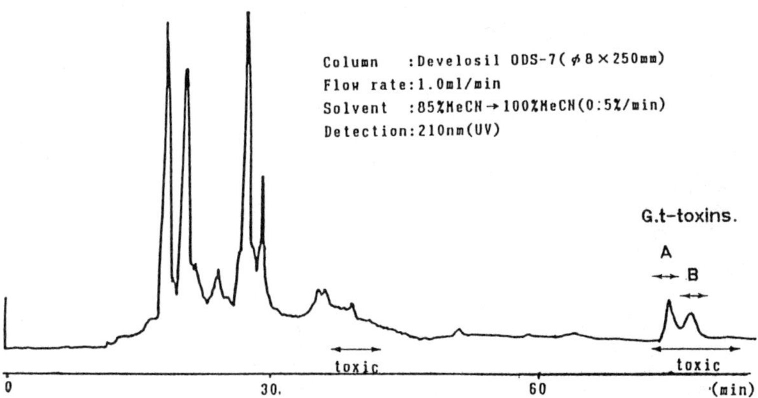

Fig. 5 HPLC chromatogram of G.t-toxins.

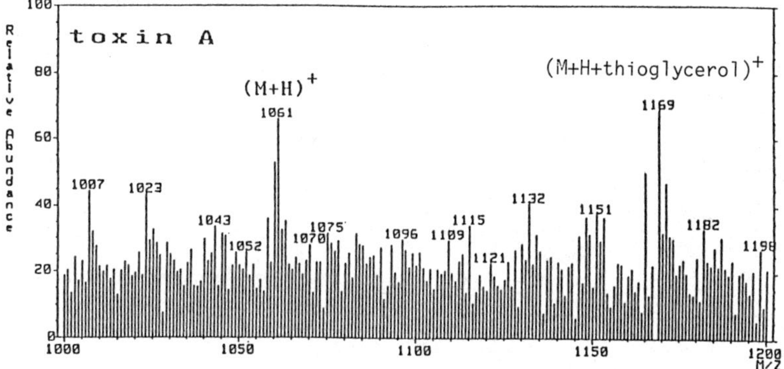

Fig. 6 FAB mass spectrum of G.t-toxin A.

1 from the LiChrosorb RP-18 column with acetonitrile-water (65:35) was 14.7 min, while that of ciguatoxin under the same condition was 5.5 min (Fig. 3). FAB mass spectra of the minor-1 showed an MH^+ ion at m/z 1057 (Fig. 4). The difference of mass unit between the MH^+ ion of ciguatoxin (m/z 1111) and that of the minor-1 (m/z 1057) corresponds to three molecules of water. The presence of four minor toxins were noticed in the less polar fraction of the moray eel extracts. However, characterization of these toxins were unsuccessful due to the smallness of samples.

Toxins in *G. toxicus* were tentatively named G.t-toxins. As shown in Fig. 5, mouse assays of the eluates indicated the presence of four toxins, but none of them coincided with ciguatoxin in retention times. Two components tentatively labeled as G.t-toxin **a** and G.t-toxin **b** gave MH^+ ions at m/z 1061 in FAB mass spectra. Fig. 6 shows the mass spectrum of G.t-toxin **a**. The difference of mass unit between ciguatoxin and G.t-toxins **a** and **b** was 50 mass unit, which suggested loss of two hydroxyls and one oxygen from ciguatoxin. Therefore, it is likely that the G.t-toxins are less oxidized forms of ciguatoxin, though further confirmation awaits comparison by ^1H NMR measurements.

Although we started extraction from a large number of moray eels, the extremely low concentration of toxins did not allow us to characterize all toxins by spectral data. Nevertheless, it is clear from the HPLC data that a number of minor toxins exist in moray eels in addition to ciguatoxin. It will be interesting to see whether the toxin profile observed in the viscera is the same in the flesh. It will be also interesting to test other fish for their toxin profiles.

G. toxicus did not contain ciguatoxin but contained less polar toxins. The mass spectral data suggested that they could be less oxidized forms of ciguatoxin. Chromatographic properties of G.t-toxins also resembled those of minor toxins in moray eel viscera. Efforts are being directed toward accumulation of all toxins for spectral comparison.

Acknowledgments

Authors are most grateful to Professor J. Roux, Director of the Louis Malarde Institute, for his continued support to the ciguatera program. Thanks are also due to Messrs. J. Bennett and M. Barsinas for collecting fish and algal samples, and to Mmes C. Lotte and J. Teore for technical assistances. This work was supported by the French Ministry of Research and Superior Education, Pasteur Institute in Paris, French Polynesia Government, and by a Grant-in-Aid for Oversea Research, the Ministry of Education, Science and Culture, Japan.

REFERENCE

1 K. Tachibana, "Structural Studies on Marine Toxins", Ph.D. Thesis, University of Hawaii, 1980.
2 M. Nukina, L.M. Koyanagi and P.J. Scheuer, Toxicon, 22, 169-176 (1984).
3 K. Tachibana, M. Nukina, Y-G. Joh and P.J. Scheuer, Biol. Bull., 172, 122-127 (1987).
4 M. Murata, A.M. Legrand and T. Yasumoto, Tetrahedron Lett., in press (1989).
5 T. Yasumoto, I. Nakajima, R. Bagnis and R. Adachi, Nippon Suisan Gakkaishi, 43, 1021-1026 (1977).
6 R. Bagnis, S. Chanteau and T. Yasumoto, C.R. Acad. Sc. Paris, t.285, Serie D-105 (1977).
7 A.M. Legrand, M. Litaudon, J.N. Genthon, R. Bagnis and T. Yasumoto, J. Appl. Phycol., in press (1989).

ANATOMICAL DISTRIBUTION OF PARALYTIC SHELLFISH TOXINS IN SOFT-SHELL CLAMS

JENNIFER L. MARTIN,* ALAN W. WHITE,** AND JOHN J. SULLIVAN***
*Department of Fisheries and Oceans, Aquaculture and Invertebrate Fisheries, Biological Station, St. Andrews, New Brunswick E0G 2X0 Canada; **Sea Grant Program, Woods Hole Oceanographic Institution, Woods Hole, Massachusetts 02543 U.S.A.; ***U.S. Food and Drug Administration, Seafood Products Research Center, Seattle, Washington 98174 U.S.A.

ABSTRACT

Soft-shell clams, *Mya arenaria*, were collected from two sites in the southwestern Bay of Fundy in January, April, May, July, September and November, 1986. The sites were Crow Harbour, where clams exceed the quarantine toxin level year-round, and Lepreau Harbour, where clams exceed the quarantine level only during the annual, summer *Alexandrium* bloom. Clams were dissected into digestive gland, gonad, gills, muscles (siphon, foot, pallial muscle and adductor muscle), and "remainder" which consisted of a small amount of material near the digestive gland and included the kidney, heart and brown gland. The concentration of total toxins (μg STX equiv./100 g) in the "remainder" was up to 10 times greater than in the other tissues throughout the year except in July. During the *Alexandrium* bloom in July, the digestive gland also contained high toxin concentrations; much lower levels occurred in the gills and gonads. In the other months, the gills contained about the same toxin concentrations as the digestive gland.

The patterns of abundance of the individual toxins in the various anatomical parts were similar within and between collection sites. In July, the sequence of abundance of the individual toxins in the clams was GTX I, GTX IV and GTX III, with STX usually the lowest of the six or seven detectable toxins. The sequence in July plankton dominated by *Alexandrium* was GTX IV, GTX III, NEO, and GTX I. Clam samples from the other months showed a striking difference from the July pattern; STX was nearly always present at the highest level, often followed by NEO, and GTX IV was always the lowest. Results indicate that soft-shell clams contain high toxin concentrations in parts other than the digestive gland. Further, results suggest that interconversion or selective retention of the toxins occurs within these animals.

INTRODUCTION

Blooms of *Alexandrium fundyense* are an annual summer occurrence in the Bay of Fundy and are responsible for soft-shell clams, *Mya arenaria*, accumulating paralytic shellfish toxins while filter feeding which result in the closure of most areas to harvesting. Shellfish toxicities for the southern Bay of Fundy behave as a unit and follow the *Alexandrium* bloom closely due to the tremendous turbulence and mixing within the system. Since 1980, several soft-shell clam sites have remained closed year round due to low, but unacceptable levels of PSP toxins. Although prior reports suggested the use of ozone for detoxifying clams [1], our efforts to do so from a permanently closed area such as this, with toxins stored for long periods, proved unsuccessful [2]. This study was undertaken to determine whether clams from an area where natural detoxification does not occur behave in a similar manner to those from an area where detoxification occurs, and to determine how the toxins are transformed or retained.

Published 1990 by Elsevier Science Publishing Co., Inc.
Toxic Marine Phytoplankton
Edna Graneli et al., Editors

MATERIALS AND METHODS

During January, April, May, July, September and November of 1986, soft-shell clams, *Mya arenaria,* were collected from Crow Harbour and Lepreau Harbour, New Brunswick. Clams were kept moist by covering with seaweed during transport (1-2 hours) to the laboratory and refrigerated at 2°C overnight. Since preliminary tests showed a decreased range in shellfish toxicity in clams of uniform size, only those with shell lengths from 4.0 to 6.0 cm were used for these experiments. For each site and each sampling, 20 clams were analyzed for whole tissue concentrations, and 55 clams for tissue and organ concentrations (digestive glands, gonads (including visceral mass), gills (plus mantle), muscles (pallial and adductor muscles plus siphon and foot) and remainder (heart, kidney and brown or Keber's gland) (Fig. 1)). The whole clams or parts were processed for "6-mouse" bioassays [3] and subsamples were removed from each to be centrifuged. The supernatant was passed through YM-10 ultrafiltration membranes, frozen and shipped to Seattle, Washington for HPLC analysis [4].

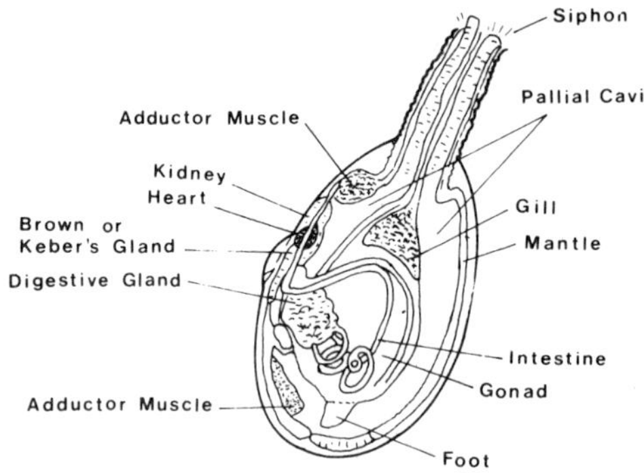

FIG. 1. Internal anatomy of the soft-shell clam.

To determine the occurrence of the PSP toxins in nature, plankton was collected with a 20-μm mesh net of 0.5 m diameter during the bloom of *A. fundyense* from three locations in southwestern Bay of Fundy: Station Prince 5 (44°57'N, 66°51'W), near the Wolves Islands (45°00'50"N, 66°47'00"W), and northeast of Grand Manan Island (44°51'97"N, 66°41'65"W) (Fig. 2). Net contents were kept on ice during the 2- to 3-hour return trip to the laboratory where the plankton was processed for toxin concentration and cell counts were determined according to the method described by White [5].

RESULTS AND DISCUSSION

Comparison between the mouse bioassay and the HPLC technique showed a good correlation between the two methods. The HPLC results in most cases were higher than those for the mouse bioassay, except at toxicities less than 30 μg STX equiv./100 g and toxicities greater than 1000 μg STX equiv./100 g where the HPLC values tended to slightly underestimate the total toxicity. Figure 3 shows the only exception to the latter from all our results - the total toxicity at Crow Harbour determined by HPLC during the *Alexandrium* bloom (July) was 78% greater than that determined by mouse bioassay. Sullivan et al. [6] observed similar results when comparing the two methods using a variety of species.

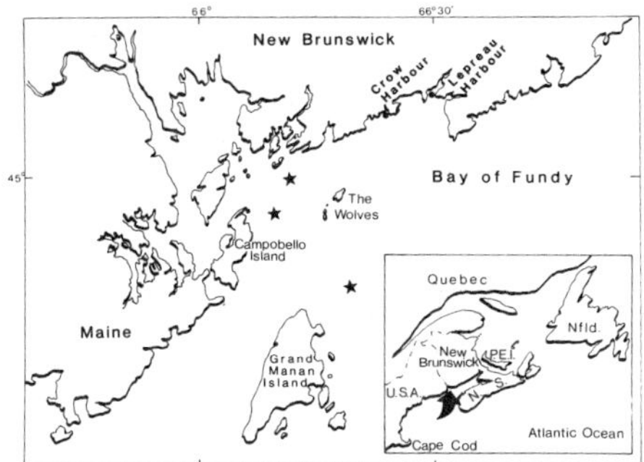

FIG. 2. Map of sampling area.

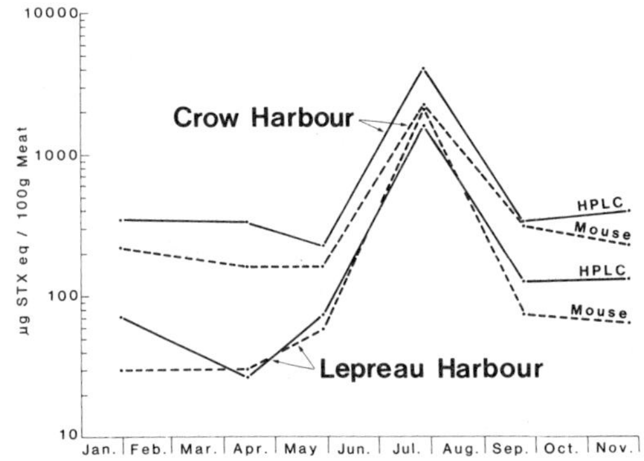

FIG. 3. Variation in PSP toxin levels in whole soft-shell clam extracts.

In general, the toxicities from whole tissue extracts show similar patterns for clams from Crow Harbour and Lepreau Harbour throughout the year (Fig. 3). Total toxin levels for Crow Harbour were from 160-310 (mouse bioassay) and 229-397 (HPLC) µg STX equiv./100 g during non-bloom periods (September-May), with a high in July of 2260 (mouse) and 4039 (HPLC) µg STX equiv./100 g meat, causing this particular area to remain closed to harvesting throughout the year. Similarly, at Lepreau Harbour, shellfish toxicities peaked in July with toxicities of 2103 (mouse) or 1602 (HPLC) µg STX equiv./100 g meat but dropped to acceptable levels by mouse bioassay (<30-75 µg STX equiv./100 g) to permit harvesting between September and May. Total toxicities for the individual organs from Crow Harbour and Lepreau Harbour also showed similar patterns. The remainders contained the most toxins throughout the year, except at Lepreau Harbour in July, and were followed by digestive glands, gills or gonads and muscles (Fig. 4). These results differ from those of Medcof et al. where toxins tended to migrate to the gills in the off-season [7]. Toxin levels increased to maximum levels for all tissues in July, one week after concentrations of *Alexandrium* reached 9.3 x 10^4 cells/L northeast of Grand Manan (Fig. 2). Earlier studies in the Bay of Fundy indicate highest concentrations of *A. fundyense* cells tend to be located offshore (northeast of Grand Manan) [8] from where they are dispersed and result in inshore shellfish toxicities. Thus, the rise in toxicity closely follows the bloom. PSP toxins were found above the level of 30 µg STX equiv./100 g in all tissues throughout the year from Crow Harbour, while the gills, muscles and gonad from Lepreau Harbour were less during April and May, resulting in a lower overall toxicity from these clams (Fig. 4).

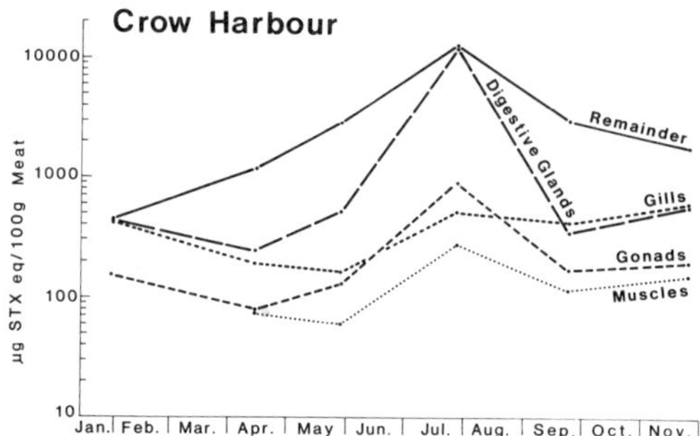

FIG. 4. Total toxin content for gills, gonads, digestive glands, muscles and remainder.

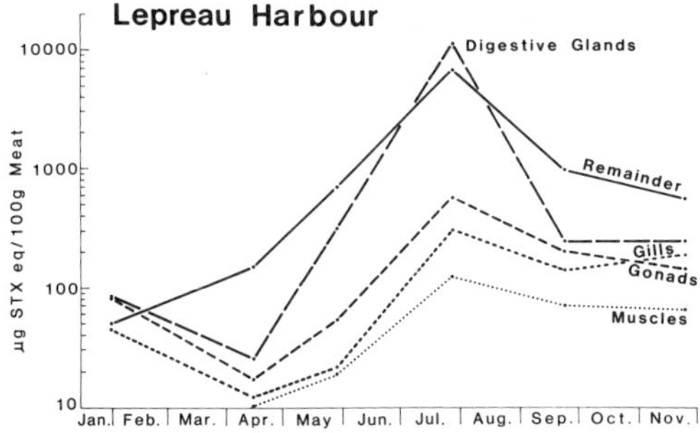

Amount of toxins in each tissue can be calculated as a percentage of the total toxins. Highest percentages were concentrated in the remainder throughout the year in Crow Harbour clams and during September and November in Lepreau Harbour clams. July was the exception for both sites when most of the toxins were located in the digestive glands (Fig. 5). This agrees with Medcof et al.'s [6] work in 1945 on toxin distribution in *Mya* from the Bay of Fundy where they also found the digestive glands to be the most toxic when the shellfish are acquiring the toxins from their food. A rise in percentage for digestive gland toxin content in May coincides with the annual early *Alexandrium* "mini" bloom that is the precursor to the July bloom (Martin, unpublished). All other tissues (gonads, gills, muscles) retained consistently low percentages of total toxins throughout the year. In July, clam toxicity values reached levels of 4039 and 1602 (HPLC) µg STX equiv./100 g at Lepreau Harbour and Crow Harbour, respectively, with clams at Lepreau Harbour able to eliminate the toxins sufficiently to allow harvesting, whereas those from Crow Harbour retained 65-70% of toxins in the remainder, and resulting in the prohibit of harvesting throughout the year. Although high percentages of toxins were retained in the remainder, when toxins from this source were combined with the rest of the clam during homogenization, the overall toxin value ranged from 160-397 µg STX equiv./100 g. These results suggest either that there may be strain variation or difference in selective retention in clams between the two areas where high percentages of toxins are stored in the kidney or brown gland of Crow Harbour clams during the non-bloom periods or the clams from Crow Harbour contained 2.5 times the levels of toxins from Lepreau Harbour (by HPLC analysis) in July and were unable to detoxify sufficiently during the non-bloom months to permit harvesting.

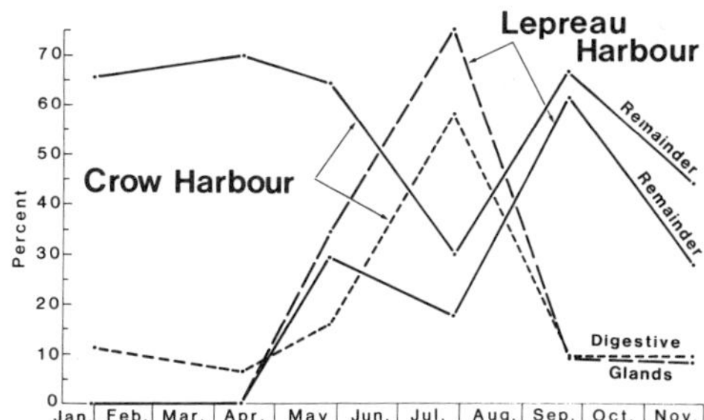

FIG. 5. Percent total body burden of toxins in digestive glands and remainder

The toxin content of *A. fundyense* cells collected from Prince 5, The Wolves and Northeast Grand Manan (Fig. 2) ranged from 7.37 x 10^{-5} to 8.87 x 10^{-5} µg STX equiv./cell. Although most analyses pool GTX I and GTX II with their respective isomers, GTX IV and III, due to difficulties in establishing equilibrium ratios, the toxin data from this experiment have been left in the raw form and the toxins treated individually. Eight of the PSP toxins were detected by HPLC analysis from the plankton in the following order of abundance: GTX (gonyautoxin) IV, GTX III, NEO (neosaxitoxin), GTX I, STX (saxitoxin), GTX II and C1/C2. Results indicate that transformation of toxins from the swimming cells occurs fairly rapidly upon ingestion by the clams. Soft-shell clams sampled in July during the bloom showed highest concentrations of GTX I, GTX IV and GTX III (with STX, GTX II, NEO, C1/C2 present at lower levels) in all tissues from Lepreau Harbour. Similar profiles were observed in the remainder, muscles and digestive glands from Crow Harbour whereas dominant toxins from the other tissues were: gills - GTX I and IV followed by STX; gonads - GTX I, GTX III and NEO; and whole clams - GTX IV, GTX III and NEO. The toxin composition through the rest of the year did not alter as markedly. STX was the main component for Crow Harbour clams from all extractions whereas at Lepreau Harbour this applied only for the remainder and muscles. In gills, gonads, digestive gland and whole clams from Lepreau Harbour, STX, NEO and GTX I or III tended to interchange as the most abundant.

Previous studies also show STX to be the major component in toxins in the butter clam (*Saxidomus giganteus*) [9] and scallops (*Placopectin magellanicus*) [10] and suggest that toxins may be enzymatically (metabolically) converted to STX or selectively retained as STX. The ability of Lepreau Harbour clams to convert major toxins from STX to NEO or GTX I or III and lower toxin levels to acceptable levels for harvesting may be either due to a variation in strains or the presence of different enzymes creating different toxin transformation than those from Crow Harbour. Kotaki et al. have demonstrated the ability of bacteria to transform GTX I, II or III to STX [11]. This further suggests the possibility that Crow Harbour and Lepreau Harbour may possess different bacteria - either within the clams or in the environment.

Results of this investigation indicate that although highest concentrations of PSP toxins occur in the digestive gland of the soft-shell clam during the bloom, the toxins tend to migrate to the remainder during the off-season. Included in the remainder were the heart, kidney and brown or Keber's gland. Although the exact function of the brown gland is uncertain, it is thought to function as a part of the excretory system [12]. Results suggest the PSP toxins are transferred within a short time from the digestive gland to the excretory system and can be retained within this system for an extended period.

This study suggests the possibility that the toxins may be either metabolically transformed by either bacteria, clams or the cells themselves or chemically altered. This may relate to variation in the ability of clams to detoxify from different areas and the storage of high concentrations or selective retention of STX in the remainder of the Crow Harbour clam during non-bloom periods.

ACKNOWLEDGMENTS

We thank B. Best, S. Bellis, F. Cunningham and B. McMullon for preparation and drafting of the manuscript. Support for A.W. White was in part by NOAA National Sea Grant College Program Office, Department of Commerce, under Grant No. NA86-D-SG090, WHOI Sea Grant Project No. M/0-2.

REFERENCES

1. W.J. Blogoslawski, M.E. Stewart, J.W. Hurst and F.G. Kern, Toxicon 17, 650-654 (1979).
2. A.W. White, J.L. Martin, M. LeGresley and W.J. Blogoslawski in: Toxic Dinoflagellates, D.M. Anderson, A.W. White and D.G. Baden, eds. (Elsevier, New York, 1985) pp. 473-478.
3. Official Methods of Analysis, 14th Edition, Procedure 18.086.092 (Association of Official Analytical Chemists, Washington, D.C. 1984).
4. J.J. Sullivan and M.M. Wekell in: Seafood Quality Determination, D.E. Kramer and J. Liston, eds., Sea Grant College Program, Anchorage, AK, pp. 357-371 (1988).
5. A.W. White, Toxicon 24, 605-610 (1986).
6 J.J. Sullivan, M.G. Simon, W.T. Iwaoka, J. Food Sci. 48, 1312-1314 (1983).
7. J.C. Medcof, A.H. Leim, A.B. Needler, A.W.H. Needler, J. Gibbard and J. Naubert, Bull. Fish. Res. Board Can. LXXV, 1-32 (1947).
8. J.L. Martin and A.W. White, Can. J. Fish. Aquat. Sci. 45, 1968-1975 (1988).
9. J.J. Sullivan, Paralytic shellfish poisoning Analytical and biochemical investigations, Ph.D. Thesis, University of Washington, 232 pp. (1982).
10. Y. Shimizu and M. Yoshioka, Science 212, 547-549 (1981).
11. Y. Kotaki, Y. Oshima and T. Yasumoto in: Toxic Dinoflagellates, D.M. Anderson, A.W. White and D.G. Baden, eds.(Elsevier, New York, 1985) pp. 287-292.
12. A.W. Martin and F.M. Harrison in: Physiology of Mollusca, Vol. II, K.M. Wilbur and C.M. Yonge, eds., pp. 353-386 (1966).

SCREENING OF MARINE PHYTOPLANKTON FOR ANTIFUNGAL SUBSTANCES

HIROSHI NAGAI*, MASAYUKI SATAKE**, MICHIO MURATA** AND TAKESHI YASUMOTO**
*Taiyo Central R&D Institute, Taiyo Fishery Co., LTD. 3-2-9, Tsukishima, Chuo-ku, Tokyo; **Faculty of Agriculture, Tohoku University, 1-1 Tsutsumidori-Amiamiyamachi, Sendai 981, Japan

ABSTRACT

Production of antifungal substances by *Chattonella antiqua, Olisthodiscus luteus, Alexandrium tamarense, Gambierdiscus toxicus, Gymnodinium sanguineum, Amphidinium carteri,* and *Amphidinium klebsii,* was confirmed. The active compound of *A. klebsii* inhibited the growth of *Aspergillus niger* at 6.2 μg/disk and was characterized by the presence of a sulfate group and a probable molecular weight of 1510. The active component of *G. toxicus* was extremely potent (10 ng/disk) and was presumed to be a polyether having the molecular weight of 1184.

INTRODUCTION

Marine phytoplankton may produce various types of bioactive substances. However, past studies have been devoted to find toxins involved in either food poisonings or massive fish kills. A number of toxins thus discovered can be classified as polyether compounds. During the course of testing their biological activities, we found many of the polyether toxins to have potent antifungal activities. Representative examples are okadaic acid (1,2), dinophysistoxin-1 (3), desulfated yessotoxin (4), and ciguatoxin (5). The results suggest the possibility that antifungal substances are rather common products of marine phytoplankton but have remained undetected because mouse bioassays or ichthyotoxicity tests have been the routine practice in screening phytoplankton. Recent discovery of antifungal compounds, goniodomins, from *Alexandrium hiranoi (Goniodoma pseudogonyaulax)* (6) and potent antineoplastic macrolides, amphidinolide A, B and C (7,8,9), from *Amhidinium* sp. further proved phytoplankton to be a promising source of the anti-eucaryotic compounds. As these substances may have potential medicinal value and also play some roles in the marine ecosystem, we initiated a search for antimicrobial substances among phytoplanlkton. In the present paper we report isolation and characterization of two potent antifungal compounds from dinoflagellates.

MATERIALS AND METHODS

Materials

Eutreptiella sp. was collected from a natural bloom. Other species were cultured in 3-liter Fernbach flasks. Code numbers in parentheses denote culture strains of the National Institute for Environmental Studies, The Environmental Agency, Japan. SW-2 medium (10) was used for *Prymnesium parvum;* and f-1 culture medium (11) for *Chattonella antiqua* (NIES-1), *Olisthodiscus luteus* (NIES-15), *Heterosigma akasiwo* (NIES-6), *Alexandrium tamarense, Prorocentrum micans* (NIES-12), *Scrippsiella trochoidea* (NIES-369), and *Gymnodinium sanguineum* (NIES-11); ES-1 medium (12) for *Gambierdiscus toxicus, Amphidinium carteri* and *Amphidinium klebsii.* Cultures were maintained at 25°C, except for *A. tamarense* and *G. sanguineum* which were cultured at 15°C, under illumination of 2000-4000 Lx with 18h light and 6h dark cycles.

Toxic Marine Phytoplankton
Edna Graneli et al., Editors

Screening methods

Cultured phytoplankton were harvested by filtration after the cells had reached stationary phases. Harvested cells were homogenized with one hundred times volume of methanol thrice at room temperature and filtered. After removing methanol by evaporation, the residues were partitioned between water and diethyl ether. The aqueous layer was then extracted with 1-butanol. The algal residues after methanol extraction was further extracted with acetone. The filtered culture medium after harvesting cells was applied to an Amberlite XAD-2 column. The column was washed with water, and then the adsorbed materials were recovered by methanol. Each fraction thus obtained was evaporated to dryness, and the residue was redissolved in methanol or chloroform. Aliquots of the solutions corresponding to 100 ml of the cultures were subjected to antifungal activity tests against *A. niger*.

Antimicrobial tests

Antimicrobial tests were performed against two fungi *Aspergillus niger* and *Penicillium funiculosum*; one Gram-positive bacterium *Bacillus subtilis*; and one Gram-negative bacterium *Escherichia coli*. Test solutions were prepared by dissolving respective extracts into methanol or chloroform. Appropriate amounts of the test solutions were applied on filter-paper disks of an 8 mm diameter. The paper disks were placed on agar plates seeded with the test microorganisms. Assays for the fungi were carried out with an yeast agar medium, and those for the bacteria with a beef extracts medium. The plates were incubated at 30°C and inhibition circles formed around the disks were measured.

Hemolysis tests

Immunoplate wells were filled with 100 μl portions of 0.5% mouse blood cell suspension in saline. The blood cell suspensions were mixed with 10-50 μl of methanol solutions of the test materials, incubated for 30 min at 30°C, and examined for hemolysis.

Isolation of the active compound from *A. klebsii*

A. klebsii cultured for three weeks were harvested by filtration and extracted three times with methanol at room temperature. The combined methanol extract was evaporated to dryness, and partitioned between water and diethyl ether. The aqueous layer was further extracted with 1-butanol. The active component extracted in the 1-butanol fraction was purified by gel-permeation chromatography on Toyopearl HW-40 (methanol-water, 1:1) and Sephadex LH-20 (methanol). Eluates were checked by antifungal tests against *A.niger* and also by hemolysis tests.

Isolation of the active compound from *G. toxicus*

G. toxicus cultured for 38 days was harvested by filtration. Tests on the culture filtrate and methanolic cell extracts indicated that the larger amount of antifungal activity was in the culture filtrate. The culture filtrate was passed through an Amberlite XAD-2 column. The column was washed with water and the active compound was recovered by washing the column with methanol. The residue obtained after evaporation of methanol was suspended in water, extracted first with diethyl ether and then with 1-butanol. The antifungal compound obtained in the 1-butanol fraction was chromatographed first on an HW-40 column (methanol-water 1:1) and next on a Develosil ODS 15/30 column (methanol-water 1:1, 7:3, and methanol). The active compound obtained in the methanol eluate from the second column was further purified on reversed phase Develosil ODS columns (acetonitril-water 9:1) and on a

TABLE I. Antifungal activity of extracts from phytoplankton.

Phytoplankton	Fraction				
	H_2O layer	1-BuOH layer	Et_2O layer	Acetone extracts	XAD-2 MeOH
(Haptophyceae)					
Prymnesium parvum	-	-	-		-
(Euglenophyceae)					
Eutriptiella sp.	-	-	-		-
(Raphidophyceae)					
Chattonella antiqua (NIES-1)	-	+	-		-
Olisthodiscus luteus (NIES-15)	-	-	-	+	-
Heterosigma akashiwo (NIES-6)	-	-	-	-	-
(Dinophyceae)					
Alexandrium tamarense	-	+	-	-	-
Prorocentrum micans (NIES-12)	-	-	-	-	-
Scrippsiella trochoidea (NIES-369)	-	-	-	-	-
Gambierdiscus toxicus	-	+	-	-	+
Gymnodinium sanguineum (NIES-11)	-	+	-	-	-
Amphidinium carteri	-	+	-	-	-
Amphidinium klebsii	-	+	-	-	-

Test: Paper Disk Method against Aspergillus niger
Aliquots of the test solutions are corresponding to 100ml of the cultures
+ :Positive - :Not detected Blank :Not tested

Develosil silica 60-5 column (chloroform-methanol-water 200:10:1). Elution of the compound was monitored by the antifungal tests.

Spectral measurements

The UV spectra were recorded on a Hitachi 200-10 spectrophotometer, and the IR on a JASCO A-202 spectrometer. FAB mass spectra were taken on a JEOL JMS DX-300HF spectrometer. ^1H NMR spectra were measured with JEOL GX-500 (500 MHz) and JEOL JNM GSX-400 (400 MHz) spectrometers, and ^{13}C NMR spectra with the JEOL JNM GSX-400 spectrometer.

RESULTS

Screening tests

Results of the screening for antifungal activity are shown in Table I. The activity was detected in butanol solubles of six species: *C. antiqua, A. tamarense, G. toxicus, G. sanguineum, A. carteri,* and *A. klebsii*. Acetone extracts of *O. luteus* and the XAD-2 retentate of *G.toxicus* culture filtrate also had the activity.

The active compound of *A. klebsii*

From 75 l culture, 3.1 mg of pure antifungal compound was obtained. It produced growth inhibition circles against *A. niger* at a concentration of 6.2 μg/disk. The potency was eight times that of amphotericin B, an antifungal drug for treatment of systemic mycoses. The compound hemolyzed mouse blood cells at 84 ng/ml, indicating that it was 120 times more potent than commercial saponin (Merck). The UV spectrum showed absorption maxima at 226, 259, 270 and 281 nm (in methanol). The IR spectrum showed characteristic absorptions at 3400, 1620 and 1220 nm, which suggested the presence of hydroxyl group, olefinic bonds and sulfate ester, respectively. FAB mass spectra (positive) showed an MH^+ ion at m/z 1511. A fragment ion at

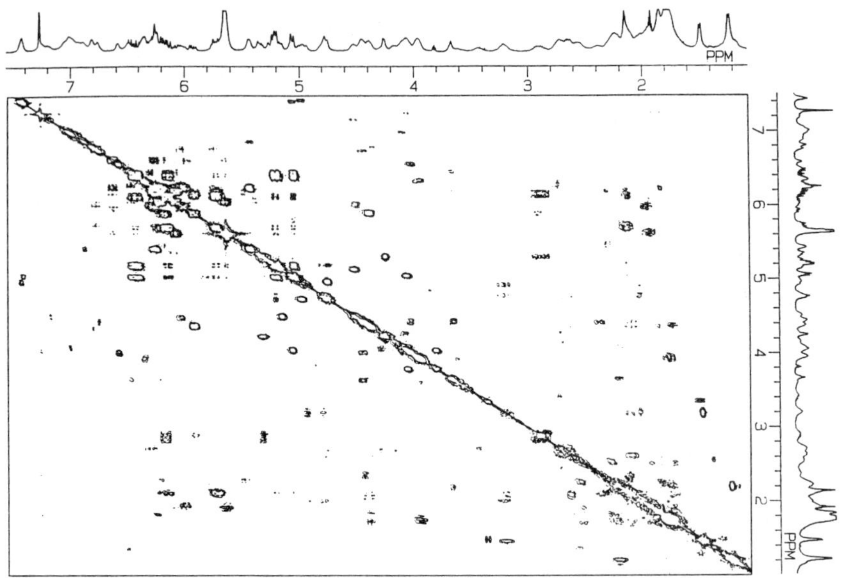

FIG. 1. ^{1}H-^{1}H COSY spectrum of the antifungal substance from Amphidinium klebsii in pyridine-d5 at -20oC (JEOL, GSX-400, 400 MHz).

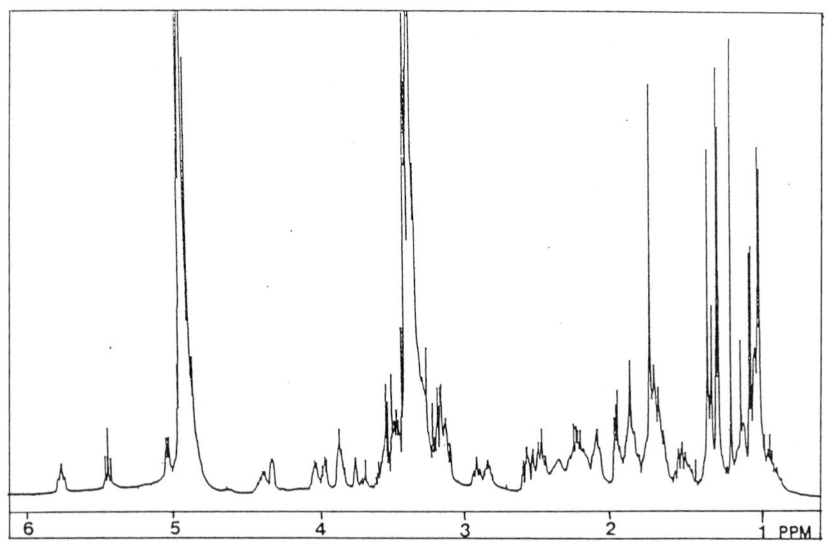

FIG. 2. ^{1}H NMR spectrum of the antifungal substance from Gambierdiscus toxicus in methanol-d4 (JEOL, GSX-500, 500 MHz)

m/z 1409 indicated desulfonation ($-SO_3Na+H$,102). FAB mass negative ion spectra showed an $(M-H)^-$ ion at m/z 1509 and two prominent ions at m/z 1487 $(M-Na)^-$ and 1465 $(M-2Na+H)^-$. The spectral data suggested the presence of at least one sulfate group in the molecule and the molecular weight as a sodium salt to be 1510. ^1H NMR spectral data were obtained in a benzene-d_6 - methanol-d_4 - deuterium oxide (10:10:1) solution. Signals observed included three methyls (two doublets, and one singlet on an olefinic carbon), oxymethine and oxymethylene protons between 3.2 and 4.8 ppm, and olefinic protons between 4.9 and 6.4 ppm. ^{13}C NMR revealed 18 olefine carbons. Probable partial structures deduced from ^1H-^1H COSY spectra (Fig. 1) suggests the compound is a polyoxy compound. The molecule seems to have an exovinyl at one terminal and a glycol at the other end.

The active compound from *G. toxicus*

The pure active compound (2.5 mg) obtained from 1000 l of *G. toxicus* inhibited the growth of *A. niger* and *P. funiculosum* at a concentration of 10 ng/disk. The potency is 5000 times that of amphotericin B. No growth-inhibiting activity against *B. subtilis* and *E. coli* was observed at 1 µg/disk. Maitotoxin, another highly potent product of *G. toxicus,* did not inhibit the growth of *A. niger* at 10 µg/disk. No absorbtion maxima was observed over 220 nm in the UV spectrum (in methanol). FAB mass spectra showed an MH^+ ion at m/z 1185. ^1H NMR spectra taken in methanol-d_3 (Fig. 3) showed the presence of eight methyls, protons on oxycarbons from 2.8 to 4.5 ppm, and olefinic protons from 5.0 to 5.8 ppm. The distribution of oxymethine and oxymethylene protons over a wide range inferred the polyether nature of the compound.

DISCUSSION

Out of 12 phytoplankton species tested, seven species showed antifungal activity. Among seven dinoflagellate species, five produced antifungals. The result suggest that marine phytoplankton, especially dinoflagellates, are a rich source of antifungal compounds, probably comparable with *Actinomyces.* The high ratio of active species coincides with the fact that many dinoflagellates produce highly bioactive polyethers such as okadaic acid (1,2), dinophysistoxin-1 (3), pectenotoxin-2 (13,14), maitotoxin (15), ciguatoxin (5), and brevetoxins (16,17). The spectral features of the two antifungals obtained in this study suggest that they also belong to the same class. The high antifungal activity of the two compounds, especially the one from *G. toxicus,* is worth special attention. If proper measures are developed for mass culture, dinoflagellates could serve as a source of useful drugs. The production of potent antifungals by dinoflagellates is interesting from the point of marine ecology. The amount ot antifungal compound of *G. toxicus* was ten times higher in the culture water than in the cells. The solubility and the chromatographic properties of the compunds suggests that they may not diffuse quickly but stay near the cells, perhaps adsorbed on organic or inorganic surfaces. It is likely, therefore, that these compounds exert an alleopathic function in the ecosystems. Effects of the compounds on organisms other than fungi will be tested after accumulation of enough materials.

REFERENCES

1. T. Tachibana, P.J. Scheuer, Y. Tsukitani, H. Kikuchi, D.V. Engen, J. Clardy, Y. Gopichand, and F.J. Schmitz, J. Am. Chem. Soc. 103, 2469-2471 (1981).

2. Y. Murakami, Y. Oshima, and T. Yasumoto, Nippon Suisan Gakkaishi 48, 69-72 (1982).
3. M. Murata,M. Shimatani, H. Sugitani, Y. Oshima, T. Yasumoto, Nippon Suisan Gakkaishi 48, 549-552 (1982).
4. M. Murata, M. Kumagai, J.S. Lee and T. Yasumoto, Tetrahedron Lett. 28, 5869-5872 (1987).
5. P.J. Scheuer, W. Takahashi, J. Tsutsumi and T. Yoshida, Science 155, 1267-1268 (1967).
6. M. Murakami, K. Makabe, K. Yamaguchi, S. Konosu, and M.R. Wälchli, Tetrahedron Lett. 29, 1149-1152 (1988).
7. J. Kobayashi, M. Ishibashi, H. Nakamura, Y. Ohizumi, T. Yamasu, T. Sasaki, and Y. Hirata, Tetrahedron Lett. 27, 5755-5758 (1986).
8. M. Ishibashi, Y. Ohizumi, M. Hamashimma, H. Nakamura, Y. Hirata, T. Sasaki and J. Kobayashi, J. Chem. Soc., Chem. Commun. 1987, 1127-1129.
9. J. Kobayashi, M. Ishibashi, M.R. Wälchli, H. Nakamura, Y Hirata, T. Sasaki and T. Ohizumi, J. Am. Chem. Soc. 110, 490-494 (1988).
10. H. Iwasaki, Biol. Bull. 121, 173-187 (1961).
11. R.R.L. Guillard and J.H.Ryther, Can. J. Microbiol. 8, 229-239 (1962).
12. L. Provasoli, Proceedings of the U.S.-Japan Conference held at Hakone, September 12-15, A. Watanabe and A. Hattori eds (Jpn Soc. Plant Physiol. 1966) pp. 63-75.
13. T. Yasumoto, M. Murata, Y. Oshima, G.K. Matsumoto and J. Clardy, Tetrahedron 41, 1019-1025 (1985).
14. A.M. Legrand, M. Litaudon, J.N. Genthon, R. Bagnis and T. Yasumoto, J. Appl. Phycol. (1989)in press.
15. A. Yokoyama, N. Murata, Y. Oshima, T. Iwashita, and T. Yasumoto, J. Biochem. 104, 730-733 (1988).
16. Y-Y. Lin, M. Risk, S.M. Ray, D.V. Engen, J. Clardy, J. Golik, J.C. James, K. Nakanishi, J. Am. Chem. Soc. 103, 6773-6775 (1981).
17. Y. Shimizu, H-N. Chou, H. Bando, G.V. Duyne,.J.C. Clardy, J. Am. Chem. Soc. 108, 514-515 (1986).

COMPARATIVE STUDIES ON PARALYTIC SHELLFISH TOXIN PROFILE OF DINOFLAGELLATES AND BIVALVES

YASUKATSU OSHIMA, KIYOKO SUGINO, HIJIRI ITAKURA, MAYUMI HIROTA AND TAKESHI YASUMOTO
Faculty of Agriculture, Tohoku University, 1-1 Tsutsumidori Amamiyamachi, Sendai 981, Japan

ABSTRACT

Paralytic shellfish toxins in wild and cultured dinoflagellates as well as infested shellfish from various countries were determined by means of an improved HPLC technique. Gymnodinium catenatum from Australia, Spain and Japan contained N-sulfocarbamoyl derivatives in large proportion, although relative abundance of each toxin differed among isolates. As minor components, unusual decarbamoyl toxins and a new toxin, 13-deoxycarbamoylsaxitoxin, were detected in the same species. Pyrodinium bahamense var. compressa and toxic shellfish in tropical Pacific were characterized by complete absence of 11-hydroxysulfate derivatives. Great variation of toxin profile was observed among Alexandrium tamarense and A. catenella. Total toxin content in one cell varied significantly depending on culture conditions while toxin profile remained fairly constant. Comparison of toxin profiles between bloom organisms and contaminated shellfish indicated rapid degradation of N-1 hydroxy toxins and conversion of 11-β-hydroxysulfate toxins to 11-α epimers during accumulation process in scallops, mussels, and oysters.

INTRODUCTION

Up to now dinoflagellates belonging to three genera are known to produce paralytic shellfish toxins (Fig. 1). Alexandrium (Protogonyaulax) spp. cause problems mainly in cold and temperate waters in the northern hemisphere, while the tropical species Pyrodinium bahamense var. compressa is found in Southeast Asia (1,2) and the Pacific coast of Guatemala (3). Recently shellfish toxicity related to the unarmored species Gymnodinium catenatum has been reported from many countries such as Spain (4), Portugal (5), Mexico (6), Australia (7) and Japan (8). In contrast with Alexandrium spp., reports on toxins in the latter two species are still scarce.

As the analytical method for paralytic shellfish toxins, sensitive and specific high performance liquid chromatography (HPLC) has been developed (9,10) and utilized not only for shellfish toxicity monitoring but also for the investigation of toxins in dinoflagellates (11,12). However, resolution for some toxins, especially N-sulfocarbamoyl-11-hydroxysulfate derivatives (CI-C4), was insufficient by the previous systems. Our effort to improve the chromatographic separation by using different column and mobile phases enabled us to detect most of the 18 naturally occurring saxitoxin analogs (13).

In the present study, intra- and inter-species variation of toxin profiles of dinoflagellates was investigated by means of the improved HPLC technique. Toxins in natural dinoflagellates were also compared to those of bivalves to estimate catabolism of toxins in shellfish.

MATERIALS AND METHODS

Specimens

G. catenatum from Tasmania (Australia), Senzaki Bay (Japan) and Galicia (Spain); A. tamarense (10 isolates) from Adelaide (Australia), Galicia, Hokkaido, Iwate, and Wakayama (Japan); and A. catenella (14 isolates) from Ibaraki, Wakayama, Kagawa, Miyazaki and Kagoshima (Japan) were cultured in enriched seawater media. P. bahamense var. compressa was collected at Palau. A. catenella and scallops (Patinopecten

Toxic Marine Phytoplankton
Edna Graneli et al., Editors

yessoensis) collected at Funka Bay and A. catenella and mussels (Mytilus edulis) collected at Ibaraki, Japan were also used for toxin profile studies.

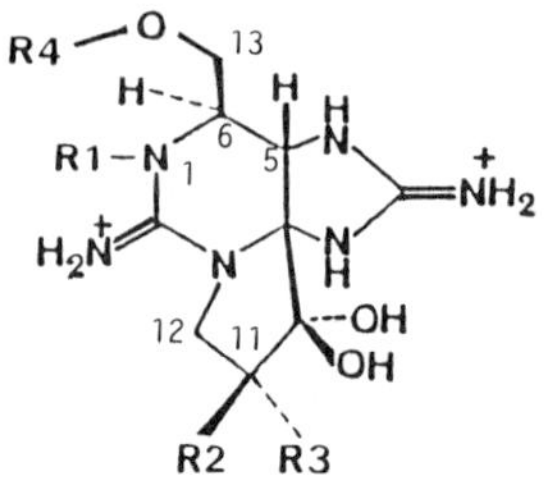

R1	R2	R3	R4:$CONH_2$	R4:$CONHSO_3^-$	R4: H
H	H	H	STX	GTX5(B1)	dcSTX
OH	H	H	neoSTX	GTX6(B2)	dcneoSTX
OH	H	OSO_3^-	GTX1	C3	dcGTX1
H	H	OSO_3^-	GTX2	C1(epiGTX8)	dcGTX2
H	OSO_3^-	H	GTX3	C2(GTX8)	dcGTX3
OH	OSO_3^-	H	GTX4	C4	dcGTX4

FIG. 1. Structures of paralytic shellfish toxins.

HPLC analysis of toxins

Details of analytical conditions were described in a previous report (13). Briefly, toxins were separated on a column of C8 bonded silica column (Develosil C8 by isocratic elution with three different mobile phases: (**a**) tetrabutylammonium in acetate buffer for analysis of C1-C4, (**b**) 1-heptanesulfonate in ammonium phosphate buffer for gonyautoxins 1-6 (GTX1-6), and (**c**) mobile phase **b** plus acetonitrile for the saxitoxin (STX), neosaxitoxin (neoSTX) group. Toxins in the eluate from the column were continuously oxidized by heating with periodate, and the resultant fluorescent compounds were detected by a fluoromonitor after acidifying the reaction mixture with acetic acid. Dinoflagellates collected by centrifugation were suspended in 0.1 N acetic acid and sonicated for 1 min. Supernatant obtained by centrifugation at 10.000 **g** for 5 min was directly injected to HPLC. Shellfish extracts were prepared according to the standard method for mouse bioassay (14) and applied for analysis after being passed through a Sep-Pak C18 cartridge column (Waters) and ultrafiltration filter (Millipore, UFC3TGCOO).

Purification of toxins

Toxins extracted from shellfish with dilute HCl were purified successively by column chromatographies on activated charcoal, Bio-Gel P-2, Bio-Rex 70 (Bio-Rad), Hitachi gel 3011C and Toyopearl HW-40 (Toso).

Spectral measurements

Fast atom bombardment (FAB) mass spectra were taken on a JEOL JMS-DX330 mass spectrometer with glycerol as matrix. PMR spectra were taken in a 4% CD_3COOD in D_2O solution on a JEOL GSX-400 spectrometer.

RESULTS AND DISCUSSION

Unusual toxin profiles and a new toxin of Gymnodinium catenatum

In the previous study (7), Gymnodinium catenatum from Tasmania showed a characteristic toxin profile consisting of a large proportion of C1-C4. However, several peaks in the chromatograms were left unidentified (Fig. 2 A: b,c). Comparison with the newly prepared standards indicated that two peaks (*) (Fig. 2 A: b) were decarbamoyl-gonyautoxins 2 and 3 (dcGTX2, dcGTX3). Firm evidence to support the identification was obtained by FAB-mass spectra (m/z 353 $(M+H)^+$, 375 $(M+Na)^+$) and PMR spectra (3.7 ppm, m, 3H for 13-CH_2 and 6-CH). The two toxins have never been detected in dinoflagellates and only reported as enzymatic hydrolysis products by a clam (15).

Another new component which was eluted after STX in HPLC (UK, Fig. 2 A: c) showed ion peaks at m/z 223 $(M-H_2O+H)^+$ and 241 $(M+H)^+$. PMR spectra indicated the toxin to be 13-deoxycarbamoylsaxitoxin, as shown in Figure 3. In a biosynthetic study, Shimizu et al. (16) indicated introduction of 13-carbon of neoSTX through transmethylation from S-adenosylmethionine. The new toxin might be an important precursor to other paralytic shellfish toxins.

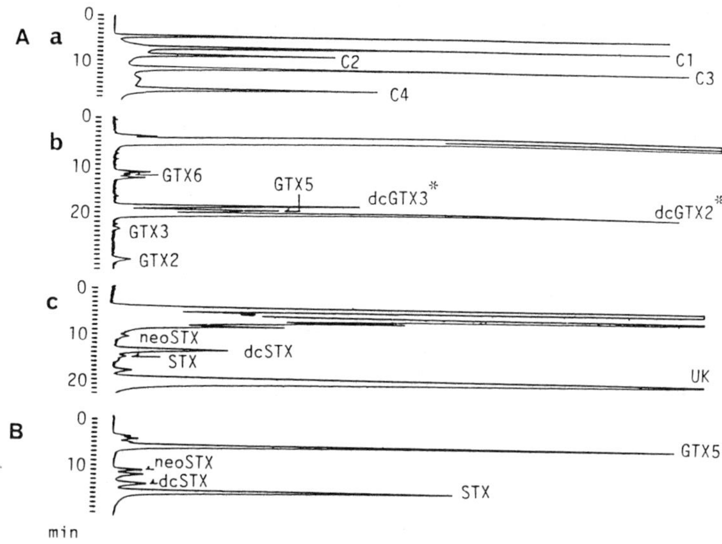

FIG. 2. Chromatograms of toxins in G. catenatum from Tasmania (**A**: a-c indicate mobile phases used for analys), and P. bahamense var. compressa from Palau (**B**).

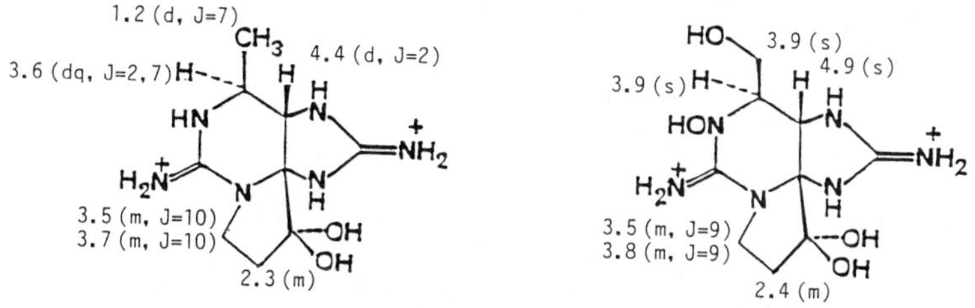

FIG. 3. Structures of a new toxin from G. catenatum (left) and decarbamoylneosaxitoxin from P. bahamense var. compressa. Numbers indicate chemical shifts in ppm measured at 400 MHz in 4% CD_3COOD in D_2O.

Toxin composition of G. catenatum are summarized in Table I. N-sulfocarbamoyl toxins comprised more than 90% of total toxins in all samples. However, Spanish and Japanese isolates contained larger proportions of GTX6 and GTX5 and smaller C3 and C4 than the Tasmanian one. DcGTX2, dcGTX3, dcSTX and the new toxin were detected in all the isolates.

Toxins in Pyrodinium bahamense var. compressa

In HPLC of Palauan P. bahamense var. compressa (Fig. 2 B), STX, neoSTX, GTX5, GTX6 and dcSTX were detected as reported previously (17). However, careful reinvestigation of toxins in infested shellfish led to the isolation of another component of weak toxicity (30 MU/μmole). Mass spectra (m/z 273 $(M+H)^+$, 237 $(M-H_2O+H)^+$) and PMR spectra indicated it being decarbamoylneosaxitoxin (Fig. 3). HPLC analyses of shellfish contaminated with the same species at the Philippines and Solomon Island and gut contents of fish (Sardinella sp.) from Brunei showed similar toxin profiles compared with the Palauan sample. In contrast to Alexandrium spp., the Pyrodinium species, including Guatemala specimens (2), showed less diversity in toxin profile and was characterized by complete absence of 11-hydroxysulfate toxins.

TABLE 1. Toxin composition of Gymnodinium catenatum (mole%)

Origin	Tasmania* Australia	Tasmania** Australia	Vigo** Spain	Senzaki** Japan
neoSTX	0.3	0.1	3.8	< 0.1
dcSTX	0.5	0.1	1.8	0.5
GTX5	0.1	< 0.1	3.5	1.5
GTX6	< 0.1	< 0.1	31.4	36.0
dcGTX2+dcGTX3	0.9	0.6	0.8	0.8
C1+C2	9.9	70.9	31.1	61.3
C3+C4	88.3	28.3	27.5	< 0.1

13-Deoxycarbamoylsaxitoxin is not included. * Bloom, ** Culture.

Toxin profiles and toxin content of Alexandrium spp.

Analysis of toxins in 24 Alexandrium isolates, including three nontoxic A. tamarense from Wakayama showed a great variation in toxin profiles as reported previously (12, 18). Typical results are shown in Table II. No significant relationship was observed between toxin profiles and two morphotypes classified according to Fukuyo's systematics (19). Interestingly, the isolates from Australia and Spain, which were identified as "A. minutum" by other taxonomists, showed comparable toxin profiles, consisting of only GTX1-4, in spite of their remote origins.

As for the configuration of 11-hydroxysulfate, the toxins exist almost exclusively as β epimers (GTX4, GTX3, dcGTX3, C2, C4) in freshly prepared samples, indicating stereospecific introduction of the moiety in biosynthesis. DcGTX3 was observed in some isolates, though in trace amounts.

Toxin content in A. tamarense (Funka Bay isolate, Table II) varied significantly during culture (35-280 fmole/cell) and showed the highest toxicity in the induction period (11,20). To determine the toxin production at a certain stage of a cell cycle, a synchronized culture is essential. Due to the difficulty to synchronize cell division by controlling physical conditions, several kinds of inhibitory reagents were tested. Addition of mitomycin c (1 μM) to the medium suppressed the cell division, causing cell volumes to become almost twice those of controls.
Under this condition short-term fluctuation of toxin content, more than 50% within 24 hrs, was observed, indicating a fairly rapid turnover of toxins in the dinoflagellate.

In contrast to toxin contents, the toxin ratios remained fairly constant under the various culture conditions, as shown in Table II. The minimum-maximum values observed for neoSTX, GTX4 and C2 were 19-51, 24-48 and 14-34 mole%, respectively. However, larger discrepancies were often observed between natural populations and cultured cells (Table I., (7, 21)).

TABLE II. Toxin composition of Alexandrium spp. culture (mole%).

Species	A. tamarense			A. catenella		
Origin	Funka Bay* Japan	Vigo Spain	Adelaide Australia	Ibaraki Japan	Wakayama Japan	Miyazaki Japan
STX	1.6 ± 1.9	-	-	-	-	-
neoSTX	30.1 ± 8.6	-	-	70.8	2.0	4.8
dcSTX	0.1 ± 0.1	-	-	-	-	-
GTX1+GTX4	39.5 ± 6.7	97.3	93.5	1.4	27.5	6.0
GTX2+GTX3	5.0 ± 2.8	2.7	6.5	0.1	-	39.6
GTX5	-	-	-	0.5	-	6.2
GTX6	-	-	-	-	10.7	1.7
dcGTX3	0.1 ± 0.1	-	-	0.1	0.1	-
C1 + C2	23.4 ± 6.9	-	-	27.1	7.4	41.7
C3 + C4	0.2 ± 0.1	-	-	-	52.3	-

* Average and standard deviation of 86 samples.

Comparison between bloom organisms and infested bivalves

Although complicated processes might be involved in the accumulation and elimination of toxins by shellfish, direct comparison of toxin profiles between bloom organisms and shellfish was expected to give a clue to the fate of toxins in bivalves. Table III shows toxin composition of natural A. catenella and two samples of scallop collected at the same point in Funka Bay with a 5 days interval. Toxicity of scallops declined from 240 to 79 MU/g during the period, reflecting disappearance of the dinoflagellate in the water. C2, the main toxic component in A. catenella, was epimerized to C1 in scallops, and both epimers were eliminated from shellfish faster than other toxins. The concentration of neoSTX in scallops also decreased from 74 to 49 nmole/g while GTX2,3 and dcGTX2,3 remained constant. A similar experiment was conducted at Ibaraki, and mussels showed fairly similar toxin profiles compared to those of the causative algae A. catenella, except for epimerization of 11-ß-hydroxysulfate and a slightly lower ratio of N-1 hydroxy toxins (data not shown). The same tendency was also observed in mussels and oysters during blooms of G. catenatum at Tasmania (7) and A. tamarense at Adelaide (21).

TABLE III. Toxin content of A. catenella and infested scallops.

Specimens Date	A. catenella Dec.1, 1988 fmole/cell(mole%)		Scallops Dec.1, 1988 nmole/g (mole%)		Scallops Dec.6, 1988 nmole/g (mole%)	
STX	1.1	(0.5)	9.5	(2.1)	3.9	(2.7)
neoSTX	11.7	(5.2)	74.3	(16.5)	49.4	(34.2)
dcSTX	< 0.2	(-)	0.3	(0.1)	0.4	(0.3)
GTX2	< 0.1	(-)	7.3	(1.6)	7.8	(5.4)
GTX3	0.1	(-)	2.0	(0.4)	0.8	(0.6)
GTX4	0.6	(0.2)	< 0.4	(-)	< 0.4	(-)
GTX5	0.2	(0.1)	< 0.1	(-)	0.6	(0.4)
dcGTX2	< 0.5	(-)	6.8	(1.5)	7.0	(4.8)
dcGTX3	< 0.1	(-)	4.7	(1.0)	3.3	(2.3)
C1	5.1	(2.3)	279.9	(62.0)	64.4	(44.5)
C2	201.1	(91.7)	66.9	(14.8)	7.0	(4.8)
Total	220.0		451.6		144.7	

ACKNOWLEDGEMENTS

The authors wish to express their sincere thanks to Dr. G. Hallegraeff of CSIRO, Australia, Ms. B. Reguera of the Spanish Institute of Oceanography, and Dr. M. Kodama of Kitasato University, Japan, for the donation of dinoflagellate samples.

REFERENCES

1. J. L. Maclean in: Toxic Dinoflagellate Blooms, D. L. Taylor and H. H. Seliger, eds. (Elsevier, New York 1979) pp. 173-178.
2. C.L. Gonzales, J.A. Ordonez and A.M. Maala in: Red Tides, T. Okaichi, D.M. Anderson and T. Nemoto, eds.(Elsevier, New York 1989) pp. 45-48.
3. F.R. Loessener, E.D. Porras and M.W. Dix in: Red Rides, T. Okaichi, D.M. Anderson and T. Nemoto, eds.(Elsevier, New York 1989) pp.113-116.
4. J.A. Rodriguez-Vazquez, Y. Oshima, J.S. Lee and T. Yasumoto in: Mycotoxins and Phycotoxins 1988, S. Natori, K. Hashimoto and Y. Ueno, eds. (Elsevier, New York) in press.
5. S. Franca and J.F. Almeida in: Red Tides, T. Okaichi, D.M. Anderson and T. Nemoto, eds. (Elsevier, New York 1989) pp. 93-96.
6. L.D. Mee, M. Espinosa and G. Diaz, Mar. envir. Res. 19, 77-92 (1986).
7. Y. Oshima, M. Hasegawa, T. Yasumoto, G. Hallegraeff and S. Blackburn, Toxicon 25, 1105-1111 (1987).
8. T. Ikeda, S. Matsuno, S. Sato, T. Ogata, M. Kodama, Y. Fukuyo and, M. Takayama in: Red Tides, T. Okaichi, D.M. Anderson and T. Nemoto, eds. (Elsevier, New York 1989) pp. 411-414.
9. Y. Oshima, M. Machida, K. Sasaki, Y. Tamaoki and T. Yasumoto, Agric. Biol. Chem. 48, 1707-1711 (1984).
10. J.J. Sullivan, J. Jonas-Davis and L.L. Kentala in: Toxic Dinolagellates, D.M. Anderson, A.W. White and D.G. Baden, eds. (Elsevier, New York 1985) pp. 275-280.
11. G.L. Boyer, J.J. Sullivan, R.J. Anderson, F.J.R. Taylor, P.J. Harrison and A. D. Cembella, Mar. Biol. 93, 361-367 (1986).
12. A.D. Cembella, J.L. Sullivan G.L. Boyer, F.J.R. Taylor and R.J. Andersen, Biochem. System. Ecol. 15, 171-186 (1987).
13. Y. Oshima, K. Sugino and T. Yasumoto in: Mycotoxins and Phycotoxins 1988, S. Natori, K. Hashimoto and Y. Ueno, eds.(Elsevier, New York) in press.
14. S. Williams (ed.) in: Official Methods of Analysis (Association of Official Analytical Chemists, Arligton, 1984) pp. 344-345.
15. J.J. Sullivan, M. Iwaoka and J. Liston, Biochem. Biophys. Res. Commun. 29, 465-472 (1983).
16. Y. Shimizu, S. Gupta, M. Norte, A. Hori, A. Genenah and M. Kobayashi in: Toxic Dinoflagellates, D.M. Anderson, A.W. White and D.G.Baden, eds.(Elsevier, New York 1985) pp. 271-274.
17. Y. Oshima, Y. Kotaki, T. Harada and T. Yasumoto in: Seafood Toxins, E.P. Ragelis ed.(American Chemical Society, Washington, D.C., 1984) pp.161-170.
18. Y. Oshima, T. Hayakawa, M. Hashimoto, Y. Kotaki and T. Yasumoto, Nipponm Suisan Gakkaishi 48, 851-854 (1982).
19. Y. Fukuyo, K. Yoshida and H. Inoue in: Toxic Dinoflagellates, D.M. Anderson, A.W. White and D.G.Baden, eds.(Elsevier, New York 1985) pp. 27-32.
20. H.T. Singh, Y. Oshima and T. Yasumoto, Nippon Suisan Gakkaishi 48, 1341-1343 (1982).
21. Y. Oshima, M. Hirota, T. Yasumoto, G. Hallegraeff, S. Blackburn and D. Steffensen, Nippon Suisan Gakkaishi 55, 925 (1989).

RED TIDE TOXIN (BREVETOXIN) ENRICHMENT IN MARINE AEROSOL

RICHARD H. PIERCE, MICHAEL S. HENRY, L. SCOTT PROFFITT AND PATRICIA A. HASBROUCK
Mote Marine Laboratory, 1600 City Island Park, Sarasota, Florida 34236, USA

ABSTRACT

This investigation addressed the bubble-mediated enrichment of brevetoxins in marine aerosol. Previous studies have shown that toxins recovered from seawater during blooms of the dinoflagellate, Ptychodiscus brevis, were also recovered with marine aerosol particles collected along the shore. To study the toxin aerosolization process an aerosol production-collection chamber was constructed. Aerosol particles were produced by bubbling air through seawater samples that contained brevetoxins. The aerosols produced by jet and film drops ejected from bursting bubbles were collected using a high volume air sampler. Toxin enrichment was calculated from the ratio of toxin concentration in aerosol relative to that in the water sample. The volume of water transported as aerosol was determined by the amount of Na^+ on the filter. The toxin content of the water and of the aerosol impacting the filter was determined by high performance liquid chromatography. The six toxins recovered from water and in the aerosol exhibited enrichment factors from 5 to 50 times the concentration in the original culture, showing a definite enrichment in aerosol with apparent selection for some toxins over others in the aerosol phase.

INTRODUCTION

Human respiratory irritation has been reported in association with phytoplankton blooms in the Gulf of Mexico since the early 1900's [1]. Correlation of respiratory irritation with bubble-produced aerosols from dinoflagellate-containing seawater was first suggested by Woodcock [2] during his investigation of a severe red tide bloom off Venice, Florida, in 1947. The causative organism from this bloom was identified as Gymnodinium breve (G. breve) by Davis [3], renamed Ptychodiscus brevis (P. brevis) by Steidinger in 1979 [3], with recent reconsideration for changing back to G. breve (Steidinger, personal communication). The purpose of this investigation was to study processes effecting transport of the toxins (brevetoxins) from bulk seawater to the sea surface with subsequent incorporation and enrichment in marine aerosol.

Although toxic phytoplankton blooms are noted worldwide for causing fish kills and the contamination of fish and shellfish, the production of airborne toxins (toxin aerosols) appears to be uniquely associated with episodes of the dinoflagellate, P. brevis (G. breve). These unarmored cells are susceptible to lysing from physical stress within breaking waves and the movement of fish gills, as well as possible chemical lysing agents [4], [5], [6], [7], [8], [9]. When the cell wall breaks, the endotoxins, as many as nine identified so far, are released to the water [10], [11], [12], [13], [14]. In the presence of breaking waves and on-shore winds, toxins associated with marine aerosol cause severe respiratory irritation to humans along the shore [2], [15], [16].

Toxic Marine Phytoplankton
Edna Graneli et al., Editors

The hypothesis tested was that entrained air bubbles thrust downward into the ocean by breaking waves, scavenge the hydrophobic toxin molecules from water transporting them to the sea surface [16]. Observations with dissolved salts, organic molecules and bacteria have shown that when bubbles burst at the sea surface, collapse of the bubble cavity provides the kinetic energy to jettison minute droplets of water into the air carrying with them dissolved salts and other substances associated with the bubbles' surfaces [17], [18], [19], [20], [21], [22], [23]. The airborne droplets (jet drops and film drops) contain a higher concentration of substances scavenged by the bubbles than in an equal volume of seawater. Brevetoxins scavenged by entrained bubbles should exhibit enrichment in bubble-produced aerosol.

Isolation and identification of brevetoxins by different researchers has resulted in various nomenclature schemes [24], [25], [10], [26], [12]. The revised nomenclature proposed by Poli, et al. [11] representing the eight P. brevis toxins by PbTx-1, -2, -3, etc., was used in this study, with a newly identified toxin (PbTx-9) also included [14].

Previous studies of toxin aerosols collected during red tide blooms by Pierce [16], [27] have shown that the toxins were associated with marine aerosol particles. Field studies of toxins in marine aerosol recovered three of the toxins (PbTx-2, -3, and -5) during natural blooms that occurred along the Florida Gulf coast as well as during an unusual red tide bloom along the North Carolina Atlantic coast [16], [27]. Preliminary studies of toxin enrichment in bubble-generated foam, jet drops and aerosol, using natural red tide as well as lab cultures in an aerosol production-collection chamber, recovered four toxins (PbTx-2, -3, -5, -6) in sufficient quantity to provide quantitative results [13]. Toxin concentrations in surface foam were 3 to 5 times greater than that in the original culture, whereas toxin enrichment in jet drops and aerosol particles was 20 to 50 times that in the cultures, showing that bubble-generated toxin aerosols were produced from jet drops and not from wind-blown surface foam [13]. A problem with the above enrichment studies was uncertainty regarding the aerosol volume. The experiments reported here used the conservative property of sodium ion (Na^+) in water and aerosol [20] to provide a more accurate determination of aerosol volume.

METHODS AND MATERIALS

Brevetoxin Source

The source of the toxins for aerosol enrichment studies was laboratory cultures of the toxic dinoflagellate, P. brevis. These cultures were initiated in 1984 at MML by inoculation with samples of the Wilson isolate (W53DB) [28] provided by the State of Florida Department of Natural Resources, Florida Marine Research Institute. These cultures are grown in two different media: a natural seawater media, Ff-10 [29], and an artificial media, NH-15 [30]. Recent comparisons of toxins from these cultures with toxins recovered from natural red tide blooms show that the same toxins are produced, although the relative amounts vary due to different growth stages, environmental conditions and different strains of P. brevis [10]; a conclusion also reported by Baden and Tomas for toxins from various P. brevis isolates [31]. Cultures used for experiments reported herein were all in the exponential growth phase containing between 25 x 10^6 and 35 x 10^6 cells/liter.

Brevetoxin Aerosol Production and Collection

Toxin enrichment in aerosol was investigated using an aerosol production-collection chamber as described by Pierce et al. [13]. Air bubbles averaging 1 mm diameter emerged from a fritted glass aerator and passed through a 4-liter volume of P. brevis culture, ejecting jet and film drops into the air as the bubbles burst at the surface. A high-volume air sampler motor pulled air across the surface of the culture at a velocity representing moderate sea breezes (approximately 5 m/sec) carrying the droplets a minimum distance of 40 cm to impact a glass-fiber filter attached to the air sampler [13].

Several experiments were performed to establish proper conditions for aerosol production and collection. Surface foam and air velocity across the water surface had a considerable effect on toxin aerosol production efficiency. Studies reported here included four sets of aerosol experiments performed using two cultures in NH-15 media and two in Ff-10 media. Experiments included control samples collected from culture media not inoculated with _P. brevis_. Prior to aerosolization, all _P. brevis_ cells were lysed with ultrasound, using a Vibra-Cell model VC-500 (Sonics and Materials, Danbury, CT) to ensure release of all endotoxins for impact with bubble surfaces, while leaving cellular debris available for interaction as in natural seawater. The original culture was analyzed in triplicate (described below) prior to each test to determine the initial toxin composition of the sample. The aerosol was collected for periods of three hours, the exposed portion of the filter was then removed, weighed and cut into small pieces, and a representative sample (approximately 10%) set aside for Na^+ analysis. Toxins also were analyzed in the culture sample remaining after aerosolization (post-aerosol sample) and in the residue adhering to the chamber sides resulting from impact by foam, splash and jet drops. The chamber sides were wiped with a glass-fiber filter paper which was then processed in a manner similar to the aerosol filters described below. The room temperature was maintained at $24^oC \pm 2^oC$. The culture samples ranged from 25^oC initially, to 20^oC after three hours of bubbling due to evaporative cooling.

The volume of culture represented by the aerosol particles collected on the filter was calculated by determining the amount of Na^+ impacting the filter. The representative sample of the filter was extracted via sonication with 20 ml of water and the sodium content of the extract was then analyzed by atomic absorption spectroscopy using an IL model 251 (Instrument Laboratories, Wilmington, MA). The volume of aerosolized culture was then calculated by comparison to the sodium content of the original culture, following the procedure described by Duce and Hoffman [32] for chemical fractionation at the sea surface.

$$E_f = \frac{\mu g\ \text{toxin} / \text{mg}\ Na^+\ \text{aerosol}}{\mu g\ \text{toxin} / \text{mg}\ Na^+\ \text{culture}} \quad \alpha \quad \frac{\mu g\ \text{toxin} / \text{liter aerosol}}{\mu g\ \text{toxin} / \text{liter culture}}$$

Brevetoxin Analysis

During preliminary tests, toxins were recovered from the original culture samples by both liquid/liquid (l/l) extraction into dichloromethane (CH_2Cl_2) and by liquid/solid (l/s) adsorption onto 100-200 µm size particles of silica bonded with octadecane (SiO_4-C-18 reverse phase chromatography column packing, Universal Scientific, Inc., Atlanta, GA) to evaluate a solid sorbent technique recently reported by Proffitt et al [33]. The l/s recovery was obtained by passing a prefiltered (0.45 µ glass fiber filter) sample of culture through a 10 cm x 2 cm column of the above sorbent at a rate of 1 liter/hour, toxins were eluted from the column with 50 ml of 85/15, methanol/ water solution and back extracted into CH_2Cl_2. For both l/s and l/l, the CH_2Cl_2 extract was evaporated in vacuo, the residue dissolved in methanol and processed for high performance liquid chromatographic (HPLC) analysis as described by Pierce et al. [26]. Toxins were recovered from the filters by replicate homogenization in methanol followed by filtration. The methanol was then reduced in volume for HPLC analysis. All reagents were analytical grade and solvents were high purity HPLC grade (Burdick and Jackson, Muskegon, MI).

Toxin analysis was performed with a Varian 5020 HPLC (Varian Instruments, Palo Alto, CA) using a 25 cm x .46 cm OD-5 spherical C-18 column (Burdick and Jackson, Muskegon, MI), with a mobile phase of 85/15, methanol/water at 1 ml/min, and a UV-50 detector at 215 nm. Verification of qualitative analysis of the six toxins recovered was provided by retention time relative to PbTx-2 as the reference toxin. Qualitative interlaboratory comparison for these toxins was provided previously in cooperation with Dr. Daniel Baden, of the University of Miami. Recent evidence from Dr. Baden's laboratory, however, indicates that the toxin designated as PbTx-6 may be a newly identified toxin, PbTx-9 [14]. This toxin is designated as PbTx-6(-9) to indicate the uncertain identity. Quantitative analysis was based on primary standards prepared from purified isolates of PbTx-2 and PbTx-3. Quantitation of the other four toxins was obtained from estimated uv response factors based on their chemical structures in relation to PbTx-2 and -3. These factors, listed in Table I, may change as primary standards for each toxin become available.

Table I. Parameters for quantitative and qualitative analysis of brevetoxins with reverse-phase HPLC.

Parameter	PbTx					
	-3	-6(-9)[1]	-2[2]	-7	-5	-1
Retention time (min.)	7.8	8.8	9.6	10.6	11.6	12.5
Relative Retention/PbTx-2	0.82	0.92	1.0	1.1	1.2	1.3
UV Detector Response [3] Conversion factor, µg/cm	0.25	0.13	0.06	0.5	0.06	0.25

[1] Identity of toxin uncertain: PbTx-6 or -9
[2] Relative Retention Reference Toxin
[3] UV Detector Response in µg toxin/cm peak ht: actual for PbTx-2 and -3, estimated for PbTx-1, -7, -5, -6(-9)

RESULTS AND DISCUSSION

Results of a representative aerosol experiment are given in Table II, showing the distribution of the six toxins recovered in the original culture before aerosolization and after 3 hours of passing air bubbles through the culture, collecting the aerosol produced. As shown here, all experiments recovered very little or no PbTx-7 in the aerosol, showing PbTx-7 to be the only toxin not enriched in aerosol. Also, PbTx-5 and -1 were present in low concentrations in water as well as aerosol, indicating the need for experiments using larger volumes of *P. brevis* culture to provide larger amounts of these less abundant toxins.

Table II. Brevetoxin enrichment (µg toxin/l) in bubble-generated aerosol NH-15 culture, 35 x 10^6 cells/l.

SAMPLE	Pbtx - (µg TOXIN/LITER)						Σ TOXINS µg/L
	-3	-6(-9)	-2	-7	-5	-1	
OC[1]	134	122	720	104	13	16	1,109
PAC[2]	40	20	170	123	--	--	353
CS[3]	600	67	3,190	22	22	380	4,281
AEROSOL	2,630	1,830	21,580	--	210	610	26,860

[1] Original culture [2] Post-aerosol culture [3] Chamber sides

The aerosol enrichment factors (E_f) calculated for toxins from all four experiments showed good agreement between duplicate experiments of the same media, and between the two types of media for PbTx-3, -2, and -5. PbTx-1 showed highly variable enrichment factors whereas PbTx-6(-9) and -7 indicated low enrichment. The mean and standard deviation of E_f values for all toxins was 25 $\pm$ 7, with three subsets: a) PbTx-2, -3 and -5 exhibiting an E_f = 26 $\pm$ 5; b) PbTx-6(-9) and -7 averaging an E_f = 10 $\pm$ 5; and c) PbTx-1 E_f = 33 $\pm$ 18.

Previous studies have used ratios of the major toxin concentrations, PbTx-2, -3, for comparative purposes. For NH-15 media, Baden [34] reported the ratio of PbTx-2/-3 to be about 3.7, and Pierce et al. [13], reported ratios of about 5.5. More recently, Baden and Tomas [31] found that the relative amounts of the major toxins vary with the particular strain of _P. brevis_ and the growth phase of the culture. Results of these studies indicated no preference of PbTx-3 over -2, which is contrary to an earlier study where the amount of PbTx-3 increased over PbTx-2 during aerosolization, attributed primarily to chemical reduction. Also notable is that PbTx-1 was not recovered in large concentrations whereas other researchers have found PbTx-1 to be one of the more abundant toxins [11], [12], [31].

SUMMARY

These experiments show that brevetoxins were enriched in marine aerosol produced by entrained bubbles, supporting the hypothesis of bubble-mediated transport to the sea surface with subsequent ejection into the air in association with jet drops and film drops from bursting bubbles. Similar mechanisms for enrichment of bacteria and marine organic substances have been reported by Blanchard [19], [35], Blanchard and Syzdek [23], Duce [20], and Gershey [21], emphasizing the need to investigate bacteria and other organics in marine aerosol produced during red tide blooms.

The use of sodium content of collected aerosol provided an accurate means to determine the volume of the aerosol for calculating toxin enrichment relative to the original culture sample. These results indicate an apparent selection for PbTx-1, -2, -3 and -5 in aerosol over PbTx-6(-9) and -7. Further studies are underway to assess physical and chemical parameters that may affect the affinity of toxins for bubble surfaces and toxin enrichment in jet drops and marine aerosol.

ACKNOWLEDGEMENTS

This investigation was funded by the State of Florida Department of Natural Resources, Florida Marine Research Institute. Special thanks to Dr. Karen Steidinger, Dr. Carmelo Tomas and Ms. Beverly Roberts of the FL DNR, for their cooperation and assistance with phytoplankton culture, and to Dr. Daniel Baden and Ms. Laurie Roszell, of the University of Miami, for assistance with brevetoxin identification and interlaboratory calibration.

REFERENCES

1. H.F. Taylor, Report of the U.S. Commissioner of Fisheries, Doc. 848, 5 (1917).
2. A.H. Woodcock, J. Mar. Res. <u>56</u> (1948).
3. C.C. Davis, Bot Gaz. 109, <u>358</u> (1948).
4. Y. Shimizu, in: Marine Natural Products, Chemical and Biological Perspectives, J.P. Scheuer, ed. (Academic Press, New York) (1978) p. 1.
5. Y. Shimizu, Pure Appl. Chem., <u>54</u>, 1,973-1,980 (1982).

6. D.G. Baden, Int. Rev. Cytol., 82, 99 (1983).
7. S. Ellis, ed., Toxicon 23, 469 (1985).
8. K.A. Steidinger, Prog. Phycol. Res. 2, 148 (1983).
9. R.E. Moon and D.F. Martin, in: The Chemistry of Allelopathy Biochemical Interactions Among Plants, ACS Series No. 268, A.C. Thompson, ed. (1985) pp. 372-380.
10. K. Nakanishi, in: Brevetoxins: Chemistry and Pharmacology of "Red Tide" Toxins from Ptychodiscus brevis (formerly Gymnodinium breve), Toxicon 23 (1985) pp. 473-479.
11. M.A. Poli, T.J. Mende and D.G. Baden, Molec. Pharmac. 30, 129-135 (1986).
12. Y, Shimizu, H.N. Chou, H. Banda, G. Van Duyne and J.C. Clardy, J. Am. Chem. Soc. 108, 515-516 (1986).
13. R.H. Pierce, M.S. Henry, S. Boggess and A. Rule, in: Climate and Health Implications of Bubble-Mediated Sea-Air Exchange, E. Monahan and M. Van Patten, eds. (in press).
14. L.S. Schulman, L.E. Roszell, T.J. Mende and D.G. Baden, Proceedings of the Fourth International Conference on Toxic Phytoplankton (in press).
15. R.M. Ingle, Dept. of Nat. Res. Spec. Bull. No. 9 (1954).
16. R.H. Pierce, Toxicon 24, 955-965 (1986).
17. F. MacIntyre, Sci. Am. 230, 62 (1974).
18. D.C. Blanchard, in: Applied Chemistry at Protein Interfaces, R. Baier, ed., Adv. Chem. Ser. 145 (1975) p. 360.
19. D.C. Blanchard, J. Geophys. Res 90, 161-163 (1985).
20. R.A. Duce, in: The Major Biogeochemical Cycles and Their Interactions B. Bolin and R.B. Cook, eds. (Wiley-Interscience, 1983) p. 427.
21. R.M. Gershey, Limnol. Oceanogr. 28, 309 (1983).
22. E.C. Monohan, in: The Role of Air-Sea Exchange in Geochemical Cycling, P. Bugt-Menard, ed. (D. Reidel, Hingham, MA 1986) pp. 129-163.
23. D.C. Blanchard and L.D. Syzdek, J. Geophys. Res. 93, 3,649-3,654 (1988).
24. G.M. Padilla, Y.S. Kim and D.F. Martin, in: Proceedings of the First International Conference of Toxic Dinoflagellate Blooms, V.R. Lo Cicero, ed. (Wakefield, Mass. Sci. and Tech. Fdn. 1975) p. 299.
25. N.-N Chou, Y. Shimizu, G.D. van Duyne and J. Clardy, in: Toxic Dinoflagellates, D.M. Anderson, A.M. White and D.G. Baden, eds. (Elsevier, New York 1985), p. 305.
26. R.H. Pierce, R.C. Brown and J.R. Kucklick, in: Toxic Dinoflagellates, D.M. Anderson, A.M. White and D.G. Baden, eds. (Elsevier, New York 1985), p. 309.
27. R.H. Pierce, Environs X (4), 7-12 (1987).
28. K.A. Steidinger and D.G. Baden, in: Dinoflagellates, D. Spector, ed. (Academic Press, New York), p. 201.
29. R.R.L. Guillard and J.H. Ryther, Detonulaconfervacae (Cleve), 8, 239-239 (1962).
30. D.V. Aldrich and W.B. Wilson, Biol. Bull., 119, 57-64 (1960).
31. D.G. Baden and C.R. Tomas, Toxicon, 6 (10), 961-963 (1988).
32. E.J. Hoffman and R.A. Duce, J. Geophys. Res. 79, 4474 (1974).
33. L.S. Proffitt, R.H. Pierce, M.S. Henry and A.J. deRosset, Florida Scientist, 52, Supp. 1 (1989) p. 28.
34. D.G. Baden and T.J. Mende, Toxicon, 20 (2) 457-461 (1982).
35. D.C. Blanchard, in: Applied Chemistry at Protein Interfaces, Adv. Chem. Ser. (1975) pp. 145-360.

TOXIN PROFILES ARE DEPENDENT ON GROWTH STAGES IN CULTURED *PTYCHODISCUS BREVIS*

LAURIE E. ROSZELL, LLOYD S. SCHULMAN, and DANIEL G. BADEN. University of Miami Rosenstiel School of Marine and Atmospheric Science, Division of Biology and Living Resources, 4600 Rickenbacker Causeway, Miami, FL 33149 U.S.A.

ABSTRACT

Ptychodiscus brevis, Florida's red tide dinoflagellate, produces at least eight polyether toxins based on two structural backbone types. Toxin yields from over 3000 liters of batch-cultured cells have been compiled over the past three years. In the logarithmic phase of growth, cultures have greater amounts of PbTx-2, the α,β-unsaturated aldehyde toxin of Type-1. As cultures progress into the stationary and declining phases, there is an increase in PbTx-3, the C-42 primary alcohol of PbTx-2, and a concurrent decline in PbTx-2. The abundance of PbTx-1 , the α,β-unsaturated aldehyde toxin of Type-2, fluctuated during all culture phases. As cultures leave logarithimic growth stages, other derivative toxins begin to appear, namely reduced PbTx-3 (PbTx-9), the o-acetate of PbTx-2, and PbTx-7, the C-43 primary alcohol of PbTx-1. We believe that these changes in toxin profile are due to metabolic events occurring in the dinoflagellate.

INTRODUCTION

Ptychodiscus brevis is the marine dinoflagellate responsible for Floridas red tide blooms and extensive fish kills [1]. From laboratory cultures at least 8 different toxins have been found to be produced by the dinoflagellates. These toxins are referred to as the PbTx series [2], and the individual toxins as PbTx 1-9. These toxins possess one of two polyether backbones (fig.1).

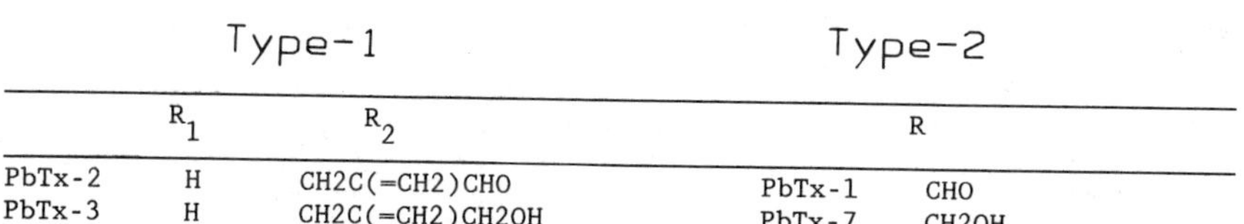

	R_1	R_2		R
PbTx-2	H	CH2C(=CH2)CHO	PbTx-1	CHO
PbTx-3	H	CH2C(=CH2)CH2OH	PbTx-7	CH2OH
PbTx-9	H	CH2CH(CH3)CH2OH		

FIGURE 1: Structures of the brevetoxins

Published 1990 by Elsevier Science Publishing Co., Inc.
Toxic Marine Phytoplankton
Edna Graneli et al., Editors

From our cultures five of these toxins are obtained, three of which are in great enough quantities to collect from each extraction of 10 liters or more. These are PbTx-1 (the α-β unsaturated aldehyde of backbone type 2), PbTx-2 (the α-β unsaturated aldehyde of backbone type 1), and PbTx-3 (the C-42 primary alcohol of type 1)(Table 1).

Both Hall [3], and Boyer *et al.*[4] have shown that in *Protogonyaulax*, toxicity can vary with different growth stages or culture conditions. Depending on the age and growth stage of a *P. brevis* culture when it is extracted, varying amounts and ratios of the three major toxins are obtained. In addition, in the older growth stages of the cultures, other derivitives of the toxins begin to appear in trace quantities. It has also been shown in *Protogonyaulax* that all toxic materials are derivatives of one parent compound, saxitoxin (Hall, 1982)[3]. An analogous process may occur in *P. brevis*; all toxins produced are derivitives of one of the two alpha-beta unsaturated aldehyde toxins. From more than 3000 liters extracted, data was collected and quantitated. Correlations were made between growth phase and toxin yield, as well as age of extracted cultures and toxin yield.

TABLE I. Toxin yields from cultures in logarithmic growth

Toxin	Yield (pg/cell)	Note
PbTx-1	1.7	(1)
PbTx-2	8.7	(1)
PbTx-3	0.42	(2)
PbTx-7	0.026	(2)
PbTx-9	0.062	(3)

Notes: (1) alpha-beta unsaturated aldehyde; (2) alpha-beta unsaturated primary alcohol; (3) PbTx-2 double reduction product.

EXPERIMENTAL

Dinoflagellates were cultured in Wilson's NH-15 media in Pyrex 10 liter carboys under continuous Gro-lux wide spectrum flourescent lights. Immediatly prior to extraction ten milliliter samples were taken for cell counts using a hemocytometer. Cultures were extracted using previously described techniques [5]. PbTx-1 and 3 were separated using C-18 reverse phase HPLC (isocratic 85% methanol 15% water, uv detection at 215 nm). PbTx-2 was crystallized in cold ethanol, yielding pure toxin. Pure toxins were weighed out on a Cahn gram electrobalance.

RESULTS

Immediatly after a dilution, amounts of all toxins were low, ranging from 0.040 to 0.060 mg/l for Pbtx-2 and 0.0037 to 0.005 mg/l for PbTx-3. This is attributed to the lag phase. As the culture moved into and through logarithmic growth, amounts of PbTx-2 were the highest, from 0.070 to 0.194 mg toxin per liter, while amounts of PbTx-3 were the lowest, from 0.006 to .013 mg per liter. As the culture became stationary, there was a decrease in the amount of PbTx-2, and an increase in the PbTx-3, though the overall amounts of PbTx-2 remained greater than those of PbTx-3. Milligram amounts of the toxins ranged from 0.005 to 0.1 mg/l for the PbTx-2 and 0.010 to 0.016 for the PbTx-3. This increase/decrease became more pronounced as the culture moved into decline; there was a further drop inthe amount of PbTx-2 (0.010 to 0.04 mg/l) and increase in the amount of PbTx-3 (0.015 to 0.040 mg/l). After a culture had been crashed for more than 10 days, there was a decrease in all toxin amounts. PbTx-1 showed no definite trends through the three phases; mg/l amounts fluctuated between 0.015 and 0.030 mg/l (Fig. 2).

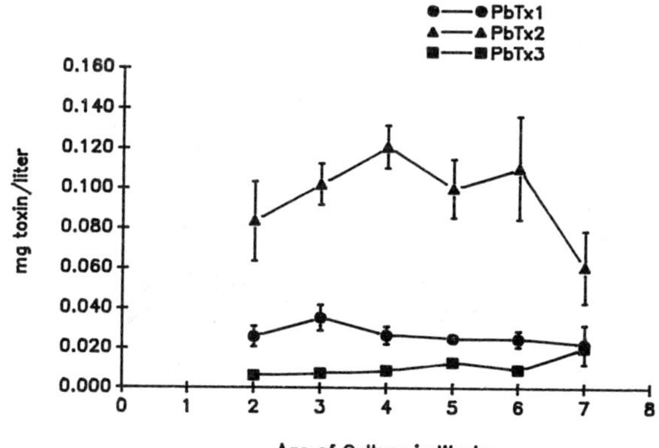

FIGURE 2: Age of extracted culture vs. toxin yield
Extraction volumes ranged from 10 to 80 L.; n=119

In addition to these findings, it was noted that in the early stages of growth extracted cultures showed significant amounts only of the three toxins discussed. However, as a culture moved into the stationary phase, other toxins began to be produced in quantities great enough to be detected on HPLC, including PbTx-9 and PbTx-7, the double reduction product of PbTx-2 and the primary alcohol of PbTx-1, respectivly. Furthermore, if a series of cultures were diluted while in the late log or stationary phase, and the product of those dilutions extracted in the late log or stationary phase of growth, these products began to appear in greater quantities. Likewise, if a series of cultures were diluted in the mid log phase , and the product of the dilutions extracted in the mid log phase, these other products were frequently not detectable by HPLC, though if fractions were pooled over a period of time enough could be accumulated for HPLC detection.

DISCUSSION

It has been shown, by Shimizu *et al.* [6] and others that PbTx-1 and PbTx-2 are the predominant toxins in *P. brevis* cultures. Our cultures also produce these toxins in predominance. From over 3000 liters of cultures extracted, we have found that the profile of the relative amounts of the toxins obtained changes through the growth stages of the cultures. The most significant change is in the ratios of PbTx-2 and PbTx-3; there is a decrease in the former and a concomitant increase in the latter (Fig.3). The conversion of PbTx-2 to PbTx-3 by the dinoflagellate has not yet been demonstrated; we have demonstrated this conversion *in vitro* using alcohol dehydrogenase from yeast. There are also increases detected in relative amounts of the double reduction product of PbTx-2 (PbTx9), and the primary alcohol of PbTx1 (PbTx7).

These findings indicate that, in any new culture, the aldehyde is the natural form of the toxin found, and changes in toxin profile can be explained by detoxification mechanisms of the dinoflagellate. As the culture ages, other toxins are derived from the aldehyde function, beginning with the alcohol, and as the culture moves into the older growth stages the double reduction product of the aldehydes are formed. These are more water soluble and hence more easily excreted by the dinoflagellate.

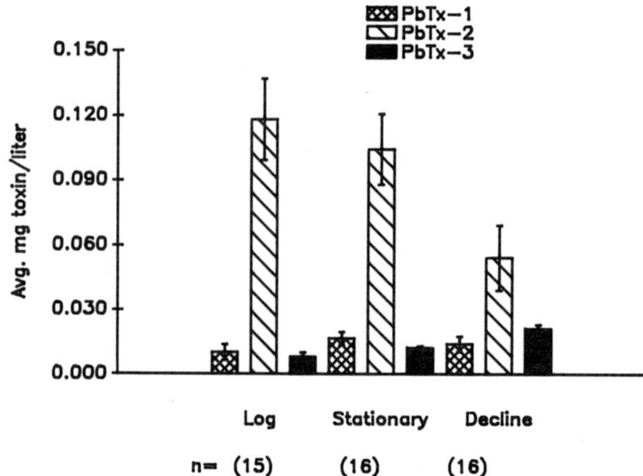

FIGURE 3: Growth stage of extracted culture vs. toxin yield. Extraction volumes ranged from 10 to 80 L.;n=52

Future research into this topic should include more attempts to quantitate all toxins found in the different growth stages, as well as more controlled experiments on the toxin variations with age. Uniformly labeled C^{14} or C^{13} PbTx-1 and PbTx-2 would be ideal for carrying out these studies, beginning with logarithmic phase cultures and analyzing toxin profiles through the cultures decline.

ACKNOWLEDGEMENTS

This work was supported in part by U.S. Army Medical Research and Development Command Contracts DAMD17-85-C-5171, DAMD17-87-C-7001, and DAMD17-88-C-8148. Opinions, interpretations, conclusions and recommendations are those of the authors and are not necessarily endorsed by the U.S. Army.

REFERENCES

1. Steidinger,K.A. Collection, enumeration, and identification of free-living marine dinoflagellates, in *Toxic Dinoflagellate Blooms* (D.L.Taylor and H.H. Seliger, eds.). Elsevier/North-Holland, Amsterdam, 435-442 (1979).
2. M.A. Poli, ,T.J. Mende and ,D.G. Baden. Brevetoxins, unique activators of voltage sensitive sodium channels, bind to specific sites in rat brain synaptosomes. *Molec. Pharmac.* 30,129. (1986).
3. S.Hall. Toxins and toxicity of *Protogonyaulax* from the northeast Pacific, Ph.D. thesis, University of Alaska, Fairbanks, 196 pp. (1982).
4. G.L. Boyer, J.J. Sullivan, R.J. Anderson, P.J. Harrison, and F.J.R. Taylor. Toxin Production in Three Isolates of *Protogonyaulax_Sp.*, in *Toxic Dinoflagellates* (Anderson, White & Baden, eds.). Elsevier, New York, 281-6 (1985).
5. D.G. Baden, T.J. Mende, M.A, Poli and R.E. Block in *Seafood Toxins*, E.P. Ragelis, ed. (ACS Symp. Ser. 262, Washington D.C. 1984) pp.359-368.
6. Y. Shimizu, H.N. Chou, H. Bando, G. VanDuyne and J. Clarddy, *J. Am. Chem. Soc.* 108, 514 (1986).

A NEW POLYETHER TOXIN FROM FLORIDA'S RED TIDE DINOFLAGELLATE *PTYCHODISCUS BREVIS*

LLOYD S. SCHULMAN*,**, LAURIE E. ROSZELL*, THOMAS J. MENDE**, ROY W. KING***, and DANIEL G. BADEN*,**.
*University of Miami Rosenstiel School of Marine and Atmospheric Science, Division of Biology and Living Resources, 4600 Rickenbacker Causeway, Miami, Fl. 33149, U.S.A.;**University of Miami School of Medicine, Department of Biochemistry, P.O. Box 016129, Miami, Florida 33101, U.S.A.;***University of Florida, Department of Chemistry, Gainesville, Florida 32611, U.S.A.

ABSTRACT

Florida's red tide dinoflagellate *Ptychodiscus brevis* produces at least 8 different polyether neurotoxins, based on two structural backbones. A new polyether toxin, designated PbTx-9 has been isolated from stationary cultures of *Ptychodiscus brevis*. This new toxin, the alpha-methylene- and aldehyde- reduced product of PbTx-2, has been confirmed by comparison studies with synthetic PbTx-9, prepared by $NaBH_4$ or NaB^3H_4 reduction of PbTx-2. Using thin layer chromatography, high performance liquid chromatography, Fourier transform infrared spectroscopy, and nuclear magnetic resonance spectroscopy, we have deduced the mechanism by which PbTx-3 and PbTx-9 arise during hydrogenation. PbTx-9 has the potential to be a new probe for sodium channel studies.

INTRODUCTION

Ptychodiscus brevis, the marine dinoflagellate that is accountable for Florida's red tides, produces at least 8 lipid-soluble neurotoxins called "brevetoxins" in culture [1]. These toxins consist of one of two different backbones which are based on polyether trans-fused lactone-containing ring structures (Fig.1) [2]. The many functional groups that result distal to the lactone ring account for the multiple derivatives that occur naturally and synthetically (Fig.1).

In culture, the quantity of each toxin fluctuates according to age and growth stage [3]. In general, PbTx-9 is not detected in logarithmic phase cultures, but is present in the stationary to declining growth stages. Even then, it is produced in very small quantities compared to PbTx-2 or PbTx-3. Toxin from several hundred liters of stationary cultures were pooled to obtain enough PbTx-9 to complete analysis.

PbTx-9, a type-1 toxin, was first characterized as "peak 2" observed during sodium borohydride reduction of PbTx-2 to produce PbTx-3 [4]. When substituting tritiated borohydride for unlabeled borohydride during reduction, PbTx-2 was reduced to tritiated PbTx-3 and tritiated PbTx-9, the observed specific activity of PbTx-9 being higher than the observed specific activity of PbTx-3 [4]. By modification of the sodium borohydride reduction of PbTx-2, yields of PbTx-9 and PbTx-3 have been optimized, and by examining the products arising from reductions performed in the presence of PbTx-3 or shift reagents, a mechanism for the production of each reduced toxin is proposed.

Published 1990 by Elsevier Science Publishing Co., Inc.
Toxic Marine Phytoplankton
Edna Graneli et al., Editors

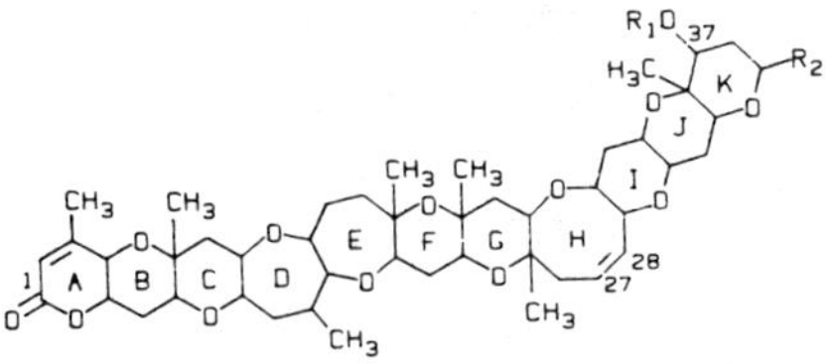

Type-1

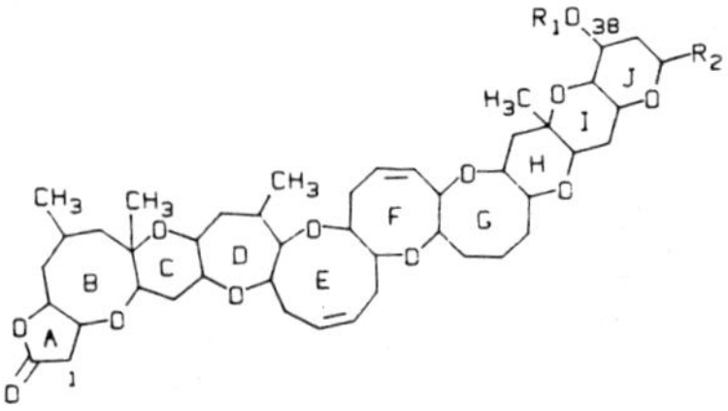

Type-2

FIG. 1. The brevetoxins are based on two different backbone structures, as indicated.

Toxin	Type	R_1	R_2
PbTx-1	2	H	$CH_2C(=CH_2)CHO$
PbTx-2	1	H	$CH_2C(=CH_2)CHO$
PbTx-3	1	H	$CH_2C(=CH_2)CH_2OH$
PbTx-5	1	$COCH_3$	$CH_2C(=CH_2)CHO$
PbTx-6	1	H	$CH_2C(=CH_2)CHO$ (27,28 epoxide)
PbTx-7	2	H	$CH_2C(=CH_2)CH_2OH$
PbTx-8	1	H	CH_2COCH_2Cl
PbTx-9	1	H	$CH_2CH(CH_3)CH_2OH$

MATERIALS AND METHODS

Extraction and Purification of Toxins

Ptychodiscus brevis was cultured and extracted as described previously [5,6]. The toxic fractions were purified by a combination of column, thin-layer, and C-18 reverse phase high performance liquid chromatography using isocratic 85% aqueous methanol as the mobile phase, and monitoring effluent at 215nm.

Chromatography

Thin-layer chromatographic analysis of natural PbTx-9 and PbTx-3 and synthetic PbTx-9 and the PbTx-3 reduced in the presence of $CeCl_3$ proceeded by applying 5 ug of each toxin to silica gel mini-plates (EM Reagents, 0.25 mm thickness, 2.5 x 5 cm) and developing in three different solvent systems as indicated in Table I. High performance liquid chromatographic analysis of each toxin was carried out on C-18 reverse phase columns employing 85% aqueous methanol as mobile phase at a flow rate of 1.4 ml/min. The retention times and homogeneity of each individual toxin, and of mixtures of natural and synthetic toxins, were monitored by ultraviolet absorbance at 215 nm.

Sodium Borohydride Reduction

Mixed products. PbTx-3 and PbTx-9 was produced synthetically in the following manner. PbTx-2 (12.400mg) was dried over P_2O_5 and was dissolved in 2.0ml of acetonitrile under constant stirring. 0.165ml aliquots of 0.01M $NaBH_4$ (in acetonitrile) solution were added slowly over a period of three minutes every 20 minutes until the PbTx-2 was completely reduced, as monitored by HPLC. One half total volume of H_2O was then added, the toxins were extracted with ether three to four times, and the combined ether fractions were dried down with nitrogen. The toxins were dissolved in methanol and purified by HPLC. The resulting yields were 5.110mg PbTx-3 (42% yield) and 2.502mg PbTx-9 (20.1% yield). By a similar method, purified PbTx-3 was subjected to sodium borohydride reduction, and the products were extracted and purified.

PbTx-2 to PbTx-3 specific (α-methylene protected). PbTx-2 (4.913mg) was dried over P_2O_5 and 147ul of a 0.4M $CeCl_3$ solution in methanol [7] was added to the toxin under constant stirring. An equal volume of methanol was then added, followed by slow addition of 17.1μl of 0.01M $NaBH_4$ solution in dimethyl formamide (DMF). This was repeated every 20 minutes until the PbTx-2 was completely reduced. During the reaction the solution became clear as PbTx-2 was converted to PbTx-3. One half total volume of H_2O was added, and the toxin extracted with ether 3 to 4 times. The combined ether fractions were dried down with nitrogen, dissolved in minimal methanol, and purified by C-18 reverse phase HPLC. The resulting yield was 4.294mg PbTx-3 (87.4% yield).

Fourier Transform Infrared Spectroscopy. Fourier transform infrared spectra of synthetic and natural PbTx-3 and PbTx-9 were obtained in KBr pellets using a Mattson Instruments Cygnus 100 FTIR; 32 scans were accumulated. Natural and synthetic toxin spectra were compared by computer, and an overlay of the fingerprint region was also performed .

Nuclear Magnetic Resonance Spectroscopy. Samples were purified by HPLC and dried over P_2O_5. The proton NMR spectrum of each toxin was obtained in 5 mm tubes on a Varian VXR-300 spectrometer, using $CDCl_3$ (99.96% enrichment) as solvent and TMS as standard. 1024 transients were accumulated for the natural PbTx-9 and 256 for the remaining toxins; no apodization was used.

RESULTS AND DISCUSSION

We and others have previously shown, by comparison of nmr spectra, that the material brevetoxin B isolated by Lin et al. [8] is identical to PbTx-2. Similarly, the material PbTx-2 has been previously reported to be chemically reduced to form PbTx-3, with the proton nmr signal at 9.5 ppm for the aldehyde disappearing and the concomitant shift of the methylene signal located α to the reduced aldehyde [1,2,4]. We now report that the reduction of PbTx-2 to its products occurs by two pathways. One pathway occurs when the PbTx-2 aldehyde function is reduced to the primary alcohol function, the product being PbTx-3. This reaction is terminal and PbTx-3 cannot be further reduced chemically using $NaBH_4$. A second competing reaction occurs when the α-methylene function of PbTx-2 is reduced first, and the aldehyde is subsequently reduced to form PbTx-9 (Fig. 2). The intermediate α-methylene-reduced aldehyde-containing toxin is proposed, but we have not been able to isolate it in purified form.

Synthetic PbTx-9 is identical to PbTx-9 isolated from cultures of *Ptychodiscus brevis*. Thin-layer, and high performance liquid chromatographic migration patterns were identical for natural and synthetic forms of the toxin (Table I). Likewise, the FTIR (Table II) and proton

TABLE I. Chromatographic parameters.

Toxin	TLC(Rf) I	II	III	HPLC (Retention time, min)
PbTx-3 natural	.65	.27	.56	8.13
PbTx-3 synthetic	.65	.27	.56	8.13
PbTx-9 natural	.53	.21	.59	9.12
PbTx-9 synthetic	.53	.21	.59	9.12
PbTx-2 natural	.73	.49	.65	9.95

I =Hexane:Tetrahydrofuran (50:50); II =Ethyl acetate:pet.ether (70:30); III =Ethyl acetate:methanol:Pet.ether (10:10:80)

NMR spectra (Table III) were virtually identical for synthetic and natural materials, with the NMR spectrum indicating a diasteromeric nature for the synthetic PbTx-9, which was expected [9].

FTIR spectra (Table II) shows that PbTx-9 is of the Type-1 backbone by its lactone signal in the 1720-1735 cm^{-1} region, as compared with type-2 toxins and their characteristic absorbance at 1760-1780 cm^{-1}. Expansion and overlay of the fingerprint region indicated that natural PbTx-9 and synthetic PbTx-9 were identical. In confirmation of these findings, the methyl signals for Rings A and D in PbTx-3 and PbTx-9 occur at 1.97 and 1.03 ppm, again indicative of Type-1 toxins [8].

Apart from some minor peaks that may be reasonably assigned to artifacts of impurities, the spectra of natural and synthetic PbTx-3 are identical. NMR spectra of natural and synthetic PbTx-9 are superimposable except for the methyl region. Reduction of the terminal methylene group creates a chiral center which, with the chiral ring system, gives rise to a mixture of diastereomers, as indicated by the methyl doublet at 0.91 and 0.96 ppm. This was expected from the non-stereospecific nature of the reduction reaction, and resulted in a ratio for the isomers of about 2:1. Interestingly, natural PbTx-9 presents a ratio of about 7:1. Comparison of PbTx-3 spectra with those of PbTx-9 indicate a loss of the methylene signal at 4.94 and 5.11 ppm upon reduction (Fig. 3).

TABLE II. Comparison of natural & synthetic PbTx-9 by FTIR wavenumber of the characteristic peaks.

PbTx-9 Natural		PbTx-9 synthetic	
3460.908	1648.32	3461.030	1647.84
1735.78	1638.01	1735.37	1638.04
1724.69	1630.77	1724.51	1630.28
1719.30	1619.12	1718.81	1618.93
1707.67	1578.29	1707.50	1578.25
1701.72	1560.79	1701.59	1560.72
1697.41	1466.73	1697.26	1466.11
1686.00	1459.79	1685.85	1458.96
1676.27	1382.16	1675.87	1382.80
1670.68	1127.79	1670.55	1127.67
1664.05	1072.42	1662.54	1069.81
1655.23	1056.70	1655.20	1056.04

Our initial conjecture was that PbTx-9, the double reduction product of PbTx-2, arose from the sequential reduction of PbTx-2 to PbTx-3 to PbTx-9. To ascertain PbTx-9 was a product of PbTx-3, a $NaBH_4$ reduction of PbTx-3 was performed under the conditions described above. PbTx-9 was not present in extracts following reaction; PbTx-3 was recovered unaltered.

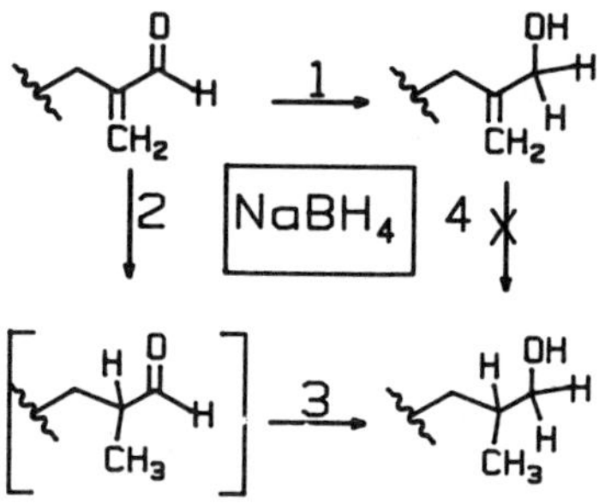

FIG. 2. Sodium borohydride reduction of PbTx-2. Reduction of the aldehyde via (1) yields PbTx-3. A competing reaction, i.e. reduction of the α-methylene first via (2) yields an intermediate. This intermediate aldehyde (which we have yet to isolate in pure form) is now reduced via (3) yielding racemic PbTx-9. Reduction of PbTx-3 via (4) cannot occur owing to deactivation of the methylene following loss of conjugation. When cerium chloride is added to block reaction (2) pathway (1) reduction becomes virtually quantitative.

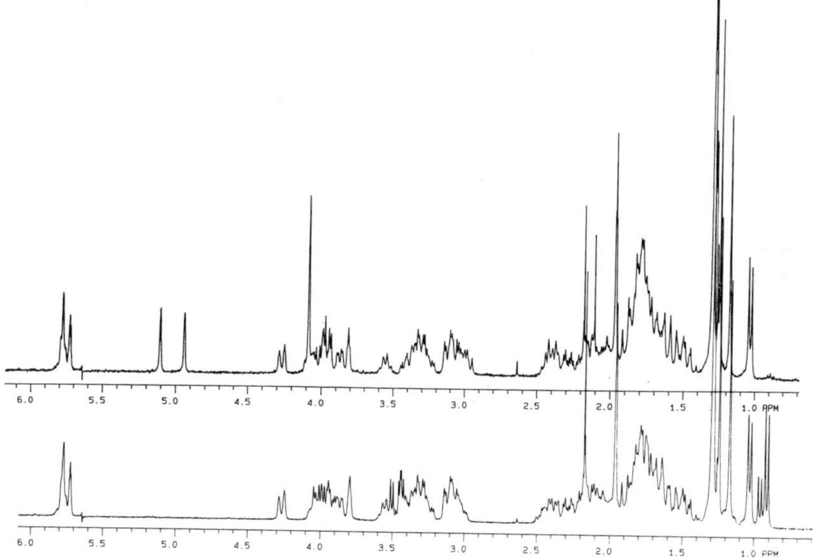

FIG. 3. Proton nuclear magnetic resonance spectra of PbTx-3 (upper) and PbTx-9 (lower). Prominent differences between the two spectra include the methylene protons in PbTx-3 at 4.94 and 5.11 ppm, which are reduced in PbTx-9 to yield diastereomeric methyl functions with chemical shifts at 0.91 and 0.96 ppm. This is accompanied by a splitting of the singlet proton signal at 4.09 in PbTx-3 (the protons on C-42) by the proton on the reduced methylene function on C-41 in PbTx-9. All remaining chemical shifts, by comparison with published spectroscopic data, indicate that both PbTx-3 and PbTx-9 are reduction products of PbTx-2.

TABLE III. Selected proton nmr shifts which indicate the reduction of the C-41, C-43 methylene function in the spectrum of PbTx-3 to yield PbTx-9 (ppm).

Proton Group	PbTx-3	PbTx-9
$=CH_2$ proton *cis* to C-42	5.11	----
$=CH_2$ proton *trans* to C-42	4.94	----
CH_3 (C-43) diastereomers	----	0.96
CH_3 (C-43)	----	0.91

Thus the reaction scheme could not proceed via. sequential reactions #1 and #4. To prove that PbTx-9 was a product of PbTx-2 (actually a negative proof), we performed a $NaBH_4$ reduction in the presence of cerium chloride [7] to protect the α-methylene group from reduction (reaction #2 in Fig. 2). The result was a nearly quantitative conversion of PbTx-2 to PbTx-3 by reaction #1.

These results are useful in two ways: First, we have proven that PbTx-9 is a reduction product of PbTx-2 and that the α-methylene of PbTx-2 is reduced prior to aldehyde reduction; and second, a method has been developed to obtain a reduction of PbTx-2 to PbTx-3 with an average yield of 84% One problem with the use of natural derivative toxins as probes of membrane function, is the small quantities in which they are produced. This problem has been overcome in the case of PbTx-9 and PbTx-3 by the reduction of PbTx-2 using $NaBH_4$. By substituting tritiated borohydride for unlabeled borohydride, each PbTx-9 & PbTx-3 can be tritiated to high specific activities produced in a suitable quantity for use as specific probes of voltage-dependent sodium channel and for radioimmuno-assays.

ACKNOWLEDGEMENTS

This work was supported in part by U.S. Army Medical Research and Development Command Contracts DAMD17-85-C-5171, DAMD17-87-C-7001, and DAMD17-88-C-8148. Opinions, interpretations, conclusions and recommendations are those of the authors and are not necessarily endorsed by the U.S. Army.

REFERENCES

1. Baden, D.G. (1989) FASEB Journal 3, 1807.
2. Shimizu, Y. Chou, H.N., Bando, H., VanDuyne, G. and Clardy, J. (1986) JACS 108, 514.
3. Roszell *et al.*, this volume.
4. Poli, M.A., Mende, T.J., and Baden, D.G. (1986) Molec. Pharmacol. 30, 129.
5. Baden, D.G., Mende,T.J., Walling, J. and Schultz, D.R. (1984) Toxicon 22, 783.
6. Baden, D.G., Mende, T.J., Lichter, W. and Wellham, L. (1981) Toxicon 19, 455.
7. Luche, J.L. (1978) JACS 100, 2226.
8. Lin, Y., Risk, M., Ray, S.M., VanEngen, D., Clardy, J., Golik, J., James, J., and Nakanishi, K. (1981) JACS 103, 6773.
9. Complete spectral data are available upon request.

RATES OF PRODUCTION OF DOMOIC ACID, A NEUROTOXIC AMINO ACID IN THE PENNATE MARINE DIATOM *NITZSCHIA PUNGENS*

D.V. Subba Rao*, A.S.W. de Freitas**, M.A. Quilliam**, R. Pocklington*, and S.S. Bates***
*Department of Fisheries and Oceans, Science Directorate, Bedford Institute of Oceanography, PO Box 1006, Dartmouth, N.S, Canada B2Y 4A2; **Atlantic Research Laboratory, National Research Council of Canada, 1411 Oxford Street, Halifax, N.S, Canada B3J 2S7; ***Dept. Fish. and Oceans, Science Branch, 343 Archibald St., Moncton, N.B., Canada E1C 5K5.

ABSTRACT

Recent work has demonstrated that blooms of the diatom *Nitzschia pungens* were the major source of a novel shellfish toxin, domoic acid, in cultured blue mussels *Mytilus edulis* from the Cardigan Bay region of eastern Prince Edward Island, (P.E.I.) Canada. This is the first example of a diatom producing a neurotoxic amino acid. The factors controlling toxin production by natural blooms of this diatom are unknown. To investigate toxin production, we employed a two-way factorial design experiment with two isolates of *N. pungens* grown in two media. Variations in growth and domoic acid levels were determined over 49 days. Division rates per day ranged between 0.5 and 0.9; cells grew exponentially until the eleventh day, reaching a plateau cell density that ranged between 39-41.5 $\cdot 10^6$ cells l^{-1}. Neither fresh sterile media nor cells in the exponential phase yielded any domoic acid. Production of domoic acid was observed only in post-exponential cultures. Maximum concentrations of domoic acid ranged from 0.3 to 2.0 $pg \cdot cell^{-1}$. The loading limit for intracellular domoic acid of 6-10 $pg \cdot cell^{-1}$ is maintained during the continuing production and excretion of domoic acid by *N. pungens* cultures during the post-exponential phase.

INTRODUCTION

In late fall of 1987, an outbreak of shellfish poisoning resulting from consumption of blue mussels, *Mytilus edulis,* cultured in the Cardigan Bay, Prince Edward Island, (P.E.I.) was associated with massive blooms of the pennate diatom *Nitzschia pungens*. The causative toxin was identified as domoic acid, a neurotoxic amino acid, novel to shellfish (1). The fact that cultures of the diatom *N. pungens* produce domoic acid de novo, the first example for a diatom, was established (2). The monospecific blooms of *N. pungens* which recurred in November 1988 seem to be the primary source of domoic acid (3). Although factors that control production of this toxin in natural blooms must be understood to permit proper management of shellfish, they remain unknown. The presence of domoic acid in Cardigan Bay mussels and its production by cultures of a small pennate benthic diatom *Amphora coffaeiformis* Cl. has been reported (4). However, data on the concentration of domoic acid and on the factors that regulate domoic acid production are not given.

In our current study we investigated how the age of the culture, source of the isolate and the nature of medium affect the concentrations of domoic acid in *N. pungens* cells and in the medium. We present observations on the variability of production of domoic acid in cultures, grown under different conditions, which could serve as analogues of natural blooms.

Toxic Marine Phytoplankton
Edna Graneli et al., Editors

METHODS

We employed a two-way factorial design using two isolates of *N. pungens* (NPBIO,and NPARL) grown in medium FE (2) and Woods Hole medium F/2 (5). Sea water from Cardigan Bay, P.E.I. and Sandy Cove near Halifax, Nova Scotia was used for enrichment. The following 4 cultures that served as inocula were set up:

Isolate	Source of sea water	Medium
NPBIO	Sandy Cove	NPBIO FE
NPBIO	Cardigan	NPBIO F/2
NPARL	Cardigan	NPARL FE
NPARL	Sandy Cove	NPARL F/2

The inocula were conditioned in the above media by successive subculturing into fresh sterile media. The first transfer was made after 3 days of growth, the second and third transfers after 7 day growth so that the inocula would be in the log phase. The final inocula had 34.2, 39.4, 29.3 and $40.1\cdot10^6$ cells l^{-1} corresponding to NPBIO FE, NPBIO F/2, NPARL FE and NPARL F/2 media, and were used to seed 20 l carboys each containing 8 l of sterile medium. Cultures were grown at 10 °C under 530 microeinsteins/m^2/s continuous fluorescent light. Samples were taken for enumeration under an inverted plankton microscope and for determination of domoic acid after 0, 6, 11, 18, 21, 28, 35, 42 and 49 days of growth. Growth constants were calculated using the formula:

$$k \text{ (divisions day)} = \{3.322/(t2-t1)\}\cdot\{\log N2/N1\}$$

where N2 and N1 correspond to cell numbers on days t2 and t1 respectively.

For domoic acid analyses, 20 ml samples of whole culture were transferred to plastic vials and frozen for subsequent analysis of total domoic acid in culture. Additionally, 200 ml samples of culture were filtered through 47 mm GF/F filters at ~kPa vacuum. The filters were tranferred to numbered petri dishes and frozen. The filtrates were similarly transferred to clean plastic bottles and refrigerated. Analytical procedures for the determination of domoic acid followed, using previously described procedures (2,6). Domoic acid from the cells retained on the filters was extracted with acetonitrile-water (1:1), reduced in volume and interfering substances removed using C-18 solid phase extraction. A high pressure liquid chromatography (HPLC) with UV detection at 242 µm and a high sensitivity modification of the 9-fluorenylmethoxycarbonyl chloride (FMOC) precolumn derivitization method for amino acids (7) were employed. The identity of domoic acid was confirmed by matching HPLC retention times and ultraviolet spectra.

RESULTS AND DISCUSSION

The initial cell densities in the four treatments ranged between 1.72 and $2.45\cdot10^6$ l^{-1}; these grew exponentially until about the 11th day, and attained a plateau with maximum density that ranged between $39\text{-}41.5\cdot10^6$ $cells^{-1}$ (TABLE I). Division rates per day ranged between 0.5 and 0.9 and were higher for those in FE (0.69 BIO, 0.87 ARL) than those for F/2 (0.47 ARL, 0.62 BIO). The division rates compare favourably with those reported by us earlier (2,8).

Neither fresh sterile media nor cells in the exponential phase yielded any domoic acid (Fig.1). Production of domoic acid was variable and was

TABLE I. Domoic acid levels in two isolates of *Nitzschia pungens*, at various ages of growth, in two media.

Culture	Growth (days)	Cell nos. ($10^3 ml^{-1}$)	DOMOIC ACID (D.A.) IN CULTURE Whole culture ($ng\ ml^{-1}$)	Proportion of D.A. in cells (%)
NPBIO FE	0	2.45	-	-
	6	21.56	-	-
	11	29.24	0.14	100
	18	37.0	0.98	99.3
	21	36.75	0.67	87.9
	28	32.99	0.87	99.3
	35	32.35	1.25	100
	42	38.38	1.92	100
	49	39.20	1.44	94.4
NPBIO F/2	0	1.88	-	-
	6	11.84	-	-
	11	26.13	0	-
	18	37.16	0.06	96.9
	21	41.49	0.25	100
	28	33.24	0.25	100
	35	32.01	0.29	97.7
	42	37.16	0.34	97.9
	49	41.48	0.21	100
NPARL FE	0	1.72	-	-
	6	20.91	-	-
	11	24.99	0.09	97.8
	18	31.12	0.46	99.4
	21	32.01	0.84	100
	28	37.24	0.76	99.7
	35	36.09	1.11	100
	42	38.22	1.74	100
	49	41.32	2.08	99.9
NPARL F/2	0	1.96	-	-
	6	5.88	-	-
	11	19.03	0.12	99.3
	18	29.97	0.20	98.3
	21	35.52	0.23	98.4
	28	32.18	0.49	99.2
	35	33.65	0.70	100
	42	37.49	0.85	100
	49	49.59	0.72	100

observed only in post-exponential cultures (Fig.1). Generally the intracellular domoic acid increased with the age of the culture. Throughout

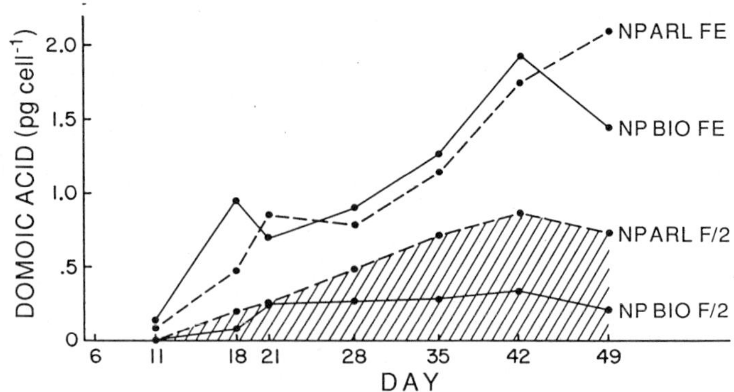

FIG. 1 DOMOIC ACID LEVELS IN *NITZSCHIA PUNGENS* CELLS

the post-exponential phase, essentially all (>88%) of the domoic acid was associated with the cells (Table I) with the maximum cellular concentrations ranging between 0.3 and 2.0 pg cell^{-1}. There were slight departures from the value expected in the cellular domoic acid levels for cultures of 18, 21 and 49 days growth in BIO FE (Fig.1) but not in their cell numbers; the departure from the expected are artifacts or minor aberrations in the measurements. Cells in FE media which usually had higher division rates, also had consistently higher domoic acid (0.4 - 1.92 pg·cell^{-1}; 0.9 - 2.08 pg·cell^{-1} ARL) than those in F/2 (0 - 0.34 pg·cell^{-1} BIO; 0.12 - 0.85 pg·cell^{-1} ARL). The cell density achieved was not a result of limitation of the major nutrients i.e. phosphate, silicate and nitrate.

Although most of the domoic acid in the cultures was intracellular and ranged between 88 and 100% (TABLE I), some excretion or release of domoic acid by *N. pungens* cultures occurred during the post-exponential phase. Data from other experiments demonstrate significant amounts of domoic acid were released to the medium by late post-exponential cells. Changes in intracellular and extracellular domoic acid levels and their temporal sequence parallel those of methionine, in the cultures of the marine xanthophycean alga *Monallantus salina,* isolated from the Gulf of Marseille (9). In this alga, the intracellular methionine appeared only in trace amounts in the exponentially growing cells (6 days); it increased to a maximum by the 11th day and remained fairly constant till the 20th day i.e. in the post-exponential cells. No extra-cellular methionine appeared in cultures between the 9th and 20th days of growth.

Our data from other ongoing experiments are consistent with these findings on domoic acid levels but with production rates as high as 2 pg·cell^{-1} day^{-1}. It appears that the loading limit for intracellular domoic acid was 6-10 pg·cell^{-1} which was maintained during the postexponential phase by the continuing production and excretion of domoic acid. Our results indicate that if domoic acid production by *N. pungens* in natural blooms follows the same pattern as our laboratory cultures, then analyses of rapidly growing blooms would not necessarily show the presence of domoic acid since the concentrations appear to be significant only in post-exponential cells.

SUMMARY

(1) Domoic acid production by *N. pungens* is confined to the post-exponential phase.

(2) Rates of domoic acid production are highly variable with values ranging up to 2 $pg \cdot cell^{-1} \cdot day^{-1}$

(3) The loading limit of intracellular domoic acid was about 6-10 $pg \cdot cell^{-1}$ with the "surplus" production presumably released into the medium.

ACKNOWLEDGEMENTS

We are grateful to Drs. J.E. Stewart and R.F. Addison, Dept. Fish. Oceans, Bedford Inst. Oceanography for their reviews and constructive criticism of the manuscript.

REFERENCES

1. R.F. Addison, and J.E. Stewart, Aquaculture 77, 263-269 (1989)
2. D.V. Subba Rao, M.A. Quilliam and R. Pocklington, Can. J. Fish. Aquat. Sci. 45, 2076 - 2079 (1988).
3. S.S. Bates et al. Can. J. Fish. Aquat. Sci. (in press)
4. Y. Shimizu, S. Gupta, K. Masuda, L. Maranda, C.K. Walker and R. Wang, Pure & Appl. Chem. 61(3), 513-516 (1989).
5. R.R.L. Guillard and J.H. Ryther, Can. J. Microbiol. 8, 229-239 (1962).
6. R. Pocklington, J.E. Milley, S.S. Bates, J.C. Bird, S.W. de Freitas, and M.A. Quilliam, Int. J. Environ. Anal. Chem. (communicated).
7. S. Einarsson, S. Folestad, B. Josefsson, and S. Langerkvist, Anal. Che. 58, 1638-1643 (1986).
8. D.V. Subba Rao, Bot.Mar. 24, 369-379 (1981).
9. B.R. Berland, D.J. Bonin, R.A. Daumas, P.L. Laborde and S.Y. Maestrini, Mar. Biol. 7, 82-92 (1970).

ENTEROTOXIC, HEPATOTOXIC AND IMMUNOTOXIC EFFECTS OF DINOFLAGELLATE TOXINS ON MICE

KIYOSHI TERAO*, EMIKO ITO*, TAKESHI YASUMOTO**, and KATSUMI YAMAGUCHI***

*Research Center for Pathogenic Fungi and Microbial Toxicoses, Chiba University, 1-8-1 Inohana, Chiba, 280, Japan; **Faculty of Agriculture, Tohoku University, Tsutsumidori Amamiya, 1-1, Sendai, 980, Japan; ***Faculty of Agriculture,The University of Tokyo, Yayoicho, Hongo, Tokyo, 113, Japan.

ABSTRACT

Toxins isolated from several species of dinoflagellates cause a diverse array of morphological injuries in organs of mice and rats. Dinophysistoxin-1 (DTX-1) from Dinophysis fortii and maitotoxin(MTX) from Gambierdiscus toxicus have enterotoxic effects. Goniodomin A (GA) from Goniodoma pseudogoniaulax causes hepatic damage. MTX and GA also induced severe lymphocytic necrosis in the thymus and marked reduction in circulating lymphocytes. These injuries are not observed in mice pretreated with bilateral adrenalectomy or with $CoCl_2$, a potent inhibitor of Ca-influx. The Ca content in the adrenal glands and the concentration of plasma cortisol increases shortly after administration of MTX. Therefore, the immunotoxic effects of MTX are due to the increased influx of Ca in the zona fasciculata of the adrenal glands. This causes an excess of cortisol in the blood and involution of the thymus follows. A similar mechanism may exist in the case of GA intoxication.

INTRODUCTION

It is now widely recognized that certain species of marine phytoplankton produce poisonous substances and transmit them to humans through the food chain. Toxins isolated from several marine dinoflagellates have been intensively studied [1-5]. Through a series of experimental studies on the biological activities of these toxins we have found that these toxins cause a diverse array of morphological changes in the organs of experimental animals[2 - 5]. The target organs of the toxins examined so far are the heart, blood vessels, liver, digestive tract, kidney, nerve system and lymphoid tissues including the thymus, spleen, and lymph nodes. We wish to summarize here the histopathological changes induced by toxins isolated from these dinoflagellates.

Toxic Marine Phytoplankton
Edna Graneli et al., Editors

RESULTS and DISCUSSION

Enterotoxic phycotoxin

Dinophysistoxin-1(DTX-1) is produced by the dinoflagellate *Dinophysis fortii* and is a diarrheagenic phycotoxin[6,7]. DTX-1 contains a C_{38} polyether fatty acid skeleton (Fig. 1). DTX-1 causes excessive fluid accumulation in the intestines of suckling mice[6]. Within 15 min the duodenum and upper portion of the small intestine becomes distended. Light and electron microscopic examinations show that three sequential stages occur in the intestinal villi after the intraperitoneal administration of more than 300 ug DTX-1 per kg body weight. The stages are as follows: extravasation, degeneration of the intestinal absorptive epithelium, and desquamation of the degenerated mucous epithelium from the villous surface[2].

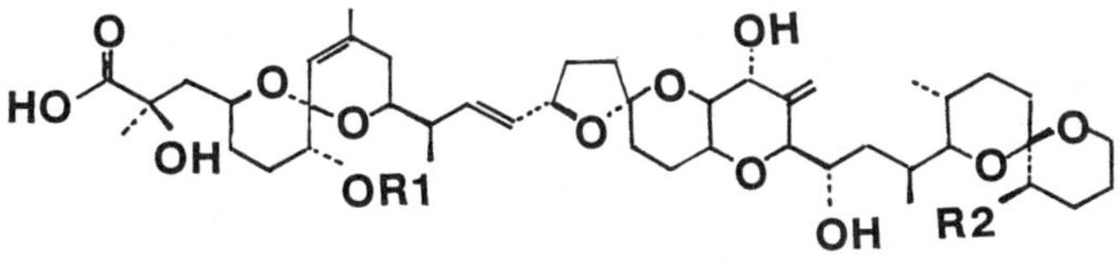

Fig. 1. Chemical Structure of Dinophysistoxin-1.

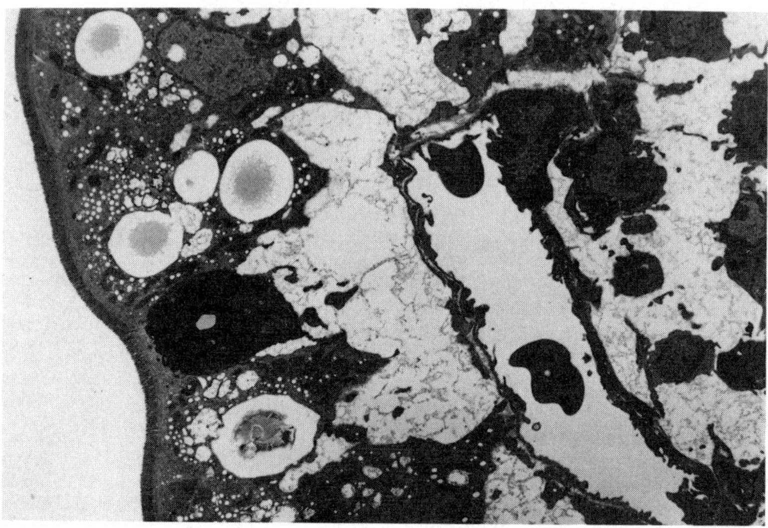

Fig. 2. An electron micrograph of mouse intestine 2hr after i. p. injection of DTX-1.

In the first stage, there is marked extravasation of serum into the lamina propria of villi. Abnormal separation is seen occasionally between the epithelial cells and the basal membrane.

The second stage, in which mucosal cells degenerate, is characterized by a marked dilation of the cisternal portion of the Golgi apparatus in the absorptive epithelium. The dilated cisternae become filled with flocculent proteinous materials. Numerous vesicles up to 0.3 um in diameter occur in the apical portions of the intestinal absorptive cells. In contrast to the inner side of the plasma membrane, no discernible changes can be seen on the surface microvilli during the very early stage. However, between 10 and 30 min after injection, degeneration of the microvilli can be seen in the epithelium. The plasma membrane along the luminal surface of several cells has extensive areas lacking any microvilli. The synapses of the myentric plexus of Auerbach, as well as those of the plexus of Meisner, become slightly swollen.

During the last stage, almost all mucosal epithelial cells become detached from the basal membrane. At the basal portion of the mucosal cells large vacuoles occur. The severity of the changes in the small intestine increases progressively with time after injection of DTX-1.

Throughout the three stages there are only slight changes in cells in the glands of Lieberkuehn.

Studies on the pathogenesis of diarrheal syndromes reveal that there are two histopathologically different responses of intestinal mucosa to diarreheagenic toxins, i.e. some diarrheagenic toxins produce no morphological changes and others produce severe morphological changes. Choleragen induces severe fluid production, but no morphological changes in the epithelium and ileal lamina propria[8,9]. In contrast, a toxin produced by <u>Clostridium difficile</u> causes hemorrhagic fluid accumulation and mucosal tissue destruction in ligated intestinal loops of hamsters[10]. Our results with DTX-1 show clearly that the phycotoxin causes diarrheal syndrome with tissue destruction. The mode of action of DTX-1 at the molecular level remains unclear.

Maitotoxin (MTX), a water-soluble toxin, is produced by the marine dinoflagellate <u>Gambierdiscus toxicus</u>[11]. Mice given 200 ng MTX/kg and rats given 400 ng MTX/kg show within 4 hr greatly distended stomachs and small intestines, often accompanied by moderate ascites. After treatment with MTX, multiple erosions and ulcers occur in the gastric mucosa, accompanied by a marked increase in total Ca content[3].

<u>Hepatotoxic Phycotoxin</u>

Goniodomin A(GA), a unique polyether macrolide, is a metabolite of the dinoflagellate <u>Goniodoma pseudogoniaulax</u>(Fig 3). The phycotoxin has an antifungal effect[12]. The 24-hr and 48-hr LD_{50} values for i.p. injection into adult mice are 1.2 and 0.7 mg GA per kg body weight, respectively [3]. GA injected intraperitoneally results in severe peritonitis with purulent effusion. Also, severe exudation with massive accumulation of neutrophiles is often seen at the sites of GA-injection, such as in subcutaneous or muscle tissues. It is not yet clear what part of the structure of GA is responsible for the local irritability.

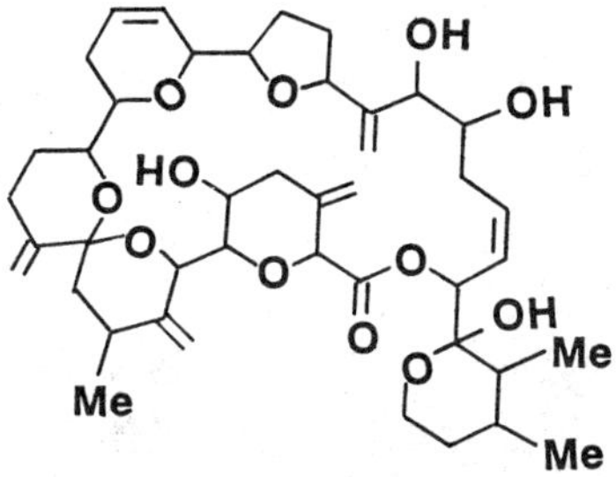

Fig. 3. Chemical Structure of Goniodomin A.

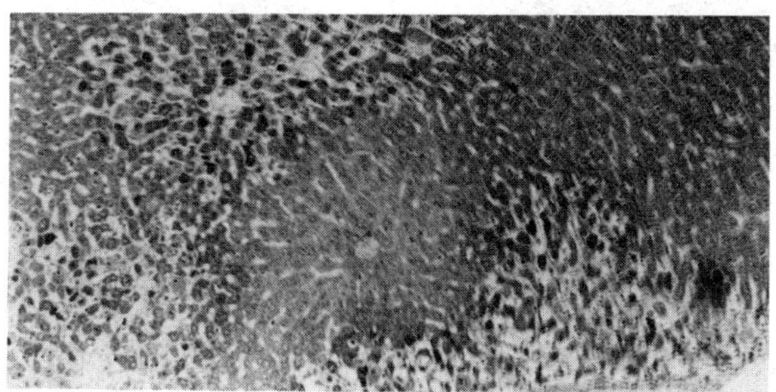

Fig. 4. Mouse liver 24 hr after the i. p. administration of GA.

The target organs of GA are the liver and thymus. Histologically, marked necrosis of the centrilobular region of the liver is seen after either i. p. injection of 0.5 mg/kg, subcutaneous injection of 2 mg/kg, or oral administration of 10 mg GA/kg (Fig. 4.). In addition, massive hemorrhage is observed in the subcapsular region of the liver after intraperitoneal or subcutaneous administration. Prominent dilation of rough-surfaced endoplasmic reticulum is apparant in hepatocytes around the necrotic lesion. The nucleoli of the hepatocytes decreases in size and round up ; the normal configuration of the nuceolus disappears. This ultrastructural feature of the affected hepatocytes suggests that GA influence RNA synthesis in the hepatocytes. Occasionally, non-fatty vacuoles can be seen in the cytoplasm of the hepatocytes in the centrilobular regions. The induction of such non-fatty vacuoles by several microbial toxins, such as cyclochlorotine and phalloidine, has been reported by several investigators.

Immunotoxic Phycotoxin

Severe damage of the thymus is also the characteristic of GA-intoxication. Twenty-four hr after administration of GA, massive necrosis of the lymphocytes in the cortical layer of the thymus occurs. In contrast, epithelial reticular cells are not affected. The weight of the spleen is reduced to about 50% of control. There is a reduction in the number of lymphocytes in the periarterial lymphoid sheath, as well as in the red pulp [4,5].

To clarify the effects of GA on lymphoid tissue, we have investigated adrenalectomized mice. In spite of GA injection, no discernible pathological changes are observed in the thymus and spleen of the adrenalectomized mice. Similar lesions in lymphoid tissues are also induced by MTX[4,5]. Four hr after a single injection of 200 or 400 ng MTX per kg body weight, the thymuses of mice and rats show degeneration of lymphocytes(Fig. 5).

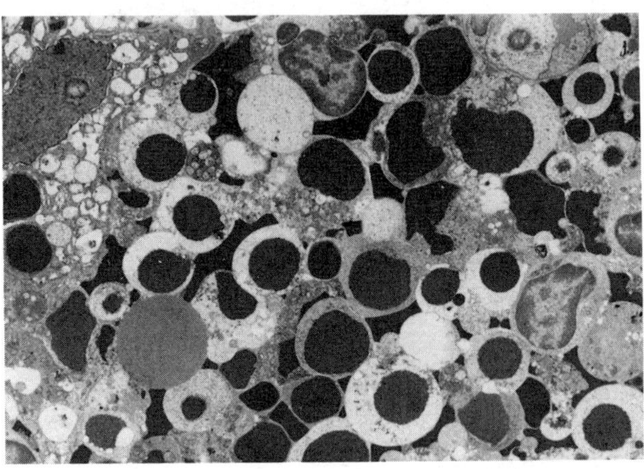

Fig.5. An electron micrograph of the mouse thymus treated with MTX 24 hr prior to sacrifice.

Within 1 hr a reduction of immunoglobulin M in serum occurs, as well as an increase in Ca content in the adrenal glands and in plasma cortisol concentration. Repeated injections of 45 ng MTX per kg body weight results in a marked reduction of lymphocytes in blood and lymphoid tissues. In contrast, mice treated with $CoCl_2$, a Ca channel inhibitor, or bilateral adrenalectomized mice show no discernible changes in the lymphoid tissues after repeated MTX injections. It has been reported that MTX results in Ca-dependent release of various hormones [13,14] and Ca-dependent contraction of various types of muscles, including smooth[15], skeletal[16] and cardiac muscles[17].

From the above-mentioned evidence we conclude that the lesion of the thymus is not a result of the direct effect of MTX. We suggest that MTX first stimulates Ca influx in cells of the zona fasciculata of the adrenal glands, which then causes the release of cortisol into the blood. The excess amount of cortisol in the blood produces acute involution of the thymus and other lymphoid tissues. The involution of the thymus induced by GA may be caused in a similar manner, although the precise mechanism remains unknown.

<u>Cardiotoxic Phycotoxin</u>

MTX often produced heart damage[17].Within 30 min, mice receiving 200 or 400 ng MTX per kg body weight show scattered coagulative necrosis in the cardiac muscle of the left ventricle. Marked swelling of the endothelial lining cells of the capillaries occurs. Therefore, the lumen becomes very constructed, or even closed. Muscle cells around the closed capillaries have degenerated. It is not yet clear whether the lesions are directly caused by the increased influx of Ca into the muscle tissue or by some other indirect effects of MTX.

REFERENCES

1. D.G. Baden in:Int. Rev. Cytol.82 G. H. Boutne, J. F. Danelli, and K. W. Jeon eds. (Academic Press, N.Y. 1983) pp 99-150.
2. K.Terao, E.Ito, T. Yanagi and T. Yasumoto, Toxicon 24, 1141-1151,(1986).
3. K. Terao, E.Ito, Y. Sakamaki, K. Igarashi, A. Yokoyama and T. Yasumoto, Toxicon, 26, 395-402, (1988).
4. K. Terao, E. Ito, M. Murakami and K. Yamaguchi,Toxicon, 27, 269-271, (1989).
5. K. Terao, E. Ito, Y. Kakinuma, K. Igarashi, Y. Ohizumi and T. Yasumoto, Toxicon, in press. (1989).
6. Y. Hamano, Y. Kinoshita and T. Yasumoto, in: Toxic Dinoflagellates. D. M. Anderson, A. W. White and D. G. Baden eds. (Elsevier ,N.Y. 1985), pp 383-388.
7. M. Murata, M. Shimatani, H. Sugitani, Y. Oshima and T. Yasumoto. Bull.Jpn. Soc. Sci. Fish. 48, 549-551,(1982).
8. G.J.Leitch, M. E. Iwert and W. Burrows, J.infect. Dis. 116, 303-312, (1966).
9. H.W.Moon, S.C. Whipp and A. L. Baetz, Lab. Invest. 25, 133-140,(1971).
10. N. S.Taylor, G. M.Thorne and J.G. Bartlett, Infect. Immun. 34, 1036-1043, (1981).
11. T. Yasumoto, Igaku no Tomo 112,886, (1980) in Japanese.
12. M.Murakami, K. Makabe, K. Yamaguchi, S. Konosu and M. R. Waelchli, Tetrahed. lett. 29, 1149-1152, (1988).
13. M. Takahashi, Y. Ohizumi and T. Yasumoto, J. biol Chem. 257, 7287-7289, (1982).
14. G. Schettini, Koike,K. I. S. Login, A. M. Judo, M. J. Cronin and T. Yasumoto, Am. J. Physiol. 247, E520-E525, (1984).
15. Y.Ohizumi and T. Yasumoto, J. Physiol., London 337,711 -721,(1983).
16. T. Miyamoto, Y. Ohizumi, H. and T. Yasumoto, Pfluegers Arch. ges. Physiol. 400, 439-441, (1984).
17. M. Kobayashi, Y. Ohizumi and T. Yamamoto, Br. J. Pharmacol. 86, 385-391, (1985).

TOXICITY OF *OSTREOPSIS LENTICULARIS* FROM THE BRITISH AND UNITED STATES VIRGIN ISLANDS

DONALD R. TINDALL,* DONALD M. MILLER,** and PATRICIA M. TINDALL***
*Department ot Botany, **Department of Physiology, and ***Protein Research Facility
Southern Illinois University, Carbondale, IL 62901, USA

ABSTRACT

Ostreopsis lenticularis is a major component of the epiphytic flora on macroalgae in near-shore coral reef environments in the Virgin Islands. Since this region is characterized by a high incidence of ciguatera, several clones were isolated, grown in large-scale culture, and examined for toxicity. Two clones of *O. ovata* were examined following the same procedures. Mouse bioassay of crude methanol extracts from *O.lenticularis* revealed LD_{50}'s from 0.45-2.17 mg/kg mouse (12.96-51.14 MU/mg cells). Liquid-liquid partitioning of methanol extracts yielded potent water soluble fractions (LD_{50}=0.10-0.51 mg/kg mouse) and ether soluble fractions (LD_{50}=0.33-0.77 mg/kg mouse). Reverse phase liquid chromatography resulted in a 3.94-fold increase in purity of the water soluble toxin (LD_{50}=0.032 mg/kg mouse). This fraction (OTX) caused greater than 50%. inhibition of the acetylcholine response in the guinea pig ileum at 1.16 pg/ml of bathing solution. Methanol extracts from cultures of *O. ovata* were not toxic to mice at doses up to 1 mg /mouse.

INTRODUCTION

Ostreopsis lenticularis Fukuyo is a principal component of the epiphytic flora on macroalgae in shallow, kinetic, near-shore coral reef environments in the Virgin Islands [1, 2]. High density populations of this species also have been reported from similar habitats in the coastal waters of Puerto Rico [2, 3]. Comparative toxicity studies of cultured cells and natural populations have shown *O. lenticularis* to surpass *Gambierdiscus toxicus* Adachi and Fukuyo as a potential progenitor of ciguatera toxins in a limited number of habitats in the Caribbean [2, 4, 5]. Other studies of *G. toxicus* from the Caribbean region have shown the species to produce significant quantities of extremely potent toxins [6, 7, 8, 9]. The objective of the present study was to obtain additional information on types and potencies of toxins produced by *O. lenticularis* from the Virgin Islands. The water soluble toxin from *O. lenticularis* was compared to maitotoxin from *G. toxicus*. An easily overlooked species, *O. ovata* Fukuyo from the Virgin Islands also was examined for toxicity.

MATERIALS AND METHODS

Clones of *Ostreopsis* examined during this study were representative of two species, *O. ovata* and *O. lenticularis*. Only specimens from culture were examined. Stock cultures were maintained side-by-side under identical environmental conditions and were transferred every 10-14 days for the past three years (some for 9 years). General characteristics of clones and their respective collection sites are shown in Table I. Our two clones of *O. ovata* conform well with the type description [10] and specimens reported from St. Barthelemy [11]. However, questions do arise regarding the designation of our remaining clones as *O. lenticularis*. All of these clones display consistent body dimensions and plate tabulations which fall within the range ot both *O. siamensis* Schmidt and *O. lenticularis*. Also, all clones are consistent with regard to their undulating cingulum (or body) when viewed from the side and all have plates with one type of pore. Pores are scattered over entire plates and are centered in a slight depression (Fig. 1). According to Fukuyo's [10] interpretation, these characteristics best fit *O. siamensis*. Since *O. siamensis* has not been reported from the Caribbean and since our current observations were limited to cultured specimens, we have continued to use the name *O. lenticularis*.

All clones were grown in 20 liter carboys containing 15 liters of filtered (0.2 μm) and

Toxic Marine Phytoplankton
Edna Graneli et al., Editors

autoclaved seawater enriched with ES nutrients [12] and soil extract [13]. Cultures were maintained under a 16:8 hour light/dark cycle (4300 lux, cool-white fluorescent] at 28° C. The various clones yielded 1.26-1.63 x 10^6 cells per liter at late log/early stationary phase (25-28 days). Cells were harvested using a Pellicon Cassette System with peristaltic pump head, 0.45 Durapore membranes, and linear channel separators (Millipore Corp.). Harvested cells were freeze-dried, weighed, and stored at -20° C.

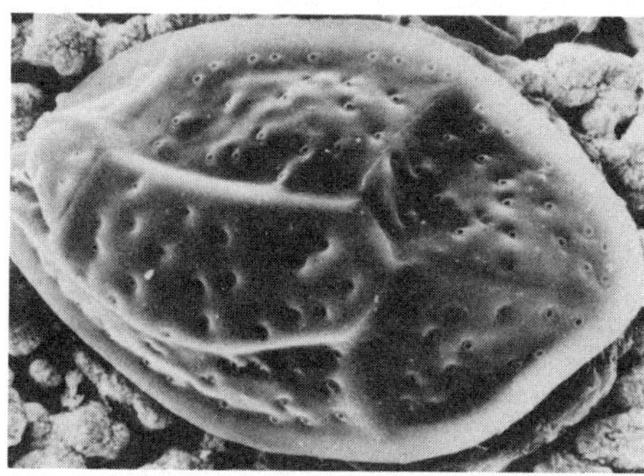

FIG. 1. Scanning electron micrograph of *O. lenticularis* (clone 872); antapical view of hypotheca.

TABLE I. Cell dimensions (ranges and means) and collection sites for *Ostreopsis* clones examined for toxicity.

Species	Clone No.	Dorso-vent. Diam. (μm)	Transdiam. (μm)	length (μm)	Collection Site
O. ovata	702	39- 51 -55	26- 32 -35	22- 25 -26	St.Thomas Lagoon, St. Thomas, U.S.V.I.
O. ovata	841	33- 43 -51	26- 31 -33	25- 26 -28	North Sound, Virgin Gorda, B.V.I.
O. lenticularis	876B	37- 61 -67	31- 44 -53	26- 33 -41	Little Lameshur Bay, St. John, U.S.V.I.
O. lenticularis	877	53- 60 -65	35- 43 -51	26- 32 -37	Greater Lameshur Bay, St. John
O. lenticularis	874	49- 61 -69	35- 41 -51	26- 31 -41	Biras Creek, Virgin Gorda
O. lenticularis	872	49- 63 -75	37- 45 -53	26- 33 -41	South Sound, Virgin Gorda
O. lenticularis	875	45- 65 -73	33- 45 -51	28- 34 -41	South Sound, Virgin Gorda

Extraction and initial phases of purification of toxic fractions from each clone were accomplished as shown in Figure 2. Crude methanol extracts and the six products of the partitioning procedure were assayed using mice. Water soluble toxic fractions were further purified by crystalization of non-toxic materials in methanol followed by liquid chromatogaphy. Toxic fractions were applied to a C18 column which was developed with acetone followed by methanol. Four fractions were collected with acetone (1A, 1B, 2, and 3). The final fraction (4) was collected using methanol. Degree of purity of toxic fractions was determined using HPLC (C8 column and methanol) [14]. LD_{50} determinations were completed using female mice (Harlan Sprague Dawley ICR 'BR') weighing 18-22 g. Doses of toxic fractions were suspended in 0.5 ml of 0.15M NaCl containing 1% Tween 60 and administered by intraperitoneal injection. Control mice were injected with carrier only. Four mice were injected with each of four doses of geometrically increasing concentrations. All mice were monitored for 48 hours. LD_{50}'s were calculated using moving average interpolation tables [15]. Toxic fractions also were assayed using the guinea pig ileum preparation as described for assaying maitotoxin from *Gambierdiscus toxicus* [7, 8, 14].

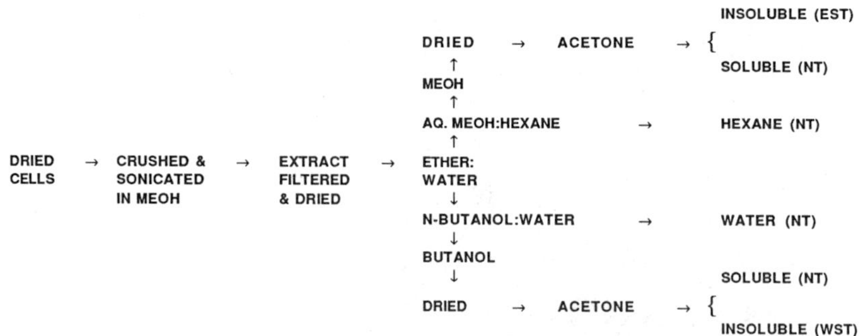

FIG. 2. Procedure for extraction and separation of toxic fractions from *O. lenticularis*. Fractions assayed with mice are noted as non-toxic (NT) or as toxic (EST and WST).

RESULTS AND DISCUSSION

Large-scale cultures of clones of *O. ovata* (702 and 841) were harvested and extracted in methanol. Methanol extracts of each clone were assayed using 36 mice (including 4 mice each at doses of 1-1.2 mg/mouse). None of the mice died nor did any show adverse signs prior to sacrifice at 72 hours. These findings concur with those reported by Nakajima et al. [16] and Besada et al. [11]. They also support Nakajima's view that the lack of notable toxicity in this species may serve as a useful taxonomic character. However, consideration should be given to the report of a collection of non-toxic *O. lenticularis* from the Caribbean [2].

TABLE II. Potency of methanol extracts from five clones of *Ostreopsis lenticularis*. One mouse unit (MU) is the LD_{50} for single 20 g mice. Two samples of clones 877 and 872 were examined.

Clone No.	Cell d.wt. (g)	MeOH Extract (mg)	LD_{50} (mg/kg)	MU (μg)	MU/mg Cells
876B	1.090	607.5	2.167	43.38	12.96
877	0.882	452.0	0.767	15.33	33.43
	0.662	339.0	0.527	10.54	48.62
874	0.872	424.0	0.498	9.96	48.82
872	0.500	292.5	0.919	16.00	36.56
	6.290	2686.0	0.527	10.54	40.51
875	1.095	504.0	0.450	9.00	51.14

The results of assays for toxicity in crude methanol extracts from cultures of *O. lenticularis* are shown in Table II. Extracts of all clones examined were lethal to mice. Although somewhat variable, all were considerably more potent that those previously reported. LD_{50}'s of extracts from our clones ranged from 0.45 to 2.17 mg/kg mouse as compared to 6.5 to 72.5 mg/kg mouse reported by Tosteson et al. [4]. These differences could have resulted from variations in culture, harvest, extraction, assay procedures and/or individual genotypes. The significant difference in potency of our clone 876B in contrast to that of other clones examined under very consistent growth conditions and cell processin procedures suggest possible genotype differences as observed in *Gambierdiscus toxicus* [9]. However, determination of the true nature of these variations will require more precise growth/toxicity studies. Interestingly, Ballantine et al. [2] reported considerable variation in potencies of wild populations of *O. lenticularis* from localities near our original collecting sites.

Liquid-liquid partitioning of methanol extracts from each clone of *O. lenticularis* resulted in a consistent distribution of cellular constituents to the six end products of the procedure. This conclusion was drawn from relative weights, HPLC analysis, and bioassay of each product. Toxicity was observed in two of the six products from each clone: (1) the acetone insoluble fraction from the water partition and (2) the acetone insoluble fraction from the ether partition [Table III).

TABLE III. Potencies of water soluble toxin (WST) and ether soluble toxin (EST) from five clones of *Ostreopsis lenticularis.*

Clone No.	Fraction	Fraction d. wt. (mg)	LD_{50} (mg/kg)	MU (µg)	MU/mg Fraction
876B	WST	34.3	0.507	10.13	98.7
	EST	17.7	0.330	6.60	151.5
877	WST	23.0	0.230	4.60	217.4
	EST	17.0	0.765	15.30	65.4
874	WST	27.5	0.100	1.99	502.5
	EST	29.5	0.442	8.84	113.1
872	WST	13.5	0.240	4.80	208.0
	EST	12.9	0.550	11.00	90.9
875	WST	39.6	0.126	2.52	396.8
	EST	31.0	0.698	13.95	71.7

WST was 3.3-5.0 times more potent than the crude methanol extract, whereas EST usually was about equal to or less potent than the crude extract. The one notable exception to this pattern was clone 876B. The same clone showed the least potency in the methanol extract. Again, these variations may denote a trait unique to this clone. Although the partitioning procedure accomplished significant improvement in potency, only about 40% of the total mouse units present in the methanol extracts were recovered in the EST and WST fractions. Except for a fairly consistent distribution between ether and water, we are not able to report a clear distinction between EST and WST. Both appear to have similar effects on mice and both cause irreversible inhibition of the acetylcholine response in guinea pig ileum preparations.

WST from clone 875 (LD_{50} =0.126 mg/kg mouse) was further purified utilizing a C18 column. Prior to application on the column, 17.6 mg of WST was dried and brought up in methanol, whereupon two types of crystals formed. The crystals were removed (total weight 2.64 mg) and assayed with mice. These crystals were not toxic to mice at doses up to 5.2 µg/mouse. The remaining 14.96 mg of WST was applied to the column which was then developed with acetone. Fraction 1B containad 3.48 mg of purified toxin which we designated ostreotoxin (OTX). The LD_{50} of OTX was 0.032 mg/kg mouse. Thus, the procedure achieved a 3.94-fold increase in potency (to mice) over that of the original WST and a 14.1-fold increase over that of the crude methanol extract. In addition to this purification step resulting in a significant improvement in potency, the OTX fraction contained 77.5% of the total mouse units detected in the WST. An additional 8.6% of the total mouse units in the WST was detected in eluent fraction 2. Subsequent chromatographic treatments did not always achieve this level of efficiency.

HPLC (C8 column with methanol) of OTX and all eluent fractions revealed that less than 3% of the total 210 nm absorbing substances were present in the OTX fraction. This confirmed the good efficiency of the C18/acetone purification step. HPLC analysis revealed no absorbance peaks for OTX that were not also present and much more apparent in samples of fractions 1A and 2 (Fig. 3). This observation may indicate a lack of UV absorbance (at 210 nm) by OTX or that the quantities of actual toxin were below the level ot detection.

OTX (LD_{50}=32.1 µg/kg mouse) at a concentration ot 1.16 pg/ml of bathing solution caused 70.9%. (n=4, SD=13.8) inhibition of the acetylcholine response in the guinea pig ileum preparation at 90 minutes after the incubation period (Fig. 4). A similar level of

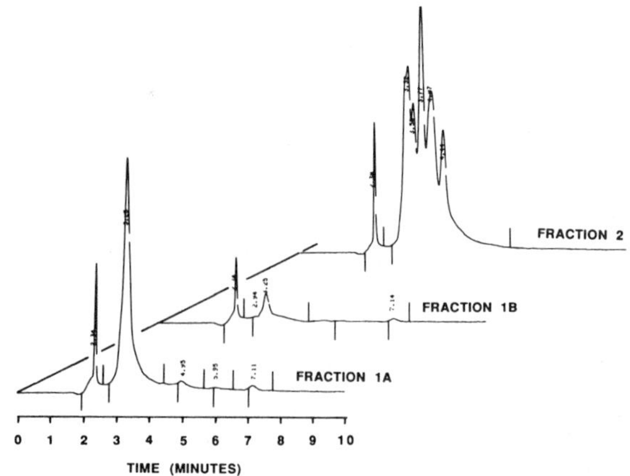

FIG. 3. Chromatograms of fractions 1A, 1B, and 2 of WST (clone 875) run on a C8 column with MeOH (A at 210 nm) at a Flow rate of 1 ml/min. All fractions were brought to equal volumes and the injection volume for each fraction was 10 µl.

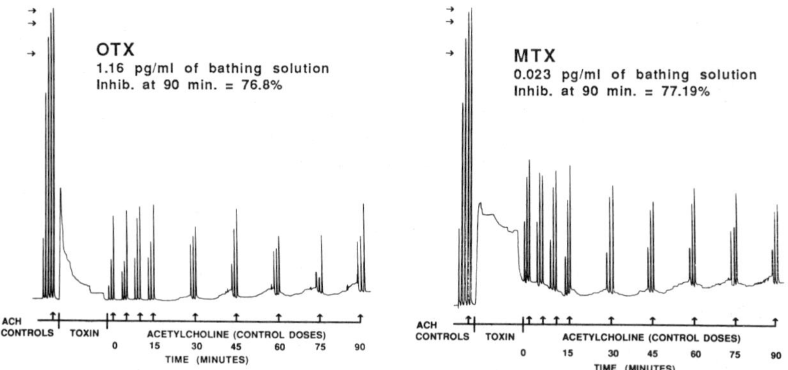

FIG. 4. Effects of OTX and MTX on the acetylcholine response in the guinea pig ileum. Ileum preparations were incubated with toxin in the bathing solution for 15 min.

inhibition was achieved with maitotoxin (MTX) (LD_{50} = 1.5 µg/kg mouse, ca. 10%. pure) from *Gambierdiscus toxicus* at 15 pg/ml of bathing solution. Although, both toxins caused an initial positive inotropic effect and irreversible inhibition, the disparity in the response of the ileum to these two toxins in relation to their respective LD_{50}'s in mice (IU:MU ratio about 2000:1 for MTX and 5.5×10^5:1 for OTX) suggested that the ileum preparation was much more sensitive to OTX. However, a second preparation of MTX (LD_{50} = 0.7 µg/kg mouse, ca. 20 %. pure) at a concentration of 0.023 pg/ml caused 77.19%. inhibition (Fig. 4). The latter results with MTX revealed a IU:MU ratio of 6.1×10^5:1 which is comparable to that of OTX. The enhanced sensitivity of the ileum preparation to the more pure sample of MTX suggests that less pure toxic fractions may contain a compound(s) which reduces or suppresses the effects of the actual antagonist. The sensitivity of the ileum to OTX in face of a relatively high LD_{50} (mouse) suggests that fewer such compounds are produced by *O. lenticularis* or that our sample of OTX is at least 20%. pure. The possible presence of a compound which reduces or suppresses the effects of neurotoxins is intriguing because such compounds could serve as a membrane protecting mechanism in the dinoflagellate cell. Also, such a compound would be an extremely valuable research tool in our quest for a clinical treatment of ciguatera and other similar syndromes.

The similar responses of OTX and MTX to various solvents and chromatographic treatments and their effects on the ileum preparation (initial positive inotropic effect and irreversible inhibition of the acetylcholine response) indicate that they are closely related compounds. However, if the differential in their respective LD_{50}'s in mice holds for pure material they will doubtlessly display structural differences. Differences in responses of mice to median lethal doses of these two toxins suggest that they are different compounds. Mice which survive the LD_{50} of OTX usually showed adverse signs throughout (and after) the 48 hour observation period. Mice which survive the LD_{50} of MTX usually were without adverse signs after 24 hours.

Clearly, *O. lenticularis* has the potential to contribute large quantities of potent toxins to the fish food web in the Virgin Islands. However, *G. toxicus* should not be discounted. Bomber et al. [9] have reported 18-150 MU/mg cells for acclimated clones of *G. toxicus* from the Caribbean region. Our estimations of potencies of *O. lenticularis* from the Virgin Islands was 12.96-51.14 MU/mg cells.

When examining toxicity of natural populations of *Ostreopsis*, care must be taken to assure that a clear distinction is made between *O. lenticularis* and *O. ovata*. The latter species is virtually non-toxic.

ACKNOWLEDGEMENTS

We extend special thanks to Jeff Bomber for technical and editorial assistance. We also thank Kevin Aikman, Faiqa Hassan, Sue Kohler, and Claude Rakatoniaina for their invaluable contributions. This research was supported by the Illinois-Indiana Sea Grant Program (COMMNA85AADSG0830N) and the U.S. Army Institute of Infectious Diseases (DAMD1787-C-7002). Mr. Myron Hokin provided research facilities at the Bitter End Yacht Club, Virgin Gorda, B.V.I.

REFERENCES

1. R. D. Carlson and D. R. Tindall in: Toxic Dinoflagellates, D. M. Anderson, A. W. White, and D. G. Baden, eds. (Elsevier, Amsterdam 1985) pp. 171-176.
2. D. L. Ballantine, T. R. Tosteson, and A. T. Bardales, J. Exp. Mar. Biol. Ecol. 4, 201-212 (1966).
3. D. L. Ballantine, A. T. Bardales, T. R. Tosteson, and H. D. Durst, Fifth Int. Coral Reef Congr.4, 417-422 [1985).
4. T. R. Tosteson, D. L. Ballantine, C. G. Tosteson, A. T. Bardales, H. D. Durst, and T. B. Higerd, Mar. Fish. Rev. 48, 57-59. (1986).
5. T. R. Tosteson, D. L. Ballantine, C. G. Tosteson, V. Hensley, and A. T. Bardales, Appl. Environ. Microbiol. 55, 137-141 (1989).
6. D. R. Tindall, R. W. Dickey, R. D. Carlson, and G. Morey-Gaines in: Seafood Toxins, ACS Symposium Series, 262, E. Ragelis, ed. (1984) pp. 225-240.
7. R. W. Dickey, D. M. Miller, and D. Tindall in: Seafood Toxins, ACS Symposium Series, 262, E. Ragelis, ed. (1984) pp. 257-269.
8. D. M. Miller, R. W. Dickey, and D. R. Tindall in. Seafood Toxins, ACS Symposium Series, 262, E. Ragelis, ed. (1984) pp. 241-255.
9. J. W. Bomber, D. R. Tindall, and D. M. Miller, J. Phycol. (accepted for publication).
10. T. Fukuyo, Bull. Jpn. Soc. Sci. Fish. 47, 967-976 (1981).
11. E. G. Besada, L. A. Loeblich, and A. R. Loeblich, III, Bull. Mar. Sci. 32,723-735(1982).
12. L. Provasoli in: Culture and Collection of Algae, A. Watanabe and A. Hattor, eds. pp. 63-75 (Jpn. J. Plant Physiol. 1968).
13. R. D. Carlson, G. Morey-Gaines, D. R. Tindall, R. W. Dickey in: Seafood Toxins, ACS Symposium Series, 262, E. Ragelis, ed. (1984) pp. 271-287.
14. D. R. Tindall and D. M. Miller, in: Toxic Dinoflagellates, D. M. Anderson, A. W.White, and D. G. Baden, eds. (Elsevier, Amsterdam 1985) pp. 321-326.
15. C. S. Weil, Biometrics 8, 249-263 (1952).
16. I. Nakajima, Bull. Jpn. Soc. Sci. Fish. 47, 1029-1033 (1981).

ENZYME IMMUNOASSAY OF BREVETOXINS

Vera L. Trainer[1] **and Daniel G. Baden**[2]. [1]Department of Biochemistry and Molecular Biology, University of Miami, Miami, FL 33149 U.S.A. [2]Rosenstiel School of Marine and Atmospheric Science, University of Miami, Miami, FL 33149 U.S.A.

ABSTRACT

The marine dinoflagellate, *Ptychodiscus brevis*, is the organism responsible for Florida's red tide. This alga produces potent toxins called brevetoxins which often remain undetected in contaminated shellfish resulting in neurological disorders in humans upon consumption. At present, the only assays which are used to detect the presence of brevetoxins in seafood are the intraperitoneal mouse bioassay and radioimmunoassay (RIA), tests which are too impractical for field use. The development of a more precise method of toxin detection has begun with the synthesis of both a brevetoxin-urease and a brevetoxin-peroxidase conjugate, each synthesized by carbodiimide linkage of the enzyme to a toxin-succinate derivative. These conjugates retain brevetoxin antigenicity and enzyme activity in enzyme-linked immunosorbent assays (ELISAs). Optimization of storage conditions and conjugation ratios is necessary for successful adaptation of these ELISAs to field studies.

INTRODUCTION

The detection and control of Florida's red tide remain serious problems faced by both the Public Health and fishing industries. Red tide blooms are caused by the marine dinoflagellate, *Ptychodiscus brevis* [1], which produces potent toxins called brevetoxins which are known as the PbTx series [2]. The only assays presently available for detection of brevetoxins are the intraperitoneal mouse bioassay and the radioimmunoassay. These tests are impractical for use in the field due to their utilization of radioisotopes and yields of false positives [3,4]. Immunoassays have been developed for detection of other marine toxins in contaminated fish and shellfish [5,6,7]. An assay which shows promise for use in field assays is the ELISA since it can be read visually without the need for expensive apparatus. Microwell microtiter plates allow large numbers of samples to be assayed simultaneously. The potential sensitivity of ELISA is greater than RIA since many molecules of product can be generated by a single molecule of enzyme.

An example of an ELISA which has been adapted for use in the field is the "poke stick" enzyme immunoassay which utilizes an anti-ciguatoxin monoclonal antibody linked to horseradish peroxidase (HRP) to detect ciguatoxin in contaminated fish [8]. Our laboratory utilizes an anti-brevetoxin antibody raised in goats for standard RIA procedures [9]. This same antiserum has been used to develop two separate ELISA protocols, one utilizing a brevetoxin-urease conjugate and the other a toxin-HRP complex. These ELISAs show high specificity through displacement of the brevetoxin-enzyme complexes by purified toxin alone. The assays can be used as a means to detect the presence of either brevetoxin antibody or brevetoxin alone (competition assay) and show promise for use in field studies.

Published 1990 by Elsevier Science Publishing Co., Inc.
Toxic Marine Phytoplankton
Edna Graneli et al., Editors

METHODS

PbTx-3 Linkage to Urease or HRP

PbTx-3 was isolated from laboratory cultures and purified by HPLC [10]. This toxin was linked to the appropriate enzyme as described previously for linkage to BSA [11] with the following changes. Rather than adding protein to toxin derivative, the carbodiimide-toxin conjugate was added at once to HRP of urease dissolved in PBS, pH 7.4, and allowed to react at $0^{o}C$ for 12 hours in the dark. The conjugate was dialyzed for 48 hours against four changes of PBS.

Toxin-urease Assay

Ninety six-well microtiter plates were incubated overnight in a humid chamber at $4^{o}C$ with polyclonal goat antibody in PBS, pH 7.4. Wells were aspirated, rinsed three times with PBS, and remaining non-specific sites were blocked with 1% gelatin in PBS. Blocking agent was aspirated and wells were incubated with toxin-urease for two hours at $37^{o}C$. After aspiration and three rinses with PBS, urease substrate [12] was added. Color development was monitored at 595 nm.

Toxin-HRP Assay

Polyclonal goat antibody dissolved in bicarbonate buffer, pH 9.6, was added to 96 well plates and incubated overnight at $4^{o}C$ in a humid chamber. Wells were then aspirated and rinsed three times with PBS containing 0.5% Tween-20 (PBS-T). Toxin-HRP dissolved in PBS-T containing 0.5% preimmune goat serum was added to wells and incubated at 37 ^{o}C for two hours. The plate was aspirated and rinsed three times with PBS-T before addition of 2,2'-azino-di-(3-ethyl-benzthiazoline sulfonate-6; ABTS) substrate containing 30% H_2O_2 in 0.1 M sodium citrate, pH 4.8. Change in absorbance was monitored at 405 nm.

Competition Assays

Competition assays were done in an identical manner as described above for the two separate toxin-enzyme assays. In addition, dilutions of PbTx-3 were added at the same time as a constant concentration of the toxin-enzyme conjugate.

Mouse Immunization

Since PbTx-3 is not antigenic due to its molecular size of 895 daltons, it was coupled as a hapten to bovine serum albumen (BSA), an effective antigenic carrier, using methods described in Baden *et al*. [11]. The toxin-BSA antigen (100 μg PbTx-3 toxin equivalents of complete conjugate) was injected into 6-8 week old Balb-c mice. The first immunization utilized an equal volume of toxin-BSA mixed thoroughly with Freund's complete adjuvant. Subsequent immunizations were done at biweekly intervals utilizing the toxin-BSA complex and Freund's incomplete adjuvant.

Radioimmunoassay (RIA)

RIAs were performed as described previously [9,11].

RESULTS AND DISCUSSION

Urease was initially selected as the enzyme for linkage to PbTx-3 for several reasons. It is absent from mammalian cells which eliminates background activity resulting from enzymatic reactions in biological fluids. Additionally, urease possesses a high turnover number which increases its sensitivity. The presence of urease results in a rise in pH and a color change of the bromocresol purple indicator dye from yellow to purple.

Checkerboard assays were initially performed with the toxin-urease complex. In these assays, decreasing dilutions of goat antibody were incubated with dilutions of the conjugate to determine the optimal conditions for simultaneously detecting the smallest amounts of antibody and balancing the tendency of antigen to leech off the plate. Adsorption of immunoglobulin to polystyrene was found to increase with increasing protein concentration up to 2 mg per cm^2 (Fig. 1). A decrease in absorbance at 595 nm was observed at high antibody concentrations (4 mg/cm^2) which is attributed to a "hook" effect [13], i.e. blockage of the antigen recognition site due to changes in antibody affinity when bound to the solid phase in a closely packed arrangement.

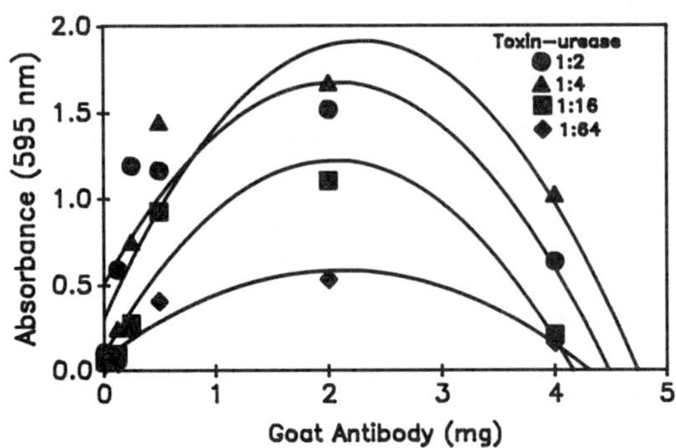

FIG. 1. Checkerboard assay of toxin-urease binding to goat antibody in 96 well microtiter plates.

The specificity of this ELISA utilizing the toxin-urease conjugate was tested in a competition assay. At an antibody concentration where absorbance values were found to be relatively high using a minimal amount of the toxin-urease complex (due to cost of toxin), increasing amounts of competitor PbTx-3 were added to separate microtiter wells. Concentrations of PbTx-3 were chosen to bracket the previously determined Kd of approximately 2.9 nM [2]. Displacement of the hapten-enzyme from its antibody recognition site by PbTx-3 was found to be logarithmic up to 10 nM competitor (Fig. 2).

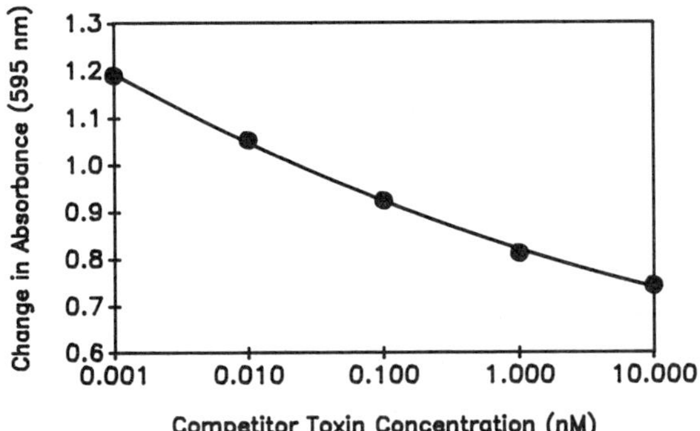

FIG. 2. Competition assay of brevetoxins using goat anti-brevetoxin antibody and a toxin-urease conjugate in 96 well microtiter plates. Color development is measured after a 3 hour incubation with indicator dye. Concentrations: 0.3 mg IgG per well, 90 nM PbTx-3 and 100 ng urease per well (toxin-urease), brevetoxin competitor of 1 pM to 10 nM. Values are averages of duplicate incubations.

Anti-brevetoxin antibody response of blood serum from mice which had been immunized seven times with PbTx-3-linked BSA antigen was compared using both RIA and toxin-urease ELISA to goat anti-brevetoxin antibody of known specificity. RIA results demonstrated the similar affinity of blood serum in two separate mice for brevetoxin antigen (Fig. 3). In terms of $[^3H]$PbTx-3 bound per mg serum antibody, the mice showed a greater antigen binding capacity than the polyclonal goat serum. This strong indication of anti-brevetoxin antibody production in both mice compares favorably with ELISA results. A net absorbance change of 1.55 ± 0.16 (mouse 1) and 1.74 ± 0.16 (mouse 2) was observed at the same serum dilutions (1:10) used for the RIA after a 30 minute incubation with substrate and indicator dye. A positive net absorbance change (above blank values of 0.13 ± 0.07) was observed at serum dilutions up to 1:1000.

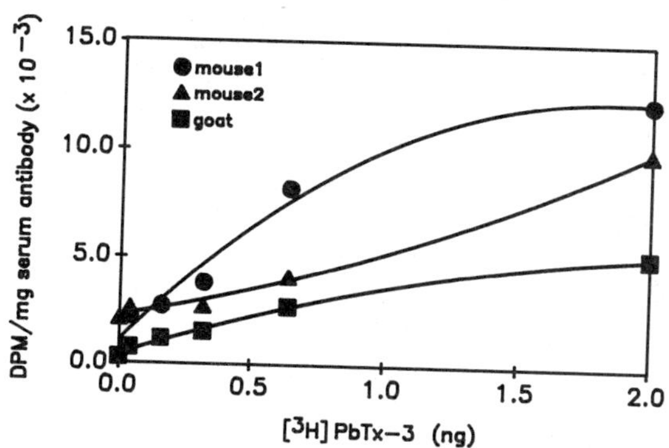

FIG. 3. Radioimmunoassay of goat and mouse serum antibodies.

Although ELISAs utilizing the brevetoxin-urease conjugate appeared to be successful, considerable difficulty was encountered in obtaining a stable enzyme preparation. Enzyme activity was lost after one week under various storage conditions, therefore subsequent assays were performed using a toxin-HRP conjugate. HRP was chosen due to its relatively low cost and its conjugation efficiency, however several precautions were taken when using this enzyme since it is inactivated by polystrene unless Tween-20 is added to the reaction buffer [12]. In addition, HRP is sensitive to bacteriostatic agents such as NaN_3. Care was taken in ELISAs utilizing the brevetoxin-HRP conjugate to add 0.5% Tween-20 to the PBS not only to prevent enzyme inactivation by polystyrene, but also to reduce background by discouraging nonspecific hydrophobic interactions. The best storage method for the complex was found to be lyophilized in glass vials in the presence of 0.1% BSA as a stabilizer (data not shown).

The molar ratio of HRP to brevetoxin during conjugation was maintained at 2 or less since activity of the enzyme label prepared with molar ratios of three or more was found to decrease markedly when coupled with antibody in a study by O'Sullivan and Marks [14]. ABTS was used as a chromogen for the HRP reaction since it is second in activity only to o-Phenylenediamine (OPD), which is unstable when exposed to light [15].

A series of blockers of nonspecific binding were tested for their effectiveness in reducing background. These included 1% BSA, 5% nonfat dry milk, 1% gelatin, and 0.5% preimmune serum (PIS). The most successful blocker was PIS dissolved in PBS probably because it is an irrelevant antibody which conforms to the size and charge of the specific goat antibody.

As with the toxin-urease conjugate, a checkerboard assay was performed to determine both activity of the enzyme and optimal dilutions of the antibody and antigen complex resulting in high absorbance values and low nonspecific binding. A "hook" effect was observed again using this conjugate, so competition assays were performed at concentrations of antibody and toxin-HRP where absorbance at 405 nm was maximal. Figure 4 shows that both full strength and 1:10 dilute antibody incubated with a constant concentration of toxin-HRP in the presence of decreasing dilutions of PbTx-3 competitor resulted in decreasing absorbance values. Antibody diluted 100-fold was beyond the detection limits of this assay as shown by the straight baseline (solid squares).

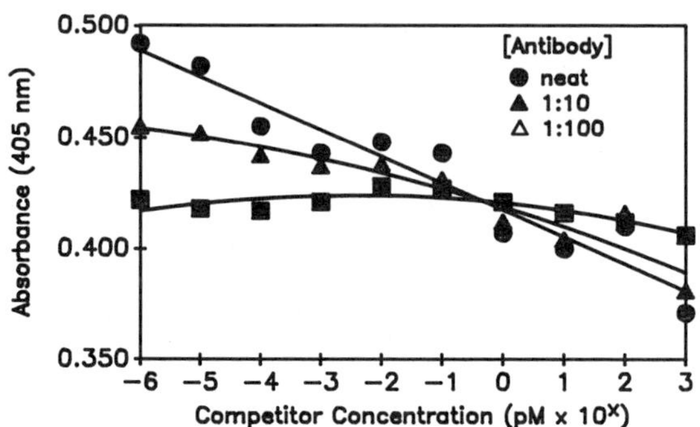

FIG. 4. Competition assay for detection of brevetoxins using goat anti-brevetoxin antibody and a toxin-peroxidase conjugate in 96 well microtiter plates. Color development is measured after 1 hour incubation with indicator dye. Concentrations: 70 μg to 0.7 mg goat antibody per well, 600 nM PbTx-3 and 35 μg (12 units) HRP per well (toxin-HRP), and brevetoxin competitors of 1 aM to 1 nM. Values are averages of duplicate incubations.

Initial experiments utilizing both enzyme conjugates were performed by diluting goat antibody in PBS, pH 7.4. Subsequent assays used bicarbonate buffer, pH 9.6, as the diluent in the initial antibody incubation which resulted in tighter binding of the antibody. This is due to the pH and ionic strength of the coating buffer which change the antibody conformation. Greater binding to solid support is observed when proteins are at their isoelectric point due to the decrease in electrostatic repulsion of closely packed proteins. Increasing pH of the buffer from 4 to 9 has been shown to increase protein binding by 10 to 25% [15].

This study analyzes the usefulness of two separate brevetoxin-enzyme conjugates in ELISAs measuring known brevetoxin concentrations. These tests can now be standardized for analysis of brevetoxin in unknown samples. The stability of the toxin-HRP conjugate under field conditions as well as endogenous peroxidase activity in biological samples must be assessed. Further optimization of the anti-brevetoxin ELISA is necessary before it can be successfully adapted for use in field studies.

ACKNOWLEDGEMENTS

This work was supported in part by U.S. Army Medical Research and Development Command Contracts DAMD17-85-C-5171, DAMD17-87-C-7001, and DAMD17-88-C-8148. The investigator adhered to the Guide for the Care and Use of Laboratory Animals, prepared by the Committee on Care and Use of Laboratory Animal Resources, National Research Council (NIH Pub. #86-23, revised 1985). Opinions, interpretations, conclusions and recommendations are those of the authors and are not necessarily indorsed by the U.S. Army. This work is in partial fulfillment of the requirements for the Ph.D. in Biochemistry and Molecular Biology at the University of Miami (V.L.T.).

REFERENCES

1. K.A. Steidinger, In: Toxic Dinoflagellate Blooms, D.L. Taylor and H. Seliger, eds. (Elsevier, Amsterdam, 1979) pp. 435-442.
2. M.A. Poli, T.J. Mende, and D.G. Baden, Mol. Pharm. 30, 129-135 (1986).
3. K.A. Steidinger and D.G. Baden, In: Dinoflagellates, D.L. Spector, ed. (Academic Press, New York, 1984) pp. 201-261.
4. D.G. Baden, Int. Rev. Cytol. 82, 99-150 (1983).
5. Y. Hokama, L.H. Kimura, M.A. Abad, L. Yokochi, P.J. Scheuer, M. Nukina, T. Yasumoto, D.G. Baden, and Y. Shimizu, In: Seafood Toxins, ACS Symposium Series 262, E.P. Ragelis, ed. (American Chemical Society, Washington, D.C., 1984) pp. 304-320.
6. R.E. Carlson, M.L. Lever, B.W. Lee, and P.E. Guire, In: Seafood Toxins, ASC Symposium Series 262, E.P. Ragelis, ed. (American Chemical Society, Washington, D.C., 1984) pp. 181-192.
7. L. Levine, H. Fujiki, H.B. Gjika, and H. Van Vunakis, Toxicon 26, 1115-1121 (1988).
8. Y. Hokama, L.K. Shirai, M. Iwamoto, M.N. Kobayashi, C.S. Goto, and L.K. Hakahawa, Biol. Bull. 172, 144 (1987).
9. D.G. Baden, T.J. Mende, A.M. Szmant, V.L. Trainer, R.A. Edwards, and L.E. Roszell, Toxicon 26, 97-103 (1988).
10. D.G. Baden, T.J. Mende, W. Lichter, and L. Wellham, Toxicon 19, 455 (1981).
11. D.G. Baden, T.J. Mende, J. Walling, and D.R. Schultz, Toxicon 22, 783 (1984).
12. G. Anido, Clin. Chem. 30, 500 (1984).
13. P. Tijssen, Laboratory Techniques in Biochemistry and Molecular Biology: Practice and Theory of Enzyme Immunoassays, R.H. Burdon and P.H. van Knippenberg, eds. (Elsevier, Amsterdam, 1985).
14. M.J. O'Sullivan and V. Marks, In: Methods in Enzymology, vol. 74, J.J. Langon and H. Van Vunakis, eds. (Academic Press, New York, 1981) pp. 147-166.
15. J.C. Standefer In: Enzyme-Mediated Immunoassay, T.T. Ngo and H.M. Lenhoff, eds. (Plenum Press, New York, 1985) p. 203.

SCREENING FOR HEMOLYTIC AND ICHTHYOTOXIC COMPONENTS OF CHRYSOCHROMULINA POLYLEPIS AND GYRODINIUM AUREOLUM FROM NORWEGIAN COASTAL WATERS

T. YASUMOTO*, B. UNDERDAL**, T. AUNE**, V. HORMAZABAL**, O. M. SKULBERG*** AND Y. OSHIMA*
*Faculty of Agriculture, Tohoku University, 1-1 Tsutsumidori Amamiyamachi, Sendai 981, Japan; **Department of Food Hygiene, Norwegian College of Veterinary Medicine, P.O. 8146 Dep., 0033 Oslo 1, Norway; ***Norwegian Institute for Water Research, P.O. Box 33, Blindern N-0313 Oslo 3, Norway.

ABSTRACT

Two species of phytoflagellates known to cause massive fish kills, Chryschromulina polylepis (Prymnesiophyceae) and Gyrodinium aureolum (Dinophyceae), were screened for bioactive compounds by hemolytic tests. Chromatographic and mass spectral data suggest occurrence of 1-acyl-3-digalactosylglycerol and octadecapentaenoic acid in the two species. These two compounds are both hemolytic and ichthyotoxic and thus likely to be implicated in the fish kills.

INTRODUCTION

Massive fish kills and damage to marine ecosystems associated with red-tides are phenomena of global occurrence. Nevertheless, the mechanisms of fish kills have not been fully elucidated. Chrysochromulina polylepis and Gyrodinium aureolum are among the important nuisance species with mass occurrence on the Norwegian coast of the North Sea, the Norwegian and Swedish coasts of the Skagerrak and the Kattegat [1]. Causes and effects of the nuisance blooms are under investigation. The difficulty in elucidating ichtyotoxins, allegedly present in these organisms, seems to derive from the lack of rapid and sensitive bioassay methods. Bioassays using fish usually requires relatively large amounts of samples, as the test materials should be dissolved in relatively large volumes of water to keep test fish. Dose-survival time responses of fish are often fluctuant. In order to surmount difficulty we applied a hemolytic test for screening ichthyotoxins, because many ichthyotoxins such as prymnesins of Prymnesium parvum [2], hemolysins of Amphidinium carteri [3], and maitotoxin of Gambierdiscus toxicus [4], are also potent hemolysins. The test is sensitive, rapid, and quantitative. In the present study we report the occurrence and chemical natures of hemolysisns in C. polylepis and G. aureolum.

MATERIALS AND METHODS

Algal samples

Water samples were collected during the phytoplankton bloom of C. polylepis (June 1988), representing Skagerrak population of the flagellate. The corresponding samples of G. aureolum representing Oslofjord population were collected during August 1988. The nuisance species accounted for more than 95% of the resident phytoplankton biomass. The water samples with algae were stored frozen (-20°C) untill used in the extraction procedure.

Extraction

Thawed algal water was mixed with an equal volume of methanol, and hemolysisns were extracted twice with chloroform. One sample of C.

Toxic Marine Phytoplankton
Edna Graneli et al., Editors

polylepis water was extracted with a quarter volume of 1-butanol for comparison with the chloroform extraction. Mussels (250 g) collected during the bloom period of C. polylepis were steamed to open shells. The whole meats were homogenized and extracted with 750 ml of 80% methanol. The aqueous methanol solution was washed with hexane twice and evaporated.

Purification

Extracts obtained from the algal waters or mussels were dissolved in chloroform and loaded on a Bond Elute silica cartridge (Separalyte). The cartridge was washed in a step wise manner with chloroform, chloroform-methanol (9:1, 1:1), and methanol (5 column volumes for each solvent). Aliquots of each eluate were used for hemolytic tests. Hemolytic components were dissolved in the minimum quantities of methanol and purified by gel permeation chromatography on an HW-40 column (Toso Co., superfine, 1 x 40 cm) with methanol as mobile phase. Eluates were monitored with a UV-flow monitor and by hemolytic tests. Subsequent purification was performed on a Develosil ODS-5 (Nomura Chemicals, 4.6 x 250 mm) column with acetonitrile-water (35:65) or on a Develosil ODS-7 column (10 x 250 mm) with methanol-water (4:1). Eluates were checked by hemolysis tests and by thin layer chromatography (TLC) on precoated silica gel plates (Merck) with chloroform-methanol-water (75:35:4). The plates were dried and sprayed with 2% mouse blood cell suspension in saline. Hemolysins were detected as white spots against a pale-red background.

Hemolysis tests

Mouse blood (1 ml) was centrifuged. The precipitated blood cells were washed twice with saline solution (0.85% NaCl) and made up to 50 ml. This 2% suspension was used for spraying TLC plates, while diluted blood cell suspension (0.4%) was used for a more elaborated hemolysis test. The diluted blood cell suspension (0.5 ml each) was put into small test tubes, mixed with a varying volume of test solution (aqueous methanol or methanol solution, between 5 and 50 µl each), and left for 30 min to observe hemolysis. As a reference, 0.1% saline solution of commercial saponin (BDH Chemicals Ltd.) was prepared, and 8, 10, 12, and 15 µl portions of the solution were added to 0.5 ml of 0.4% blood cell suspensions, respectively. Normally, complete hemolysis took place with 10 µl of the saponin solution, allowing us to define the hemolytic potency of 10 µg of the saponin as one hemolytic unit (HU). The number of hemolytic units of test materials were calculated from the volumes and concentrations of their solution to cause complete hemolysis in 30 min. Micro-scale hemolytic tests were carried out similarly using micro-well plates, instead of test tubes, filled with 100 µl of the 0.4% blood cell suspension in each well. The maximum amount of a test solution added to one well was 30 µl.

Spectral measurements

Fast atom bombardment (FAB) negative ion mass spectra of purified materials were taken on a JEOL JMS-DX330 mass spectrometer with glycerol as matrix. ^{1}H NMR spectra were taken in a CD_3OD solution on a JEOL GSX-400 spectrometer.

RESULTS AND DISCUSSION

Chloroform extracts from 22.5 L of C. polylepis water yielded 90.0 mg of residue. The result of differnt fractions from silica gel column chromatography is presented in TABLE I. The chloroform-methanol (1:1) eluate showed the highest hemolytic activity, followed by the chloroform-methanol (9:1) eluate. About 83% of the initial hemolytic activity was recovered in these two eluates. Hemolytic activities of other eluates were

TABLE I. Silica gel chromatography of C. polylepis extracts.

Solvent (20 ml each)	Residue (mg)	Hemolytic units
Chloroform	42.7	<< 20
Chloroform-Methanol (9:1)	6.1	66
Chloroform-Methanol (1:1)	43.9	100
Methanol	13.3	< 20

Sample: Chloroform extracts of algal water, 90 mg, 200 HU.
Column: Bond Elute Silica, 3.0 g.

insignificant. As prymnesins of P. parvum and hemolysin-3 and -4 of A. carteri gave poor recoveries under similar chromatographic conditions and were eluted with methanol, the occurrence of related compounds in C. polylepis seemed unlikely. In TLC the chloroform-methanol (9:1) eluate revealed three spots. Two spots with high Rfs (0.70, 0.60) corresponded to reference stearic acid and octadecapentenoic acid ($C_{18:5}$), respectively. The major spot had a low Rf (0.25), corresponding to that of hemolysin-2 (Fig. 1) of A. carteri and was tentatively code-named H2LC (hemolysin-2-like compound). H2LC was the dominant compound in the chloroform-methanol (1:1) eluate. Hemolytic spots corresponding to prymnesins or hemolysin-1, -3 and -4 of A. carteri were undetectable. On a reversed phase C_{18} column, H2LC was further separated into two broad peaks, H2LC-1 (23-30 min) and H2LC-2 (49-52 min). A FAB mass negative ion spectrum of H2LC-1 (Fig. 2A) showed a prominent ion at m/z 673, which corresponded to an $(M-H)^-$ ion of 1-octadecapentaenoyl-3-digalactosylglycerol. It was thus suggested that H2LC-1 was an analogue of hemolysin-2 differing in an acyl moiety (Fig. 1). The FAB mass spectrum showed a series of less prominent ions differing by 14 mass units, suggesting the presence of analogues of different acyls. As the occurrence of odd numbered acyls are relatively rare, the origin of these ions will be investigated in future. A FAB mass spectrum of H2LC-2 showed an ion at m/z 701, suggesting that H2LC-2 could be a homologue of H2LC having alonger acyl moiety ($C_{20:5}$, Fig. 1). After purification of the chloroform-methanol (9:1) eluate on a Sep-pak C_{18} with 80% methanol, the presence of a trace amount of a hemolytic compound

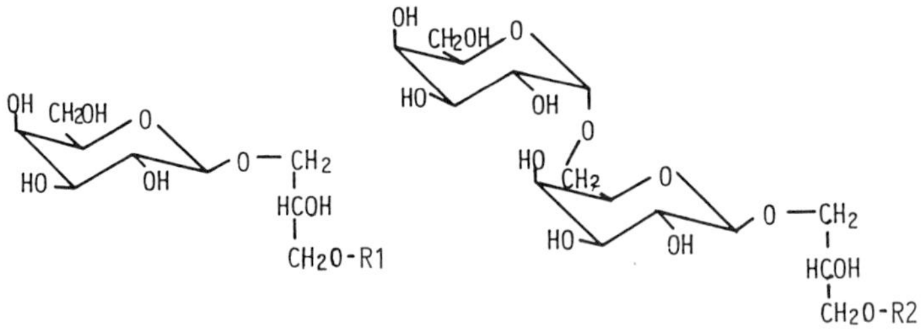

hemolysins-1 and 2 of A. carteri: **R1**= acyl($C_{18:4\omega3}$), R2= acyl($C_{18:4\omega3}$)
hemolysins of C. polylepis: **R2**= acyl($C_{18:5}$, $C_{20:5}$)
hemolysins of G. aureolum: **R2**= acyl($C_{20:5}$)

FIG. 1. Structures of hemolysin-1 and hemolysin-2 from A. carteri and proposed structures of hemolysins from C. polylepis and G. aureolum.

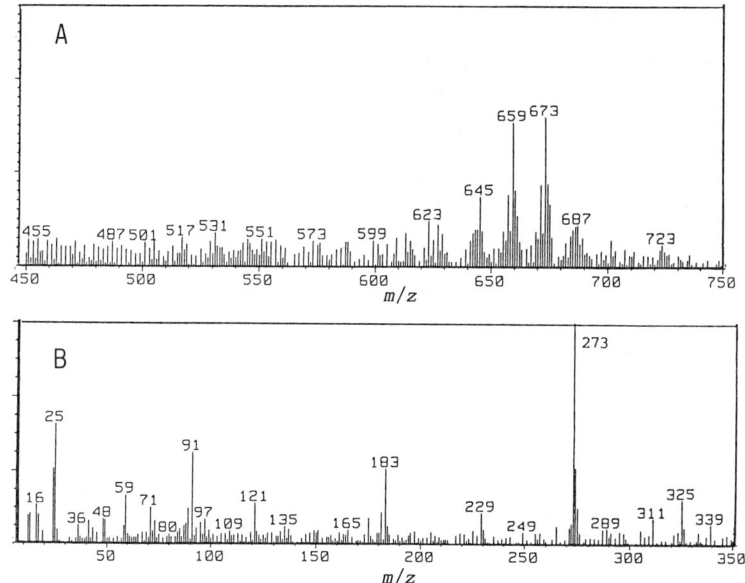

FIG. 2. Negative FAB mass spectra of hemolysin-2-like compound (H2LC-1) of C. polylepis(A) and 1-octadecapentaenoic acid isolated from G. aureolum(B).

resembling hemolysin-1 of A. carteri (Fig. 1) was observed by TLC. Mass spectral analysis of this spot was unsuccesful due to the small sample size.

The chloroform extract prepared from 4.8 L of G. aureolum bloom water was partitioned between aqueous 90% methanol and hexane. The aqueous methanol fraction was evaporated and the residue was chromatographed on a silica gel 60 (5 g) column with 50 ml each of chloroform, chloroform-methanol (95:5, 90:10, 70:30, 50:50), and methanol in succession. The major hemolytic activity was observed in the chloroform-methanol (95:5) eluate, and the minor in the chloroform-methanol (70:30) eluate. TLC data suggested that the hemolytic activity of the former eluate was due to fatty acids with varying degree of unsaturation. The highly unsaturated fatty acid fraction was purified on a DEAE cellulose column with methanol and chloroform-acetic acid (3:1). The acidic fraction was then chromatographed on a C_{18} column with methanol-water (4:1). The purified acid was identified as a $C_{18:5}$ fatty acid by a FAB mass spectrum (Fig. 2B), $(M-H)^-$ at m/z 273]. The ^1H NMR spectrum of the purified acid also agreed with that of 3Z,6Z,9Z,12Z,15Z-octadecapentaenoic acid in having characteristic signals at 0.98 (t, CH_3), 2.1 ($C{=}CH{-}CH_2$), 3.25 ($=CHCH_2{-}COO$), 2.8 ($=CH{-}CH_2{-}CH=$), and 5.4 ppm ($-CH{=}CH-$). Some extra peaks of artifacts appeared in the spectrum due to the vulnerability of this highly unsaturated acid to oxidation. The hemolysin in the chloroform-methanol (70:30) eluate was indistinguishable by TLC from H2LC of C. polylepis. The higher field of a FAB mass spectrum also resembled that of H2LC in having ions at m/z 673, 687, 701. These results suggest that the polar hemolysin of G. aureolum is the same with that of C. polylepis, and is likely to be a homologue of hemolysin-2 of A. carteri. Hemolysins corresponding to prymnesins of P. parvum or hemolysin-3 and -4 of A. carteri were not confirmed in G.

aureolum.

The present study indicated that hemolytic compounds in C. plolylepis and G. aureolum were polyunsaturated fatty acids and monoacyl-digalactosylglycerol, which resembled hemolysin-2 of A. carteri. However, the relative abundance of the two compounds was significantly different between the two species. In C. polylepis the galactolipid was largely responsible for hemolytic activity, while polyunsaturated fatty acids were the major hemolytic compounds in G. aureolum. Because both compounds were unstable in chloroform solution, it was possible that the actual concentration of these hemolysins in the phytoplankters were higher than that found in the present study. The fact that the butanol extract of C.polylepis was 10 times more potent in hemolytic activity than chloroform extracts supports this hypothesis.

Mussels exposed to a C. polylepis bloom displayed a higher hemolytic activity (80 HU/g) than non-contaminated mussels (16 HU/g). The results indicated accumulation of hemolytic substances by the mussels. When mussel extracts were chromatographed on a silica gel column, hemolytic and cytotoxic activity was observed mainly in chloroform-methanol (1:1) eluate, suggesting that the accumulated hemolysin was H2LC. Polyunsaturated fatty acids are known to be abundant in phytoplankton [5] and reach relatively high concentrations in the hepatopancreas of shellfish [6]. Reports on monoacylglycerolipids such as hemolysin-1 and -2 in marine organisms are rare [3,7], but the occurrence of diacyl analogues in algae seems to be common [8,9]. Elucidation of the biological and ecological role of these compounds will be an interesting subject of future research. Search for other hemolytic and cytotoxic compounds in plankton and mussels should be also continued for better understanding of red-tide related phenomena.

In our previous study (T.Y. and Y.O.), hemolytic activities of $C_{18:4}$ fatty acid, hemolysin-1, and hemolysin-2 of A. carteri have been estimated to be 1.9, 0.8, and 0.24 SU/mg, respectively, where one SU (saponin unit) stands for hemolytic potency of 1 mg of Merck's saponin. The minimum concentrations of $C_{18:4}$ fatty acid and hemolysin-1 required to kill killifish were 10 and 7-18 ppm, respectively [10]. Ichthyotoxicity of hemolysin-2 was not tested. In killed fish immersed in water containing 50 ppm of hemolysin-1, congestion in gills was observed first, the operculum stayed open, and the scales began to peel off. The fish died within 24-50 min following vigorous convulsive movements. These results suggest implication of poly-unsaturated fatty acids and hemolysins in fish kills and other deleterious effects of C. polylepis and G. aureolum. Further experiments to support identification of hemolysins and a search for other toxins will be continued.

REFERENCES

1. B. Underdal, O.M. Skulberg, E. Dahl and T. Aune, Ambio 18, 265-270 (1989).
2. M. Shilo, Bacteriol. Rev. 31, 180-193 (1967).
3. T. Yasumoto, N. Seino, Y. Oshima and M. Murata, Biol. Bull. 172, 128-131 (1987).
4. I. Nakajima, Y. Oshima and T. Yasumoto, Bull. Jpn. Soc. Sci. Fish., 47, 1029-1033 (1981).
5. G.W. Harrington, D.H. Beach, J.E. Dunham and G.G. Holz,Jr., Protozool. 17, 213-219 (1970).
6. K. Hayashi, Nippon Suisan Gakkaishi, 52, 1559-1563 (1986).
7. H. Kozakai, Y. Oshima and T. Yasumoto, Agric. Biol. Chem., 46, 233-236 (1982).
8. N. Fusetani and Y. Hashimoto, Agric. Biol. Chem., 39 2021-2025 (1975).
9. K. Sakata and K. Ina, Agric. Biol. Chem., 47, 2957-60 (1983).
10. M. Watanabe, M.S. Thesis, Department of Food Chemistry, Faculty of Agriculture, Tohoku University, 1986.

VII REGIONAL OUTBREAKS AND MANAGEMENT ISSUES

DIARRHETIC SHELLFISH TOXINS IN BIVALVE MOLLUSCS ALONG THE COAST OF PORTUGAL

ALVITO, P.,* SOUSA, I.,* FRANCA, S.,* AND SAMPAYO, M. A. DE M.**
*Lab. Microbiologia Exp., Instituto Nacional de Saúde (INSA), 1699 Lisboa Codex; **Instituto Nacional de Investigação das Pescas (INIP), 1400 Lisboa, Portugal.

ABSTRACT

This report concerns the first studies on diarrhetic shellfish toxins in bivalve molluscs along the Portuguese coast. It focusses on the occurrence of these toxins during 1987 and its correlation with the presence of *Dinophysis sacculus* Stein and *Dinophysis acuta* Ehr. in the plankton. Samples showing positive results by the adult mouse bioassay (Yasumoto's method) were examined using the infant mouse test to detect the diarrhetic effect and to eliminate false positive results associated occasionally with hepatopancreas extracts of contaminated shellfish. Reproducible results from both tests confirm the presence of diarrhetic toxins on the Portuguese coast. This study points to the need to review the safe limit of DSP toxins for human consumption established in Portugal.

INTRODUCTION

Diarrhetic shellfish poisoning (DSP) is a gastroenteritis caused by ingestion of shellfishes contaminated with dinoflagellate toxins [1,2]. The occurrence of the poisoning was first reported in Japan [1] but DSP was soon recognized in many parts of the world. The most affected areas are Japan and Europe [3]. Until now, the presence of DSP toxins has been correlated with the occurrence of *Dinophysis fortii*, *D. norvegica*, *D. mitra*, *D. acuta*, *D. acuminata* and *Prorocentrum lima* [4]. It has been noted that in the presence of *D. fortii* at a cell density as low as 200 cells/litre, shellfish become toxic enough to affect man [1] and even lower concentrations of *D. acuminata* may maintain toxicity [5].

The purpose of the present work was to relate the occurrence of DSP toxins in bivalve molluscs and the presence of *Dinophysis* spp. in the plankton of the Portuguese coast. The diarrhetic nature of the toxins detected was confirmed by comparative studies using adult mouse and infant mouse bioassays.

MATERIALS AND METHODS

Materials

All bioassays were run with hepatopancreas of bivalve molluscs (*Mytilus* sp., *Cerastoderma* sp., *Spisula* sp.) collected along the Portuguese coast between July and November 1987. Phytoplankton samples were collected from surface or as an integrated 5 meters water column depending on the depth of the sampled area at the same times and places as were shellfish. Mice for bioassays, Charles River CD-1, were supplied by Instituto Gulbenkian de Ciência, Oeiras, Portugal.

Methods

Adult mouse test. Lethal potencies of shellfish samples were determined by Yasumoto's method [6] according to the modification used in France [7].

Toxic Marine Phytoplankton
Edna Graneli et al., Editors

Mice weighing from 16 to 20 g were used (Yasumoto, personal communication). Observations of intraperitoneally injected mice were made over a 24 h period. As in France and Spain were considered a period of 5 h after i.p. injection as the minimum survival time to allow public consumption of the corresponding shellfish. As a standard toxin was not available, it was not possible to estimate the lethal potencies of shellfish samples in terms of toxin concentration. When 1 ml Tween 60 solution was injected as a control, no abnormal behaviour was observed.

Infant mouse test. The suckling mouse test [8] is usually used in human pathology for detection of bacterial enterotoxin. It was carried out according to the method described by Marcaillou-Le Baut [7]. Toxins effects were estimated as a fluid accumulation ratio calculated as the ratio of intestinal weight to remaining body weight (R). Diarrhetic effects were estimated as follows: $R<0.07$ test was considered negative, $R>0.09$ was positive and $0.07<R<0.09$ questionably positive. Six infant mice were used for each assay to prevent data lost due to injured mouse, but only four were considered. A reference extract was obtained by mixing three of the most toxic extracts (after adult mouse test). Serial dilutions of it were made and the expression of R for different hepatopancreas concentrations in Tween 60 solution was estimated.

RESULTS AND DISCUSSION

Comparative analysis of the DSP and phytoplankton data (Table I) reveals an association between shellfish toxicity and occurrence of *Dinophysis sacculus* and *Dinophysis acuta* in the plankton. When these species were observed at concentrations higher than 200 cells/l, the bivalves were toxic either at the same time or after a brief delay. During 1987, DSP toxins were detected along the Portuguese west coast from July to November, as shown in Table I. The toxicity detected in the shellfish was persistent in Aveiro region (coastal zone and lagoon) and in Figueira da Foz (Mondego estuary) and limited to a very short period of time in Obidos lagoon and in Espinho. Of the several bivalve species collected and analyzed for DSP toxins, *Mytilus* sp. showed the highest levels of toxicity. Thus, it would appear to be a good toxicity indicator.

The effects of i.p. injection of toxic extracts on mice were inactivation and weakness, with death occurring aproximately 60 min to 24 h after injection. However, some injected mice died within 1 to 5 min after some excitement, followed by a loss of balance and abrupt jumping before death. To clarify this different symptomatology and eliminate false positive results occasionally obtained by the adult mouse test (of the hepatopancreas extracts of contaminated shellfish), a more specific test for detection of diarrhetic effects was conducted. In this assay method there was no interference from the free fatty acids which are lethal to adult mice by i.p. injection [9]. Results of a typical infant mouse test are shown in Fig. 1.

The test conditions were checked by gavage administration to infant mice with serial dilutions of the reference extract. R values are reported in Table II. The mean value of R was determined and the bioassay was considered clearly positive at the concentration of 2.5 g toxic hepatopancreas/ml.

A comparison of the results obtained using the infant mouse test and the lethality test (Fig. 2.) shows good reproducibility between the tests. Of 40 positive lethal samples assayed by the infant test, 34 (85%) had $R>0.07$ and 20 had $R>0.09$. It is important that a longer surveillance time (24 to 48 h) be used for mice after i.p. injection using Yasumoto's method. As shown in Fig. 2., it was observed that for toxicity levels represented by + (survival time between 5 and 24 h) the corresponding R indicated positive diarrhetic effects.

In spite of the correlation obtained between the results of the two

TABLE I. Occurrence of diarrhetic shellfish toxins in bivalve molluscs (Mytilus sp., Cerastoderma sp. and Spisula sp.) and Dinophysis spp. in plankton (Dinophysis sacculus and Dinophysis acuta) along the Portuguese coast. DSP toxicity was detected by mouse lethality bioassay and is represented here by +++, ++, + and a which correspond, respectively, (+++) to the presence of toxins causing two mice death between 0 and 5 h, (++) one mouse death between 0 and 5 h, (+) one or two mice death between 5 and 24 h and (a) to the absence of toxins. Dinophysis spp. are recorded as cell number/l. (*) refers to the presence of Dinophysis in the region of shellfish collecting area and (-) refers to no Dinophysis cells recorded.

PORTUGUESE NW COAST — ATLANTIC OCEAN

LAT. 39° — 40° — 41°

STATION	OBIDOS LAG. (com. w.sea)				FIGUEIRA da FOZ			AVEIRO: Costa Nova					AVEIRO: S.Jacinto				ESPINHO			
	Biv.		Dinophysis		Biv.	Dinophysis		Biv.			Dinophysis		Biv.		Dinophysis		Biv.		Dinophysis	
1987 Week	M	C	D.sac.	D. ac.	M	D.sac.	D. ac.	M	C	S	D.sac.	D. ac.	M	C	D.sac.	D. ac.	M	S	D.sac.	D. ac.
JUL. 30	+	a	24000	—		—	—				—	—			—	—			—	—
32	+++		—	—		—	—				—	—			—	—			—	—
AGT. 33	+++	++	2000*	1000		—	—				1000	1000	+++		4000	2000			—	—
34		a	1000	2000*		—	—	+++			4000	6000	+++		1000	6000			500	28000
35			—	—		—	—	+++		+	1500	6000	+++		4000	6000		+++	—	6000
36			500*	—	+++	—	500	+++	+++		1000	500	+++	+++	1000	—	+++	+++	200	200
37	a		1000*	250	+++	—	—	+++	++	+	—	16000	+++	++	—	5000	++	a	200	—
SEPT. 38		a	—	—		—	500	+++		++	—	2000			100	200	a	+	—	—
39			—	—	+++	—	—	+++	+++	+	—	—	+++	+++	—	500		+++	2000	—
40			—	—	+++	—	—	+++		+	—	500	+++		—	1000			—	—
41		a	—	—	+++	—	—	+++			—	—	+++		—	1000		a	—	—
OCT. 42			—	—	+++	—	—	+++	a	+	—	—	+++		—	—			—	—
43		a	—	—	+++	—	—	+++	a		—	—	+++		—	—		a	—	—
44			—	—	++	—	—	+++			—	—	++		—	—			—	—
45			—	—	+++	—	—	+++	a	+	100	—	+++	+++	1000	—			—	—
NOV. 46			—	—	+	—	—	+++	a	a	—	—	+++	+++	—	—			—	—
47			—	—		—	—	+++	++	+	—	—	+++	++	—	—		a	—	—
48			—	—		—	—	a	a		—	—			—	—			—	—

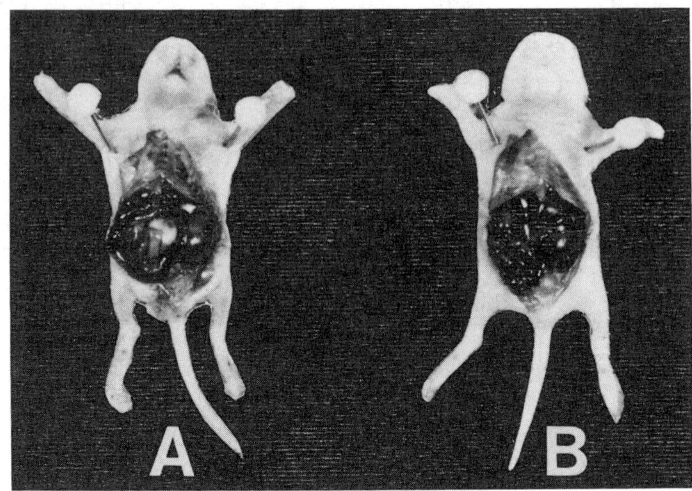

FIG. 1. Infant mice 4 h after gavage administration of hepatopancreas extract. A. Positive result showing fluid accumulation in the intestine. B. Negative result. The black colour of the intestine indicates total circulation of the administered extract.

TABLE II. Expression of R (Intestinal weight/Body weight) for different hepatopancreas concentrations in Tween 60 solutions.

Hepatopancreas g/ml	R				Arithmetic mean	Standard deviation
Controls	0.0537	0.0609	0.0547	0.0552	0.0561	0.0028
	0.0607	0.0565	0.0629	0.0653	0.0613	0.0032
0.32	0.0657	0.0678	0.0741	0.0738	0.0703	0.0036
0.63	0.0596	0.0620	0.0606	0.0605	0.0606	0.0008
1.25	0.0627	0.0566	0.0660	0.0666	0.0629	0.0039
2.50	0.1034	0.0970	0.1119	0.1024	0.1036	0.0053
5.00	0.0967	0.1042	0.1341	0.0907	0.1064	0.0166

bioassays, there were additional behaviours in adult mouse tests probably related to the presence of other unknown toxic compounds. The analysis of these toxins by means of High-Performance Liquid Chromatography is a priority in the development of further research on the toxicity originating from dinoflagellates.

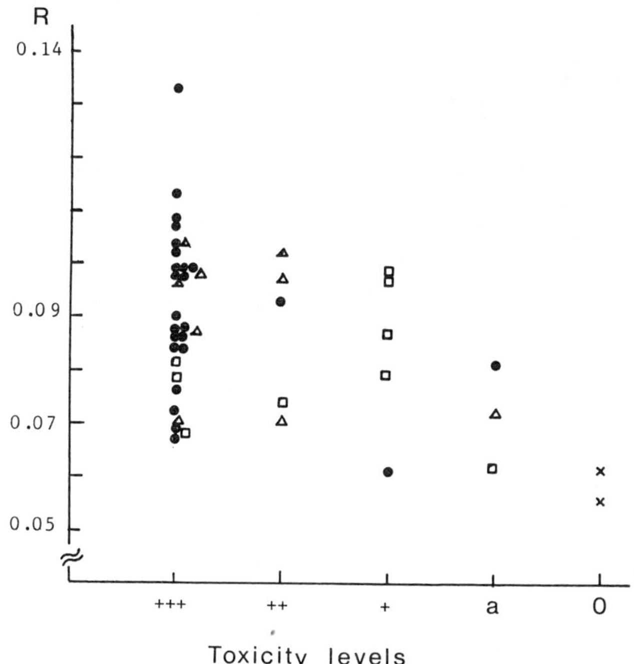

FIG. 2. Comparison between infant mouse and lethality bioassays. The ordinate represents the values of R corresponding to the different shellfish assayed: ● - Mytilus sp., Δ - Cerastoderma sp., □ - Spisula sp.. The abcissa represents the corresponding results of the lethality test. a - refers to the absence of toxins and 0 - refers to Tween 60 solution administered by gavage (controls).

ACKNOWLEDGEMENTS

This work was supported in part by JNICT (Portugal). Thanks are due to the Local Committee of the Fourth International Conference on Toxic Marine Phytoplankton for support to attend this conference. S. F. is indebted to Dr. P. Lassus and Dr. C. Marcaillou-Le Baut, IFREMER-Nantes, for invaluable advice and for help in the improvement of DSP studies in Portugal. The financial support received from the European Science Foundation and the European Medical Research Councils for a Short-term visiting fellowship with those scientists is gratefully acknowledge. We thank J. Romão for providing us with infant mice and E. Domingues for preparing the photographs.

REFERENCES

1. T. Yasumoto, Y. Oshima and M. Yamaguchi. Bull. Japan Soc. Sci. Fish. 44, 1249-1255 (1978).
2. T. Yasumoto, Y. Oshima, W. Sugawara, Y. Fukuyo, H. Oguri, T. Igarashi and N. Fujita. Bull. Japan Soc. Sci. Fish. 46, 1405-1411 (1980).
3. T. Yasumoto in: Progress in Venom and Toxin Research, P. Gopalakrishnakone and C. K. Tan, eds. Publ. by Nat. Univ. of Singapora,

(1987), pp. 348-355.
4. Jong-Soo Lee, M. Murata and T. Yasumoto in: 7th International Symposium on Mycotoxins and Phycotoxins, Abs., (1988), p. 13.
5. M. Kat in: Red Tides. Biology, Environmental Science, and Toxicology, Okaichi, Anderson and Nemoto, eds. (Elsevier, Amsterdam 1989) pp. 73-76.
6. T. Yasumoto, Y. Oshima and M. Yamaguchi. Bull. Japan Soc. Sci. Fish. 43, 207-211 (1978).
7. C. Marcaillou-Le Baut, D. Lucas and L. Le Dean in: Toxic Dinoflagellates, Anderson, White and Baden, eds. (Elsevier, Amsterdam 1985), pp. 485-488.
8. A. G. Dean, Y-C. Ching, R. G. Williams and L. B. Hardan. J. Infect. Dis. 125, 407-411 (1971).
9. Y. Hamano, Y. Kinoshita and T. Yasumoto in: Toxic Dinoflagellates, Anderson, White and Baden, eds. (Elsevier, Amsterdam 1985) pp. 383-388.

FIRST REPORT OF GYMNODINIUM CATENATUM GRAHAM ON THE SPANISH MEDITERRANEAN COAST

I. BRAVO,* B. REGUERA,* A. MARTINEZ** AND S. FRAGA*
*Instituto Español de Oceanografia. Centro Oceanografico de Vigo. Apdo. 1552. 36280 Vigo. Spain.
**Ministerio de Sanidad y Consumo. Direccion Comisionada de Galicia. Apdo. 90. 36280 Vigo. Spain.

ABSTRACT

Routine toxicity testing of Venus verrucosa of Andalucian (southern Spain) origin revealed rising levels of PSP toxins during January 1989. Toxin levels in the clam Callista chione reached 200 µg saxitoxin equiv./ 100 g meat in February and harvesting of shellfish was forbidden. Plankton samples taken along the coast between Malaga and Bahia de Algeciras just north of Gibraltar (the northern shore of the Alboran Sea), an area subject to continuous inflow of Atlantic waters, revealed Gymnodinium catenatum in concentrations up to 3 x 10^3 cells/l in mid February. No other toxic species were found.

INTRODUCTION

The naked dinoflagellate Gymnodinium catenatum Graham has been held responsible for PSP toxicity in molluscs from Mexico (1), northwest Spain (2, 3), Portugal (4,5), Tasmania (6,7) and Japan (8). Here we report the first appearance of G. catenatum from the Spanish Mediterranean coast associated with a PSP outbreak.

The Alboran sea is a partially isolated region of the Western Mediterranean connected to the Atlantic Ocean through the Straits of Gibraltar (Fig.1). Atlantic surface water flows into the Mediterranean through these Straits, and high salinity Mediterranean water flows into the Atlantic below it. Thus the Alboran Sea is an important mixing zone for these two water bodies (9,10). The coastal region of interest here (coast of Malaga) lies on the northern limb of the western gyre of the Alboran Sea (Fig.1). Shellfish beds at depths of less than 50 m in this zone are exploited to yield about 3000 tonnes a year of bivalves, among which the Mediterranean cockle (Cerastoderma tuberculata), the warty Venus (Venus verrucosa), and a clam (Callista chione) have the highest commercial value (Camiñas, internal report). The Costa del Sol is an important and luxurious tourist area.

RESULTS AND DISCUSSION

In November 1987, PSP toxins were detected in samples of Venus verrucosa from the Mediterranean coast of southern Spain. A maximum value of 54 µg saxitoxin equiv./100 g meat was recorded. At that time, there was no monitoring of phytoplankton in the region, so no link with the flora was established. On 5 January 1989, PSP toxins were detected by mouse bioassay in molluscs samples from the Costa del Sol (Fig.1), between the Bay of Algeciras and Fuengirola, with levels around 40 µg/100 g meat of saxitoxin equivalent. These values increased progressively until early February when they reached values of 200 µg/100 meat. The collection and sale of molluscs was forbidden by the health authorities in mid-January when toxin levels reached 80 µg/100g meat.

In early February, phytoplankton sampling began, and G. catenatum was

Toxic Marine Phytoplankton
Edna Graneli et al., Editors

detected. Between 7 February and 10 March, samples were collected at 0, 5, and 10 m at the locations shown in Fig. 1. The highest concentration of G. catenatum observed was 3 x 10^3 cells/l at Linea de la Concepcion in Algeciras Bay on 7 February. Subsequently, all samples revealed a decline in the abundance of G. catenatum. Table I shows the concentrations of G. catenatum and Table II the toxicity values determined by mouse bioassay. The dominant species was Gonyaulax polygramma. The rest of the phytoplankton composition was diatoms (Stephanopyxis turris, Skeletonema costatum, Thalassiosira cf. rotula, Eucampia zoodiacus, Hemiaulus sinensis, Chaetoceros socialis, Rhizosolenia alata, Coscinodiscus sp., and Bacteriastrum sp.).

Although cells of G. catenatum were not isolated and cultured in order to unequivocally identify the species responsible for the toxicity, no other species were found which have been responsible for causing PSP toxicity elsewhere in the world. Despite the absence of information prior to 7 February, it can be assumed from the toxicity data that the concentrations of G. catenatum reached higher values in the preceding weeks (see Tables I and II).

TABLE I. Concentrations of G. catenatum (cells/l) at differents locations

Date	7-9 Feb.			15-16 Feb.			1-22 Feb.			9-10 March		
Depth (m)	0	5	10	0	5	10	0	5	10	0	5	10
Fuengirola	1000			520		1440	ND[a)]	ND	ND	ND	ND	320
Estepona				40			ND	ND	ND	ND	ND	ND
Marbella						320	ND	ND	560	ND	ND	640
La Linea	3000			2320			60			120	160[b)]	

a) ND = Not detected   b) Sample taken at 3 m

TABLE II. Toxicity of shellfish (µg saxitoxin equiv./ 100 g meat)

Date	5 Jan.	12 Jan.	18 Jan.	20 Jan.	6 Feb.
Toxicity	40	87	109	94	200

G. catenatum was recorded for the first time in the Galician rias (NW Spain) in 1976 after a major PSP episode (2). It was found again in 1981 in the rias (11) a month after having been detected in Sines (Portugal) in August the same year (2). The definitive identification of G. catenatum as the agent responsible for PSP on the Galician coast was made after a PSP episode in 1985 using HPLC analysis of the organisms and of affected shellfish (12). In 1986, this species was identified for the first time as the cause of PSP in Portugal (4,5). It is very likely that the slight toxicity found in molluscs from the southern Mediterranean coast of Spain in 1987 and the toxicity found in C. tuberculata imported from Marrocco in February 1989 (A. Martinez, unpublished data) was caused by the same species as in the 1989 episode.

G. catenatum has also been detected on the Southern Tyrrhenian coast of Italy, in a eutrophic euhaline coastal lagoon with no aquacultural developments (13). It seems possible that G.catenatum is autochthonous in European waters, but none of the four regions from which it is known (Italy, Malaga, Portugal and Galicia) has been intensively investigated over long time periods. It may have been present in undetectable concentrations prior to 1976 when blooms were first noticed. Cysts of G. catenatum, first identified as Varia type A are known from cores taken in the Kattegat, and date approximately 2000 to 300 yr BP (14,15) If the species is an introduction, then it may have spread from Galicia to the Mediterranean, or in the reverse direction, or it may have been introduced more than once. Genetic studies might be used to resolve these problems.

Growth experiments with G.catenatum in our culture collection in Vigo show maximal growth rates at temperatures between 20 ºC and 25 ºC (Bravo, Reguera, unpublished data). But recent blooms in nature have occurred over a much wider range of temperatures, from 14 ºC in 1988 in Galicia (Fraga et al, this symposium) to 28 ºC in the Italian bloom (13).

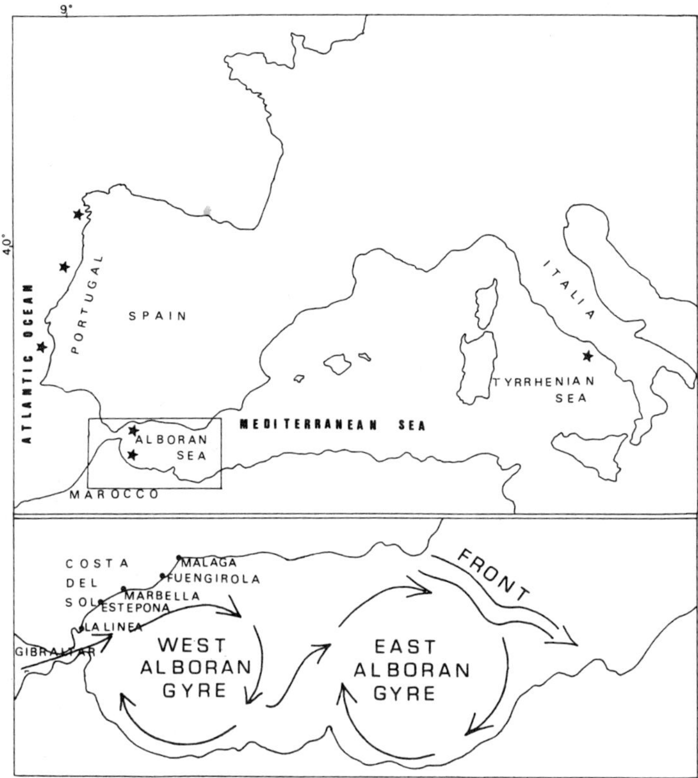

FIG. 1. Locations off the Iberian Peninsula and Mediterranean countries where G. catenatum has been reported and stations where samples were taken.

ACKNOWLEDGEMENTS

This work would not have been possible without the valuable help of J.A. Camiñas, D. Cortes, A. Carpena, J. Perez Rubin y D. Godoy from the Centro Costero de Fuengirola (Instituto Español de Oceanografia).

REFERENCES

1. G. Morey-Gaines, Phycologia 21(2), 154-163 (1982).
2. M. Estrada, F.J. Sanchez and S. Fraga, Inv.Pesq 48(1), 31-40 (1984).
3. S. Fraga, D.M. Anderson, I. Bravo, B. Reguera, K.A. Steidinger, C.M. Yentsch. Estuar. Coast. Shelf Sci. 27,349-361 (1988).
4. M.A. de M. Sampayo in: Red Tides: Biology, Environmental Science, and Toxicology. T. Okaichi, D.M. Anderson and T. Nemoto eds. (Elsevier

Science Publishing. 1989) pp. 89-92.

5. S. Franca and J.F. Almeida in: Red Tides: Biology, Environmental Science, and Toxicology. T. Okaichi, D.M. Anderson and T. Nemoto eds. (Elsevier Science Publishing. 1989) pp. 93-96.
6. G. Hallegraeff and C. Sumner, Aust.Fish. 45, 15-18 (1986).
7. Y. Oshima, M. Hasegawa, T. Yasumoto, G. Hallegraeff and S. Blackburn, Toxicon 25(10), 1105-1111 (1987).
8. T. Ikeda, S. Matsuno, S. Sato, T. Ogata, M. Kodama, Y. Fukuyo and H. Takayama. in: Red Tides: Biology, Environmental Science, and Toxicology. T. Okaichi, D.M. Anderson and T. Nemoto eds. (Elsevier Science Publishing, 1989) pp. 411-414.
9. G. Parrilla and T.H. Kinder, Bol.Inst.Esp.Oceanogr. 4(1), 135-165 (1987).
10. J. Tintore, P.E. La Violette, I. Blade and A. Cruzado, J.Phys.Oceanogr. 18, 1384 (1988).
11. M.J. Campos, S. Fraga, J. Mariño and F.J. Sanchez, ICES CM 1982/L:27 (1982) (mimeo).
12. D.M. Anderson, J.J. Sullivan and B. Reguera, Toxicon (in press).
13. G.C. Carrada, R. Cassotti and V. Saggiorno, Rapp.Comm.int.Mer Medit. 31, 2 (1988).
14. Nordberg, K. and H. Bergsten. Mar.Geol. 83, 135-158 (1988).
15. Dale, B. (This Conference)

RED TIDE OUTBREAKS AND THEIR MANAGEMENT IN THE PHILIPPINES

RHODORA A. CORRALES AND EDGARDO D. GOMEZ
Marine Science Institute
University of the Philippines
Diliman, Quezon City

ABSTRACT

In the Philippines, health and socio-economic problems associated with red tide in Manila and the Visayas have escalated. The most recent widespread outbreak (1988-89) of Pyrodinium bahamense var. compressum could have been the worst example of this phenomenon the country had experienced to date. Records of paralytic shellfish poisoning (PSP) increased as the public became "aware" of the "red tide".

The paper presents a brief account of the country's red tide experiences, the government's actions and researches so far made. Activities and plans for the development and implementation of a red tide management program are discussed. Areas which need most urgent attention are highlighted.

INTRODUCTION

The Philippines (Fig. 1) with a territorial area of about 1,965,700 sq. km. and more than 7,200 islands, lies between 21° 25' and 40° 23'N Lat. and 116°00' and 127° 00'E Long. It is bounded on the east by the Pacific Ocean, on the south by the Celebes Sea and the coastal waters of Borneo and on the west and the north by the South China Sea.

Blooms of Pyrodinium bahamense var. compressum in important fishing areas have been experienced since 1983. The recent outbreaks, totalling six, occurred in a series between July 13, 1988 to February 14, 1989 [1]. Epidemics of paralytic shellfish poisoning have caused grave human health problems and negative socio-economic impacts.

This is a brief account of the country's toxic red tide experiences, their impacts and the activities and plans for the development of a national red tide program.

OUTBREAKS AND IMPACTS OF Pyrodinium RED TIDES IN THE PHILIPPINES

Figure 1 and Table 1 show when and where Pyrodinium red tides have been reported. These are mostly sites of mussel (Perna viridis) industries, i.e. culture and/or collection from the wild, hence, where people derive their daily sustenance. Mangroves grow along or near the coasts of these areas.

Records of fatalities and illnesses due to PSP from 1983 to the present (Table 2) are believed to be far below the actual numbers because of unreported and unconfirmed cases. Pastor et al. [1] report that in the Philippines, PSP illnesses may occur below the AOAC standard of 80 micro-grams per 100 grams meat (as measured by the mouse bioassay). Deaths due to PSP now reach about 34 and reported cases more than a thousand.

Published 1990 by Elsevier Science Publishing Co., Inc.
Toxic Marine Phytoplankton
Edna Graneli et al., Editors

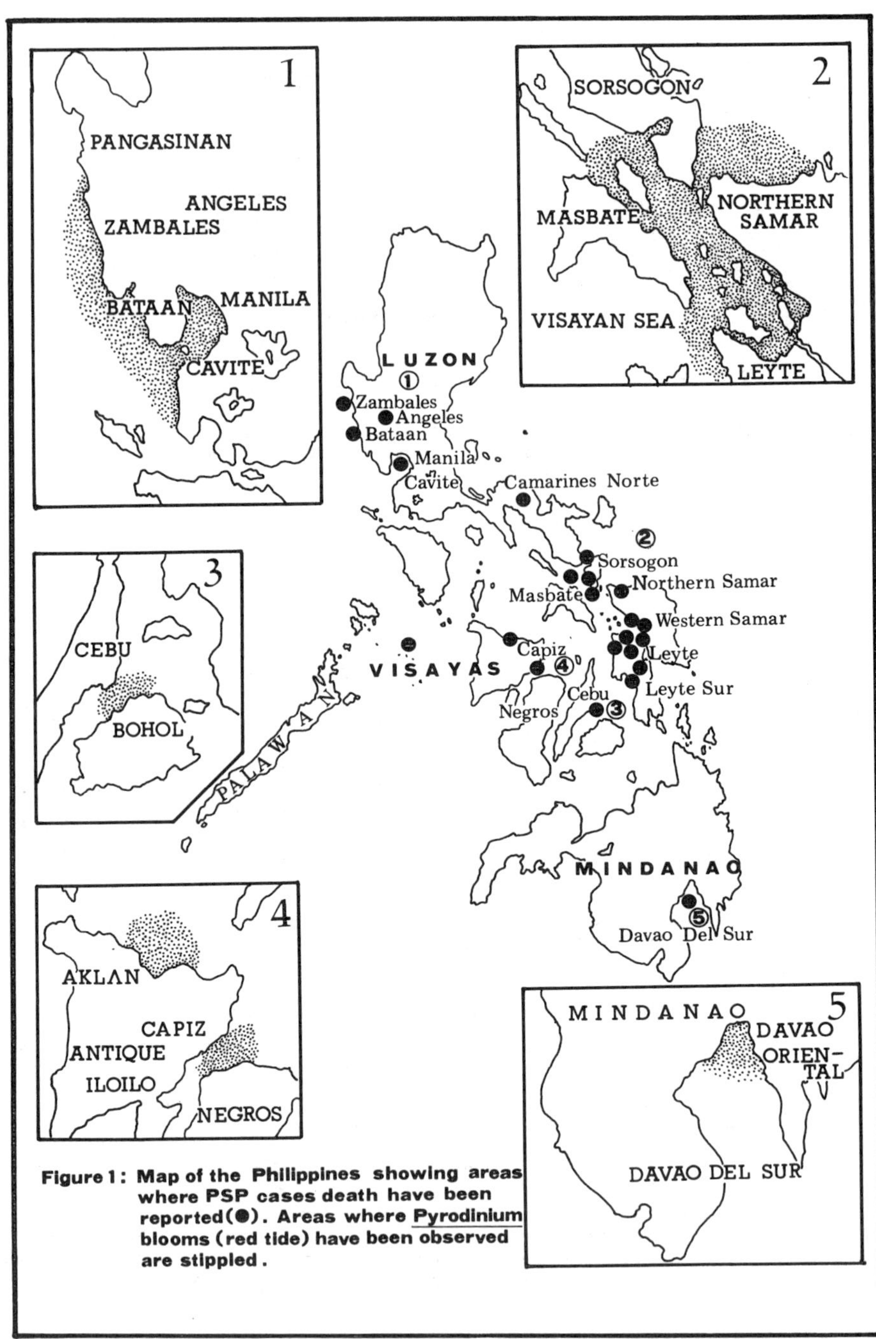

Figure 1: Map of the Philippines showing areas where PSP cases death have been reported(●). Areas where Pyrodinium blooms (red tide) have been observed are stippled.

TABLE 1. Summary of the dates and places of *Pyrodinium* Red Tide Outbreaks in the Philippines

Dates	Places
1983, June	Samar Sea, Maqueda Bay and Vicinity
1983, July	Balete Bay, Davao
1987, June	Samar Sea, Maqueda Bay, and Zambales Area
1988, July-Sept.	Samar/Leyte Area
1988, Aug.-Sept.	Bataan Area
1988, Sept.-Oct.	Manila Bay Area
1988, December	Negros Island
1989, February	Cebu Island

Pastor *et al*. [1], Gacutan *et al*. [2], Gonzales *et al*. [3]

TABLE 2. Philippine Records of Paralytic Shellfish Poisoning (PSP) Cases and Deaths from 1983-1989*

Province	Cases	Deaths	Period Recorded
Bataan	44	1	Aug.-Sept. 1988
Zambales	?	?	June 1987
Manila	14	2	Sept. 1988
Cavite	8	1	Sept. 1988
Sorsogon	?	?	Sept. 1988
Samar/Leyte	211	6	June 1983
	691	14	June 1987
	22	0	Jul.-Sept. 1988
Capiz	3	1	December 1988
Negros	109	4	December 1988
Cebu	24	5	February 1989
Davao	1	0	August 1983
Total	1,127	34	

Pastor *et al*. [1]   ? Cases/reports unconfirmed

The 1983 first Samar outbreak must have cost the local mussel industry about $5 million [3]. Maclean [4] estimates that the same level of loss was incurred in the subsequent 1987 and 1988 outbreaks.

The first Manila Bay area experience (1988) and the subsequent ones in other areas were widely covered by the media. The whole seafood industry (fishermen, mussel gatherers, dealers and restaurateurs) was negatively affected as a result of the "red tide scare" which spread in the National Capital Region and almost the entire country. For many months, many people refused to buy seafoods. Japan and Singapore banned their Philippine imports of shrimps, adding more millions of dollars to the country's economic loss.

Socio-political impacts of the recent red tides could be viewed from the reactions of the various sectors of society. Legislators and administrators showed concern and reviewed present systems of red tide management. For a time, anti-government rallies and demonstrations were staged, some indicating different interpretations of this natural phenomenon.

PRESENT RED TIDE MANAGEMENT SCHEME

Aimed primarily to safeguard public health and minimize or control economic losses, the Philippines' present red tide management scheme (Fig. 2) involves: 1) Monitoring and surveillance and 2) Regulatory Measures. The latter includes a) Public Awareness, b) Ban on harvesting and selling during outbreaks, c) Health Service, and d) Support for Affected Fishermen.

Plankton counts are regularly done by the Bureau of Fisheries and Aquatic Resources (BFAR) once monthly during periods of non-occurrence and at least once weekly during blooms in several stations in red tide affected areas. Mussels or oyster samples are sent to the Bureau of Food and Drug (BFAD) in Manila for mouse bioassay. During outbreaks, regulatory measures include the imposition of bans on the harvesting and selling of shellfish from affected areas and the subsequent lifting of the same. Public health services for PSP victims are provided by local centers and doctors dispatched from the Manila office of the Department of Health (DOH). Red tide bulletins are issued jointly by the BFAR and the BFAD-DOH to give the public more accurate and updated information on the outbreaks. Recently (in 1988), the Office of the President through the Department of Agriculture, the Land Bank, and with the participation of the Philippine Partnership for the Development of Human Resources in Rural Areas (PHIPDHRRA) [5] provided soft loans at 12% rate from a Natural Livelihood Support Fund (NLSF) of 200 million pesos to shellfish gatherers adversely affected by the red tide.

A PROPOSED RED TIDE MANAGEMENT SCHEME

An Inter-agency Committee on Environmental Health (IACEH) established by the Office of the President, formed a Technical Working Group (TWG) on red tide, composed of representatives from the Department of Health (DOH), Department of Agriculture, Bureau of Fisheries and Aquatic Resources (DA-BFAR), and the University of the Philippines Marine Science Institute (UPMSI). The group, together with a consultant, reviewed the present management scheme and visited selected monitoring areas. To shorten the time between detection of blooms or toxicity and the flow of information and appropriate responses, a management scheme shown in Fig. 3 has been devised and will be proposed for consideration. Important features include: a) performance of mouse bioassay at the municipal or regional level by the local monitoring teams which is presently being done only at the national level by BFAD in Manila, b) increased plankton sampling frequency (weekly during non-occurrence and daily during blooms), c) active involvement of local government and responsible citizens in a local red tide committee specially in public information and in the implementation of bans, d) confirmatory toxicological tests to be done at the BFAD, and e) creation of a National Red Tide Committee to act as an advisory or regulatory body. The National Committee will also be responsible on what, when, and how red tide announcements will be made to ensure public safety, reduce panic, and instill confidence in the system.

RESEARCH COMPONENT OF A NATIONAL RED TIDE PROGRAM

Broad range research programme have accompanied red tides in developed countries whereas in developing countries where funds, trained personnel, and equipment are insufficient or lacking, red tide research is wanting [6]. Knowledge on Philippine red tides is limited to data and information gathered from monitoring efforts since 1983 which could be further analysed. Epidemiological studies have been initiated [1] and the use of some locally available detoxification materials [7] have been tried. There has been no attempt, however, to undertake more in depth oceanographic and ecological studies that would support the monitoring program, enhance its efficiency

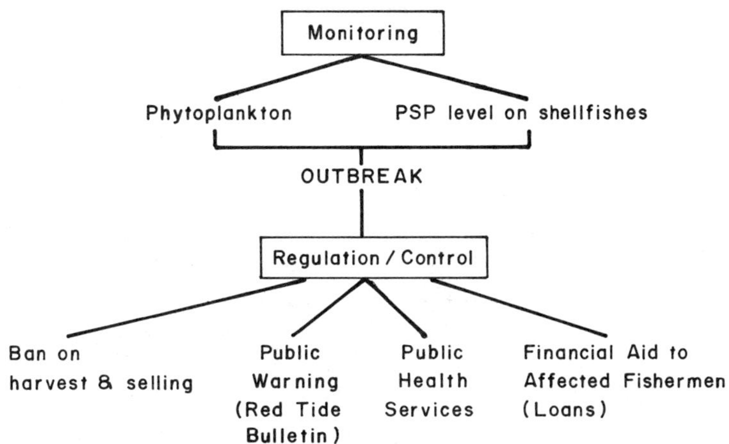

Fig. 2. Diagram of the Present Red Tide Management Scheme in the Philippines.

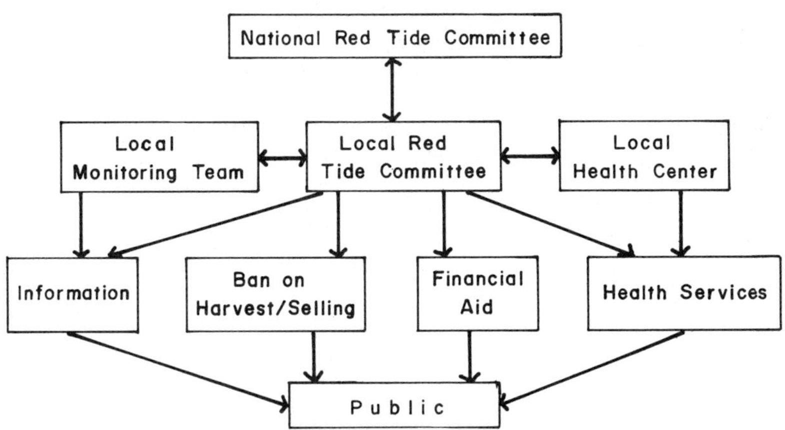

Fig. 3. Diagram of A Proposed Red Tide Management Scheme for the Philippines.

and help develop predictisng systems for imminent blooms. This could also contribute to the compilation and analyses of data sets on the red tides in the region. It is therefore, strongly suggested that the research component of a National Red Tide Program be supported vigorously.

REFERENCES

1. N.Y. Pastor, I. Gopuz, M.C. Quizon, N. Bautista, M. White, and M. Dayrit, Management and Training Workshop on Pyrodinium Red Tides, Brunei, Darussalam, May 1989 (unpub).
2. R.M. Gacutan, M. Tabbu, E. Aujero, F. Icatlo, Jr. Marine Biology 87 (1985).
3. C. Gonzales, J.A. Ordonez, and A.M. Maala in: Red Tides Biology, Environmental Science and Toxicology, Okaichi et al. eds. (Elsevier, Amsterdam 1989) pp. 45-48.
4. J. Maclean. Management and Training Workshop on Pyrodinium Red Tides, Brunei Darussalam May 1989 (unpub.)
5. A. Pamintuan, A. Abrematea, and D. Gomez. The Philippine Star, October (1988).
6. D. Anderson in: Red Tides Biology, Environmental Science, and Toxicology, Okaichi et al., eds. (Elsevier, Amsterdam 1989) pp. 11-16.
7. R. Gacutan, M.Y. Tabbu, T. de Castro, H. Gallego, M. Bulalacao, L. Arafiles, and F. Icatlo. Proc. First Asian Fish. Forum (1986).

ACKNOWLEDGEMENTS

The senior author is grateful to the organizing committee for the travel grant which enabled her to attend the conference. Help of P. Sa-a and A. Gentiles in the preparation of the manuscript is also acknowledged.

PSP TOXICITY OF WILD AND CULTURED BLUE MUSSELS INDUCED BY ALEXANDRIUM EXCAVATUM IN GASPÉ BAY (CANADA) : IMPLICATIONS FOR AQUACULTURE

DESBIENS M. *, COULOMBE F.*, GAUDREAULT J.*, CEMBELLA A.D.**, LAROCQUE R.**
*Ministère de l'Agriculture du Québec, Direction de la recherche, 96, Montée Sandy Beach, Gaspé, Québec, Canada G0C 1R0; **Institut Maurice Lamontagne, Ministère des Pêches et Océans du Canada, 850, route de la Mer, Mont-Joli, Québec, Canada G5H 3Z4

ABSTRACT

Wild and cultured mussels were immersed in a PSP affected zone at different sites and depths. Wild mussels became less toxic than cultured mussels. Samples placed at the bottom of the water column were more toxic than those near the surface; both were more toxic than mussels immersed near shore. The impacts of these findings on the implantation of a mussel culture in the area is briefly discussed.

INTRODUCTION

Eastern Canada is periodically affected by PSP problems which seriously limit bivalve mollusc culture [1], in particular that of the blue mussel. Historical data collected by supervising authorities concerning toxicity in mollusc zones take into account only bio-assays carried out on bivalves harvested near shore. The behaviour of cultured mussels at large in the water column which come into contact with the toxic dinoflagellate, *Alexandrium excavatum*, is poorly known. The main objectives of this study are to describe the spatial and temporal evolution of the toxicity of mussels under different growing conditions and to obtain solutions which would permit growers to circumvent this problem and thereby use the potential offered by Gaspé Bay, a known toxic area, for the culture of mussels.

In particular, a description of the dynamics of toxicity induced in mussels suspended in the water column as well as the establishment of the time of year when there is a slighter tendency for paralysing toxins to accumulate would constitute valuable advantages for aquaculture.

MATERIAL AND METHODS

In 1987 and 1988 cultured mussels (*Mytilus edulis*) and wild mussels from zones free of toxicity were placed in holding containers immersed at four different sites in Gaspe Bay in order to compare the spatial and temporal dynamics of toxicity induced by *Alexandrium excavatum*. At the first three sites, the structures were placed immediately below the lowest limit of the average low tide (at the shore in such a way so as to be immersed at all times, as well as at 200-600 m offshore at the surface (1 m) and at the bottom at (15 m) of the water colum. At the fourth site, which was selected in 1988 in addition, the mussels were placed at two intermediate levels (4 and 8 m) in order to determine the evolution of this phenomenon according to depth level.

After inspection, all the lots of mussels were, to begin with, below the detection level of 42 µg STXeq/loo g of meat. Samples were taken biweekly from mid-June to mid-November, 1987. In order to better describe the kinetics of the intoxication-detoxication process in time, the frequency and the period of sampling were increased in certain cases in 1988 to a weekly basis from mid-April to mid-November; a supplementary

Toxic Marine Phytoplankton
Edna Graneli et al., Editors

sample was also taken from under the ice in January, 1989. The extraction of toxins was carried out according to the official method approved by the A.O.A.C. [2] and the hydrolysates were sent to a government laboratory for bio-assays on mice.

RESULTS

Regardless of the type of mussel or of the depth, increases in toxicity were recorded during the first half of June in 1987 and 1988, attaining up to 6 900 µg at some places. Although we do not have data for the beginning of the 1987 increase, we have noticed a remarkable synchronism in the periods of spring toxicity for both years. No secondary increase took place in 1987 (Fig. 1) while in 1988, a noticeable increase (more than 3 500 µg at some sites) took place at the end of August which later disappeared rapidly at the end of September (Fig. 2). The absence of toxicity then continued at least until the end of January, 1989, as shown by the sample taken from under the ice.

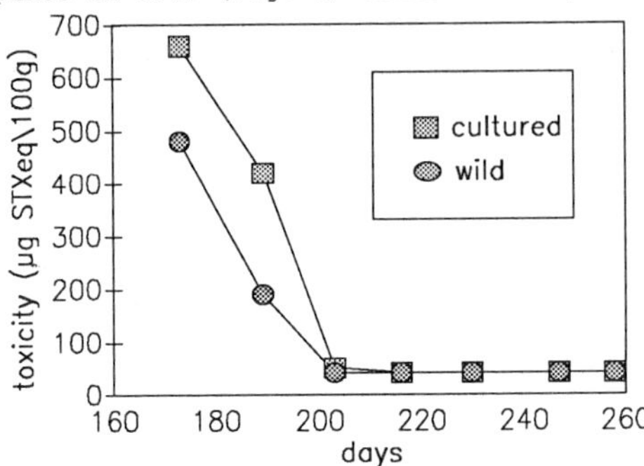

FIG. 1 Toxicity of mussels offshore (1987)

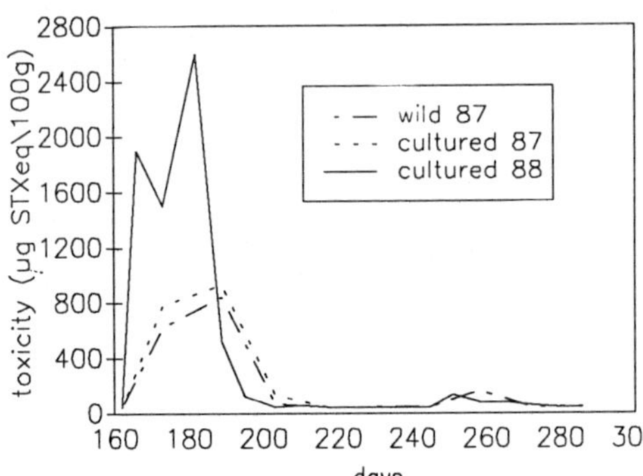

FIG. 2 Toxicity of mussels offshore (1988)

Toxicity distribution in the water column was not uniform. The spring toxicity increase was characterized by important toxic concentrations first at the surface (Fig. 3a); then gradually they were found closer to the bottom (Fig. 3b, c) before being eliminated (Fig. 3d). However, in September, 1988, the mussels placed at the bottom became more toxic than those at the surface (Fig.3e, f); this pattern was maintained during the entire length of this toxic phase, simultaneously at all four sites.

Cultured mussels and wild mussels placed near the shore were systematically less toxic (less than 300 µg) than those placed from 300 to 600 m further offshore (up to 4 900 µg), regardless of whether they were at the surface or at the bottom (Fig. 4). Cultured mussels attained higher toxicity levels than wild mussels in 1987 (Fig. 1), although the difference was relatively slight.

DISCUSSION

Overall observation of the data as it related to time over the two consecutive years of study allows us to establish two critical periods during which increases in toxicity might be particularly prevalent, that is

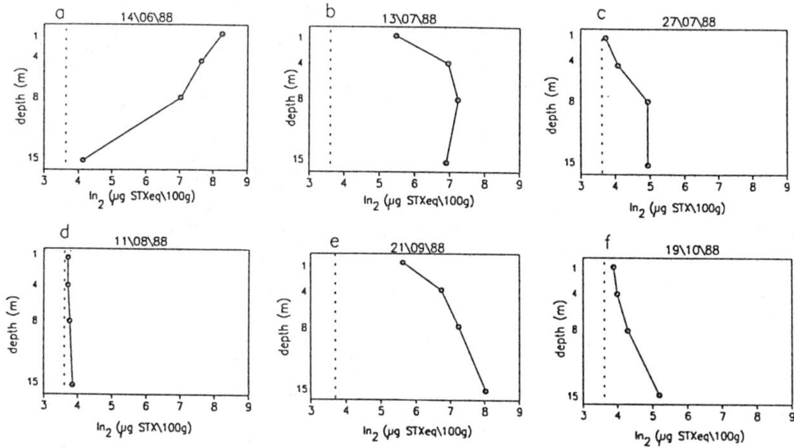

FIG. 3. Evolution of toxicity levels at different depths (cultured mussels)

at mid-June and at the very end of the month of August. The absence of toxicity during the entire winter from the end of September on, in spite of the high incidence of toxicity during the summer, indicates that the dynamics of intoxication and detoxication are sufficiently rapid to permit harvesting the mussels from the month of October on while at the same time respecting innocuity standards of 80 µg STXeq/100 g. However, winter toxicity caused by putting A. excavatum cysts in resuspension is possible [3] and might endanger the extension of the harvest period.

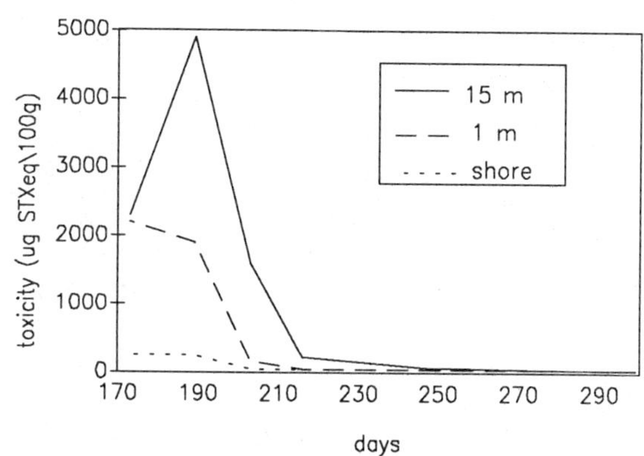

FIG. 4 Toxicity of mussels offshore (depths 1m, 15m) and near shore (1987)

On the other hand, ice cover during the entire winter period can further reduce water movement; thus there is a high probability that mussels harvested during the winter at such a site are exempt of toxicity, as long as fall detoxication is sufficient.

At the beginning of the study, we expected to find that mussels placed at the surface would attain higher rates of toxicity than those placed at deeper levels, since the mobile cells of dinoflagellates tend to proliferate in the first 5 m of the water column [4, 5, 6]. The results show this to be the case at the beginning of the toxic period in mid-June, 1988, but the vertical gradient of toxicity is rapidly inverted during this same period. During the entire Septemher increase, the mussels near the bottom became the most toxic. Data for 1987 are similar. Therefore, the entire water column provides a favourable environment for the accumulation of toxins by mussels in coastal, semi-open waters such as Gaspé Bay. The immersion of mussel-growing structures at specific depths in order to move them out of reach of summer increases in toxicity is not only inefficient but can also bring about, at some point during the critical period, toxicity levels that increase with water depth. The causes of this situation are still unknown; in fact, the data concerning samples of phytoplankton and cyst counts are still being studied. Only plankton net drags carried out 6 to 7 days before the appearance of toxicity at the end of August reveal qualitatively important concentrations of mobile cells of Alexandrium excavatum.

Based on the observed differences in the toxicity levels of mussels placed offshore and those near the shore as well as between cultured mussels and wild mussels, we have come to the conclusion that it is not possible to extrapolate data based on mussels collected near the shore of a mollusc zone by government inspection authorities to the toxicity rates in cultured mussels outside this zone. This will complicate the search for favourable sites which are free of toxicity. In fact, if monitoring of the intertidal zone of a specific sector reveals non-existant or low levels of toxicity, this does not positively indicate that mussels grown in suspension outside this zone will be non-toxic.

In conclusion, the production of cultured mussels free of toxicity in a semi-open bay where the toxic dinoflagellate Alexandrium excavatum proliferates would be possible by harvesting in the fall in open waters or in the winter under the ice cover. The distribution of toxicity throughout the water column does not allow a grower to avoid intoxication of mussels simply by modifying the depth of holding structures.

ACKNOWLEDGEMENTS

The authors would like to thank Sharon Thibault, Lucie Barriault, Louise Therrien, Gilles Lapointe and Danielle Packwood for their technical assistance as well as John Reid of Health and Welfare Canada for carrying out the bio-assays.

REFERENCES

1. A. Prakash, J.C. Medcof, A.D. Tennant, J. Fish. Res. Bd. Can. Bull. 177, 87p. (1971).
2. Official Methods of Analysis, 13th ed. (A.O.A.C., Arlington VA), secs 18.089-18.090 (1984).
3. C.M. Yentsch, L.C. Incze, Proc. 6th FDA Science Symposium on Aquaculture, Feb. 12-14, New Orleans, (1980) pp. 49-62.
4. T. Braarud, Sarsia 60, 41-62 (1975).
5. A.D. Cembella, J.C. Therriault, Red Tides: Biology, Environmental Science and Toxicology, Okaichi, Anderson, Nemoto eds. (EIsevier, Amsterdam) ,pp.81-84, (1989) .
6. L. Nishitani, K.K. Chew , Aquaculture 39, 317-329 (1984).

PUBLIC HEALTH ASPECTS OF CIGUATERA POISONING CONTRACTED ON TROPICAL VACATIONS BY NORTH AMERICAN TOURISTS

Anita R. Freudenthal, Nassau County Department of Health, Mineola, Long Island, New York 11501.

ABSTRACT

Since 1978, snapper (*Lutjanus* spp.), grouper (*Epinephelus* spp.) and other fish have caused ciguatera poisoning on at least 13 occasions, involving 24 New York area tourists of the Bahamian and Caribbean islands. Case histories indicate a need for increased clinical recognition, reporting, and education of consumers for this emerging northern public health problem.

INTRODUCTION

Ciguatera poisoning is an illness characterized by its variability (1). Various combinations of more than 175 symptoms appear in minutes, hours or several days after consumption of any one of more than 425 species of fish (2,3,4,5). Gastrointestinal distress usually initiates the syndrome and an array of cardiovascular and neurological symptoms usually follows, all of varying duration and intensity. The acute phase can result in death, but rarely does. A chronic phase can persist for weeks, months or several years. The greatest management problem is the inability to predict, both in time and space, the intermittent nature of the occurrence, duration, and type of affected fish and level of toxicity. In the last decade, great advances have been made in the understanding of ciguatera fish poisoning, starting with the discovery of a causative agent, the dinoflagellate *Gambierdiscus toxicus* Adachi and Fukuyo (6,7,8) and in confirming its presence in endemic areas (9,10).

This disease is endemic to the tropics and of considerable importance to the public health and economy of the Pacific and Caribbean islands (13). It is reported to be on the increase, but the true incidence is not known (14,15). Many natives have experienced multiple bouts of the illness. In the United States, ciguatera is endemic only to mainland Florida and the Hawaiian islands. The American Virgin Islands, Puerto Rico and the South Pacific territories also have fish which can cause the intoxication (16). In non-endemic regions there are two patterns for exposure: 1) when tropical fish are imported and purchased for home or restaurant consumption; and 2) when visitors to endemic regions eat fish at a restaurant, hotel or local home (17,18). Increased numbers of tourists, especially from northern regions such as New York, are frequenting ciguatera-endemic areas. In spite of numerous reports, recommendations and precautionary advice, neither the average northern tourist nor physician is well informed about ciguatera poisoning (19,20). All cases reported here have occurred in the tourist population.

EPIDEMIOLOGICAL OBSERVATIONS

Starting with the first reported case in 1978, 13 episodes (12 "confirmed",1 suspect) were investigated by The Nassau County Department of Health (Fig. 1). The age of the 10 males and 14 females ranged from the 20s to the 50s. All were active/athletic professional/business people. Nineteen of the 24 were interrogated about both the particulars of the incident and

Published 1990 by Elsevier Science Publishing Co., Inc.
Toxic Marine Phytoplankton
Edna Graneli et al., Editors

specifics regarding the victims' condition (Table I). Some indicated that, prior to the current debriefing, no health professional had asked about fish consumption. Many specifically noted that the fish had a normal

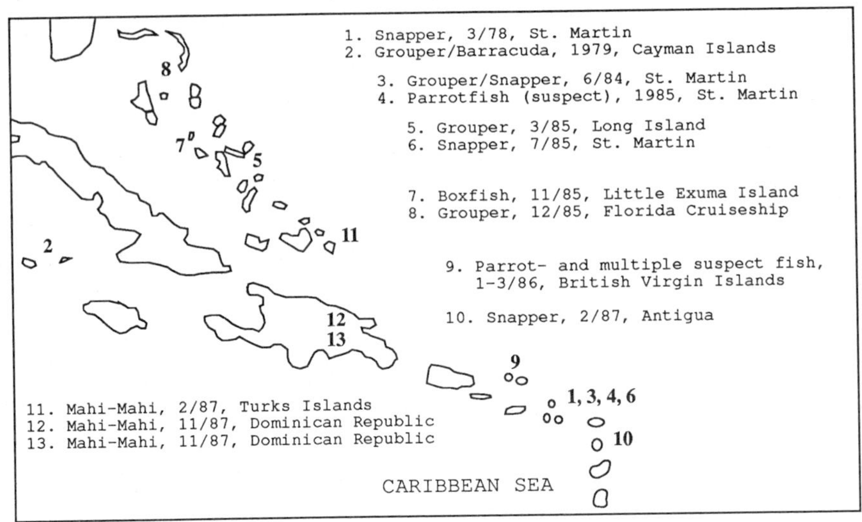

FIG. 1. Geographical distribution and date of 13 occurrences based upon the location where the implicated fish was consumed.

appearance, odor and taste, and "It was the best piece of fish I ever ate". No implicated fish could be recovered for supportive diagnosis. In these cases, consumption history and clinical findings were suggestive of ciguatera poisoning, although only 3 obtained this diagnosis during the early acute stages. Because of this lack of awareness by both patient and physician, some saw up to 7 specialists (Table II). Cost of the illness in these episodes ranged from several thousand to ten thousand dollars including lost vacation and work time, physician consultations, assorted medical tests and medications. There is no single definitive test for ciguatera poisoning, nor is there a single medication. In such an illness, with multiple symptoms, where treatment can be only symptomatic, a multiplicity of tests, diagnoses and medication is frequently encountered (Table III) (21).

Ciguatera is currently a non-reportable illness. Almost all victims met others whose cases had never been reported. Unquestionably, these are only a small number of the actual incidents of this poisoning occurring among this type of consumer of tropical reef fish. The low number of reported episodes, reflects the lack of clinical recognition and reporting. Although the symptoms are variable, there was a typical profile in this study group (Table IV). The temperature sensation reversal phenomenon, typical of this intoxication, was often particularly vivid, e.g., stepping on a cold floor caused feet to feel burned. Extreme fatigue, often lasting 4 to 8 months, coupled with heightening chronic symptoms, caused mental anguish worse than the physical disability. No deaths occurred, however, in one episode, a cat was given the leftover boxfish. A veterinarian diagnosed "fish poisoning", and constant diarrhea caused death in several weeks.

TABLE I. A typical case history: mid-March, 1985

7:30 p.m.	Female, mid-40s, ate grouper (rock hind) in tropical island restaurant; "tasted fresh, meaty, delicious".
8:30 p.m.	Left restaurant "feeling odd, something happening".
9:00 p.m.	Very tired, went right to sleep.
10:00 p.m.	Awakened by violent vomiting, continued 12 hours.
11:00 p.m.	"Terrible" diarrhea started, lasted 24 hours; could not move, carried to bathroom.
Next a.m.	Very nauseous, waves of nausea even after vomiting stopped, dry heaves; waves of abdominal cramps and pain; "unbelievable weakness"; couldn't lift arms or legs; slight headache; felt cold, used blanket but shivering; "bones were hot and flesh cold or bones freezing and flesh hot".
2nd day	"Felt as if having heart attack/stroke, as if blood couldn't flow on left side in leg, "experienced vascular irregularities, heartbeat fluttering"; numbness on left side, pins and needles up to head and down to toes.
3rd day	On plane to Long Island, New York home.
4th day	To physician who "didn't know what I was talking about, he was astonished". In wheel chair.
April	Put hand under cold water and would burn like a match, wore gloves; itch "so bad I could lose my mind"; this intensity lasted 4 weeks.
May	In bed 2 months, couldn't lift arms, or straighten legs; numb on left side then right side.
6+ months	Still bothered by foods, if eat pickling mixtures, wine, vinaigrettes then red marks from nose to ear; still tired, very upset, nervous, "no settling down", thought from pressures of business.
8 months	Still has fatigue; noticed thinning of hair.
11 months	Itch on palms and soles of feet then ended; woke up one morning and "felt like myself"
3-4 years	Still occasional pins and needles, some numbness

TABLE II. Institutions/Specialists consulted by phone and/or visit.

Internal medicine	Emergency Rms	Local island MD
Infectious Diseases	Homeopathic	Marine Biologist
Tropical Diseases	Cardiologist	Health Depts.:
Otolaryngologist (E,N,T)	Neurologist	Hawaii, Conn.,
Poison Control Center	Audiologist	Nassau County, NY.
Allergist	Psychiatrist	Epidemiologist
General Practitioner	Toxicologist	MAYO Clinic

In one episode, a family lives several months each year at a Caribbean island vacation home eating 3-4 fish meals per week, usually as soup. The fish cooked include: boxfish/trunkfish (*Lactophrys* spp.), yellowtail snapper (*Ocyurus chrysurus*), angel fish (*Holacanthus* spp.), queen trigger (*Balistes vetula*), porgies (*Archosargus, Diplodus, Calamus* spp.), blue tang (*Acanthurus coeruleus*), grunt (*Haemulon* spp.), groupers (rock hind, *Epinephelus adscensionis*, Nassau, *E. striatus*), and especially parrot fish (*Scarus* spp.); most of these are potentially toxic. Tingling and numbness have plagued one family victim, now in his 60s, for 10-20 years.

The recent increase in the export of possible ciguatera-containing fish to northern markets stimulates the appetite for these fish, which is usually acquired during tropical vacations. However, one victim who had contracted the disease from red snapper in the Caribbean and was symptom-free for a year, again ate red snapper in a restaurant near his Long Island, New York home, and the tingling in his fingertips resumed.

TABLE III. Diagnoses and medical tests prior to eventual ciguatera diagnosis, and medications used before and after correct diagnosis.

Self/Physician(s) "Diagnosis"	Tests	"Medications" taken/prescribed
Multiple Sclerosis	EMG, EKG	I.V.Glucose
Coronary Disease	MRI, NMR	Kaopectate
Brain Tumor	CATscan	Milk of Magnesia
Intestinal Flu/Virus	Urine Sample	Benedryl
Depression/Emotional	Blood Work	Cortisone
Food Allergy Attack	Allergy Workup	Indocin
Stress	Stoolsample:	Rose Petals on tongue
	for parasites	Sea water soak
Neurologic Illness	andbacteria	Atarax
Going Crazy	(staphyloccus	Seldane
Losing Mind	(and salmonella)	Vitamins C,B1,B6,B12

DISCUSSION

The traditional ciguatera victims have been inhabitants of the islands where the illness is endemic. Newer consumers need to learn to exercise judgment to avoid implicated fish. This preventive strategy depends to a large extent on the utilization of knowledge to identify and forewarn persons of the potential risk of acquiring ciguatera poisoning, if fish management and detection strategies fail to prevent marketing and consumption.

RECOMMENDATIONS

- Collect epidemiolcgic data from those previously afflicted.
- Acquire more precise and expansive data (personal, episodic and ecological) in future episodes.
- Standardize questionaires for use worldwide.
- Propose a 2-day storage requirement for fish carcasses from hotels and restaurants in endemic areas to be available for toxicologic study and confirmation of diagnosis.
- Promote use of "stick test" immunoassay toxin detection on suspect fish for frequent providers/users in endemic areas, (restaurants/hotels and owners of vacation homes (20)).
- Convince travel personnel to alert prospective tourists, especially pregnant and nursing women who are particularly susceptible.
- Provide further education to improve diagnosis in non-endemic regions, targeting pharmacists and specialists most frequented by victims.
- Stimulate official reporting of incidents.
- Develop consumer and physician awareness programs in health departments (i.e.) in the use of mannitol as treatment (21).

- o Organize a self-help group, hot-line/clearing-house in localities, similar to emotional support provided by counseling aid groups (i.e., accident or disease victims).
- o Coordinate management efforts which involve: the habitat, the fishery, the scientist, the consumer, and the health care provider.

TABLE IV. Typical progression of symptoms in these episodes.

	ACUTE PHASE
<u>Gastrointestinal</u> beginning/lasting hours to first days after meal	Nausea, Vomiting Abdominal Pain/Cramps Diarrhea (up to 3 days).
<u>Subjective</u> accompanying GI during first/few hours, days	Weight/Vise/Elephant on chest Felt: "funny"/"terrible"/"weird sinking feeling"; Felt like: shock/dead/stroke/heart attack; Thought would passout; Dizzy; Faint; "Drunk".
<u>Cardiovascular</u> beginning hours/few days and lasting a few days after meal	Bradycardia (low pulse rate ≈40-50/min) Hypotension (low blood pressure ≈90/40-90/60 mm Hg) Cardiogram with skipped beat. "Heart palpitations"/ fluttering. High white cellcount.
	POST-ACUTE PHASE
<u>Neuromuscular</u> usually by day 2, and lasting days, weeks, months	Muscle aches (myalgia)/asthenia weakness Malaise/fatigue/prostration Afternoon and early eve tiredness. Mild to severe headaches. Strange facial sensation. Fingers curl up.
<u>Neurological</u> within 24 hours usually succeeding cessation of GI symptoms and lasting days, weeks, months, few years	Pruritis (lips, extremities, genitals itchy). Tingling (perioral, extremities). Paresthesia (extremities, circumoral numbness) Metallic taste Chills/feeling cold - heating/burning Temperature reversal (hot felt cold, cold felt hot). Catatonic.
	CHRONIC PHASE
<u>Dermatological</u>	Papular rashes/ spots on body (brown) Red marks on face Hair loss/thinning on head.
<u>Eye and Ear</u>	Trouble focusing, eyes strained Uncomfortable to look up then down Ataxia (feeling off balance/light-headed).
<u>Psychological</u> lingering months to few years	Anxious/ "unsettled"/ depression "Going crazy"/"can't cope with usual things" Fear of "losing mind"/"can't deal with life" Fear of losing memory. Write or update wills.
<u>Exacerbated by</u> liquor, vinegar wine, pickles, etc., for several years	Itching Tingling (toes and fingers after 1 drink) Numbness Red streaks across cheek

REFERENCES

1. R. Bagnis, Hawaii Med. J., 28, 25 (1968).
2. R. Bagnis, T. Kuberski and S. Laugier, Ann. J., Trop. Med. Hyg., 28, 1067-1073 (1979).
3. B. S. Anderson, J. K. Sims, N. H. Wiebenga, and M. Sugi, Hawaii Med. J., 42, 1975-1981 (1982).
4. J. K. Sims, Ann. Emerg. Med. 16, 1006-1015 (1987).
5. B. W. Halstead, Poisons and Venomous Marine Animals of the World (Darwin Press, Inc., Princeton, NJ, 1978).
6. T. Yasumoto, I. Nakajima, R.A. Bagnis, and R. Adachi, Bull. Jap. Soc. Sci. Fish 43, 1021-1026 (1977).
7. R. Adachi and Y. Fukuyo, Bull. Jap. Soc. Sci. Fish 45, 67-71 (1979).
8. R. Bagnis, S. Chanteau, E. Chungue, J. M. H., T. Yasumoto and A. Inoue, Toxicon 18, 199-200 (1980).
9. Y. Shimizu, H. Shimizu, P.J. Scheuer, Y. Hok, M. Oyama, and J.T. Miyahara, Bull. Jap. Soc. Sci. Fish 48, 811-813 (1982).
10. D. R. Tindall, R. W. Dickey, R. D. Carlson, and G. Morey-Gaines in: Seafood Toxins, E. P. Regalis, ed. (Am. Chem. Soc. Syn. Ser. No. 2, Washington, D.C.) pp. 225-240.
11. A. H. Banner in: Biology and Geology of Coral Reefs, Vol.III, O. A. Jones and R. Endean, eds. (Academic Press, NY 1976) pp. 177-213.12.
12. N. D. Lewis in: Seafood Toxins, E. P. Regalis, ed., Am. Chem. Soc. Syn. Ser. No. 262, Wash. D.C. (1984).
13. L. V. Sick, personal communication, NOAA, Charleston, SC (1986).
14. J. G. Morris, Jr., L. P. Smith, P. A. Blake and R. Schneider, Am. J. Trop. Med. Hyg. 31, 574-578, (1982).
15. C.D.C., MMWR, 35, 263-265, (1986).
16. E. C. D. Todd in: Proceedings of the Third International Conference on Toxic Dinoflagellates, D. M. Anderson, A. W. White, D. G. Baden, eds. (Elsevier, New York, 1985) pp. 505-510.
17. E. M. Grant, Guide to Fishes (Dept. of Harbours and Marine, Brisbane, Australia, 1982).
18. A. M. H. Ho, R. M. Fraser and E. D. C. Todd, Ann. Emerg. Med., 15, 1225-1228, (1986).
19. D. P. de Sylva and M. Poli in: World Record Game Fishes. (International Game Fish Association, 1982).
20. Y. Hokama, L. K. Shirai, M. N. Kurosawa, L. M. Iwan, C.S. Goto and A. M. Osugi, Biol. Bull. 172, 144-153 (1987).
21. N. Palafox, L.G. Jain, A. Z. Pinano, T. M. Gulick, R. K. Williams and I. J. Schatz, JAMA, 259, 2740-2742 (1988).

DISTRIBUTION AND MAGNITUDE OF DOMOIC ACID CONTAMINATION OF SHELLFISH IN ATLANTIC CANADA DURING 1988

M.W. GILGAN, B.G. BURNS AND G.J. LANDRY
Chemistry Section, Scientific and Technical Services Laboratory, Inspection Services Branch, Department of Fisheries and Oceans, P.O. Box 550, Halifax, N.S. B3J 2S7, Canada

ABSTRACT

During the late fall of 1987, domoic acid contamination of shellfish was responsible for severe illness in many Canadians. The contamination of the shellfish has subsequently been attributed to the occurrence of a toxic variety of the diatom, *Nitzschia pungens*. The presence and abundance of domoic acid in shellfish throughout Atlantic Canada was monitored by the HPLC analysis of extracts prepared by the AOAC procedure for PSP toxins. Several outbreaks of severely contaminated shellfish occurred characterized by long, low-level preliminary contamination followed by an abrupt rise to severe levels and an equally abrupt decline in the domoic acid content. The toxin was first detected in mussels (*Mytilus edulis*) and soft-shell clams (*Mya arenaria*) collected from the shore in late July-early August in the Bay of Fundy and vicinity. In this area, peak toxin accumulation occurred in these species in September-October, then declined to not detectable. The toxin accumulation in clams was apparently about one-half that in mussels. In the area which produced the toxic shellfish of the previous winter, Cardigan River, P.E.I. and environs, the toxin accumulation did not become significant until late November and peaked in December. The amount of domoic acid accumulated in Bay of Fundy mussels was *ca.* 75 ug/g, whereas that of the vicinity of Cardigan Bay (Brudenell River) was *ca.* 350 ug/g.

INTRODUCTION

In December 1987, a group of government employees working in the laboratories of the National Research Council, in Halifax (NRC, ARL), Nova Scotia determined that the cause of a rash of poisonings attributed to cultured blue mussels (*Mytilus edulis*) from Prince Edward Island (P.E.I.), Canada (fig. 1) was domoic acid [1]. The source of the toxin was not immediately apparent, but it was soon attributed to a diatom, *Nitzschia pungens* [2]. The domoic acid was detected in the shellfish by evaluating extracts of the soft tissues prepared for detecting PSP with the mouse bioassay. At that time Health and Welfare Canada (HWC), an agency responsible for the general food safety in Canada, proclaimed that shellfish would be considered safe if toxins, including domoic acid, could not be detected by the mouse bioassay.

During the work to isolate and identify domoic acid, workers at the NRC, ARL laboratory established an assay for the toxin based on reversed-phase high performance liquid chromatography (HPLC) with an ultra-violet light detector [1]. The method was accurate, rapid and had very wide linear range. Upon implementation of an enhanced shellfish monitoring program in Canada in

Toxic Marine Phytoplankton
Edna Graneli et al., Editors

1988, the method was adapted for use on a routine basis by the Halifax Fish Inspection Laboratory of the department of Fisheries and Oceans. It permits monitoring of levels of domoic acid in shellfish harvested at key sampling sites so that when serious contamination was detected the public could be quickly informed and harvesting stopped to safeguard consumer safety and protect the reputation of the Canadian shellfish industry. As a result some interesting data was collected which has been reported here.

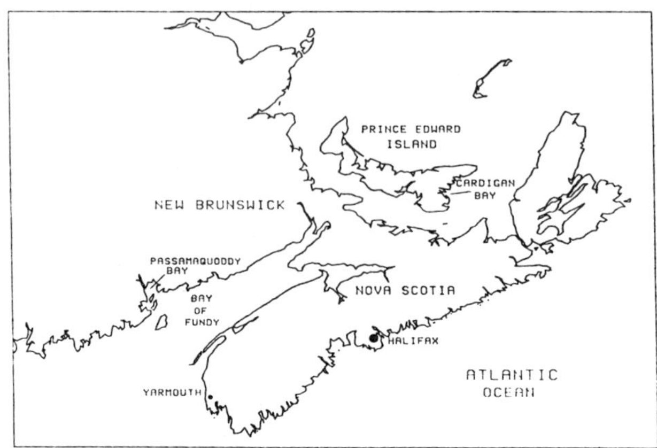

FIG. 1. Map of Maritime Provinces of Canada

METHOD

Extract Preparation

The procedure used to prepare extracts for bioassay or HPLC evaluation was that specified by the Association of Official Analytical Chemists [3], with minor modifications in the various Fisheries Inspection Laboratories. Samples of the filtrate were dispensed to 20 mL vials for storage (short term: refrigerator, 4 C; long term: freezer, -20 or -30 C) or delivery at ambient temperature by courier or mail to the bioassay or HPLC testing laboratory.

Chromatographic Sample Preparation

An aliquot of the crude PSP extract was mixed with an equal volume of methanol (A.C.S. or HPLC grade), centrifuged to clarify the mixture and a sample of the supernatant fluid forced through a 0.2 u filter (PTFE or nylon). An aliquot of the filtrate was diluted with an equal volume of glass distilled water in an autosampler vial.

Chromatographic Evaluation

The conditions for HPLC were essentially those first used by the NRC laboratory [1], with minor modifications to suit the particular column and apparatus available. The apparatus was a Beckman/Altex gradient pumping system, a Shimadzu autosampler and column oven, a Vydac 201TP54 (25 cm X 4.6 mm ID, C-18) column, a Kratos model SF-769Z monitor set at 242 nm and a Spectra-Physics model SP4270 integrator. The solvent system was degassed, filtered (0.2 u), acetonitrile:glass redistilled water:trifluoroacetic acid (TFA) (10:90:0.1) pumped at 1.0 mL/min (*ca.* 900 psi).

The retention time and system response was checked at the beginning and end of a set of samples with a quantitative standard injection. The retention time was checked after every ten samples by an injection of dilute standard (*ca.* 5 ng). When used, a check-sample consisting of a previously evaluated shellfish extract was included in the routine at the beginning, middle and end of the sample train to ensure that resolution and response remained adequate. Results are reported as ug domoic acid/g shellfish soft tissue, uncorrected for recovery.

RESULTS AND DISCUSSION

The percent recovery was difficult to determine since the absolute value of domoic acid content was not known and insufficient pure standard was available for spike recovery trials. Comparison of the recovery by the AOAC PSP extraction procedure and the best recovery by other methods (hot distilled water extract [1]) indicated that the former extraction produced about 75-80 % of the available toxin. This is quite adequate for the mouse bioassay. It was accepted as adequate for analytical purposes by HWC. Thus,PSP evaluation could be used for domoic acid estimation. This laboratory determined that normal extracts could be transported to Yarmouth, N.S. and back (3 days) without significantly affecting the domoic acid estimation in the HPLC monitoring program. Samples prepared for the HPLC were stable for several weeks at 4 C when properly sealed to prevent methanol evaporation. For reasons not yet clear, crude extracts show variable stability to all conditions of storage, but in general little change occurs with frozen (-20 or -3 C) storage. The variable storage stability has limited the usefulness of check-samples.

Initially, domoic acid was very scarce, particularly in the purity necessary for analytical work. The NRC, ARL generously supplied the scientific community with samples of pure domoic acid standard in small amounts. The samples were stabilized by the addition of TFA to prevent bacterial action, but this caused the samples to be unstable to freezing. None of the standards used in the Inspection laboratory were frozen except for the very dilute material used for frequent retention time checks. Despite the dilution of the TFA of the standard the retention time standard was still not stable to freezing.

To evaluate precision, repeated samples of standard were evaluated at high levels (24 ng/injection) by peak height determination to determine the standard deviation (example; mean peak height: 5766, number of estimates (n): 5, standard deviation (sd): 19, percent relative sd (% rsd): 0.3) and similarly at a low level, pure domoic acid (6.0 ng) evaluated on the basis of the high level standardization (n: 8, mean: 5.9 ng, sd: 0.15, % rsd: 3) or a low level shellfish extract (n: 10, mean: 8.2 ug/g, sd: 0.12, % rsd: 1). The long term evaluation of a check-sample extract, stored either frozen or in the refrigerator (n: 310, mean:51.2 ug/g, sd: 2.73, % rsd: 5.33) gave acceptable precision.

For the interpretation of results of the monitoring, it was also necessary to know the variability of the domoic acid in the individual shellfish and the stability of the domoic in stored (dry) shellfish. Shellfish were stored at *ca.* 4 C for up to a week with no loss of domoic acid. The variability of shellfish within a sample was examined in detail, first with a sample of cultured blue mussels contaminated by domoic acid during the fall 1987 episode and stored frozen and the second, with fresh digestive glands of Atlantic scallop (*Placopecten magellanicus*). The evaluation of individual mussels gave a mean of 150 ug domoic acid/g (240, 100, 20, 250, 170, 90, 100, 290, 60, 70, 240, 110) (n: 12, sd of 89, % rsd 59). The individual digestive glands gave an average of 112 ug/g (67, 83, 120, 92, 200) (n: 5, sd: 52.6, % rsd: 45.0). It is apparent from both of these samples, and in conformity with expectations, that the potential variation in the sample domoic acid contents due to small differences in the selection of the sample (including random chance) is much greater than expected analytical errors, even with the bioassay. Hence a large sample size is important.

At the inception of the intensified shellfish monitoring which followed the poisonings of late 1987, the acceptance level for domoic acid contamination was the detection limit of the mouse bioassay. This is of course not a precise value. However, domoic acid could be detected by bioassay at least as low as 70 ug/g in the shellfish. Since the HPLC assay can readily detect levels lower than 1 ug/g, it was used for the first several months as an early warning for the presence of domoic acid. As a result, extracts of many species from various locations in Canada and elsewhere were examined. The first domoic acid found in animals not from the Cardigan Bay area of Prince Edward Island was detected in red (horse) mussels, *Volsella modiolus* and digestive glands of Atlantic scallop from the Bay of Fundy (fig. 1). The presence of domoic acid was confirmed chromatographically by the NRC, ARL. The red mussels, while sometimes consumed by individuals is not considered to be a commercial species and was of lesser concern. The presence of domoic acid in scallops was of very major concern. However, after the determination of many hundreds of samples of scallop tissues for which domoic acid was detected in the digestive gland, none was detected in the principal edible part, the adductor muscle or "meat", and only rarely small amounts were detected in the less frequently eaten gonadal tissue. Since other major species in the area (blue mussels and soft-shell clam, *Mya arenaria*), did not show detectable levels of domoic acid, it was presumed that the former two species could retain domoic acid for at least most of a year. As a result of these detections, the monitoring was intensified and changed to include offshore species. It is intended that these will be the subject of a separate report.

When it became evident that domoic acid was widespread and that the levels detected by the mouse bioassay might still represent some hazard to humans, the acceptance level was reset in July 1988 by HWC to 20 ug/g, where it remains at present. At this stage the quantitation became more critical but suitable alternate methods or confirmation procedures were still not available. The largest concerns were to not misidentify domoic acid in samples and to detect it at as low a level as possible with the normal rapid procedure. Confirmations were done under conditions which gave much higher resolution system than was used by the NRC laboratory and the detection limit was maintained by keeping the background level of the system as low as possible by flushing the system with methanol. As a result domoic acid could be detected with reasonable certainty at as low as 0.2 ug/g (reported as trace). Results reported have not been corrected for recovery since the precise recovery was not known nor could be determined. Similarly since the reliability and precision of the standard was not always sure, generally only two significant figures were reported. Chromatographic confirmations were done on all samples from a new site or area, or from a new species suspected to contain domoic acid.

A summary of the results obtained from characteristic sites and species from Nova Scotia, New Brunswick and Prince Edward Island are shown in fig. 2. The domoic acid detected in inshore species from Nova Scotia, primarily from the vicinity of Yarmouth north into the Bay of Fundy, was infrequent and generally very low level (<2 ug/g). Except for the data shown in fig. 2, no consistent trend developed; the highest value was found in a blue mussel (8.0 ug/g), well below the acceptance level. The red mussels which gave the results shown in fig. 2 were obtained from two sites north of Yarmouth on the Bay of Fundy. It is apparent that low levels of domoic acid were present in the red mussel before the sampling was started. Samples of red mussels from the same area contained domoic acid in April, 1988, which may indicate they were usually contaminated. This was not true for soft-shell clams or blue mussels in the same general area. Samples of these animals did not show any domoic acid, or at most traces, prior to late July, 1988.

Soft-shell clams and blue mussels from the New Brunswick side of the Bay of Fundy in the general area of Passamaquoddy Bay, did not show measurable amounts of domoic acid until late July (Bar Road, fig. 2). After the initial appearance, little change occurred until near the end of August, when at the

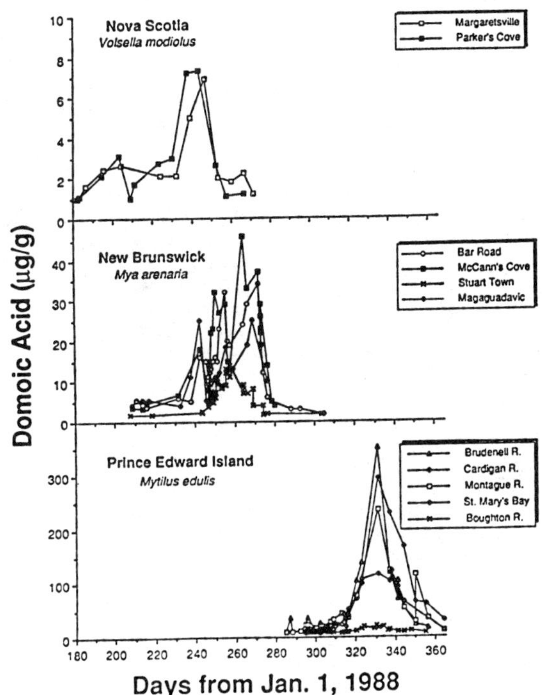

FIG. 2. Domoic acid content of the soft tissues of major commercial shellfish from New Brunswick and Prince Edward Island and an indicator species from Nova Scotia.

same time as the small, transient increase in domoic acid in red mussels in Nova Scotia, there began a more dramatic increase in the levels in Passamaquoddy Bay sites (all sites indicated for New Brunswick, fig. 2). The exposure in the Passamaquoddy Bay sites appear to have been in three episodes, each resulting in slightly greater accumulation, until it abruptly dropped to low levels near the end of October. These peaks of domoic acid contamination presumably reflect the abundance of toxic specie(s) in the water on which the soft-shell clams and blue mussels were feeding. While not shown here, locally non-commercial, blue mussels from Bar Road became more toxic than the soft-shell clams (max. 74 *vs.* 33 ug/g) but otherwise followed the same pattern as the clams from the same site. There is a real possibility that the clam values are significantly underestimated using the AOAC PSP extraction procedure.

It would appear that the intoxication which occurred in P.E.I. was due to one episode, since all sites sampled showed the same time for domoic acid increases (start: Nov 11, peak: Nov 25), even though the sites are not immediately adjacent, nor even in the same local embayment.

The behaviour of the domoic acid in red mussels suggests that they tend to retain it more than soft-shell clams and blue mussels. It was presumed that the half-life of domoic acid in the red mussel was perhaps one-half to one year, hence the occurrence of domoic acid in the red mussels in April. The appearance of the curve in fig. 2 does not indicate such a long half-life. Hence perhaps what this really shows is that there is a relatively continuous occurrence of organism(s) which contain domoic acid but at levels too low to accumulate in the rapidly depurating soft-shell clams and blue mussels, but which can accumulate in the slowly depurating red mussels and Atlantic scallops. The domoic acid in the digestive gland of the Atlantic scallop is strongly retained. It was not cleared from contaminated animals held in flowing sea water at ambient Halifax temperatures (*ca.* 8-13 C) for over four months.

In conclusion, the inshore shellfish of Atlantic Canada from three areas were significantly affected by domoic acid intoxication during the late summer to early winter of 1988; South-West Nova Scotia, Passamaquoddy Bay New Brunswick and Cardigan Bay area of Prince Edward Island. The latter two area experienced domoic acid accumulations which exceeded the Canadian tolerance for domoic acid contamination and required that specific areas be closed to the commercial shellfishery. The former two episodes may have been initiated by shoreline contamination from deeper water since they occurred at the same time, suggesting a common origin. Rapid, large changes in the contamination of the shellfish occurred during each episode which indicates that monitoring requires a rapid response time.

ACKNOWLEDGEMENTS

The authors acknowledge the assistance of Myrna Gillis, Winston Manuel, Denice Cowan in the Hallifax Fisheries Inspection Laboratory in preparing samples and performing analyses, and to all of the laboratory technicians of the Fisheries Inspection Laboratories in Yarmouth, N.S., Sydney, N.S., Black's Harbour, N.B. and Charlottetown P.E.I. for diligently extracting shellfish samples. We thank Roland Cormier, Fisheries Inspection, Moncton, N.B. for supplying analytical results of the last half of December, 1988 for which this laboratory did not perform the analyses. They thank Dan Casavechia for supplying the flowing sea water facility.

REFERENCES

1. C.J. Bird, R.K. Boyd, D. Brewer, C.A. Craft, A.S.W. de Freitas, E.W. Dyer, D.J. Embree, M. Falk, M.G. Flack, R.A. Foxall, C. Glllis, M. Greenwell, W.R. Hardstaff, W.D. Jamieson, M.V. Laycock, P. Leblanc, N.I. Lewis, A.W. McCulloch, G.K. McCully, M. McInerney-Northcott, A.G. McInnes, J.L. McLachlan, P. Odense, D. O'Neil, V. Pathak, M.A. Quilliam, M.A. Ragan, P.F. Seto, P.G. Sim, D. Tappen, P. Thibault, J.A. Walter, J.L.C. Wright; A.M. Baokman, A.R. Taylor, D. Dewar (CISTI); M. Gilgan and D.J.A. Richard (Dept. of Fisheries & Oceans), Atlantic Research Laboratory Technical Report 56., NRCC No. 29083, July, 1988. pp. 1-86.
2. S.S. Bates, C.J. Bird, R.K. Boyd, A.S.W. de Freitas, M. Falk, R.A. Foxall, L.A. Hanic, W.D. Jamieson, A.W. McCulloch, P. Odense, M.A. Quilliam, P.G. Sim, P. Thibault, J.A. Walter and J.L.C. Wright, Atlantlc Research Laboratory Technical Report 57, NRCC No. 29086, September,1988.
3. S. Williams, ed., Official Methods of Analysis of the Association of Official Analytical Chemists, 14th Edition, 1984, 18.090.

MICROALGAL SPORES IN SHIP'S BALLAST WATER : A DANGER TO AQUACULTURE

HALLEGRAEFF,G.M.[1], C.J.BOLCH[1], J.BRYAN[2] and B.KOERBIN[2]

[1]Division of Fisheries,CSIRO Marine Laboratories, GPO Box 1538, Hobart, Tasmania 7001, Australia

[2]Australian Quarantine Inspection Service, Department of Primary Industries and Energy, GPO Box 573F, Hobart, Tasmania 7001, Australia

ABSTRACT

Resting spores of diatoms and dinoflagellates were found in the ballast tank sediments of Japanese cargo ships servicing Tasmanian ports. While most phytoplankton cells do not survive the long, dark voyage in ballast tanks and subsequent discharge into different temperature, salinity and nutrient regimes, phytoplankton resting spores are well capable of surviving such unfavourable conditions. Ballast tank sediments, even after being stored in the dark at 4°C for 6 months, were found to be capable of producing viable cultures of diatoms (e.g. *Chaetoceros*, *Odontella*) and dinoflagellates (e.g. *Protoperidinium*, *Scrippsiella*). Of special concern was the finding of the cyst of the toxic dinoflagellate *Alexandrium* cf.*tamarense* . Precautionary measures on ballast water discharge are called for, especially where ballast water and/or sediments may be carried into important aquaculture areas.

INTRODUCTION

Ever since humans first took to the sea in ships, they have inadvertently transferred marine organisms. These may have been discharged with the water or sediment contained in ship's ballast tanks or formed part of the fouling community on the vessel's hulls. Modern antifouling paints, the increasing speed of seagoing vessels and rapid port turnaround times have decreased the chances of dispersal of the fouling species. However, since the 1970s the greater use of specialised bulk container ships, which carry cargo in only one direction and use water as ballast for stability when unladen, has significantly increased the likelihood of marine organisms being dispersed in ballast water (1). Some 60 million tonnes of ballast water are discharged in Australian ports every year, with Japanese vessels accounting for over a third of this amount (2). Woodchip carriers servicing Triabunna, Tasmania, take up 20,000 to 25,000 tonnes of ballast water in Japanese harbours when the woodchips are offloaded. Phyto- and zooplankton species (including larval fish and invertebrates (3)) are sucked up with the water as it is being pumped into the ballast tanks of the ships. If weather conditions stir up the water column during pumping, fine sediment (including spores of diatoms and dinoflagellates) will be included in the ballast water. On arrival at Triabunna, ballast water is pumped straight into the harbour and mud accumulated on the bottom of the ballast tanks is shovelled into buckets and dumped over the ship's sides.

Toxic Marine Phytoplankton
Edna Graneli et al., Editors

The present study was prompted by the recent appearance in Tasmanian waters of the dinoflagellate *Gymnodinium catenatum*, the causative organism of paralytic shellfish toxins detected in shellfish (4). This conspicuous chainforming organism was first noticed in the Tasmanian plankton in 1980; it was not reported in surveys of the region in 1945-1950 and 1975-1978 (5). Extensive studies of Tasmanian sediments have indicated that *G.catenatum* cysts are confined to the area around the main shipping port of Hobart and the woodchip harbour of Triabunna, and they have never been seen in surveys of other Tasmanian, Australian or New Zealand waters (6). Furthermore, surveys on ^{137}Cs-dated sediment depth cores have confirmed the absence of *G.catenatum* cysts in sediments deposited in Hobart Harbour prior to 1954 (unpublished data). Previously, this dinoflagellate species was known only from Mexico, Argentina, Spain and Japan (including Yatsushiro, which has been a port of origin of Japanese woodchip carriers servicing Tasmania since 1971).

On a global scale, the number and intensity of noxious plankton blooms seem to be on the rise and their geographic extent seems to be spreading. To some degree, this may be simply attributed to an increased awareness of toxic species, increased utilisation of coastal waters for aquaculture and an increase in plankton blooms caused by coastal eutrophication. The present work suggests that, in addition, transport of these plankton organisms by human activity may also be contributing (7).

EXPERIMENTAL

Ballast tank sediments (duplicate 500 mL samples) were collected from twelve Japanese woodchip container ships visiting Triabunna, Tasmania (Australia) between December 1987 and August 1988 (Table 1). These ships were selected because of their ports of origin were known (Kushiro, Ishinomaki, Yatsushiro or Yura), their voyages are short (16 to 21 days) and they have no stopovers in other ports. Sediments were stored in a dark refrigerator at 4°C until further examination.

Samples were sonicated for 2 min (Braun Labsonic homogeniser) to dislodge detritus particles, screened through a 90µm sieve, collected onto a 20 µm sieve and examined by light microscopy. Some cysts were collected on Nuclepore filters (2 µm), air-dried, sputter coated with platinum, and examined with a Philips 515 scanning electron microscope. Selected cyst types were also isolated by micropipette, washed twice in filtered seawater and incubated in a nutrient medium (filtered Triabunna seawater with nutrients added according to medium G) at 15°C and 150 $\mu E^{-2}s^{-1}$ white light. These cysts were examined regularly for germination.

RESULTS

Four out of twelve ships contained dense concentrations of the diatoms *Actinoptychus senarius* (from the port of Kushiro), *Bacteriastrum furcatum* (from Yura), *Thalassiosira* species (from Kushiro; including *T.anguste-lineata* and *T.pacifica*, neither of which is endemic to Australian waters (8)) or *Odontella aurita* (from Yatsushiro).

Table 1. Ballast tank samples collected from Japanese woodchip carriers arriving in Triabunna, Tasmania.

name of vessel	port of origin	departure Japan	arrival Tasmania	midocean exchange	ballast tank microplankton		
					diatoms	diatom spores	dinoflagellate cysts
Shearwater	Yatsushiro	4 Mar 1988	25 Mar 1988	no	+++	+	+++
Meridian	Kushiro	21 Nov 1987	11 Dec 1987	no	+++	-	++
Forest Trader	Yura	?	16 Aug 1988	no	+++	+	++
Jujo Maru	Yatsushiro	13 Nov 1987	2 Dec 1987	no	+++	-	+
Jujo Maru	Kushiro	26 Mar 1988	13 Apr 1988	no	++	-	+
Forest Trader	Ishinomaki	30 Jan 1988	20 Feb 1988	no	+	-	+
Forest Trader	Yatsushiro	19 Dec 1987	6 Jan 1988	no	+	-	-
Jujo Maru	Kushiro	10 Feb 1988	26 Feb 1988	no	-	-	+
Meridian	Ishinomaki	?	30 Jan 1988	no	-	-	-
Meridian	Ishinomaki	23 Feb 1988	14 Mar 1988	yes	-	-	-
Meridian	Ishinomaki	8 Apr 1988	29 Apr 1988	yes	-	-	-
Jujo Maru	Kushiro	25 Apr 1988	11 May 1988	yes	-	-	-

+++ very abundant; ++ abundant; + present; - absent

Sediments from the four tanks were stored in a dark refrigerator at 4°C for 6 months. When incubated in the nutrient medium, viable diatom cultures of *Odontella aurita* and *Chaetoceros socialis* were produced. These cells germinated from resting spores (9) in the ballast tank sediments (Figs 1-3). Some small pennate diatoms (*Navicula, Neodelphineus*), which do not have resting stages, were also cultured from the sediments (Table 2).

Dinoflagellate cells were rare in ballast tank samples, but seven out of the twelve ships contained dinoflagellate resting spores (cysts). Viable dinoflagellate cysts (Figs 4-8) were isolated by micropipette and germinated into *Protoperidinium* and *Scrippsiella* cultures. Of considerable concern was the find of the oval mucoid resting cysts, thought to belong to the toxic dinoflagellate *Alexandrium tamarense* (Fig.7), but which failed to germinate for confirmation of identification.

DISCUSSION

While the present work shows conclusively that microalgal species can be transferred via ship's ballast water, it is very difficult to assess how often introduced species have actually established themselves in their new environments. The diatom *Odontella (Biddulphia) sinensis*, now one of the most common diatoms in the North Sea, English Channel and Irish Sea, was unknown in European waters before 1903. It is thought that this species may have been introduced from the Indo-Pacific via ballast water (10).

Table 2 Microalgal species in ballast tank waters and sediments (from all ships listed in Table 1); + dominant species.

Diatom cells	**Diatom resting spores**
+ *Actinoptychus senarius*	*Chaetoceros affine*
Asteromphalus sp.	*Chaetoceros laciniosus*
+ *Bacteriastrum furcatum*	*Chaetoceros* cf. *seiracanthus*
Chaetoceros lorenzianum	*Chaetoceros* cf. *similis*
Chaetoceros pseudocurvisetus	*Odontella aurita*
Chaetoceros spp.	
Cocconeis sp.	**Diatoms cultured**
Coscinodiscus cf.*gigas*	
Coscinodiscus wailesii	*Chaetoceros socialis*
Coscinodiscus spp.	*Odontella aurita*
Cyclotella striata	*Neodelphineus* sp.
Diploneis sp.	*Navicula* sp.
Ditylum brightwelli	
Nitzschia spp.	**Dinoflagellate cysts**
Paralia sulcata	
Pleurosigma sp.	*Alexandrium* cf. *tamarense*
+ *Odontella aurita*	cf.*Diplopelta parva*
Rhizosolenia setigera	*Gonyaulax grindleyi*
Skeletonema costatum	*Protoperidinium conicum*
Stephanopyxis sp.	*Protoperidinium conicoides*
Surirella sp.	*Protoperidinium leonis*
Thalassiosira allenii	*Protoperidinium oblongum*
+ *Thalassiosira anguste-lineata*	*Protoperidinium* cf. *punctulatum*
Thalassiosira eccentrica	*Protoperidinium* sp.
Thalassiosira nordenskioldii	*Scrippsiella* cf. *trochoidea*
Thalassiosira oestrupii	cf.*Zygabikodinium lenticulatum*
Thalassiosira pacifica	
Thalassiosira punctigera	**Dinoflagellates cultured**
Thalassiosira rotula	
Thalassiosira tenera	*Gymnodinium* cf. *simplex*
Thalassiosira spp.	*Katodinium* cf. *rotundatum*
Thalassionema frauenfeldii	*Protoperidinium claudicans*
Thalassionema nitzschioides	*Protoperidinium leonis*
	Protoperidinium oblongum
	Protoperidinium spp.(2)
	Scrippsiella precaria
	Scrippsiella cf. *trochoidea*

Other records of exotic diatoms appearing in European waters include *Coscinodiscus wailesii* and *Pleurosigma planctonicum* (11), while *Rhizosolenia calcar-avis* may be an introduction into the Caspian Sea (12).

We believe that the existing evidence is strong enough to call for precautionary measures, especially where ballast water may be carried into important aquaculture areas. Some of the ships bound for Triabunna have adopted the practice of exchanging their ballast water en route in mid-ocean to avoid prosecution under coastal pollution regulations (e.g.MARPOL Marine Pollution Convention).Under some weather conditions, however, mid-ocean exchange may not be carried out for safety reasons. Furthermore, it may not be fully effective with bottom-dwelling species (including microalgal spores) unless the sediments are removed at the same time as the water is exchanged. Nevertheless, as a first step, this appears to be a simple and effective way of reducing

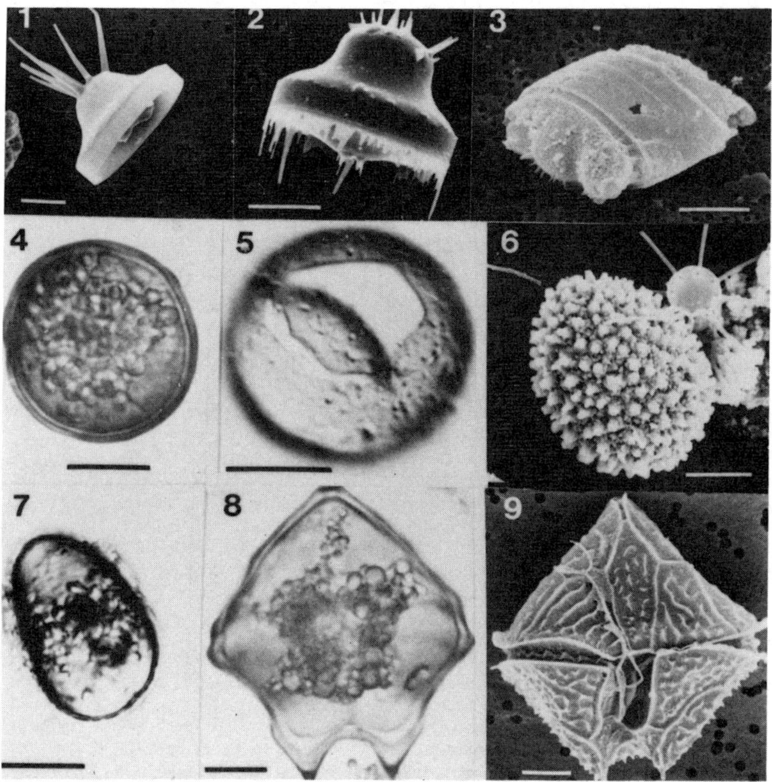

Microalgal spores from ship's ballast water. Figs 1-3, Diatom resting spores; 1. *Chaetoceros affine* ; 2. *Chaetoceros* sp.; 3. *Odontella aurita* ; Figs 4-8. Dinoflagellate resting cysts; 4. *Zygabikodinium lenticulatum* ; 5. *Protoperidinium* cf. *punctulatum* ; 6. *Scrippsiella trochoidea* ; 7. *Alexandrium tamarense* group; 8. *Protoperidinium leonis* ; 9. Motile cell of the dinoflagellate *Protoperidinium leonis* , germinated from the cyst shown in Fig. 8; All scale bars 10 μm.

planktonic introductions (Table 1, ref.3), since it replaces coastal, plankton-rich Japanese waters with oceanic, plankton-poor Central West-Pacific water. After extensive discussions between Australian Government Departments responsible for Fisheries, the Environment, Transport and Communications, and Quarantine, the problem of ballast water has been acknowledged as a quarantine issue of national significance. The Australian Quarantine Act 1908 (Proclamation 6(g)) is now being amended to prohibit the introduction via ballast water of "all disease germs, microbes and disease agents ". Control options available to ships' masters range from certification by foreign authorities that the ballast water has been taken on board in a place free from toxic dinoflagellates, evidence that re-ballasting has taken place en route to Australia,to evidence for effective water treatment in ballast tanks or on shore. Penalties for knowingly importing into Australia anything in contravention of this Act range from Aus $50,000 (natural person) to Aus $ 200,000 (body corporate).It is hoped that other countries sensitive to introductions of foreign species will soon follow this example. When unwanted species are introduced into an open, unexploited niche, the results could be disastrous for endemic flora and fauna, and for the commercial operations that depend upon them.

REFERENCES

1. Carlton,J.T.,Oceanogr.Mar.Biol.Ann.Rev.23, 313 (1985)
2. Hutchings,P.A., Van der Velde,J.T. and Keable,S.J.,Occ.Rep. Aust.Mus.no.3, 147pp (1987)
3. Williams,R.J.,Griffiths,F.B.,Van der Wall,E.J. and Kelly,J., Est.Coast.Mar.Sci.26,409 (1988)
4. Oshima,Y.,Hasegawa,M.,Yasumoto,T.,Hallegraeff,G.M. and Blackburn,S.I., Toxicon 25, 1105 (1987)
5. Hallegraeff,G.M.,Steffensen,D.A. and Wetherbee,R., J.Plankton Res.10, 533 (1988)
6. Bolch,C.J. and Hallegraeff,G.M.,Bot.Mar.(1990)(in press)
7. Anderson,D.M.,in:T.Okaichi,D.M.Anderson and T.Nemoto (eds), Red Tides:Biology,Environmental Science and Toxicology,pp.11-16 (1989)
8. Hallegraeff,G.M.,Bot.Mar.27,495 (1984)
9. Garrison,D.L.,in:K.A.Steidinger and L.M.Walker (eds),Marine Plankton Life Cycle strategies,pp.1-14, CRC press (1984)
10. Ostenfeld,C.H.,Medd.Komm.Havunders.,Ser.Plankton 1,no.6, 44pp (1908)
11. Boalch,G.T. and Harbour,D.S.,Nature 269, 687 (1977)
12. Zenkevitch,L.,in:Biology of the Seas of the USSR.Interscience Publ,NY (1963)

---

* This study was supported by Fishing Industry Research Trust Account grants 86/84 and 89/39

A THREE-MONTH RED TIDE EVENT IN HONG KONG

CATHERINE W.Y. LAM AND SUZANNE S.Y. YIP
Environmental Protection Department, 28/F Southorn Centre, 130 Hennessy Road, Wanchai, Hong Kong.

ABSTRACT

A persistent bloom of *Gonyaulax polygramma* occurred in Tolo Harbour, a marine bay in the northeastern part of Hong Kong, from early February to mid-May 1988. The collapse of the bloom in early May was coupled to the onset of the summer thermal stratification and resulted in a severe anoxic condition throughout the bay. A massive fish kill occurred on 5 May. The dynamics of the bloom development and decline are analysed. The causes of fish kills are examined. We speculate that *G. polygramma* is oceanic in origin and that the cells were carried by tides and oceanic intrusion into Tolo Harbour. The eutrophic conditions of Tolo Harbour, the uniform meteorological conditions, with cool temperatures, overcast skies and low rainfall from February to April, are believed to have been favourable for the development of the bloom. The fish kill is mainly attributed to anoxia. An abnormal condition of high pH and high ammonia concentrations in the surface water during the decline of the bloom could also have been a contributing cause. Furthermore, it is evident that *G. polygramma* can exert further negative effects on fish in stimulating the production of a thick mucous layer around the gill filaments, reducing the oxygen uptake efficiency.

INTRODUCTION

Tolo Harbour, an eutrophic [1-6] marine bay in the northeastern part of Hong Kong, has suffered from frequent red tide occurrences in the 1980's [2,6]. A red tide, caused mainly by *Gonyaulax polygramma,* occurred in Tolo Harbour continuously for three and a half months from early February to mid May 1988. None of the previous red tide incidents had such a long duration. Furthermore, *G. polygramma* has not been reported before as a dominant red tide causative organism in the area. During the whole sequence of the bloom, the distribution of *G. polygramma* and the hydrological conditions in Tolo Barbour were closely monitored. Based on these information, the dynamics of the developmet and decline of the bloom are described in this paper. The adverse impacts of the bloom are reported and the probable causes for fish kills examined.

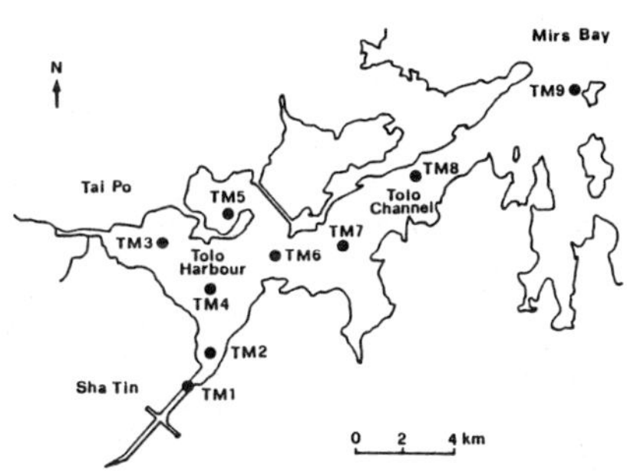

Fig.1. Map of Tolo Harbour showing location of monitoring stations

METHODS

Samples for phytoplankton and water analyses were collected with a Van Dorn water sampler at nine stations (TM1-9) in Tolo Harbour and Tolo Channel (Fig. 1)

Published 1990 by Elsevier Science Publishing Co., Inc.
Toxic Marine Phytoplankton
Edna Graneli et al., Editors

at biweekly intervals. Phytoplankton samples were collected from the 0.5 m depth and were preserved by Lugol's solution for species identification and counting in a Sedgewick Rafter cell by light microscopy. Water samples were collected at three depths; 0.5 m below surface, middle, 0.5 m above the seabed. The depth in Tolo Harbour and Channel ranges from 6 m in the inner harbour to over 20 m in the channel. A wide range of water quality parameters were analysed. For the purposes of this paper, only ammonium nitrogen, chlorophyll-a and phaeopigments data are presented. In situ field measurements at every meter in the water column included temperature, salinity, dissolved oxygen and pH.

An intensive survey involving daily measurements of phytoplankton concentrations and the most important hydrological parametes was carried out at 4 stations (TM3, 4, 6, 7) from April 25 to May 6 1988.

Fish gills were examined under a stereo light microscope. Both healthy and sick fish from the same fish cultures were examined. A fish from the open sea, bought at the market, was used as a control.

RESULTS

Bloom Development and Distribution

Relatively low concentrations (10-20 cells/ml) of *Gonyaulax polygramma* was first detected at the mouth of the Tolo Channel in early January. Red tide conditions were apparent in early February, with the sea outside the entrance of the Tolo Channel turning reddish brown. The bloom spread gradually into the Channel and eventually into the inner Harbour.

The fluctuations of the concentration of *G. polygramma* in the Tolo Channel (at station TM8) and inner Harbour (at station TM3) are presented in Fig. 2. In mid February, the concentration in the channel (> 1,000 cells per ml) was high enough to form a visible red tide, whilst only low concen-trations was detected in the harbour. By early March, red tide conditions were observed both in the harbour and the channel. Concentrations ranged from 1,250 to 6,700 cells/ml. The species continued to bloom and peak abundances (14,550 cells/ml at TM5 and 23,000 cells/ml at TM8) were reached by mid April. During this period the bloom was almost monospecific.

The bloom began to decline towards the end of April and concentrations dropped below 10,000 cells/ml. However, concentrations of *G. polygramma* again increased in the inner Harbour at TM3 in early May, but was then co-dominated by *Skeletonema costatum,* which reached a high concentration of

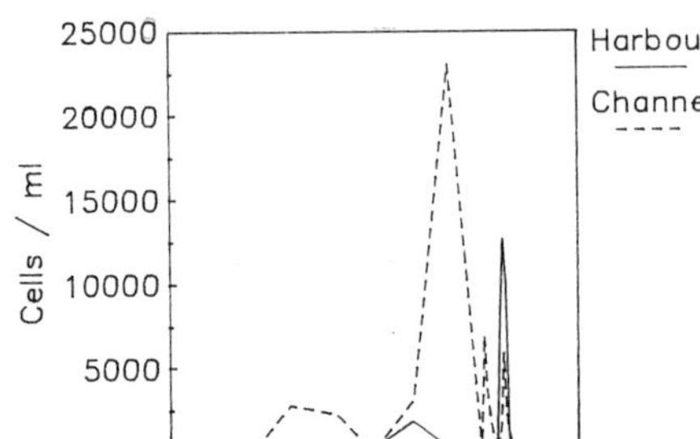

Fig. 2. Variation of cell concentration of *G. polygramma* in Tolo Harbour during the bloom period

40,000 cells/ml and soon dominated over *G. polygramma,* whose concentration dropped below 500 cells/ml on 6 May and further decreased to <100 cells/ml on 18 May. During this period *Thalassiosira mala* also co-dominated and *Prorocentrmm minimum* emerged to later form another red tide event.

G. polygramma discoloured Tolo Harbour and Channel to a greyish green with reddish-brown patches. A massive quantity of scum was sighted floating in large areas of Tolo Harbour and Channel towards the end of April. A foul smell was reported along the shores.

Meteorological Conditions

From February to mid April, the air temperature was relatively cool and constant, with monthly means ranging from 16.3 to 18.1 °C. During this period the sky was unusullay overcast; the monthly mean of total bright sunshine, ranging from 6.1 to 76.6 h, was lower than the normal long term range of 52.1–108.7 h. There was little rainfall (11.9-41.1 mm as compared to the normal range of 41.9-60.4 mm). The weather condition changed around mid April. The air temperature surged rapidly to a mean of 22.9 °C in the latter half of April and to 26.9 °C in May. Rainfall began to increase gradualy (61.8 mm in May) and the sunshine period was longer (70.1 hours in May). During this perlod all the daily scalar mean wind speeds recorded (4.5–14.8 km/h) were lower than the long term mean of 18.5 km/h.

Hydrologial Conditions

Temperature and Thermal Stratification: The temporal variation of sea temperature in surface and bottom waters of the inner Tolo Harbour (station TM3) is presented in Fig.3. The surface temperature rose sharply from a mid April value of 18 °C to 28 °C in early May, parallel to a rise in air temperature. The bottom water was heated up more slowly, resulting in a pronounced thermal stratification during this period. A similar develpoment was observed in other parts of the Harbour as well as in the Channel.

Dissolved Oxygen: The general temporal variation of dissolved oxygen in the surface and bottom waters of Tolo Harbour and Channel is shown in Fig.4. The dissolved oxygen levels remained within the normal range, with higher levels at the surface due to algal photosynthesis. A distinct oxygen stratification was formed in April; while surface oxygen levels surged to a high of over 20 mg/l the

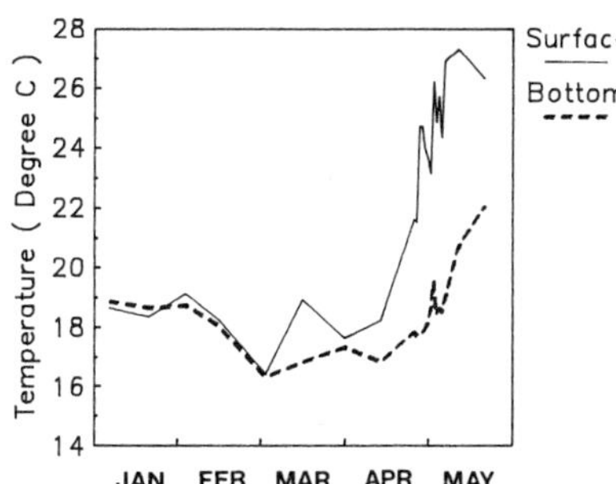

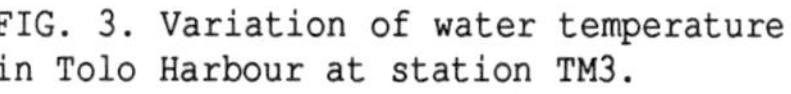

FIG. 3. Variation of water temperature in Tolo Harbour at station TM3.

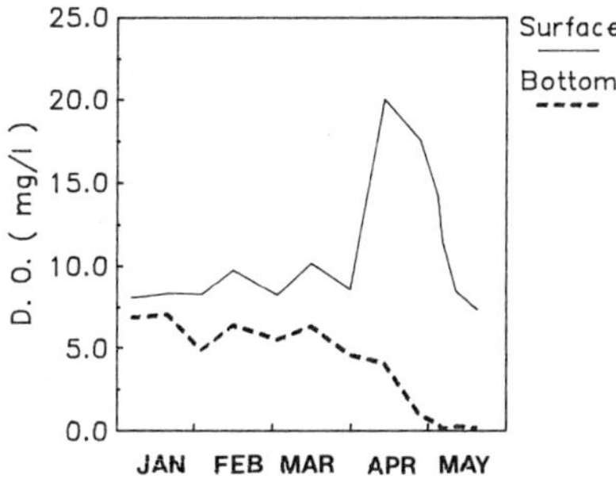

FIG. 4. Variation of dissolved oxygen in Tolo Channel at station TM8 .

levels at the bottom decreased rapidly. From April 27 to the end of May anoxic conditions expanded upwards from the bottom and eventually the whole water layer below the 3 m depth became anoxic (Fig.5)

pH: The temporal variation of pH values in the surface and middle waters in Tolo Harbour and Channel is presented in Fig. 6. From January through March, pH remained constant within the normal range of 8.2 - 8.5. From April and onwards, surface pH began to rise to a high value of 8.85, while that in the middle water fell to a minimum of 7.95.

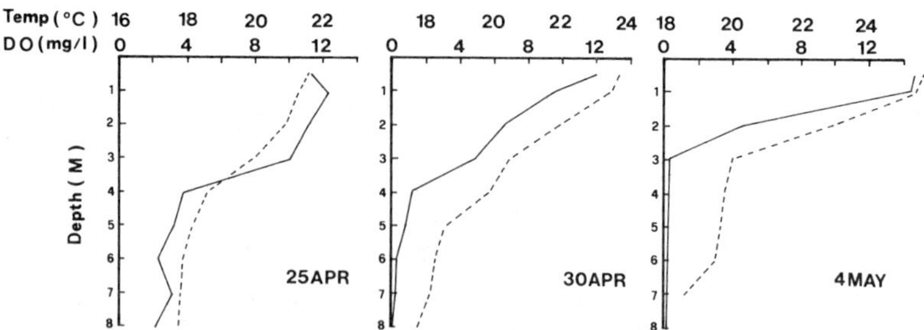

FIG. 5. Stratification of temperature and dissolved oxygen in Tolo Harbour at station TM3.

Ammonium Nitrogen: From January to March, the ammonium nitrogen level ranged from below 0.01 to 0.3 mg/l and usually remained below 0.1 mg/l. A high level of 1.2 mg/l was observed in the surface water on 3 May but then dropped below 0.3 mg/l in the following days. The concentration in the lower water column remained below 0.1 mg/l for most of the time. The algal decay (most pronounced on May 3) naturally produced large amounts of ammonium compounds. Fish crowding at the surface water prior to the major fish kill (May 5) probably also contributed to the high ammonium concentration.

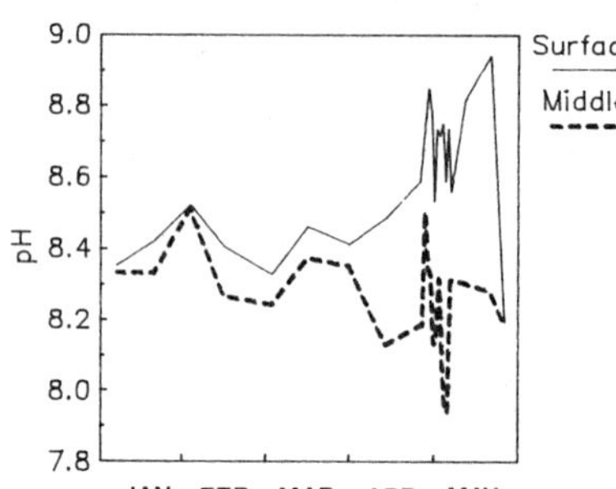

FIG 6. Variation in pH in Tolo Harbour and Channel.

Fish Kills and other Environmental Impacts

During the red tide, three events of fish kills and two incidents of shellfish mortality were reported (unpubl.results), most of these occurred in early May. The most serious fish kill occurred on 5 May with a loss of 35 tonnes fish, representing a value of 7.0 million HK$. When the gills of dead fish were examined under the microscope, it was found that a thick mucous layer, not present in the control sample, had developed around the gills.

In the surface water, the development and decline of the *G. polygramma* bloom also led to other hydrological changes. The water became more alkaline (average 8.85). When total ammonia was recorded at 1.2 mg/l, the un-ionised ammonium concentration was around 0.15 mg/l (the safe concentration for fish is 0.025 mg/l [7]). Various factors contributing to fish mortality are presented in Fig.7.

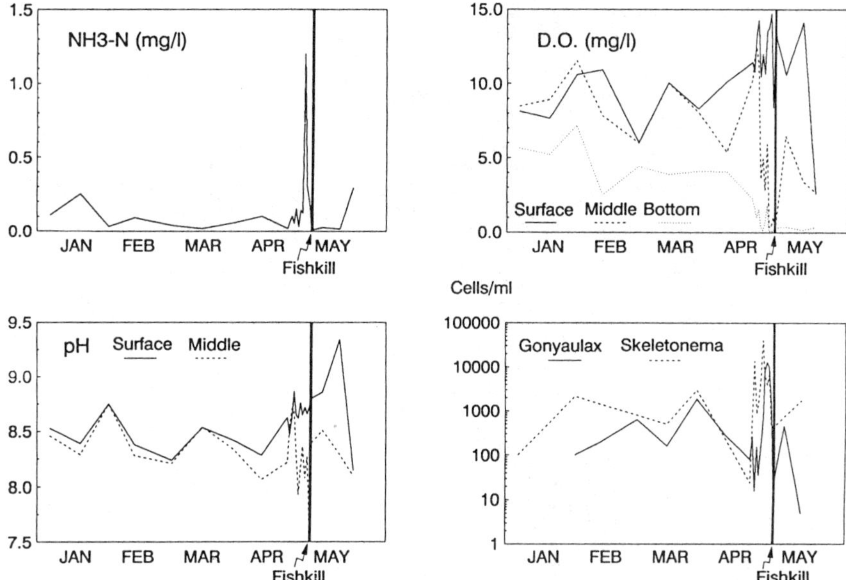

Fig.7. Hydrochemical and biological data from station TM3 prior to and after the fishkill

DISCUSSION

Dynamics of the Development and Decline of the Bloom

Gonyaulax polygramma has not been reported before as a red tide causative organism in Hong Kong. Since the red tide caused by this species was first observed outside Tolo Channel it is probable that the population originated from outside the area rather than from an inside seeding population. After the population had spread into the Channel and eventually into the Harbour, eutrophic conditions [6,8] probably provided the basis for it to flourish and to form an intense red tide. We also believe that the prevailing meteorological conditions; cool temperatures, overcast skies and low preciptation from February to mid April favoured the development of the bloom. Note that red tides of *G. polygramma* also affected extensive areas of other Hong Kong coastal waters during the same period [unpubl. results] . However, these areas are more open and much less eutrophic and there the red tides were just fleeting events of short duration.

The rapid decline of the bloom in Tolo Harbour and Channel towards the end of April was probably triggered by the sudden change in climatic conditions and subsequent changes in hydrological conditions. Thus, with the onset of summer, a sudden increase in air temperature resulting in a pronounced water thermocline, and a calm atmospheric condition helped to maintain the situation. The higher temperatures and the bright sunlight was evidently not favourable for the growth of *G. polygramma*. Furthermore, thermal stratification must have prevented the vertical transfer of gases and ions and the surface layer must have become depleted of at least one essential nutrient through algal uptake. Because of this, further algal growth was evidently hampered.

Factors Contributing to Fish Kills

Since massive fish kills occurred only when the anoxic condition in Tolo Harbour was at its most severe it is likely that this was the major cause of the mortality. However, we believe that a combination of factors could have contributed to the crisis. Thus, the fish must have been severely stressed in the presence of the red tide since *G. polygramma* cells evidently caused irritation to the gills, which must have reduced the oxygen uptake efficiency. The stress was probably also enhanced by the gradual upward movement of the oxygen depleted layer. Furthermore, since the collapse of the red tide resulted in masses of dead algae in the surface water it is highly likely that the decay of these consumed large amounts of oxygen [9].

The water became more alkaline through algal photosynthetic activity and the resulting high pH (average 8.85) probably stressed the fish, which normally require a pH range of 7.8 – 8.3 [10]. At the higher pH range and with the onset of summer, with an increase in temperature, the formation of toxic un-ionised ammonium was probably also favoured [11,12]. Ammonium intoxication therefore could be a contributing factor to the fish kills [7].

ACKNOWLEDGEMENTS

We wish to thank the Director of Environmental Protection for his permission to publish this work. We are also grateful to the staff of the Marine Section for their contribution to field work and technical assistance.

REFENENCES

1. R.G. Wear, G.B. Thompson and H.P. Stirling, Asian Marine Biology 1, 59-75 (1984).
2. P.R. Holmes and C.W.Y. Lam, Asian Marine Biology 2, 1-10 (1985).
3. D.J.H. Phillips, Water Quality in Tolo Harbour and Channel: A Review (Report No. EPA/TM6/86, Environm. Protect. Dept., Hong Kong Government 1986).
4. B.S.S. Chan and I.J. Hodgkiss, Asian Marine Biology 4, 79-90 (1987).
5. P.R. Holmes, J. Inst. of Water and Environmental Management 2 (2), 171-179 (1988).
6. G W Y. Lam and K.C. Ho in : Red Tides – Biology, Environmental Science and Toxicology, T. Okaichi, D.M. Anderson and T. Nemato, eds. (Elsevier 1989) pp. 49 – 52.
7. J.S. Alabaster and R. Loyd, Water Criteria for Freshwater Fish (Butterworth, London 1980).
8. C.W.Y. Lam and K.C. Ho in: Red Tides-Biology, Environmental Science and Toxicology, T. Okaichi, D.M. Anderson and T. Nemato eds (Elsevier 1989) pp. 49-52.
9. P.G. Soulsby, M. Mollowney, G. Marsh and D. Lowthion, Water Science Technology 17, 745 – 756 (1984).
10. A.D. McVicar and R.H. Richards in: Aquarium Systems, A.D. Hawkins, ed. (Academic Press, London 1981) pp. 279 - 301.
11. M. Whitefield, J. Mar. Biol. Ass. U.K. 54, 565 – 580 (1974).
12. J.P. Wickens and M.M. Helm in: Aquarium Systems, A.D. Hawkins, ed. (Academic Press, London 1981) pp. 63 – 128.

RECENT DATA ON DIARRHETIC SHELLFISH POISONING IN FRANCE

CLAIRE MARCAILLOU-LE BAUT AND PIERRE MASSELIN
Institut Français de Recherche pour l'Exploitation de la Mer (IFREMER)
B.P. n° 1049, 44037 Nantes cedex 01 (France)

ABSTRACT

First of all, the validity of the mouse test method used in France is discussed versus other methods used in Europe. The french procedure is assumed to be suitable for the purpose of a monitoring network and to warrant a safe level of 0,5 M.U. g^{-1} hepatopancreas or less in shellfishes checked for sale.

Then, we have summarized here the results of a several years survey of toxin levels in shellfish using the data collected by the national program. It is thus possible to demonstrate from a one year monthly survey that no toxicity occurs in shellfish in winter and that during late April-summer period the toxic levels are coïncident with observations of *Dinophysis* cells in seawater. These results were corroborated by the HPLC analyses of okadaïc acid, though it was not used as a routine method.

INTRODUCTION

In France, the toxin-detection method performed routinely by the French national monitoring system (created in 1984) consists in a biological test on the adult mouse. This test, first described in Japan by Yasumoto *et al.* [7, 8,], has been slightly modified in France to meet the requirements of a systematic control in which toxicity is expressed differently than by the official Japanese method [6]. Nevertheless, it is possible to express these results relative to a generally accepted threshold and thus to consider the advantages and limitations of a test now used routinely in France.

In addition, to have a better understanding of the changes in and extent of toxicity, monthly (and sometimes semimonthly) monitoring was carried out during the critical period (May to August) in 1988. Quantitative evaluation of toxicity was regularly determined by two methods : the official Japanese biological test and high-performance liquid chromatography (HPLC) of okadaïc acid. Monitoring was carried out in 3 mussel production areas : Antifer in Normandy and Douarnenez Bay and Vilaine Bay in southern Brittany.

After describing the method used, we shall consider successively the validity of the biological test in routine practice and the results of toxicologic monitoring during a year's time.

METHODS

Mouse test

Three successive extractions from 30 g of mussel hepatopancreas were obtained using acetone. After evaporation of the solvent, the residue was resuspended in 6 ml of a Tween 60 solution. One milliliter of this extract was then injected intraperitoneally into 3 male mice (18 to 20 g).

Toxic Marine Phytoplankton
Edna Graneli et al., Editors

For monthly monitoring, the extract was diluted when the sample was very toxic until approximately 24 h lethality was obtained. This made it possible to express the toxicity of these samples in terms of a mouse unit comparable to that defined by the Japanese, *i.e.*, the minimum quantity of extract needed to kill 2 out of 3 mice in 24 h [7, 8].

Within the context of the monitoring system, the observation period for determining survival time was reduced to 5 h. In fact, the adult mouse test was calibrated on a test using young mice which showed the specific diarrhetic toxins involved. It was thus determined that survival of adult mice beyond 5 h corresponded to a toxicity lower than that revealed by the young mouse test [5]. Consequently, the survival of 3 mice for more than 5 h was considered to be a negative response ; however, as a precautionary measure, mice were kept under observation for at least 24 h.

Dinophysis sp. count

Water samples were obtained by means of a Niskin bottle. After fixation in formol or Lugol's solution and decantation in wells suitable for inverted microscopy, cells belonging to the genus *Dinophysis* were counted first and then individual species were determined. On the Antifer site, daily sampling enabled mean cell concentration to be evaluated over a 10 day period. For the Douarnenez Bay and Vilaine Bay sites, only isolated values were available.

Chemical analysis

The laboratory method described by Lee *et al.* [3] was used for chemical analysis.

Two points are especially noteworthy concerning the use of the proposed method. First, synthesis of the fluorescence reagent, which is unavailable on the French market, was performed in our laboratory ; secondly, the composition of the elution solvent for HPLC is acetonitrile 80 %/H_2O 20 %. Thanks to the use of an okadaïc acid standard (Fig. 1)

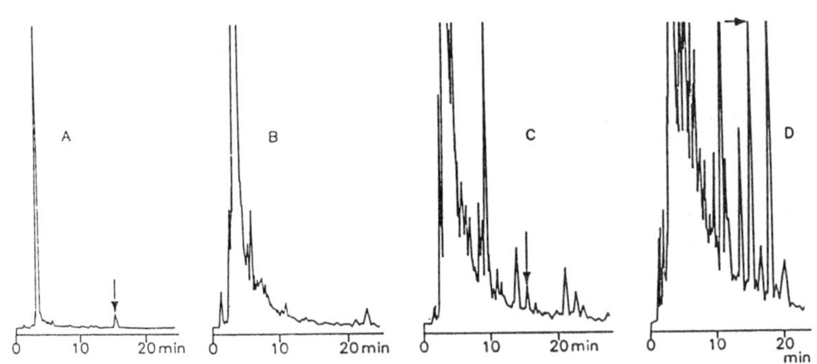

FIG. 1. Chromatograms of 9-Anthrylmethyl Esters of :
A - Okadaic Acid
B - Uncontaminated Mussels - Antifer - January 1988
C - Contaminated Mussels - Antifer - August 1988
D - Plankton sample -Antifer - August 1988.

kindly provided by Prof. Yasumoto, it was possible to perfect the assay. The linearity of detection response was checked on the 10 to 50 ng interval (r = 0.996), the detection threshold being 1 ng for pure acid. Analysis variability for a batch of hepatopancreas was 14 %. An example of the chromatogram obtained is given in figure 1.

RESULTS AND DISCUSSION

Evaluation of toxicity by the monitoring system

During the last two years (1987 and 1988), the monitoring system performed 750 biological tests. The breakdown of the results according to the length of survival time is shown in Table 1.

TABLE I. Percentage of results obtained as a function of survival time.

Survival time (T) (in hours)	Test results (in percentage)
T < 5	18 positive
5 < T < 24 h	20 negative
T > 24 h	62 negative

Positive results led to the closing of the areas indicated in figure 2.

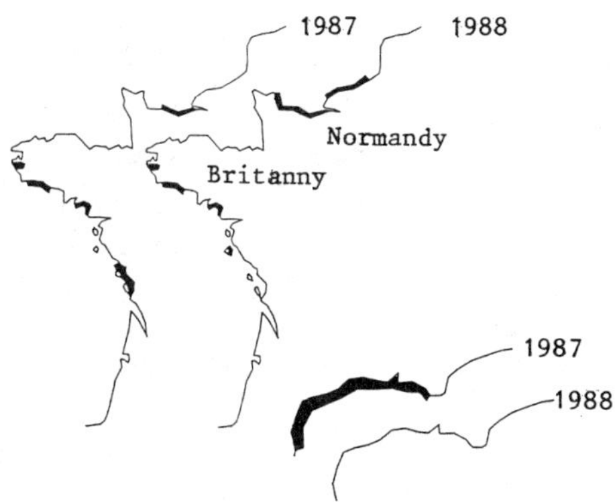

FIG. 2. Sites of closed areas in 1987 and 1988.

For samples considered nontoxic but having survival times of between 5 and 24 h (20 % of total results), it is debatable whether the designation "nontoxic" really applies. Two cases may be considered :

(1) Samples associated with survival time "close to 24 h". The quantity of toxic material injected (5 g) in this case corresponds to the minimum quantity defining the mouse unit (MU). The sample thus contains about 0.2 $MU.g^{-1}$ of hepatopancreas. The threshold in use in Japan and accepted by some European countries is : 0.5 $MU.g^{-1}$ of hepatopancreas.

(2) Samples associated with survival times of more than 5 h but much shorter than 24 h. To be strictly exact, the extract should be diluted to quantify the toxicity in mouse units. Previous

studies have shown that these survival times correspond to a toxicity close to the sublethal level [5].

A half-dilution of the extract would thus result either in survival or lethality at 24 h or later, which would give the sample a toxicity of around 0.4 MU.g^{-1} of hepatopancreas.

The threshold can be situated approximately at 5 h relative to a quantity of okadaïc acid. In effect, a relation between the survival time of 3 mice and okadaïc acid concentration was determined for observations of lethality time greater than 5 h (Fig. 3). This relation shows that less than 4 µg of okadaïc acid are necessary to observe a lethality time greater than 5 hours. However, expression of toxicity in equivalents of okadaïc acid should be employed with caution. Our preliminary results obtained during monthly monitoring indicate that this toxin alone does not account for total toxicity in certain cases.

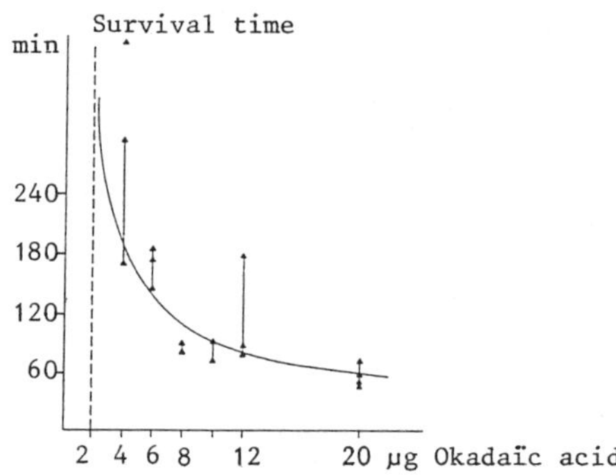

FIG. 3. Dose-effect relation in adult mice of okadaic acid.

Toxicologic shellfish monitoring during a year's time

From February to December 1988, mussel and water samples were obtained monthly (and more frequently in June and July)) in 3 coastal sites. Each shellfish sample was subjected to a mouse test, with evaluation in mouse units and an okadaïc acid assay. Cells of the genus *Dinophysis* were counted in each water sample.

This monitoring confirmed that from September to April mussels were not toxic and that no okadaïc acid could be found, while few or no *Dinophysis* cells were detected in corresponding water samples (Fig. 1). Toxicity appeared at the beginning of May or later, depending on the site, after a more or less rapid increase in the number of toxic algal cells.

For the Antifer site (Fig. 4), a progression in the count, associated with an increase in toxicity, could be noted from the end of May to July, even though there was considerable variation in *Dinophysis* concentrations in the water. Contrary to what might have been expected, this tendency was not confirmed by chemical analysis. In effect, analysis of a sample from this area sent to Japan in 1984 had shown an 80 % concentration of okadaïc acid [2].

For Douarnenez Bay and Vilaine Bay, the lesser frequency of water sampling provided a rough estimation of changes in *Dinophysis* counts during the summer period, although it was evident that concentrations were markedly lower than those observed at Antifer (Fig. 5). However, toxicity levels for Douarnenez Bay were higher than 2 MU.g^{-1}, that is, more than double the toxicity measured at Antifer. In Vilaine Bay, maximum toxicity occurred before the cell peak was reached. For both sites, comparison of

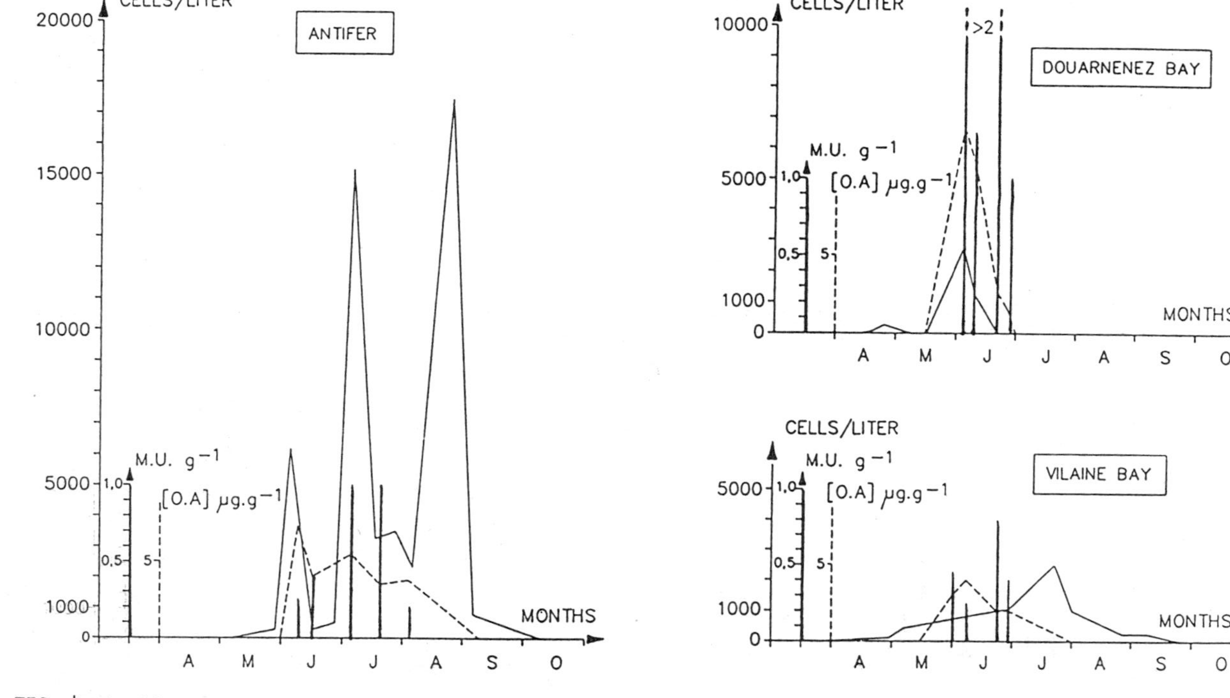

FIG. 4. Antifer site : Toxicity monitoring expressed in mouse units (diagram) and okadaic acid concentration (- - -) parallel to changes in *Dinophysis* concentrations in water.

FIG. 5. Vilaine Bay and Douarnenez Bay : Toxicity monitoring expressed in mouse units (diagram) and okadaic acid concentration (- - -) parallel changes in *Dinophysis* concentrations in water.

overall toxicity and okadaïc acid concentration showed that there was no coherence between these two measurements : okadaïc acid concentration could not account for the toxicity level.

These observations obtained in three different sites raise the problem of the type of toxins involved. The presence of other toxins than okadaïc acid could be one explanation, but the role played by the different *Dinophysis* species in toxicity cannot be excluded. In fact, the dominant species at Antifer was *D. acuminata*. During a particularly large proliferation of this species, we were able to obtain an extract from a phytoplankton concentrate containing 10 species of dinoflagellates, with *D. acuminata* representing 7 % of the total. HPLC analysis was carried out, and the sample was found to have 4.8 µg of okadaïc acid per 500,000 cells (*i.e.*, 10 pg/cell), which was markedly higher than in previous analyses [4] and confirmed the majority presence of this toxin in mussels. However, for the other two areas the species involved were different. *D. sacculus* in Douarnenez Bay and *D. norvegica, D. acuminata* and *D. sacculus* in Vilaine Bay. Unfortunately, the presence of many other species and the fact that the cell concentration of toxic species was too low did not allow us to obtain a representative sample for analysis.

CONCLUSION

The methodology described here for detection of shellfish toxicity is generally satisfactory for a monitoring system. It is relatively easy and rapid to use for purposes of consumer protection. Nonetheless, it would be advisable to provide correspondences with other methods to be able to compare results. Chemical analysis, though far more satisfactory than the biological test, cannot be used routinely at the present time. Its development is hindered by the unavailability of toxin standards. Moreover, toxicologic monitoring has shown that the okadaïc acid assay alone is not a sufficient method.

REFERENCES

1. Y. Hamano, Y. Kinoshita, and T. Yasumoto in : Toxic Dinoflagellates, D. Anderson, A. White and D. Baden, eds (Elsevier, Amsterdam 1985) pp. 383-388.
2. M. Kumagai, T. Yanagi, M. Murata, T. Yasumoto, M. Kat, P. Lassus, and J. A. Rodriguez-Vasquez, Agric. Biol. Chem. 50 (11), 2853-2857 (1986).
3. J.S. Lee, T. Yanagi, R. Kenma, and T. Yasumoto, Agric. Biol. Chem. 51 (3), 877-881 (1987).
4. J.S. Lee, T. Igarashi, S. Fraga, E. Dahl, P. Hovgaard, and T. Yasumoto, J. Appl. Physiol.
5. C. Marcaillou-Le Baut, D. Lucas, and L. Le Déan in : Toxic Dinoflagellates, D. Anderson, A. White and D. Baden, eds (Elsevier, Amsterdam 1985) pp. 485-488.
6. B. Underdal, M. Yndestad, and T. Aune in : Toxic Dinoflagellates, D. Anderson, A. White and D. Baden, eds (Elsevier Amsterdam 1985) pp. 489-494.
7. T. Yasumoto, Y. oshima, and M. Sugawara, Y. Fukuyo, H. Oguri, T. Igarashi, and N. Fujita, Bull. Japan. Soc. Sci. Fish 46 (11), 1405-1411 (1980).
8. T. Yasumoto, Y. Oshima, W. Sugawara, Y. Fukuyo, H. Oguri, T. Igarashi, and N. Fujita, Bull. Japan. Soc. Sci. Fish 46 (11), 1405-1411 (1980).

FIRST OCCURRENCE OF ALEXANDRIUM CATENELLA IN FUNKA BAY, HOKKAIDO, ALONG WITH ITS UNIQUE TOXIN COMPOSITION

TAMAO NOGUCHI,* MANABU ASAKAWA,* OSAMU ARAKAWA,* YASUWO FUKUYO,* SACHIO NISHIO,** KENJI TANNO,*** AND KANEHISA HASHIMOTO****
*Department of Fisheries, Faculty of Agriculture, University of Tokyo, Bunkyo, Tokyo 113; **Shikoku Women's University, Tokushima 771-11; ***Japan Food Research Laboratories, Shibuya, Tokyo 151; ****Food Science Laboratory, Faculty of Education, Ibaraki University, Mito 310, Japan

ABSTRACT

In late autumn of 1988, the scallop Patinopecten yessoensis was suddenly infested with paralytic shellfish poison (PSP) in Funka Bay, causing a serious damage to fisheries industry. In this bay, scallops have extensively been infested with Alexandrium tamarensis in late spring to early summer in almost every year, but never in autumn or winter.

The plankton was isolated and identified as A. catenella, based on morphological characters. Dinoflagellates cultured in ES medium at 15°C for 21 days showed that $2.5x10^4$ cells were roughly equivalent to one mouse unit. HPLC analysis of the extract demonstrated that the toxin was composed mainly of neosaxitoxin (neoSTX) and protogonyautoxin 2(PX_2 or GTX_8). The toxin was partially purified from the infested scallop digestive gland by column chromatography on Bio-Gel P-2. It exhibited a more complicated composition, with GTX_1 and GTX_3 as the major components, and PX_2, GTX_4 and neoSTX as the minor, suggesting a bioconversion of PSP in scallops.

INTRODUCTION

Funka Bay, Hokkaido, is one of the representative scallop culture areas in Japan. The first infestation of paralytic shellfish poison (PSP) in bivalves occurred in this region in June 1978. Several people were poisoned by ingesting mussels collected from this bay, and one death was recorded. The phytoplankton species was determined to be Alexandrium tamarensis [1]. Since then, this species has infested scallops in this bay from late spring to early summer in almost every year, posing a tremendous damage to fisheries and related industries.

In 1988, scallops became toxic only in April alone in this bay. Unlike in previous years, scallops became toxic again in late autumn, with the highest toxicity score exceeding 400 MU/g digestive gland.

The purposes of the present study were to isolate and identify the responsible phytoplankton, and to analyze the toxin composition.

MATERIALS and METHODS

Scallop

Five individual specimens of the scallop Patinopecten yessoensis were collected from the upper, middle and lower layers of a floating scallop-culture raft, at Otoshibe and Shikabe in Funka Bay in August through December 1988 (TABLE I). These specimens were immediately transported to the Marine

Toxic Marine Phytoplankton
Edna Graneli et al., Editors

Biochemistry Laboratory, University of Tokyo, on ice, and the digestive glands were excised.

Plankton

During the above-mentioned period, attempts were made to isolate the dominant motile phytoplankton strains from the seawater near the three layers of the scallop raft. A total of seven strains were isolated and cultivated successfully (TABLE II).

The medium used was modified ES medium [2](12 mg of $NaNO_3$ and 0.5 mg of K_2HPO_4, 10 µg of vitamin B_1, 1 µg of B_{12}, 0.1 µg of biotin, 26 µg of EDTA-Fe, 33 µg of EDTA-Mn, 100 mg of Tris and 5 ml soil extract in 100 ml of seawater), under either 12 L/12 D or 24 L/ 0 D, at 15°C for 21 days. After cultivation, cells were harvested by centrifuging at 2,000 x g for 20 min, and used for toxicity and toxin composition assays.

As described below, only one of the seven strains was found to be toxic. This strain was designated No.3151 for convenience.

METHODS of ANALYSIS

Toxicity

Digestive glands were assayed for PSP toxicity by an official Japanese method using mouse bioassay [3]. In assay for toxicity of phytoplankton, cells were ultrasonicated in an equal volume of 0.1 M acetic acid. The lyzate was centrifuged at 2,000 x g for 20 min, and the supernatant obtained was assayed for toxicity similarly.

Toxin composition

Toxins were partially purified from toxic digestive glands and used for toxin composition analysis, essentially according to the procedure previously reported [4]. Three volumes of 80% ethanol (pH 2) was added to the digestive glands and homogenized. The homogenate was centrifuged at 2,000 x g for 20 min. The pellet was similarly treated two more times. The supernatants were combined, concentrated under reduced pressure, and defatted with dichloromethane. The water layer obtained was concentrated under reduced pressure, and ultrafiltered through a Diaflo YM-2 membrane (Amicon) to eliminate substances of more than 1,000-dalton. The filtrate was loaded onto a Bio-Gel P-2 column (2.5 x 94 cm), which was washed with water, followed by 0.1 M acetic acid. Toxic fractions, as monitored by the mouse bioassay, were combined, concentrated and freeze-dried. The toxin thus partially purified was dissolved in a small amount of water and examined for toxin composition by the HPLC method of Nagashima *et al.* [5], in which a reversed-phase silica YMC-gel ODS-5 column was used, with heptanesulfonate-containing phosphate buffer/methanol mixtures as mobile phase. The toxins separated were finally detected by fluorescence, following oxidation by periodate reagent. A toxin extract was prepared from *A. catenella* strain No.3151 and analyzed for composition.

RESULTS and DISCUSSION

Toxicity of scallop

As seen in TABLE I, scallops collected from Otoshibe in August and September showed toxicity scores less than 6 MU/g digestive gland, but the toxicity tended to increase in October. November specimens exhibited higher scores, 101, 69 and 46 MU/g digestive gland at the upper, middle and lower layers, respectively. In November, Shikabe specimens showed as high a toxicity as 432 MU/g digestive gland. After ten days, however, the toxicity sharply decreased down to 133-140 MU/g.

TABLE I. Toxicity of the digestive gland of scallop *Patinopecten yessoensis* collected from two sites in Funka Bay, Hokkaido

Site and date of collection		Toxicity in MU/g (mean ± S.D.)		
		Upper layer	Middle layer	Lower layer
Otoshibe	Aug.12, 1988	5.5 ± 0.1	5.3 ± 0.3	5.4 ± 0.3
	Sept.20, 1988	< 2	3.9 ± 1.9	5.3 ± 0.2
	Oct.20, 1988	6.7 ± 5.5	17 ± 16	13 ± 5.8
	Nov.17, 1988	101 ± 12.2	69 ± 4.6	46 ± 3.2
Shikabe	Nov.20, 1988		432 ± 73.7	
	Dec. 1, 1988	140 ± 26.5	133 ± 25.2	133 ± 5.8

Toxicity and identification of responsible plankton

The seawater (*ca*. 3 l) was monthly collected at Otoshibe, at each of three depths corresponding to the three layers of the scallop raft, and screened for phytoplankton. No plankton was isolated from the August seawater. Two plankton cells were isolated from the lower layer seawater in September, but not cultivated successfully.

In October, the seawater from the middle layer was found to contain scores of phytoplankton cells. A dominant, motile strain (designated No. 3151) was isolated and cultivated successfully. In November, one strain each was isolated from the upper and middle layers, and cultivated (designated No. 4001 and 4151).

TABLE II. Details of plankton strains isolated from Funka Bay

Strain No.	Site and date of collection			Depth(m)	Days of culture
3151	Otoshibe	Oct.27,	1988	15	21~35
4001	〃	Nov.19,	〃	0	25
4151	〃	〃	〃	15	25
5001	〃	Dec. 6,	〃	0	25
5002	Shikabe	〃	〃	0	25
5201	〃	〃	〃	20	25
5202	〃	〃	〃	20	25

On the other hand, two strains (No. 5001 and 5002) were isolated and cultivated from the upper layer, and another two strains (No. 5201 and 5202) from the middle layer of seawater at Shikabe in December 1988.

Details of these strains are summarized in TABLE II. Among these strains, only strain 3151 was toxic, with a toxicity of $2.5x10^4$cells/MU. This value is in a toxicity range of A. catenella and A. tamarensis so far reported, $2.5-140x10^3$ cells/MU [6,7].

Strain 3151 was unambiguously identified as A. catenella, on the basis of morphological characters such as the shape of thecal plates [8].

Toxin composition of scallop

HPLC profiles of the partially purified toxins from Otoshibe and Shikabe scallops are shown in FIG. 1. The gonyautoxin (GTX) fraction from Otoshibe scallops gave rise to a large peak regarded as protogonyautoxin (PX) near the void

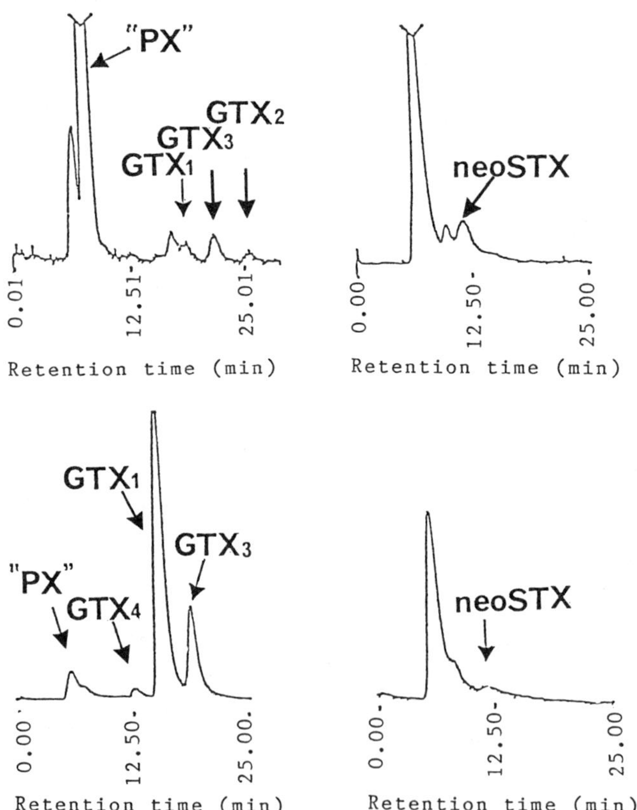

FIG. 1. HPLC of the toxin from scallops collected at Otoshibe (above) and Shikabe(bottom).

volume, along with small peaks of gonyautoxin 1-3 (GTX_{1-3}). The saxitoxin (STX) fraction showed three peaks, one was neoSTX. In contrast, the GTX fraction from Shikabe scallops elicited large peaks of GTX_1 and GTX_3, along with small ones of GTX_4 and "PX". The STX fraction gave rise to a small peak of neoSTX. Each component in a sample was identified, referring to the retention time of authentic toxins run before and after HPLC analysis of the test sample or by co-chromatography with authentic toxins.

Toxin composition in *A. catenella*

As FIG. 2 shows, GTX fraction from *A. catenella* strain 3151 which was cultivated in ES medium, was characterized by the presence of "PX" along with GTX_1, GTX_3 and GTX_4 (12L / 12D) whereas the STX fraction by neoSTX. Toxicity ratios of "PX" : GTXs : neoSTX were roughly 1:2:20. On hydrolysis with 0.1 N HCl for 10 min, "PX" peak completely disappeared and instead a sharp peak of GTX_3 appeared [9], indicating that "PX" was PX_2 [5](FIG.3).

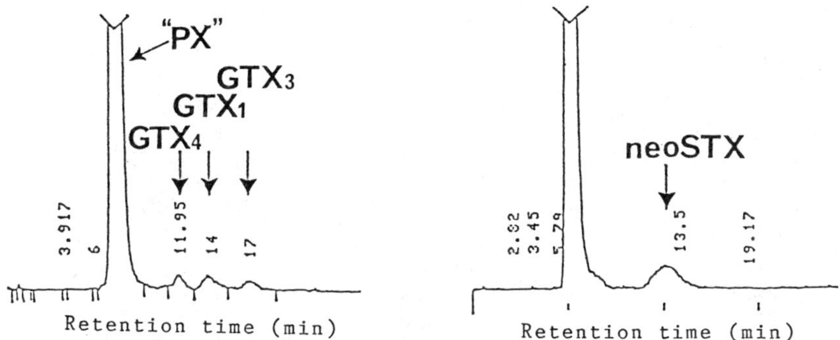

FIG. 2. HPLC of the toxin of *A. catenella* strain 3151 which was cultivated in ES medium under 12 L/12 D. Refer to the text for further details.

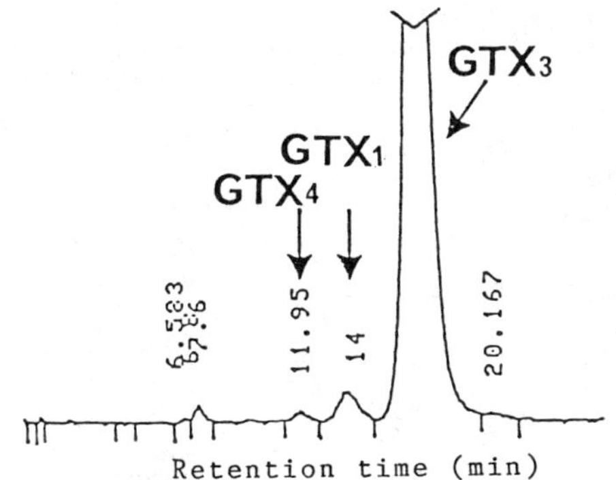

FIG. 3. HPLC of the toxin of *A. catenella* strain 3151 after acid hydrolysis. The same toxin as used in FIG.2 was hydrolyzed in 0.1N HCl for 10 min, and subjcted to HPLC. Refer to the text for further details.

As described above, Otoshibe scallops contained a component, supposedly PX_2, as the major toxin (FIG. 1), reminding us of the toxin composition of *A. catenella* which was cultured in ES medium. On the other hand, the toxin of Shikabe scallops was featured by the presence of GTX_{1-4} (FIG. 1), suggesting an *in vivo* conversion of PSP after ingestion.

Several investigators examined the toxin composition of *A. catenella*, whether of wild or culture origin, detecting GTX_{1-4} as the major components, and neoSTX as the minor. Our strain 3151, which contained PX_2 and neoSTX as the major toxins, was considered to be unique in this respect.

Prior to this study, it was exclusively *A. tamarensis* that had infested bivalves in Funka Bay. The recent occurrence of *A. catenella*, may be regarded as an example of surprise infestation of PSP-producing species to new areas.

SUMMARY

1. In Funka Bay, Hokkaido, the scallop *Patinopecten yessoensis* was intoxicated with PSP unexpectedly in late autumn of 1988.
2. The responsible plankton was identified as *Alexandrium catenella*, whose lethal potency was estimated to be 2.5×10^4 cells / mouse unit.
3. HPLC analysis showed that the toxin of *A. catenella* was composed mainly of neoSTX and PX_2 (GTX_8), and that of infested scallops mainly of GTX_1 and GTX_3, suggesting a bioconversion of PSP components in this bivalve.

ACKNOWLEDGEMENT

The expenses of the present study were supported in part by a research fund from the Ministry of Agriculture, Forestry and Fisheries of Japan.

REFERENCES

1. Y. Nishihama in: Toxic Dinoflagellate - Implication in Shellfish Poisoning, Fisheries Science Series No. 56, Y. Fukuyo, ed. (Koseisha-Koseikaku, Tokyo 1985) pp. 47-58.
2. T. Okaichi, S. Nishio and Y. Imatomi in: Toxic Phytoplankton Occurrence, Mode of Action and Toxins, Fisheries Science Series No.42, Japn. Soc. Sci. Fish., ed. (Koseisha-Koseikaku, Tokyo 1982) pp. 22-34.
3. T. Kawabata: "Food Hygiene Examination Manual", Vol. 2, Japn. Food Hyg. Assoc., Tokyo, 1978, pp. 240-244.
4. D.-F. Hwang, T. Noguchi, Y. Nagashima, I.-C. Liao and K. Hashimoto: Nippon Suisan Gakkaishi, **53**, 623-626 (1987).
5. Y. Nagashima, Y. Sato, T. Noguchi, Y. Fuchi, K. Hayashi and K. Hashimoto: Mar. Biol., **98**, 243-246 (1988).
6. Y. Oshima in: Toxic Phytoplankton - Occurrence, Mode of Action, and Toxins, Fisheries Science Series, No. 42, Japn. Soc. Sci. Fish., ed. (Koseisha-Koseikaku, Tokyo 1982) pp. 73-87.
7. T. Noguchi, J. Maruyama, Y. Onoue, K. Hashimoto and T. Ikeda: Nippon Suisan Gakkaishi, **49**, 499 (1983).
8. Y. Fukuyo: Bull. Mar. Sci., **37**, 529-537 (1987).
9. N.H. Proctor, Ph.D.Thesis, Univ. of California, San Francisco (1973).

BREVETOXIN CONTAMINATION OF MERCENARIA MERCENARIA AND CRASSOSTREA VIRGINICA: A MANAGEMENT ISSUE

PATRICIA A. TESTER[1] AND PATRICIA K. FOWLER[2]
[1]National Marine Fisheries Service, NOAA, Beaufort, NC 28516 USA
[2]North Carolina Department of Human Resources, Shellfish Sanitation Program, Morehead City, NC 28557 USA

ABSTRACT

The first red tide, Ptychodiscus brevis, bloom ever recorded along the North Carolina coast was both massive and persistent. Between 2 November 1987 and 21 January 1988, 145,280 hectares of shellfish harvesting areas were closed. This caused severe economic loss to coastal communities (>$24.7 million) and was especially devastating to the clam (Mercenaria mercenaria) fishery. We examine factors affecting the toxicity of Crassostrea virginica and M. mercenaria in the field and comment on existing guidelines which govern the reopening of the shellfish harvesting areas after a P. brevis bloom.

INTRODUCTION

On 2 November 1987, P. brevis (Davis) Steidinger, 1979 (formerly Gymnodinium breve Davis) cells were identified from water samples taken off Emerald Isle, NC (34°41'N, 77°00'W) after people at the beach complained of eye and respiratory irritation. This was the first recorded occurrence of P. brevis north of Jacksonville, FL and the initial water samples containing P. brevis along the North Carolina coast were collected 29 to 39 days after a bloom was reported between Sarasota and Charlotte Harbor, FL (10-29 September 1987). The North Carolina red tide occurred 13 days after a strong shoreward intrusion of Gulf Stream water onto the continental shelf between Capes Hatteras and Lookout (Fig. 1, inset). Tester et al. [1] suggest that the Gulf Loop Current-Florida Current-Gulf Stream system was the transport mechanism for these cells. By the time reports of respiratory irritation were received from the North Carolina coast on 29 and 30 October, 1987, there were 6 million P. brevis cells L^{-1} in nearshore water samples. The harvesting of oysters (C. virginica), which had started in all parts of the State by 15 October, was halted when affected waters were closed on 2 November 1987. Clam (M. mercenaria) harvesting in North Carolina is allowed all year but most commercial harvesting occurs after mid-November [2]. During the fall of 1987, clam harvesting in the affected areas was delayed until shellfish waters were reopened between 19 February and 6 May 1988. A total of 145,280 hectares of approved shellfish harvesting waters were closed during this red tide and this affected 50% of the oyster and 98% of the clam harvesting areas [2]. The dollar value of the North Carolina shellfish resource was reduced by almost 50% from the previous season (Nov.-Dec. 1986 and Jan.-Mar. 1987).

P. brevis has a requirement for high salinity (>24 ppt), and so it is not frequently found in Florida bays where the majority of clam and oyster resources exist. For example, the harvesting of oysters from Apalachicola Bay (about 80% of the Florida harvest) has never been significantly impacted by a red tide [3,4]. Perhaps because of this, there is little information on shellfish depuration rates of brevetoxins, and especially on depuration rates that would apply to winter field conditions off the North Carolina coast. This study documents the occurrence and effects of the P. brevis bloom in North Carolina and evaluates the impact of current guidelines related to toxicity and depuration of this dinoflagellate. Based on our findings, we recommend a review of the US Food and Drug Administration's (FDA) interpretation of the American Public Health Association (APHA) guidelines concerning the acceptable amount of NSP toxin allowed in harvestable shellfish.

Toxic Marine Phytoplankton
Edna Graneli et al., Editors

METHODS

In accordance with the recommendations of the National Shellfish Sanitation Program Manual of Operations [5] and procedures followed by the Florida Department of Natural Resources (FDNR) [6], North Carolina shellfish waters were closed to the harvesting of shellfish when P. brevis cells exceeded 5,000 L^{-1}. Nearshore, surface water samples from selected locations were taken either daily or every other day during the course of this bloom from 2 November 1987 to 22 February 1988. Overall 117 stations were sampled for P. brevis. Water samples were collected from open water areas rather than from shore and were preserved in the field using Utermöhl's solution [7]. Cell counts were made using an inverted microscope [8] on each of the 990 water samples collected. Shellfish samples were taken from selected sites within approved harvesting areas.

P. brevis toxins (brevetoxins) are detected in shellfish meats by a standardized mouse bioassay [9]; all reopenings of shellfish harvesting areas after the bloom were based on bioassay test results from the Environmental Sciences Branch of the North Carolina Division of Health Services (NCDHS). Toxicity is expressed in mouse units (MU), one MU being that amount of crude toxin contained in 100 g shellfish tissue that, on average, will kill 50% of the test mice (20 g weight) in 15.5 hours. The standard observation time is 24 hours. The bioassay is based on the relationship of dose to death time of mice injected intraperitoneally with lipid-soluble crude toxin residues extracted from shellfish with diethyl ether [10]. The assay's lowest, reproducible sensitivity is 10 MU 100 g^{-1} of shellfish meat. If no mice died within 24 hours the results were reported as ND (toxin not detected). Current guidelines consider any detectable level of NSP toxin, using the standard bioassay, as potentially unsafe for human consumption.

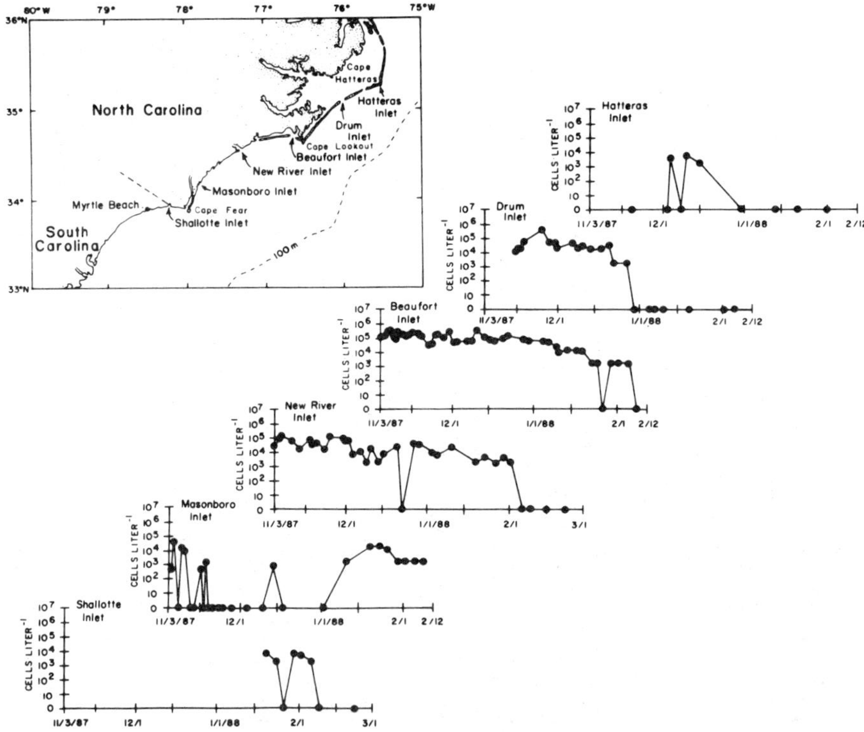

FIG. 1. Numbers of Ptychodiscus brevis cells from samples taken 4 November 1987 through 22 February 1988, from selected locations along the North Carolina coast.

RESULTS AND DISCUSSION

P. brevis cells are both motile and positively phototactic, so they concentrate near the surface during the day [11]. Consequently, physical factors such as insolation, wind, currents and tides affected the distribution of this bloom. During the early stages, cell concentrations from nearshore were highest in waters immediately adjacent to Beaufort Inlet. (Fig. 1). As would be expected, the shellfish meats taken from areas with the highest and most persistent P. brevis cell counts consistently exhibited the highest toxicity. All samples with >30 MU were collected within about 75 km of Beaufort Inlet. As P. brevis cells were moved into different areas along the North Carolina coast by winds and water movements, the cell concentrations fluctuated with environmental conditions. Because of this, sampling efforts were extended along the coast and additional closures of shellfishing waters were done.

The initial closure of shellfish waters on 2 November 1987, included a total of 23,857 hectares. By 5 November waters were closed southward to the Cape Fear area. From 19-21 January 1988 a ban on shellfish harvesting from Cape Fear south to the South Carolina state line was implemented bringing the total area closed to 145,280 hectares. The last samples with motile P. brevis cells were collected from North Carolina nearshore waters on 11 February, and by mid-February, the bloom had dissipated off Myrtle Beach, SC. Morris and Campbell conducted the epidemiological investigation of neuro-toxic shellfish poisoning (NSP) in North Carolina and found that 48 persons became ill from consumption of shellfish contaminated by the P. brevis toxin [12]. Thirty-five of the 48 cases (73%) occurred before the first ban on shellfish harvesting was instituted on 2 November, and all but six cases occurred between 27 October to 5 November 1987. One other case occurred in December and was traced to shellfish harvested on 31 October and frozen until the time of the meal.

Shellfish samples were not collected or assayed routinely when the P. brevis cell counts in the area fell below 5,000 cells L^{-1} and remained so for at least two weeks. Although routine analyses were not performed until the late phase of this bloom, some oysters were collected near Bogue Inlet during the early weeks of the bloom and were assayed by the FDA. They contained as much as 180 MU 100 g^{-1} [13]. In comparison the highest level of brevetoxin ever reported was 550 MU 100 g^{-1} determined by Cummins et al. for coquinas (Donax variabilis) in the surf zone [14]. In that same study, P. brevis cell counts of 820,000 L^{-1} corresponded to toxicity levels of 270 MU 100 g^{-1} in M. campechiensis and 75 MU 100 g^{-1} in C. virginica obtained concurrently. McFarren et al. [10] suggest that a total of 400-500 MU of brevetoxins will produce a mild form of illness in humans and that as little as 50-80 MU can produce some symptoms of poisoning.

Generally, both accumulation and depuration of brevetoxins will depend on the species of shellfish, the persistence of the bloom, and the temperature of the surrounding water. Routine bioassays for brevetoxins in North Carolina shellfish started in January 1988. For the 163 shellfish samples assayed (N=775 mice injected), during the NC bloom, 39 were >20 MU 100 g^{-1} (range=21 to 70), 43 were <20 MU 100 g^{-1} (range=11 to 20), 46 were <10 MU 100 g^{-1}, and 35 were listed as nondetectable (ND). The bioassay data also suggested that the toxicity of clams collected from the same area and about the same time were not as toxic as oysters (Table I). Of eight specific comparisons, clams were less in six and equally toxic in two.

All openings of shellfish waters prior to 31 March 1988, were for the harvesting of clams only. The first of these openings occurred on 19 February, some 3.5 months after the red tide first appeared. Results of shellfish bioassays from this area were positive for oysters but clams had no detectable toxin. The shellfish areas near Beaufort Inlet, where P. brevis cell counts had been highest, were the last areas to be reopened to harvesting (between 8 April and 6 May 1988).

Economic Impact and Recommended Changes to APHA Guidelines

The impact of this toxic dinoflagellate bloom had its greatest effect on the clam fishermen. Calm landings were less than half of the previous season's and this represented a reduction of almost $2 million dockside. The management decisions to close and reopen shellfish harvesting areas made during this first red tide bloom in North Carolina were conservative ones, as was appropriate. Those decisions were based upon the FDA's interpretation of APHA procedural guidelines [15]. Current guidelines consider any detectable level of NSP toxin per 100 g shellfish meats as potentially unsafe for human consumption and they require a 24 hour observation period for the injected mice. A request has been made to the FDA to change the interpretation of the APHA guidelines for the mouse bioassay for determination of brevetoxins in shellfish [16]. The recommended change is for a 6 to 15.5 hour observation period (6 hours continuous observation), corresponding to <20 MU 100 g^{-1} shellfish meat. Several lines of evidence argue in favor of this change.

Table I. Direct comparisons of toxicity of M. mercenaria and C. virginica collected from the same areas at or about the same time along the North Carolina coast. Toxicity is the amount of brevetoxins and is measured as a mouse unit (MU).

Distance from Beaufort Inlet Date	Sample	Ave MU 100 g^{-1}	Range	# Dead # Tested	Time to death (h) <6	6-24	>24
74 km North							
2-02-88	Clams	ND	-	0/5			
2-02-88	Oysters	<20	<10-<20	5/5		5	
63 km North							
2-02-88	Clams	ND	-	0/5			
2-02-88	Oysters	<10	<10	2/5		2	
45 km North							
1-26-88	Clams	ND	-	0/5			
1-26-88	Oysters	24	<10-100	2/5	1	1	
2-16-88	Clams	ND	-	0/5			
2-10-88	Oysters	<20	<10-<20	3/5		3	
20 km North							
1-26-88	Clams	<20	<10-40	3/5	1	2	
1-26-88	Oysters	<20	<10-22	4/4	2	2	
2-16-88	Clams	<10	<10-43	1/5	1		
2-10-88	Oysters	<20	<20-23	5/5	1	4	
15 km North							
2-23-88	Clams	12	<10-34	2/5	1		1
2-09-88	Oysters	33	32-33	5/5	5		
69 km South							
2-16-88	Clams	30	25-34	5/5	3	2	
2-16-88	Oysters	29	<20-38	5/5	5		

McFarren et al. [10 conducted the original research upon which the APHA guidelines for detection NSP toxins was based. According to them [10, page 114],

> "With the test proposed herein, 10 mouse units is the least amount that can be detected, and under ordinary circumstances, only 20 to 30 mouse units will be detected, **since, for the sake of expediency and accuracy, the mice will not be watched for longer than 3 to 6 hr.** Under these circumstances, if any toxin is detected by mouse bioassay, then presumably, the shellfish should be considered unsafe for human consumption, and the area should be closed to harvesting." (emphasis added)

We agree. If mouse deaths occur within the 6-15.5 hour observation period the shellfish harvesting areas should remain closed; if no mouse deaths occur within the observation period the area should be considered safe. Serious problems with false positive results can occur with the long observation period. Shellfish fatty acids are lipid-soluble, bioactive compounds which can be extracted during the bioassay procedure and, which can kill mice within 24 hours [17].

Another rationale exists for the proposed change in the guidelines. Saxitoxins, the toxins responsible for paralytic shellfish poisoning (PSP), are potent and can be fatal. Current guidelines allow shellfish containing measurable amounts of PSP toxin (detectable in 6 hours with standard mouse bioassay) to be marketed [9,18], yet "no detectable" amount of NSP is permitted. This represents an inconsistency in the application of the results of the mouse bioassays for NSP and PSP. In all areas of North Carolina where NSP could be detected the shellfish harvesting waters remained closed.

Since 1970, the State of Florida has successfully used the criteria of <20 MU 100 g^{-1} and 6 hour continuous observation period to reopen shellfish harvesting waters after a P. brevis bloom. If these same criteria, now used by the State of Florida, had been applied to the reopening of the North

Carolina shellfish beds, the economic impact to the fishery would have been considerably less. We urge the reevaluation of the FDA interpretation of APHA requirements and consideration of the newly suggested changes. These changes could greatly mitigate the effects of any subsequent P. brevis blooms beyond Florida waters.

ACKNOWLEDGEMENTS

We would like to thank J. Neal for the mouse bioassay results. Drs. Morris and Campbell kindly allowed us access to their unpublished epidemiological report. The staff of the NC Shellfish Sanitation Program deserves special thanks (most especially D. Mason) and The North Carolina Division of Marine Fisheries is credited for collecting water and shellfish samples. We thank B. Watkins for reviewing the manuscript and J. Turner for drafting the figure. PKF wishes to thank Dr. R. Levine, C. Hoke, R. Rowe, S. Covil and R. Benton for their support and approval of travel funds to participate in this conference.

REFERENCES

1. P.A. Tester, P.K. Fowler, and J.T. Turner in: Novel Phytoplankton Blooms: Causes and Impacts of Recurrent Brown Tides and Other Unusual Blooms, E.M. Cosper, E.J. Carpenter and V.M. Bricelj eds. (Springer-Verlag, Berlin in press).
2. NC Div. Mar. Fish., Morehead City, NC, unpublished landing statistics.
3. R.A. Lutz and L.S. Incze in: Toxic Dinoflagellate Blooms, D.L. Taylor and H.H. Seliger eds. (Elsevier, Amsterdam 1979) pp. 476-483.
4. S.M. Ray and D.V. Aldrich in: Animal Toxins, F.E. Russell and P.R. Saunders eds. (Pergamon Press, NY 1967) pp. 75-83.
5. National Shellfish Sanitation Program in: Manual of Operations, Part 1 (U.S. Dept. Health Human Serv., FDA, Wash., DC, 1986). pp. C20-C25.
6. Florida Department of Natural Resources, Contingency Plan for Control of Shellfish Potentially Contaminated by Marine Biotoxins (Bureau Mar. Res., St. Petersburg, FL 1985) pp. 1-10.
7. R.R.L. Guillard in: Handbook of Phycological Methods, J.R. Stein ed. (Cambridge Univ. Press 1973) pp. 289-311.
8. J.W.C. Lund, C. Kipling, and D.E. LeCren, Hydrobiologia 11,143-170 (1979).
9. J.E. Delaney in: Laboratory Procedures for the Examination of Seawater & Shellfish, A.E. Greenberg and D.A. Hunt eds. (Am. Publ. Health Assoc., Wash., DC 1985) pp. 64-80.
10. E.B. McFarren, H.Tanabe, F.J.Silva, W.B. Wilson, J.E. Campbell, and K.H. Lewis, Toxicon 3,111-123 (1965).
11. K.A. Steidinger and RD.M. Ingle, Environ. Lett. 3,271-277 (1972).
12. P.M. Morris and D.S. Campbell, NC Dept. Human Res., Env. Epidemiology Br., Raleigh, NC, unpublished report.
13. S. Hall, USFDA, Wash., DC, personal communication.
14. J.M. Cummins, A.C. Jones, and A.A. Stevens, Trans. Am. Fish. Soc. 100,112-116 (1971).
15. J.J. Miescier, USFDA, Wash., DC, personal communication.
16. K.A. Steidinger, Florida Dept. Nat. Res., St. Petersburg, Fl, personal communication.
17. J.S. Lee, Agri. Biol. Chem. 51,877-881 (1987).
18. D.G. Baden, T.J. Mende, M. A. Poli, and R.E. Block in: Seafood Toxins, E.P. Ragelis ed. (Am. Chem. Soc., Wash., DC 1984) pp. 359-367.

AMNESIC SHELLFISH POISONING - A NEW SEAFOOD TOXIN SYNDROME

E.C.D. Todd
Bureau of Microbial Hazards, Health Protection Branch, Tunney's Pasture, Ottawa, Ontario K1A 0L2

ABSTRACT

An intoxication not previously documented occurred in November and December, 1987, in 107 persons in Canada who developed gastroenteritis within 24 hours or neurologic symptoms within 48 hours of eating cultivated mussels from Prince Edward Island (P.E.I.). The symptoms included vomiting, abdominal cramps, diarrhea, disorientation and memory loss. Short term memory loss was the most persistent symptom and lasted for over 18 months in several cases. Three deaths were directly associated with the outbreak and brain examination showed lesions. The mussels consumed were harvested from river estuaries in eastern P.E.I.. Domoic acid, an excitatory amino-acid, was identified as the toxic agent based on the mouse bioassay of the mussel extracts and chemical identification of the compound. In ten cases where mussel consumption data were available there was a close correlation between amount of domoic acid ingested (40 to 290 mg) and the severity of the symptoms. Monkeys fed toxic mussel extracts developed gastrointestinal and neurologic symptoms and possible brain lesions. The domoic acid originated from a bloom of the diatom *Nitzschia pungens*. Harvesting of all shellfish was banned until stocks were tested and found to be negative for domoic acid.

INTRODUCTION

Although occasional illnesses have been related to shellfish in Canada, the only poisoning to be reported with regularity is paralytic shellfish poison. Monitoring programs have been established for many years in areas where the *Alexandrium* dinoflagellate occurs in sufficient numbers to poison shellfish, ie.,the British Columbia seaboard, Gulf of St. Lawrence, and Bay of Fundy and more recently the coast of Newfoundland. Most illnesses in recent years have been traced to consumption of wild shellfish harvested in areas known to be toxic and where warning notices have been posted. Up till 1987 no illnesses are known to have occurred from shellfish cultivated in Canada.

In November and December of 1987 reports of gastrointestinal and neurologic illness associated with mussels from eastern Prince Edward Island (PEI) were received from across Canada. The syndrome was different from PSP in that the main neurologic symptoms were memory loss, confusion and disorientation. These mussels were cultivated in sheltered small river estuaries in 23-29 ppt salinity with water temperatures varying from -2° in January to 23°C in the late summer. The first commercial operation on PEI began in 1975 and demand continued to exceed production until the 1987 problem. With the development of this industry, restaurants in Quebec and other provinces began selling meals with the main course as mussels and retail stores sold bags of these shellfish to shoppers.

The mussels are grown in 3m plastic mesh socks suspended from 100m

Published 1990 by Elsevier Science Publishing Co., Inc.
Toxic Marine Phytoplankton
Edna Graneli et al., Editors

polypropylene lines anchored at either end and kept at the surface by buoys. The main harvesting period is between November and April and winter collection is done either by retrieving socks through holes in the ice or by employing divers. This paper describes the outbreak, toxicity of the implicated mussels and the origin of the toxic compound isolated from the mussels.

METHODS

Case Finding

According to Perl et al [1] a case of mussel-associated intoxication was any person who consumed mussels from PEI on or after November 1 1987 and who developed vomiting, diarrhea and/or abdominal cramps within 24 hours, or confusion, memory loss, disorientation, seizures, coma and/or cranial nerve palsies within 48 hours. A standardised illness questionnaire was developed for Quebec and a similar one for the other Canadian provinces. Information collected included each patient's name, age, sex, address, time and place of mussel consumption, illness onset time, symptoms occurring, duration time and medical care received. Supplementary information came from patients' hospital charts and post mortem examination. Local, provincial and federal health officials were involved in collecting and collating the data.

Laboratory Analysis

Mussels associated with cases and with wild and cultivated shellfish from PEI and other parts of Canada were analyzed for pesticides, heavy metals, pathogenic bacteria and viruses. These and plankton gathered from PEI river estuaries were also subjected to mouse bioassay and HPLC analysis for domoic acid.

Monkey Feeding Studies

Cymologous monkeys weighing between 3.4 and 5.1 kg were fed between 16 and 50 ml of blended mussel digestive glands to give doses of 19,868-28,470 µg domoic acid. Symptoms were observed and blood, urine and feces specimens taken for domoic acid analysis. One monkey was given 25,000µg incompletely purified domoic acid in polybuffer.

RESULTS AND DISCUSSION

The Disease

From 229 preliminary reports of illness from November 4 to December 4, 1987, 107 met the case definition [1]. These were received from PEI (5.6 cases/100kg mussels shipped), British Columbia (3.2 cases/100kg),Quebec (3.1 cases/100kg), Ontario (1.5 cases/100kg), and New Brunswick (1.1 cases/100kg). Symptoms included nausea (76%), vomiting (75%), abdominal cramps (49%), diarrhea (42%), headache (41%), anorexia (41%), loss of balance or dizziness (39%) and memory loss (24%). There was a close association between memory loss and age with 47% of cases 70 or over suffering from this condition. Short-term memory was the main persistent symptom with some cases unable to recognize their families and home environments and unable to carry out simple tasks. Those with mild symptoms consumed <75mg whereas those with neurological problems ingested between 115 and 290 mg. Twelve of 19 hospitalized cases were in intensive care. These were either aged over 60 or had chronic renal failure. These experienced additional symptoms of coma, mutism, seizures, chewing, grimacing, hiccoughs, profuse respiratory secretions (requiring intubation) and unstable blood pressure. Three deaths occurred within three weeks of mussel ingestion. The death of another patient, not exactly fitting the case definition, was also linked to eating mussels; and a fifth death occurred in a case who died of a cardiac arrest unrelated to the poisoning three months earlier. Brain tissue from four of these persons was examined for neuronal necrosis in the hippocampus, thalamus, amygdala, nucleus accumens and claustrum (Dr. Stirling Carpenter, Montreal Neurological Institute, Quebec, personal communication). Most of the severely affected elderly patients were still seriously incapacitated especially with loss of memory, 18 months after ingestion.

Mouse Bioassay

The first confirmatory indication of a toxin in the mussels was determined by mouse bioassay on November 27, 1987. PSP- type acid extracts caused mice to die 15-45 minutes after injection but usually preceded by a unique scratching syndrome of the shoulders by the hind leg followed by convulsions. This procedure was used to monitor shellfish from the Atlantic seaboard until harvesting and sale was resumed in the spring.

The AOAC PSP bioassay procedure [2] was modified by using 3 instead of 2 mice and the time of observation was extended from 15 minutes to 4 hours after which time the mice were periodically examined for up to 18-24 hours, usually after overnight. Deaths associated with the contaminated mussels containing the domoic acid were never observed after 135 minutes.

The toxin in the mussels was identified as domoic acid by high-performance liquid chromatography, high-voltage paper electrophoresis and ion-exchange chromatography and characterized by spectroscopical techniques [3,4]. Domoic acid is a 311 mol wt excitatory amino acid,containing the structure of glutamic acid and resembling kainic acid (Figure 1). Domoic acid is an agonist of the kainic acid subtype of receptor [5]. A practical HPLC method for identifying domoic acid in shellfish was developed [6,7,8]. During most of the 1988 season this procedure was used to monitor shellfish and identify toxic areas (>20μg domoic acid/g) so that harvesting could be stopped.

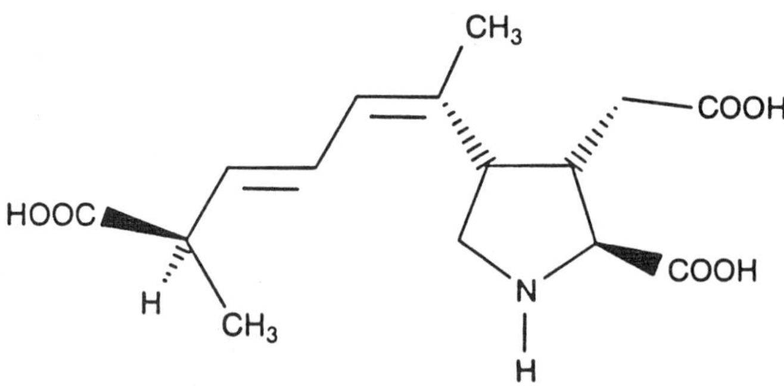

FIG. 1. Structure of domoic acid

Monkey Studies

The monkeys dosed orally with the contaminated mussel extracts developed gastroenteritis such as vomiting, anorexia, diarrhea (3/5) and some neurologic effects e.g. withdrawal and wet-dog shakes (3/5), disorientation, such as glassy-eye stares and prostration or weakness (2/5) and trembling (1/5). The gastrointestinal symptoms began between 2.75 and 6.0 hours and the neurologic ones between 0.25 and 6.0 hours. These last endured for up to 120 hours. Salivation, a feature noticed in severe human cases, occurred in 2 of the 5 animals. The monkey given 25,000μg of domoic acid in polybuffer showed similar effects.

In other studies with pure domoic acid given to monkeys, the effects became stronger as the level was increased. The animal given an intraperitoneal injection of 4 mg/kg within two minutes developed mastication, salivation, projectile vomiting and later had retching, weakness, teeth grinding, fixed gaze and lethargy. It died 3.5 hours after injection. The animals with the most severe effects showed brain lesions on autopsy particularly in the hippocampus, hypothalamus, the medulla oblonga and the retina. [9,10]. The orally dosed monkeys did not have significant lesions. Some lesions may have occurred but been transient as the animals were not sacrificed until between 16 and 25 days after ingestion.

Rodent studies have also shown that domoic acid gives excitatory effects if injected intraperitoneally, intravenously or directly into the brain [9,10,11,12,13]. Absorption of domoic acid by the oral route in rats and mice is limited; most is excreted in the feces. The toxic mussels had the same effect as pure domoic acid [11,12].

Source of Domoic Acid

At first, the origin of the domoic acid in the Cardigan area was postulated to be the red alga *Chondria baileyana*, a species found in PEI waters. This is because the only known source of domoic acid until then were seaweeds *Chondria armata* [14,15] and *Alsidium corallinum* [16], both red algae in the family Rhodomelaceae from warm waters (Japan and the Mediterranean). However, even though a herbarium specimen of *Chondria baileyana* collected from Prince Edward Island was found to contain domoic acid, the low mass of red seaweed in the estuaries would not be sufficient to account for the toxicity found in mussels. It was calculated that 6kg of domoic acid was present in the 63,000 kg of mussels growing in the contaminated estuaries [3,17]. Another potential source of the acid were dinoflagellates, e.g.,*Prorocentrum* spp. since these are known to produce some known seafood toxins. However, levels of these in the waters or in the mussel digestive glands were too low to explain the amount of the toxin. A plankton bloom at the time of the outbreak was sampled and extracts were found to affect mice in the same way that the toxic mussels did. This bloom consisting almost entirely of the pennate diatom *Nitzschia pungens* f. *multiseries* which was later determined to contain domoic acid, and a positive correlation was found between the number of *N. pungens* cells and the concentration of domoic acid in the plankton sampled in December and January. Three isolates of this *Nitzschia* strain obtained from Cardigan Bay produced domoic acid in culture at levels (1 to 20 pg/cell) comparable to values estimated for *N. pungens* in the plankton samples [17,18]. Other phytoplankton from the same area including the closely related species *Nitzschia seriata*, failed to produce domoic acid. The conclusion was drawn that this strain of *Nitzschia* was the main, if not sole, source of the toxic compound.

REFERENCES

1. T.M. Perl, L. Bedard, T. Kosatsky, J.C. Hockin, E.C.D. Todd, and R.S. Remis. New Engl. J. Med. (submitted)
2. Anon. Official Methods of Analysis Assoc. Official Analty. Chem., Arlington, Virginia. 18.086-18.092 (1984).
3. J.L.C. Wright, R.K. Boyd, A.S.W. de Freitas, M. Falk, R.A. Foxall, W.D. Jamieson, M.V. Laycock, A.W. McCulloch, A.G. McInnes, P. Odense, V.P. Pathak, M.A. Quilliam, M.A. Ragan, P.G. Sim, P. Thibault, M. Gilgan, D.J.A. Richard, and D. Dewar. Can. J. Chem., 67, 481-490 (1989).
4. M. Falk. Can. J. Spec. 33, 117-121 (1988).
5. B. Meldrum in: Neurotoxins and Their Pharmacological Implications, P. Jenner, ed. (Raven Press, New York (1987) pp. 33-53.
6. J.F. Lawrence, C.F. Charbonneau, C. Ménard, M.A. Quilliam, and P.G. Sim. J. Chromatorgr. 462, 349-356 (1989).
7. J.F. Lawrence, C.F. Charbonneau, B.D. Page, and G.M.A. Lacroix. J. Chromatogr. 462, 419-425 (1989).
8. M.A. Quilliam, P.G. Sim, A.W. McCullough, and A.G. McInness. Int. J. Environ. Anal. Chem. (in press).
9. L. Tryphonas, J. Truelove, F. Iverson, E.C.D. Todd, E. Nera. F. Proc. Symposium on Domoic Acid Toxicity, Ottawa, Ontario, April 10-11, 1989.
10. F. Iverson, J. Truelove, L. Tryphonas and E.A. Nera. Proc. Symposium on Domoic Acid Toxicity, Ottawa, Ontario April 10-11, 1989.
11. F. Iverson, J. Truelove, E. Nera, L. Tryphonas, J.S. Campbell, and E. Lok. Food Chem. Toxicol.(in press).
12. L. Tryphonas, J. Truelove, E. Nera, and F. Iverson. Toxicol. Pathol. (in press).
13. G. Debonnel, L. Beauchesne, and C. de Montigny. Can. J. Physiol. Pharmacol. (in press).
14. T. Takemoto and K. Daigo. Chem. Pharm. Bull. 6., 578- (1958).
15. K. Daigo. J. Jap. Pharm. Assn. 79, 353 (1959).
16. G. Impellizzeri, S. Mangiafico, G. Oriente, M. Piattelli, S. Scuito, E. Fattorusso, S. Magno, C. Santacroce, and D. Sica. Phytochemistry 14, 1549-1557 (1975).
17. S.S. Bates, C.J. Bird, A.S.W. de Freitas, R. Foxall, M. Gilgan, L.A. Hanic, G.A. Johnson, A.W. McCulloch, P. Odense, R. Pocklington, M.A. Quilliam, P.G. Sim, J.C. Smith, D.V. Subba Rao, E.C.D. Todd, J.A. Walter, and J.L.C. Wright. Can. J. Fish Aquat. Sci. 47,1203-1215(1989).
18. D.V. Subba Rao, M.A. Quilliam, and R. Pocklington. Can. J. Fish. Aquat. Sci. 45, 2076-2079 (1988).

INTERNATIONAL RED TIDE INFORMATION AND ASSISTANCE SERVICE (IRTIAS)

ALAN W. WHITE
Sea Grant Program, Woods Hole Oceanographic Institution, Woods Hole, MA, U.S.A. 02543

ABSTRACT

Toxic algal blooms and red tides represent a global problem for fisheries, mariculture and public health. Developing countries in particular can be severely affected by red tide outbreaks because they lack the expertise, information and infrastructure to respond quickly and effectively to the resulting fisheries and public health issues. When confronted with sudden red tide emergencies, developing countries are in need of rapid, international assistance from experienced scientists and fisheries and public health managers. With this in mind, the Woods Hole Oceanographic Institution Sea Grant Program is developing an International Red Tide Information and Assistance Service (IRTIAS). The primary goals are to establish an international database on red tide activities and to disseminate red tide information worldwide, especially to developing countries. The first project is the preparation of an international directory of scientific and managerial expertise on red tides. The directory can be used for rapid access to information and international assistance when red tide catastrophes occur. A questionnaire will be distributed in 1989 to acquire material for the information base and the directory.

INTRODUCTION

It is clear that toxic algal blooms and red tides now constitute a problem of global dimensions for fisheries, mariculture and public health [1]. Over the past twenty years or so toxic algal blooms and red tides have increased around the world in their frequency, magnitude, and geographic extent, as well as in their resulting effects [2]. Fish kills, shellfish contamination and poisonings of humans and other animals associated with well-recognized, toxic algal phenomena, such as PSP, DSP, and ciguatera, now occur nearly worldwide. In addition, new (or at least previously unknown) kinds of toxic algal outbreaks have recently appeared, such as brown tide (Aureococcus anophagefferens) in the northeastern United States, domoic acid poisoning (probably caused by a species of Nitzschia) in eastern Canada in 1987, the Chrysochromulina polylepis bloom in Scandinavia in 1988, and others.

RED TIDE PREPAREDNESS

Developed countries that have a long history of experience with toxic red tides (e.g., United States, Canada, Norway and Japan) conduct surveillance, monitoring, and management programs for red tides and red tide toxins to protect public health and minimize the impact on commercial fisheries. Some of these countries are expanding their efforts to include sophisticated, computerized remote sensing of the environment for red tides. However, even with sound management practices in place, developed countries can still have difficulties handling sudden, new events in terms of protecting fisheries and public health, as for example in eastern Canada in 1987 when 3 people were killed and about 100 were poisoned from eating mussels containing domoic acid.

Developing countries are especially vulnerable to the effects of toxic bloom and red tide outbreaks because of lack of information on how to cope with red tides, insufficient scientific and managerial expertise in this field, and the absence of an effective and responsive fisheries and public health infrastructure. A case in point is the first reported instance of PSP in Guatemala, which occurred in 1987 during a Pyrodinium bahamense var. compressa red tide, and which caused 26 human deaths and more than 175 serious illnesses [3]. Similarly, the first reported instance of PSP in the Philippines in 1983, also caused by a Pyrodinium red tide, caused 21 deaths and nearly 300 serious illnesses, and resulted in the banning of all shellfish in

Published 1990 by Elsevier Science Publishing Co., Inc.
Toxic Marine Phytoplankton
Edna Graneli et al., Editors

the central region for 8 months [4]. In these cases, rapid international assistance may well have saved lives and could have minimized the impact on fisheries by defining the seafood safety risks and indicating which fisheries would be affected and which would not.

INTERNATIONAL COOPERATION

The need for international cooperation and assistance with toxic algal bloom and red tide problems has been recognized and discussed at a number of international red tide conferences since 1974 [5-10]. Yet, in general, over this period no group or agency has been willing to "push" this concept through to implementation. International agencies, such as FAO and WHO, have been presented with the problem but have not taken up the challenge. There are two notable exceptions. One is an effort of the International Council for the Exploration of the Sea (ICES) to form a group of National Coordinating Centers for Exchange of Information on Exceptional Plankton Blooms. A summary report of the major toxic algal events in the 18 countries (most in Europe) that are members of ICES has been prepared annually since 1985. The second is an effort initiated by the Intergovernmental Oceanographic Commission (IOC) through its Western Pacific Group (WESTPAC) to hold red tide training sessions in Southeast Asia. The first training session was held in Australia in 1984, the second in Thailand in 1985, and the third in Brunei in 1989 under the cooperative sponsorship of a number of organizations with interests in Southeast Asia. Further, IOC sponsored a research-oriented Workshop on International Cooperation in the Study of Ocean Blooms and Red Tides in 1987 [9].

INTERNATIONAL RED TIDE INFORMATION AND ASSISTANCE SERVICE

In response to the need for addressing the global, red tide problem on an international level, the Woods Hole Oceanographic Institution Sea Grant Program has initiated an International Red Tide Information and Assistance Service (IRTIAS). The primary goals of this service, which is part of the International Marine Science Cooperation Program, are to develop and maintain an international information base on red tide activities and sources of expertise and to prepare this and related information on red tides for dissemination to institutions and agencies worldwide, especially to countries with little or no experience in dealing with red tide situations. The service is intended to improve and accelerate international information exchange and cooperation in this field and to help expedite rapid international assistance during toxic bloom and red tide emergencies. In connection with the ICES effort mentioned above and in cooperation with Dr. D.M. Anderson at the Woods Hole Oceanographic Institution, the service has been designated as the U.S. national coordinating center for red tide information.

International Red Tide Directory

The first project of IRTIAS is the preparation of an international directory of scientific and managerial expertise on toxic algal blooms and red tides. The compilation of such a red tide directory was recommended at the IOC workshop in 1987 [9] and at an international, red tide/aquaculture workshop in Ireland in 1987 [10]. The directory will list scientists, fisheries managers and public health officials with experience in dealing with red tide outbreaks. The directory will be especially useful to developing countries in providing sources of expertise for rapid assistance during red tide emergencies, and will thereby help to minimize the effects on fisheries and public health. Material for the red tide information base and red tide directory will be obtained in 1989 by questionnaire. Publication of the directory is expected in 1990.

The International Red Tide Information and Assistance Service is aimed at taking the first steps toward providing information exchange and outreach on toxic algal blooms and red tides on a much-needed international level. Subsequent projects of the service will include preparation and distribution of information sheets on various red tide topics, such as identification of causative organisms, fisheries risks, preventative measures for aquaculturists, and assessment of public health risks.

ACKNOWLEDGMENT

This work was sponsored by NOAA National Sea Grant College Program Office, Department of Commerce, under Grant No. NA86-AA-D-SG090, WHOI Sea Grant Projects No. E/L-1 and M/O-2. The U.S. Government is authorized to produce and distribute reprints for governmental purposes notwithstanding any copyright notation that may appear hereon.

REFERENCES

1. A.W. White, in: Proc. Int'l. Conf. on the Impact of Toxic Algae on Mariculture. Aqua-Nor '87 Int'l. Fish Farming Exhibition, Trondheim, Norway (1988) pp. 9-14.
2. D.M. Anderson, in: Red Tides, T. Okaichi, D.M. Anderson and T. Nemoto, eds. (Elsevier, New York 1989) pp. 11-16.
3. F. Rosales-Loessener, E. De Porras and M.W. Dix, in: Red Tides, T. Okaichi, D.M. Anderson and T. Nemoto, eds. (Elsevier, New York 1989) pp. 47-62.
4. R.A. Estudillo and C.L. Gonzales, in: Toxic Red Tides and Shellfish Toxicity in Southeast Asia, A.W. White, M. Anraku and K.K. Hooi, eds. (Southeast Asian Fisheries Development Center and International Development Research Centre, Singapore 1984) pp. 52-79.
5. V.R. LoCicero, ed. Proc. First Int'l. Conf. on Toxic Dinoflagellate Blooms (Massachusetts Science and Technology Foundation, Wakefield, MA 1975).
6. D.L. Taylor and H.H. Seliger, eds. Toxic Dinoflagellate Blooms. Proc. Second Int'l. Conf. on Toxic Dinoflagellate Blooms (Elsevier, New York 1979).
7. D.M. Anderson, A.W. White and D.G. Baden, eds. Toxic Dinoflagellates. Proc. Third Int'l Conf. on Toxic Dinoflagellates (Elsevier, New York 1985).
8. T. Okaichi, D.M. Anderson and T. Nemoto, eds. Red Tides. Proc. First Int'l. Symp. on Red Tides (Elsevier, New York 1989).
9. Workshop on International Cooperation in the Study of Ocean Blooms and Red Tides (Intergovernmental Oceanographic Commission, Takamatsu, Japan, 16-17 November 1987).
10. B. Dale, D.G. Baden, B. Mck. Bary, L. Edler, S. Fraga, I.R. Jenkinson, G.M. Hallegraeff, T. Okaichi, K. Tangen, F.J.R. Taylor, A.W. White, C.M. Yentsch and C.S. Yentsch. Proc. Int'l. Conf. and Workshop on the Problems of Toxic Dinoflagellate Blooms in Aquaculture (Sherkin Island Marine Station, Sherkin Island, Co. Cork, Ireland 1987).

VIII WORKSHOPS

EVALUATION OF HEALTH RISKS DUE TO PHYTOPLANKTON BLOOMS

Chairperson
L. EDEBO
Dept. of Clinical Bacteriology, University of Göteborg, Guldhedsg. 10, S-413 46 Göteborg, Sweden

Rapporteur
J. HAAMER
Aschebergsg. 21, S-411 27 Göteborg, Sweden

The chairman emphasized that international cooperation and consensus are important since the knowledge about known toxic algae and toxins is expanding and new kinds of poisoning are expected to appear. Blooms have a tendency to catch the eye of the public and to create a sense of catastrophy. When such occasions occur internationally based advisory groups should be put into action to accumulate knowledge and enhance objectivity. Preferably also internationally accepted analysis procedures should be employed.

Because of patchiness and uneveness of the distribution of toxic algae in the water, mussels should be analyzed when the risk of mussels are assessed. Since the concentration of toxin in mussels is an expression of the concentration in the water, mussels may also be used to assess the health risks of plankton-associated substances in the sea-water. Where there is a macroscopic bloom sieving of sea water may be used for sampling.

Dr. Horstmann proposed the use of satellite control for searching of blooms. This procedure was available before, but has been abandoned. Dr. Horstmann suggested the meeting to express its opinion in favor of reestablishing the satellite control. Dr. Lindahl questioned the possibility of using satellite control in several cases of serious disturbances when rather low concentrations of phytoplankton had been encountered. Dr. Edebo advocated the use of mussels rather than plankton sampling. Mr. Hansen pointed out that such a procedure was operating in the "mussel watch" program for environmental toxins.

Mr Tangen has six years' experience of using phytoplankton sampling as an early warning of DSP in mussels. This procedure is particulary useful since toxicity testing in mice may take a week. The plankton tests are used in combination with mouse i.p. tests. Ms. Doyle described that with the aid of phytoplankton analysis and consideration of the geography of the mussel farms it was possible to predict toxicity of mussels two weeks ahead. Dr. MacKenzie told that in New Zeeland *Dinophysis* spp. were not found without the mussels becoming toxic. Dr. Stabell replied that the toxicity differed between different populations.

Dr. Edebo emphasized the importance of sampling in the analysis of mussels since the difference between individual mussels may be great. With respect to specific information there is a ranking in the information obtained increasing from animal < cell < immunochemical < chemical tests. It would be preferred to use the same extraction procedure for different modes of testing. In Sweden there is a public opinion against unnecessary animal tests. Dr. Stabell reported that in Norway the i.p. mouse test was still used since some DST (yessotoxin, pectenotoxin) will not be demonstrated by HPLC.

It was agreed that methanol extraction should be used for the different modes of testing DSP toxins.

Dr. Lange emphasized that bioassays should be correlated to HPLC analysis. He furthermore asked for epidemological data particulary from Japan on the long-time effect of small doses of marine biotoxins. Mr. Tangen pointed out the difficulties in demonstrating low concentrations of toxins present in Norway with HPLC and Dr. Lassus commented that only for okadaic acid and DTX-1 is a practical HPLC-method available.

Dr. Edebo called for a new way of grouping the old DST-group. A new group tentatively called "Lipid soluble toxins" would contain "Diarrheic shellfish toxins" (sensu strictu) with okadaic acid, DTX-1 and DTX-3 and "Hepatotoxic shellfish toxins" with the pectenotoxins and yessotoxin. Furthermore, purified standards were asked for. So far standards are only available for okadaic acid and domoic acid.

Published 1990 by Elsevier Science Publishing Co., Inc.
Toxic Marine Phytoplankton
Edna Graneli et al., Editors

HAS THERE BEEN A GLOBAL EXPANSION OF ALGAL BLOOMS? IF SO, IS THERE A CONNECTION WITH HUMAN ACTIVITIES?

Chairperson

THEODORE J. SMAYDA
Graduate School of Oceanography
University of Rhode Island
Kingston, RI, U.S.A. 02881-0809

Rapporteur

ALAN W. WHITE
Sea Grant Program
Woods Hole Oceanographic Institution
Woods Hole, MA, U.S.A. 02543

Bloom Definition

Opinions varied as to what is the most important aspect to stress in developing a definition of an "algal bloom." The need was emphasized for two types of definitions -- a definition for scientific purposes based on environmental and ecological grounds and a definition for the purposes of human affairs based on the impact of blooms on public health, fisheries and aquaculture. The ecological definition would consider population variance, i.e., differences from "normal" levels of species abundance. This would require long-term monitoring to define "normal" levels. In regard to considering the harmful effects of algal blooms as part of a definition, it was noted that there are of course good blooms as well as bad. It was sugggested that toxic algal blooms fall into three categories: eutrophication dependent, toxicity associated with low cell numbers (as for *Dinophysis* blooms), and toxicity associated with large numbers of cells but not obviously with nutrient enrichment of coastal waters.

Global Increase

The question was considered whether it is possible to conclude at this time if there has been a global increase in the frequency and intensity of algal blooms around the world. Although it is generally suspected that toxic and noxious algal blooms have been increasing in frequency and intensity worldwide over the past 20 years or so, it was agreed that it is not possible to conclude this with certainty on a global level because the long-term data on the abundance of algae in the sea are insufficient in scales of both time and space. It will take a long time to be able to answer this question. An appropriate time frame for monitoring algal blooms appears to be in the range of 5 to 10 years at a minimum. The need for internationally coordinated efforts was emphasized. The obvious links with present plans for internationally coordinated studies of global climate change were noted. The usefulness of reviewing historical data in terms of delineating past trends was mentioned by several people. Past meteorological records can also provide a considerable amount of information relative to algal abundance.

The need was identified for the scientific community to reach agreement on how to approach the problem of algal blooms on an international level. Efforts can be directed at two aspects of the problem -- long-term monitoring and short-term mechanisms. There was considerable feeling that we should direct attention to solving bloom problems (short-term mechanisms) because of the tremendous impact of blooms on public health, fisheries and aquaculture. The causes and mechanisms of blooms may differ as there seem to be two major types of blooms, those in which nutrient additions to coastal systems are obviously implicated (for example, in Tolo Harbour (Hong Kong), in the Seto Inland Sea (Japan), and in the Aegean Sea in the vicinity of sewage outfalls) and those blooms that are not obviously associated with coastal enrichment (for example, *Alexandrium, Pyrodinium, Dinophysis*, etc.). It was suggested that although our knowledge of specific features of algal blooms is increasing (such as nutrient uptake and association with hydrographic fronts) we should focus interdisciplinary attention (biology, chemistry, physics) on particular blooms to develop a full understanding of the short-

Published 1990 by Elsevier Science Publishing Co., Inc.
Toxic Marine Phytoplankton
Edna Graneli et al., Editors

term mechanisms of bloom formation, transport and dissipation. It was stressed that there is an immediate need to coalesce the best available information around the world to help in managing bloom events and their effects.

Discussion followed on how to develop funding for long-term projects on algal blooms. International organizations may be of assistance here. Perhaps half of any such funding could be used for conducting long-term monitoring work and half for conducting short-term studies on bloom mechanisms and amelioration of bloom effects. Establishing capital investments and using the interest accruing from them for conducting long-term monitoring of blooms was also suggested.

Global Spreading

The question was posed whether a global spreading of algal blooms is now occurring. The circumstantial evidence is building to suggest strongly that some algal organisms have recently been introduced from certain parts of the world to others, either within the same region or in different regions suggesting trans-oceanic spreading. Ocean currents are viewed as the chief natural cause of spreading. Regarding man's involvement in the spreading of marine microalgae, it appears feasible that ballast water may be a prime mechanism for long-distance transport of algae, including those responsible for toxic and noxious blooms. Evidence is inceasing that marine microalgae can survive long-distance transport in the ballast tanks of ships. Controls and regulations pertaining to the discharge of ballast water by commercial shipping are now being adopted in Australia. Another possible mechanism of algal spreading related to man's activities is the relay of shellfish stock from one region to another. Viable vegetative cells and cysts in and on the shellfish may be introduced to new waters in this manner.

Man's Involvement

Circumstantial evidence points to man's involvement in the increase in frequency, intensity and global distribution of some algal bloom events. This is especially so in the case of coastal blooms associated with nutrient enrichment from municipal and industrial activities. The association of other human activities with increased bloom occurrence is suspected, as for example with ballast water, shellfish transport, aquaculture activities themselves (self-pollution), and with more general environmental alterations on a global scale.

Statement of Concern

It was agreed that the workshop prepare a statement of concern to bring to the attention of international organizations. The following general statement was prepared and approved. It was transmitted to the Intergovernmental Oceanographic Commission (IOC) in Paris.

"Scientists attending the 4th International Conference on Toxic Marine Phytoplankton in Lund, Sweden, conclude that there is a potential that global expansion of algal bloom problems is now occurring.

"While in the traditional scientific sense, the term 'algal bloom' refers to a deviation from 'normal' abundance, this term is increasingly being equated with harmful algal phenomena affecting public health, fisheries and aquaculture.

"The conference participants reached a consensus that some human activities may be involved in increasing the intensity and global distribution of blooms and recommended that international research efforts be undertaken to evaluate the possibility of global expansion of algal blooms and man's involvement in this phenomenon."

STANDARDIZING EXTRACTION AND ANALYSIS TECHNIQUES FOR MARINE PHYTOPLANKTON TOXINS

Chairperson
OLE B. STABELL
Norwegian College of Veterinary Medicine, Dept. of Food Hygiene, Oslo, Norway

Rapporteur
ALLAN D. CEMBELLA
Maurice Lamontagne Institute, Biological Oceanography Division, Dept. of Fisheries and Oceans, Mont-Joli, Quebeck, Canada

This workshop was convened in order to compare analytical techniques and extraction methods for the determination of marine phycotoxins. Although most of the discussion was devoted to methods for diarrheic shellfish poisons (DSP), which are perhaps the toxins of greatest public health significance in northern Europe and Japan, the general approach was equally valid for the analysis of other phycotoxins, including paralytic shellfish poisons (PSP), brevetoxins, ciguatera and domoic acid. The chairman initiated the discussion by focusing on problems related to a recent inter-laboratory collaborative study to test the efficacity, reproducibility, and robustness of the fluorescence-based HPLC method developed in Dr. Yasumoto's laboratory for the determination of DSP toxins. In this particular pre-column derivatization method, okadaic acid, as well as other carboxylic acid metabolites, are esterified with the fluorochrome anthanyldiazomethane (ADAM) to yield a fluorescent product. The technique has not been subjected to the rigorous comparisons necessary for official international acceptance as a standard method. However, its use in several laboratories in various countries suggests that it represents the nearest approach to a routine technique presently available.

In general, the results of the collaborative study of the DSP method were rather discouraging. The intra-laboratory replication (precision) of sample toxin values, and the ability of each laboratory to accurately reproduce a gradient of varying proportions of okadaic acid/DTX_1, was usually acceptable, but quantitative toxicity values for the same sample often differed several-fold among the various laboratories. Using this study as a model to illustrate the difficulties encountered in the standardization of techniques for the analysis of marine phycotoxins, the workshop participants were able to identify several key areas which require further investigation, before future inter-calibration studies are attempted.

Extraction and sample preparation

A consensus was reached among the participants that, at least for the determination of DSP toxins, sample extraction and preparation pose a greater problem than factors which are purely analytical. A conventional 80% aqueous methanol solvent was deemed to be acceptable for the extraction of DSP toxins from contaminated shellfish. Most inconsistencies in the method appear to arise in the fluorescent-derivatization and sample-rinsing steps. The fluorochrome reagent ADAM is notoriously unstable, therefore, the resulting degradation products may yield artifact peaks on HPLC chromatograms. Deterioration of the fluorochrome may also lead to incomplete derivatization of the toxin components. Recommendations were made that this reagent be stored at low temperature (-70 ^{o}C), preferably in complete darkness in a dessicator, and that it be sub-divided into small aliquots to be prepared in solvent immediately before use. Opinions were mixed regarding the potential shelf-life of the reagent when stored under these conditions; estimates generally ranged from several months to years.

Published 1990 by Elsevier Science Publishing Co., Inc.
Toxic Marine Phytoplankton
Edna Graneli et al., Editors

Since the diazo-reaction involved in the derivatization procedure is expected to be rather pH sensitive, it was suggested that the technique could be made more reliable by proper pH control during the reaction. The pH-buffering effect in biological material can likely account for the discrepancies in reaction time compared to purified toxins.

The importance of improving the rinsing procedure during sample clean-up was heavily emphasized. In some cases, as much as half of the total toxin was reported to have been lost in the solid-phase extraction step. The magnitude of this loss was highly dependent upon both the technical expertise of the particular individual performing the procedure, and the specific technique adopted. When the rinsing step is properly carried out, a better recovery of DSP standard from spiked sample mixtures, a more stable baseline on the HPLC chromatograms, and a reduction in the number of artifact peaks can be expected.

A suggestion was made that since ADAM-derivatives of okadaic acid have a tendency to adhere to glassware, acetone rather than methanol should be used to dissolve the ADAM-okadaic acid ester. It was also recommended that methanol be avoided as a storage solvent for DSP toxin standards, in favor of acetonitrile, to reduce the possibility of methyl-ester formation.

Use of internal standards

The value of internal standards, specifically in the DSP technique, was considered and actively debated. The use of an internal standard offers several advantages, particularly if the purified compound is available in quantity at a reasonable price, while toxin standards are scarce. Ideally, the internal standard should react with the fluorescence reagent in a manner analogous to the purified toxin standard. The fluorescent derivative should exhibit stable chromatographic behavior which is similar, but not identical, to that of the derivatized toxin. In this way, reaction kinetics, extraction efficiency, recovery of material, and chromatographic stability can be evaluated without sacrificing precious toxins.

Deoxycholate has been exhaustively tested as a potential analogue of okadaic acid, as it fulfills some of the basic criteria. Unfortunately, in the opinion of Norwegian workers, who have pursued this most intensely, the use of deoxycholate as an internal standard is appropriate for verifying retention time by HPLC, but its introduction into the technique for quantitative determination is premature. It was suggested that one of the problems with the use of deoxycholate as an internal standard for okadaic acid arises from the difference in pKa, thereby resulting in unpredictable reaction kinetics during derivatization.

Availability of reference standards and analytical reagents

All participants stressed the urgent need for reference standards for the calibration of analytical instruments, particularly for HPLC systems. In the case of DSP reference toxins, several current or potential new sources were reported. A group in Tenerife, Canary Islands, is apparently involved in the mass culture of *Prorocentrum lima* for the production of okadaic acid. Similar efforts are also underway at the IFREMER laboratory at Nantes, France. A Hawaiian company, Moana Bioproducts, is now offering purified okadaic acid for reference material as a commercial product. Small quantities may also be available through exchange from Japanese colleagues.

The fluorochrome reagent ADAM has often been synthesized in small batches by individual laboratories performing the analysis of DSP toxins by the fluorescence-HPLC technique. Given the range of competence among these laboratories, this "home-made" ADAM has often been of questionable purity and stability. This has undoubtedly led to difficulties in applying the method successfully. The participants of the workshop were informed that

Merck Corp. in the United States is now able to supply this reagent in purified form.

High-quality domoic acid, the primary toxin involved in amnesiac shellfish poisoning (ASP), has been prepared from contaminated mussels from Prince Edward Island, Canada. Reference domoic acid (designated DACS-1) is available from the Atlantic Research Laboratory, National Research Council of Canada, Halifax, Canada, as a marine analytical standard, or, in a different preparation, from Diagnostic Chemicals, Prince Edward Island.

Some of the brevetoxins have been purified by Dr. D. Baden, University of Miami, Florida, and at least one is commercially available in the United States.

The lack of readily available reference PSP toxins has severely restricted the introduction of the Sullivan HPLC method into regulatory laboratories, and has even hampered attempts to conduct inter-laboratory collaborative studies to verify the technique. With the exception of saxitoxin, the other components in the suite of PSP toxins are extremely difficult to obtain in purified form. One potential source of reference standard mixtures for PSP, the Food and Drug Administration, Washington, is either unwilling or unable to satisfy the high demand for these calibration materials. A project involving the Atlantic Research Laboratory, NRC, Halifax, Canada, which had as a secondary objective the production of sufficient purified PSP toxins to supply reference standards for instrument calibration, has sufferred from technical difficulties, and will not be able to furnish these compounds in the near future. A biotechnology group in California, directed by Dr. M. Neuschul, has also proposed the production of these toxins, but the current status of this project is unknown.

Conclusions

The workshop participants were generally in agreement that the successful introduction of standardized methods, verified by proper inter-laboratory calibrations, requires that the method development be substantially completed, and not merely in progress. Thus, there must be a consensus on sample extraction and preparation techniques, chromatographic conditions, and the type of analytical column employed for the separation. There was a further recommendation that the laboratories engaged in these studies attempt to obtain the same primary standards, internal standards, and reagents of certified purity to minimize inconsistencies which are unrelated to the appropriateness of the particular method. For example, there were major technical flaws in the collaborative study of the Lee and Yasumoto HPLC method for DSP, in that mixed samples of shellfish hepatopancreas containing various proportions of okadaic acid and DTX_1, rather than pure toxins, were employed for intercalibration. Each laboratory provided its own reference standards and fluorescence reagent. Finally, there was no objective independent reference for the toxin concentrations obtained, since the "true values" were provided by one of the laboratories participating in the study.

The problem of the availability of reference materials and high quality reagents for the analysis of marine phycotoxins can only be rectified by a concerted international effort to produce and freely distribute these vital compounds. Initial advances in toxin identification and methods development have often been pioneered by chemists working on marine natural-products, and those involved in pure analytical research. Such individuals are often disinclined or unable to engage in the mass production and distribution of biotoxins. Several participants noted that private corporations may be reluctant to market such potent biotoxins for fear of product-liability lawsuits. Government-funded research institutions may also be constrained from mass production of such materials, due to the negative public connotations associated with their potential use as biological and chemical warfare agents.

A final plea was made by all participants that information on the modification of existing methods should be freely shared and rapidly disseminated in the scientific community. This is perhaps justification for the preparation of a technical manual on methods for toxin analysis, with a clear indication of the potential pitfalls of each technique, and which could be up-dated at periodic intervals. Practical workshops on toxin methodology could also serve a valuable role in the training of less experienced analysts, particularly those working in regulatory and inspection laboratories.

THE TAXONOMY OF GONYAULAX, PYRODINIUM, ALEXANDRIUM, GESSNERIUM, PROTOGONYAULAX AND GONIODOMA

Chairman
K. A. STEIDINGER, Marine Research Institute, Florida Dept. of Natural Resources, 100 8th Ave. S.E., St. Petersburg, FL 33701, USA

Rapporteur
Ø. MOESTRUP, Institut for Sporeplanter, Ø. Farimagsgade 2 D, DK-1353 Copenhagen K, Denmark

The workshop discussed only one issue: could a consensus be reached on a generic name for the tamarensis/catenella group of Gonyaulax/Protogonyaulax/Alexandrium/Gessnerium/etc.

Max Taylor began the discussion by stating that all genera are artificial and often overlapping. He considered Gessnerium to be ill founded and a synonym of Alexandrium. The main problem is whether Protogonyaulax and Alexandrium should be kept separate or united under the oldest name, Alexandrium. Was there a risk that future studies might show the correct name to be Goniodoma? Could/should Protogonyaulax be conserved?

K. Steidinger pointed out that there was no real difference between Alexandrium and Protogonyaulax and referred to SEM micrographs in one of the posters at the Congress where in the type species of Alexandrium the first apical plate in some cells reaches the pore plate, in others the plate rim does, and in others again growth of the plate rims apparently separates the first apical plate from the pore plate.

E. Balech then explained his conclusions based on a detailed study of material of the type species of Alexandrium, A. minutum Halim, from the type locality, Alexandria Harbour. The material was confirmed by Halim to represent the type species of Alexandrium. Balech concluded that Alexandrium is validly published; the lack of a Latin diagnosis is irrelevant (it was described under the Zoological Code of Nomenclature). The description is very good for the time (1960) and Alexandrium is a well established genus. Balech's studies had shown, that the measurements agreed with Halim's description, it is the only small species of Alexandrium.

In the first description of Alexandrium one drawing shows 1' separate from Po, in another the 2 plates almost touch each other. Balech had dissected several hundred individuals from Halim's material and found all intermediates between touching and non-touching Po and 1', i.e. the feature used to separate Protogonyaulax and Alexandrium. A pore was always present in 1'. All early Alexandrium/Gonyaulax species, incl. A. minutum, were described without a ventral pore. No importance was ascribed to this discrepancy, it was not illustrated in Braarud's article of A. ostenfeldii, but it was found by Tangen in some of Braarud's unpublished notes on the same organism. A. minutum is a widely distributed species, whose morphology is rather variable. Italian material is much sculptured in the hypotheca, while cells from Australia and Alexandria Harbour are smooth. French material is intermediate. Is this caused by environmental conditions?

Toxic Marine Phytoplankton
Edna Graneli et al., Editors

The percentage of cells with 1' - Po apparently disconnected varies geographically. The cell shape also varies geographically and at different stages of the life cycle. Balech concluded that Alexandrium and Protogonyaulax species should all be grouped under the oldest generic name Alexandrium. He is now preparing a monograph on Alexandrium.

Y. Fukuyu was not convinced that Balech's findings agreed with the original description of Alexandrium and suggested that 2 species might be involved. Can another species be found which fits Halim's description better? Balech again pointed out that Halim's original figures illustrated Po and 1' almost in contact in one drawing and separated in another and that this suggested variability within a population.

B. Dale stated that we should only create a new name if we are convinced that a new name is necessary. Very small discrepancies between older published descriptions and new findings should not be used to establish new genera. New observations will always supply details which are not in the original description. B. Dale praised E. Balech for going back to the material from the type locality.

G. Hallegraeff also praised Balech's work. He agreed that Alexandrium minimum is closely related to the tamerensis/catenella group, it even has the same type of cyst and produces the same toxins. There is no need for 2 separate genera. He considered it very important that a consensus could be reached now regarding which generic name to use.

E. Balech further stated that Goniodoma is not an acceptable name for the tamarensis/catenella complex, it is a distinct but related genus. In Goniodoma the pore plate is arranged transversally, in Alexandrium longitudinally. Cells of Goniodoma have distinct lists along the sulcus. Three species of Goniodoma are presently recognized by Balech, all with the same general structure. Goniodoma is a legitimate name under the Botanical Code of Nomenclature. Gessnerium may be used as a subgenus of Alexandrium.

Y. Shimizu concluded that it is time for a consensus to be reached, it is confusing to chemists, biologists, and the public alike to use so many names.

K. Steidinger finished by stating that we should take advantage of Balech's extensive work. She suggested that the generic description of Alexandrium should be emended according to Balech's ideas.

Max Taylor accepted this, i.e., to include also the tamarensis/catenella group in Alexandrium, but stressed that the emendation should be done in such a way that Goniodoma is clearly excluded. Only then can the risk of future name changes be reduced. There were no objections to this, i.e. a general consensus was reached on using the name Alexandrium for the tamarensis/catenella group, as emended by Balech. The emendation should be published as soon as possible, preferably in the Congress proceedings (see below).

Ca. 80 persons attended the workshop.

IX CONFERENCE OVERVIEW

Conference Overview

RED TIDES, BROWN TIDES AND OTHER HARMFUL ALGAL BLOOMS: THE VIEW INTO THE 1990's.

F.J.R. TAYLOR

Departments of Oceanography and Botany, University of British Columbia, Vancouver, B.C., Canada V6T 1W5

The increases in attendance and country representation in this series of conferences has accompanied a substantially growing awareness of the problems associated with phytoplankton blooms, even by non-specialists, in many parts of the world. It has taken too long for the general recognition that these are global problems concerning human health and economic growth. This recognition seems to have been due largely to the great expansion of aquaculture in recent years, with increased fish farming and shellfish consumption providing an increased "bioassay" of the environment. At the same time, the expanded marketing of such products, both nationally and internationally, has resulted in much more widespread epidemics of human intoxications rather than the low-key local outbreaks of before.

There is a great array of material included in this volume and it is plainly not possible to cover more than some major points. This overview attempts to put them in perspective , particularly in the light of what has gone before and to identify some major outstanding problems. Papers will be referred to by the presenting author only, for brevity.

SCOPE

It is evident from the difference in title of this conference compared to others in the series that the scope of such meetings has greatly widened. This has been to accommodate signicant problem algal blooms outside of the dinoflagellate "red tides" that the earlier conferences were primarily concerned with. For example, the economically important fish-killing blooms produced by chloromonad flagellates in the Inland Sea of Japan have been added to recent conferences. At the present one a major item was the 1988 *Chrysochromulina polylepis* bloom (papers by Lindahl, Granelli and others) but there were also papers on the chrysophyte *Aureococcus* "Brown Tide" that has starved scallops and caused other types of harm in bays in the eastern United States (Cosper), Amnesic Shellfish Poison resulting from domoic acid produced by several diatoms (Todd, Shimizu) and physical damage to the gills of farmed fish by certain species of the diatom *Chaetoceros* (Horner) in addition to those topics covered previously. The report, by Lam, of fish death due to oxygen depletion associated with a *Gonyaulax polygramma* bloom in Hong Kong waters was personally nostalgic, since it was a similar bloom that led to my first paper in this field in 1962 [1]. Table I indicates the range of phenomena now recognised.

TABLE I. Types of harmful consequences associated with marine phytoplankton blooms

Marine Fauna Mortalities		
due to physical damage (usually gills)		
oxygen depletion		
chemical - direct action		
chemical - indirect food-chain?		
Human Intoxications		
Paralytic Shellfish Poisoning (PSP)	known since	1793
Diarrheic Shellfish Poisoning (DSP)		1978
Neurotoxic Shellfish Poisoning (NSP)		1970s
Amnesic Shellfish Poisoning (ASP)		1987
Ciguatera Fish Poisoning (CFP)		16th.century

Published 1990 by Elsevier Science Publishing Co., Inc.
Toxic Marine Phytoplankton
Edna Graneli et al., Editors

Multiple effects can be associated with a single bloom, *Gymnodinium breve* being a classic example. Its blooms kill wild fish in large numbers while simultaneously making oysters toxic (NSP) and causing eye, throat and nasal irritation due to its aerosol effects (Pierce). Although human intoxications are obviously a major concern it should be recalled that other animals can also be harmed by oral ingestion of contaminated shellfish or fish. For example, PSP and CFP outbreaks are frequently accompanied by the death of domestic animals and at this conference we heard of the possible death of whales due to their ingestion of mackeral. It has become evident that mackerels in the western Atlantic frequently have saxitoxin family toxins in their livers (Anderson), presumably acquired through the zooplankton food chain. There is also a possible link, although more tenuous, between the recent death of dolphins and brevetoxins (Steidinger). In passing it can be noted that caged or penned fish are usually much more vulnerable to direct noxious phytoplankton since they are unable to escape the bloom "clouds".

The list of organisms that have been implicated in such outbreaks has also grown dramatically. Table II lists the twenty genera or so (depending on taxonomic practice), from six algal groups, that are now known to be have harmful species. To keep perspective it is important to remember that there are 2000 living dinoflagellate species and 10,000 diatoms; consequently the harmful members constitute only a very small proportion of them.

TABLE II. A minimal list of genera of marine phytoplankton with well established harmful species (approximate number in parentheses). Taxa that have been associated with oxygen depletion or "gill clogging" are not included.

DINOFLAGELLATES
Alexandrium (incl. *Protogonyaulax*; 7), *Amphidinium* (2), *Cochlodinium* (2), *Dinophysis* (5), *Gambierdiscus* (1), *Gymnodinium* (5), *Ostreopsis* (2), *Prorocentrum* (2), *Pyrodinium* (1)
CHLOROMONADS
Chattonella (3), *Fibrocapsa* (1), *Heterosigma* (1)
CHRYSOMONADS
Aureococcus (1)
PRYMNESIOMONADS
Chrysochromulina (1), *Phaeocystis* (1), *Prymnesium* (3)
SILICOFLAGELLATES
Dictyocha (= *Distephanus*)
DIATOMS
Chaetoceros (2), *Nitzschia* (1), *Amphora* (1)

TOXINS

Much of the conference concerned the nature of the toxins and their production. The greatly increased understanding of their structures has cleared a lot of the confusion in the earlier literature resulting from the use of different methodologies and names. The Brevetoxin family is a classic example. Many bacterial and animal toxins are large molecules (e.g. the phospholipases of bee venom, etc.) whereas phytoplankton toxins are generally smaller, with molecular weights less than 1000..

Functionally they fall into three main groups: *transmembrane blockers*, *transmembrane facilitators* or *membrane lysers*. Not much was said of their pharmacology at this conference although this is an important area (workshop on human health) and those with very specific actions on cells can be useful research tools. The mode of action of some is still not known.

The plenary talk by Yasumoto (honoured for his work at the conference banquet) provided a concise outline of the structural types now known. The newly discovered polyether structure of ciguatoxin was announced by him at the conference.

TABLE III. Basic types of phytoplankton toxins

Water soluble toxins consist of the much studied
saxitoxin family (causing PSP,at least 18 forms: see the contribution by Oshima)
some amino acids (domoic acid, mycosporine-like aa's)

Fat soluble toxins
okadaic acid and its derivatives (dinophysis toxins 1 and 3; producing DSP),
macrolides (pectenotoxins - also DSP; amphidinolide, prorocentrolide)
polyethers (brevetoxins falling into two types, A and B - fish kills, NSP; ciguatoxin - ciguatera; yessotoxin - DSP, formerly dinophysistoxin 2),
fatty acids (fish killers)
glycolipids (fish killers).
peptides (hepatotoxins from fresh water cyanobacteria)

In addition to these two peculiar small, phosphorus-containing ichthyotoxins (PB-1, Gb-4) have been previously described from *Gymnodium breve* [2]. The structure or even general nature of several others, such as maitotoxin and ostreopsistoxin (both probably contributing to the ciguatera syndrome), those from *Chyroschromulina*, *Aureococcus* and *Heterosigma*, are only partially known or unknown.

The synthetic pathways of both the saxitoxins and the brevetoxins have been elucidated by the painstaking studies of Shimizu and co-workers over the past decade. At this conference a summary of the synthesis of the former (in a strain of the cyanobacterium *Aphanizomenon*) from arginine, acetate and methionine was provided. Its primarily intranuclear location is indicated by preliminary immunochemical fluorescence. Brevetoxins appear to be formed as polyketides from dicarboxylic acids.

Toxin content can vary considerably, partly due to genetic differences, but it is also an inverse function of growth rate and can be influenced by several environmental factors, studied by several groups in the past and summarised here by Anderson. Higher concentrations in cysts were confirmed by Okaichi. Phosphorus limitation was mentioned several times as an increaser of both PSP toxins (earlier work by Cembella, Taylor and others) and *Chrysochromulina* toxicity.

Multiple forms of the toxin are often present simultaneously in the synthesiser, i.e. a characteristic toxin spectrum is present. This is also usually true in the organism which concencentrates the toxin, the transvector. Early confusion arose from differences found in different strains and between the producer and the transvector. The recognition of strain specific spectra and of bioconversion within the transvector (to more toxic forms such as saxitoxin) was a product of research in recent years. Yasumoto raised the possibility that ciguatoxin may be an oxidation product in the herbivorous fish transvectors. Several papers (Anderson, Cembella), reported spectral shifts as well as toxin content under different growing conditions. This could be taken, together with possible bacterial involvement (Kodama, Ogata), to cast doubt on earlier reports of the predictability of toxin spectra within strains. However, since different strains grown under identical conditions do produce different spectra it appears that there is an underlying genetic component on which the other influences are superimposed. There is an interesting latitudinal effect with PSP, with diminishing toxicity from the Gulf of St. Lawrence to south of Cape Cod (earlier work by Miranda et al., cited by Anderson), but there are tropical PSP producers, such as *Pyrodinium* (which is such a serious problem in the Philippines and other parts of the western Pacific that it was the sole subject of a workshop in Brunei this year).

Bacterial involvement?

It has been 10 years since Silva first presented data that seemed to indicate that bacteria may be the source of PSP toxins, using cultures that would now be identified as *Alexandrium lusitanicum* (see taxonomy). This is an extremely important fact to establish since it would change the fundamental approaches used to examine toxin source and variability. The synthesis of tetrodotoxin by species of *Vibrio* and *Pseudomonas* is now well established and so this seems like a feasible possibility. Unfortunately, although there are further indications that bacteria, and in particular a species of *Moraxella* (Kodama, Ogata) and *Vibrio vulnificus* (Yasumoto) isolated from toxic cultures, *Alteromonas* (from crabs) may be sources, there are difficulties in linking the presence of toxin in the bacteria with the intracellular production in the dinoflagellates. Intracellular bacteria are usually readily visible with transmission electron microscopy, although their appearance can vary substantially. Taking that into account, intensive examination by several groups, including those of Moestrup and our own (Doucette), have been unable to see a trace of bacteria within cells from the most toxic isolates of *Alexandrium*, even though they can be found quite commonly in the nuclei of other dinoflagellates. As noted by Moestrup at the conference and myself at the Takamatsu conference, the objects figured by Kodama in TEM sections of nuclei are not clearly bacteria. This very important question needs rapid resolution.

Toxin in bivalve molluscs

It has long been known through the work of Twarog, Shumway and others reported in previous volumes of this series, that different species of bivalves react differently to the presence of toxic phytoplankton, acquiring toxicity and losing it (depurating) at different rates, while some are actually harmed by it. Scallops won't eat *Aureococcus*, stopping filtration in its presence and starving to death during the recent prolonged blooms in Long Island embayments. At the conference Bricelj and coworkers reported the ingestion and toxin acquisition rates of *Mytilus edulis* under controlled conditions and showed that *Mercenaria mercenaria* will consume toxic dinoflagellates even though it does not usually become toxic during PSP outbreaks. The digestive gland of *Mytilus* contained by far the greatest concentration of toxin, confirming earlier observations. Martin reported that this was also true for *Mya arenaria* during a toxic bloom but that the highest concentration at other times were in a group of tissues near the digestive gland that includes the kidneys and heart. The classic example of this persistence outside the digestive gland is the black siphon tip of *Saxidomus giganteus*.

Toxin functions

At several points in the conference there were clues as to the possible functions of these compounds in the producing organisms. It has already been shown that the saxitoxins may reduce, although not prevent predation by some zooplankton. Indeed, the concentration by zooplankton can lead to fish kills (reported by White at previous conferences) or be concentrated in the liver of mackerals (this conference). A dramatic demonstration of changed behaviour by a zooplankter by high concentrations of toxin exudate was provided by the videotape of Hansen. It showed the tintinnid *Favella ehrenbergii*, which has long been known to ingest toxic dinoflagellates, swimming continuously backwards, and finally lysing, in the presence of exudate from *Alexandrium tamarense*. Other functions include allelopathy (inhibition of other potentially competitive species), with antibacterial and antifungal effects offering the possibility of practical applications.

Defence is also a possibility in the transvectors, offering an explanation for their accumulation of toxin and conversion of less toxic to more toxic forms. Recent studies published elsewhere have shown that sea otters are very orally sensitive to saxitoxins and won't eat shellfish contaminated by them. However, the presence of ciguatoxin does not protect herbivorous fish from the predators. More work is needed on this interesting topic, as well as the mechanisms used to compensate for the effects of the toxins on their nerves.

NEW METHODOLOGIES

In several papers and a workshop new, or relatively recent methods were outlined and another workshop concerned itself with standardisation of currently used toxicity assays. The value of HPLC as a tool in toxicological research, pioneered by Sullivan for PSP toxins, is now unquestionable and the conclusions of numerous papers depended on its use.

There was an emphasis at this meeting on the use of immunochemical techniques, particularly the enzyme-linked type (ELISA) for the detection of saxitoxins (a Canadian kit was described by Cembella), a Japanese-American kit for DSP was exhibited, and the development of methods for brevetoxins was described by Trainer. Others, such as a stick test for ciguatoxins, have been described at previous meetings.

Immunochemical methods are sensitive and relative inexpensive but are still somewhat limited since their specificity is determined by only parts (often unknown) of the test molecules, with false positives an ongoing problem. A current flourishing of immunotaxonomic applications, particularly in the fluorescent labelling (FITC) of tiny, coccoid picoplankton species, was described in the case of *Aureococcus* by Anderson. Once again, the significance of the lack of reactivity between two strains has not yet been evaluated in terms of "species".

TAXONOMY

The use of different names for the same organisms has been an ongoing problem, particularly for non-specialists. The use of different generic names has been annoying and potentially confusing but, once the commonality is recognised, the literature can be readily combined. For example, at previous conferences it was pointed out that the name *Olisthodiscus luteus* had been wrongly applied to the flagellate *Heterosigma akashiwo* outside of Japan [3] and the literatures (except for the original description of the former and recent papers that recognise the problem) dealing with them can be combined. In view of the great similarity between *Gymnodinium breve* and other toxic species of *Gymnodinium* its placement in *Ptychodiscus* by some authors seems to be inappropriate. Hence, the designation of its toxins as "GB" rather than "PB" seems to be more appropriate.

At this conference the organisers wanted a resolution of the *Alexandrium/ Protogonyaulax/Gonyaulax* applicability problem (discussed at several conferences in the series), with only one name to be used in the included papers. At a workshop session Steidinger reviewed the history of the problem. It has become increasingly evident that members of *Alexandrium* and *Protogonyaulax* are very closely related and that the criteria which have been used to try to keep them separate (and thus cause the least disturbance to the literature) have been weakened by observations of their variability. If they are combined *Alexandrium* has priority, being the older name. *Protogonyaulax* has been used preferentially in more than 100 papers since its proposal in 1975 (computer-aided search) and it is unfortunate that this action will require an upheaval of some magnitude. It was pointed out that the insistence that there be no overlap between genera sets a rather drastic precedent for dinoflagellate taxonomy since there are several other large, well-known genera which overlap (hence the odd-looking possibility that *Gymnodinium nagasakiense* and *Gyrodinium aureolum* may be the same species) and it is difficult to see the utility of combining them. When *Alexandrium* was created its author did not indicate what the essential features were which separate it from others, such as *Goniodoma* (= *Triadinium*), and so the need for a new, more exclusive diagnosis was recognised and created for this volume.

While the choice of which generic name to use is a significant problem it is minor compared with the much more fundamental and separate question of whether particular organisms under study are the same species or not. While the former is largely a matter of convenience, the latter is supposed to be a scientifically verifiable reality. This topic was not featured at this conference although information bearing on it was.

On the one hand, many species are being named based on small, albeit consistent morphological differences recognised by the painstaking study of field material. On the other hand, there is the view, based largely on laboratory studies, that many of these "morphospecies" are parts of large continua, species complexes. It is becoming increasingly evident that there is considerable genetic variability within particular morphospecies. Furthermore, studies comparing enzymes and toxins in species differing in small morphological features, such as those of the *Alexandrium* group (one data set reported here by Anderson), have shown that they are not concordant. Sexual fusion has been possible between some of these (Fukuyo). Plainly much work is needed to clarify these important questions.

ECOLOGY

In the past decade or so there has been much progress in this area. The old approach of seeing all phytoplankton blooms as being much the same, with the organisms envisioned as nutrient-requiring packets of chlorophyll, permitted only the most elementary understanding of such phenomena. Red Tides and similar monospecific blooms require autecological information, i.e. the tolerances and preferences of the species in question. An important conceptual development was the recognition that there are distinct phases to such blooms, with different factors of importance in different phases. Factors which are limiting in one location may not be so in another in which the same species blooms. The ability to migrate vertically is a key factor in many, but not all, cases. Most occur in the summer after two weeks of calms, but not all. There are indications of reduced grazing pressure in several species, as noted in the section on toxin functions. Some blooms are plainly in response to eutrophication but others are equally plainly not. One cannot simply apply findings derived from one organism in one location to another species and place although they do provide clues as to what to look for.

For these, and other reasons, accurate prediction still remains an elusive goal in most regions. Models are useful for sensitivity testing but may not be essential for prediction. In most cases where prediction has been possible it has been based rather on indices, the measurement of some feature, such as temperature at a particular location, that has been found to be a good indicator. Long term field studies are essential for the determination of such indicators but these are usually difficult to fund in non-crisis years. There are logistic problems since blooms can develop and vanish in two weeks although the *Aureococcus* blooms have proved to be remarkably long-lived, with fatal consequences to scallops (Cosper). First-time blooms are usually studied too late in their development, although some data may be available from other concurrent studies, as was the case in the recent *Chrysochromulina* bloom. Examination of the chemical composition of the cells can yield insights into probable nutrient limitation, as was also done in this case.

Two major concerns at this conference, which were also considered at one of the workshops introduced by Smayda, were whether harmful blooms are increasing in frequency and if they are spreading. A cursory inspection of the literature certainly suggests these to be true. Many more blooms are being reported each year. Until 1986 *Heterosigma akashiwo* was known as a fish killer only in Japan. In that year farmed salmon were killed in British Columbia. More recently salmon have been killed by the same organism in Chile and New Zealand (both reported here) and, more surprisingly, in Singapore. It is difficult to conclude much from areas that have received little previous study although the high consumption of seafood by local populations, such as in Japan, can serve as indicators of the occurrence of toxicity over long time periods. Caged or penned fish are much more vulnerable to noxious blooms than wild stock, since the latter can avoid the blooms, and so the recent rise in fish farming is likely to lead to many more reports of these types of blooms.

A major difficulty is the distinction between natural fluctuations on a variety of time scales and progressive anthropogenic changes. Unusual oceanographic events can lead to incursions of species into areas that they have not been known from in recent years, such as the spread of *Gymnodinium breve* up the east coast of the United States in 1987-88 (Tester), but how cyclic are these? *G. breve* blooms are now annual events within the Gulf of Mexico: have they always been and not noted earlier? The well documented, gradual eutrophication of the Inland Sea of Japan undoubtedly led to increased harmful blooms in the 1970s, but blooms of other organisms occur in remote, pristine areas. Have humans contributed to the spread of harmful organisms through the transport of cysts by transplanting shellfish or other means? Hallegraeff provided evidence of the feasibility of this in the ballast water of large ships, particularly in the context of Japan and Tasmania. While such observations indicate a great need for awareness and prevention, if it is possible, they must be seen against a background of extreme latitudinal cosmopolitanism in the natural communities (including mirror communities in the northern and southern hemispheres). It is possible that artificial dispersal is not significant, when compared with natural dispersal factors but the question deserves further investigation.

The geological record can give insights into these questions. Cyst-forming species leave a signal in the sediments that can give information on a longer time scale, as stressed by Dale. There are examples of species being more abundant in regions where they are now sparse or absent. He pointed to the abundance of cysts of *Gymnodinium catenatum* in southern Norwegian sediments with a lack of curent planktonic records. I drew attention to the dominance of *Pyrodinium bahamense* cysts in Recent sediments off Israel and the Arabian Gulf. These areas do not seem to regularly support large blooms of these organisms at present.

CONCLUSION

These are exciting times for the study of harmful blooms, with knowledge progressing rapidly as new scientists are attracted to the field and economic and human health concerns stimulate more support for research. It still appears, however, that many Governments have to learn of the negative impacts the hard way. Every coastal country is likely to be negatively affected by plankton blooms to a greater or lesser extent due to the great cosmopolitanism of the harmful species; absence of records is usually a reflection of underdevelopment or lack of awareness. It is a pity that it seems to take catastrophes or near-catastrophes to get adequate research resources assigned.

The management of harmful blooms was not discussed to any great extent at this conference although it is obviously a critical area. Blooms are of such great extent that there is little hope of eliminating them once they have developed but, if human activities are contributing to their occurrence, these can be identified and changed. Accurate prediction is a primary research goal which can allow steps to be taken to ameliorate the effects, such as the towing of salmon pens or increased shellfish toxin monitoring. Methods of protecting marine life need to be improved as they are still very primitive at present. Antidotes don't exist for several of the toxins harming humans.

Every field of research seems to pass through three phases: an initial phase in which the problem appears simple because little is known about it; a middle phase when knowledge accumulates and it gets much more complicated; and a clarifying phase in which it is determined how much of the complication is irrelevant. In my estimation we are nearing the end of the middle phase in this field of study but this will probably look naive twenty years from now. A great deal remains to be answered, with difficult challenges ahead. Thanks to conferences like this we can approach them with a more realistic appreciation of the factors involved.

REFERENCES

1. J.R. Grindley & F.J.R. Taylor, Nature 195, 1324 (1962)

2. T. Yasumoto, in Toxic Dinoflagellates, D.M. Anderson, A.W. White & D.G. Baden, eds (Elsevier, New York 1985) pp.259-270.

3. F.J.R. Taylor, in Toxic Dinoflagellates, D.M. Anderson, A.W. White & D.G. Baden, eds (Elsevier, New York 1985) pp.11-26

APPENDIX

THE GLOBAL DISTRIBUTION OF HARMFUL EFFECTS OF PHYTOPLANKTON

During the "First International Symposium on Red Tides", in Takamatsu, Japan, November 1987, the participants were asked to indicate on maps all the records of PSP, DSP, NSP and fishkills known to them. The results were kindly made available to us by Prof. T. Okaichi and have been put together in Figs 1 and 2.

During the Lund conference we thought it would be a good idea to try to bring these records up to date. We thus asked the participants to indicate - on maps displayed by the main lecture room - new records of PSP, DSP, ASP, harmful effects of diatoms, harmful effects of other phytoplankton and fishkills. Figs 3 and 4 summarize these indications and, thus, only show cases that were known to the participants and that have taken place between November 1987 and June 1989.

Although both conferences were attended by more than 200 scientists from all over the world, it is highly probable that several records of harmful effects have not been indicated due to lack of information. The way of indicating "records" probably also differ from person to person, and one mark on the maps should be seen as either indicating an isolated occurrence or a general area affected.

The distribution maps we present here are perhaps of little scientific accurancy but give a general, and somewhat frightening, picture of the worldwide problem that harmful phytoplankton now comprise. It is clear that the problems in no way have diminished over the last two years and that new areas/localities have been affected by both PSP and DSP since 1987. Of particular concern is the high number of records of harmful effects by other phytoplankton (including diatoms), and if the tendency of a continuous addition of new genera and species to the list of harmful taxa persists, phytoplankton monitoring programs all over the world should find their important task of warning the public of the presence of potentially dangerous phytoplankton increasingly difficult.

Bo Sundström Lars Edler Edna Granéli

Toxic Marine Phytoplankton
Edna Graneli et al., Editors

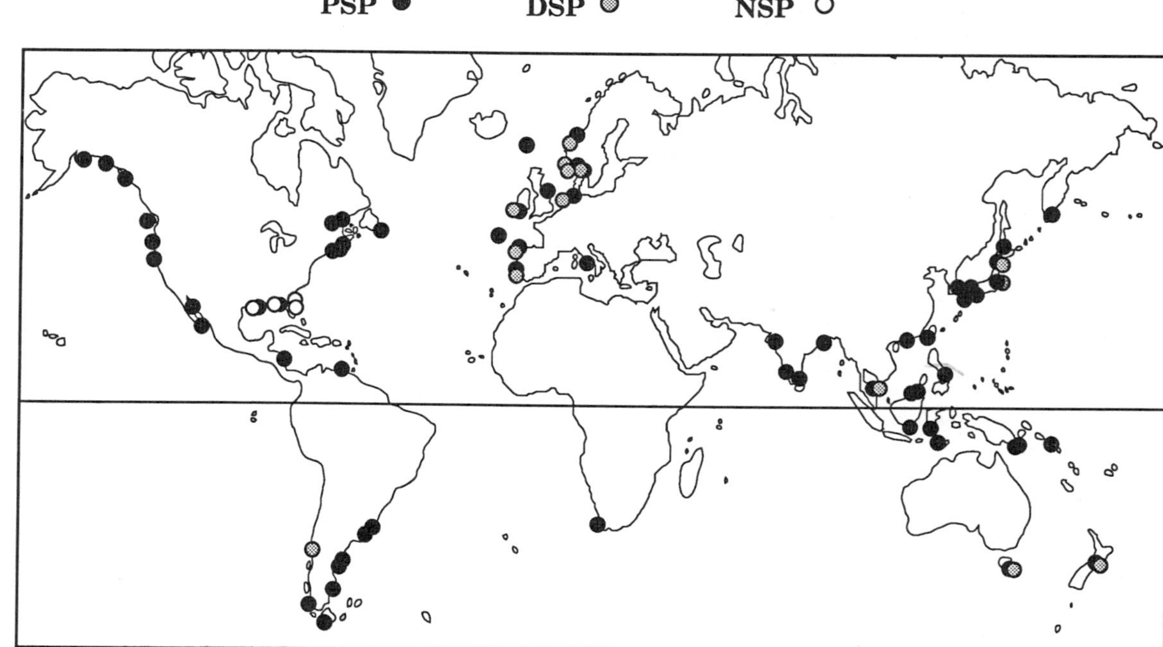

FIG. 1. Global distribution of human intoxication by Paralytic Shellfish Poison (PSP), Diarrheic Shellfish Poison (DSP) and Neurological Shellfish Poison (NSP). (Prior to the International Symposium on Red Tides, Takamatsu, Japan, November 10-14, 1987)

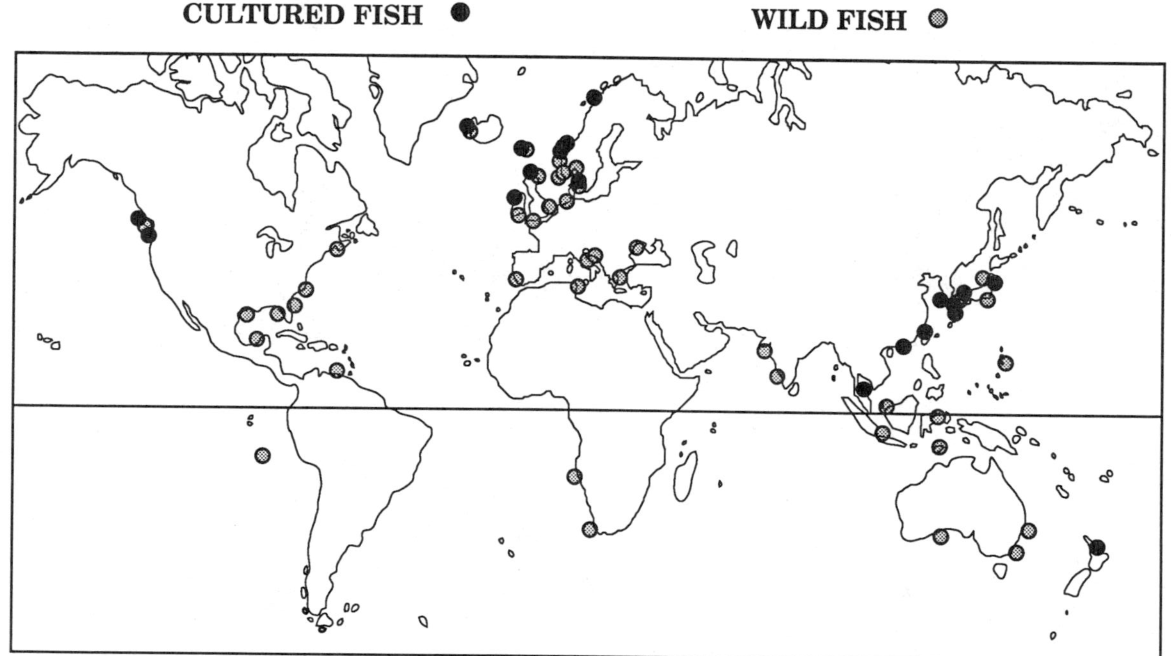

FIG. 2. Global distribution of wild and cultured fishkills (various causes) (Prior to the International Symposium on Red Tides, Takamatsu, Nov. 10-14, 1987)

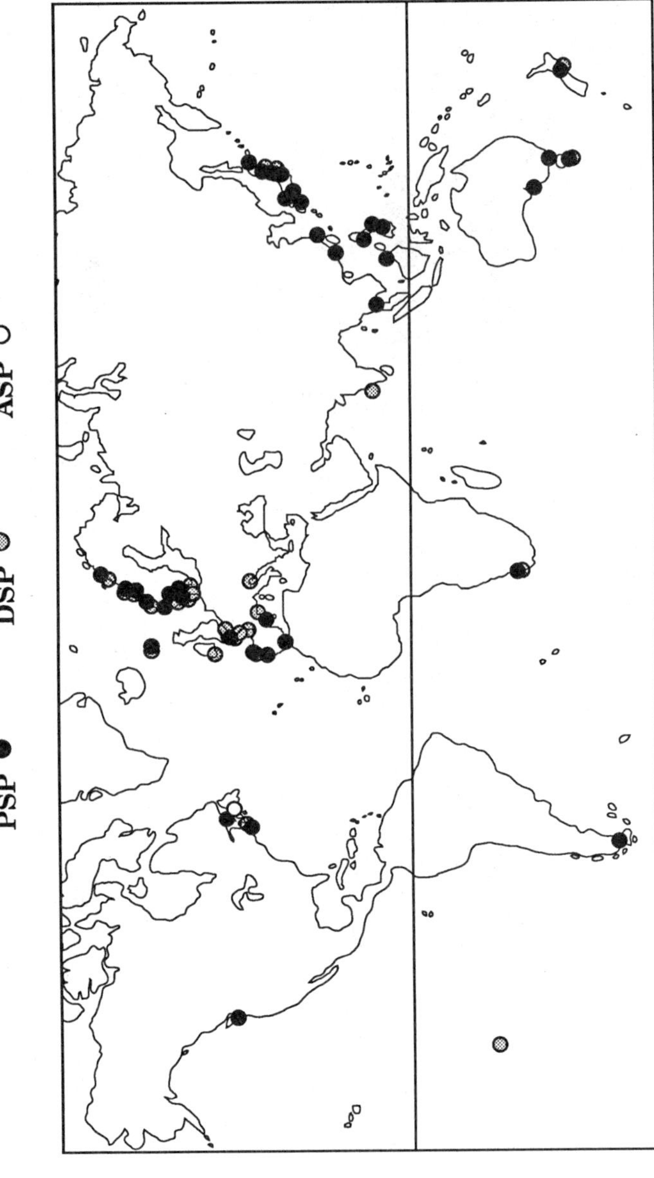

FIG. 3. Global distribution of human intoxication by Paralytic Shellfish Poison (PSP), Diarrheic Shellfish Poison (DSP) and Amnesic Shellfish Poison (ASP) (New records from Nov. 1987 to June 26-30, 1989)

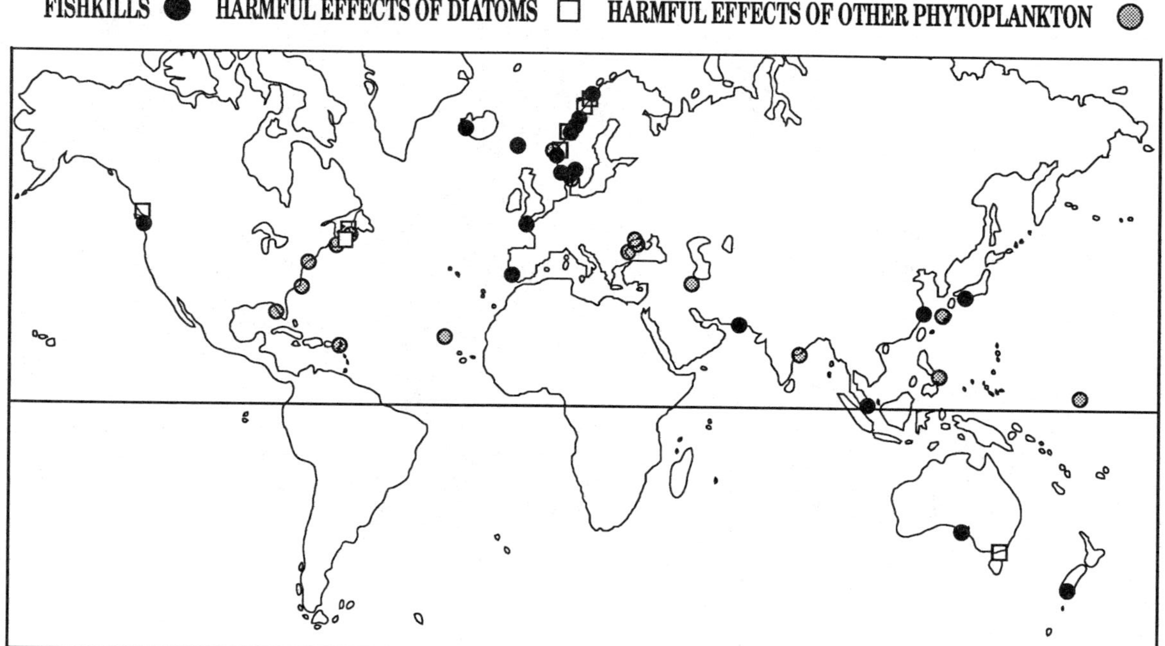

FIG. 4. Global distribution of fishkills and harmful effects of diatoms and other phytoplankton except dinoflagellates (New records from Nov. 1987 to June 26-30, 1989)

AUTHOR INDEX

Index